NATO ASI Series

Advanced Science Institutes Series

A series presenting the results of activities sponsored by the NATO Science Committee, which aims at the dissemination of advanced scientific and technological knowledge, with a view to strengthening links between scientific communities.

The Series is published by an international board of publishers in conjunction with the NATO Scientific Affairs Division

A Life Sciences B Physics	Plenum Publishing Corporation London and New York
C Mathematical and Physical Sciences D Behavioural and Social Sciences E Applied Sciences	Kluwer Academic Publishers Dordrecht, Boston and London
F Computer and Systems Sciences G Ecological Sciences H Cell Biology I Global Environmental Change	Springer-Verlag Berlin Heidelberg New York London Paris Tokyo Hong Kong Barcelona Budapest

PARTNERSHIP SUB-SERIES

1. Disarmament Technologies	Kluwer Academic Publishers
2. Environment	Springer-Verlag/Kluwer Acad. Publishers
3. High Technology	Kluwer Academic Publishers
4. Science and Technology Policy	Kluwer Academic Publishers
5. Computer Networking	Kluwer Academic Publishers

The Partnership Sub-Series incorporates activities undertaken in collaboration with NATO's Cooperation Partners, the countries of the CIS and Central and Eastern Europe, in Priority Areas of concern to those countries.

NATO-PCO DATABASE

The electronic index to the NATO ASI Series provides full bibliographical references (with keywords and/or abstracts) to about 50000 contributions from international scientists published in all sections of the NATO ASI Series. Access to the NATO-PCO DATABASE compiled by the NATO Publication Coordination Office is possible in two ways:

- via online FILE 128 (NATO-PCO DATABASE) hosted by ESRIN, Via Galileo Galilei, I-00044 Frascati, Italy.
- via CD-ROM "NATO Science & Technology Disk" with user-friendly retrieval software in English, French and German (© WTV GmbH and DATAWARE Technologies Inc. 1992).

The CD-ROM can be ordered through any member of the Board of Publishers or through NATO-PCO, Overijse, Belgium.

Series I: Global Environmental Change, Vol. 45

Springer
Berlin
Heidelberg
New York
Barcelona
Budapest
Hong Kong
London
Milan
Paris
Santa Clara
Singapore
Tokyo

Radiation and Water in the Climate System

Remote Measurements

Edited by

Ehrhard Raschke
Institut für Atmosphärenphysik
GKSS-Forschungszentrum
D-21502 Geesthacht, Germany

Springer

Published in cooperation with NATO Scientific Affairs Division

Proceedings of the NATO Advanced Science Institute "Remote Sensing of Processes Governing Energy and Water Cycles in the Climate System", held in Plön, Germany, May 1–12, 1995

Library of Congress Cataloging-in-Publication Data

Radiation and water in the climate system : remote measurements / edited by Ehrhard Raschke.
p. cm. -- (NATO ASI series. Series I, Global environmental change ; vol. 45)
"Published in cooperation with NATO Scientific Affairs Division."
"Proceedings of the NATO Advanced Science Institute: "Remote Sensing of Processes Governing Energy and Water Cycles in the Climate System", held in Plön, Germany, May 1-12, 1995"--T.p. verso.
Includes bibliographical references and index.
ISBN 3-540-61470-2 (hard cover)
1. Dynamic meteorology--Remote sensing--Congresse. 2. Atmospheric physics--Remote sensing--Congresses. 3. Cloud physics--Remote sensing--Congresses. 4. Atmospheric radiation--Remote sensing--Congresses. 5. Hydrologic cycle--Remote sensing--Congresses. I. Raschke, Ehrhard. II. North Atlantic Treaty Organization. Scientific Affairs Division. III. NATO Advanced Science Institute: "Remote Sensing of Processes Governing Energy and Water Cycles in the Climate System" (1995 : Plön, Germany) IV. Series.
QC880.R33 1996
551.57'028--dc20 96-27902
CIP

ISBN 3-540-61470-2 Springer-Verlag Berlin Heidelberg New York

Printed in Germany

Typesetting: Camera ready by the editors
Printed on acid-free paper
SPIN: 10477259 31/3137 - 5 4 3 2 1 0

Foreword

The search for an accurate knowledge of the transfer of energy and water in our climate system and of all participating processes is still a burning issue within environmental and climate research. More reliable ways should be found to model them in their complex interplay with cascades of processes in that system, and their impact on all our resources on earth. Such modelling needs also validation with accurate and complete observations which in turn should also enable to observe variations of climate components.

These are the goals of the Global Energy and Water Experiment (GEWEX), which began its work about 8 years ago. Many advanced remote sensing capabilities have now been developed, of which only a few will enter into satellite based platforms. These new developments should be brought to the attention of a broader community of younger scientists, in particular of those countries which quite recently could open their borders for free exchanges.

Therefore, after the very successful Advanced Study Institute on the "Energy and Water Cycles in the Climate System", which was held from 29. September to 13. October 1991 in Glücksburg, Germany, another one has been felt to be required to bring this *need for, but also the limitations of remote sensing techniques* for many components of those cycles to the attention of young scientists. The invited co-directors (Y. Fouquart and V. V. Salomonson) and all speakers agreed immediately and constructively to cooperate in its preparations; and they did their best during the courses and responded as far as possible to all wishes of the auditorium.

We had a large number of applicants of which finally 74 could come into the Kurhaus Plön, Germany, from 1 to 12 May, 1995. They came from as far as Japan, Taiwan, South Africa, Brazil, Bulgaria, Ukraine, from all NATO-Member Countries and from other parts of Europe.

As for the earlier ASI in Glücksburg, all participants were asked to report in a permanent poster exhibition and during a few oral sessions on their own research, which in some cases was not at all related to the theme of our Summer School. Many long night sessions were required to bring the more than 40 posters into an order, that allowed to write the summarizing reports, which are reproduced at the end of this book.

This book contains somewhat modified and shortened versions of the original presentations, where most illustrations, which were shown during the lectures, have been left away in favour of an enhanced quotation of literature for deeper information. It is not perfect, and the reader may find here and there inconsistencies or even (what she / he may interpret as) an error or incompleteness. We left it entirely to the responsibility of each author to prepare the text and content according to her / his own lines, since we felt an urgent need to bring these lectures as fast as possible to the attention of the open scientific community for further use. We asked all authors to provide as much as possible information on related reviews and books.

At the end of this work we want to thank in particular all donators, whose financial contributions were fundamental for this event. These were the NATO Division for Scientific Affairs, the Directorate of the World Climate Research Programme in Geneva, the GKSS Research Centre in Geesthacht and its associated "Club of Friends and Donators" and also many of our colleagues, who managed to support fully or in part their collaborators or students for their stay in the encouraging environment of the lake area in Schleswig-Holstein. We acknowledge also remarks of two anonymous reviewers of our proposal.

During the preparations for this ASI and later on we appreciated the helpful assistance of Dr. R. Nolte-Holube, Mrs. S. Hartmann, Dr. M. Serwazi, Mrs. W. Jansen, Mrs. H. Lemke and Mr. J. Oldag of the GKSS Research Center in Geesthacht.

The hotel staff in Plön has been very patient, helpful and friendly to all of us. At the end of the school, a disco-evening transformed the students and speakers back to the real world.

Finally, we must in particular thank Mr. M. Ikonomou for his skill and willingness to work carefully his way through all manuscripts, which arrived in often nonstandardized forms.

Ehrhard Raschke (Director of this NATO-ASI)

Contents

Kurhotel

(1) Dr. G. Kiely, IRELAND (2) Dr. Marcus Serwazi, GERMANY (3) Iván Csiszár, HUNGARY (4) Joern Oldag, GERMANY (5) Koen de Ridder, BELGIUM (6) Michael Lieder, GERMANY (7) Dr. Cheng-Ta Chen, GERMANY (8) H. Klein Baltink, THE NETHERLANDS (9) Xiaofeng Li, USA (10) Sigrun Tahl, GERMANY (11) Boyd Tolton, CANADA (12) Jason Burford, CANADA (13) Dr. Bernd Zbell, GERMANY (14) Aashish Clerk, CANADA (15) Daniel B. Michelson, SWEDEN (16) Dr. O. Udovik, UKRAINE (17) Hajime Okamoto, JAPAN (18) Patrick Simpson, UK (19) Frederic Raison, FRANCE (20) Dr. Gerald Mace, USA (21) Hans Thonfeld, GERMANY (22) Colin Jones, USA (23) Stéphane Oury, FRANCE (24) Michael Frech, GERMANY (25) Ralf Bennartz, GERMANY (26) Alev Temel Tahmaz, TURKEY (27) Dr. Josef Noll, THE NETHERLANDS (28) Dr. Selim Seker, TURKEY (29) Susanne Mecklenburg, GERMANY (30) Silvio Petaccia, ITALY (31) Prof. Dr. E. Raschke, GERMANY (32) Christina Köpken, GERMANY (33) Charles Drummond, USA (34) Dr. A.D. Del Genio, USA (35) Antonio Serrano Pérez, SPAIN (36) Dr. Julia Abbate, ITALY (37) Dr. W. B. Rossow, USA (38) AntonV. Volkov, RUSSIA (39) R.G. Booysen, SOUTH AFRICA (40) Dieter Jasbinschek, GERMANY (41) Marthinus Melck Truter, SOUTH AFRICA (42) Dr. C. Weitkamp, GERMANY (43) Dr. Piia Post, ESTONIA (44) Claudia Schröder, GERMANY (45) Barbara Bartsch, GERMANY (46) Dr. Ana Maria Silva, PORTUGAL (47) Henriette Lemke, GERMANY (48) Hela Mehrtens, GERMANY (49) Hélène Chepfer, FRANCE (50) Àgnes Merza, HUNGARY (51) Kerstin Lippert, GERMANY (52) Dr. Elisabetta Pierazzo, USA (53) Halina Kowalewska, POLAND (54) Dr. Kristinka Ivanova, BULGARIA (55) Nilgün Kubilay, TURKEY (56) Márta Diószeghy, HUNGARY (57) Patricia de Rosnay, FRANCE (58) Ayfer Serap Sögüt, TURKEY (59) Marie Doutriaux, FRANCE (60) Deniz Okcu, TURKEY

PART A

SOME BACKGROUND ON OBSERVATIONS, DATA ASSIMILATION AND MODELING

Chapter 1

Energy and Water Cycles in the Climate System: What can we measure from space?

Ehrhard Raschke
GKSS-Forschungszentrum
D-21502 Geesthacht
Germany

Abstract

This first paper of a lecture series on various specific problems, which are related to spaceborne and ground-based remote sensing of several components of our climate system describing energy and water cycles over the entire globe and specific regions of it, intends to provide a first brief overview on our present knowledge of them. It further will introduce into the principles, limitations and problems of their direct and remote measurements from space, within the atmosphere and from ground where the latter ones are often the first step towards a space experiment. More recent details on our present knowledge on them and their observations can be found in the many individual papers, following thereafter.

We concentrate here, following the major research strategy of the Global Energy and Water Cycles Experiment (GEWEX, see WCRP 1990), on the major components of the water budget, on the radiation budget and on water transports in the atmosphere and at ground. This brief review cannot be complete nor exhausting. It purposely quotes some work which has been performed some 20 years ago and earlier.

1.1 Introduction

Our climate system, consisting of the atmosphere, oceans with varying sea-ice cover, land-surfaces with their complex vegetative cover and orography and of the cryosphere (ice sheets over Greenland and Antarctica, continental glaciers), is steadily in motion within a wide spectrum of spatial and temporal scales which are ranging from millimeters to 10.000 km and from milliseconds to several or hundreds of years, respectively. These dynamical processes and their complex interactions and the resulting exchanges of energy, momentum and material must completely be understood and then also be reconstructed to a certain degree of detail

NATO ASI Series, Vol. I 45
Radiation and Water in the Climate System:
Remote Measurements
Edited by Ehrhard Raschke

(and accuracy!) with numerical models. Such global numerical models are now used for the simulation of the present and prediction of future states of weather over 3 to 10 days, or the weather statistics during the next season, the ocean circulations or even the development of the climate in general during the next decades (e.g.: Cubasch et al., 1995; Kumar and Hoerling, 1995). Research is underway to use them also for the reconstruction of the climate during past historical periods considering different external forcing scenarios.

More demands than ever before are now occuring to this field of science from the open public, due to the issue of increasing concentrations of the greenhouse gases carbon dioxide and methane, and also of others. The world's population increases further and will in future need more food, more water and also more space to live and to dump and recycle its waste and other disposals.

But also many immediate applications benefit already at present from an improved forecasting of atmospheric and surface properties and also ocean currents, depending on the space and time scale. In fact, one estimates an annual global saving of the world's economy by about 30–40 Bill. US-$, due to the present days use of informations to be derived from weather forecasts at different scales.

The nation's decisionmakers need reliable and believable information on future developments over the next decades not only of their own economies and the future developments in their countries but also on the climate and its impact on the use of the earth's surface and related resources. This information enables them to reach national and international agreements on possibly drastic and expensive measures to keep our climate system in an acceptable state, which does not cause too strong social unrest, provide such a state can be defined and reliably predicted.

The World Meteorological Organization (WMO) and other bodies have, therefore, invited many scientists to advice them in these matters within the framework of the Intergovernmental Panel on Climate and Climate Changes (IPCC, 1996, to appear). Their reports reflect the state of art of the present understanding of the earth's climate. In it also some conclusions are made on the development of the climate during the upcoming 50 years.

As one possible and quite straightforward way to meet these demands one has chosen to built up systems of numerical simulation models, which are capable to consider all relevant processes in the climate system, i.e. within the atmosphere and coupled to it the oceans, continents and the cryosphere. This strategy arose from the practices of numerical weather forecast. It is the basis of the World Climate Research Programme (WCRP). The performance of such models needs detailed and careful validation of results which have been calculated for the present climate. It must be based on reliable observations.

But many ground-based observational systems degrade in quantity and quality and need increasingly be supplemented by space borne and other automatic observations. Furthermore, we still do not know the regional and temporal distribution of most climate variables well enough over areas where almost no "ground-truth" is yet available, such as over all oceans and other continental areas, and the many global atmospheric models compute still quite different results for the same input. Very few direct measurements are available over the oceans.

The many careful intercomparisons of model results within the Atmospheric Model Intercomparison Project (AMIP, see Gates, 1992) and also of various model components, such as the routines for radiation having been intercompared during the International Comparison of Radiation Codes in Climate Models (ICRCCM; Ellingson and Fouquart, 1993) or the PILPS-project concentrating on land-surface processes over vegetated areas (Henderson-Sellers, 1995) have demonstrated the need for a better understanding of the various processes involved in energy and water transfers.

In the following we will, as it had been planned for this summer school, concentrate only on the description of the present knowledge of the components of the radiation budget at the top and bottom of the atmosphere and of the availability and transport of water in the atmosphere, at ground and in the uppermost soil layers. Here, we will describe briefly how these quantities can be inferred from remote satellite measurements and to some degree also from those at ground.

More details on the modelling (Del Genio, this volume), on data analysis and assimilation schemes (Courtier, this volume) and in particular on special space-borne measurements and some principle arguments of the radiative transfer theory (Fouquart and Vesperini, this volume) are given in the papers to follow. The same holds also for some of the more recent "climatologies" on the radiation budget components and clouds, and the precipitation and evaporation over the oceans. We also will stress the need for accurate and reliable ground-based informations, which in many cases are urgently required to validate the quantities which are estimated from the satellite data.

It is one of the major goals of the Global Energy and Water Cycle Experiment (GEWEX; WCRP, 1988) to develop and use operationally spaceborne techniques and appropriate analysis schemes to measure these quantities with high spatial detail over the entire globe.

Very detailed accounts of our present knowledge on energy and water cycles are given in more recent textbooks, e.g. by Hartmann (1995) or Peixoto and Oort (1992), in some other papers of these and many other authors, and also in the publications of earlier NATO-Advanced Study Institutes (e.g. Raschke and Jacob, 1992; Oliver and Oliver, 1994). They, therefore, will not be repeated in this book, although they have been presented and discussed during the various lectures.

1.2 Water in the Climate System

Our climate system (see e.g. summary by Chahine, 1992) contains more than 1.4 Mill. $\times 10^{12}$ t water, of which only a very small fraction (about 0.0001%) is abundant within the atmosphere, and another but larger fraction (about 4%) at the land surfaces. Fig. 1.1, from Chahine (1992), displays these distributions and also the annual mean fluxes between the different reservoirs which are formed by the atmosphere, the continents and the oceans. This schematic shows that the atmosphere in general transports, in general, water from the oceans over the continents, from where it flows back into the oceans.

This freshwater supply into the oceans is believed to obey a strong influence on the formation of deep sea currents in the Arctics, which in turn may effect the mean position of warmer near-surface branches of the world-wide conveyor belt and lead to changes of the continental climate (e.g. Schmitz, 1995).

Over the oceans the atmosphere recycles its water content about 40 times during each year (precipitation divided by atmospheric storage), over continents only about 24 times. Thus the mean residence time of water within the atmosphere is between about 9 to 15 days, while in the oceans it may remain more than 3000 years. This recycling over land becomes important in precipitation forecasts, when frontal systems move inward over a continent and the soil moisture might be predicted with too low values, then also too small amounts of precipitation are often predicted.

These immensely large water reservoirs slow down all climate processes. A "fast" regime is found in the system atmosphere, land surfaces and upper ocean layers, which directly controls the amplitude and regional patterns of climate variations. A "slow" regime is caused by large

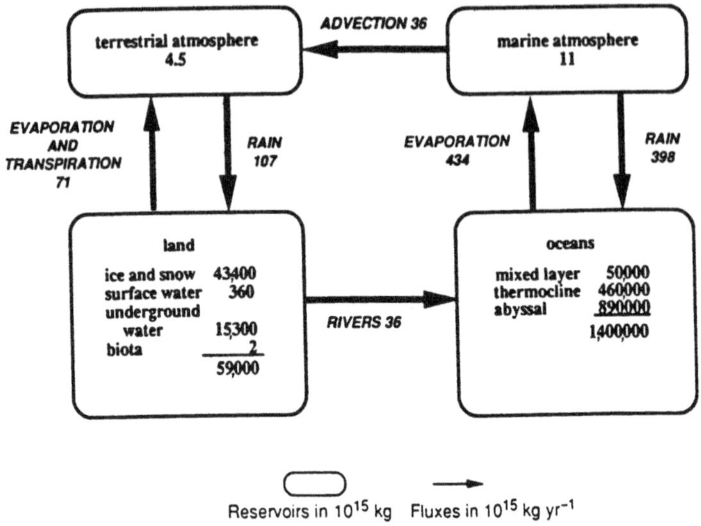

Figure 1.1: *Estimates of the global water reservoirs in the atmosphere, on the continents and within the oceans and of the transports between them (from Chahine, 1992).*

heat capacities of the continental ice shelfs and the major (deeper) ocean layers. It modulates all longterm transient processes.

Water in such large amounts is still available on our planet, due to the high mass of our earth and the moderate temperatures caused by the distance of the Sun. Its availability during almost the whole life of the Earth is responsible for many geochemical processes which led to the present state of our climate system. Further details have recently been discussed by Webster (1994), and can also be found in many textbooks on planetary sciences. Fig. 1.2 shows the estimated phase transition curves for water and hypothetical climate trajectories for Mars, Earth and Venus.

In the following sections 1.3 and 1.4 more details are provided on the abundance of water in its gaseous and liquid and solid phases in the atmosphere and on their possible retrievals with passive and active remote sensing techniques. Section 1.5 reports on the radiation budget components, while the remaining chapters describe the precipitation, evaporation and runoff.

1.3 Atmospheric water vapor
(see also contributions by Simmer, Weitkamp)

Water vapor is the major greenhouse gas, contributing dominantly to the major radiative warming in the lower troposphere, and some cooling in the upper layers. This is illustrated in Fig. 1.3 from Manabe and Möller (1961), where also the contributions of other greenhouse gases (CO_2 and O_3) are shown for some annual and global average. Its abundance in at least the lower layers of the troposphere is closely but not completely related to the mean air temperature (e.g.: Gaffen et al., 1992, Stephens et al., 1990), thus the water vapor concentration will exert a positive greenhouse effect. When the surface temperatures will increase, then also the mixing ratio of the water vapor will increase, and more absorption of radiative energy can occur.

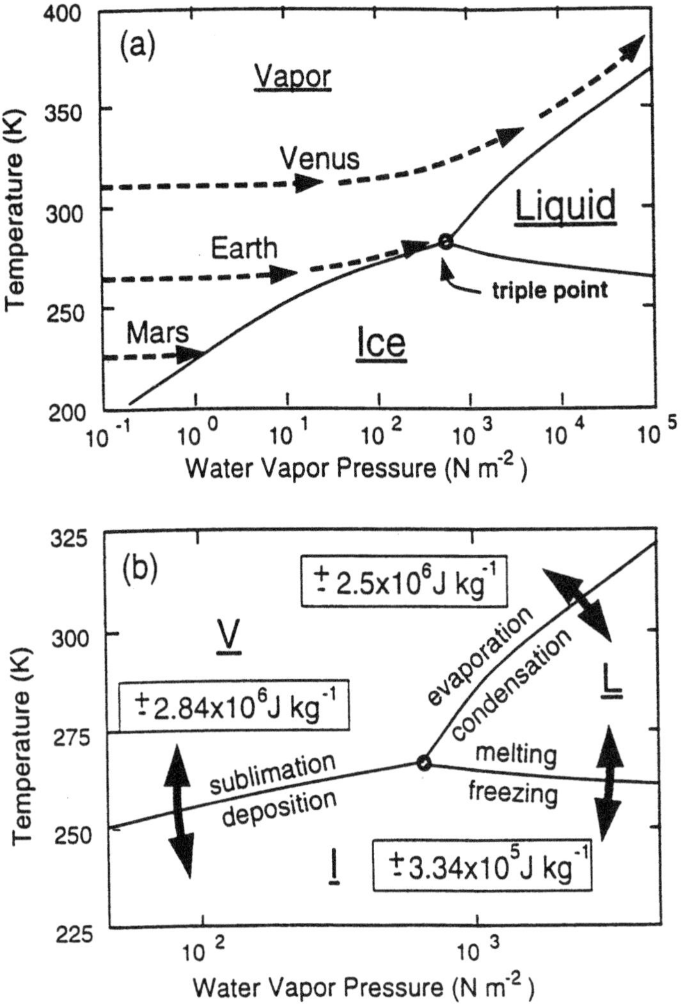

Figure 1.2: *Phases of water: Phase transition (solid curves) and hypothetical climate trajectories for Venus, Earth and Mars (from Webster, 1994). This figure indicates that on Earth all three phases can conveniently coexist.*

Drastic temporary variations of the mixing ratio can, however, also be observed in the lower troposphere due to the formation of different air masses on their journey over the globe.

The higher layers of the troposhere are much dryer due to lower temperatures, but as the images of Meteosat and other satellites each day visualize, if made from measurements of the emission of water vapor in its band 6.3 μm, there are larger dryer and moister regions, mostly due to larger-scale atmospheric dynamics. These moister regions are important for the formation and life time of higher clouds systems, such as cirrus and mid-level stratiform layers. Schmetz et al. (1988) use such measurements to determine the upper tropospheric humidity and also the horizontal water vapor transports, where winds are estimated from motions of cloud elements but also other observable structures in the upper-tropospheric water vapor fields. Their analysis is based on principles which were developed already 25 years ago (e.g. Möller, 1961).

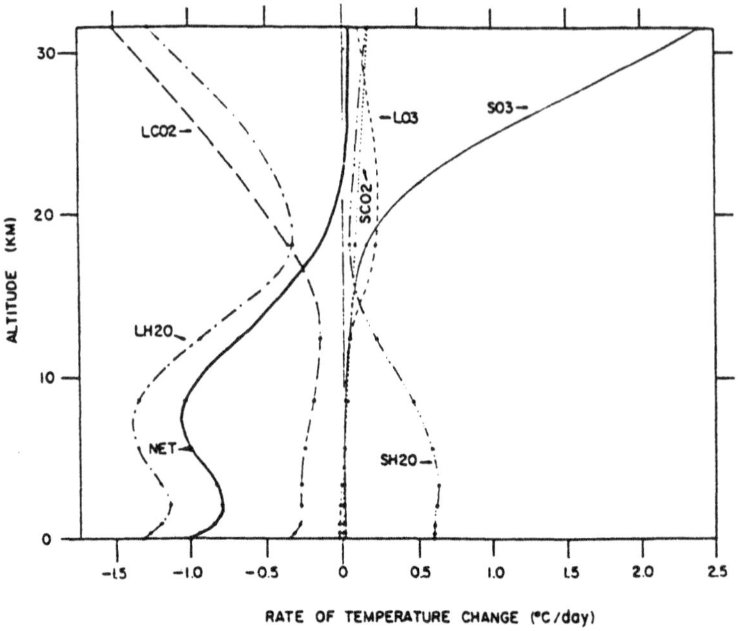

Figure 1.3: *Radiative heating and cooling of the atmosphere, computed for the greenhouse gases water vapor, carbon dioxide and ozone (from Manabe and Möller, 1961). This figure arose in one of the earliest studies of the effects of radiative energy transfer on dynamical processes computed in numerical models for the atmosphere.*

In the lower layers of the stratosphere the water vapor mixing ratio assumes already low values around 2 to 5 $\times$ 10^{-6} g_{H_2O}/g_{air}. Read et al. (1995) confirmed the many direct soundings with analyses of microwave limbsounding data from the UARS (Upper Atmosphere Research Satellite). Convection from below causes that most portions of the water vapor freeze out, when air penetrates through the tropopause into the stratosphere (see also details in Webster, 1995).

In the upper stratosphere and even more in the , the water vapor participates also in many photochemical processes, which are responsible for the maintenance of the ozone layer. It originates there to some part from the methane, which is steadily penetrating from the troposphere below into the middle atmosphere. Presently observed increases of the number of annual occurances of noctilucent clouds, and also an apparent increase of their areal extent, are attributed to this water vapor source.

Direct measurements of the atmospheric water vapor with operational sondes show large and often also systematic errors in all layers above about 400 hPa, where mostly too dry values are obtained. Other sensors for direct measurements, such as the complex Lyman-α probes or cryogenically cooled mirrors, measure higher moistures, but they are too expensive for world wide operational use in the operational "throw-away sondes".

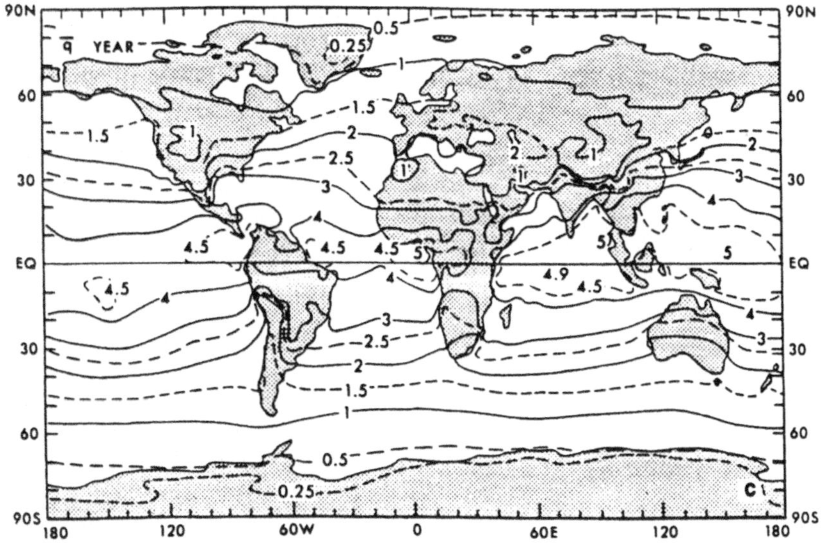

Figure 1.4: *Total atmospheric precipitable water in vertical columns (in units of 10 kg m^{-2}), from Peixoto and Oort (1992). These values are primarily based on conventional analyses. About 80% of the total vertical water column is contained in the lowest 2000 meters of the atmosphere, and less than 1% in the strato- and mesosphere.*

Recent airborne comparisons of various water vapor sensors have been described by Ström et al. (1993). Their results will urge for improvements of the present water vapor sensors used in the operational radio-wind sondes.

Space-borne soundings can make use of various principle which are described below:

- Measurements of the upward spectral emission to space within strong absorption bands provide global coverage but very low vertical resolution due to the half-width of its weighting functions in the troposphere of about 5 km. They have firstly been done from the very first American meteorological satellites of the TIROS-series (e.g. Möller, 1961)

- The limb emission (as discussed by Read et al., 1995) or absorption (occultation measurements), provides an excellently high vertical resolution, but only relatively few data points are obtained over the globe during each day.

- The attenuation of radio signals which is transmitted from the Global Positioning Systems (GPS) in earth satellites and had been considered often as noise for geodetic purposes (e.g.: Ware et al., 1996). The attenuation of these signals is proportional to changes of the index of refraction of air, and thus of the temperature and humidity, where the latter in the lowest 5 km-layers can be determined.

- Attenuation occurs also along the path of lidar rays, where however the technology is not yet available for use in space. Measurements with Raman-Lidar, proven as an excellent tool for water vapor soundings (see Weitkamp, this volume), require relatively long integration times.

The GEWEX Water Vapor Project (GVaP) within GEWEX enhances our present activities to approach higher accuracies in water vapor measurements and also spatially denser informations. Several commercial aircraft in Europe and in the US carry already now automatic instrumental packages to record atmospheric water vapor from take-off to the landing. These additional airborne reports (AIREP) on profiles near the major airports will immediately be fed into the operational analyses as also those measurements of the occasional research field campaigns.

1.4 Clouds in the atmosphere

(see also contributions by Ackerman, Rossow, Simmer, Stewart, Stuhlmann, Weitkamp)

As many careful global analyses of, however, limited data sets have shown, clouds at present occur daily over about 50 to 55% of the world, where highest ocurrences can be found in the "belts" of stronger convective and synoptic-scale activities. On the other side the subtropical branches of the Hadley-cell and, moreover, several regions over the continents contain very dry air and consequently only very few clouds. Due to importance of clouds on the weather at the surface, on the transport of precipitation and within the huge sequences of air-chemical processes, world-wide observations began very early, being now complemented by space-borne data. Fig. 1.5 displays results on the day-night difference of the fractional coverage by clouds which was obtained from ground-based observations by Hahn et al. (1995). They analysed most available groundbased cloud observations and obtained global cloud distributions which are quite similar to those determined from satellite data (see Rossow, this volume).

Analyses of satellite data began as early as they could be visualized in images and manipulated in computers. Such early overlay statistics led e.g. to the "discovery" of the "Double-ITC" over the Pacific Ocean. Operational meteorologists also developed manual analysis techniques (Nephanalyses).

The more modern analyses of satellite measurements can now make use

- of contrasts of cloud fields against the ambient background, the Earth's surface, in their intensities of emitted and reflected radiation, respectively, and
- of the spectral properties of these quantities. For instance, ice clouds reflect slightly less radiation than water clouds, thus "differences" between albedoes of simulataneous measurements can be used to identify cirrus over bright lower level clouds, or low level stratus over snow and ice fields.
- of up to 24 measurements per day from most geostationary satellites to meet the diurnal variations.

Such data form the major source of informations within the quite straightforward algorithms which at present are used in most cloud analysis schemes, also within the International Satellite Cloud Climatology Project (ISCCP; see papers by Rossow in this volume, and Rossow and Schiffer, 1991). Here the radiative transfer properties are estimated by comparison of measured with calculated radiance data where additional ground-based observations of snow cover and of the surface temperatures are used to "enhance" the contrasts. The ISCCP has at present analysed worldwide data sets from more than 10 complete years, beginning in 1983. The diurnal variation of cloud fields is observable over most regions of the world in the data of geostationary satellites, which - with the exception of the INSAT-series - enter also the ISCCP analyses.

There are also several attempts to identify cloud species with pattern recognition methods (e.g.: Lovejoy et al., 1992); but those methods have until present not reached an operational status

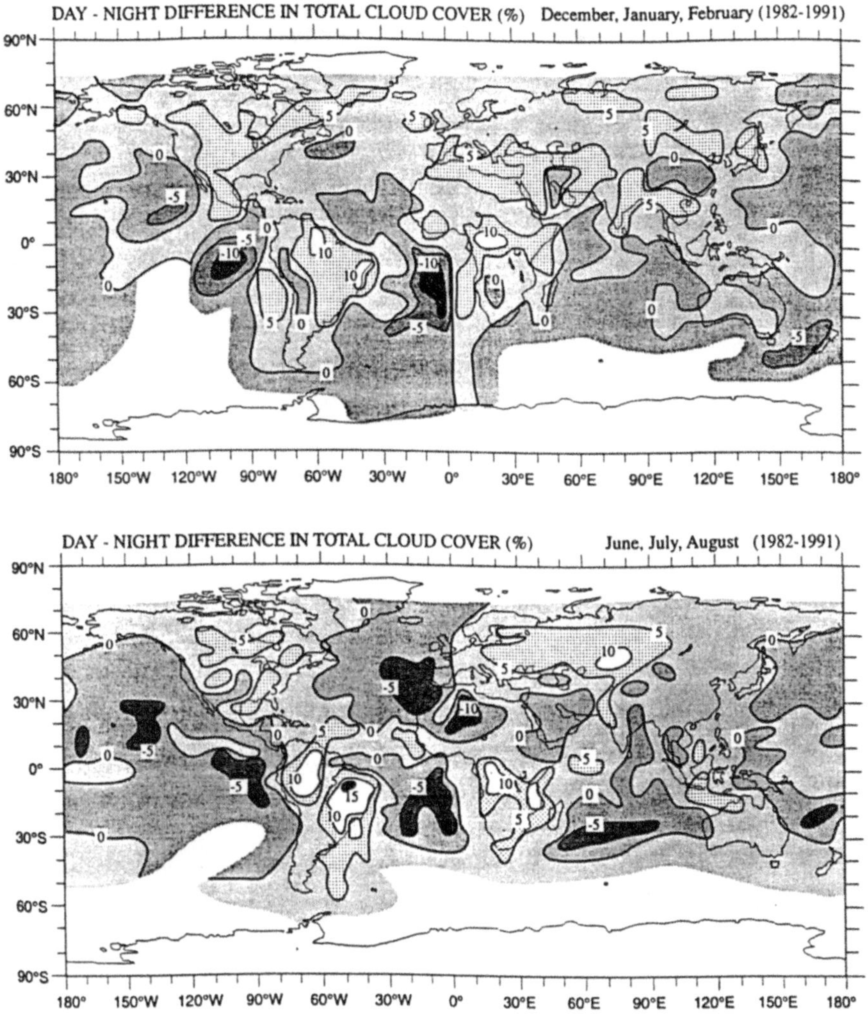

Figure 1.5: *Difference in cloud cover between day (06:00–18:00 LT) and night (18:00–06:00 LT). All values in percent (Hahn et al., 1995).*

due to their high demands in computational requirements. We also should not overlook the very early manual analyses ("nephanalyses") of satellite images, of at that time often poor quality, to identify major storm systems over the oceans.

The high effects of cloud fields on the transfer of radiative energy within the atmosphere and our still not adequately good knowledge on the relations of their individual properties (such as horizontal structures, particle number densities and size distributions) to the larger-scale circulation has led to the establishment of several regional projects, where numerical modellers and experimenters joint for research for the final goal to develop parametrization schemes for the description of clouds and their impacts in larger-scale circulation models. These experiments are known under the names FIRE (First ISCCP Regional Experiment; Randall et al. 1996), EUCREX (European Cloud and Radiation Experiment; Raschke et al., 1989) and JACCS (Japanese Cloud and Climate Studies; Asano et al., 1992). They provided much new information on microphysical properties of clouds, in particular of the high cirrus. There were also other related investigations. Another initiative will start over the Arctics sea ice fields within the ACSYS (Arctic Climate Systems Study; WCRP 1994).

As an example, some most recent measurements on the vertical distribution of particle size spectra within a cirrus layer are shown in Fig. 1.6. They were obtained within an "elderly" frontal cirrus layer over the polar seas near the island of Svalbord and show a bimodal distribution, which also can be reconstructed with models for more explicit cloud microphysical processes.

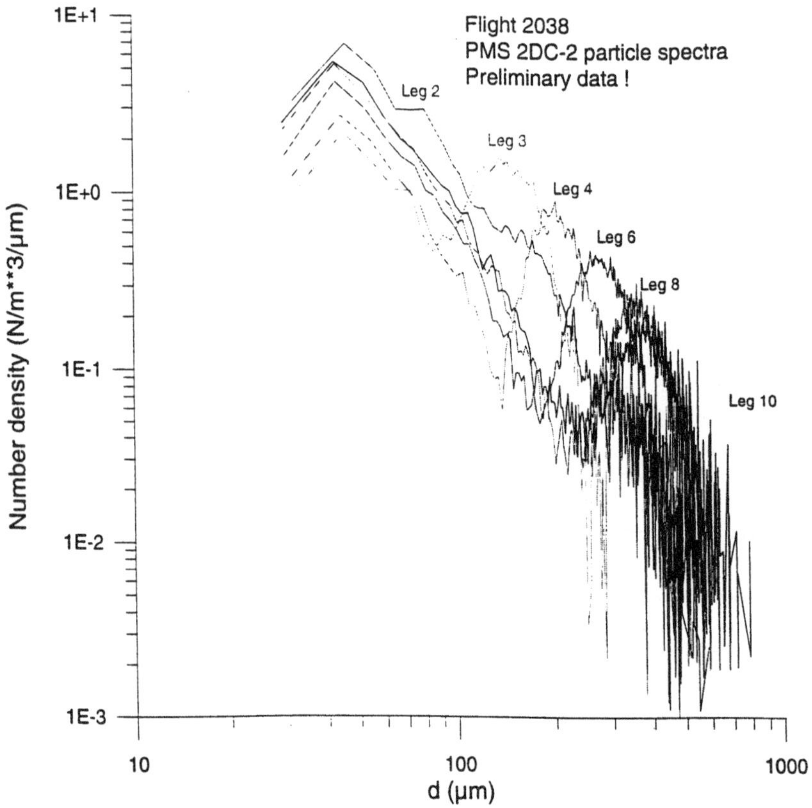

Figure 1.6: *Particle size distributions measured at different altitudes with aircraft within a cirrus layer (from W. Koch, 1995, private communication). The respective flight levels are from top downward at 7910, 7300, 6690, 5970, 4250 and 3040 meters above the Arctic ocean, which below the lowest leg was covered with cumulus fractus.*

The ISCCP makes so far use only of simultaneous measurements within the solar and thermal infrared portions of the spectrum and added some a-priori informations on the states of the atmosphere and ground. In the future, also the emission in some microwave bands will be added to obtain information on the water and ice content of clouds, which amounts to only a few percent of the total water content of the atmosphere. The horizontal resolution will enhanced from the present 250 km to about 30 km to show more regional detail (ISCCP-II).

The liquid water contained in clouds can be estimated from measurements of the emission in the microwave region. But they often contain larger errors due to the structure of cloud fields and the relatively wide footprint of about 30 to 50 km of such measurements, causing in many cases "beam-filling" problems.

All these passive satellite measurements failed so far to provide sufficiently accurate informations on the often very thin cirrus and its characteristics, in particular when it is located on top of lower cloud fields. Further, no information is directly obtained from them on the lower cloud boundaries and thus cloud thicknesses. Therefore, the space agencies are now investigating the feasibility for new satellites with backscattering and cloud profiling radar (working in the millimeter range of the spectrum) onboard. Once such data can be included into analysis schemes, the lower cloud boundary altitudes can be estimated within uncertainty ranges of about 300 m, which corresponds to a temperature uncertainty of 3 to 4 K, or in the downward longwave radiation of 5 Wm^{-2} (at a lower cloud boundary temperature of 0°C). The measurements from the recent Shuttle mission with a lidar onboard (LITE) has demonstrated that spaceborne backscattering lidar may even be able to identify the upper boundary of the planetary boundary layer, when moderately thick clouds are above.

1.5 Radiation Budget Components

(see also contributions by Del Genio, Rossow, Stuhlmann)

It is the meridionally nonuniform distribution of solar radiative energy, which forces all circulation processes within our climate system. Its average radiative power of 343 Wm^{-2} (i.e. one-fourth of the solar constant) shows some small variations of up to 1 to 2% due to dynamical processes within the Sun's photosphere and interior, of which the so-called 11-year cycle might be strong enough to effect processes within our climate system. Statistical analyses show some evidence in various connections (e.g.: Currie, 1995). The amplitude of the annual variation of insolation is due to the ellipticity of the earth's orbit around the Sun and amounts to 7% or about 23 Wm^{-2} between December and June.

This incident radiation, as schematically shown in Fig. 1.7, is only partly reflected back to space by clouds, the atmosphere and ground (in total about 30%, of which one half is solely due to clouds), while an amount of about 20% of it heats directly the atmosphere and clouds (see also the radiation flux divergence figure in Section 1.3 of this paper), and the remainder penetrates, however then spectrally considerably modified to ground. Fig. 1.7 relates these partitions to those of the terrestrial heat radiation and the exchange of sensible and latent heat between the atmosphere and ground, where the latter two amount to about 5% and 24%, respectively, of the solar radiation incident at the top of the atmosphere.

1.5.1 Radiation budget at top of the atmosphere

Regional distributions of all radiation budget components at the top of the atmosphere and at ground had been estimated in the pre-satellite age from the at that time known climatologies of clouds and their properties and the temperatures and other quantities as well (e.g.: London, 1957). Since the first meteorological satellites were launched in the sixties of this century, estimates of the radiation budget components have been made with different degrees of complexity

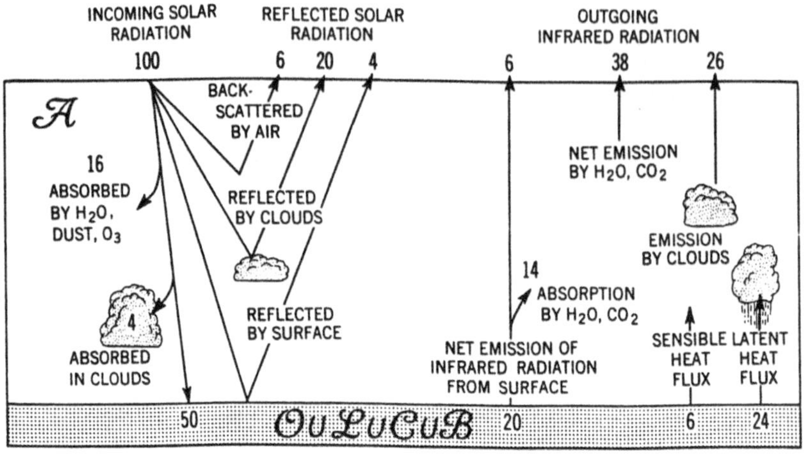

Figure 1.7: *Schematic of the annual global radiation budget components and the partitioning of cloud fields on their redistribution (from Peixoto and Oort, 1992).*

depending on the available data sets (e.g. Raschke et al., 1973, for a summary of all earlier measurements; Kandel et al., 1989; Stephens, 1979; Vonder Haar and Suomi, 1969). More recently Stephens et al. (1991 a and b; 1993) analysed large data sets from the ERBE and brought them into connections with components of the hydrological cycle in the atmosphere.

These spaceborne data already showed, that the earth is "darker" with albedo values around 29 to 30% and warmer with a mean emission temperature of about 252 K than previously thought on the basis of the above-mentioned conventional estimates (35%, 250 K, respectively). Such values for the globe, but also the regional distributions must be computed in simulations of the present climates (control runs).

The analysis schemes, developed sofar, have to perform the following three major operations with measured radiance data:

- conversion of spectrally limited to total radiances, i.e. correction for instrumental filtering.
- "integration" to convert radiance into outgoing radiative flux densities, taking into account the angular distribution of emitted / reflected radiation.
- consideration of diurnal variations, dependent on the number of daily available measurements.

Both of them depend on the measured "scene" (i.e. clouds, oceans, land surfaces with different radiative properties). Further "corrections" are required to overcome the diurnal variations of the outgoing radiation fields. They may be minimized increasing the number of satellites measuring simultaneously radiation budget components, as done for the Earth Radiation Budget Experiment (ERBE; Barkstrom et al., 1989). Several schemes add to the measurements of a single polar orbiting satellite, which may see each region of the earth only 2 to 3 times, the informations on cloud and temperature variations which are contained in the operational data of geostationary satellites. Various sampling studies showed that such methodologies could reduce systematic uncertainties to values of about 5 to 10 Wm^{-2} for daily averages (e.g.: Dieckmann and Smith, 1989).

Fig. 1.8 shows purposely as an example results which had already been derived from the measurements of the very early Nimbus satellites (from Raschke et al., 1973) instead of the many very recent results which are discussed by Stuhlmann (this volume). The geographical distribution and also amounts of the annual net radiation, resulting from the incident radiation and the global cloud and temperature distributions, and surface albedoes as well, show the same patterns as observed now with radiative surplus areas between about 40° northern and southern latitude and the known deficit regions poleward. However, at the first time, these data revealed also regional deficits over the subtropical bright and hot deserts over both hemispheres, which enforce compensating circulations (see their discussions by Sohn and Smith, 1992).

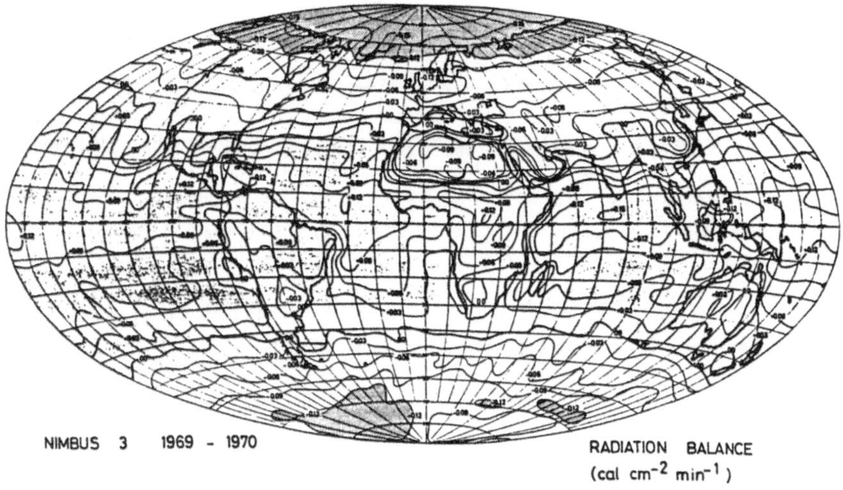

Figure 1.8: *Global distribution of the annual net-radiation budget at the top of the atmosphere, derived from the measurements of the satellite Nimbus 3, made in 1969 and 1970 (from Raschke et al., 1973).*

The energy transports which will be required to balance the deficits in the planetary radiation budget have frequently been estimated. Fig. 1.9 (from Sohn and Smith, 1992) shows an annual average, based on the ERB data derived from Nimbus 7 wide field-of-view measurements. It clarly indicates the dominating "source" over the tropical Pacific and "sinks" over the Sahara and central South America.

Various analyses of such data demonstrated also the magnitude and sign of the effect of clouds (often called "cloud forcing": see Stuhlmann, in this volume) on the outgoing radiation fields. In such estimates, in principle monthly averages of the albedo or longwave emission to space are subtracted from "clear-sky values" measured over each area during the same month but in many cases only during one or two days. These estimates bear many error sources. They however revealed, that high-level cirrus - when no other lower clouds are present - may slightly increase the radiative energy gain of the planet ("heating" it) while in the average all clouds reduce the net gain of radiative energy by about 20 to 25 Wm^{-2}. Thus , in general, clouds "cool" the planet Earth. Such cloud-effect analyses are also an excellent tool to validate model

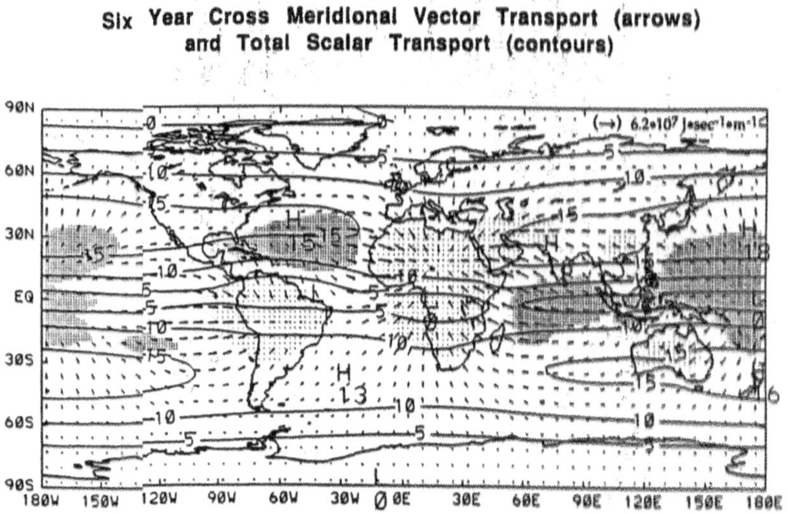

Figure 1.9: *Residual vectorial horizontal transport from the required zonal-mean transport (arrows) and the scalar horizontal total energy transport (contours) for a 6 year interval (April 1979 to March 1985). Units are in 10^7 J s^{-1} m^{-1} (from Sohn and Smith, 1992).*

results, simulated for the present climate. Wielicki et al. (1995) have summarized most recent investigations. Raschke et al. (1973) were possibly the first estimating the effect of clouds on the planetary albedo; they also could identify the drastic decrease of the surface albedo over the Arctic sea ice during the summer season.

1.5.2 Radiation budget at the surface

An accurate knowledge of these components of the heat budget of the earth is necessary to validate numerical model output for the present climate, but to estimate also the effect of related processes, such as the evapotranspiration over continents and evaporation over oceans. Since the networks of reliable stations of the different countries are very sparse, and do not meet the required spatial density which could be estimated from spatial variabilities of cloud and temperature informations (see the pioneering work by Gandin, 1963), and since no stations at all exist over the oceans, whose temperature may react sensitively to trends in the radiation budget components at the ocean surface, there is a need to derive such information from the available satellite based data sets. Control of such new information is possible at least over continental areas with data from operational networks and the international Baseline Station Radiation Network (BSRN; Whitlock et al., 1995), which has been established within the framework of the WCRP. Automatic stations for the downward radiation components should be installed onboard of "ships of opportunity".

During the past years, several methods have been developed to estimate the components of the radiation budget at ground with satellite and other ancillar data. We can scetch here their principle only and refer the reader to the papers by Stuhlmann and Rossow in this volume.

The surface albedo and temperatures can only be estimated from "clear-sky" measurements, with specific corrections for atmospheric effects. Over continents both depend also on the cloudiness and are, therefore, not very accurate. Over the oceans, both properties are very stable within small uncertainty limits. More recent operational analyses of the sea surface temperatures, making use of the multi-spectral capabilities of the NOAA-AVHRR, reach error limits of about 1K and even less. These are excellent accuracies to monitor anomalies in the surface temperature, which are related to oceanic circulation phenomena.

Thus greater care has been given to the derivation of downward fluxes of short- and longwave radiative energy which according to some sensitivity studies should be known within limits of less than 10 Wm^{-2} for monthly averages over areas of about 100×100 km^2.

The downward atmospheric radiation has been shown to be determined mainly by the temperature and humidity (water content!) of the lowest layers of the troposphere (up to about 1.000 to 2.000 m), and on the altitude and temperature of cloud bottoms (e.g.: Schmetz, 1989). Their relative influence increases with decreasing water content; thus it is very large over polar regions and almost negligible in the tropics. The more operational methods of the work within the WCRP use therefore operational analyses of the temperature and moisture fields as "predictors". Kornblueh (1995, private communication) tries an additional estimate of the cloud bottom height using a simplified assimilation scheme and obtained in case studies an agreement within 50 m comparing his data with simultaneous lidar measurements (over the Arctic ice). Fig. 1.10 shows some of Kornblueh's results.

It is believed, that only active remote sensing with lidar and cloud profiling radar, the latter measuring at 3 or 8 mm wavelength, will improve the situation. The GEWEX Science Steering Group has therefore established a working group to support and critically review ongoing activities in Europe, USA and Japan. An example of such measurements, made from ground is shown in Fig. 1.11.

The downward solar radiation can in principle be estimated using the cloud albedo as an indicator for the cloud optical thickness (e.g.: Möser and Raschke, 1983). Various procedures have been developed in the past years and applied to regional and also global data sets (see contributions of Stuhlmann and Rossow, this volume). They yield uncertainties - as compared to the often quite inaccurate ground-based data - of 5 to 10% of monthly averages, i.e. about 10 to 20 Wm^{-2}. Another, and possibly simpler approach makes use of the vertical profiles of solar net radiation in each layer of the atmosphere (Li et al., 1993), to be estimated with the cloud information from satellite data. But here also some information on the surface albedo is required (see comments made above). Such retrievals require steady control by measurements from stations, which are representative for areas of about 10 to 50 km extent, dependent on the spatial scale of available satellite data. The BSRN-data (e.g. Whitlock et al., 1995) might be able to meet these requirements, although the location of the individual stations has not been chosen according to the representativeness for larger areas.

1.6 Precipitation
(see also sections by DelGenio, Hallikainen, Joe, Noll, Stewart, Simmer)

Precipitation in form of rain or snow falls out of mostly unstable clouds; thus most precipitation events should be expected in regions of the earth with convective activity and synoptic-scale activities. But also orography plays a major role to destabilize the flow approaching it and causes precipitation up to a certain altitude, where clouds cannot carry large amounts of water. Various global precipitation climatologies have been established earlier on such informations, on ground-based measurements in inhabitated areas and a variety of ship reports (e.g.: Hulme, 1995; Jäger, 1976; Legates, 1995). Precipitation maps show patterns which are closely related to the instabilities mentioned above. An example is shown in Fig. 1.12. Such still "conventional"

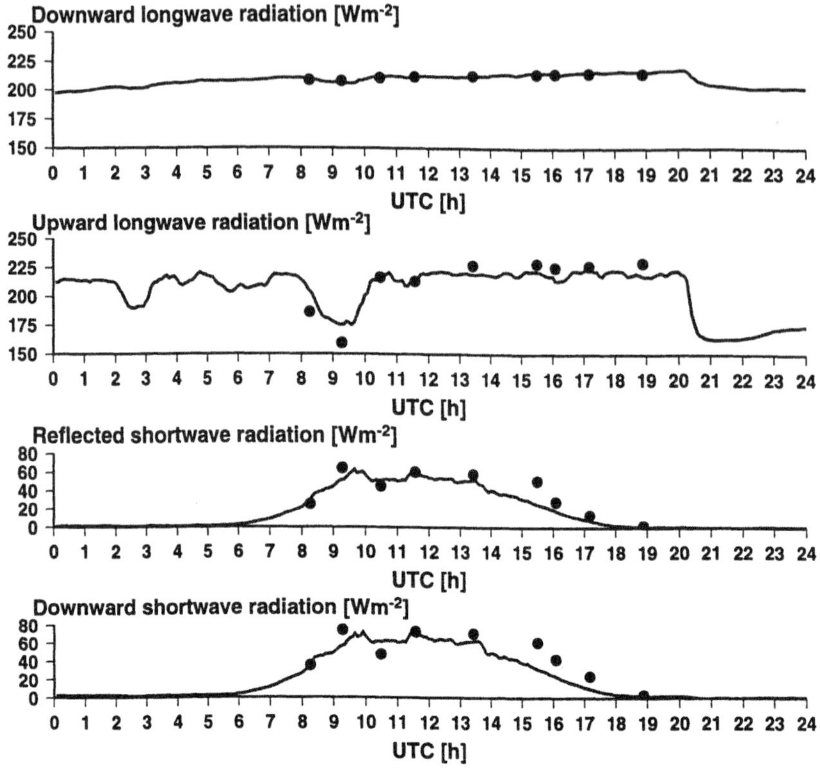

Figure 1.10: *Comparison of surface radiation fluxes which have been retrieved from the conventional analyses and from passive satellite data with measured values over the Arctic ice (from Kornblueh, 1995, private communication), where cloud bases have additionally been measured with lidar.*

analyses of the precipitation over oceans are not worse than the estimates made from the very recent satellite measurements (e.g.: Huffmann et al., 1995), since the latter are based on many assumptions and require steady validation.

1.6.1 Ground-based gauge measurements

The various operational and often also community-owned or commercial networks over continental areas are in many cases not very reliable due to systematic errors of the different rain gauge types and their exposition, which have been investigated and some corrections (amounting in some cases up to recommended 20% increase of the measured values!) are suggested (e.g.: Sevruk, 1982). Legates' (1995) and Hulme's (1995) summaries of such analyses are quite discouraging in this respect, since the present ground-based networks might be the only means to "calibrate" other precipitation data derived from radar and satellite measurements. Despite of such uncertainties, they are at present the only means to search for expected trends or variations in the amount and patterns of precipitation. Some recent trend analyses by Hulme (1995)

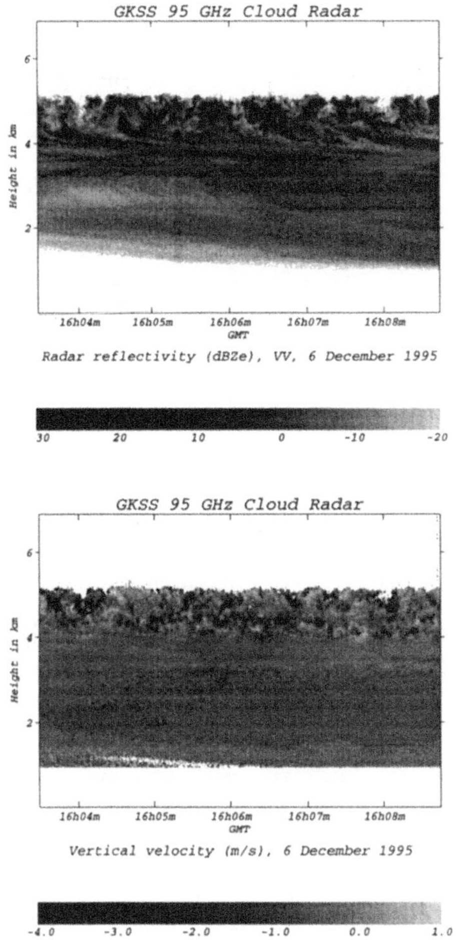

Figure 1.11: *Time-series of returns of radar signals (3 mm), scattered downward from a dense cloud layer reaching from ground to about 5 km altitude (from Quante et al., 1996). Note the convective mixing in its top layers.*

are shown in Fig. 1.13. They indicate even a small increase within increasing temperatures near ground (Fig. 1.14).

Various possibilities of shipborne precipitation measurements and related problems are discussed by Petty (1995). They mostly arise from the fact that a ship changes the atmospheric flow. There are indeed now developments underway of automatic precipitation recorders, whose principles are based on the backscattering of optical, radar and sound waves by precipitation

Figure 1.12: *Annual global distribution of precipitation as estimated from ground-based measurements (from Legates, 1995).*

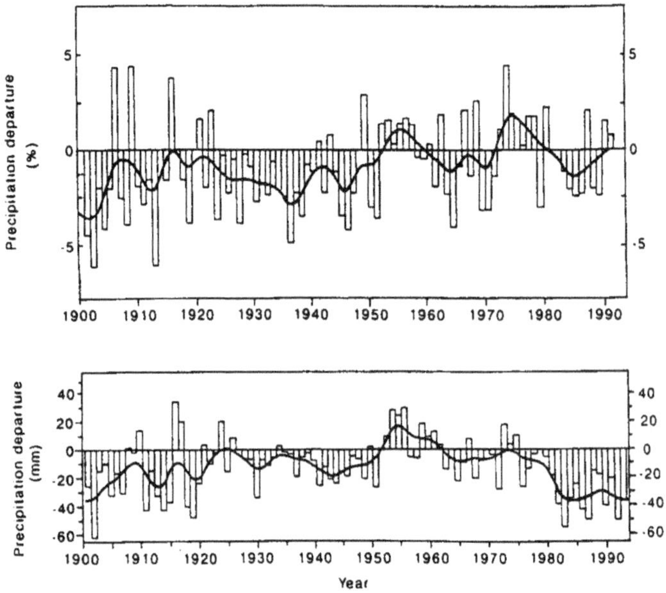

Figure 1.13: *Precipitation trends between 1900 and 1992 (from Hulme, 1995).*

particles (e.g. Sheppard, 1990; using speedometers), which even might be a few meters away from the instrument and thus be only moderately effected by the ship's effect on the atmospheric wind field. A large test series is now made within the framework of the Baltic Sea Experiment (BALTEX), a regional subprogramme of the GEWEX, where several optical probes are mounted on ferry boats which are operationally crossing the Baltic sea.

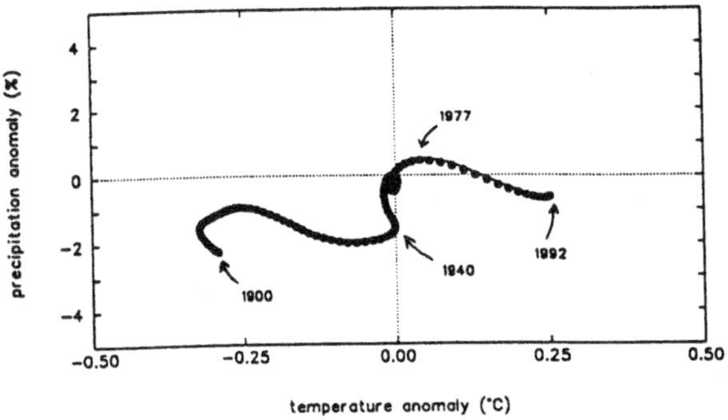

Figure 1.14: *Possible dependence of the precipitation trends shown in Fig. 1.13 on the mean global temperature increase (Hulme, 1995).*

Snow fall measurements are made also operationally at many surface stations with various means, but they require corrections for both cases, thowing over weakly frozen ground below and blowing snow. For instance, it appears impossible to measure the snow falling over the Arctic sea ice. Thus, within the ACSYS (Arctic Climate System Study) an improvement of the present analyses of the divergence of net-water transports into and within this region is encouraged. Such values provide estimates of the "net precipitation" - snow fall minus evaporation.

1.6.2 Measurements from satellites

At all, there is a need to establish reliable global and regional statistics on the precipitation, where also space-borne measurements play an important role, since they can be obtained over almost all portions of the globe. During the past decades various possibilities have been investigated.

One of them is based on the fact, that in most regions with convective instability the amount of precipitation is quite proportional to the vertical mass flow. This led to relevant estimates using consecutive infrared imagery, which is available from geostationary satellites. Others "calibrated" measured infrared window-radiances (i.e. measurements of the emission at 11 μm), below a theshold value, versus some observations (e.g. from weather ships) or other estimates. Such precipitation indices are still operationally calculated over the whole earth between about 60° northern and southern latitude.

Passive multispectral radiometry in the spectral region of microwaves has been demonstrated already by K.J.K. Buettner in 1964 to provide useful information on the precipitation rate (or better: the potential of it, since the radiometer cannot dicriminate between precipitation inside and outside of clouds) of cloud systems over the oceans. This method is based on the fact, that the clear ocean reflects back to space a large fraction of the "cold" space, while precipitating clouds with large and in particular warm droplets appear then above this cold area as "warmer" spots, whose radiance to space can directly be related to the water content on the basis of detailed radiative transfer calculations. Some comparison with shipborne observations of precipitation are made to "calibrate" the retrieved values. Serious errors are due to the temporal sampling and the "beam filling" by precipitation cells, which are in many cases smaller in size than the beam of sensors in space, which at present amount to about 30 km. Further details are described by Simmer (this volume).

Since at present such observations are made only from one single polar orbiting satellite, such precipitation values contain larger sampling errors. The Global Precipitation Data Center (GPDC) of the WCRP collects all such data and merges them with continental based gauge data and the results of numerical forecasts, for the use of comparison with model results. Fig. 1.15 shows one of their products.

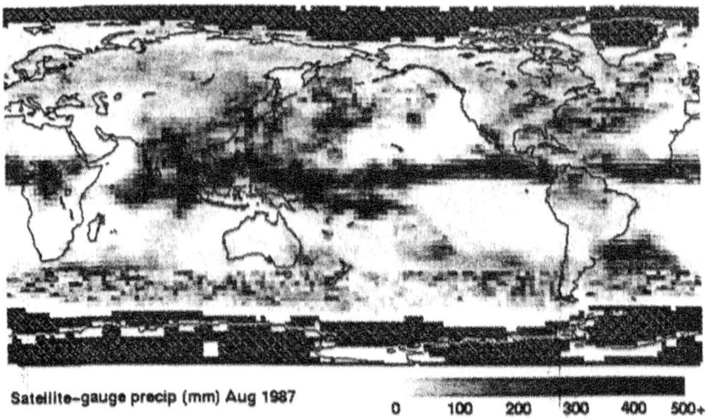

Figure 1.15: *Annual precipitation over the globe, as derived from satellite and ground-based measurements by the Global Precipitation Data Center in Offenbach (Huffman et al., 1995) for the validation of model simulations. This map shows all significant precipitation belts which are known. However, intensive validation - in particular over oceans - is required.*

Snow fall cannot be measured nor observed yet from satellites, but later, when the snow has fallen, fresh snow fields may be identified by the albedo and temperatures. Groundbased data on the snow field thickness and satellite data on the extent help the water authorities in remote areas to estimate the water equivalent, which later may be available for floods and groud water restoration. Snow thickness estimates from microwave radiometry bears many difficulties due to effects from the ground below (see Hallikainen in this volume).

1.6.3 Remote sensing at ground

Intensive ground truth is also required over the oceans, as it is envisaged also for other weather elements. Therefore some developments are underway to monitor precipitation amounts or intensities automatically with remote techniques which use the visible light of simple but robust lidar (e.g.: Hasse et al. (1995) developed such an instrument which is now installed onboard of several ferry boats crossing the Baltic Sea). Others used inexpensive radar used by the police to control the speed of traffic or even of a miniature sodar. These and possibly other developments should be tested to establish as soon as possible a "network of ships of opportunity" providing at least some control data for a validation of analyses from satellite measurements.

At the end of this section we also should recognize the challenging opportunities which ground-based, airborne and space-borne radar can provide. Ground-based rain radar is now the back-bone of precipitation observations in many countries. The USA installed during the last years an almost nationwide network of digital Doppler-radar which allows a continentalwide picture on precipitation events and intensities, their convective activities and horizontal movements. Such information is of enormous economic importance, for instance to minimize the damage by strong events, such as thunderstorms or tornadoes or to manage the water resources. A similar network already exists also over Western Europe and the Nordic countries.

But does even such a dense network, even over plain and uniform land "measure" the amount of water reaching the ground?? Collier and Knowles (1995), Joss and Waldvogel (1990), Sauvageot (1994) and Zawadski (1982) and others point out that the inversion of even such multi-informative signals is dependent on various assumptions on the properties of the falling rain, in particular on the droplet size distributions and on the falling speed. Ground glitter and other sources may deteriate the signals and their information contents. Further, more distant rain events can only be seen several hundred meters above ground, where the falling rain has still some chance to evaporate while falling to the ground.

Nevertheless such information is of vital importance for weather analyses and shortrange forecasts (in particular of hazardeous events) and also for many hydrological applications. To interpolate between the ranges of validity of the various radar stations, there are also attempts to "blend" the radar data with measurements of the geostationary satellite Meteosat. Such products and their values are discussed by Browning and Collier (1989).

Despite of these still existing uncertainties a new satellite (TRMM: Tropical Rainfall Monitoring Mission) will soon be launched into a near-equatorial orbit to measure with radar and passive multispectral radiometers precipitation events and their intensity over the near equatorial regions. It is an urgent challenge for the numerical modellers to assimilate such information into their analysis schemes and forecast models to improve our knowledge on the quantitative precipitation over oceans and continents. Further details on data assimilation and modelling problems are given by Courtier and DelGenio, respectively (this volume).

1.7 Evaporation and Evapotranspiration

(see also contributions by Bastiaansen, Hallikainen, Katsaros, Kiely, Paloscia, Schlüssel)

Regional values of the evaporation over sea and the evapotranspiration over continental surfaces are required to estimate and to validate model results on the energy and water cycles over regions or to derive within dedicated process studies parametrization schemes for use in global models. A quite complete account of the complexity of such research is given in a recent review by Parlange et al. (1995).

The evaporation from water surfaces or even just from flat and bare soil cannot be measured directly, rather it is derived from vertical flux (of momentum, sensible and latent heat) measurements, and then related to easier measurable quantities like the "moisture deficit" above ground, wind speed and the net radiation budget at ground levels. These principles led to reliable values over many areas, provided the soil moisture over the continents was known. Valuable assistance came here from modelling of the atmospheric boundary layer over various surface types.

The evaporation estimates over oceans, as reported in details by Schlüssel (this volume) make use of such relations and derive the required input information from satelite measurements of the sea surface temperature, near surface winds (from roughness of the sea) and moisture. A simplified formulation of the boundary layer dynamics helps to estimate the vertical water vapour fluxes. Unfortunately no recent "ground-truth" experiment is known to validate such data, although urgently required. Many studies are made with most available shipborne measurements and summarized in the "Bunker-Atlas from the Northern Atlantic Ocean" by Isemer and Hasse (1987). Most recent analyses are summarized by Da Silva et al. (1994). An example of evaporation data as derived from such ship observations is shown in Fig. 1.16.

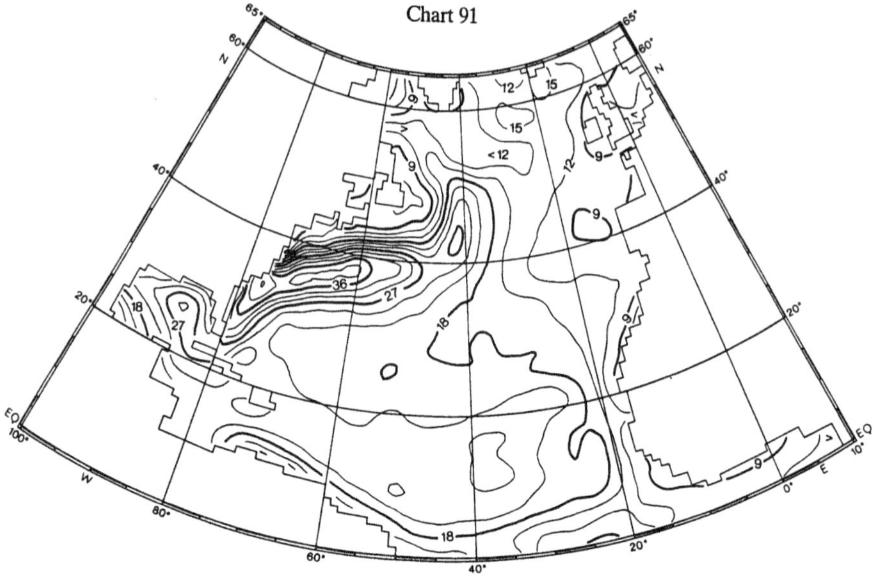

Figure 1.16: *Evaporation over the North Atlantic Ocean as derived from shipborne measurements (from Isemer and Hasse, 1987).*

Evapotranspiration over vegetated continental surfaces, consisting in many cases of two components: the evaporation from soil surfaces and the transpiration from the vegetation, is now estimated from models simulating the vertical water flow in vegetation and the soil. There are many complex formulations now published and studied in a comparative way within the PILPS project (Henderson-Sellers, 1995). Data sets have already been collected within larger field projects such as the BOREAS (Sellers et al., 1995), EFEDA (Bolle et al., 1993), HAPEX (André et al., 1989), FIFE (Sellers et al., 1992) and others and used for the development of parametrization schemes to be used in climate and weather forecast models. More will be done within the framework of the GEWEX regional experiments, to improve the coupling between atmospheric and hydrological models.

Are there opportunities to independently map the evapotranspiration over continental areas? Possibly there are no such opportunities, since no measurement principle is known to provide under all weather conditions vertical profiles of atmospheric moisture or temperature and of the momentum fluxes in the layers in close neighborhood to the ground. Nor do we often not know the moisture of the soil in the near surface layer (which may lead to a rapid recycling of water and a precipitation enhancement downstream) of the water supply within the root zones. Only a few components are sometimes, and not at all weather conditions, observable, and need then fed into a model formulation.

It is hoped, that these deficits for the soil moisture can be overcome by multispectral SAR-measurements in the X- and L-bands at 3 cm and 15 cm wavelength, respectively, where the reflection is strongly effected by the water content of soil. But we need to control the model results steadily. This may only be possible installing instrumentally "controlled" areas of the size of one or several grid element of an operational weather forecast model, where steady and quality-controlled measurements are performed. One of them is now in development at the Meteorological Observatory of the German Weather Service, near Lindenberg (about 50 km south-east of Berlin).

1.8 Runoff

The runoff in rivers cannot be directly monitored with instruments onboard of satellites. But satellites can very efficiently be used as relay stations from many automatic river gauges. Such networks are now in preparation for the major river systems in developing countries and may become even more common to centralize such data streams and control their quality from the beginning, since they are of global interest.

Occasional observations of flooded river beds have been reported (and illustrated with often fine color images) using the imaging data of earth resources satellites, such as Landsat or the European ERS-1, which are made in the visible and microwave portions of the spectrum (e.g. Gunley and King, 1995). They provided immediate and areal informations for the management of such disasters.

1.9 Some Conclusive Remarks

This brief overview on our present knowledge on the various radiation and water flows in our climate system, which are components of energy and water cycles, and on the present possibilities of spaceborne observations has shown that despite of the already advanced developments of such techniques the backbone of such observations may in some case still remain with surface-based and also airborne direct measurements.

As a consequence, ground-based networks must be maintained and kept at highest technological levels to provide best data quality. It will remain to the data assimilation techniques, which have successfully been developed and applied for several purposes in meteorology and oceanography (see contribution by Courtier in this volume), to combine the various data streams into a useful frame. On the other side many operational ground-based networks begin to erode often due to the false belief, that they can (in part) be replaced by space-borne observations.

Furthermore, the need for a steady control of quantities, such as the precipitation or radiation budget over the vast ocean areas has in most publications on these quantities been overlooked, due to the fact that related measurement techniques need to be developed and installed onboard ships of opportunity. And one also may not overlook the more political fact, that the space engineers and politicians neglected for many reasons the fact that the development of space technology requires also adequate investments into the scientific background and ground-based networks.

1.10 References

André J-C et al. (1989) Evaporation over land-surfaces: First results from the HAPEX-MOBILHY special observing period. *Ann Geophys* 6: 477–492

Asano S and the JACCS/MRI Research Groups (1994) Japanese Cloud and Climate Study (JACCS): Research Plan and Preliminary Results. *Proceed. 8 th Conference Atmosph. Radiation, Nashville.* Available from Am. Meteor. Soc., Boston, Mass

Barkstrom B, Harrison EF, Smith GL, Green RN, Kibler J, Cess RD and the ERBE Science Team (1989) Earth Radiation Budget Experiment (ERBE) archival and April 1985 results. *Bull Amer Meteor Soc* 70: 1254–1256

Bolle HJ et al. (1993) EFEDA-European field experiment in a desertification-threatened area. *Ann Geophys* 11: 173–193

Browning KA, Collier CG (1989) Nowcasting of precipitation systems. *Rev Geophys* 27: 345–370

Chahine MT (1992) The hydrological cycle and its influence on climate. *Nature* 359: 373–380

Collier CG, Knowles JM (1986) Accuracy of rainfall estimates by radar, Part III: Application to short term forecasting. *Journ Hydrol* 83: 235–249

Cubasch U, Waszkewitz J, Hegerl G, Perlwitz J (1995) Regional Climate hange as simulated in time-slice experiments. *Climatic Change* 31: 273–304

Currie RG (1995) Luni-Solar and Solar Cycle 10-11 Year Signals on Chinese Dryness and Wettness Indices. *Int Journ Climatol* 15: 497–516

Darnell WL, Staylor WF, Gupta SK, Ritchey NA, Wilber AC (1992) Seasonal variation of surface radiation budget derived from International Satellite Cloud Climatology Project C1 data. *J Geophys Res* 97: 15.741–15.760

Da Silva A, Young CC, Levitus S (1994) Atlas of Surface Marine Data. Vol. 4: Anomalies of Fresh Water Fluxes. NOAA-Atlas NESDIS 9, 308 pp

Desbois M, Désalmant F (1995) Past and Present Precipitation Related to Climate Changes. Methodologies and Perspectives. *Bull Am Meteor Soc* 76: 1173–1176

Dieckmann F, Smith GL (1989) Investigation of Scene Identification Algorithms for Radiation Budegt Measurements. *Journ Geophys Res* 94 D3: 3395–3412

Ellingson RG, Fouquart Y (1991) The Intercomparison of Radiation Codes in Climate Models: An Overview. *Journ Geophys Res* 96 D5: 8925–8927

Gaffen DJ, Elliot WP, Robock A (1992) Relationships between tropospheric water vapor and surface temperature as observed by radiosondes. *Geophys Res Letters* 19: 1839–1842

Gandin LS (1963) Objective Analysis of Meteorol. Fields. Gidrometeorolichesko Isdatelst'vo. Tranl. Israel Programme for Scientific Translations. Jerusalem. 242 pp

Gates WL (1992) AMIP: The Atmospheric Model Intercomparison Project. *Bull Amer Meteor Soc* 73: 1962–1970

Gunley LE, King MD (1995) Remote Sensing of Flooding in the US. Upper Midwest During the Summer 1993. *Bull Am Meteor Soc* 76: 933–944

Hahn CJ, Warren SG, London J (1995) The effect of moonlight on observation of cloud cover at night, and application to cloud climatology. *Journal of Climate* 2: 1429–1446

Hartmann DL (1994) Global Physical Climatology. Int. Geophys. Series, Vol. 56, *Academic Press*, 408 pp

Hasse L, Grossklaus U, Isemer HJ, Uhlig R (1994) New ship rain gauge. In Instruments and Observing Methods. Report No. 57. WMO, Geneva, WMO/TD 588, p 97–101

Hastenrath S (1982) On meridional heat transports in the world ocean. *Journ Phys Oceanogr* 10: 159–170

Henderson-Sellers A, Pitman AJ, Love PK, Ivannejard P, Chen TH (1995) The Project for Intercomparison of Land Surface Parametrization Schemes (PILPS): Phase 2 and 3. *Bull Am Meteor Soc* 76: 489–504

Huffmann GJ, Adler RF, Rudolf B, Schneider U, Keehn PR (1995) Global Precipitation Estimates Based on a Technique for Combining Satellite Based Estimates, Rain Gauge Analysis and NWP Model Precipitation Information. *Jour of Climate* 8: 1284–1295

Hulme M (1995) Estimating global changes in precipitation. *Weather* 50: 34–42

IPCC (1996) Intergovernmental Project on Climate Change - Report to be issued by the WMO in Geneva

Isemer HJ, Hasse L (1987) The Bunker Atlas for the North Atlantic Ocean. Vol. 2: Airsea interactions. *Springer Verlag*, 252 pp

Jäger L (1976) Monatskarten des Niederschlags für die ganze Erde. Ber. D. Deutschen Wetterdienstes, 18, 139, DWD Offenbach, 38 pp

Joss J, Waldvogel A (1990) Precipitation measurements and hydrology. *Radar in meteorology: Battan Memorial and 40th Anniversary of Radar Meteorol. Conf., ed. by D Atlas, Amer. Meteor. Soc.*, 577–606

Kandel RS, Monge J-L, Viollier M, Pakhomov LA, Adasko VI, Reitenbach RG, Raschke E, Stuhlmann R (1994) The ScaRaB project: Earth radiation budget observations from the meteor satellite. *Adv Space Res* 14: (1)47-(1)54

Kumar A, Hoerling MP (1995) Prospects and Limitations of Seasonal Atmospheric GCM Predictions. *Bull Am Meteor Soc* 76: 335–345

Legates DR (1995) Global and regional precipitation; A comparative asessment of existing climatologies. *Intern Journ Climatol* 15: 237–258

Li Z, Leighton HG, Masuda K, Takashima T (1993) Estimation of SW Flux Absorbed at the Surface from TOA Reflected Flux. *Journ Climate* 6: 317–330

London J (1957) A study of the atmospheric heat balance. *Final Rept*, Contr. AF 19 (122-165), Res. Div. Coll. Eng., New York University, 99pp

Lovejoy S, Schertzer D, Watson B (1992) Radiative Transfer and Multifractal Clouds: Theory and Applications. *In Int Radiat Sympos*, 1992, A. Deepak Publ

Manabe S, Möller, F (1961) On the radiative equilibrium and heat balance of the atmosphere. *Monthly Wea Rev* 89: 503–532

Möller F (1961) Atmospheric Water Vapor Measurements at 6-7 Microns from a Satellite. *Planet Space Science Rev* 5: 202–206

Möser W, Raschke E (1983) Mapping of global radiation and cloudiness from METEOSAT imaging data. *Meteor Rundsch* 36: 33–41

Oliver HR, Oliver SA (1993) *The Role of Water and the Hydrological Cycle in Global Change.* NATO-ASI Series I, Vol 31, Springer Verlag, 468 pp

Parlange MB, Eichinger WE, Albertson JD (1995) Regional Scale Evaporation and the Atmospheric Boundary Layer. *Rev Geophys* 33: 99–124

Peixoto JP, Oort AH (1992) *Physics of Climate.* Am. Institute of Physics. 520 pp

Petty GW (1995) Frequency and Characteristics of Global Oceanic Precipitation from Shipboard Present Weather Reports. *Bull Am Meteor Soc* 76: 1593–1616

Randall DA, Albrecht BA, Cox SK, Johnson D, Minnis P, Rossow WB, Starr D O'C (1996) On FIRE at Ten. *Advances in Geophys*, in press

Raschke E, Vonder Haar TH, Bandeen W, Pasternak M (1973) The annual radiation balance of the earth atmosphere system during 1969–1970 from Nimbus 3 measurements. *Journ Atmos Sci* 30: 341–364

Raschke E, Schmetz J, Heintzenberg J, Kandel R, Saunders R (1990) The International Cirus Experiment (ICE) - A Joint European Effort. *ESA Journal* 14: 193–199

Raschke E, Jacob D (ed.) (1993) *Energy and Water Cycles in the Climate system.* NATO-ASI I, Vol. 5, Springer Verlag, 467 pp

Read WG, Watres JW, Flower DA, Froidvaux L, Jacob RF, Hartmann DL, Harwood RS, Rood RB (1995) Upper-Tropospheric Water Vapor from UARS/MLS. *Bull Am Meteor Soc* 76: 2381–2389

Rosenfeld D, Wolff DB, Amitai E (1995) Classification of rain regimes by three- dimensional properties of reflectivity fields. *Journ Appl Meteor* 33: 682–693

Rossow WB, Schiffer RA (1991) ISCCP cloud data products. *Bull Am Meteor Soc* 72: 1–20

Sadler JC, Oda L, Kilonsky BJ (1976) *Pacific Ocean Cloudiness from Satellite Observations.* Univ. Hawaii Tech. Rept. Meteorol., Honolulu

Sauvageot H (1994) Rainfall estimates by radar: a review. *Atmospheric Research* 35: 27–54

Schmetz J (1989) Towards a surface radiation climatology: Retrieval of downward irradiances from satellites. *Atmos Res* 23: 287–321

Schmetz J et al. (1995) Monthly Mean Large-Scale Analyses of Upper-Tropospheric Humidity and Winf Fields from Three Geospatioary Satellites. *Bull Am Meteor Soc* 76: 1578–1584

Schmitz WJ Jr (1995) On the interbasin-scale thermohaline circulation. *Rev Geoph* 33: 151–174

Sellers PJ, Hall FG, Asrar G, Strebel DE, Murphy RE (1992) An overview of the First International Satellite Land Surface Project (ISLSCP) Field Experiment (FIFE). *Journ Geophys Res* 97: 18.354-18.371

Sellers P et al. (1995) The Boreal Ecosystem-Atmosphere Study (BOREAS): An Overview and Preliminary results from the 1994 Field Year. *Bull Am Meteor Soc* 76: 1549–1577

Sheppard B (1990) Measurements of Raindrop Size Distributions using a Small Doppler Radar. *Journ Atm Ocean Technol* 7: 255–268

Smith EA, Sohn B-J (1990) Surface Forcing of Interannual Variations in the Radiation Balance over North Africa. Part 1: Partitioning the Surface and Cloud Forcing. *Climatic Change* 17: 147–192

Sohn B-J, Smith EA (1992) Global Energy Transports and the Influence of Clouds on Transport Requirements - A Satellite Analysis. *Journ of Climate* 5: 717–735

Stephens GL (1990) On the relationship between water vapor over the oceans and the sea surface temperature. *J Climate* 3: 634–645

Stephens GL, Greenwald TJ (1991) (a,b): Observations of the Earth's radiation budget in relation to the atmospheric hydrology (I, II). *J Geophys Res* 96: 15.311–15.324 , 15325–15340

Stephens GL, Randall DA, Wittmeyer IL, Dazlich DA (1993) ... (III), Comparison of Observations over the Oceans with a GCM. *J Geophys Res* 98: 4931–4950

Ström J, Busen R, Quante M, Guillemet B, Brown PRA, Heintzenberg J Pre-EUCREX intercomparison of airborne humidity measuring instruments. *Journ Atmos Ocean Tech* 11: 1392–1399

Stull RB (1988) An Introduction to Boundary Layer Meteorology, Acad. Science Library, *Kluwer Academic Publ*, 536 pp

Vonder Haar TH, Suomi VE (1969) Satellite observations of the Earth's radiation budget. *Science* 163: 667–669

Ware R, Exner M, Feng D, Gorbunov M, Hardy K, Herman B, Kuo Y, Meehan T, Melbourne W, Rocken C, Schreiner W, Sokolovsky S, Solheim F, Zhou X, Anthes R, Businger S, Trenberth K (1996) GPS Sounding of the Atmosphere from Low Earth Orbit: Preliminary Results. *Bull Am Meteor Soc* 77: 19–40

WCRP (1990) Scientific Plan for the Global Energy and Water Cycle Experiment. WCRP-40. Available through WMO in Geneva.

WCRP (1994) ACSYS

Webster PJ (1994) The role of hydrological processes in ocean-atmosphere interactions. *Rev Geophys* 32: 427–476

Wielicki BA, Cess RD, King MD, Randall DA, Harrison EF (1995) Mission to Planet Earth: Role of Clouds and Radiation in Climate. *Bull Amer Meteo Soc* 76: 2125–2153

Whitlock CH et al. (1995) First Global WCRP Shortwave Surface Radiation Budget Datasets. *Bull Amer Meteo Soc* 76: 905–922

Zawadski I (1982) The quantitative interpretation of weather radar measurements. *Atmos-Ocean* 20: 158–180

Chapter 2

Observational Requirements for Modeling of Global and Regional Climate Change

Anthony D. Del Genio
NASA Goddard Institute for Space Studies
2880 Broadway
New York, NY 10025
USA

2.1 Introduction

Predicting the rate and geographic distribution of future climate changes caused by anthropogenic perturbations such as increasing greenhouse gas and aerosol concentrations is perhaps the most important single problem in the earth sciences today. General circulation models (GCMs) are the tools of choice for making such projections. Simulating climate change with GCMs has become a worldwide activity; 12 modeling groups thus far have performed equilibrium doubled CO_2 calculations with atmosphere-only GCMs, and 8 are simulating transient greenhouse gas increase scenarios with coupled ocean-atmosphere models for the 1995 IPCC scientific assessment.

Unfortunately, the predictions of the various modeling groups differ widely: equilibrium warmings range from 1.7 to 5.3°C, and the rate and geographic distribution of warming depends on the type of ocean representation used and the type of climate forcing assumed (e.g., whether effects of tropospheric aerosols are included in addition to greenhouse gases). In fact, it has not been proven that decadal climate change is predictable; in simple models of the coupled system, multiple equilibria are easily produced with slightly different initial conditions (Marotzke and Willebrand, 1991). This latter aspect of climate change requires a fundamental breakthrough in our understanding of the coupled climate system, and it does not appear to be particularly amenable to insight using observations of the current climate. But the issues of how to specify accurate climate forcings and simulate realistic feedbacks as responses to these forcings can definitely benefit from a thoughtfully designed observing strategy.

In this paper we address the question of how data sets and models can be used together to ultimately produce more accurate predictions of climate change (within the constraints set by the unknown predictability of the system). Section 2.2 describes the current status of climate models and model-data comparisons as they pertain to climate change prediction. Section

NATO ASI Series, Vol. I 45
Radiation and Water in the Climate System:
Remote Measurements
Edited by Ehrhard Raschke

2.3 discusses several physics-based approaches to the question of predicting regional aspects of climate change. Section 2.4 then outlines a multi-tiered observational strategy for the eventual accurate modeling of climate change based on validation, physical process, and monitoring components. Section 2.5 concludes with some thoughts on the future assessment of whether climate models are good enough for their intended applications.

2.2 Myths and Methods of Climate Model Diagnosis

Satellite data sets currently in existence provide good integrated measures of several important climate parameters such as top-of-the-atmosphere (TOA) shortwave and longwave radiation fluxes, total cloud cover, and vertically integrated precipitable water vapor. These serve in principle as zeroth order validation tools for climate GCMs. It is possible with appropriate selection of free parameters in climate model parameterizations to match the globally integrated values of such quantities as well as some features of their geographical distributions, but this by itself tells us nothing about the utility of the models for predicting climate change. Evidence of the degree of tuning inherent in climate models is provided by their wildly varying estimates of hydrologic quantities: In a recent intercomparison of 19 models, July cloud cover ranged from 40 to 72% (Cess et al., 1990), while column precipitable water varied from 17 to 27 mm (Randall et al., 1992).

Much of this tuning is designed to provide realistic looking simulations of top-of-the-atmosphere radiation budgets, because of the perception that this is the most fundamental and well-observed climate quantity. Nonetheless, some models are out of global radiation balance by more than the 5–7 Wm^{-2} apparent uncertainty of existing satellite estimates. Furthermore, individual components of the TOA and surface energy budgets may vary by as much as 50-60 Wm^{-2} among models (Table 2.1).

Parameter	Current climate	Climate change
Cloud cover (%)	40/72	-4.4/-0.2
TOA radiation flux (Wm^{-2})		
Absorbed solar	203/258	-3.3/+8.3
Outgoing longwave	222/255	6.6/12.5
Shortwave cloud forcing	-33/-70	-5.9/+7.4
Longwave cloud forcing	13/48	-4.2/+3.0
Net cloud forcing	-45/-2	-2.9/+5.1
Surface latent heat flux (Wm^{-2})	55/107	8/20
Surface sensible heat flux (Wm^{-2})	3/36	0/4
Precipitable water (mm)	17/27	5/11
Precipitation (mm d^{-1})	2.1/3.7	0.3/0.7

Table 2.1: *Range of global mean perpetual July values of selected climate parameters simulated by current climate GCMs in a recent intercomparison (adapted from Cess et al. (1990) and Randall et al. (1992)). The second column indicates the range of magnitudes of simulated climate change in the parameter in response to a 4° C warming of sea surface temperature.*

Even for models that agree well with globally averaged estimates of fluxes, cloud cover, etc., large discrepancies in the regional distributions of these quantities often exist. Consider, for example, the GISS GCM, whose global TOA radiation field is within 3.6 Wm^{-2} of balance, with small global errors in both shortwave and longwave components. Figure 2.1 shows differences in absorbed shortwave radiation between the GCM and ERBE; it can be seen that this global agreement is actually the result of compensating errors in the tropics and midlatitudes. The nature of the errors in Figure 2.1 is characteristic of the behavior of many climate models. One can combine the simulated surface energy budget with the prescribed sea surface temperature (SST) field to derive an implied meridional ocean heat flux. By underestimating the energy received by the tropical ocean and overestimating that received by the midlatitude ocean, as is suggested by Figure 2.1, one underestimates the need for poleward ocean heat transport.

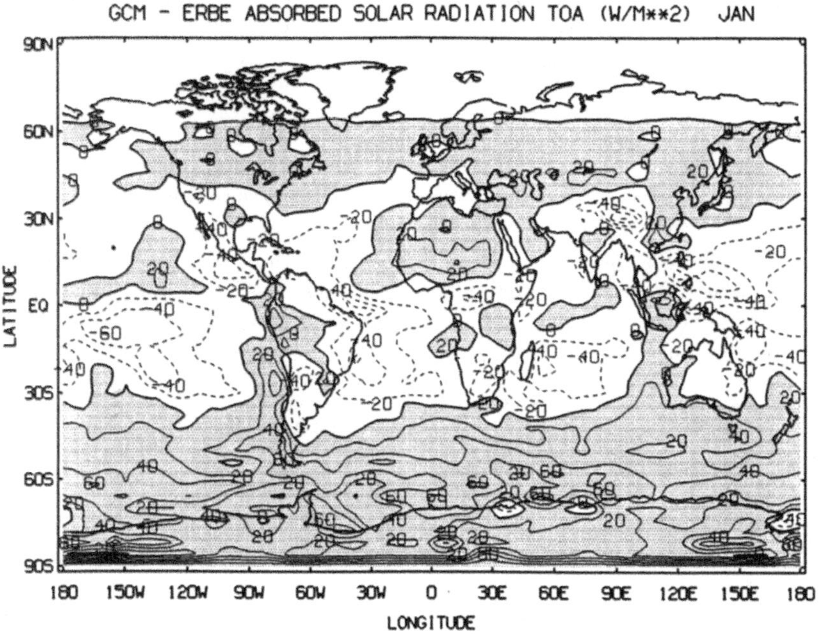

Figure 2.1: *January differences in TOA absorbed shortwave radiation between the GISS GCM and ERBE (Del Genio et al., 1996).*

Indeed, in a comparison of a large number of GCMs participating in the Atmospheric Model Intercomparison Project (AMIP), most models severely underestimate this transport, especially in the Southern Hemisphere (Gleckler et al., 1995).

Examination of the causes of these regional errors illustrates the difficulty in finding simple solutions to model-data discrepancies. First, the nature of the errors (opposite signs in different regions) means that they most likely cannot simply be "tuned" out by varying a free parameter, because such changes usually affect the given field in the same direction everywhere. This is perhaps the most common myth about climate modeling. Consider first the overestimate of solar absorption in midlatitudes, which is insensitive to large changes in cloud microphysical parameters (Del Genio et al., 1996). Analysis of the model shows that the eddy kinetic energy of transient midlatitude storms is underestimated by 30%. Since synoptic-scale storms account for the bulk of upward moisture transport in midlatitudes (Del Genio et al., 1994), the midlatitude atmosphere in the GCM is too dry, and this suppresses cloud formation. Examination of individual storms suggests that the GCM is especially deficient in cloudiness along cold fronts, perhaps because the model cannot resolve the tilted frontal structure and associated vertical motions. But going to higher resolution (4° × 5° to 2° × 2.5°) does not improve the situation, nor do higher order differencing schemes. Thus, an apparent problem in a different part of the GCM affects the performance of the cloud parameterization, with no clear-cut solution.

A different situation exists in the tropics, where too little sunlight reaches the surface in the GCM. This problem illustrates another common misperception in the climate community. It has become popular to assume that model climate can be validated simply by ensuring correct TOA broadband radiation fluxes or monochromatic radiances corresponding to particular satellite instruments. Furthermore it is asserted that errors can be diagnosed simply by comparing the individual shortwave and longwave contributions, without making use of information on physical parameters because estimates of the latter are too dependent on retrieval algorithm to be useful. This is based on the observation that high clouds have a distinct longwave signature, while low clouds do not. But clouds reach unit longwave emittance at fairly low visible optical depths ($\tau \approx 4$), while shortwave reflectance continues to increase with increasing optical depth. Thus, for the optically thick cumulus anvil clouds that pervade the tropics, an error in cloud cover will be manifested as an error in both the shortwave and longwave, but an error in optical thickness will show up only in the shortwave. The latter is indistinguishable from an error in low cloud cover or optical thickness using fluxes alone.

We can diagnose the tropical shortwave error in the GISS GCM only by combining satellite and surface-based data sets (Fig. 2.2). First, comparison of the GCM with ERBE outgoing longwave radiation indicates smaller tropical errors, both positive and negative, ruling out anvil cloud cover errors. On the other hand, the GCM simulates much more low cloud cover in these regions than is indicated by the surface-observed cloud climatology. Furthermore, ENSO variations in tropical Pacific cloud forcing (Fig. 2.3) are consistent with ERBE (Ramanathan and Collins, 1991), while the distribution of tropical high cloud optical thicknesses is consistent with that deduced by ISCCP (Fig. 2.4).

The comparison of model and observed anvil cloud behavior illustrates two more points about the validation of GCMs. The first is that variability reveals model performance in ways that are impossible using information about the mean state alone. In the above example, ENSO variability allows the anvil cloud effect to be isolated (because the primary tropical effect of ENSO is a shift in the location of convection centers), while the mean state is a composite of high and low cloud effects. Furthermore, a quantity and its derivative are not the same thing, and validating a model's mean state gives no information about whether it will respond realistically to climate perturbations.

Climate modelers are only now beginning to test their GCMs against observed variability. Some types of variability are more useful than others in understanding climate change; what is crucial is to determine the mechanism of variability in the current climate and whether it is relevant to that which causes the climate feedback (cf. Del Genio, 1996a). For example, the seasonal cycle is often said to be useless as a proxy for climate change because it is dominated by seasonal shifts in the location of the Hadley cell, which are irrelevant to long-term climate change. But poleward of ±30° latitude, baroclinic waves dominate the dynamics, and winter-to-summer weakening of the meridional temperature gradient, which affects baroclinic instability, is a good proxy for the polar amplification of warming that occurs in decadal climate change simulations. Thus, observations of seasonal water vapor variations (Del Genio et al., 1994) in midlatitudes may contain useful information about regional long-term water vapor feedback.

On the other hand, ENSO variations in the tropics may or may not be a very good indicator of long-term climate feedbacks: In ENSO, net equatorial warming strengthens the Hadley cell (Pan and Oort, 1983) and thus dries the subtropics due to enhanced subsidence. In decadal

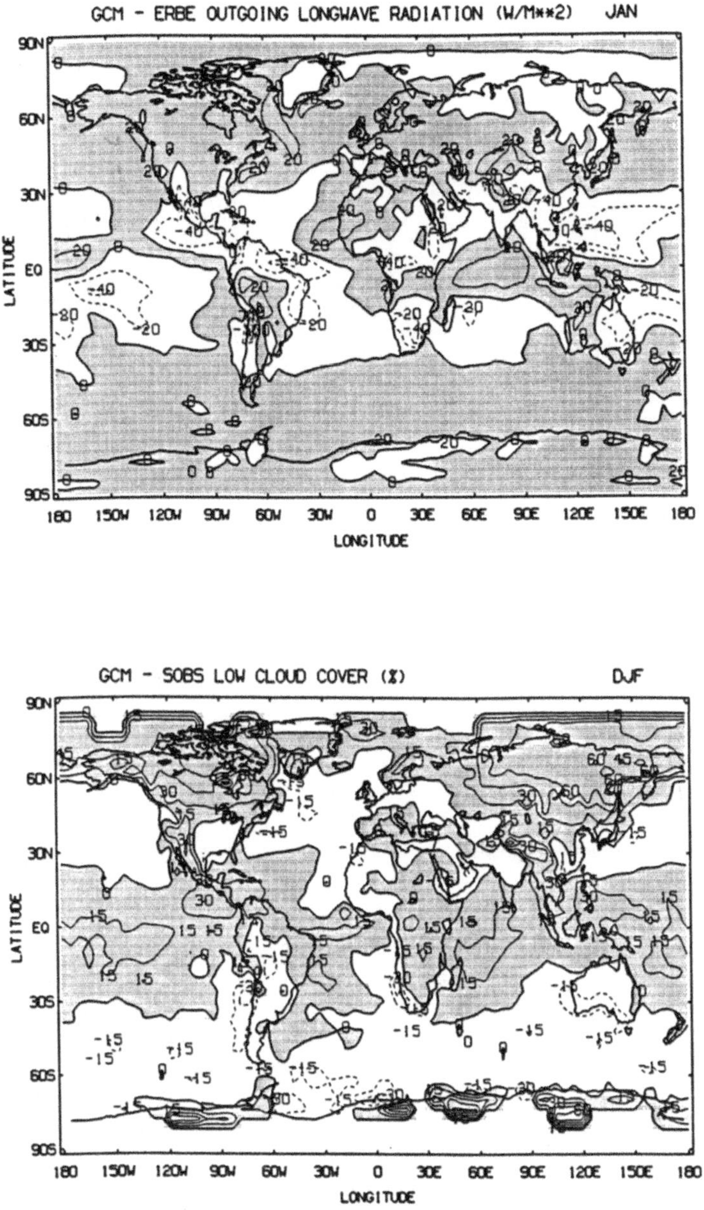

Figure 2.2: *(Top) January differences in outgoing longwave radiation between the GISS GCM and ERBE; (bottom) differences in boreal winter low cloud cover between the GISS GCM and the atlas of surface cloud observations (Del Genio et al., 1996).*

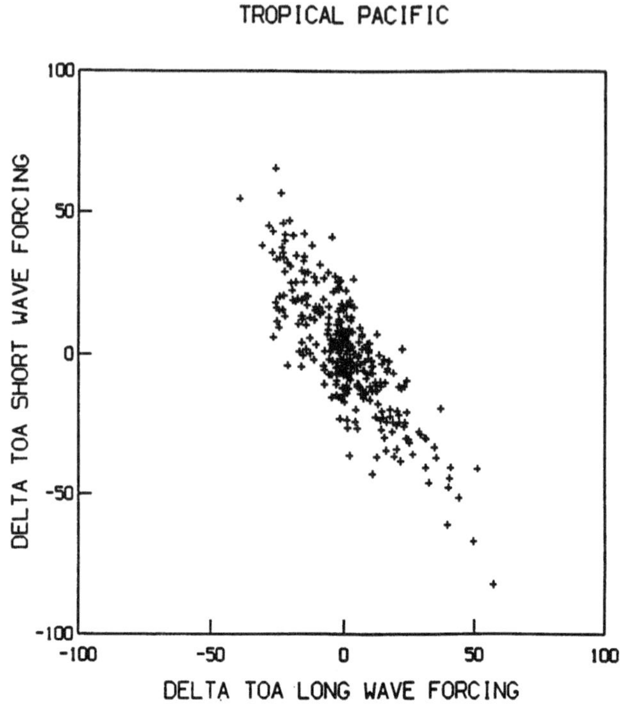

Figure 2.3: *GCM monthly mean differences in TOA shortwave cloud forcing vs. TOA longwave cloud forcing in the tropical Pacific between ENSO and non-ENSO months (Del Genio et al., 1996).*

climate change simulations, however, the Hadley cell slightly weakens in many models, and subtropical water vapor feedback is controlled instead by eddies and the change in the vertical gradient of specific humidity.

All of this does not mean that observations of the mean state are unimportant. Errors in the mean state, such as those in Figure 2.1, may cause climate drift when an atmospheric GCM is coupled to an ocean GCM, thus necessitating the dreaded "flux correction" and complicating interpretations of any coupled GCM simulation. But it is misguided for modelers to focus too heavily on the mean state - observations of variability are the most direct means of assessing whether a climate GCM has the appropriate physics for its intended applications.

The second point to be made about the anvil cloud example is that diagnosis of GCM errors requires observations of the fluxes as well as all the physical parameters that determine those fluxes. For the anvil case, we suspect that errors in radiative properties (e.g., assuming equivalent spheres for ice clouds) cause artificial agreement with ERBE cloud forcing variations, but to be certain, we require global data on cloud ice water path, which do not currently exist. Thus, only certain kinds of diagnosis are possible with current data sets. It is fair to say, however, that climate modelers have generally not exploited the data that are available.

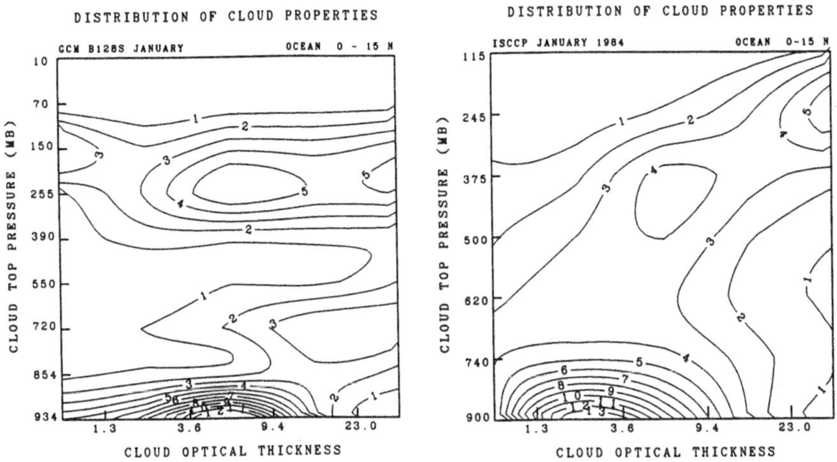

Figure 2.4: *Two-dimensional January frequency distributions of cloud top pressure and visible optical thickness for oceans at 0–15°N latitude. Left: GISS GCM; right: ISCCP (Del Genio et al., 1996).*

2.3 Approaches to Understanding Regional Climate Change

Almost all the effort that has been put into understanding long-term climate change thus far has been directed toward assessing the global sensitivity of the climate to perturbations. The only definitive statements that have been made about scales smaller than global are that the polar regions are expected to warm to a greater degree than the tropics, and that land responds to a transient climate forcing sooner than does the ocean (cf. Hansen et al., 1988). The former occurs because of snow/ice- albedo feedback in the polar regions and the relative absence of convection there to remove surface heating to higher altitudes; the latter occurs because of the small heat capacity of the land surface relative to the ocean.

It is important to discuss the spatial scales we are referring to when discussing regional climate change. The highest resolution GCMs used to predict climate change resolve scales of about 200 km. But there is likely to be significant aliasing of effects from smaller unresolved scales to the grid scale. Thus, although Schleswig-Holstein and the Schwarzwald occupy different gridboxes in a GCM, one cannot interpret any differences in projected climate changes between these two regions with any confidence. The situation is not likely to improve in the near future; at higher resolution, some of the scale separation assumptions that underlie many parameterizations begin to break down, and the benefits of higher resolution may be quite modest as a result. Thus, prediction of regional differences in climate change is probably limited to scales in excess of about 1000 km for the foreseeable future.

We propose instead a physics-based approach to regional climate change, one in which "regions" are defined according to characteristic variations in one or more central aspects of the climate system. At least three different means of such regional classification of climate change can be imagined. These are not mutually exclusive, so net climate change may be the product of the convolution of all three effects.

a) Regional cloud regimes.
The globe can be divided into regions in which different cloud types dominate. This is really a segregation into different climate regimes, because cloud types typically follow properties of the temperature and water vapor profiles and the dynamics. Cloud regimes are conveniently identified by two-dimensional frequency distributions of cloud top pressure and optical thickness, such as those shown in Figure 2.4. Each regime defines a particular set of controlling physical processes, so if we can understand why a particular cloud type prevails in a given regime, we may be able to anticipate how it should change as the climate is perturbed. Such regions need not be geographically contiguous; they merely must share common important physical processes. At a minimum, probably 5 different cloud regimes can be identified in this way (see Figures 2.4 and 2.5):

- Tropical convection zones.
This regime includes the tropical West Pacific, South Pacific Convergence Zone, Amazon Basin, central Africa, and the maritime continent. As indicated by the right panel in Figure 2.4, it is characterized by high, optically thick clouds (deep convective towers and associated anvils) and low-level boundary layer stratus. The anvil clouds generally control the TOA radiative fluxes in this regime. The central question for climate change is whether the strength of convection will increase in a warmer climate, and if so, how it will affect anvil radiative properties (cf. Ye et al., 1996). Some insight might be gained by comparing the radiative properties of maritime and continental convective clusters, which probably differ in the strength of cumulus updrafts. Anvil properties depend on detrainment of liquid and ice from convective updrafts, but the dynamics and microphysics of this process is represented crudely if at all in current GCMs. Another issue for these clouds is the correct particle size and phase function for ice crystals (Minnis et al., 1993), and whether the microphysical properties of these ice clouds differ from thin cirrus.

-Subtropical/tropical ocean subsidence regions.
These areas, off the west coasts of North America, South America, Africa, and Australia, are dominated by low-level marine stratus, capped by a strong trade inversion. The radiative and microphysical properties of these clouds have been relatively well-defined by several field programs (FIRE, ASTEX), and their typical single layer, pure liquid nature allows them to be monitored more easily from space than the other regimes. The great uncertainty is the question of the mechanisms that cause the transition from these almost unbroken cloud decks to the more scattered, deeper fields of trade cumuli that prevail equatorward and west of these zones. Possible contributors to marine stratus breakup include detachment of the cloud layer from the surface due to stabilization of the subcloud layer by solar heating or drizzle evaporation, and cloud top entrainment instability (cf. Wang and Albrecht, 1994). A related question is the cause of the ISCCP-observed tendency for the optical thickness of these clouds to decrease with temperature (Tselioudis et al., 1992). If marine stratus breakup and thinning at higher temperatures are indicative of what to expect in a climate change, then a region of positive cloud feedback may exist. Most GCMs underpredict cloudiness in these regions (cf. Fig. 2.2, bottom).

-Subtropical/midlatitude continents.
This may actually be several separate regimes, encompassing both the relatively cloud-free subtropical deserts and the cloudier midlatitude continents, which themselves vary with the seasons. In winter, midlevel optically thick and moderate optical thickness clouds (e.g., nimbostratus and altostratus) are common, along with thin frontal cirrus and low stratus, in the United States and Europe. In summer, both shallow and deep cumulus increase in importance, but the other cloud types are still evident. Some continental regions such as Siberia in winter have more in common with the subtropical deserts than with the more populated

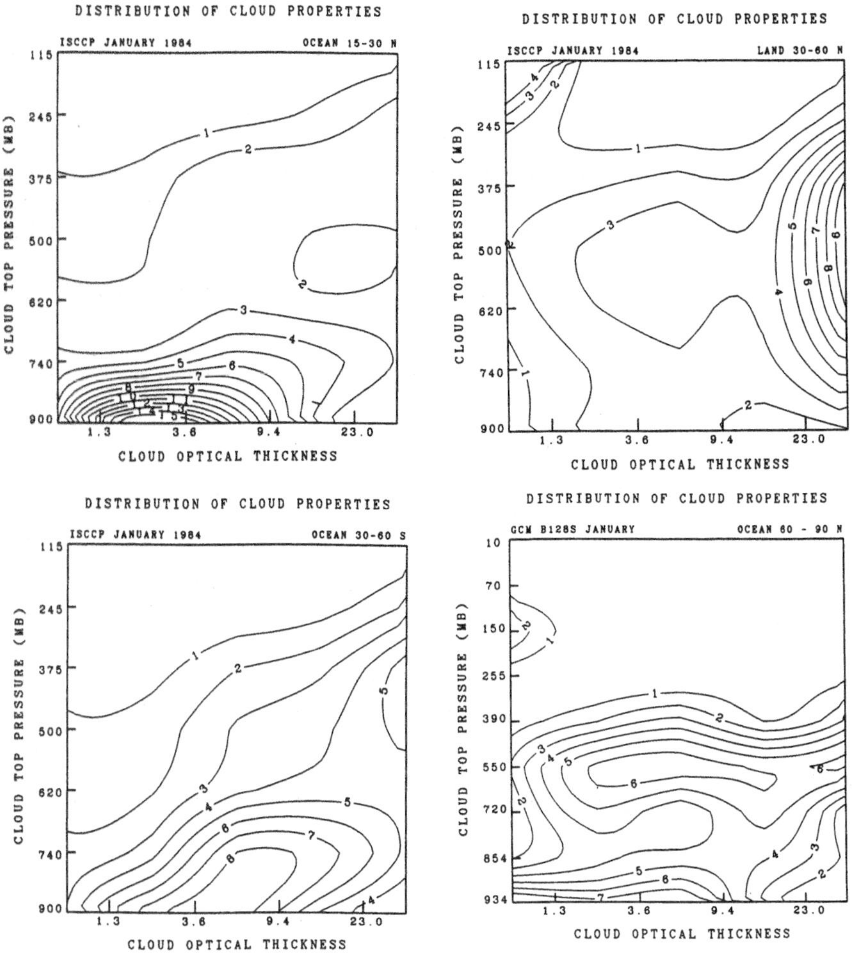

Figure 2.5: *As in Fig. 2.4 but for other cloud regimes. Upper left: oceans, 15-30°N; upper right: land, 30-60°N; lower left: oceans, 30-60°S; lower right: oceans, 60-90°N. The first three histograms are from ISCCP data; that for the polar region is from the GISS GCM, since there are almost no ISCCP optical thickness data there in January (cf. Del Genio et al., 1996).*

midlatitude continental regions. Thin midlatitude cirrus have been analyzed by field programs (FIRE, ICE), but midlatitude storm clouds are poorly documented, even though storms themselves have been the subject of many field studies geared toward improving weather prediction. These clouds involve complex mixed-phase microphysics, and it is not known, for example, whether a low-level midlatitude winter ice cloud has similar or different particle sizes and phase functions from the more widely observed cirrus cloud type. In the United States, Europe and Australia, a long-term trend of increasing cloudiness has been observed (cf. Henderson-Sellers, 1992), possibly the only direct observation of cloud feedback that will exist for quite some time.

-Midlatitude ocean storm tracks.
Off the east coasts of North America and Asia, and throughout the "roaring 40's" latitude band of the Southern Hemisphere, baroclinic wave activity is more persistent than over the midlatitude continents. The cloud regime in these locations thus also consists of midlevel optically thick nimbostratus, but with a higher proportion of moderate optical thickness low stratus than is found over the continents. Because of their almost ubiquitous cloudiness and their lower cloud tops relative to the tropical convection zones, these are the most negative net cloud forcing regions on the planet. It is this cloud regime that causes most of the error in implied ocean heat transport in GCMs. Polar amplification of climate warming, combined with the dependence of baroclinic instability on meridional temperature gradient, suggests that cloudiness here should decrease in a greenhouse climate change, but changes in the optical properties of these clouds associated with a more vigorous hydrologic cycle are harder to anticipate. Despite their possible climatic importance, little attention has been paid to this cloud regime by either the climate modeling or observation community.

-Polar regions.
Little is known about the prevailing cloud types in the polar regions. Current wisdom suggests that the cloud distribution is bimodal, with boundary layer cloudiness controlled by surface heat and moisture fluxes and midlevel cloudiness due to poleward moisture transport by midlatitude storms (Curry and Ebert, 1992). Multilayer cloudiness may be common, as well as "cloudless" ice crystal precipitation that complicates attempts to compare data sets. Because of the difficulty in distinguishing snow/ice-covered surfaces from clouds, traditional satellite detection techniques have large uncertainties. The ISCCP seasonal cycle of polar cloudiness disagrees completely with that inferred by surface observers, and the ERBE algorithm gets the incorrect sign of cloud forcing at these latitudes. Consequently, most GCM modelers pay no attention to their polar cloudiness, absent an accepted validation standard. But snow/ice-albedo feedback and polar amplification of climate warming, central to the overall question of climate sensitivity, may well be determined by climate changes in polar clouds.

b) Regional dynamic variability effects.
Since clouds are to a great extent tracers of the general circulation, it is logical to approach the question of regional climate feedbacks from the standpoint of dynamics. Systematic trends in specific components of the seasonal or annual mean dynamics, such as the baroclinic wave issue discussed above, are one example. But the current climate also varies considerably on interannual time scales, as a result of ocean-atmosphere interactions. Might the regional pattern of climate changes in a warmer world bear any resemblance to the preferred modes of interannual variability of the current climate?

The most direct example of observed interannual variability is the eastward shift in tropical Pacific SST that occurs during ENSO. Under non-ENSO conditions, rising motion, deep convection, and heavy precipitation characterize the west Pacific warm pool region and the maritime continent, while subsidence, low stratus, and drier conditions prevail in the east Pacific and along the west coast of South America. ENSO warming of the central and east Pacific shifts the Walker circulation eastward and causes regional climate anomalies on either side of the Pacific, e.g., drought in Australia and Indonesia and flood conditions over South America.

It is not clear how the tropical ocean circulation will change in response to greenhouse gas forcing. But there are purely energetic reasons for anticipating that longitudinal Pacific SST gradients might weaken as climate warms, causing the regional pattern of climate change in the tropics to resemble that for the warm phase of ENSO. First, because of the prevailing trade winds near the equator and attendant ocean upwelling off the west coast of South America, the ocean mixed layer and thermocline are shallower in the east Pacific than in the west. Thus, the effective ocean heat capacity is lower in the east, and the eastern ocean surface should therefore warm more quickly than the western warm pool as greenhouse gas concentrations increase. Furthermore, to the extent that highly reflective anvil clouds in the west dampen the rate of warming of the ocean surface there, any increase in anvil ice water content might result in a smaller perturbation of the surface energy budget in the west than in the east. Finally, preferential increases in evaporative cooling in the warmer west Pacific and in downward longwave heating in the less opaque atmospheric boundary layer over the east Pacific might also lead to a stronger warming trend in the eastern ocean.

Several of these effects are nonlinear: If the longitudinal SST gradient decreases, the Walker cell will weaken, thus reducing both the trade winds that maintain the ocean mixed layer depth gradient and the convergence that supplies water to the cumulus anvils. The latter implies that the climate change in tropical cloud forcing will depend not only on the thermodynamics of clouds and convection but also on the regional pattern of SST change. In idealized experiments with prescribed SST changes in the GISS GCM, the global climate sensitivity is largely determined by changes in tropical anvil clouds (Del Genio et al., 1996). When the SST changes are applied uniformly, maintaining the current SST gradient, large increases in anvil cloud cover and ice water content occur, substantially reducing the global mean sunlight absorbed by the planet and generating a net negative cloud feedback and very low climate sensitivity. However, if the same basin-average SST change is applied so as to reduce the longitudinal gradient in the warmer climate, the increase in anvil cloudiness and reflectance is smaller, and net positive cloud feedback and relatively high global climate sensitivity result.

The latter scenario appears to be more relevant to actual greenhouse climate change: In a doubled CO_2 experiment with the most recent version of the GISS atmospheric GCM coupled to a mixed layer ocean with fixed ocean heat transports, the tropical Pacific SST gradient weakens by 0.5–1.0°C across the basin, and the deep convection pattern shifts eastward in a manner reminiscent of ENSO (Fig. 2.6). The climate sensitivity of this model is 3.5°C, despite parameterized detrainment of cumulus condensate into anvil clouds and interactive cloud radiative properties. Thus, the concept of a tropical cirrus "thermostat" limiting the magnitude of climate change does not appear to be valid, because changes in clouds and dynamics interact to lessen the impact of local SST increases.

One direct implication of ocean dynamics concerns the latitudinal profile of greenhouse warming. In equilibrium, polar amplification is predicted by all mixed-layer models. But if high latitude rainfall and ice melting were to increase sufficiently to slow down the North Atlantic thermohaline circulation, the climate system might make a transition to a different state (Broecker, 1991), and the North Atlantic could conceivably cool for several decades. Thus, while the rest of the globe was beginning to experience warming, western Europe, which depends on the Gulf Stream to maintain a comfortable climate, might actually cool temporarily instead. Even without a transition, if land warms faster than ocean, a tendency for oceanic high pressure anomalies might result. The associated clockwise flow might cause western continental regions to warm more slowly than their eastern counterparts (Hansen et al., 1988).

Another aspect of ENSO is the teleconnection patterns that propagate thousands of kilometers from the forcing region in the tropical Pacific (Horel and Wallace, 1981). These are known to create regional climate anomalies in the Northern midlatitudes. The specific teleconnection pattern differs for each ENSO depending on the magnitude and location of the peak SST anomaly. If the spatial pattern of SST change in a greenhouse warming resembles ENSO, then the remote effect in midlatitudes might well be partly determined by the relevant teleconnection pattern, i.e., one manifestation of regional climate change might be an increase in the frequency of occurrence of spatial anomalies characteristic of the warm phase of ENSO. Since clouds as a general rule prefer low pressure, one might expect regional cloud cover increases and decreases to correlate to some extent with regional low and high pressure anomalies, respectively.

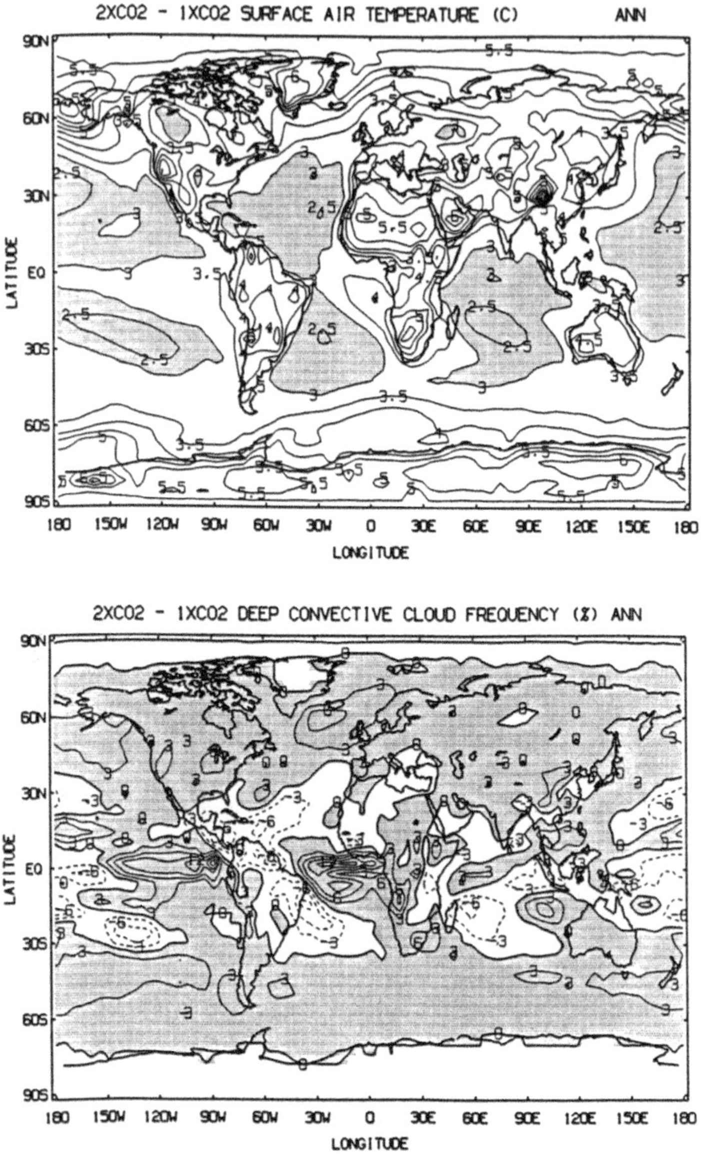

Figure 2.6: *Doubled CO_2 minus current climate annual mean differences in surface air temperature (top) and deep convection occurrence frequency (bottom) simulated by the GISS GCM.*

The GISS GCM shows evidence of the dynamical and ocean mixed layer influences described above in its regional patterns of cloud cover and sea level pressure change for a doubling of CO_2 (Fig. 2.7). First, cloud cover decreases in the Atlantic and Pacific midlatitude storm tracks, as expected from the baroclinic instability argument (but not consistently in the southern midlatitudes, because snow/ice-albedo feedback and thus polar amplification are less in the Antarctic than in the Arctic). In the tropical east Pacific, cloud cover increases as convection shifts eastward, but it decreases over South America and Australia/Indonesia, which experience anomalous subsidence. A high pressure anomaly forms over the west and central Pacific, and a possible teleconnection pattern emanates poleward from there, with a low pressure anomaly (and increased cloud) along the west coast of the United States, high pressure (and decreased cloud) in the North Pacific south of Alaska, and low pressure again over eastern Canada (but without as obvious a cloud change). A similar pattern may propagate into the Southern Hemisphere.

c) Regional climate forcing.

Although greenhouse gas concentration changes are fairly uniform globally (with the possible exception of ozone), the competing negative climate forcing associated with aerosols is quite inhomogeneous spatially. Thus, a third approach to understanding regional climate change might be tied to an assessment of the magnitude of regional aerosol impacts. Aerosol climate forcing consists of three components (Charlson et al., 1992): (1) A direct radiative effect associated with reflection of sunlight by aerosols in otherwise clear air; (2) An indirect radiative effect due to the fact that higher aerosol concentration implies more cloud condensation nuclei and thus smaller and more reflective cloud droplets for a given cloud water content; (3) A related indirect microphysical effect because smaller droplets precipitate less efficiently than larger ones, thus implying higher cloud water content.

Even if there were no climate forcing due to increasing aerosol concentrations, an indirect aerosol effect on climate sensitivity might still exist because of the climatological difference between continental and marine air. Continental air has several orders of magnitude greater concentration of cloud condensation nuclei (CCN) than does marine air, and consequently cloud droplets are smaller over land and precipitate less easily for a given water content (Twomey, 1977; Han et al., 1994). As climate changes, the different microphysical settings of land and ocean clouds might cause them to react differently to changes in the hydrologic cycle. In the GISS GCM, which has higher effective precipitation thresholds over land than ocean, midlatitude cloud cover generally tends to decrease over ocean but increase over land for a doubling of CO_2 (Fig. 2.7). As climate warms, there is less opportunity for cloud water content to increase over ocean before precipitation sets in than is the case over land. This might contribute to longer cloud lifetimes over land in the warmer climate, since clouds with larger cloud water content take longer to dissipate by evaporation.

But in reality, tropospheric aerosol concentrations appear to be increasing at a fairly rapid rate due to anthropogenic activity, including combustion of fossil fuels and biomass burning (Charlson et al., 1991; Penner et al., 1992). There is significant uncertainty in the total aerosol direct climate forcing, because ground-based data exist over only a small fraction of the Earth, satellite monitoring of the relatively small aerosol optical depth is difficult, and the radiative properties of different aerosols vary with particle size, shape and composition. Best estimates of the direct forcing suggest that it is of order -1 to -2 Wm^{-2} globally, with the signal much larger in the Northern Hemisphere and especially near and downwind of industrial regions. If correct, this is a substantial fraction of the current radiative imbalance due to increased greenhouse gas concentrations, and it may explain why climate has warmed to a lesser extent thus far than many GCM predictions (Charlson et al., 1992). But unlike greenhouse forcing, which is fairly uniform globally, aerosol forcing is quite inhomogeneous, so net climate forcing to date may be near zero or negative in some regions while being strongly positive in others.

The indirect radiative effect is even more uncertain. One complication is that cloud droplet number and size do not simply respond linearly to an increase in CCN concentration. Continen-

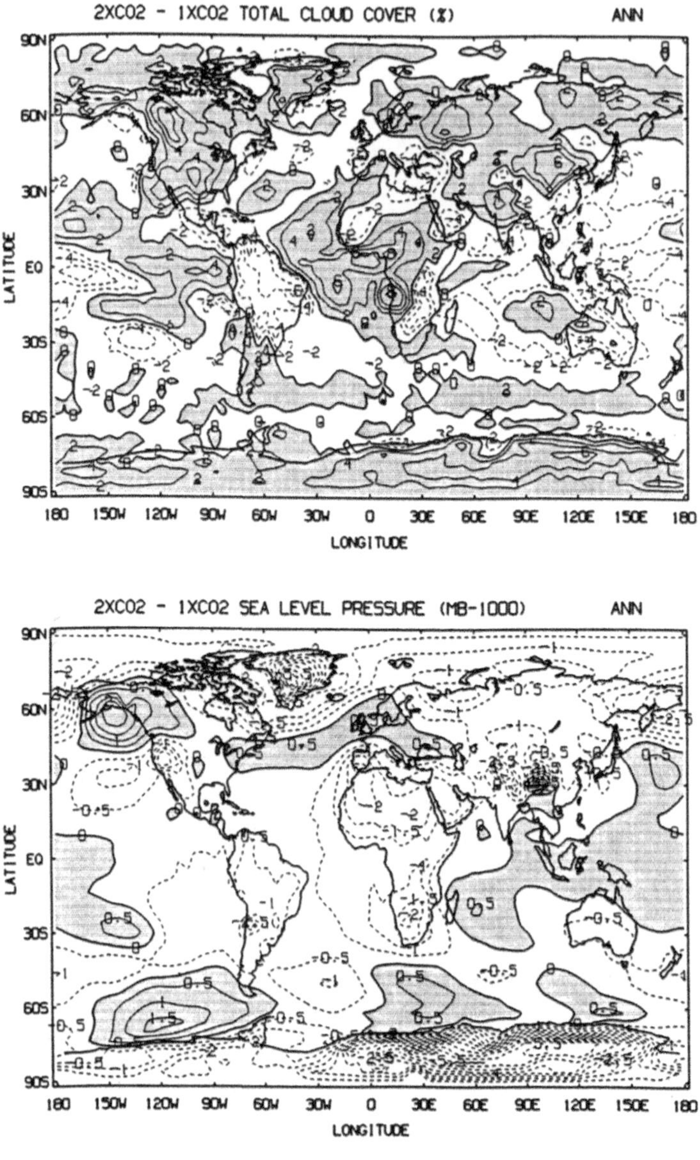

Figure 2.7: *Doubled CO_2 minus current climate annual mean differences in total cloud cover (top) and sea level pressure (bottom) simulated by the GISS GCM.*

tal air may often have sufficient numbers of CCN to nucleate clouds with maximum efficiency, so that an increase in aerosol has no effect on cloud droplet size. Marine air, on the other hand, is deficient in CCN, and thus oceanic clouds are more susceptible to aerosol changes (Platnick and Twomey, 1994). Thus, the indirect effect may not have the same spatial pattern as the direct effect. A recent GCM estimate suggests that the global indirect radiative effect may be comparable in magnitude to the direct effect (Jones et al., 1994). The magnitude of the indirect microphysical effect is an even greater unknown. A final complicating factor is that biogenic sources of sulfate in the oceans may contribute their own indirect effect far from industrial regions (Charlson et al., 1987), but climate changes in plankton emissions into the atmosphere are completely unknown.

One possible hint of a regional indirect climate effect is the observation that the diurnal temperature range of surface air has decreased by about 0.5°C over the past four decades over much of the Earth's land area (Karl et al., 1993) with most of the change having occurred at night. Greenhouse gas increases and the direct effect of aerosols seem to be insufficient to explain the observed narrowing. Hansen et al. (1995a) use GCM simulations to argue that either low or midlevel cloud increases over this time are the only plausible candidates to produce the observed magnitude of the effect. They suggest that anthropogenic aerosol increases might have provided the required stimulus for a low cloud cover increase. Indeed, cloud cover has increased over both the United States and Europe during the past 50 years (Henderson-Sellers, 1992). But whether this can be traced to aerosols is as yet unknown; the GISS GCM predicts increasing cloud cover over midlatitude land in a warming climate without anthropogenic aerosol climate forcing (Fig. 2.7), and climatic increases in the vigor of the hydrologic cycle might also explain the observed weakening of the diurnal cycle (Dai, 1995).

2.4 Observational Strategies for Improving Climate Model Predictions

With this background, we can now assess the suitability of existing observations and suggest strategies for future data acquisition that would maximize the impact on climate model performance. Needed data fall into one of three categories, according to how they are to be used by the climate model. First, there are validation data required to determine how well GCMs simulate the mean state and variability of parameters in the current climate. Second, we need physical process data to determine why the GCM performs incorrectly in a given situation and thus to provide insight into parameterization improvements. Third, we require monitoring data to ultimately detect the elements of decadal climate change. We discuss specific data requirements for each class below.

a) Validation data sets.
The basic requirement for validation data are that they be global and document both the mean state and variability on diurnal, seasonal, and interannual time scales. A non-rigorous but practically useful accuracy standard for such data sets is that uncertainties be significantly less than the range of differences in a given parameter among various climate models. Since almost every GCM in the world is participating in the AMIP intercomparison, required accuracy estimates of this kind will be available in the near future. A sample of the range of model predictions for several parameters from an earlier intercomparison is given in Table 2.1. The status of data sets for particular parameters follows.

-TOA radiation budget.
ERBE (Barkstrom, 1984) has become the standard of comparison for many GCMs, but the data set covers only 4 complete years, with a current gap until CERES flies on the TRMM satellite in 1997. The angle and diurnal models used to derive the fluxes are crude, but errors are probably < 10 Wm^{-2} regionally. The net annual imbalance of 6 Wm^{-2} presumably reflects

a bias rather than a real imbalance. Seasonal and interannual variations in fluxes can be determined, but the diurnal cycle is probably useless. Cloud forcing derived from ERBE has larger errors, associated with cloud detection and temperature/humidity differences between clear and cloudy skies. Cloud forcing estimates are unreliable over snow and ice.

-Surface radiation budget.
Direct information is not available globally; TOA information is combined with radiation models to derive surface radiation estimates with 10–20 Wm^{-2} accuracy (cf. Zhang et al., 1995). Surface longwave fluxes may not be sufficiently accurate at high latitudes, where humidity is low enough for cloud base errors to have an impact. Surface incident shortwave fluxes had been thought to be fairly reliable, but recent controversy over 25 Wm^{-2} of apparent unexplained shortwave absorption by the atmosphere (Cess et al., 1995) has temporarily called this into question. Surface albedo errors in GCMs may well be significant, but few GCMs have attempted to validate their surface radiation budget.

-Turbulent surface fluxes.
Evaporation algorithms for the oceans only, based on microwave precipitable water and SST estimates (Liu, 1988), provide crude estimates on monthly and longer time scales and allow interannual variability to be documented, but evapotranspiration over land is not observed globally. Sensible heat flux estimates are lacking over most of the globe but are important only at higher latitudes. Precipitation estimates on monthly and longer time scales are acceptable over some land areas but existing satellite algorithms for ocean are validated only regionally and of unknown quality globally. The TRMM satellite mission will address this problem for the tropics and subtropics beginning in 1997 (Del Genio, 1996b, this volume, chapter 20). Although precipitation is much larger in the tropics and of first-order importance for the general circulation there, the more demanding requirement driven by climate modeling is the accuracy of 0.5-1.0 mm d^{-1} required to ensure accurate driving of the thermohaline circulation in the North Atlantic (Zaucker et al., 1994).

-Cloud cover, top height, optical thickness.
ISCCP (Rossow and Schiffer, 1991) has provided a useful climatology of these cloud properties spanning almost a decade, with information on the mean and all three important time scales of variability. Like ERBE, it is extremely uncertain over snow and ice. Inter-satellite calibration problems cause some spurious interannual signal. ISCCP misses or misidentifies very thin cirrus, and there are errors due to the use of Mie phase functions and similar particle size assumptions for all clouds (Minnis et al., 1993; Liao et al., 1995), but a re-analysis of the data set is in the process of rectifying this situation. The 5 km pixel size causes errors in cloud cover of no more than 10%, and much less in many places (Wielicki and Parker, 1992); subpixel cloudiness effects on optical thickness estimates have not been quantified. Climate modelers have underutilized these data, particularly the optical thickness information. Satellite data cannot uniquely determine low and middle cloud amounts due to shielding by higher clouds. The former can be obtained from the atlas of surface observations (Warren et al., 1986, 1988), but coverage varies over the globe. The surface data set also performs separation into frequencies and amounts-when-present, and attempts to distinguish cloud types morphologically. SAGE II (Liao et al., 1995) provides a multi-year record of cirrus frequency and/or cover.

-Cloud base height.
A global data set does not exist, although crude global patterns have recently been derived from radiosonde profiles (Wang and Rossow, 1995). This parameter is almost never examined by climate modelers, but accuracy sufficient to produce surface longwave flux errors of 10 Wm^{-2} or less is desireable for surface energy budget studies. For some climate process problems (see discussion below), accuracy higher than that attainable globally may be required.

-Cloud particle size.
Satellite near-infrared retrievals of droplet effective radius for liquid clouds outside the polar regions have recently become available (Han et al., 1994). These refer primarily to cloud top conditions; the vertical profile of droplet size through clouds is unknown except for isolated in situ measurements. Particle size for ice clouds is much more uncertain; comparable global surveys are only beginning to be produced, and remote retrievals from field programs typically differ markedly from in situ ice particle size estimates. Ice scattering properties are not well-defined by assuming "equivalent spheres" for Mie scattering calculations, the standard approach in GCMs. More realistic ice phase functions have been defined (Macke, 1993); it remains to be seen whether a universal phase function for all ice particle shapes is adequate for GCM purposes. Only GCMs with prognostic cloud water budgets can currently make use of such information.

-Cloud liquid and ice water path.
Microwave retrievals of vertically integrated liquid water path are available over ocean, but not over land. But the algorithms used by different workers produce dramatically different results, even disagreeing about whether liquid water path is higher in midlatitudes or the tropics (Lin and Rossow, 1994). Such data are thus virtually useless for GCM validation, a point missed by many climate modelers who have nonetheless compared their prognostic models to one of the published estimates. The problems involve whether clear points are included in the average, the threshold assumed for detecting cloud water, and errors in the assumed water vapor profile and cloud temperature. Ice water path has not yet been retrieved from satellite; this is perhaps the major gap in validation data for GCMs with prognostic cloud water schemes. ISCCP can indirectly estimate ice water path by combining microwave liquid water path and near-infrared particle size estimates with its own column optical thickness retrieval (Lin and Rossow, 1994), but this depends on their assumptions about ice particle size and phase function. A major issue for GCMs is where the transition from liquid to ice occurs as a function of temperature, cloud depth and age, and other parameters (cf. Mitchell et al., 1989; Li and LeTreut, 1992; Del Genio et al., 1996).

-Water vapor.
Rossow (1996, this volume, chapter 8) provides a comprehensive discussion of existing water vapor observation capabilities. Global precipitable water estimates from a combination of microwave and radiosonde data will be produced by the GEWEX Water Vapor Project, but these are of limited use for climate models, because the real climate issue is the vertical redistribution of water by dynamical processes. For the lower troposphere, radiosondes provide a useful if incomplete global picture, but large differences exist in radiosondes used in different parts of the world (Elliott and Gaffen, 1991). Upper troposphere water vapor has been observed by SAGE II (Rind et al., 1991), but its sparse coverage allows only a multi-year average to be obtained with any useful accuracy; there is almost no tropical water vapor information, and what exists is biased toward non-convecting regions. The GOES 6.7 μm channel has been used to create a decade-long climatology of upper troposphere relative humidity (Soden and Bretherton, 1993). However, this diagnostic and that derived from SAGE II disagree in the sense of their Northern midlatitude seasonal cycles (as well as the magnitude of their tropical seasonal cycles), creating a quandary for modelers wishing to use such data to diagnose feedback processes (cf. Del Genio et al., 1994). The 183 GHz channel on SSM/T-2 is beginning to provide upper and middle troposphere water vapor profiling on a variety of time scales over oceans.

-Temperature.
Climatological temperature distributions and interannual variations are available from radiosonde data (Oort, 1983), although some modelers validate against analysis products instead. TOVS gives some temperature information on shorter time scales. GCMs typically exhibit

temperature errors in the tropical and polar upper troposphere. SST climatologies generally differ from each other by 0.5–1.0°C, but there are important differences in interannual variation among various data sets as a result of different weighting of in situ and satellite observations.

-General circulation.
Conventional climatologies exist (cf. Oort, 1983), but information on synoptic time scales is restricted to analysis products provided by ECMWF, NMC, etc. (Raschke, 1996, chapter 1). These are fairly accurate in northern midlatitudes, where observations strongly influence the analysis, but problems exist in the tropics, where the parent GCM's cumulus parameterization is more likely to determine the nature of the flow, and in the southern midlatitudes, where virtually no observations constrain the analysis. Thus, such analyses must be used judiciously as validation tools.

-Sea ice.
Estimates of sea ice extent vary significantly among different data sets; the ten-year record compiled by AMIP to force GCMs contains a sharp discontinuity early in the 1980's which is probably not real. Other important properties of sea ice, such as thickness and areal coverage of leads, which dramatically affect the polar surface energy balance, are even more uncertain. This subject has not received the attention it deserves, given the extreme sensitivity of GCMs in general to sea ice variations.

b) Physical process data sets.
The validation data described above can only tell a modeler whether a particular parameter is being simulated correctly or not in the GCM. To correct flawed or incomplete parameterizations, information is required on the physical processes that control various climate parameters.

Some processes are not understood fundamentally. A good example is cloud top entrainment instability, which was originally touted as the mechanism that controls the breakup of marine stratocumulus decks (Randall, 1980). Recent observational and theoretical evidence suggests that this mechanism occurs if at all under more restrictive instability conditions than originally thought, and may or may not be important relative to other processes that can detach stratus from the surface moisture source (Kuo and Schubert, 1988; Wang and Albrecht, 1994).

Other processes are understood on the cloud scale but not on the GCM grid scale. The most obvious example is the relationship between cloud cover and relative humidity. On the cloud scale, condensation occurs very close to 100% relative humidity, at least for liquid clouds. But an area the size of a GCM gridbox is generally not saturated, yet some fraction may be occupied by clouds because of subgrid-scale variations in temperature and moisture. Some GCMs assume a correlation between cloud cover and relative humidity on the grid scale; observations suggest that this is true statistically but with a considerable degree of scatter (Soden and Bretherton, 1993; Walcek, 1994). Other GCMs assume a subgrid-scale distribution of some thermodynamic quantity to diagnose cloud cover, but what controls this distribution is not known, and arbitrary assumptions can sometimes have significant unintended feedback effects (Miller and Del Genio, 1994).

Process studies can involve either actual physical mechanisms that control a variable or diagnosis of the component parameters that determine an integrated measure of radiative effect. Process understanding is best accomplished by combining satellite data, in situ measurements, theory, and model sensitivity studies. An instructive ongoing case in point is the question of the factors controlling optical thickness feedback, a major aspect of disagreement among climate models. Basic thermodynamics indicates that the adiabatic liquid water content of a lifted cloud should increase with temperature (Betts and Harshvardhan, 1987), and aircraft data for low and midlevel clouds generally seem to verify this (Feigelson, 1978). This implies a negative optical thickness feedback if other factors are held fixed (Somerville and Remer, 1984).

But ISCCP data suggest that this behavior occurs only at cold temperatures, and more so over land than ocean; at warm temperatures, optical thickness decreases with temperature instead (Tselioudis et al., 1992). To understand this, consider first that optical thickness is a function not only of liquid water content, but also cloud physical thickness and droplet effective radius. Sensitivity tests with the GISS GCM show that cloud thickness variations alone can simulate the ISCCP behavior at most latitudes. In the GCM, cloud physical thickness is a function of stability (Del Genio et al., 1996), as suggested by analyses of ISCCP and temperature profile observations for several marine stratus regions (Klein and Hartmann, 1993). Surface-based remote sensing data from the DOE Atmospheric Radiation Measurement (ARM) Program in the central U.S. suggest that liquid water path is correlated both with cloud top pressure (Del Genio, 1996a) and cloud base pressure (Fig. 2.8), suggesting that the physical thickness explanation is viable. And finally, a doubled CO_2 simulation with the GCM produces a low cloud optical thickness feedback that is consistent with the observed temperature dependence for current climate variability (Fig. 2.9). This is an encouraging, but unfortunately rare, example of a climate feedback mechanism being understood and validated by a combination of models with several independent sources and types of data, from local to global.

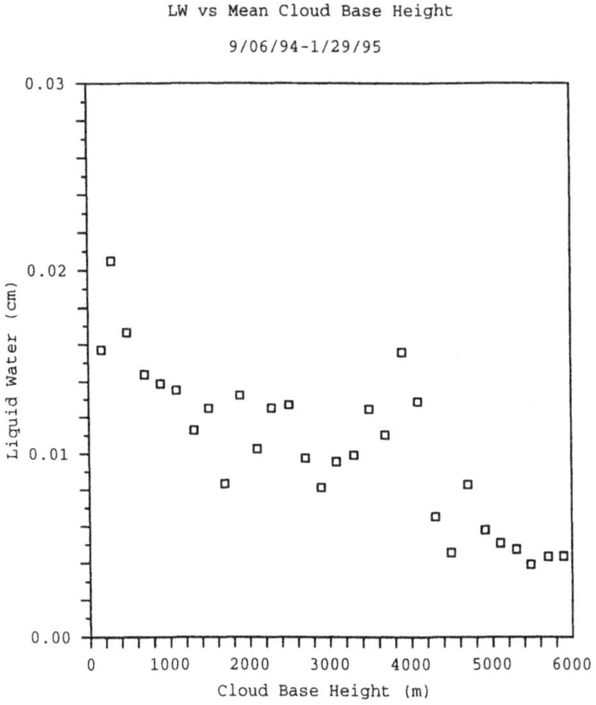

Figure 2.8: *Microwave water radiometer liquid water path vs. Belfort ceilometer cloud base height for a 5-month period at the DOE ARM Southern Great Plains Cloud and Radiation Testbed site.*

A useful strategy for future field programs targeted at questions of regional decadal climate change might be to prioritize them according to one or more of the key regional forcing or response mechanisms discussed in the previous section. Considering first the different regional

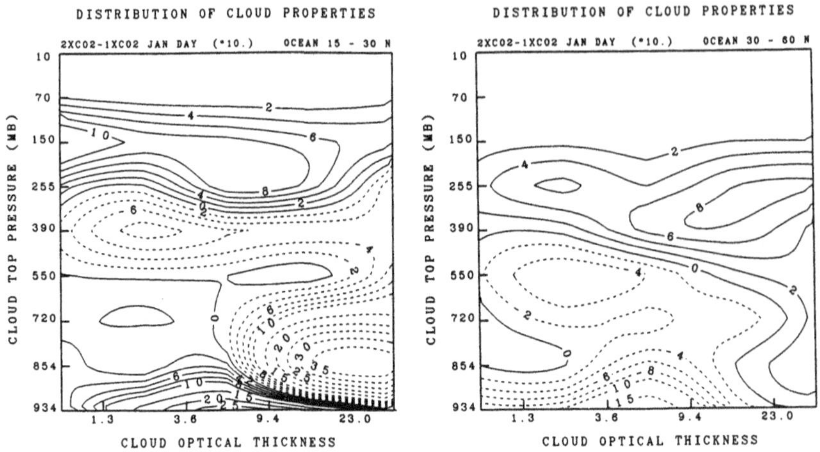

Figure 2.9: *Doubled CO_2 minus current climate January differences in cloud top pressure-optical thickness histograms simulated by the GISS GCM over oceans at 15–30°N (left) and 30–60°N (right), showing the latitudinally changing character of low cloud optical thickness feedback.*

cloud regimes, 2 of the 5 have already been the subject of intensive field programs: FIRE and ASTEX for the marine stratus regime, and ARM, FIRE and ICE for the midlatitude continental cloudiness regime. The Arctic stratus regime will be studied intensively by ARM, SHEBA, and FIRE III in several years. ARM will also have a site in the tropical west Pacific, and FIRE III is interested in a tropical cirrus field program, thus covering the tropical convection regime.

This leaves only the midlatitude oceanic storm track regime as unobserved in detail from the climatic standpoint. Of course, considerable attention has been paid to storm tracks by field programs such as GALE and STORM, but these have been aimed at understanding storm dynamics and cyclogenesis rather than clouds, water vapor and radiative properties. Considering that these clouds have the largest net cloud forcing of any on Earth, that climate models consistently underpredict them, and that changes in ocean heat transport as climate changes may depend in part on how well we simulate the effects of these storms on the surface energy and freshwater budgets, there is a clear need for a targeted observation program designed to study cloud-radiation-precipitation-dynamics interactions in the midlatitude storm tracks.

Designing a regional field program to understand the effects of global dynamical normal modes of variability on regional climate change is more difficult. First, climate models would have to determine that such variability was important and agree on the specific regions in which the dynamical signal was greatest. Intercomparisons of coupled ocean-atmosphere simulations of transient climate change are only beginning, so the prospects of such a consensus in the near term are slim. But this problem is best studied with long time series of data rather than limited duration field studies, i.e., via global climate monitoring, which we discuss in the next section.

Finally, an obvious strategy for regional field programs would be to focus on areas of potentially large aerosol climate effects. This would necessarily include one of the Northern Hemisphere industrialized areas, but should also include tropical regions of active deforestation and a more remote ocean area in which cloud susceptibility is highest and biogenic sources of aerosol are important. But this too is a potential focus for a global satellite climate monitoring effort, if suitably precise remote measurements of aerosol and cloud properties can be made.

It is also true that existing field programs have not yet fully exploited their potential to contribute to climate research. In part this is due to the surprising lack of interaction among climate modeling, satellite retrieval, and aircraft observation experts to date. This is a problem of spatial scale, from the cloud microphysical scale to the satellite pixel scale to the GCM grid scale, a range of many orders of magnitude. It is also a problem of temporal scale, from the several hour duration of a typical aircraft flight to the monthly time scale on which most GCM parameters are studied. To bridge the scale gaps, all parties must move in directions they traditionally resist. Modelers must understand the behavior of their GCMs on synoptic time scales. Data from field programs, which do not represent climatologies, can nonetheless be compared by modelers to the frequency of occurrence of similar phenomena in the GCM, and those instances that are relevant can be analyzed in more detail to see how well they correspond to what occurs in the real world. Field program workers in turn need to consider the domain-averaged characteristics of their observations and statistical relationships between observed parameters, to link to the satellite and GCM scales. Availability of field experiment data in useful, readable form is also a chronic problem causing these data to be underutilized by modelers.

Satellite data sets must in turn provide enough information to permit a modeler to diagnose the same variable in the same way in the model (and modelers in turn must more often "fly the satellite" over their GCMs to create the most instructive comparisons possible). For some satellite data, information may exist at a variety of spatial scales, but resolution is degraded to limit the volume of the data set. A better approach would be to create several regional data sets at full resolution for limited time periods (e.g., the canonical seasonal months) to supplement global data at lower resolution. For example, Landsat images at sub-kilometer resolution have proven valuable as a complement to ISCCP data at 5 km resolution (Wielicki and Parker, 1992).

c) Climate change monitoring data
Ultimately, we will not know with certainty whether GCM predictions of climate change are realistic until unambiguous evidence of long-term trends in a variety of climate parameters can be observed. To date, there is a useful record of surface temperatures over much of the globe spanning more than a century, and several regional trends of cloud cover have been noted over a period of a few decades (cf. Henderson-Sellers, 1992). But many of the data one would like to use to detect trends are not adequate for the purpose. For example, there is a fairly long record of radiosonde specific humidity at various locations around the world, but different nations use different instruments with different accuracies, and both instrument types and reporting methods have changed over time, making deduction of trends a particularly difficult undertaking (Elliott and Gaffen, 1991). Satellite data sets in principle can be used for monitoring, but in practice, for many parameters there are gaps in the record, and for others, calibration uncertainties make it impossible to detect a trend when one satellite replaces another.

If climate monitoring is to be viable, one must first ask what parameters one wishes to monitor, and whether it is possible to detect changes of the magnitude that we expect to occur over the next few decades. To understand the sensitivity of climate to external perturbations, one needs to know the dominant forcings and feedbacks. Greenhouse gas concentrations are adequately measured in situ (except for ozone, whose vertical profile is important and could be monitored with a sufficient density of ozonesondes), so the primary forcing monitoring need is for tropospheric aerosol optical depths. The relevant feedbacks we need to monitor are changes in the vertical profile of water vapor, changes in the spatial distribution of clouds and cloud properties, and changes in surface properties.

The next question is whether changes in these quantities can be monitored with current instrument capabilities. To define accuracy requirements, one can take two approaches: (1) By how much do climate models or theory predict that a given parameter is going to change over several decades? (2) By how much would a parameter have to change to cause a significant TOA radiative flux change? To answer the latter question, consider that the current atmospheric loading of greenhouse gases since the Industrial Revolution has been sufficient to cause a net radiative imbalance of 1.5-2.5 Wm^{-2} to date (IPCC, 1992). Thus, one might argue that a change in flux roughly an order of magnitude smaller than this, 0.25 Wm^{-2}, is significant and defines a useful accuracy for monitoring instruments. (By comparison, over the next 20 years the expected additional climate forcing due to increasing greenhouse gas concentrations is 1 Wm^{-2}). One can calculate the required change in a given parameter using a 1-dimensional radiative transfer model analogous to that used in a GCM.

To answer the former question, we must resort to theory and existing climate model simulations. Typical predicted warmings over the next 20 years in transient climate change GCM simulations are of order 0.7°C (Hansen et al., 1988). We can use the Clausius-Clapeyron equation to estimate associated changes in water vapor, given constant relative humidity (keeping in mind that upper troposphere changes will in general be larger than lower troposphere changes because warming increases with height in climate change scenarios). Cloud height changes can be estimated from the dependence of convective penetration depth on surface temperature (Del Genio, 1993a). The temperature dependence of adiabatic liquid water content, which can be derived from simple thermodynamic arguments (Betts and Harshvardhan, 1987) can be used to infer changes in cloud optical thickness in the absence of aerosols. Furthermore, if one assumes constant droplet number concentration, an estimate of cloud particle size changes can be made as well. In the presence of indirect aerosol effects, changes in both quantities might be much larger. Changes in cloud cover are still not understood fundamentally, but we can use the results of climate change simulations (cf. Fig. 2.7) to estimate plausible regional changes. Aerosol changes can be estimated empirically from current, albeit crude, assessments of the anthropogenic effect over the past century (Charlson et al., 1992).

Table 2.2 shows the predicted changes in each of these quantities over the next 20 years, as well as the more restrictive estimate of changes required to produce a 0.25 Wm^{-2} perturbation of the TOA radiation balance. A more complete list and discussion of the basis for the predicted changes can be found in Del Genio (1993b) and Hansen et al. (1995b). The table gives us a rough idea of how accurately we need to monitor various parameters to actually detect a climate change and attribute it to a particular combination of forcings and feedbacks.

Some of the changes are quite small and present an imposing challenge for satellite instruments and retrieval algorithms. But the numbers in the table do not refer to absolute accuracies. Instead they denote the precision with which changes in these quantities must be measured, a less formidable demand on observing systems.

How can these requirements be met? If one starts with the general statement that all climate forcings and feedbacks operate via modification of either the solar or terrestrial radiation spectrum, then the requirement is for monitoring instruments that cover the solar and thermal spectra with sufficient sensitivity to detect these small changes. Hansen et al. (1995b) have proposed a small satellite mission, Climsat, that includes a Michelson interferometer and scanning polarimeter for this purpose. Space-proven predecessors of both instruments have flown on planetary missions, and each is self-calibrating, a crucial element for monitoring of decadal trends (Hansen et al., 1995b). Combined with monitoring of stratospheric aerosols, water vapor and spectral solar irradiance by other spacecraft, as well as ground-based monitoring of ozone

Parameter	Plausible 20 year change	Change required for Δflux= 0.25
Water vapor (Δq/q)		
Upper troposphere	.10	.02
Lower troposphere	.04	.02
Cloud top pressure (mb)	12	5
Cloud optical thickness		
Cirrus	.1	.02
Stratus (without/with aerosol)	.5/1.0	.07
Cloud particle size (μm)		
Cirrus	.3	.06
Stratus (without/with aerosol)	.1/1.0	.2
Cloud cover (%, regional)		
Cirrus	.03	.004
Stratus	.03	.003
Tropospheric aerosol optical depth	.04	.01

Table 2.2: *Predicted changes in selected important climate forcing and feedback parameters over the next 20 years, and changes required for a significant radiative flux perturbation.*

profiles and more accurate estimates of aerosol forcing in key areas, virtually all the important climate forcings and feedbacks would be continuously observed. This would provide the most stringent possible test of the realism of GCMs used to predict decadal climate change.

2.5 Conclusion

As a final note, it is worth commenting on the stated IPCC uncertainties in climate sensitivity (2–5°C; IPCC, 1992) and how in practice this range will be narrowed in the future. Current estimates of uncertainty are simply based on the full range of results obtained by every climate GCM that has ever performed an equilibrium doubled CO_2 simulation. One might imagine two routes to reducing the uncertainty. In one case, only results from the most recent versions of GCMs would be included. Simplified SST perturbation sensitivity experiments indicate a dramatic convergence of recent results (Cess et al., 1996). But it would be dangerous to conclude that this is evidence of better understanding of the climate system. In part at least it is likely to be the result of the psychology of the now-ubiquitous climate model intercomparison projects, which causes outliers to try to change their models so as to move toward the middle, or "consensus", result. One saving grace is that agreement in a particular GCM intercomparison does not guarantee agreement in greenhouse gas change simulations. The different sensitivity of the current GISS GCM to prescribed uniform SST perturbations and doubled CO_2 is a case in point.

A better strategy would be to identify individual climate feedbacks and the physical processes that cause them. For at least a few examples, such as upper tropospheric water vapor (Del Genio et al., 1994), convective available potential energy (Ye et al., 1996), and low cloud optical thickness (Tselioudis et al., 1992; Del Genio et al., 1996), current observations of variability are good enough to test model parameterizations and indirectly relate climate feedbacks to processes operating on shorter time scales. The observational requirements described in this paper, if met, would provide a host of other such figures of merit. Evaluated in this way, a subset of climate models that get the same answer for the same reason, verified by data for the current climate, will eventually define the true range of uncertainty, until actual detection of decadal climate change settles the issue once and for all.

Acknowledgments

I thank William Kovari for assistance with the figures used in this paper. This research was supported by the NASA Climate Modeling, FIRE III and Earth Observing System Interdisciplinary Programs, and the DOE Atmospheric Radiation Measurement Program.

2.6 References

Barkstrom BR (1984) The Earth Radiation Budget Experiment (ERBE). *Bull Amer Meteor Soc* 67: 1170–1185

Betts AK, Harshvardhan (1987) Thermodynamic constraint on the cloud liquid water feedback in climate models. *J Geophys Res* 92: 8483–8485

Broecker WS (1991) The great ocean conveyor. *Oceanography* 4: 79–89

Cess RD, Potter GL, Blanchet JP, Boer GJ, Del Genio AD, Deque M, Dymnikov V, Galin V, Gates WL, Ghan SJ, Kiehl JT, Lacis AA, Le Treut H, Li Z-X, Liang X-Z, McAvaney BJ, Meleshko VP, Mitchell JFB, Morcrette J-J, Randall DA, Rikus L, Roeckner EE, Royer JF, Schlese U, Sheinin DA, Slingo A, Sokolov AP, Taylor KE, Washington WM, Wetherald RT, Yagai I, Zhang M-H (1990) Intercomparison and interpretation of climate feedback processes in 19 atmospheric general circulation models. *J Geophys Res* 95: 16601-16615

Cess RD, Zhang MH, Minnis P, Corsetti L, Dutton EG, Forgan BW, Garber DP, Gates WL, Hack JJ, Harrison EF, Jing X, Kiehl JT, Long CN, Morcrette J-J, Potter GL, Ramanathan V, Subasilar B, Whitlock CH, Young DF, Zhou Y (1995) Absorption of solar radiation by clouds: Observations versus models. *Science* 267: 496–499

Cess RD, Zhang MH, Potter GL, Alekseev V, Barker HW, Cohen-Solal E, Colman RA, Dazlich DA, Del Genio AD, Dix MR, Dymnikov V, Esch M, Fowler LD, Fraser JR, Galin V, Gates WL, Hack JJ, Ingram WJ, Kiehl JT, Le Treut H, Lo KK-W, McAvaney BJ, Meleshko VP, Morcrette J-J, Randall DA, Roeckner E, Royer J-F, Schlesinger ME, Sporyshev PV, Timbal B, Volodin EM, Taylor KE, Wang W, Wetherald RT (1996) Cloud feedback in atmospheric general circulation models: An update. *J Geophys Res*, submitted

Charlson RJ, Lovelock JE, Andreae MO, Warren SG (1987) Oceanic phytoplankton, atmospheric sulfur, cloud albedo and climate. *Nature* 326: 655–661

Charlson RJ, Langner J, Rodhe H, Leovy CB, Warren SG (1991) Perturbation of the northern hemisphere radiative balance by backscattering from anthropogenic sulfate aerosols. *Tellus* 43AB: 152–163

Charlson RJ, Schwartz SE, Hales JM, Cess RD, Coakley JA, Hansen JE, Hofmann DJ (1992) Climate forcing by anthropogenic aerosols. *Science* 255: 423–430

Curry JA, Ebert EE (1992) Annual cycle of radiation fluxes over the Arctic Ocean: Sensitivity to cloud optical properties. *J Climate* 5: 1267–1280

Dai A (1995) Global precipitation variability and its relationship with other climate changes. Ph.D. dissertation, Columbia University, New York

Del Genio AD (1993a) Convective and large-scale cloud processes in global climate models. Energy and Water Cycles in the Climate System (E Raschke and D Jacob, eds), *Springer-Verlag*, Berlin Heidelberg New York: 95–121

Del Genio AD (1993b) Accuracy requirements. Proceedings, Workshop on Long-Term Monitoring of Global Climate Forcings and Feedbacks (J Hansen, W Rossow and I Fung, eds), NASA CP-3234: 13–19

Del Genio AD (1996a) GCM implications for mechanisms determining cloud and water vapor feedbacks. In Climate Sensitivity to Radiative Perturbations: Physical Mechanisms and Validation (H Le Treut, ed), *Springer-Verlag*, Berlin Heidelberg New York, p. 107–125

Del Genio AD (1996b) TRMM: The Tropical Rainfall Measuring Mission: This volume, chapter 20

Del Genio AD, Kovari W, Yao M-S (1994) Climatic implications of the seasonal variation of upper troposphere water vapor. *Geophys Res Letters* 21: 2701–2704

Del Genio AD, Yao M-S, Kovari W, Lo KK-W (1996) A prognostic cloud water parameterization for global climate models. *J Climate* 9: 270–304

Elliott WP, Gaffen DJ (1991) On the utility of radiosonde humidity archives for climate studies. *Bull Amer Meteor Soc* 72: 1507–1520

Feigelson EM (1978) Preliminary radiation model of a cloudy atmosphere. Part I - Structure of clouds and solar radiation. *Beitr Phys Atmos* 51: 203–229

Gleckler PJ, Randall DA, Boer G, Colman R, Dix M, Galin V, Helfand M, Kiehl J, Kitoh A, Lau W, Liang X-Y, Lykossov V, McAvaney B, Miyakoda K, Planton S, Stern W (1995) Cloud- radiative effects on implied oceanic energy transports as simulated by atmospheric general circulation models. *Geophys Res Letters* 22: 791–794

Han Q, Rossow WB, Lacis AA (1994) Near-global survey of effective cloud droplet radii in liquid water clouds using ISCCP data. *J Climate* 7: 465–497

Hansen J, Fung I, Lacis A, Rind D, Lebedeff S, Ruedy R, Russell G, Stone P (1988) Global climate changes as forecast by Goddard Institute for Space Studies three-dimensional model. *J Geophys Res* 93: 9341–9364

Hansen J, Sato M, Ruedy R (1995a) Long-term changes of the diurnal temperature cycle: Implications about mechanisms of global climate change. *Atmos Res*, submitted

Hansen J, Rossow W, Carlson B, Lacis A, Travis L, Del Genio A, Fung I, Cairns B, Mishchenko M, Sato M (1995b) Low-cost long-term monitoring of global climate forcings and feedbacks. *Clim Change*, in press

Henderson-Sellers A (1992) Continental cloudiness changes this century. *Geo Journal* 27: 255–262

Horel JD, Wallace JM (1981) Planetary-scale atmospheric phenomena associated with the Southern Oscillation. *Mon Wea Rev* 109: 813–829

IPCC (Intergovernmental Panel on Climate Change), Varney SK (1992) Climate change 1992. In The Supplementary Report to the IPCC Scientific Assessment (Houghton JT, Callander BA, eds.), Cambridge University Press, 200 pp

Jones A, Roberts LD, Slingo A (1994) A climate model study of indirect radiative forcing by anthropogenic sulphate aerosols. *Nature* 370: 450–453

Karl TR, Jones PD, Knight RW, Kukla G, Plummer N, Razuvayev V, Gallo KP, Lindseay J, Charlson RJ, Peterson TC (1993) A new perspective on recent global warming. *Bull Amer Meteor Soc* 74: 1007–1023

Klein SA, Hartmann DL (1993) The seasonal cycle of low stratiform clouds. *J Climate* 6: 1587–1606

Kuo H, Schubert WH (1988) Stability of cloud-topped boundary layers. *Quart J Roy Meteor Soc* 114: 887–916

Li Z-X, LeTreut H (1992) Cloud-radiation feedbacks in a general circulation model and their dependence on cloud modeling assumptions. *Clim Dyn* 7: 133–139

Liao X, Rind D, Rossow WB (1995) Comparison between SAGE II and ISCCP high-level clouds. Part II: Locating cloud tops. *J Geophys Res* 100: 1137–1147

Lin B, Rossow WB (1994) Observations of cloud liquid water path over oceans: Optical and microwave remote sensing methods. *J Geophys Res* 99: 20909–20927

Liu WT (1988) Moisture and latent heat flux variabilities in the tropical Pacific derived from satellite data. *J Geophys Res* 93: 6749–6760

Macke A (1993) Scattering of light by polyhedral ice crystals. *Appl Optics* 32: 2780–2788

Marotzke J, Willebrand J (1991) Multiple equilibria of the global thermohaline circulation. *J Phys Ocean* 21: 1372–1385

Miller RL, Del Genio AD (1994) Tropical cloud feedbacks and decadal variability of climate. *J Climate* 7: 1388–1402

Minnis P, Liou K-N, Takano Y(1993) Inference of cirrus cloud properties using satellite-observed visible and infrared radiances. Part I: Parameterization of radiance fields. *J Atmos Sci* 50: 1279–1304

Mitchell JFB, Senior CA, Ingram WJ (1989) CO_2 and climate: A missing feedback? *Nature* 341: 132–134

Oort AH (1983) Global atmospheric circulation statistics, 1958–1973. NOAA Prof. Paper 14, U.S. Dept. of Commerce, Washington, DC: 180 pp

Pan YH, Oort AH (1983) Global climate variations connected with sea surface temperature anomalies in the eastern equatorial Pacific Ocean for the 1958-1973 period. *Mon Wea Rev* 111: 1244–1258

Penner JE, Dickinson RE, O'Neill CA (1992) Effects of aerosol from biomass burning on the global radiation budget. *Science* 256: 1432–1434

Platnick S, Twomey S (1994) Determining the susceptibility of cloud albedo to changes in droplet concentration with the Advanced Very High Resolution Radiometer. *J Appl Meteor* 33: 334–347

Ramanathan V, Collins W (1991) Thermodynamic regulation of ocean warming by cirrus clouds deduced from observations of the 1987 El Niño. *Nature* 351: 27–32

Randall DA (1980) Conditional instability of the first kind upside-down. *J Atmos Sci* 37: 125–130

Randall DA, Cess RD, Blanchet JP, Boer GJ, Dazlich DA, Del Genio AD, Deque M, Dymnikov V, Galin V, Ghan SJ, Lacis AA, Le Treut H, Li Z-X, Liang X-Z, McAvaney BJ, Meleshko VP, Mitchell JFB, Morcrette J-J, Potter GL, Rikus L, Roeckner E, Royer JF, Schlese U, Sheinen DA, Slingo J, Sokolov AP, Taylor KE, Washington WM, Wetherald RT, Yagai I, Zhang M-H (1992) Intercomparison and interpretation of surface energy fluxes in atmospheric general circulation models. *J Geophys Res* 97: 3711–3724

Raschke E (1996) Energy and water cycles in the climate system. This volume, chapter 1

Rind D, Chiou E-W, Chu W, Larsen J, Oltmans S, Lerner J, McCormick P, McMaster L (1991) Positive water vapor feedback in climate models confirmed by satellite data. *Nature* 349: 500–503

Rossow WB (1996) Atmospheric water vapor. This volume, chapter 8

Rossow WB, Schiffer RA (1991) ISCCP cloud data products. *Bull Amer Meteor Soc* 72: 2–20

Soden BJ, Bretherton FP (1993) Upper tropospheric relative humidity from the GOES 6.7 μm channel: Method and climatology for July 1987. *J Geophys Res* 98: 16669–16688

Somerville RCJ, Remer LA (1984) Cloud optical thickness feedbacks in the CO_2 climate problem. *J Geophys Res* 89: 9668–9673

Tselioudis G, Rossow WB, Rind D (1992) Global patterns of cloud optical thickness variation with temperature. *J Climate* 5: 1484–1495

Twomey S (1977) Atmospheric Aerosols. Elsevier Amsterdam Oxford New York: 302 pp

Walcek CJ (1994) Cloud cover and its relationship to relative humidity during a springtime midlatitude cyclone. *Mon Wea Rev* 122: 1021–1035

Wang Q, Albrecht BA (1994) Observations of cloud-top entrainment in marine stratocumulus clouds. *J Atmos Sci* 51: 1530–1547

Wang J, Rossow WB (1995) Determination of cloud vertical structure from upper-air observations. *J Appl Meteor* 34: 2243–2258

Warren SG, Hahn CJ, London J, Chervin RM, Jenne RL(1986) Global distribution of total cloud cover and cloud type amounts over land. NCAR/TN-273+STR/DOE Tech. Report ER/60085-H1: 29 pp and 200 maps

Warren SG, Hahn CJ, London J, Chervin RM, Jenne RL (1988) Global distribution of total cloud cover and cloud type amounts over ocean. NCAR/TN-317+STR/DOE Tech. Report ER-0406: 42 pp + 170 maps

Wielicki BA, Parker L (1992) On the determination of cloud cover from satellite sensors: The effect of sensor spatial resolution. *J Geophys Res* 97: 12799–12823

Ye B, Del Genio AD, Lo KK-W (1996) CAPE variations in the current climate and in a climate change. *J Climate* , submitted

Zaucker F, Stocker TF, Broecker WS(1994) Atmospheric freshwater fluxes and their effect on the global thermohaline circulation. *J Geophys Res* 99: 12443–12457

Zhang Y-C, Rossow WB, Lacis AA (1994) Calculation of surface and top of atmosphere radiative fluxes from physical quantities based on ISCCP data sets. 1. Method and sensitivity to input data uncertainties. *J Geophys Res* 100: 1149–1165

Chapter 3

Surface Fluxes of Momentum, Heat, and Water Vapor

John D. Albertson, University of California, Davis, CA
Gerard Kiely, University College Cork, Ireland
Marc B. Parlange, University of California, Davis, CA

Abstract

Advances in global scale hydrology and climatology demand an improved understanding of the interaction between the Earth's surface and its atmosphere. Such an understanding is being sought through field experiments over the ocean and the land and through the use of remote sensing. In many cases, the traditional eddy correlation measurement technique to obtain fluxes of heat and momentum is problematic and it becomes necessary to employ less direct techniques. In this paper the inertial-dissipation method is reviewed and a refined approach is presented. The new dissipation model circumvents several of the short-comings inherent to previous formulations of the dissipation approach. Experimental data are presented to support new empirical scaling forms for turbulent kinetic energy and scalar variance dissipation rates and the refined dissipation-flux model is tested. Excellent agreement with direct flux measurements is obtained.

3.1 Introduction

Momentum and scalar fluxes between the Earth's surface and its atmosphere are fundamental to issues ranging from watershed management to climate systems, as well as to our basic understanding of flow and transport in the atmospheric surface layer. Efforts to model the atmospheric boundary layer (ABL) in itself or as part of a larger climate simulation model are limited by an inadequate understanding of surface fluxes over the land and the ocean. Advances in regional and global scale hydro-meteorology must rely on observations of the exchange between the Earth's surface and its atmosphere. For these observations to be relevant, they must be made over a spatial scale appropriate to the problem being addressed. Satellite-based instruments hold promise for diagnosing and mapping surface fluxes. However, the development of reliable algorithms for relating the instrument readings to actual surface fluxes hinges on a program of careful comparison of algorithm estimates to actual measurements of the fluxes. Benchmark flux measurements may be acquired through (i) inference from vertical

NATO ASI Series, Vol. I 45
Radiation and Water in the Climate System:
Remote Measurements
Edited by Ehrhard Raschke

profiles of mean meteorological quantities using Monin-Obukhov similarity theory, (ii) remote sensing of ABL turbulence with lidar or scintillometry, or (iii) *in situ* point measurements of turbulent fluxes using eddy correlation or dissipation techniques in the ABL. For these techniques to provide measurements that represent integrated fluxes over large spatial scales, the measurements must be made above the blending height; here we rely on the thorough mixing provided by the extremely high Reynolds number turbulent flow of the ABL. For a complete discussion of the idea of a blending height and the relationship between measurement height and spatial integration of surface fluxes see the recent review paper by Parlange et al. (1995b).

Although the eddy correlation (EC) method provides a direct measurement of the vertical fluxes, there are applications for which it is inappropriate, particularly over water and for remote field experiments requiring long term unattended operation. Fluxes from the mean profile method or the bulk aerodynamic method have limitations as well. The flux-dissipation method is used widely over water surfaces, typically using power spectra to determine the dissipation rates, followed by an iterative procedure to determine fluxes. However, this approach suffers from uncertainties due to empirical inertial subrange constants, the jumpy (or noisy) nature of power spectra, and the reliance on an iterative solution. In this paper we introduce a one-step (non iterative) dissipation method for determining fluxes of momentum, sensible heat and water vapor. This new method which uses third order structure functions to determine dissipation rates, avoids many of the pitfalls previously assumed unavoidable with the dissipation method. The results presented here may also improve the reliability of flux estimates from the ABL remote sensing instruments, such as lidar and scintillometers, since these techniques typically rely on an empirical relationship between scalar fluxes and the dissipation rate of scalar variances (see Hill et al., 1992 and Eichinger et al., 1993).

3.1.1 The Atmospheric Boundary Layer

The ABL is that layer of air, directly above the Earth's surface in which the effects of the surface (friction, evaporation, heating and cooling) are felt directly on time scales less than a day. The ABL is broadly made up of two layers, the atmospheric surface layer (ASL) and the mixed layer (ML). The ASL is affected primarily by surface fluxes while the ML is affected by both surface fluxes and boundary layer entrainment of the free atmosphere air from above. Figure 3.1 is a schematic of the composition of the ABL. The ASL occupies about the lowest 10% of the fully developed daytime ABL, or approximately the first 100 m above the Earth's surface (Parlange et al., 1995b).

In the ASL, the turbulent flow is often assumed to be statistically stationary, when considering periods of 10 minutes to 1 hour. The principle mechanism for the mechanical production of turbulence in this time period is the vertical gradient of the mean wind. Figure 3.2 shows schematics of the vertical profiles of the means of velocity U, potential temperature Θ, and humidity $\overline{q}$ for the daytime convective boundary layer (CBL). It is seen from Figure 3.2, that the steepest gradients occur in the lower 10% of the boundary layer (i.e. the ASL), while in the upper 90% of the CBL strong convective mixing and entrainment of the free atmosphere air at the top of the ABL smooth out almost all vertical variations in the mean profiles.

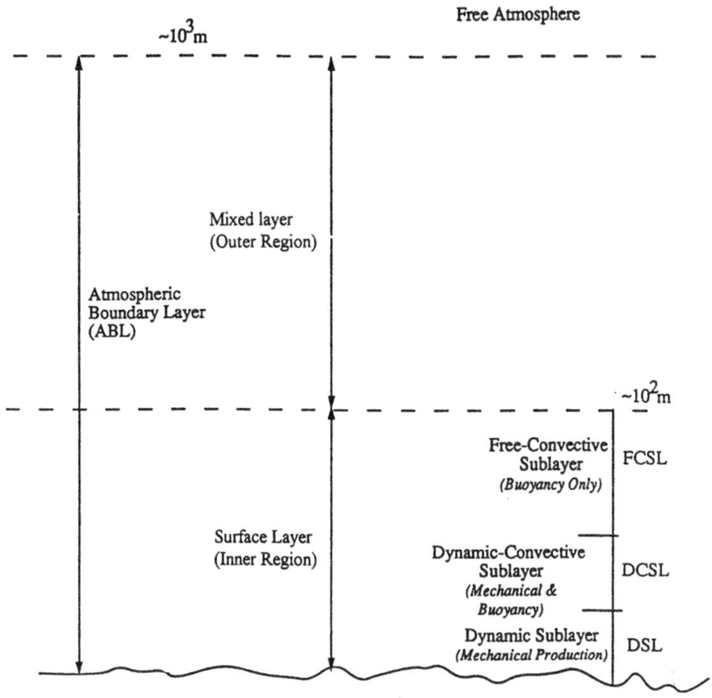

Figure 3.1: *Atmospheric boundary layer schematic.*

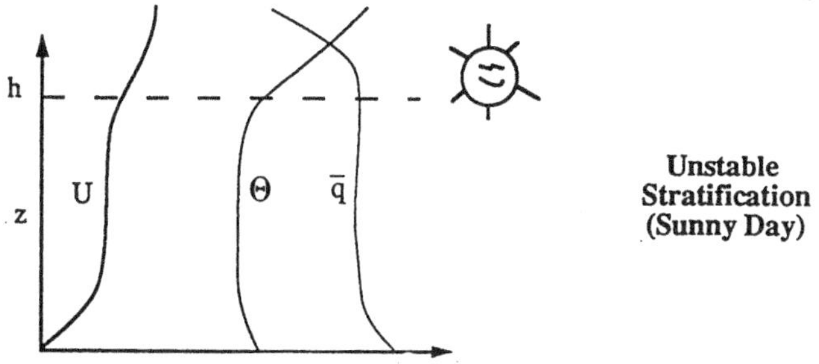

Figure 3.2: *Mean profiles of wind speed, potential temperature, and humidity in the convective atmospheric boundary layer.*

3.1.2 Surface Layer Scaling

ASL Turbulence

Before getting into the details of ASL scaling, it is instructive to review briefly the turbulence spectrum as shown schematically in Figure 3.3. The eddies carrying most of the energy are of the larger, integral scale size. The integral scale for longitudinal velocity (defined by l)

is representative of the large, turbulent energy producing eddies. At the other end of the spectrum, energy is dissipated by molecular viscosity via eddies of length scale approaching the Kolmogorov scale η (typically of order $\approx$ 1 mm). Between these two extremes lies what is known as the inertial subrange, which marks the broad range of scales between the anisotropic energy bearing eddies and the viscosity controlled dissipative eddies. This inertial range, for high Reynolds number flows as found in the ASL, spreads over several orders of magnitude of wavenumbers. Referring to Figure 3.3, the spectral density peaks at wavenumbers of the order of l^{-1} . For the mid-spectrum inertial scales, Kolmogorov (1941) postulated that the energy content is controlled simply by the eddy size (k) and the rate of energy transfer through the spectrum (taken under an assumption of steady state as ε, the average dissipation rate of turbulent kinetic energy). Consequently, straight dimensional analysis yields Kolmogorov's celebrated -5/3 spectrum for the inertial subrange, $E(k) \sim \varepsilon^{2/3}k^{-5/3}$. It is seen from Figure 3.3, that as k increases, the energy falls with a slope of -5/3 (in a log-log framework). For wavenumbers approaching η^{-1}, local dissipation induces steadily faster decay of energy content. We make much use of Kolmogorov's relationship in later sections.

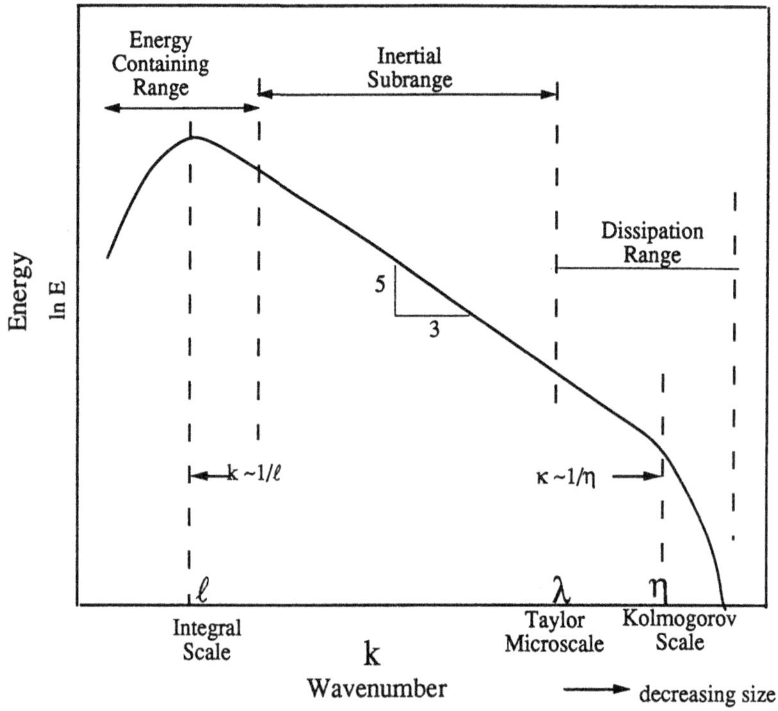

Figure 3.3: *Schematic of the power spectrum. The integral length scale is represented by l and the Kolmogorov length scale is represented by η. Wavenumbers (k) are in radians/length.*

Turbulence due to the gradient of the mean wind speed is created by mechanical (shear) turbulence. A second form of turbulence production is that due to buoyancy, induced by vertical temperature stratification. In general, turbulence is produced by a combination of both during daytime hours, with the ratio of buoyant to mechanical production of turbulence increasing with height. This ratio is represented by the stability parameter, $-z/L$, where z is the height above the surface and L is the Monin-Obukhov length

$$L = -\frac{u_*^3 \Theta}{kg < w\theta >} \tag{3.1}$$

where the friction (shear) velocity is u_*, the fluctuating vertical velocity is w, the mean potential air temperature is Θ and its fluctuating component is θ, the dimensionless parameter k is von Karman's constant ($\simeq 0.4$) and g is gravitational acceleration. The three classifications of atmospheric stability can be expressed in terms of the stability parameter as

$$\begin{aligned} &unstable\ stratification & -\frac{z}{L} &> 0 \\ &neutral & -\frac{z}{L} &\Rightarrow 0 \\ &stable & -\frac{z}{L} &< 0 \end{aligned} \tag{3.2}$$

Monin-Obukhov Scaling

Prandtl set out the concept of the logarithmic wind velocity profile, based on dimensional analysis. In plane parallel flow, an increase of the horizontal mean velocity in the z direction is evidence of a downward momentum flux. Thus the gradient of mean velocity is determined by the shear stress at the earth surface, and the distance from that surface, such that

$$\frac{dU}{dz} = \frac{u_*}{kz} \tag{3.3}$$

Upon integration from z_1 to z_2 we obtain the familiar log law

$$U_2 - U_1 = \frac{u_*}{k} \ln\left(\frac{z_2}{z_1}\right) \tag{3.4}$$

where U_1 and U_2 are the mean longitudinal velocities at elevations z_1 and z_2.

The similarity theory of Monin and Obukhov extended Prandtl's scaling to scalars and into the unstable and stable regimes. This was accomplished by noting the analogy between scalars and velocity and by admitting additional variables to the dimensional analysis to account for the effect of density stratification (i.e. (3.1)). Monin-Obukhov (MO) scaling of mean profiles in the ASL, for example, yields

$$\frac{dU}{dz} = \frac{u_*}{kz}\Phi_m\left(\frac{z}{L}\right) \tag{3.5a}$$

$$\frac{d\Theta}{dz} = \frac{\Theta_*}{kz}\Phi_H\left(\frac{z}{L}\right) \tag{3.5b}$$

$$\frac{d\overline{q}}{dz} = \frac{q_*}{kz}\Phi_q\left(\frac{z}{L}\right) \tag{3.5c}$$

with $\Theta_*(= -H/\rho c_p u_*)$ being a temperature scale, $q_*(= -E/\rho u_*)$ a humidity scale, and Φ_m, Φ_H, and Φ_q representing non-dimensional gradients of mean velocity, temperature, and humidity, respectively. A basic premise of MO scaling is that the stability functions (e.g. Φ_m, Φ_H, and Φ_q) are universal functions of the stability parameter.

Recently a three sublayer model has been proposed as an extension of MO scaling (see Kader, 1988, and Kader and Yaglom, 1990). This model represents the lowest sublayer of the ASL as the *Dynamic Sublayer* (DSL) in which buoyant action is negligible compared to mechanical production. The uppermost sublayer of the ASL, governed by buoyancy only, is termed the *Free Convection Sublayer* (FCSL). Between these two is a sublayer, driven by both mechanical and buoyant forces, termed the *Dynamic-Convective Sublayer* (DCSL). The sublayers are separated by narrow regions of transition. The $-z/L$ limits of the sublayers, based on the order of magnitude analysis of Kader (1992), are

$$\begin{aligned} DSL &: \quad -z/L < 0.04 \\ DCSL &: \quad 0.12 < -z/L < 1.2 \\ FCSL &: \quad -z/L > 2.0 \end{aligned}$$

The TKE and temperature variance budget equations are scaled in the context of the three sublayer model using directional dimensional analysis (DDA) which involves different length scales for the different directions, with L_x for the horizontal and L_z for the vertical (see Panton, 1984, p.207–209). The use of DDA implies that the vertical and horizontal motions are uncoupled. In the DCSL the energy transferred between the horizontal and vertical motions through vortex stretching is considered to be much less than that produced in each the vertical and horizontal directions. Therefore, the transfer between directions is ignored and we consider the horizontal motions to be driven solely by horizontal processes (shear) and the vertical motions to be driven solely by vertical processes (buoyancy) only. We will return to this model below as we examine the scaling of the TKE and scalar variance budget equations. For further review of DDA and its application to dissipation rates see Albertson et al. (1996).

Fluxes and methods of determination

The vertical fluxes of momentum, heat and moisture are defined as

$$u_* = [-<uw>]^{1/2} \tag{3.6a}$$

$$H = \rho c_p <w\theta> \tag{3.6b}$$

$$E = \rho <wq> \tag{3.6c}$$

where q is the fluctuation from the mean specific humidity and ρ is the mean air density. The latent heat flux (LE) is related to the evaporation rate (E) as $LE = L_v E$, where L_v is the latent heat of vaporization of water (see Brutsaert, 1982).

The four standard techniques of flux determination are eddy correlation, mean profiles, bulk aerodynamic, and inertial-dissipation (see Brutsaert, 1982). The eddy correlation (EC) method is a direct measurement of fluxes, while the other three are indirect methods relying on similarity theory. The EC method requires the measurement of the fluctuating components of the vertical velocity, and an associated fluctuating component for the term being transported, e.g.: horizontal velocity, for momentum flux; temperature, for heat flux; and humidity, for moisture flux. The use of EC on a moving platform (ship or aircraft) is prone to errors due to contamination of the w signal by motion of the platform. The mean profile method requires precise measurements of the vertical gradients of the mean meteorological variables, which is also difficult when applied from ships or aircraft. The bulk aerodynamic method is largely a conceptual approach consisting of bulk transfer equations with drag coefficients which are estimated experimentally (Brutsaert, 1982, p.201; Brutsaert, 1986). The dissipation approach is significantly less susceptible to errors from ship motion, but does require measurement of turbulent fluctuations at a frequency capable of resolving fluctuations in the inertial subrange as well as the use of similarity theory and other assumptions in its formulation. With this last approach the fluxes (covariances) are not determined directly but are derived from turbulence measurements and use of variance budget equations that relate surface fluxes to mean dissipation rates. Therefore, we begin the discussion of this method with an examination of the budget equations for turbulent kinetic energy and temperature variance (with temperature serving as a proxy for any other scalar).

3.1.3 Turbulent kinetic energy

The mean turbulent kinetic energy (TKE) dissipation rate (ε) is related to the production rate of TKE and hence the surface fluxes through the TKE budget equation. The TKE budget equation is obtained by multiplying the momentum equation for the $\alpha(=1,2,3)$ direction, by u_α, time averaging all terms and subtracting the equation governing the kinetic energy of the mean flow (Tennekes and Lumley, 1972, p63). For a stationary flow with horizontal homogeneity we write the budget equation for the time averaged TKE per unit mass with the Boussinesq approximation as

$$-<uw>\frac{\partial U}{\partial z}+\frac{g<w\theta>}{\Theta}-\frac{1}{2}\frac{\partial<ew>}{\partial z}-\frac{1}{\rho}\frac{\partial<pw>}{\partial z}=\varepsilon \tag{3.7}$$

where p is the fluctuating component of the pressure, and e is twice the TKE (i.e. TKE $=\frac{e}{2}=\frac{u^2+v^2+w^2}{2}$). The first term in (3.7) represents the mechanical (or shear) production of TKE by interaction of the Reynolds stress with the mean velocity gradient. The second term represents the TKE production by buoyancy, the third the divergence of vertical turbulent flux of TKE and the fourth the transfer term for TKE due to pressure-velocity interaction. The latter two terms on the left hand side of (3.7) are generally called the transport terms. This budget is non-dimensionalized by u_*^3/kz to arrive at

$$\Phi_m\left(\frac{z}{L}\right)-\frac{z}{L}-\frac{Fkz}{u_*^3}-\frac{Pkz}{u_*^3}=\Phi_\varepsilon\left(\frac{z}{L}\right) \tag{3.8}$$

where

$$\Phi_\varepsilon\left(\frac{z}{L}\right)=\frac{\varepsilon kz}{u_*^3} \tag{3.9}$$

and F and P are written for $\frac{1}{2}\frac{\partial<ew>}{\partial z}$ and $\frac{1}{\rho}\frac{\partial<pw>}{\partial z}$, respectively.

It is often assumed that total production (mechanical plus buoyant) equals dissipation (Deacon, 1988), regardless of the state of atmospheric stratification and position in the boundary layer, i.e.

$$\Phi_m-\frac{z}{L}\approx\Phi_\varepsilon \tag{3.10}$$

where Φ_m is the dimensionless mechanical production and $-z/L$ is the dimensionless buoyant production.

In the Kansas experiments, Wyngaard and Coté (1971; abbreviated here as WC71) made measurements of the production, turbulent transport and dissipation of TKE for near neutral and moderately unstable stratification. Their results showed dissipation exceeding production for the slightly unstable flows, with an equality approached for more strongly convective flows. Their empirical functions for non-dimensional shear production and dissipation are

$$\Phi_m\left(\frac{z}{L}\right) = \left(1+16\left|\frac{z}{L}\right|\right)^{-\frac{1}{4}} \quad (3.11a)$$

$$\Phi_\varepsilon\left(\frac{z}{L}\right) = \left(1+0.5\left|\frac{z}{L}\right|^{\frac{2}{3}}\right)^{\frac{3}{2}} \quad (3.11b)$$

WC71 suggested that the pressure-velocity term may be important but that direct measurements were needed to investigate this. Work by McBean et al. (1971) also found dissipation to exceed total production for $-z/L > 0.3$. Further work by McBean and Elliot (1975) in measurements of F and P over dry prairie land showed that these two terms were somewhat balanced, with P adding energy and F removing energy. Leavitt and Paulson (1975) in an ocean experiment concluded that dissipation equalled production. Champagne et al. (1977), in work over bare furrowed soil, found dissipation to exceed production. Frenzen and Vogel (1992), in an experiment over wheat in Wyoming, found the dissipation rate to be less than production and showed that with corrections the WC71 data show dissipation equalling production. Many of these studies represent narrow ranges of stability. Yet, the different circumstances and measurement techniques of the studies inhibit the drawing of conclusions from the studies taken as an ensemble. Therefore, we present new experimental results from dissipation measurements made over a wide range of stability (Albertson et al., 1996; Kiely et al., 1996) and we investigate the scaling of these measurements in the context of the three sublayer model.

Accepting that all one point fluctuation moments in the DSL are independent of z (Kader, 1992), then F and P vanish in this region. Therefore, in the DSL the normalized dissipation rate should equal the normalized production rate, which is know to be a constant of order 1.0. In the DCSL the F and P terms may be significant. From DDA the shear velocity u_* has dimensions of $L_x^{1/2}L_z^{1/2}t^{-1}$ and the convective velocity $w_*(=[<w\theta>gz/\Theta]^{1/3})$ has dimensions of $L_z t^{-1}$, where t is used to represent the time dimension. The convective velocity is used to scale the vertical motion and the combination u_*^2/w_* is used to scale the horizontal motions. In the FCSL the scaling is independent of u_* and thus the relevant velocity is w_*. From dimensional analysis for the TKE we obtain (Albertson et al., 1996)

$$\Phi_\varepsilon = \Phi_m = C_1 \approx 1 \qquad -\frac{z}{L} < 0.04 \quad (3.12a)$$

$$\Phi_\varepsilon = C_2\left(-\frac{z}{L}\right)^{-\frac{1}{3}} + C_3\left(-\frac{z}{L}\right) \qquad 0.12 < -\frac{z}{L} < 1.2 \quad (3.12b)$$

$$\Phi_\varepsilon = C_4\left(-\frac{z}{L}\right)^{\frac{1}{3}} + C_5\left(-\frac{z}{L}\right) \qquad -\frac{z}{L} > 2.0 \quad (3.12c)$$

In the DCSL, C_2 is a constant that describes the effects of mechanical production and the vertical transport of $< u^2 >$ and $< v^2 >$. The value of C_3 accounts for the buoyant production and the combined effects of vertical transport of $< w^2 >$ and the pressure-velocity interaction (Kader, 1992). In the FCSL, C_4 describes the contribution of shear production, and C_5 represents that due to buoyancy and the transport contributions F and P. Since shear production is negligible in the FCSL, the C_4 term may be neglected in practice (i.e. $C_4 \to 0$). The constants in (3.12) are determined below by regression fit to the experimental results of Albertson et al. (1996).

3.1.4 Temperature variance

The mean dissipation rate of the temperature variance (ε_θ) is related to the vertical temperature flux ($< w\theta >$) through the temperature variance budget equation, (actually the budget for $\frac{1}{2} < \theta^2 >$), which for steady, horizontally homogenous flow is written

$$-< w\theta > \frac{\partial \Theta}{\partial z} - \frac{1}{2}\frac{\partial < w\theta^2 >}{\partial z} = \varepsilon_\theta \tag{3.13}$$

The first term is the average rate of production of temperature variance by interaction of the vertical heat flux with the vertical gradient of mean potential temperature. The second term represents the flux divergence of the variance. Non-dimensionalizing (3.13) by $u_*\Theta_*^2/kz$ yields

$$\Phi_H\left(\frac{z}{L}\right) - \frac{kz}{u_*\Theta_*^2}T_r = \Phi_{\varepsilon_\theta}\left(\frac{z}{L}\right) \tag{3.14}$$

where T_r $(= \frac{1}{2}\frac{\partial < w\theta^2 >}{\partial z})$ is the dimensional transport, Φ_H is the dimensionless production of temperature variance from (3.5), and $\Phi_{\varepsilon_\theta}(= \frac{\varepsilon_\theta kz}{\Theta_*^2 u_*})$ represents the dimensionless dissipation rate of temperature variance. Hogstrom (1990) found measurements of the transport term to contain large scatter, as with WC71, but found no systematic deviation from zero.

The classic Businger-Dyer empirical formula for the normalized production in the near neutral and unstable region is (Businger, 1966; Dyer, 1967)

$$\Phi_H = \left(1 + 16\left|\frac{z}{L}\right|\right)^{-\frac{1}{2}} \qquad \text{for} \qquad 0 < -\frac{z}{L} < 2 \tag{3.15}$$

although the empirical constants are subject to some uncertainty and varying interpretation. However, several recent studies have found that Φ_H scales convectively (i.e. $\propto (-z/L)^{-\frac{1}{3}}$) for $-z/L \gg 0$ (e.g. Kader and Perepelkin, 1984). An important point is that under convective scaling, $\Phi_{\varepsilon_\theta}$ is also proportional to $(-z/L)^{-\frac{1}{3}}$, and hence ε_θ is independent of surface shear stress. This simplifies greatly the calculation of H from ε_θ.

Following the approach described above for the three sublayer model we obtain the following scaling form for the temperature variance dissipation

$$\Phi_{\varepsilon_\theta} = \Phi_H = B_1 \approx 1 \qquad -\frac{z}{L} < 0.04 \tag{3.16a}$$

$$\Phi_{\varepsilon_\theta} = B_2\left(-\frac{z}{L}\right)^{-\frac{1}{3}} \qquad 0.12 < -\frac{z}{L} < 1.2 \tag{3.16b}$$

$$\Phi_{\varepsilon_\theta} = B_3\left(-\frac{z}{L}\right)^{-\frac{1}{3}} \qquad -\frac{z}{L} > 2. \tag{3.16c}$$

Note that the production and dissipation rates scale with $(-z/L)^{-1/3}$ for all but the most neutral region of the convective boundary layer, and not with $(-z/L)^{-1/2}$ as was suggested in earlier research (e.g. WC71). This -1/3 scaling simplifies greatly the process by which heat fluxes are computed from inertial subrange measurements of scalar dissipation rates. The constants in (3.16) are determined below from the results of recent experiments (Kiely et al., 1996).

3.1.5 Computing fluxes from dissipation rates

From the dimensionless dissipation functions Φ_ε and $\Phi_{\varepsilon_\theta}$ and measurements of ε and ε_θ we may estimate the momentum and heat fluxes. The u_* and H estimates may be determined from

$$u_* = \left[\frac{kz\varepsilon}{\Phi_\varepsilon\left(\frac{z}{L}\right)}\right]^{\frac{1}{3}} \tag{3.17}$$

$$H = \rho c_p \left[\frac{kzu_*\varepsilon_\theta}{\Phi_{\varepsilon_\theta}\left(\frac{z}{L}\right)}\right]^{\frac{1}{2}} \tag{3.18}$$

However, the useful application of this approach rests on the accuracy of the empirical functions Φ_ε and $\Phi_{\varepsilon_\theta}$. For the classical interpolation-type models of Φ_ε and $\Phi_{\varepsilon_\theta}$, the flux estimation demands an iterative technique, as (3.17) and (3.18) are coupled in a way that does not submit to a closed form solution. Such a scheme typically starts at an assumption of neutral conditions ($-z/L=0$), thus providing estimates Φ_ε and $\Phi_{\varepsilon_\theta}$, and in turn estimates of u_* and H. These fluxes provide an improved estimate of $-z/L$, which yields new values of Φ_ε and $\Phi_{\varepsilon_\theta}$, toward revised estimates of the fluxes, and so on iteratively.

Deacon (1959) was the first to suggest that fluxes (u_* and H) could be estimated from dissipation rates, but he did not actually employ the technique and he cautioned that it may not work well for strongly unstable stratifications. He mentioned that a similar approach could be used for evaporation but that its use would be limited by the lack of instruments capable of making fast measurements of water vapor concentration fluctuations. The dissipation method has since been used, mostly, over the ocean environment, with instrumentation on ships or buoys, (e.g. Fairall and Larsen, 1986; DeLeonibus and Simpson, 1987; Skupniewicz and Davidson, 1991; Edson et al., 1991). However, aircraft based data were used by Durand et al. (1991) and land based, point instrumentation were used by Hicks and Dyer (1974), Kader and Yaglom (1990) and Marsden et al. (1993). Hill et al. (1992) used optical scintillation methods over land surface path lengths of 150 meters. Others who have used scintillation methods for fluxes include Andreas (1988) and Hill et al. (1992). In a fascinating study, Raman lidar derived dissipation rates of humidity variance were used to estimate the surface flux of water vapor by Eichinger et al. (1993).

With the new three sublayer model for the dissipation rates of TKE and scalar variance, we proceed to present a model for calculating the fluxes of momentum, sensible heat and latent heat from inertial subrange estimates of the dissipation rates. The derivation of this new method is presented only briefly here. Essentially, the approach grew out of the need to produce a more accurate determination of fluxes from dissipation rates.

Sensible heat flux

As the vertical heat flux vanishes in the neutral limit of the DSL, we focus our model development on the convectively scaled power law of (3.16b) and (3.16c). It seems reasonable to apply this single form over the full range of unstable stratification. This form is appropriate wherever the heat flux is significant (i.e. $-z/L > 0$). A test of this assumption is provided below. Using the definitions of $\Phi_{\varepsilon_\theta}$, Θ_*, and L we may write

$$\begin{aligned} \Phi_{\varepsilon_\theta} &= \frac{\varepsilon_\theta kzu_*}{< w\theta >^2} = B\left(-\frac{z}{L}\right)^{-\frac{1}{3}} \\ &= B\left(\frac{g}{\Theta}\right)^{-\frac{1}{3}}\left[\frac{u_*}{< w\theta >^{\frac{1}{3}}}\right](kz)^{-\frac{1}{3}} \end{aligned} \tag{3.19}$$

where B is a single empirical constant. This convective form provides a closed form expression for the sensible heat flux

$$H = B^{-\frac{3}{5}} \rho c_p \left(\frac{g}{\Theta}\right)^{\frac{1}{5}} (kz)^{\frac{4}{5}} \varepsilon_\theta^{\frac{3}{5}} \tag{3.20}$$

This method provides a simple and direct computation of sensible heat flux, which does not require iteration or depend on u_*. Below, we will describe how to compute ε_θ.

Latent heat flux

Assuming similarity of scalars (see Brutsaert, 1982), we can extend our temperature analysis to water vapor. In fact, it is reasonable to believe that the result obtained for $\Phi_{\varepsilon_\theta}$ would apply to any scalar. Therefore, we write the normalized dissipation rate for water vapor as

$$\begin{aligned} \Phi_{\varepsilon_q} &= \frac{\varepsilon_q kz}{q_*^2 u_*} = \frac{\varepsilon_q kz u_*}{< wq >^2} = B\left(-\frac{z}{L}\right)^{-\frac{1}{3}} \\ &= B\left(\frac{g}{\Theta}\right)^{-\frac{1}{3}} \left[\frac{u_*}{< w\theta >^{\frac{1}{3}}}\right] (kz)^{-\frac{1}{3}} \end{aligned} \tag{3.21}$$

This convective scaling form yields the following expression for evaporation

$$E = \rho \left[B^{-1} \left(\frac{g}{\Theta}\right)^{\frac{1}{3}} < w\theta >^{\frac{1}{3}} (kz)^{\frac{4}{3}} \varepsilon_q\right]^{\frac{1}{2}} \tag{3.22}$$

where $< w\theta >$ is taken in this case from the above calculation based on ε_θ. The dissipation rate for humidity variance can be calculated by the methods described below for temperature with the substitution of q for θ.

Momentum flux

For true convective scaling the dissipation rates of the TKE and scalar variances are independent of u_*, and so measurements of the dissipation rates do not contain the information necessary to estimate u_*. For neutral stratification the normalized dissipation rate of TKE is a constant. This neutral behavior for Φ_ε seems to continue up to about $-z/L = 0.1$ (Albertson et al., 1996). In this limited region we may use

$$\Phi_\varepsilon = \frac{\varepsilon kz}{u_*^3} = C$$

and on rearrangement

$$u_* = \left[C^{-1} \varepsilon kz\right]^{\frac{1}{3}} \tag{3.23}$$

The momentum flux per unit mass is simply $-u_*^2$. For larger values of $-z/L$ we can compute u_* from the empirical interpolation form of Φ_ε as used by WC71 (i.e. (3.11b)) with H taken from (3.20) for use in calculating L. Note this requires some iteration. Our main focus here is in estimating sensible and latent heat (water vapor) fluxes and, therefore, we will not extend the u_* model any further.

In the next section, we review several inertial subrange methods for computing ε and ε_θ from velocity and temperature fluctuation measurements.

3.1.6 Inertial range methods of determining dissipation rates

We review three methods for determining dissipation rates from inertial subrange scaling in the spirit of Kolmogorov (1941): power spectra, second order structure functions, and third order structure functions.

Spectral method

The dissipation rates (ε and ε_θ) have been determined most frequently from the one-dimensional power spectra in the inertial subrange using (Kolmogorov, 1941; Corrsin, 1951)

$$E_u(k) = \alpha_u \varepsilon^{\frac{2}{3}} k^{-5/3} \tag{3.24}$$

and

$$E_\theta(k) = \beta_\theta \varepsilon^{-\frac{1}{3}} \varepsilon_\theta k^{-\frac{5}{3}} \tag{3.25}$$

where α_u and β_θ are empirical constants that have been determined from experiments to be about 0.55 and 0.8, respectively (McBean et al., 1971; Antonia et al., 1979; Kaimal and Finnigan, 1994). Thus, the dissipation rates can be obtained from (3.24) and (3.25) evaluated at one or more wavenumbers (k) using measured spectral densities in the inertial subrange. This is the approach adopted by most researchers, who have made flux estimates from dissipation rates over oceans and land, (e.g. Hicks and Dyer, 1972; Fairall et al., 1990; Skupniewicz and Davidson, 1991; Kader, 1992; and Eichinger et al., 1993).

The power spectra method is subject to errors introduced by the jumpiness of the spectra, the data treatment required for Fourier analysis (e.g. windowing and tapering), and from the uncertainty of the constants α_u and β_θ. The dissipation rate for sensible heat computed from (3.25) is potentially more erroneous than that for TKE, as the former is dependent on an estimate (with all the attendant problems) of the latter.

Second order structure function

The second order structure function represents the averaged squared differences in a flow variable over spatial separation r in the direction of flow. For longitudinal velocity $D_{uu}(r) =< (u(x+r) - u(x))^2 >$ and for temperature $D_{\theta\theta}(r) =< (\theta(x+r) - \theta(x))^2 >$ (see Monin and Yaglom, 1975). These terms scale in the inertial subrange according to Kolmogorov (1941; for velocity) and Obukhov (1949; for temperature) as

$$D_{uu}(r) = C_{uu} \varepsilon^{\frac{2}{3}} r^{\frac{2}{3}} \tag{3.26}$$

$$D_{\theta\theta}(r) = C_{\theta\theta} \varepsilon_\theta \varepsilon^{-\frac{1}{3}} r^{\frac{2}{3}} \tag{3.27}$$

where $C_{uu}(= 4.0\alpha_u)$ and $C_{\theta\theta}(= 4.0\beta_\theta)$ are empirical constants (Anselmet et al., 1984). From these equations the dissipation rates can be computed, using values of the constants taken from the literature. This approach was used by Taylor (1961) in perhaps the first application of the inertial-dissipation method. However, due to the uncertainty in these empirical constants, there is some degree of imprecision in this approach as well.

Third order structure function

For velocity the third order structure function represents the averaged cubed velocity differences over lag r, $D_{uuu}(r) =< (u(x+r) - u(x))^3 >$ (Monin and Yaglom, 1975, Ch.8). For temperature the mixed third order structure function is $D_{u\theta\theta} =< (u(x+r) - u(x))(\theta(x+r) - \theta(x))^2 >$. These structure functions scale in the inertial subrange with r as

$$D_{uuu}(r) = -\frac{4}{5}\varepsilon r \tag{3.28}$$

$$D_{u\theta\theta}(r) = -\frac{4}{3}\varepsilon_\theta r \tag{3.29}$$

Note that with this approach the dissipation rates for momentum and heat can be directly computed without resort to empirical constants. This method is superior to the second order methods in that no empirical constants are used and also the dissipation rate for temperature variance is not dependent on prior numerical calculation of the dissipation rate for momentum. This approach has been used by Albertson et al. (1996) for momentum, and by Kiely et al. (1996) for heat. The application of dissipation rates derived by third order structure functions to compute fluxes is shown below.

3.2 Experiments

Surface energy balance and atmospheric turbulence measurements were carried out in the summer of '94, at two sites in California (Albertson et al., 1996; Kiely et al., 1996). One was located at the Campbell Tract research facility at the University of California at Davis in the Central Valley of California and the second was at a dry lake bed at Owens Valley in southeastern California. Eddy correlation equipment consisted of a one dimensional sonic anemometer with a fine wire (dia. = 0.0127 mm) thermocouple and a Krypton hygrometer operating at 10 Hz, with covariances taken over 20 minute averaging periods. This enabled the direct measurement of the vertical fluxes of sensible and latent heat. A three-dimensional sonic anemometer was used to record the three velocity components at 21 Hz for the Campbell Tract site and 56 Hz for the Owens Lake site. Instantaneous air temperature was also measured from the speed of sound recorded by the 3-D sonic. From the 3-D sonic we obtain direct measurements of u_* (and so the flux of momentum). The eddy correlation equipment ran continuously for the experimental duration (six weeks at Davis and two weeks at Owens Lake) with data logged on 20 minute time increments. Typically, the 3-D sonic ran for up to 12 hours per day and the data recorded to a new file every 20 minutes (i.e. 25200 data points at 21 Hz and 67200 points at 56 Hz) to match the eddy correlation and energy balance time steps.

The Davis site is a flat bare soil field of 500 m by 500 m extent. In the northeast corner of the field an irrigated portion extends 155 m in a north-south line and 115 m in an east-west line. The surface roughness length is z_0 = 2 mm. The fetch for uniform surface roughness exceeded 400 m and for surface wetness the fetch exceeded 100 m. The experiments were performed in June and July, with daytime highs of about 30°C and nighttime lows of about 15°C. The 3-D sonic anemometer was set at z=0.85 m for the initial 4 days and at z=1.5 m thereafter. Irrigations were performed at the beginning of the experiment and also three weeks later. The three week period in between was dry, with no recorded rain. By saturating the soil surface most of the available energy was forced to latent heat rather than to sensible heat, thus extending the range of near neutral flows encountered. Throughout the experiment a wide range of atmospheric stabilities were observed with low $-z/L$ values during the days immediately following irrigation and increasingly large values as the drying out of the bare soil continued and more of the available energy was forced to sensible heat.

The dry Owens Lake site, which had daytime highs of about 40°C and night-time lows of about 15°C during August 1994, enabled investigation over a wider range of convective atmospheric conditions than in Davis. Owens Lake is an arid flat landscape with uniform fetch exceeding 10 km, and a lakebed area of about 200 km^2. The surface roughness length of the lakebed has been estimated at $z_0 = 0.13$ mm (Katul et al., 1995b). The 3-D sonic anemometer was set at z=2.65 m, and data were recorded at 56 Hz.

3.3 Dissipation Results

3.3.1 Data screening

Of the collected data, 180 files (each of 20 minute duration) were selected for analysis. The selection was based on a requirement that the files support unambiguous decomposition into means and fluctuations and that they possess turbulence intensity values ($T.I. = \sigma_u / < U >$) of less than 50%, as necessary for the application of Taylor's hypothesis (Stull, 1988). The Davis experiment contributed 105 of these files and Owens Lake the remainder. The atmospheric stability range encountered was $0.004 < -z/L < 8.1$. For each file the power spectra of the longitudinal velocity and temperature were computed using square windowing of 2048 points, bell tapering the first and last 10% of the window (Stull, 1988, p.309), using an FFT to compute the spectra of the window, repeating the process on the remaining windows and averaging all windows for each wavenumber. The power spectra of w was also calculated for each file to verify local isotropy, which is necessary for inertial subrange scaling and is indicated by $\frac{E_w}{E_u} = \frac{4}{3}$ (Tennekes and Lumley, 1972).

Typical power spectra of velocity and temperature are shown in Figures 3.4a, and 4b. These spectra have been frequency smoothed for presentation; in their original form they are quite jumpy. Note that they follow the expected -5/3 scaling over a wide range of wavenumbers. Typical second order structure functions for velocity and temperature are shown in Figures 3.5a and 5b. As expected they both follow the 2/3 scaling. Figures 3.6a and 6b are typical velocity and temperature third order structure functions. They both follow the r^1 scaling for the short lag portion which is expected to scale inertially. The third order structure functions are considered to be a more stringent test of the inertial subrange (Katul et al., 1995a). The second and third order structure functions do not require smoothing as the time averaging process provides stable measures that vary smoothly with r.

3.3.2 Normalized dissipation rates for TKE

From the computed power spectra, log transformation of (3.24) allows the straightforward determination of ε from the regressed intercept of $\log(E)$ vs $\log(k)$. A similar approach to the second order structure function provides an estimate of ε. Due to the uncertain nature of the actual values of α_u and S_2 and the susceptibility of power spectra and the second order structure function to intermittency effects on the inertial subrange scaling, we place more confidence in the third order structure function. A comparison of the three methods was made by Albertson et al. (1996). Here we focus on the third order approach and compute ε from (3.28). For the third order structure function, the Φ_ε results for the 180 data files are binned in equal log increments of $-z/L$ and presented in Figure 3.7. The scaling forms of (3.12) were fit to the

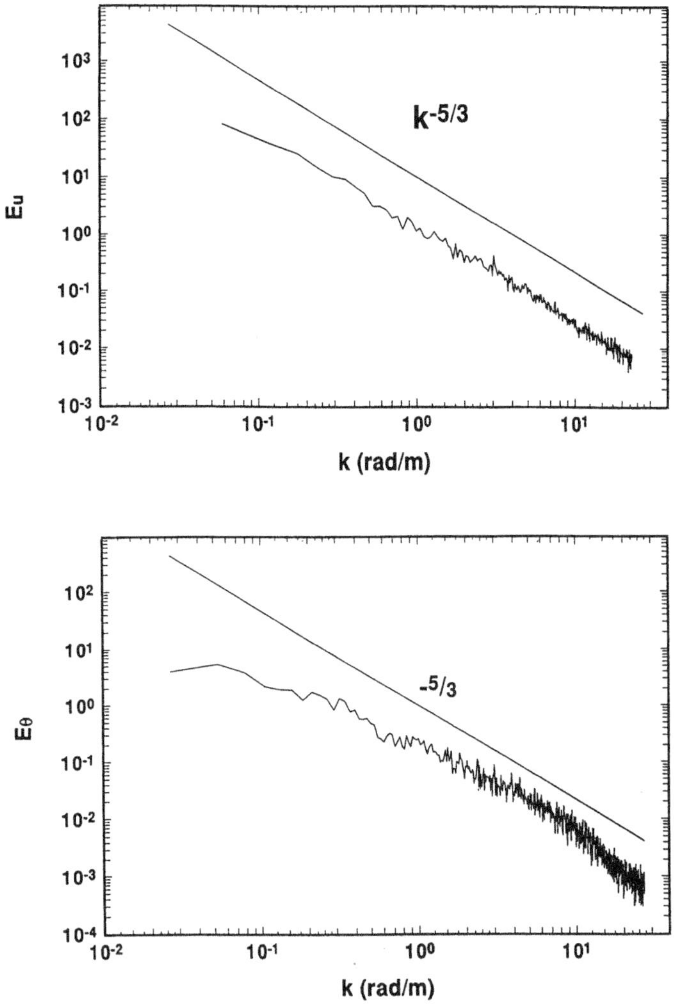

Figure 3.4: *(a) Sample power spectrum of longitudinal velocity from one of the measured 3D sonic files (top). (b) Sample power spectrum of temperature fluctuations (bottom).*

unbinned third order structure function based dissipation estimates of Albertson et al. (1996), resulting in

$$\Phi_\varepsilon = 0.61 \qquad -\frac{z}{L} < 0.04 \tag{3.30a}$$

$$\Phi_\varepsilon = 0.35\left(-\frac{z}{L}\right)^{-\frac{1}{3}} + 2.28\left(-\frac{z}{L}\right) \qquad 0.12 < -\frac{z}{L} < 1.2 \tag{3.30b}$$

$$\Phi_\varepsilon = 1.81\left(-\frac{z}{L}\right) \qquad -\frac{z}{L} > 2.0 \tag{3.30c}$$

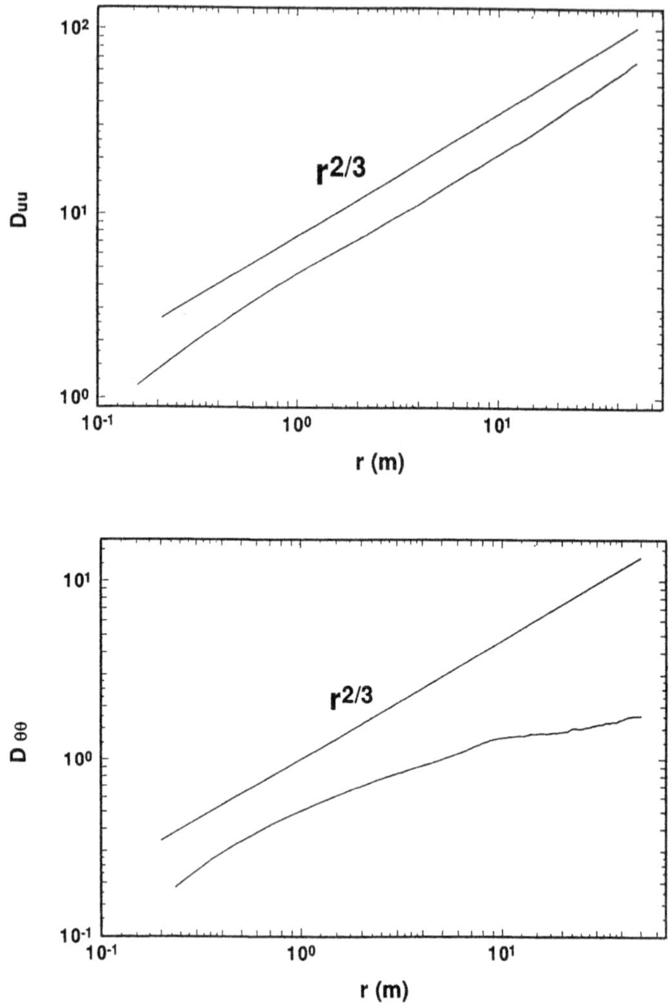

Figure 3.5: *(a) Sample second order structure function of longitudinal velocity (top). (b) Sample second order structure function of temperature fluctuations (bottom).*

These are shown by a solid line in Figure 3.7, with production represented by $[(1-16\,|z/L|)^{-1/4} - z/L]$ and the normalized dissipation model of WC71 is shown for comparison. It is clear from our results in Figure 3.7, that in the DSL the dissipation rate is significantly less than production, in the DCSL the dissipation rate is about equal to production, and in the FCSL the dissipation rate slightly exceeds production. We present (3.30) as a scaling form for TKE dissipation rates based on the three sublayer model.

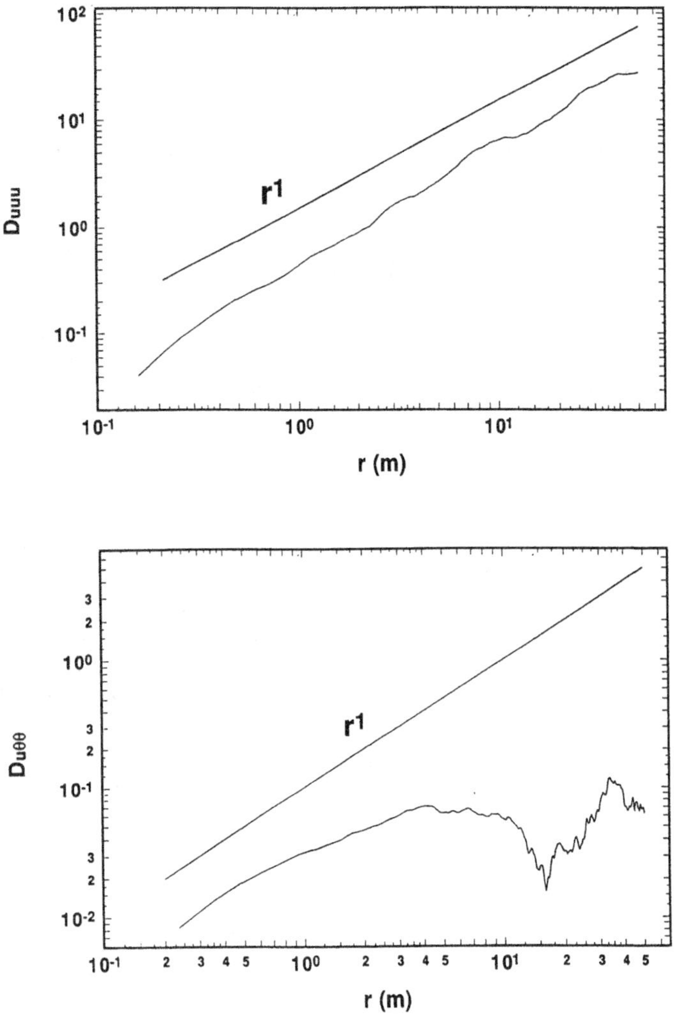

Figure 3.6: *(a) Sample third order structure function of longitudinal velocity (top). (b) Sample mixed third order structure function of velocity and temperature fluctuations (bottom).*

3.3.3 Normalized dissipation rates of temperature

From the temperature spectra, log transformation of (3.25) allows the straightforward determination of ε_θ from the regressed intercept of log (E_θ) vs log(k), with ε known from the above analysis. A similar approach to the second order structure function provides an estimate of ε_θ. Due to the uncertainty associated with the constants in the spectral and the second order structure function methods (Kiely et al., 1996), we determine the constants for (3.16) using only the third order structure function (3.29) based results. A comparison of the three inertial subrange methods was made by Kiely et al. (1996). The three sublayer model for $\Phi_{\varepsilon_\theta}$ (3.16) was

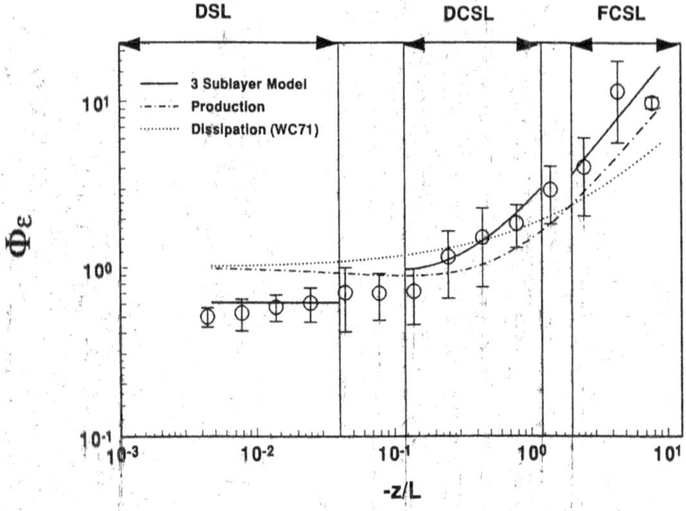

Figure 3.7: *Dimensionless TKE dissipation function plotted vs. the stability parameter -z/L. The circles represent the mean values of the data in each bin and the bars denote +/- one standard deviation of the data in the bins. The dissipation rates were computed from the third order structure functions. The fit 3 sublayer model is shown with a solid line. The normalized production is shown with dot-dash and the dissipation model of WC71 is shown with a dotted line.*

least squares fit to the unbinned third order structure function derived dissipation results, and resulted in the following model (Kiely et al., 1996)

$$\Phi_{\varepsilon_\theta} = 0.88 \qquad -\frac{z}{L} < 0.04 \tag{3.31a}$$

$$\Phi_{\varepsilon_\theta} = 0.11\left(-\frac{z}{L}\right)^{-\frac{1}{3}} \qquad 0.12 < -\frac{z}{L} < 1.2 \tag{3.31b}$$

$$\Phi_{\varepsilon_\theta} = 0.11\left(-\frac{z}{L}\right)^{-\frac{1}{3}} \qquad -\frac{z}{L} > 2.0 \tag{3.31c}$$

which is shown in Figure 3.8 along with the Businger-Dyer model of production. The dimensionless dissipation rate scales with a single convective power law over an extended range of $-z/L$, thus simplifying greatly the calculation of sensible heat flux from ε_θ and supporting the application of the single power law for flux calculations over the entire range of $-z/L > 0$. This is similar to the extended convective scaling found for the standard deviation of temperature fluctuations (e.g. Albertson et al., 1995a).

3.4 Dissipation-Flux Model

Recall that the pitfalls in the traditional spectral or second order structure function dissipation methods have been described to include: the uncertainty of the empirical constants for the second order subrange scaling of velocity and scalars; the requirement of a two step process of first estimating ε from u and then ε_θ from ε and inertial subrange scaling of θ; possible errors due to the jumpiness of the spectra when examined over a narrow range of wavenumbers; and, the need for an iterative solution of heat fluxes. We avoid all of these problems with the present approach.

With the application of third order structure functions and the use of Taylor's hypothesis we may obtain estimates of the dissipation rates from single lag values of the structure functions

$$\varepsilon = -\frac{5}{4}\frac{D_{uuu}(\tau)}{\tau U} \tag{3.32a}$$

$$\varepsilon_\theta = -\frac{3}{4}\frac{D_{u\theta\theta}(\tau)}{\tau U} \tag{3.32b}$$

$$\varepsilon_q = -\frac{3}{4}\frac{D_{uqq}(\tau)}{\tau U} \tag{3.32c}$$

where $\tau(= r/U)$ is the time lag corresponding to r. Note that τ must be chosen to correspond to spatial lags that fall in the inertial subrange. We write the flux equations for the sensible heat and evaporative fluxes by making the substitution of (3.32) into (3.20) and (3.22) to obtain

$$H = (0.11)^{-\frac{3}{5}}\rho c_p \left(\frac{g}{\Theta}\right)^{\frac{1}{5}} (kz)^{\frac{4}{5}} \left[-\frac{3}{4}\frac{D_{u\theta\theta}(\tau)}{\tau U}\right]^{\frac{3}{5}} \tag{3.33a}$$

$$E = \rho\left[(0.11)^{-1}\left(\frac{g}{\Theta}\right)^{\frac{1}{3}} < w\theta >^{\frac{1}{3}} (kz)^{\frac{4}{3}} \left[-\frac{3}{4}\frac{D_{uqq}(\tau)}{\tau U}\right]\right]^{\frac{1}{2}} \tag{3.33b}$$

where the time average in the structure functions $D_{u\theta\theta}(\tau)$ and $D_{uqq}(\tau)$ can be computed directly by a data logger for a single time lag τ. This is possible because the third order structure function is smooth enough to provide accurate estimates of the intercept from the known slope with just one point on the curve.

As described above the friction velocity may be computed for moderate values of $-z/L$ using

$$u_* = \left[C_1^{-1}\left(-\frac{5}{4}\frac{D_{uuu}(\tau)}{\tau U}\right) kz\right]^{\frac{1}{3}} \tag{3.34}$$

This approach for determining the fluxes of H, E and u_* has several advantages over the classical method of computing fluxes from dissipation rates: (i) it does not require potentially error-inducing data treatment as with the Fourier transform, (ii) it does not rely on regressions over a range of wavenumbers as needed with jumpy power spectra, (iii) it does not rely on uncertain spectral scaling constants (i.e. α_u and β_θ), and (iv) it is free of iteration.

3.5 Evaluation of Model Performance

To test the above described model we present sensible and latent heat flux estimates using the model and compare them to direct measurements from eddy correlation. The flux comparisons are for time periods not covered by the data files used to fit the dissipation rate scaling forms of Φ_ε and $\Phi_{\varepsilon_\theta}$. The sensible heat flux comparison covers a three day period during which a wide range of H was encountered. The comparison is shown in Figure 3.9. Note that the model estimates from (3.33a) match the eddy correlation measurements to within the 10% stated accuracy range for eddy correlation. The latent heat flux comparison is made for a day with fast response humidity measurements available and a wet surface, to test the model over a wide range of LE. Only one day of data is available for this comparison. The comparison of model estimates of LE using (3.33b) to eddy correlation measurements of LE is shown in Figure 3.10. This agreement is also considered excellent.

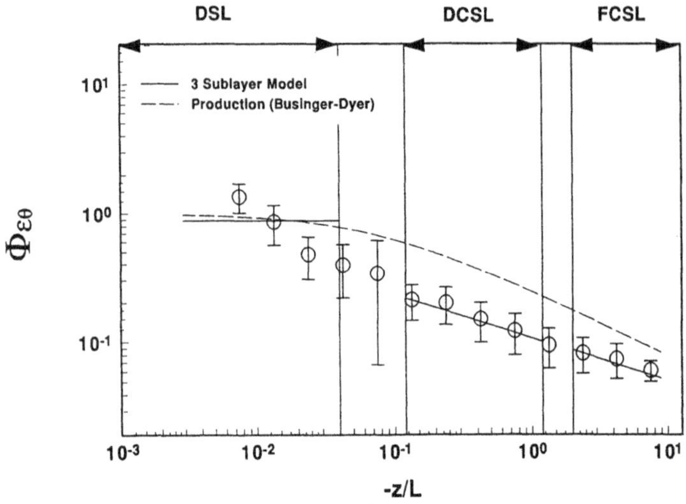

Figure 3.8: *Dimensionless temperature dissipation function plotted vs. the stability parameter -z/L. The circles represent the mean values of the data in each bin and the bars denote +/- one standard deviation of the data in the bins. The dissipation rates were computed from the mixed third order structure functions. The fit 3 sublayer model is shown with a solid line and the normalized production is shown with a dashed line.*

3.6 Discussion and Conclusions

We have presented measurements from 180 velocity/temperature files in the ASL over a wide range of atmospheric stabilities (three decade range of $-z/L$). Average dissipation rates of TKE and temperature variance were computed using third order structure functions. We place more confidence in this approach as it is devoid of empirical constants and also since the temperature dissipation rate does not require a priori knowledge of the TKE dissipation rate. These dissipation rates are evaluated in the context of the three sublayer model of Kader and Yaglom (1990). The new empirical functions developed show that in the DSL the dissipation rates of TKE are constant and therefore independent of z, and are significantly less than production. In the DCSL and in the FCSL the dissipation rates of TKE follow yet exceed production, as defined by the empirical function of Wyngaard and Coté (1971). The dissipation rate of temperature variance in our fit to the three sublayer model are shown to scale with a single convective power law over a broad range of stability.

We use these dissipation rate scaling forms to develop a simple technique for computing the fluxes of momentum, sensible heat, and water vapor, in a direct one step method. This is in contrast to the traditional, iterative dissipation approach of computing fluxes. Sensible heat flux was estimated from dissipation measurements independent of those used to fit the $\Phi_{\varepsilon_\theta}$

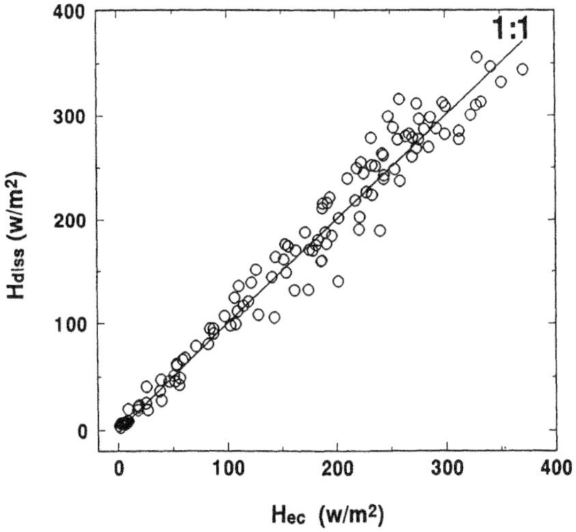

Figure 3.9: *A comparison of the sensible heat flux predictions from the presented dissipation model (H_{diss}) with direct measurements from eddy correlation (H_{ec}).*

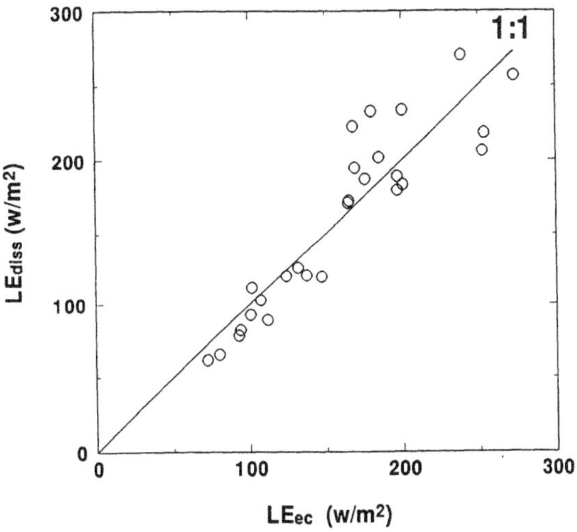

Figure 3.10: *A comparison of the latent heat flux predictions from the presented dissipation model (LE_{diss}) with direct measurements from eddy correlation (LE_{ec}).*

function. These estimates match direct measurements from eddy correlation to within the accuracy of the eddy correlation equipment. A similar comparison was made between latent heat flux estimates from the dissipation model and direct eddy correlation measurements with similarly good results.

In conclusion, we present a simple dissipation method of computing fluxes which is as accurate as the eddy correlation method, and arguably more accurate and simple than the traditional iterative dissipation approach.

Acknowledgments

The authors wish to thank Anthony Cahill and Mike Mata for their assistance in the field and Scott Tyler for his logistical help at Owens Valley. This research has been supported and financed, in part, by the National Science Foundation (EAR-93-04331), CA State Salinity Drainage Task Force, Kearney Foundation, CA Water Resources Centre (W-182), the UC Davis Superfund grant (5 P42ESO4699-07), and the NASA Graduate Student Fellowship in Global Change Research program. The support of University College Cork and the Fulbright Program is acknowledged by the Cork author.

3.7 References

Albertson JD, Parlange MB, Katul GG, Chu C-R, Stricker H, Tyler S (1995) Sensible heat flux from arid regions: A simple flux-variance method. *Water Resour Res* 31: 969–973

Albertson JD, Kiely G, Parlange MB, Eichinger WE (1996) The average dissipation rate of turbulent kinetic energy in the neutral and unstable atmospheric surface layer. *Journal of Geophysical Research*

Andreas EL (1998) Using scintillation at two wavelengths to measure path-averaged heat fluxes in free convection. *Boundary Layer Meteorol* 54: 167–182

Anselmet F, Gagne Y, Hopfinger EJ, Antonia RA (1984) Higher order velocity structure functions in turbulent shear flows. *J Fluid Mech* 140: 63–89

Antonia RA, Chambers AJ, Phong-Anant D, Rajagopalan S (1979) Properties of spatial temperature derivatives in the stmospheric surface layer. *Boundary-Layer Meteorol* 17: 101–118

Brutsaert W (1982) *Evaporation into the Atmosphere.* Kluwer Academic Publishers 299pp

Brutsaert W (1986) Catchment scale evaporation and the atmospheric boundary layer. *Water Resour Res* 22, suppl., 39S–45S

Businger JA (1966) Transfer of momentum and heat in the planetary boundary layer. *Proc Symp Arctic Heat Budget and Atmos Circulation*, Rand Corp. RM-5233-NSF, pp.305–332

Champagne FH, Friehe CA, LaRue JC, Wyngaard JC (1977) Flux measurements, flux estimation techniques, and fine-scale turbulence measurements in the unstable surface layer over land. *J Atmos Sci* 34: 515–530

Corrsin S (1951) On the spectrum of isotropic temperature fluctuations in an isotropic turbulence. *J Appl Phys* 22: 469–473

Deacon EL (1959) The measurement of turbulent transfer in the lower atmosphere. *Adv Geophys* 6: 211–228

Deacon EL (1988) The streamwise Kolmogorov constant. *Boundary Layer Meteorol* 42: 9–17

Durand P, De Sa L, Druilhet A, Said F (1991) Use of the inertial dissipation method for calculating turbulent fluxes from low-level airborne measurements. *J Atmos Ocean Tech* 8: 78–84

DeLeonibus PS, Simpson LS (1987) Dissipation observations of drag coefficients over the open ocean. *IEEE J Oceanic Eng* OE-12: 296–300

Dyer AJ (1974) A review of flux-profile relations. *Boundary-Layer Meteorol* 1: 363–372

Dyer AJ (1967) Measurements of evaporation and heat transfer in the lower atmosphere by an automatic eddy-correlation technique. *Quart J Roy Meteorol Soc* 93: 501–508

Edson JB, Fairall CW, Mestayer PG, Larssen SE (1991) A study of the inertial-dissipation method for computing air-sea fluxes. *J Geophys Res* 96C: 10689–10711

Eichinger WE, Cooper DI, Holtkamp DB, Karl RR Jr., Quick CR, Till JJ (1993) Derivation of water vapour fluxes from lidar measurements. *Boundary-Layer Meteorol* 63: 39–64

Fairall CW, Edson JB, Larsen SE, Mestayer PG (1990) Inertial-dissipation air-sea measurements: A prototype system using realtime spectral computations. *J Atmos Ocean Tech* 7: 425–453

Fairall CW, Larsen SE (1986) Inertial-dissipation methods and turbulent fluxes at the air-ocean interface. *Boundary-Layer Meteorol* 34: 287–301

Frenzen P, Vogel CA (1992) The turbulent kinetic energy budget in the atmospheric surface layer: A review and an experimental re-examination in the field. *Boundary-Layer Meteorol* 60: 49–76

Hicks BB, Dyer AJ (1972) The spectral density technique for the determination of eddy fluxes. *Quart J R Met Soc* 98: 838-844

Hill RG, Ochs GR, Wilson JJ (1992) Measuring surface layer fluxes of heat and momentum using optical scintillation. *Boundary-Layer Meteorol* 58: 391–408

Hogstrom U (1990) Analysis of Turbulence Structures in the Surface Layer with a Modified Similarity Formulation for near Neutral Conditions. *J Atmos Sci* 47: 1949–1972

Kader BA (1988) Three-level structure of an unstably stratified atmospheric surface layer. *Izv Atmos Ocean Phys* 24: 907-919 (English translation)

Kader BA (1992) Determination of momentum and heat fluxes by spectral method. *Boundary-Layer Meteorol* 61: 323–347

Kader BA, Perepelkin VG (1984) Profiles of the mean velocity and temperature in the near surface layer of the atmosphere under conditions of neutral and unstable stratification. *Izv Atm Ocean Phys* 20: 112–119. (English Translation)

Kader BA, Yaglom AM (1990) Mean fields and fluctuation moments in unstably stratified turbulent boundary layers *J Fluid Mech* 212: 637–662

Kaimal JC, Finnigan JJ (1994) *Atmospheric Boundary Layer Flows: Their Structure and Measurement*, Oxford University Press 289pp

Katul GG, Parlange MB, Albertson JD, Chu C-R (1995a) Local isotropy and anistropy in the sheared and heated atmospheric surface layer. *Boundary-Layer Meteorol* 72 123–148

Katul GG, Chu C-R, Parlange MB, Albertson JD, Ortenburger TA (1995b) The large scale spectral characteristics of stratified atmospheric surface layer flows. *J Geophys Res*, in press

Kiely G, Albertson JD, Parlange MB, Eichinger WE (1996) On the scaling of the average dissipation rate of temperature variance in the atmospheric surface layer. *Boundary-Layer Meteorol*, submitted

Kolmogorov AN (1941) The local structure of turbulence in incompressible viscous fluid for very large Reynolds number *Dokl Akad Nauk SSSR* 30: 301–305

Leavitt E, Paulson CA (1975) Statistics of surface layer turbulence over the tropical ocean. *J Phys Oceanogr* 5: 143–156

McBean GA, Stewart RW, Miyake M (1971) The turbulent energy budget near the surface. *J Geophys Res* 76: 6540–6549

McBean GA, Elliott JA (1975) The vertical transport of kinetic energy by turbulence and pressure in the boundary layer. *J Atmos Sci* 32: 753–766

Marsden RF, McBean GA, Proctor BA (1993) Momentum and sensible heat fluxes calculated by the dissipation technique during the Ocean Storms Project. *Boundary-Layer Meteorol* 63: 22–38

Monin AS, Yaglom AM (1975) *Statistical Fluid Mechanics* Vol. II, J. Lumley (Ed.), MIT Press, 874pp

Obukhov AM (1949) Structure of the temperature field in a turbulent flow. *Izv Akad Naul SSSR, Ser Georgr i Geofiz* 13: 58–69

Panton RL (1984) *Incompressible Flow*, Wiley-Interscience, 780pp

Parlange MB, Eichinger WE, Albertson JD (1995) Regional scale evaporation and the atmospheric boundary layer. *Reviews of Geophysics* 33: 99–124

Skupniewicz CE, Davidson KL (1991) Hot-film measurements from a small buoy: Surface wind estimates using the inertial dissipation method. *J Atmos Ocean Tech* 8: 309–322

Stull R (1988) *An Introduction to Boundary Layer Meteorology*, Kluwer Academic Press, 666pp

Taylor RJ (1961) A new approach to the measurement of turbulent fluxes in the lower atmosphere. *J Fluid Mech* 10: 449–458

Tennekes H, Lumley J (1972) *A First Course in Turbulence*, MIT Press, 300pp

Wyngaard JC, Coté OR (1971) The budgets of turbulent kinetic energy and temperature variance in the atmospheric surface layer. *J Atmos Sci* 28: 190–201

Further Reading

Brutsaert W (1982) *Evaporation into the Atmosphere.* Kluwer Academic Publishers, 299pp

Parlange MB, Eichinger WE, Albertson JD (1995b) Regional scale evaporation and the atmospheric boundary layer. *Reviews of Geophysics* 33: 99–124

Stull R (1988) *An Introduction to Boundary Layer Meteorology.* Kluwer Academic Press, 666pp

Wyngaard JC (1990) Scalar fluxes in the planetary boundary layer - Theory, modeling, and measurement. *Boundary-Layer Meteorol.* 50: 49–75

Chapter 4

Introduction to Numerical Weather Prediction Data Assimilation

Phillipe Courtier
ECMWF, Shinfield Park
Reading, Berkshire RG2 9AX
UK

4.1 The Problem

In numerical weather prediction data assimilation consists of the process which estimates the initial conditions of the forecast using all the available information. A description of the current observing system can be found in McGrath (1993); around 105 elementary pieces of information are currently used over 24 hours by the operational data assimilation system.

The current operational ECMWF model (Simmons, 1991) has a horizontal resolution of 90 km and covers the whole globe. On the vertical the atmosphere is sampled with 31 levels from the surface up to 10 hPa. The number of degrees of freedom of the model is then of the order of 10^7.

Over 24 hours, the estimation problem is clearly underdetermined. The time dimension is thus a critical element of any data assimilation system: it is essential to carry forward in time information from past observations using the forecast model since it is the best information propagator available.

We have identified the main difficulties of data assimilation:

- large dimension problem
- time dimension critical but non linear dynamic
- observations of variable nature and quality

The purpose of this paper is to provide the theoretical basis of the algorithms used in operational meteorology together with some recent developments. Most of the material presented here is well documented in the meteorological literature (Lorenc, 1986; Ghil and Manalotte-Rizzoli, 1991; Daley, 1991) or in other fields under the generic name of inverse problems (Tarantola, 1987).

NATO ASI Series, Vol. I 45
Radiation and Water in the Climate System:
Remote Measurements
Edited by Ehrhard Raschke

In the next section we shall introduce linear estimation theory. Section 4.3 briefly describes the Optimal Interpolation algorithm and then concentrates on the variational approach. Section 4.4 deals with the time dimension; it theoretically introduces the Kalman filter then the four-dimensional variational assimilation algorithm is described, together with some simplifications which allow an operational implementation in the foreseeable future. Finally, an algorithm is presented as an attempt to provide error bars from the variational algorithms and to improve the treatment of the time dimension.

4.2 Introduction to Linear Estimation

In the following the random variables are underlined whereas realizations of these random variables are not.

4.2.1 A simple illustration

Let us assume that we are interested in an estimation of the temperature of the ECMWF Lecture Theatre. We consider the room temperature as a random variable $\underline{T}$ of expectation $<\underline{T}>$ and variance $<(\underline{T}-<\underline{T}>)^2>$ which are not supposed to be known.

We have a thermometer of known precision σ_0 and we read T_0. The measurement T_0 is considered as one realization of a random variable $\underline{T}_0$ of expectation $<\underline{T}>$ (the observation is assumed unbiased) and variance

$$<(\underline{T}-<\underline{T}>)^2>=\sigma_0^2.$$

In the absence of any other information, the Best Linear Unbiased Estimator (BLUE) of the room temperature is $\underline{T}_a=\underline{T}_0$ and for the particular realization T_0 of the measurement we have $T_a=T_0$.

In other words, we have been able to produce a rather pedantic presentation of what you should do when you want to have an idea of the room temperature - simply read the thermometer; at least it is not against common sense!

However, if you look carefully at the ECMWF attendees of a presentation in the Lecture Theatre, most of them are wearing a jumper: they have a priori (background) information T_b on the room temperature (it's gonna be cold!). T_b is considered as one realization of a random variable $\underline{T}_b$ of expectation $<\underline{T}>$ and variance $<(\underline{T}_b-<\underline{T}>)^2>=\sigma_b^2$

Intuitively, one realizes that the background information, combined with the observation, should lead to a better estimate than the observation alone. We then look for the BLUE, a linear estimate

$$\underline{T}_a=\alpha\underline{T}_b+\beta\underline{T}_0 \tag{4.1}$$

which is unbiased and best (of minimum error variance). As the background information and the observations are assumed unbiased, we get

$$<\underline{T}>=\alpha<\underline{T}>+\beta<\underline{T}>.$$

and then $\alpha+\beta=1$.

In meterological practice (4.1) is usually rewritten as

$$\underline{T}_a=\underline{T}_b+\beta(\underline{T}_0-\underline{T}_b).$$

We look for a correction to the background $\underline{T}_b$ which is a linear function of the difference observation minus background.

β is evaluated so as to fulfil the minimum vanance criterion for the estimate

$$\begin{aligned} <(\underline{T}_a - <\underline{T}>)^2> &= (1-\beta^2) <(\underline{T}_b - <\underline{T}>)^2> + \beta^2 <(\underline{T}_0 - <\underline{T}>)^2> \\ &= (1-\beta^2)\sigma_b^2 + \beta^2\sigma_0^2. \end{aligned} \tag{4.2}$$

which is minimum for $\beta = \frac{\sigma_b^2}{\sigma_b^2+\sigma_0^2}$

and $\underline{T}_a = \underline{T}_b + \frac{\sigma_b^2}{\sigma_b^2+\sigma_0^2}(\underline{T}_0 - \underline{T}_b)$

and for the actual realizations T_0 and T_b,

$$T_a = T_b + + \frac{\sigma_b^2}{\sigma_b^2+\sigma_0^2}(T_0 - T_b)$$

both pieces of information are weighted according to their statistical quality. If we consider the limit case of a very low quality measurement ($\sigma_0 \gg \sigma_b$). then the analysis remains equal to the background. On the other hand, if the observation is of very high quality ($\sigma_0 \ll \sigma_b$), the analysis is equal to the observation.

The variance of analysis error is found by replacing β with its expression in (4.2)

$$<(\underline{T}_a - <\underline{T}>)^2> = \left(\frac{1}{\sigma_0^2} + \frac{1}{\sigma_b^2}\right)^{-1}.$$

4.2.2 General formulation of linear estimation

The state of the model considered as a random variable is denoted by $\underline{x}$. As already said, for real size problems, it is a vector of length 10^7. Observations $\underline{y}$ are available under the form

$$\underline{y} = H\underline{x} + \underline{\epsilon}. \tag{4.3}$$

H is the so-called observation operator and $\underline{\epsilon}$ the observation errors (representativeness and instrumental as explained by Lorenc, 1986). The observations are assumed unbiased ($<\underline{\epsilon}> = 0$) and of known error covariances $<\underline{\epsilon}\underline{\epsilon}^t> = O$.

A background $\underline{x}_b$ is available unbiased $<\underline{x}_b> = <\underline{x}>$ and of known error covariance

$$<(\underline{x}_b - <\underline{x}>)(\underline{x}_b - <\underline{x}>)^t> = B$$

The Best Linear Unbiased Estimator is then given by

$$\underline{x}_a = \underline{x}_b + K(\underline{y} - H\underline{x}_b) \tag{4.4}$$

with

$$K = BH^t(HBH^t + O)^{-1} = (B^{-1} + H^tO^{-1}H)^{-1}H^tO^{-1}. \tag{4.5}$$

The covariances of analysis error

$$<(\underline{x}_a - <\underline{x}>)(\underline{x}_a - <\underline{x}>)^t> = (I - KH)B = (B^{-1} + H^tO^{-1}H)^{-1}. \tag{4.6}$$

This classical result of linear estimation is clearly established in several ways in Jaszwinski (1970) or in Lorenc (1986) and by various other authors. A more general formulation where a background is not necessarily included is presented (in a very accessible way) in Tarantola (1987). The analysis value is obtained for a given realization of the background and the observations replacing the random variables in (4.4) by their values.

In meteorology the practical difficulty is that it is impossible to use (4.4) and (4.5) directly. B, for example, is a matrix of size $10^7 \times 10^7$ which is about 1000 times the total archiving capabilities of ECMWF and one million times the memory size of the current computers.

The scientific difficulty of data assimilation is then to find algorithms which simplify (4.4) and (4.5) to an affordable amount of computer resources, while preserving some of the essential characters.

4.2.3 Quasi linear case

The previous results can easily be generalised when the observation operators are weakly non linear: the tangent linear approximation H' is assumed to be valid for the order of magnitude of the background errors:

$$\begin{aligned} \underline{y} &= H(\underline{x}) + \underline{\epsilon} \\ &= H(x_b) + H(\underline{x}) - H(x_b) + \underline{\epsilon} \\ &\simeq H(x_b) + H' \cdot (\underline{x} - x_b) + \underline{\epsilon}. \end{aligned}$$

Equations (4.4), (4.5) and (4.6) then become

$$\underline{x}_a = \underline{x}_b + K(\underline{y} - H(x_b)) \tag{4.7}$$

with

$$K = BH'^t(H'BH'^t + O)^{-1} \tag{4.8}$$

and

$$<(\underline{x}_a - <x>)(\underline{x}_a - <x>)^t >= (I - KH')B. \tag{4.9}$$

4.3 Two Practical Implementations of Linear Estimation

4.3.1 Optimal Interpolation (OI)

The OI was introduced in meteorology by Gandin (1963) and significantly contributed to the progress of numerical weather forecast quality during the late 70's and the 80's. The first ECMWF implementation is described in Lorenc (1981) and a revised implementation in Shaw et al. (1987).

The basic idea to simplify (4.4) and (4.5) is that, for a given geographical location, only the neighbouring observations are useful. This argument relies on the relatively small horizontal length scale (500 km) of the height forecast error correlation (see Hollingsworth and Lönnberg, 1986 and Lönnberg and Hollingsworth, 1986).

For each location a small set of predictors is kept and the same equation as (4.4) and (4.5) is solved but with matrices of tractable size. In some operational implementations wind and mass are analysed at a given horizontal location and at a given level with as few as 15 predictors.

At ECMWF, a box of a typical size 500 km is considered and an order of magnitude of 700 predictors is kept.

The second approximation comes from the necessity of an analytical model for the forecast error covariances which have to be computed at the observation locations. This model in most (all?) implementations relies on isotropy and geostrophy on the f-plane (f is assumed instant locally) with a separability vertical/horizontal (the spatial correlations can be expressed as a product of two functions, one which depends on the horizontal distance only and one on the vertical distance only).

Generally speaking, the more local, the more noisy the analyses are (there are jumps in the data used from one location to another). A second weakness of OI is that if, in principle, any linear (or linearised) observation operator may be used, it is difficult to practically make use of observations indirectly related to the model parameters since in (4.5) all the forecast error covariances in observation space have to be computed. Durand (1985) implemented a direct use of satellite radiances but it significantly increased the PERIDOT OI code complexity.

The analysis error variances are easily computed using (4.6) for each simplified problem.

4.3.2 Variational formulation of linear estimation - 3D-Var

Coming back to the simple example of the room temperature estimation, if one introduces the function

$$J(T) = \frac{1}{2}\frac{(T-T_b)^2}{\sigma_b^2} + \frac{1}{2}\frac{(T-T_0)^2}{\sigma_0^2},$$

this function is minimal for

$$T_a = \frac{\sigma_0^2}{\sigma_0^2+\sigma_b^2}T_b + \frac{\sigma_b^2}{\sigma_0^2+\sigma_b^2}T_0$$

which is the same as the BLUE. Furthermore the second derivative of J is

$$J'' = \frac{1}{\sigma_b^2} + \frac{1}{\sigma_0^2}$$

and is equal to the inverse of the covariance matrix of estimation error.

It is thus possible to find the estimate of the statistical problems minimizing a deterministic function which measures the misfit between the estimate and the information weighted by its statistical quality.

In other words, a linear regression and a least square fit provide the same answer. In the general case, let us consider the cost function

$$J(x) = \frac{1}{2}(x-x_b)^t B^{-1}(x-x_b) + \frac{1}{2}(Hx-y)^t O^{-1}(Hx-y) = J_b + J_o. \tag{4.10}$$

Using the adjoint technique one can compute the gradient of the cost function and thus use an iterative minimization scheme suitable for large scale problems (quasi-Newton like M1QN3 from INRIA, Gilbert and LeMarechal, 1989). Here is the first approximation: for computational reasons, we shall allow 30 to 100 iterations (but not 1000!) and the convergence will not be achieved to machine accuracy. The OI scheme solved a multitude of approximate problems exactly. 3D-Var approximately solves the global problem.

However, one difficulty remains, namely the specification of the covariances of the forecast errors B: the spatial correlations are generally called "structure functions". As already said, dealing with the full matrix is intractable and one has to introduce simplifications. In both the ECMWF (Courtier et al., 1993) and the

NMC (Parrish and Derber, 1992) implementations of J_b the basic ideas are similar and already expressed in Phillips (1986). 3D isotropy and geostrophic coupling over the sphere using Hough mode separation (ECMWF) or a balance equation (NMC) are practically feasible while providing a theoretical improvement on top of the OI implementation. We now describe the key points of the ECMWF implementation while referring to Courtier et al. (1993) for a detailed presentation.

4.3.3 Structure function specification in 3D-Var

2D univariate homogeneous covariances

The basic property we are using is that, as homogeneous covariances over the sphere are by definition invariant with rotations, it becomes diagonal while expressed in the spherical harmonics basis (Boer, 1983). More precisely, the covariance between two points P and Q of the sphere is a function of only $\mu = \cos\theta$ where θ is the angle between P and Q. f may be expressed as a Legendre polynomial series

$$f(\mu) = \sum_n f_n P_n^0(\mu)\sqrt{2n+1}.$$

(The square root is here only for normalisation).

We then have the cost function expression J_b for a field $x(\lambda,\mu) = \sum_{n=0}^{N}\sum_{m=-n}^{n} x_n^m Y_n^m(\lambda,\mu)$ and a background field

$$\begin{aligned} x_b(\lambda,\mu) &= \sum_{n=0}^{N}\sum_{m=-n}^{n} x_{bn}^m Y_n^m(\lambda,\mu) \\ J_b(x) &= \sum_{n=0}^{N} f_n^{-1} \sum_{m=-n}^{n} |x_n^m - x_{bn}^m|^2 \end{aligned}$$

2D univariate, homogeneous correlations

Knowing the field $\sigma_b(\lambda,\mu)$ of the standard deviations of error, assuming the correlations homogeneous means that the covariances of $\frac{x(\lambda,\mu)}{\sigma_b(\lambda,\mu)}$ are homogeneous; we are back to the previous case f being a correlation which verifies $f(0) = 1$ and then $\sum_{n=0}^{N} f_n(2n+1) = 1$.

The algorithm to compute J_b then becomes

1. from the spectral component x_n^m and x_{bn}^m of the control variable and the background respectively, computes the difference $\delta x_n^m = x_n^m - x_{bn}^m$
2. transforms δx_n^m to grid point space and obtains the field $\delta x(\lambda,\mu)$
3. divides $\delta x(\lambda,\mu)$ by the standard deviations $\sigma_b(\lambda,\mu)$
4. transforms back to spectral space and obtains the spectral coefficients $\left(\frac{\delta x}{\sigma_b}\right)_n^m$
5. computes the cost function as $J_b(x) = \sum_{n=0}^{N} f_n^{-1} \sum_{m=-n}^{n} \left(\frac{\delta x}{\sigma_b}\right)_n^{m^2}$

3D univariate separable structure functions

As already said, assuming separable structure functions separable means that the correlation between two points can be expressed as a product of a function of the horizontal distance only and a function of the vertical distance only.

In 3D-Var, one may consider the vertical distance as a function of the hybrid vertical coordinate. After projection on the eigenvectors of the vertical correlation matrix, we are back to the 2D case. In optimal interpolation, the separability is assumed with pressure as vertical coordinate; this is different from 3D-Var where the vertical coordinate is terrain-following close to the surface. This leads to differences in the analysis increments near orography.

3D univariate, non separable

The idea is to have the vertical correlation matrix dependent on the total wavenumber n. More precisely, having obtained $\left(\frac{\delta x}{\sigma_b}\right)_n^m$ as in step iv of the 2D univariate case for all levels $\mathbf{k}$ (it is a vector of length k), one may consider a vertical covariance matrix $A(n)$ as a function of the total wavenumber n.

The implication of this formulation on the structure function in grid point space is discussed in Phillips (1986) and in the 3D-Var framework in Courtier et al. (1993). This implementation of non separability can be seen as a kind of 3D isotropy with sharper vertical structures associated to small horizontal scales. The specification of the matrices $A(n)$ is achieved using statistics of the departure between a 24 h forecast and a 48 h forecast valid for the same time. The approach proposed by Parrish and Derber (1992) is described and validated by Rabier and McNally (1993).

Multivariate

$(x - x_b)$ is separated into a Rossby and a Gravity contribution. The latter is penalized ensuring 10% of the flow ageostrophic. For more details, we refer to Courtier et al. (1993), in particular for some difficulties introduced by the vertical discretization while going back from geopotential to temperature and surface pressure. In the current implementation, it requires separability of the σ_b variation.

4.4 Dealing with the Time Dimension

We said in the introduction that the time dimension is a critical aspect for the specification of the initial condition of a numerical weather prediction. In this section we introduce an algorithm, the Kalman filter (Kalman, 1960; Ghil et al., 1981) which provides a comprehensive and rigorous framework in the linear case. It may easily be extended to the quasilinear case and is then called extended Kalman filter (Jaswinski, 1970).

4.4.1 The extended Kalman filter

Here we only present the algorithm; the theoretical results can be found in the above references and, in particular, in Jazwinski (1970). Denoting by $x(t)$ the state of the atmosphere at time t, we are able to propagate forward in time the information using the forecasting model M,

$$x(t+T) = M(t+T,t)x(t).$$

M is non-linear but the tangent linear model R remains valid, to a large extent, for the propagation of the forecast errors (Lacarra and Talagrand, 1988; Vukicevic, 1991; Rabier and Courtier, 1992)

$$\delta x(t+T) = R(t+T,t)\delta x(t) \quad + \text{random noise (model errors)}.$$

The random noise depicts the fact that the model is not perfect. Here we assume no bias and no time correlation and that we know the covariance matrix Q of the model errors. Relaxation of these assumptions is discussed by Jawinski (1970). At a given time, observations are available

$$y_i = H_i x(t_i) \quad + \text{random noise (observation errors).}$$

We assume the covariance matrix of observation errors O_i is known. Then we are in a position to apply an extended Kalman filter.

<u>The forecast step</u>

$$x^f(t_{i+1}) = M(t_{i+1}, t_i) x^a(t_i) \tag{4.11}$$

$$B^f(t_{i+1}) = R(t_{i+1}, t_i) B^a(t_i) R(t_{i+1}, t_i)^t + Q \tag{4.12}$$

<u>The analysis step</u>

$$x^a(t_i) = x^f(t_i) + K_i(y_i - H_i x^f(t_i)) \tag{4.13}$$

$$B^a(t_i) = (I - K_i H_i') B^f(t_i) \tag{4.14}$$

with

$$K_i = B^f(t_i) H_i'^t (H_i' B^f(t_i) H_i'^t + O_i)^{-1} \tag{4.15}$$

In the analysis step one recognizes the linear estimation equations (4.2), (4.8) and (4.9). What the Kalman filter brings is a way of making use of the dynamics to transport the information in time through equation (4.11) and its quality with equation (4.12).

In the current operational implementation of four-dimensional assimilation relying on OI, (4.11) is solved using the forecast model. However, solving (4.12) is an intractable task; as the dimension of the model phase space is 10^7, (4.12) would require 10^7 model integrations. (4.12) is then replaced by a simple evolution law.

The correlations are kept constant in time while the variances are assumed to follow a growth according to the typical doubling time of the forecast errors of a couple of days with a saturation toward a climatological value. As a consequence, neither the variances of the forecast errors nor the correlations depend on the meteorological situation. This has been recognized as an OI weakness for years.

4.4.2 The 4D Variational algorithm

4D-Var then consists of the minimization problem

$$\mathcal{P}_{4D} : \text{minimize } J(x(t_0)) = \frac{1}{2}(x(t_0) - x_b)^t B^{-1}(x(t_0) - x_b) + \frac{1}{2}\sum_{i=0}^{N}(H_i x(t_i) - y_i)^t O_i^{-1}(H_i x(t_i) - y_i) \tag{4.16}$$

with $x(t_i) = M(t_i, t_0)x(t_0)$. x_b is the background information valid for time to which summarises all the information used before time t_0 and B is the error covariance matrix of x_b.

A classical result, assuming a perfect model and linearity of H and M, is that if $x^*(t_0)$ is the result of $\mathcal{P}_{4D}$ then $x^*(t_N) = M(t_N, t_0)x^*(t_0)$ can also be obtained by applying the Kalman filter to the same statistical estimation problem (Jawinski, 1970; Ghil et al., 1981; Lorenc, 1986; see Thepaut and Courtier, 1991 or Rabier and Courtier, 1992 for a detailed presentation using the same notations as here). In meteorological applications, however, H and M are weakly nonlinear: assuming that the tangent-linear operators $\mathcal{R}$ and H' of M and H respectively satisfy, to acceptable accuracy for meteorological order of magnitude of the estimation error, the relations

$$\begin{aligned} M(t_i,t_0)(x(t_0)+\delta x(t_0)) &\simeq M(t_i,t_0)x(t_0)+\mathcal{R}(t_i,t_0)\delta x(t_0) \\ H_i(x(t_i)+\delta x(t_i)) &\simeq H_i x(t_i)+H'\cdot\delta x(t_i) \end{aligned} \tag{4.17}$$

for perturbation $\delta x(t_0)$, then the 4D-Var problem $\mathcal{P}_{4D}$ is equivalent to the so-called extended Kalman filter under the above-mentioned hypothesis, namely the quasi-linearity, the perfect model and the fixed lag (see previous references). This consists of two steps (^{f} and ^{a} denote forecast and analysis respectively).

4D-Var implicitly uses flow-dependent structure functions as can be seen from equation (4.12) (Thépaut et al., 1993a), so that 4D-Var is a scientific improvement on the current operational implementation. Moreover, 4D-Var is also an algorithmic improvement on the Kalman filter (4.11–4.15) where the equation (4.12) has to be solved explicitly, instead of implicitly in 4D-Var.

There are two main weaknesses in the 4D-Var implementation. First, the model is assumed to be perfect: in Eq. (4.12) no source terms Q are present (Talagrand, 1988; Cohn and Parrish, 1991; Daley, 1991; Wergen, 1992), nevertheless Derber (1989) and Zupanski (1993) demonstrated how to address a model bias in 4D-Var. Second, we do not have access to the analysis error covariance $B^a(t_N)$. Here we suggest that, if (4.12) is approximate anyway, it is not scientifically worthwhile solving it exactly. This idea has been followed by most of the NWP centres which implemented optimal interpolation; the dynamics $\mathcal{R}$ was replaced by the identity or a simple law for the temporal evolution of the variances. In the Kalman filter context, this is discussed e.g. by Dee (1991) and Cohn (1992). In other words, it may be scientifically acceptable to replace $\mathcal{R}$ by an approximate tangent-linear model in (4.2) provided this approximation is smaller than the approximation of neglecting the model error source term Q.

4.4.3 The incremental formulation of variational assimilation

Let us assume from now on that $\mathcal{R}$ is any linear operator, for which we will later stipulate the link with the model M. We define the 4D-Var problem:

$$\mathcal{P}'_{4D} : \text{minimize } J(\delta x(t_0)) = \frac{1}{2}\delta x(t_0)^t B^{-1}\delta x(t_0) + \frac{1}{2}\sum_{i=0}^{N}(H_i x(t_i)-y_i)^t O_i^{-1}(H_i x(t_i)-y_i) \tag{4.18}$$

with $x(t_i)=M(t_i,t_0)x_b+\mathcal{R}(t_i,t_0)\delta x(t_0)$ (we then have $\delta x(t_0)=x(t_0)-x_b$).

Remark 1
If $\mathcal{R}$ is the tangent-linear model, $\mathcal{P}'_{4D}$ and $\mathcal{P}_{4D}$ are equivalent to within the accuracy of the tangent-linear approximation.

Remark 2
If $\mathcal{R}$ is any linear operator which we assume would describe the forecast error evolution exactly, $\mathcal{P}'_{4D}$ leads to the same result as the Kalman filter described by equations (4.11–4.15). However, in a nonlinear problem there is no linear operator which describes the error evolution exactly; introducing $\mathcal{R}$ will then remain an approximation.

Remark 3
$\mathcal{P}'_{4D}$ is better than $\mathcal{P}_{4D}$ as far as an operational implementation is concerned since we keep the original model M for propagating in time the state of the atmosphere, but use an approximate propagation in time of the errors, thus introducing some flexibility on the cost of 4D-Var.

Remark 4

A variant of $\mathcal{P}'_{4D}$ is the quadratic problem

$$\mathcal{P}''_{4D}: \text{minimize } J(\delta x(t_0)) = \frac{1}{2}\delta x(t_0)^t B^{-1}\delta x(t_0) + \frac{1}{2}\sum_{i=0}^{N}(y_{b,i}+H'_i\delta x(t_i)-y_i)^t O_i^{-1}(y_{b,i}+H'_i\delta x(t_i)-y_i) \tag{4.19}$$

with $\delta x(t_i) = \mathcal{R}(t_i,t_0)\delta x(t_0)$ and $y_{b,i} = H[M(t_i,t_0)x_b]$.

The cost of $\mathcal{P}'_{4D}$ and $\mathcal{P}''_{4D}$ are similar but the storage requirement for the background trajectory is different: in $\mathcal{P}'_{4D}$ it is the background vertical column at the observation point and in $\mathcal{P}''_{4D}$ it is the observation equivalent of the background which have to be stored. In $\mathcal{P}'_{4D}$, $\mathcal{R}$ is an approximate linearisation of M, similarly in $\mathcal{P}''_{4D}$, H'_i is an approximate linearisation of H_i. Approximating the full problem by a quadratic one has theoretical advantages since the solution involves solving only linear equations in principle, as is guaranteed unique.

The structure functions used in the current T213 optimal interpolation operational at ECMWF have a cut-off at wave number 63 (Lönnberg, 1988). If we were to use a T106 truncation for $\mathcal{R}$, this would already be an enhancement in terms of resolution. An adiabatic version for $\mathcal{R}$ with some basic simplified diabatic processes like horizontal and vertical diffusion and surface friction would produce the same benefits in terms of implicit flow dependent structure functions as obtained by Thépaut et al. (1993).

The CPU cost of an adiabatic semi-Lagrangian T106 L31 model is typically 1/16 of the CPU cost of the T213 L31 version. The gain of the incremental approach is then of one order of magnitude.

Remark 5

Another variant of $\mathcal{P}'_{4D}$ can be obtained by replacing $H'_i\delta x(t_i)$ in Eq. 4.19 by a finite difference:

$$\mathcal{P}'''_{4D}: \text{minimize } J(\delta x(t_0)) = \frac{1}{2}\delta x(t_0)^t B^{-1}\delta x(t_0) + \frac{1}{2}\sum_{i=0}^{N}(y_{b,i}-\tilde{y}_{b,i}+\tilde{y}_i-y_i)^t O_i^{-1}(y_{b,i}-\tilde{y}_{b,i}+\tilde{y}_i-y_i)^t \tag{4.20}$$

with $\tilde{y}_{b,i} = H_i(\tilde{M}(\tilde{x}_b(t_0))$ being the model equivalent of the observation obtained from a simplified (low resolution, adiabatic) background trajectory and $\tilde{y}_i$ being the model equivalent of the observation obtained from the simplified trajectory issued from $\tilde{x}_b(t_0) + \delta\tilde{x}(t_0)$. $\tilde{x}_b(t_0)$ is the background but at lower resolution. Using the Taylor formula, one has to first order $\tilde{y}_i - \tilde{y}_{b,i} = H'_i Jx(t_i)$. The practical advantage of this formulation over $\mathcal{P}''_{4D}$ or $\mathcal{P}'_{4D}$ is that there is less technical development required once the full 4D-Var problem has already been implemented. It is with this formulation that the numerical experimentations are performed at ECMWF. These three implementations $\mathcal{P}'_{4D}$, $\mathcal{P}''_{4D}$ and $\mathcal{P}'''_{4D}$ are equivalent in the quasi-linear context. They are expected to behave differently in the presence of strong nonlinearities, however we have not seen any arguments as to why one should be superior to the others.

4.4.4 Further developments

There are two ways of improving the model $\mathcal{R}$. Firstly, one could increase the horizontal resolution: the main drawback here is the cost involved since the CPU follows a power law close to 3. In addition the trajectory storage and thus the IO also follow a cubic law (quadratic at a given time step but the number of time steps increases linearly).

Secondly, it is necessary to take into account the physics. The experiments performed so far (Thépaut et al., 1993a and Rabier and Courtier, 1992) have used only horizontal and/or vertical diffusion with a simple surface friction. Rabier et al. (1993) showed that large-scale condensation is essential in order to get reasonable humidity fields in the upper troposphere. More generally, it is expected that the important feedback loops present in the model M will have to be described to a reasonable accuracy with $\mathcal{R}$; this is expected to be of particular importance in the tropics. Zou et al. (1993) and Zupanski (1993) have performed feasibility studies using the adjoint of physical parametrizations. The automatic methods developed at INRIA will assist us in formulating a series of tangent-linear models including progressively more effects of the physics (Rostaing et al., 1993).

In terms of cost this will eventually double the CPU cost of 4D-Var (as the cost of the physical parametrizations is about 50% of the cost of the model) but it will immediately double the storage required for the trajectory (and the related IO). Currently, only t values are stored since the dynamics are nonlinear only with respect to these t values and not the $t - \Delta t$. Since the physics are nonlinear with respect to $t - \Delta t$ values, they too will have to be stored. It should be pointed out, however, that a 2 time-level semi-Lagrangian scheme would not require this extra storage.

The physics is far more nonlinear than the dynamics. As a consequence, the tangent-linear approximation is likely to be less valid for the full model than for the adiabatic version. This means that $\mathcal{P}'_{4D}$ or $\mathcal{P}''_{4D}$ are not necessarily a very good approximation of $\mathcal{P}_{4D}$. A simple way for accounting for some of the nonlinearities in the final analysis is to define a sequence $\mathcal{P}''_{4D}(n)$ of assimilation:

$$\mathcal{P}''_{4D}(n) : \text{minimize } J(\delta x^n(t_0)) \tag{4.21}$$

with

$$\delta x(t_i) = \mathcal{R}(t_i, t_0)\delta x(t_0) \tag{4.22}$$

and

$$y_i^{n-1} = H_i[M(t_i, t_0)(x^{n-1} + \delta^{*n-1}x(t_0))]. \tag{4.23}$$

$\delta^{*n-1}x(t_0)$ is the result of the (approximate) minimization of $\mathcal{P}''_{4D}(n-1)$ and $\delta^{*0}x(t_0) = 0$ and

$$x^{n-1} = x^{n-2} + \delta^{*n-1}x(t_0) \qquad x^0 = x_b \qquad x^{-1} = x_b$$

This algorithm can be seen as a pair of nested loops. The outer loop uses the complete model in Eq. (4.23) to re-define the model trajectory at each iteration of the outer loop. The inner loop uses the tangent-linear and adjoint of a simpler (e.g. adiabatic) model (Eq. 4.22) to minimize the cost function (Eq. 4.21) for the increments calculated with respect to the re-defined trajectory.

This approach allows a progressive inclusion of physical processes without dealing with large-scale non-differentiable minimization problems, of which little is known in practice. The drawback is that we have no guarantee that the sequence $\delta^{*n}x(t_0)$ will converge. Experimental work is necessary to address this issue but we have to be pragmatic. Highly non regular problems will remain intractable for a long time but we have here a reasonable approach that is probably robust.

Remark $\mathcal{R}$ does not have to be kept constant in this iterative process and one can imagine a sequence $\mathcal{R}^n$ where the resolution and the number of physical processes dealt with increase with n.

4.5 Conclusion

Data assimilation schemes allow the use of any kind of observations provided that their error statistics are known. In that respect, satellite data are not different from ground based observations. In the current operational practice, the satellite data used are twofold.

Cloud track winds
Wind vectors are derived from cloud displacement as can be observed from the imager of a geostationary satellite. Schmetz et al. (1993) describe their production from Meteosat while Kelly (1993) assesses their impact in numerical weather prediction.

TOVS soundings
Temperature and humidity information is measured by the infrared and microwave radiometers on board the NOAA polar orbiting satellites. The problems specific to these data are discussed by Eyre (1993) and their impact in numerical weather prediction is presented by McNally et al. (1993) and in the variational context by Andersson et al. (1994).

Acknowledgment

This lecture was also presented at the 1993 ECMWF seminar. Carole typed the manuscript with care, patience and a great magnifier.

4.6 References

Andersson E, Pailleux J, Thepaut J-N, Eyre JR, McNally AP, Kelly GA, Courtier P (1994) Use of cloud cleared radiances in three/four-dimensional variational data assimilation. *Q J R Meteorol Soc* 120: 627–654

Boer GJ (1983) Homogeneous and isotropic turbulence on the sphere. *J Atmos Sci* 40: 154–163

Cohn SE, Parrish DF (1991) The behaviour of forecast error covariances for a Kalman filter in two dimensions. *Mon Wea Rev* 129: 1757–1785

Cohn SE (1992) Short-term dynamics of forecast error covariances. *Proceedings of the ECMWF workshop on "variational assimilation with special emphasis on three-dimensional aspects"*, 9–12 November 1992, Reading, UK, pp 157–170

Courtier P, Thepaut J-N, Hollingsworth A (1993) A strategy for operational implementation of 4D-Var using an incremental approach. *Q J R Meteorol Soc*, submitted

Daley R (1991) *Atmospheric data analysis.* Cambridge atmospheric and space science series. Cambridge University Press

Dee DP (1991) Simplification of the Kalman filter for meteorological data assimilation. *Q J Meteorol Soc* 117: 365–384

Derber JC (1989) A variational continuous assimilation technique. *Mon Wea Rev* 117: 2437–2446

Durand Y (1985) The use of satellite data in the French high resolution analysis. *Proceedings of the 1985 ECMWF workshop on high resolution analysis*, 24–26 June 1985. pp 89–128

Eyre JR (1993) Assimilation of TOVS data: problems and possibilities. *Proceedings of the 1993 ECMWF seminar on "Developments in the use of satellite data in numerical weather prediction"*, 6–10 September 1993, pp 287–302

Gandin LS (1963) Objective analysis of meteorological fields translated from Russian by the Israeli Programme for Scientific Translations, 1965

Ghil M, Cohn S, Tavantzis J, Bube K, Isaacson E (1981) *Application of estimation theory to numerical weather prediction.* Dynamic meteorology: data assimilation methods, L Bengtsson, M Ghil and E Källen, eds, Springer Verlag, 330 pp.

Ghil M, Manalotte-Rizzoli P (1991) Data assimilation in meteorology and oceanography. *Adv Geophys* 33: 141–266

Hollingsworth A, Lonnberg P (1986) The statistical structure of short-range forecast errors as determined from radiosonde data. Part I: The wind field. *Tellus* 38A: 111–136

Jaszwinski AH (1970) *Stochastic processes and filtering theory.* Academic Press, New York

Kalman RE (1960) A new approach for linear filtering and prediction problems. *Trans ASME, Ser D, J Basic Eng* 82: 35–45

Kelly GA (1993) Assimilation and impact of satellite winds. *Proceedings of the 1993 ECMWF seminar on "Developments in the use of satellite data in numerical weather prediction"*, 6–10 September 1993, pp 331–348

Lacarra JF, Talagrand O (1988) Short-range evolution of small perturbations in a barotropic model. em Tellus 40A: 81–95

Lönnberg P, Hollingsworth A (1986) The statistical structure of short-range forecast errors as determined from radiosonde data. Part II: The covariance of height and wind errors. *Tellus* 38A: 137–161

Lönnberg P (1988) Developments in the ECMWF analysis system. 1988 ECMWF seminar on data assimilation and the use of satellite data, pp 75–119, Reading, UK

Lorenc AC (1981) A global three-dimensional multivariate statistical interpolation scheme. *Mon Wea Rev* 109: 701–721

Lorenc A (1986) Analysis methods for numerical weather prediction. *Q J R Meteorol Soc* 112: 1177–1194

McGrath R (1993) Operational monitoring of satellite data and its availability. *Proceedings of the 1993 ECMWF seminar on "Developments in the use of satellite data in Numerical Weather Prediction"*, 6–10 September 1993

McNally AP, Kelly GA, Eyre JR, Andersson E (1993) Experiments using one-dimensional variational analysis of TOVS data at ECMWF. *Proceedings of the 1993 ECMWF seminar on "Developments in the use of satellite data in numerical weather prediction"*, 6–10 September 1993, pp 221–232

Parrish DF, Derber J (1992) The National Meteorological Centre's spectral statistical interpolation analysis system. *Mon Wea Rev* 120: 1747–1763

Phillips NA (1986) The spatial statistics of random geostrophic mode and first-guess error. *Tellus* 38A: 314–322

Rabier F, Courtier P (1992) Four-dimensional assimilation in the presence of baroclinic instability. *Q J R Meteorol Soc*118: 649–672

Rabier F, Courtier P, Pailleux J, Talagrand O, Vasiljevic D (1993) A comparison between 4D variational assimilation and simplified sequential estimation relying on 3D variational analysis. *Q J R Meteorol Soc* 119: 845–880

Rabier F, McNally A (1993) Evaluation of forecast error covariance matrix. *Tech Memo 195*, ECMWF

Rostaing N, Dalmas S, Galligo A (1993) Automatic differentiation in odyssee. To appear in *Tellus*

Schmetz J, Holmlund K, Ottenbacher A (1993) Developments in operational satellite winds from Meteosat. *Proceedings of the 1993 ECMWF seminar on "Developments in the use of satellite data in numerical weather prediction"*, 6–10 September 1993, pp 143–154

Shaw DB, Lonnberg P, Hollingsworth A, Unden P (1987) The 1984/1985 revisions of the ECMWF mass and wind analysis. *Q J R Meteorol Soc* 113: 553–566

Simmons AJ (1991) Development of a high resolution, semi-Lagrangian version of the ECMWF forecast model. *Proceedings of the 1991 ECMWF seminar on "Numerical methods in atmospheric models"*, 9–13 September 1991

Talagrand O (1988) Four-dimensional variational assimilation. *Proceedings of the 1988 ECMWF seminar*, Reading, UK

Tarantola A (1987) Inverse problem theory. Methods for data fitting and model parameter estimation. Elsevier

Thépaut J-N, Courtier P (1991) Four-dimensional variational data assimilation using the adjoint of a multi-level primitive equation model. *Q J R Meteorol Soc* 117: 1225–1254

Thépaut J-N, Hoffman R, Courtier P (1993) Interactions of dynamics and observations in a four-dimensional variational assimilation. *Mon Wea Rev* 121: 3393–3414

Vukicevic T (1991) Non linear and linear evolution of initial forecast errors. *Mon Wea Rev* 119: 1602–1611

Wergen W (1992) The effect of model errors in variational assimilation. *Tellus* 44A: 297–313

Zou X, Navon IM, Sela J (1993) Variational data assimilation with moist threshold process using NMC spectral model. *Tellus*, in press

Zupanski D (1993) The effect of discontinuities in the Betts-Miller cumulus convection scheme on four-dimensional data assimilation. *Tellus*, in press

Zupanski M (1993) Regional four-dimensional variational data assimilation in a quasi-operational forecasting environment. *Mon Wea Rev* 121: 2396–2408

Chapter 5

Priciples of Active and Passive Remote Measurements of Water in the Atmosphere and at Ground

Yves Fouquart and Michèle Vesperini
Laboratoire d'Optique Atmosphérique
Université des Sciences et Technologies de Lille
59655 Villeneuve d'Ascq Cedex
France

Remote sensing consists in inferring a parameter indirectely from measurements of a dependent quantity. For earth and atmospheric parameters, remote sensing is based on measuring radiation which contains the signature of the required characteristics and constituents of the atmosphere (temperature, pressure, absorbing gases, aerosols and cloud), or of the surface (temperature, moisture, surface wind stress, canopy, etc...). Any quantitative interpretation of remote sensed measurements requires that the emission and the transfer of radiation through the atmosphere is accurately accounted for. This is a necessary condition to permit geophysical information to be retrieved from measurements. The first part describes the physical processes related to emission and transfer of radiation in the atmosphere. In the second part, the characteristics of the different spectral domains are outlined to show which type of information can be extracted from the different measurements, depending on the wavelength. A special emphasis is put on water vapour information.

5.1 Radiative Transfer Equation

Let $L_\nu(M, \vec{s})$ be the monochromatic radiance at wavenumber $\nu = 1/\lambda$, (in the interval $\nu, \nu + d\nu$) at a point M in the solid angle $d\omega$ centered on direction $\vec{s}$. $L_\nu(M, \vec{s})$ has dimension $[\mathrm{Wm^{-2}sr^{-1}\,(cm^{-1})^{-1}}]$. Consider a cylindrical volume element which axis is along $\vec{s}$, with surface dS and length dl. Let choose an axis along $\vec{s}$ and call $L_\nu(l, \vec{s})$ and $L_\nu(l + dl, \vec{s})$ the radiances at the entrance and exit of the cylinder.

The photons reaching the exit of the cylinder ($L_\nu(l + dl, \vec{s})$) may have three origins (Figure 5.1):

NATO ASI Series, Vol. I 45
Radiation and Water in the Climate System:
Remote Measurements
Edited by Ehrhard Raschke

1. incoming photons, $L_\nu(l, \vec{s})$, of which a fraction, dL_ν^{ext}, have been lost either by absorption in the medium or because of a scattering event with a change of their direction. Extinction refers to the total loss due to absorption and scattering and $dL_\nu^{ext} < 0$.

2. emitted in the volume element

3. incident in the volume element from direction $\vec{s}' \neq \vec{s}$ and scattered in direction $\vec{s}$.

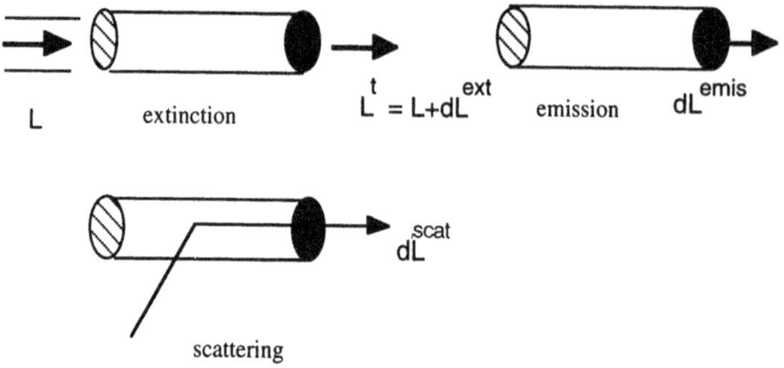

Figure 5.1: *Radiative transfer processes.*

The budget writes as

$$L_\nu(l + dl) = L_\nu(l) + dL_\nu^{ext} + dL_\nu^{sc} + dL_\nu^{em} \tag{5.1}$$

The so-called source function is in $[\mathrm{Wm^{-2}sr^{-1}\,(cm^{-1})^{-1}m^{-1}}]$,

$$J_\nu(l, \vec{s}) = \frac{dL_\nu^{sc}}{dl} + \frac{dL_\nu^{em}}{dl} \tag{5.2}$$

and combining (5.1) and (5.2) we get the general expression:

$$dL_\nu(l) = L_\nu(l + dl, \vec{s}) - L_\nu(l, \vec{s}) = dL_\nu^{ext} + J_\nu(l)dl \tag{5.3}$$

5.1.1 Extinction

Extinction is caused by two processes: absorption and scattering. Unlike scattering and reflexion, absorption is always associated with a physical change of the medium (temperature, internal energy).

The scattering and absorbing processes are linear and statistically independent, thus

$$dL_\nu^{ext} = -L_\nu(l, \vec{s})\sigma_\nu^{ext} dl \tag{5.4}$$

$$\sigma_\nu^{ext} = \sigma_\nu^{abs} + \sigma_\nu^{sc} \tag{5.5}$$

σ_ν^{ext} is the extinction coefficient $[\mathrm{m^{-1}}]$. These coefficients are proportional to the number n of absorbing molecules and/or to the number of scattering particles. One thus defines the efficiency cross sections for one molecule or one particle, s_ν, which has the dimension of a surface $[\mathrm{m^{-2}}]$.

$$\sigma_\nu = n s_\nu \tag{5.6}$$

For absorption, the density of the absorbing gas ρ [kg m^{-3}] is preferred (mass concentrations),

$$\sigma_\nu^{abs} = \rho s_\nu^{abs}, \tag{5.7}$$

s_ν^{abs} has then dimensions of [kg m^{-2}]$^{-1}$.

The quantity $du = n\,dl$ or $du = \rho\,dl$ is the total absorber amount along dl. For water vapour, the S.I. unit is [kg.m^{-2}] although [g cm^{-2}] or equivalently, precipitable water [cm] are still commonly used (1 g cm^{-2} = 1 precip. cm).

5.1.2 Beer-Lambert's law

Assuming that there is neither scattering nor emission in the volume element, then

$$L_\nu(l + dl, \vec{s}) - L_\nu(l, \vec{s}) = dL_\nu^{ext} = -\sigma_\nu^{ext} L_\nu(l, \vec{s})dl \tag{5.8}$$

so that for absorption only,

$$L_\nu(l_1, \vec{s}) = L_\nu(l_o, \vec{s})e^{-\int_{l_o}^{l_1} \sigma_\nu^{abs}(l')dl'} \tag{5.9}$$

5.1.3 Optical thickness

The absorption optical thickness is defined as

$$\delta_\nu = \int_{l_o}^{l_1} \sigma_\nu^{abs}(l')dl' = \int_{l_o}^{l_1} n(l')s^{abs}(l')dl' \tag{5.10}$$

such that

$$L_\nu(l_1, \vec{s}) = L_\nu(l_o, \vec{s})e^{-\delta_\nu} \tag{5.11}$$

5.1.4 Scattering phase function

Scattering is described by:

- the scattering coefficient σ_ν^{sc}, which can be considered as the probability for a photon to be scattered along the optical path dl, with no mention of direction.
- the scattering phase function $p_\nu(\vec{s}, \vec{s}\,')$, probability for a photon incident in direction $\vec{s}$ to be scattered in direction $\vec{s}\,'$. For non-polarized radiation, the phase function is symmetrical: $p_\nu(\vec{s}, \vec{s}\,') = p_\nu(\vec{s}\,', \vec{s})$. When considering the contributions from all directions $\vec{s}\,'$ included in solid angle $d\omega'$, the resulting radiation scattered into direction $\vec{s}$ is proportional to $p_\nu(\vec{s}, \vec{s}\,')\,d\omega'$. The phase function is normalized:

$$\frac{\iint_{4\pi} p_\nu(\vec{s}, \vec{s}\,')\,d\omega'}{\iint_{4\pi} d\omega'} = 1 = \iint_{4\pi} \frac{p_\nu(\vec{s}, \vec{s}\,')}{4\pi}\,d\omega' \tag{5.12}$$

The amount of energy incident from the element of emitting surface dS into the solid angle $d\omega$, during the time interval dt, at frequency ν (between ν and $\nu + d\nu$) is [W.s = J],

$$dE_\nu = dt d\nu d\omega L_\nu(l, \vec{s})dS \tag{5.13}$$

The amount of energy which is scattered in all directions (4π steradians) is [J].

$$dM_\nu = dE_\nu \sigma_\nu^{sc} dl \tag{5.14}$$

whereas in the solid angle $d\omega'$ centered on $\vec{s}'$, it is

$$dL_\nu^{sc}\, dt d\nu = \frac{p_\nu(\vec{s}, \vec{s}')}{4\pi} d\omega' dE_\nu \sigma_\nu^{sc} dl \tag{5.15}$$

5.2 Emission

5.2.1 Thermodynamic equilibrium

If we consider neither nuclear nor chemical processes, the internal energy of an atom or a molecule is made of kinetic energy, electronic, vibrational and rotational energy. When a photon interacts with matter, two cases are possible:

(1) The interaction takes place with no modification to the internal energy: this is a scattering process. It is approximately the case when the energy levels of the molecule are quantized and the energy of the photon is very different. Generally, there are interactions of variable strength but the process can still be considered as scattering: the absorption of a photon causes a transition to an upper energy level (excited level), with a short life time and the photon is re-emitted at the same frequency (the conversion in kinetic energy is negligible). This scattering process is called **coherent**. If the molecule comes back to its initial level through a cascade of transitions, the scattering is **incoherent**.

(2) In the case of absorption, there is also a transition to a higher level but the molecule which has absorbed the photon has a collision with other molecules before re-emitting (i.e. within the life time of the excited level). This is the case of non-radiative transitions which transform the energy of the photon into thermal (kinetic) energy.

The absorption process corresponds to the transformation of electromagnetic (radiative) energy into kinetic (thermal) energy, the emission process is the reverse. Since molecular jiggling is randomly distributed, absorption and emission processes are isotropic, whereas scattering is anisotropic.

The thermodynamic equilibrium characterizes the state of matter and radiation inside an isolated volume. In these conditions,

- the state of equilibrium is stationary: by virtue of the second principle of thermodynamics, there is no thermal change without external interaction (cold source)
- temperature is uniform throughout the volume
- the population of the energy levels is given by the Boltzmann's law

$$n_2 = n_1 e^{-h\nu/kT} \tag{5.16}$$

- the absorption coefficient equals the emission coefficient ($\varepsilon_\nu = \sigma_\nu^{abs}$)
- for a black body the emittance[1] is such that $\varepsilon_\nu = 1$ at all wavenumber ν
- the radiation crossing a surface is isotropic and depends only on temperature, due to the number of collisions per unit time, through the Planck's function $B_\nu(T)$.

$$L_\nu = \varepsilon_\nu B_\nu(T) \tag{5.17}$$

[1]the term "emittance" should be preferred to "emissivity" (IAMAP radiation commission, 1978)

5.2.2 Planck's law

$B_\nu(T)$ is the Planck function which gives the dependence of black body emission with temperature and wavenumber.

$$B_\nu(T) = \frac{c_1'\nu^3}{e^{c_2'\nu/T} - 1} \qquad B_\lambda(T) = \frac{c_1\lambda^{-5}}{e^{c_2/\lambda T} - 1}$$

with $B_\nu d\nu = B_\lambda d\lambda$

ν is in cm^{-1}	λ is in μm
$B_\nu(T)$ in $\mathrm{Wm}^{-2}\ \mathrm{ster}^{-1}\,(\mathrm{cm}^{-1})^{-1}$	$B_\lambda(T)$ in $\mathrm{Wm}^{-2}\ \mathrm{ster}^{-1}\ \mu\mathrm{m}^{-1}$
$c_1' = 2\mathrm{hc}^2 = 1.191\,10^{-8}\ \mathrm{Wm}^{-2}\ \mathrm{ster}^{-1}\,(\mathrm{cm}^{-1})^{-4}$	$c_1 = 1.191\,10^8\ \mathrm{Wm}^{-2}\ \mathrm{ster}^{-1}\,\mu\mathrm{m}^4$
$c_2' = \mathrm{hc/k} = 1.439\ \mathrm{K}(\mathrm{cm}^{-1})^{-1}$	$c_2 = 1.439\,10^4\ \mathrm{K}\ (\ \mu\mathrm{m})$

5.2.3 Stefan's law

Since $B_\lambda(T)$ is isotropic, the angular integration of the Plank function over the half space gives the exitance,

$$M(\lambda) = \int_{2\pi} B_\lambda(T)\cos\theta d\omega = \pi B_\lambda(T) \tag{5.18}$$

Integrating over all wavenumbers gives

$$M = \int_0^\infty \int_{2\pi} B_\lambda(T)\cos\theta d\omega\, d\lambda = \sigma T^4 \tag{5.19}$$

hence

$$\int_0^\infty B_\lambda(T)d\lambda = \frac{\sigma T^4}{\pi} \tag{5.20}$$

with the Stephan's constant $\sigma = 5,67\,10^{-8}\ [\mathrm{Wm}^{-2}\mathrm{K}^{-4}]$.

5.2.4 Wien's law

The maximum of the Planck's function is given by the Wien's law. Neglecting 1 in the denominator of $B_\lambda(T)$, it is easy to check that $\frac{\partial B}{\partial \lambda} = 0$ for

$$T\,\lambda_m \approx \frac{c_2}{5} = 0.29.10^4, \quad \text{or} \quad \nu_m = 1.96\,T \tag{5.21}$$

Examples:

Sun	$(T \approx 5700\mathrm{K})$	$\lambda_m \approx 0.505\ \mu$m
Earth	$(T \approx 220K)$	$\lambda_m \approx 13\ \mu$m
	$(T \approx 270K)$	$\lambda_m \approx 9\ \mu$m
South pole	$(T \approx 200K)$	$\lambda_m \approx 14\ \mu$m
Venus	$(T \approx 753K)$	$\lambda_m \approx 3.8\ \mu$m

There is thus a very clear distinction between the wavelength domains corresponding to solar and terrestrial radiation.

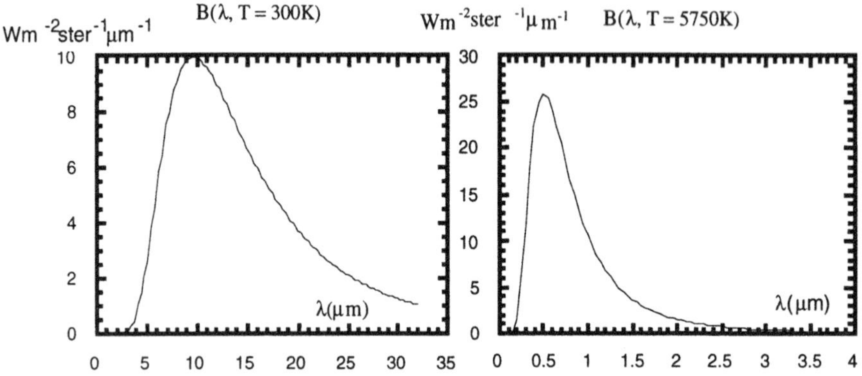

Figure 5.2: *Black body emission at terrestrial and solar temperatures*

5.2.5 Local thermodynamic equilibrium (LTE)

If the volume is no more isolated, radiation exchanges with the environment will occur. The corresponding exchange of energy is

$$h_\nu = -div F_\nu = \rho C_p \frac{\partial T}{\partial t} \tag{5.22}$$

where F_ν is the net flux [$\mathrm{Wm^{-2}}$], $F_\nu = \int_{4\pi} \cos\theta L_\nu d\theta$

Rigorously, the Boltzmann's law does not apply: for instance, the absorption of photons coming from outside can increase the population of the energy levels. However, if the rate of de-excitation by collision dominates the rate of radiative transitions, the population of excited levels is only weakly dependent on radiative processes, and the Boltzmann's law (5.16) remains an excellent approximation.

5.3 Molecular absorption

5.3.1 Energy levels and transitions

Besides the energy associated to nuclear processes and chemical links, atoms and molecules possess energy in different forms: (1) thermal (kinetic energy), (2) electronic, (3) vibrational and (4) rotational. For an isolated molecule

$$E = E_t + E_e + E_v + E_r \tag{5.23}$$

The three last forms are quantized: the molecule can only store discrete values of energy (levels) which depend upon its structure.

Usually, energy transitions can result from any combination of energy levels associated to the three forms of quantized energy. However, we are interested in wavelengths which are mostly associated with vibration-rotation transitions, (i.e. simultaneous change of rotational and vibrational energy levels) and pure rotational transitions. Vibration-rotation transitions

are responsible for optical radiation: UV to IR, respectively of wavelength $\lambda = 0.3$ to $100\ \mu$m, wavenumber $\nu = 1/\lambda = 33000$ to $100\ \mathrm{cm}^{-1}$, frequency $\tilde{\nu} = c\nu = 10^{15}$ to 10^{12} Hz. Electronic transitions correspond to higher frequencies (UV). The transitions which involve the smallest changes of energy are pure rotational (30 μm to a few cm, or a few hundreds of cm^{-1} to 0.1 cm^{-1}, $3 \cdot 10^9$ to 10^{13} Hz), and thus concern the microwave and the far infrared. Pure vibrational transitions are more energetic ($> 500\ \mathrm{cm}^{-1}$).
In fact, most transitions combine rotational and vibrational transitions. Several rotational transitions (small energy change) can be associated with each vibrational transition (large energy change) so that the absorption spectrum is made of different bands corresponding to different vibrational transitions, each of them constituted of many single lines due to rotational transitions.

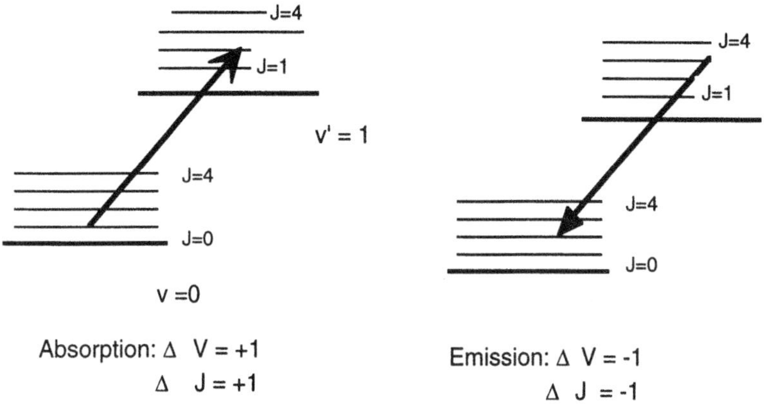

Figure 5.3: *Combination of vibrational and rotational transitions.*

Absorption and emission processes are symmetric. If $E(J'')$ represents the energy of the upper level (initial state) of the transition and $E(J')$, that of the lower state (final), the central frequency $\tilde{\nu}_o$ of the transition is such that

$$h\tilde{\nu}_o = E(J'') - E(J') = F(J) \tag{5.24}$$

and the associated wavenumber ν is ($\tilde{\nu} = c/\lambda$, c speed of light and $\nu = 1/\lambda$)

$$\frac{h\nu_o}{c} = E(J'') - E(J') = F(J) \tag{5.25}$$

Absorption or emission take place when electromagnetic radiation interacts with matter. Neutral molecules are electrically balanced, however the motion of atoms and electrons creates a dipole electrical structure (electric or magnetic dipole) or a quadrupole one. The action of electromagnetic wave on the molecule is that of the oscillating fields which constitute the wave, on this electric structure. A modification of this structure results which corresponds to a change in the energy state.

The strongest interactions are associated to the molecular electric dipole (about 10^5 stronger than magnetic dipole interactions and 10^8 stronger than electric quadripole interactions). The intensity of an electric dipole transition is proportional to the square of the matrix element of the dipole moment (Goody and Young, 1989)

$$R_{ij} = \int \Psi_i^* \mathbf{M} \Psi_j dV \tag{5.26}$$

The Ψ_j are the wave functions (solutions of the Schrödinger's equation) corresponding to the initial state i and final state j. dV is a volume element and the asterisk denotes complex conjugation. Since the wave functions are orthogonal, the intensity of a transition is zero ($R_{ij} = 0$) if the dipole moment $\mathbf{M}$ does not change during the transition. For symmetric linear molecules such as CO_2 and for diatomic molecules, the dipole moment is not changed by a rotational transition. These molecules have no pure rotational spectrum but only vibration-rotation bands which contain much more energy and thus correspond to smaller wavelengths.

5.3.2 Main absorption bands of atmospheric gases

Figure 5.4 shows the vertical transmittance computed for a given atmosphere in the longwave. The main absorbing gas in the longwave is water vapour with a vibration-rotation band centered

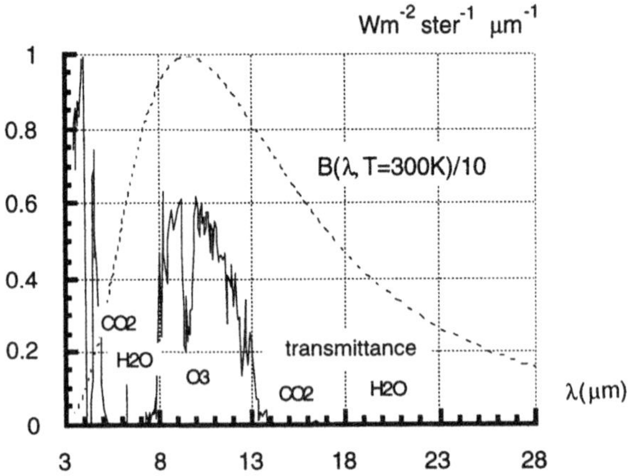

Figure 5.4: *Vertical transmittance through a tropical atmosphere: longwave.*

near 6.3 μm and ranging approximately from 4 to 8 μm. Then, many water vapour rotation bands ranges from 12–13 μm to millimetric wavelengths. Between these two water vapour bands, absorption is much weaker and the atmosphere presents a transparency which permits the remote sensing of surface temperature. This so-called "atmospheric window" stretches from 8 to 12 μm. Its role in radiative exchanges is particularly important because it corresponds to the maximum emission of the black body at usual atmospheric temperatures. The principal ozone rotation band is located at the center of the atmospheric window (9.6 μm). The CO_2 absorption near 15 μm is due to the fundamental vibration-rotation transition and to all transitions between vibration states which energies are such that their differences lead to close frequencies, that is vibrational transitions from level 1 to 2, 2 to 3 or any other combination. This is the case for isotopes and "hot bands" that is energy levels which are populated at high temperature. The other greenhouse trace gases such as CO_2 (weak bands), CFCs, CH_4 etc..., present some bands in the atmospheric window whose absorption, although weak, is very efficient because

(i) the atmosphere is not opaque, and (ii) they are located near the maximum of emission by the atmosphere and the surface. Nevertheless, the main absorber in the atmospheric window remains water vapour, with a continuum structure due to the superposition of the line wings of the bands surrounding the window.

In the shortwave (Figure 5.5), the main absorbers are atomic oxygen and ozone. Their strong absorption protects life from UV radiation. In the visible and in the near infrared, absorption is due to the ozone Chapuis' band around 0.5 μm (it is the main absorber of the channel 1 of

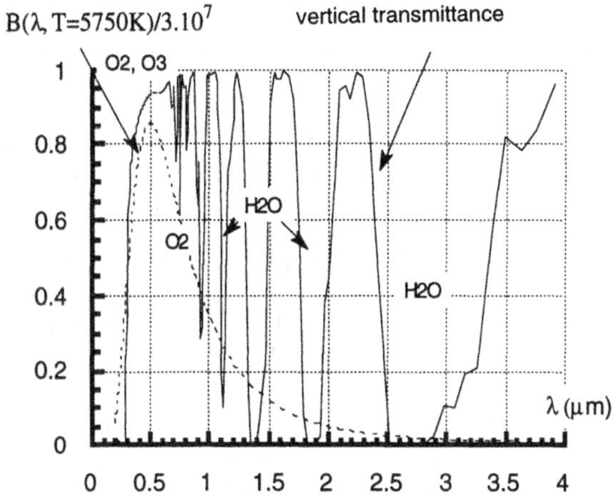

Figure 5.5: *Vertical transmittance through a tropical atmosphere: shortwave.*

AVHRR), to the oxygen band around 0.76 μm, and to many water vapour absorption bands (0.93, 1.87, 2.7 μm, etc). In the troposphere, the absorption is very weak in the visible window from 0.4 to 0.67 μm, but important and variable in the solar infrared.

5.3.3 Molecular absorption coefficient

Although radiation is also absorbed by particles in suspension (aerosols) and by drops and crystals in clouds, the main cause of absorption of electromagnetic radiation in the atmosphere is molecular absorption. Compared to these other types of absorption, the main characteristics of molecular absorption is its very sharp spectral variation. Figure 5.6 shows spectral variations of the transmittance through a tropical atmosphere at high spectral resolution.

To distinguish molecular absorption from the other types of absorption, we will call k_ν^{abs} the corresponding absorption coefficient. For the theoretical case of an isolated spectral line,

$$k_\nu^{abs} = S(J'')g(\nu - \nu_o) \tag{5.27}$$

where S is the line intensity, ν_o the frequency or more generally the wavenumber associated to the transition ($h\nu_o/c = E(J') - E(J'')$) and $g(\nu - \nu_o)$ the normalized line shape implying

$$\int_{-\infty}^{\infty} k_\nu^{abs} d\nu = S(J'') \tag{5.28}$$

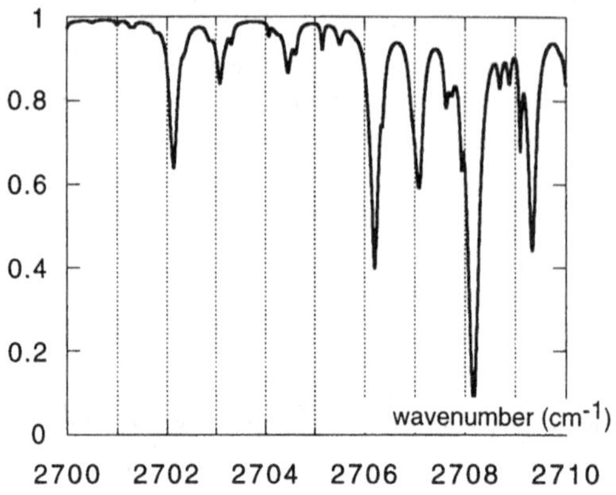

Figure 5.6: *Vertical transmittance through a tropical atmosphere: high spectral resolution.*

5.3.4 Intensity of a spectral line: temperature dependence

The line intensity is proportional to the intensity of the transition and to the number of molecules that are in the initial state J''. Since the population of the levels depends on temperature through the Boltzmann's law, the line intensity changes with temperature. For example, for linear molecules (this is the case of CO_2),

$$\frac{S_J(T)}{S_J(T_\circ)} = \frac{T}{T_\circ} \exp\{-\frac{\mathrm{hc}}{\mathrm{k}}(\frac{1}{T} - \frac{1}{T_\circ})F(J)\} \tag{5.29}$$

whereas for asymmetric top molecules (this is the case for water vapour)

$$\frac{S_J(T)}{S_J(T_\circ)} = [\frac{T}{T_\circ}]^{\frac{3}{2}} \exp\{-\frac{\mathrm{hc}}{\mathrm{k}}(\frac{1}{T} - \frac{1}{T_\circ})F(J)\} \tag{5.30}$$

with h, the Planck's constant, k the Boltzmann's constant, $F(J) = E_r(J'') - E_r(J')$, the energy of the rotational transition, $\mathrm{hc/k} = 1.439\ \mathrm{K}^{-1}\ (\mathrm{cm}^{-1})^{-1}$.

Since $F(J)$ represents the rotational energy, the temperature influence is, first, to redistribute intensities within a given band.

5.3.5 Spectral line shape

Natural width

The energy levels of molecules are not determined uniquely: by virtue of Heisenberg's principle, if an excited state has a limited life time Δt, the energy E of the level presents an uncertainty $\Delta E = \frac{h}{2\pi\Delta t}$ and then the frequency of the transition has an uncertainty $\Delta\tilde{\nu}_\circ = \frac{1}{2\pi\Delta t}$. As a consequence, all frequencies between $\tilde{\nu}_\circ - \Delta\tilde{\nu}_\circ$ and $\tilde{\nu}_\circ + \Delta\tilde{\nu}_\circ$ belong to the same transition. This broadening is very weak, about $10^{-12}\ \mathrm{cm}^{-1}$ for the CO_2 band near 15 μm.

Doppler broadening

Doppler broadening is due to thermal motion of molecules. It thus depends mainly on temperature. If the speed of an emitting (or absorbing) molecule has component (u) along the direction of propagation of the wave, the Doppler's effect leads to a frequency shift

$$\Delta\tilde{\nu} = \frac{\tilde{\nu}_o u}{c} \tag{5.31}$$

(same expression in wavenumber). In this case,

$$g(\nu - \nu_o) = \frac{1}{\alpha_D\sqrt{\pi}} \exp\{\frac{(\nu - \nu_o)^2}{\alpha_D^2}\} \tag{5.32}$$

with

$$\alpha_D = \frac{\nu_o}{c}\sqrt{\frac{2kT\ln 2}{m}} \tag{5.33}$$

For atomic oxygen at 0.5577 μm (or 17953 cm^{-1}), $\alpha_D \approx 3.3\,10^{-2}$ cm^{-1} at 300K, for H_2O rotation near 200 cm^{-1} $\alpha_D \approx 3.5\,10^{-4}$ cm^{-1}.

Collision broadening

The wavenumber associated to a transition depends on initial and final states of energy through $h\nu/c = E(J') - E(J'')$. When a collision occurs between two molecules, the levels of energy are changed, and ν changes. This cause of broadening is preponderant in the troposphere.

The line shape takes the Lorentz's form

$$g(\nu - \nu_o) = \frac{\alpha_L}{\pi[(\nu - \nu_o)^2 + \alpha_L^2]} \tag{5.34}$$

α_L is the half-width (the width of the line at half height), it is proportional to the number of collisions per unit time, it thus depends on pressure and temperature, approximately

$$\alpha_L = \alpha_o \frac{p}{p_o}\sqrt{\frac{T_o}{T}} \tag{5.35}$$

Voigt shape

When collision and Doppler broadenings are comparable and since the two types of broadening are not correlated, the line shape is the result of their convolution:

$$g_V(\nu - \nu_o) = \int_0^\infty g_D(\nu - \zeta) g_L(\zeta - \nu_o) d\zeta \tag{5.36}$$

In practice, $\alpha_L \approx \alpha_D$ near 33 km for CO_2, 31 km for H_2O, 30 km for O_3.

Actually, a Voigt line is made of a Doppler central part combined with Lorentz wings. If the line is saturated, when transmittance is $\simeq 0$, the exact description of its shape around $\tilde{\nu}_o$ is not required. Then, with a good accuracy, a Voigt line can be schematically written as (Fels, 1979):

$$g_V(\nu-\nu_o) = C \qquad \text{if } |\nu-\nu_o| \leq \Delta\nu_o$$
$$g_V(\nu-\nu_o) = \frac{\alpha_L}{\pi(\nu-\nu_o)^2} \qquad \text{elsewhere}$$

The normalization condition implies

$$\int_{-\infty}^{\infty} g_V(\nu-\nu_o)\, d\nu = 1 = 2C\Delta\nu_o + \tfrac{2\alpha_L}{\pi\Delta\nu_o} \tag{5.37}$$

that is

$$C = \frac{1}{2\Delta\nu_o} - \frac{\alpha_L}{\pi(\Delta\nu_o)^2} \tag{5.38}$$

$\Delta\nu_o$ depends on the Lorentz and Doppler widths

$$\Delta\nu_o = 2(1+\zeta)\tfrac{\alpha_L}{\pi} + \beta\alpha_D \tag{5.39}$$

ζ and β are empirical coefficients, they guarantee matching the extreme cases, namely Lorentz and Doppler. The best approximation for the average transmittance of an entire line is obtained for $\zeta = 0.25$ and $\beta = 1.25$.

5.4 Scattering

Scattering is extremely important in the solar range but its influence is generally small in thermal infrared, because of the ratio between the particle size and the wavelength.

The scattering source function (5.15) is

$$dL_\nu^{sc} = \frac{\sigma_\nu^{sc}}{4\pi}\left[\iint_{4\pi} p_\nu(\vec{s},\vec{s}\,')L_\nu(l,\vec{s}\,')d\omega'\right] dl \tag{5.40}$$

where $p_\nu(\vec{s},\vec{s}\,')$ (symmetrical) is the probability, for an incident radiance in the direction $\vec{s}\,'$ to be scattered in the direction $\vec{s}$. With cylindric coordinates and assuming horizontal homogeneity of the atmosphere (plane parallel hypothesis): $dl = dz/cos\theta = dz/\mu$, the radiative transfer equation derives from equations (5.2),(5.3),(5.4) and (5.40)

$$\mu\frac{dL_\nu}{dz} = -\sigma_\nu^{ext}L_\nu(z,\vec{s}) + \frac{\sigma_\nu^{sc}}{4\pi}\left[\iint_{4\pi} p_\nu(\vec{s},\vec{s}\,')L_\nu(l,\vec{s}\,')d\omega'\right] \tag{5.41}$$

with $\sigma_\nu^{ext}dz = -d\delta_\nu$, where δ_ν is the optical depth

$$\mu\frac{dL_\nu}{d\delta_\nu} = L_\nu(\delta_\nu,\vec{s}) - \frac{\varpi_\nu}{4\pi}\left[\iint_{4\pi} p_\nu(\vec{s},\vec{s}\,')L_\nu(\delta_\nu,\vec{s}\,')d\omega'\right] \tag{5.42}$$

where $\varpi_\nu = \sigma_\nu^{sc}/\sigma_\nu^{ext}$ is the single scattering albedo. For a pure absorbing atmosphere, $\varpi_\nu = 0$ whereas $\varpi_\nu = 1$ if the atmosphere is only scattering.

5.4.1 Molecular scattering (Rayleigh)

Molecular scattering results from interactions between the electromagnetic wave and the electric dipole of the molecule. The corresponding scattering phase function is nearly independent of λ. With angle $\Theta = (\widehat{\vec{s},\vec{s}\,'})$

$$p(\Theta) = \frac{3}{4}(1+\cos^2\Theta) \tag{5.43}$$

The scattering cross section of a molecule is

$$\sigma_\lambda^{sc} = 8\pi^3 \frac{m^2 - 1}{3N^2\lambda^4} \quad (5.44)$$

where m is the refractive index of air and N the density of molecules. Typically, the resulting optical depth is $\delta_\lambda = \delta_\lambda^o \simeq 0.01\,\lambda^{-4}$ for $p = p_o = 1013$ hPa, else $\delta_\lambda = \frac{p}{p_o}\delta_\nu^o$.

5.4.2 Scattering by spherical particles

Whereas gas molecules have very few interactions between each others, interactions of radiation with water and aerosols particles can not be described at the molecular scale, since multiple interactions between molecules can not be described simply. These particles are thus considered as a continuous medium characterized by its complex refractive index, and the Maxwell theory which describes the transfer of an electromagnetic wave through a surface is used. The Mie

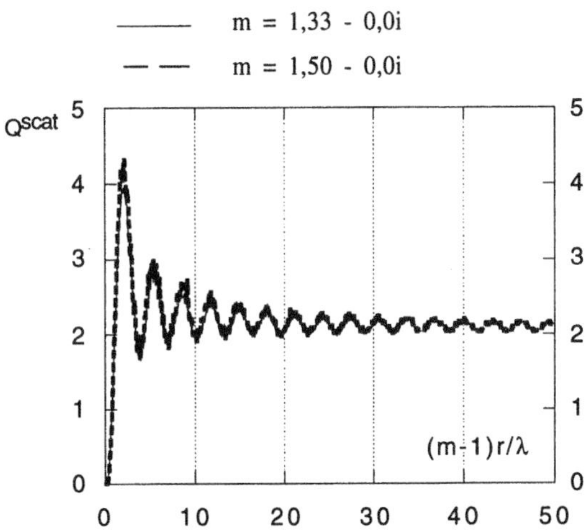

Figure 5.7: *Mie scattering efficiency.*

theory develops the electromagnetic wave in spherical harmonics to solve exactly Maxwell equations in the case of spherical surfaces. This is approximately the case for water droplets and most aerosols, where observations agree with Mie theory. The Mie scattering efficiency for one particle is the non dimensional quantity $Q^{sc}(m, r/\lambda)$ such as

$$\sigma_\lambda^{sc}(r) = \pi r^2 Q^{sc}(m, r/\lambda) \quad (5.45)$$

Similarly, we define the absorption and extinction efficiencies. In these expressions, r is the particle radius and m its complex refractive index. Figure 5.7 shows the variation of Q^{sc} as a function of the so-called Mie parameter $x = 4\pi(m-1)\,r/\lambda$ for two values of m. Note that Q^{sc} essentially depends on $(m-1)\,r/\lambda$ and tends to 2 for large particles, according to Babinet's principle (Goody and Young, 1989, p301).

Aerosol and cloud particles occur randomly with different sizes, and consequently, the scattering events are incoherent. If $n(r)$ is the particle size distribution, the mean optical coefficients for many particles are

$$\sigma_\lambda^{sc} = \frac{\int_0^\infty \sigma_\lambda^{sc}(r)\,n(r)dr}{\int_0^\infty n(r)dr} \quad (5.46)$$

and

$$p_\lambda(\Theta) = \frac{\int_0^\infty p_\lambda(\Theta, r)\sigma_\lambda^{sc}(r)n(r)dr}{\int_0^\infty n(r)dr} \quad (5.47)$$

Scattering is not very sensitive to the precise details of $n(r)$. The weight in $p(\Theta)$ of the particles of radius r, (i.e. $\sigma_\lambda^{sc}(r)n(r)$) is close to $2\pi r^2 n(r)$. As a consequence, different size distributions can be nearly equivalent provided that they have the same effective radius

$$r_e = \frac{\int_0^\infty r\,[r^2 n(r)]dr}{\int_0^\infty [r^2 n(r)]dr} \quad (5.48)$$

For low clouds, drops are mostly spherical and, in the visible, their size is large enough compared to wavelength ($r \approx 1$ to $20\ \mu$m), so that $Q^{sc} \approx 2$ and

$$\begin{aligned} \sigma_\lambda^{sc} &\approx \int_0^\infty 2\pi r^2 n(r)dr \\ &\approx \frac{4\pi}{3}\int_0^\infty [r^3 n(r)]dr \frac{\int_0^\infty [r^2 n(r)]\,dr}{\int_0^\infty r[r^2 n(r)]\,dr} = \frac{3\mathrm{w}}{2\rho\, r_e} \end{aligned}$$

where w is the liquid water content (kg m^{-3}), $\rho = 1000$ kg m^{-3} and r_e is the effective radius of the size distribution. For a cloud with constant r_e, the total optical thickness is proportional to the integrated liquid water content or liquid water path W [kg.m^{-2}]:

$$\delta \approx \frac{3\mathrm{W}}{2 < r_e >} \quad (5.49)$$

$< r_e >$ is, now, the average value of the effective radius and

$$\mathrm{W} = \int_{z_b}^{z_t} \mathrm{w}(z)dz \quad (5.50)$$

(z_b et z_t, are the altitudes of cloud base and cloud top respectively).

5.5 Solution of the RadiativeTransfer Equation

The Radiative Transfer Equation (5.3, (RTE)) established in section 5.1,

$$dL_\nu(l) = L_\nu(l + dl) - L_\nu(l) = -\sigma_\nu^{ext} L_\nu(l)dl + J_\nu(l)dl \quad (5.51)$$

can be expressed by using the **optical depth** δ_ν as coordinate rather than l. Since $d\delta_\nu = -\sigma_\nu^{ext}dl$,

$$\delta_\nu = [\delta_\nu(l')]_x^l = -\int_x^l \sigma_\nu^{ext}(l')dl' \quad (5.52)$$

$$\delta_\nu = \int_l^x \sigma_\nu^{ext}(l')dl' \quad (5.53)$$

With these conventions, the optical depth refers to the boundary x at which $\delta_\nu(x) = 0$. The RTE expressed as function of δ_ν becomes

$$\mu\frac{dL_\nu}{d\delta_\nu}(\delta_\nu, \mu, \varphi) = L_\nu(\delta_\nu, \mu, \varphi) - J_\nu(\delta_\nu, \mu, \varphi) \quad (5.54)$$

with:

$$J_\nu(\delta_\nu, \mu, \varphi) = \frac{J_\nu(l, \mu, \varphi)}{\sigma_\nu^{ext}} \quad (5.55)$$

and can be formally integrated for upwarding radiation (with the convention $\mu > 0$ and being the surface coordinate) as

$$L_\nu^\uparrow(\delta_\nu,\mu,\varphi) = L_\nu^\uparrow(,\mu,\varphi)\exp\{-\tfrac{-\delta_\nu}{\mu}\} + \int_{\delta_\nu} J_\nu(\delta'_\nu,\mu,\varphi)\exp\{-\tfrac{-\delta'_\nu}{\mu}\}d\delta'_\nu/\mu \tag{5.56}$$

and for downwarding radiation ($\mu < 0$) as

$$L_\nu^\downarrow(\delta_\nu,\mu,\varphi) = L_\nu^\downarrow(\textsc{toa},\mu,\varphi)\exp\{-\tfrac{\delta_\nu}{\mu}\} + \int_0^{\delta_\nu} J_\nu(\delta'_\nu,\mu,\varphi)\exp\{-\frac{\delta_\nu - \delta'_\nu}{\mu}\}d\delta'_\nu/\mu \tag{5.57}$$

In these expressions is the total optical thickness of the atmosphere, or equivalently, the optical depth at the surface, $L_\nu^\uparrow(,\mu,\varphi)$ is the upward radiation at the surface in direction (μ,φ) and $L_\nu^\downarrow(\textsc{toa},\mu,\varphi)$ is the downward radiation at the top of the atmosphere (TOA)

The physical interpretation of these solutions is as follows: the upward (downward) radiation emerging at coordinate δ_ν results from the radiation produced at the lower (upper) boundary, attenuated along its path (exponential term), plus the contributions $J_\nu d\delta_\nu/\mu$ of each layer of effective optical thickness $d\delta_\nu/\mu$ (in the direction (μ,φ)), attenuated along the corresponding optical paths.

5.5.1 Shortwave: successive orders of scattering

Scattering is the dominant process in the shortwave, the source function is

$$J_\nu(\delta_\nu,\mu,\varphi) = \frac{\varpi_\nu}{4\pi}[\iint_{4\pi} p(\Theta)L_\nu(\delta_\nu,\vec{s}\,')d\omega'] \tag{5.58}$$

$$\text{[reminder: } \iint_{4\pi} p_\nu(\vec{s},\vec{s}\,')L_\nu(\delta_\nu,\vec{s}\,')d\omega' = \int_{-1}^{1}\int_0^{2\pi} p_\nu(\mu,\varphi,\mu',\varphi')L_\nu(\delta_\nu,\mu',\varphi')d\mu'd\varphi'\text{]}$$

A physically simple method consists in developing the source function according to successive scatterings

$$J_\nu^{(n)}(\delta_\nu,\mu,\varphi) = \frac{\varpi_\nu}{4\pi}\int_{-1}^{1}\int_0^{2\pi} p_\nu(\mu,\varphi,\mu',\varphi')L_\nu^{(n-1)}(\delta_\nu,\mu',\varphi')\,d\mu'd\varphi' \tag{5.59}$$

where $L_\nu^{(n-1)}(\delta_\nu,\mu',\varphi')$ corresponds to the photons which have undergone $n-1$ scatterings events. The order 0 corresponds to the direct radiation (no scattering). It can be obtained with Beer's law,

$$L_\nu^o = L_\nu^{TOA}\exp\{-\delta_\nu/\mu\} = F_\nu^o\exp\{-\delta_\nu/\mu_o\} \tag{5.60}$$

where F_ν^o is the solar flux incident at the top of the atmosphere in the direction (μ_o,φ_o). Physically, the radiance at δ_ν comes from all depths δ'_ν where it is scattered from direction (μ_o,φ_o) into direction (μ,φ).

Primary scattering approximation

The primary scattering approximation ($L_\nu \simeq L_\nu^1$) can be made when multiple scattering is negligible. This approximation is valid if the amount of scatterers is small that is if the optical thickness of the total atmosphere, , is small. In this case, the upward radiation emerging at coordinate δ_ν is

$$L_\nu^1(\delta_\nu,\mu,\varphi) = \frac{\varpi_\nu}{4\pi}\int_{\delta_\nu} F_\nu^o\exp\{-\delta_\nu/\mu_o\}p_\nu(\mu,\varphi,\mu',\varphi')\exp\{-\frac{--\delta'_\nu}{\mu}\}d\delta'_\nu/\mu \tag{5.61}$$

then for $\delta_\nu = 0$ that is at the top of the atmosphere,

$$L^1_\nu(0,\mu,\varphi) = \frac{\varpi_\nu}{4\pi} F^o_\nu\, p_\nu(\mu,\varphi,\mu',\varphi')(\tfrac{1}{\mu}+\tfrac{1}{\mu_o})^{-1}\tfrac{1}{\mu}(1-\exp\{-(\tfrac{1}{\mu}+\tfrac{1}{\mu_o}))\tag{5.62}$$

and if $\ll 1$,

$$L^1_\nu(0,\mu,\varphi) \simeq \frac{\varpi_\nu}{4\pi} F^o_\nu\, p_\nu(\mu,\varphi,\mu',\varphi')d/\mu\tag{5.63}$$

5.5.2 Longwave case

Let's use again the formal integration of the beginning of section 5

$$L^\uparrow_\nu(\delta_\nu,\mu,\varphi) = L^\uparrow_\nu(,\mu,\varphi)\exp\{-\tfrac{-\delta_\nu}{\mu}\} + \int_{\delta_\nu} J_\nu(\delta'_\nu,\mu,\varphi)\exp\{-\tfrac{-\delta'_\nu}{\mu}\}d\delta'_\nu/\mu\tag{5.64}$$

and

$$L^\downarrow_\nu(\delta_\nu,\mu,\varphi) = L^\downarrow_\nu(\textsc{toa},\mu,\varphi)\exp\{-\tfrac{\delta_\nu}{\mu}\} + \int_0^{\delta_\nu} J_\nu(\delta'_\nu,\mu,\varphi)\exp\{-\frac{\delta_\nu-\delta'_\nu}{\mu}\}d\delta'_\nu/\mu\tag{5.65}$$

In the longwave, scattering is very weak with respect to absorption. It is thus common to make use of transmittances rather than optical depths.

$$\tau_\nu(z,z',\mu) = \exp\{-\frac{|\delta_\nu(z)-\delta_\nu(z')|}{\mu}\}\tag{5.66}$$

In the conditions of local thermodynamic equilibrium, the source function is given by Planck's law:

$$J_\nu(\delta'_\nu,\mu,\varphi) = B_\nu(T(z'))\tag{5.67}$$

and for an horizontally homogeneous medium, it only depends on altitude. Hence, radiation does not depend on azimuth φ anymore, in these conditions,

$$L^\uparrow_\nu(z,\mu) = L^\uparrow_\nu(0,\mu)\,\tau_\nu(0,z,\mu) + \int_0^z B_\nu(z')d\tau_\nu(z,z',\mu)\tag{5.68}$$

$$L^\downarrow_\nu(z,\mu) = L^\downarrow_\nu(\textsc{toa},\mu)\,\tau_\nu(z,\infty,\mu) - \int_z^\infty B_\nu(z')d\tau_\nu(z,z',\mu)\tag{5.69}$$

Generally, the extraterrestrial longwave radiation is negligible (fossil big bang radiation at 2K) and the radiation leaving the surface is, most of the time, the emission of a body which has a given emittance ε_ν and temperature T_s. In these conditions

$$L^\uparrow_\nu(0,\mu) = \varepsilon_\nu B_\nu(T_s) + 2\pi\int_0^1 \rho_\nu(\mu,\mu')L^\downarrow_\nu(0,\mu')d\mu'\tag{5.70}$$

The surface reflectance $\rho_\nu(\mu,\mu')$ is often very small in the IR. The emittance of water is near 0.98 on average, that of the emerged surfaces depends a lot on the quantity of water present in a very thin layer at the top of the soil (skin), for dry sand, it can be as small as 0.6 in the atmospheric window (Foot, 1988).

However,with the noticeable exception of cloudy conditions, the downward atmospheric radiation is much smaller than the surface emission itself. The term of reflectance is quite often omitted in the IR but not in the microwave where it is of the same order as the emission term. Note that it would be inconsistent to use the right emittances and to neglect the reflecting term: for example, omitting the reflexion in the case of a desert atmosphere and using experimental surface emittances leads to larger errors than taking $\varepsilon_\nu = 1$.

5.5.3 Calculation of monochromatic transmittance

If scattering is neglected, the extinction coefficient σ_ν^{ext} reduces to the absorption coefficient; except in the case of clouds, it reduces to the molecular absorption coefficient k_ν.
Then the transmittance is

$$\tau_\nu(z, z', \mu) = \exp\{-\int_{z''}^{z} k_\nu(z'')\rho(z'')dz''\} \tag{5.71}$$

This expression would be correct if the atmosphere had only one absorbing gas with density ρ. Indeed, we have to generalize:

$$\tau_\nu(z, z', \mu) = \exp\{-\int_{z''}^{z} \sum_j k_\nu^j(z'')\rho_j(z'')dz''\} \tag{5.72}$$

where k_ν^j is the absorption coefficient of gas j.

For a given gas, the absorption coefficient results from contributions of all the neighbouring lines corresponding to different transitions with quantum number i

$$k_\nu^j(z") = \sum_i S_i g_j(\nu - \nu_{oi}) \tag{5.73}$$

where ν_{oi} represents the central frequency of the line i of gas j and $g(\nu - \nu_{oi})$, the line shape. For the most general case of collision broadening,

$$k_\nu^j(z") = \sum_i S_i \frac{\alpha_i}{\pi[(\nu - \nu_{oi})^2 + \alpha_i^2]} \tag{5.74}$$

5.5.4 Spectral integration of transmittances

Line-by-line (LBL) models

Since optical sensors do not provide monochromatic measurements, transmittances have to be considered over a given bandwidth $\Delta\nu$ as

$$\tau_{\Delta\nu}(z, z', \mu) = \frac{1}{\Delta\nu}\int_{\Delta\nu} \tau_\nu(z, z', \mu)d\nu \tag{5.75}$$

Line-by-line (LBL) models perform a strict calculation of this integral by summing up the contributions of each line in each spectral interval of typically 0.1 cm^{-1} or 0.05 cm^{-1}. This calculation is quite cumbersome because the number of spectral lines to consider is very large: the wings of remote lines as far as 100 cm^{-1} or more can still contribute significantly to absorption.

Band models

The alternative to expensive computer-time LBL models are band models, which are physically based approximations of spectraly averaged transmittances. With the rapid development of computational capacity, this somewhat traditional approach tends to be replaced line-by-line models (LBL), in which the full complexity of molecular absorption can be taken into account. Nevertheless, line-by-line models remain time consuming and faster methods are still required in many cases. Moreover, band models are still a useful concept to understand some of the physical processes that are associated with absorption, particularly in the case of saturation processes.

a) Homogeneous path, equivalent width of an absorbing line

First, we consider the homogeneous case (p and T constant) for a unique gas with density ρ, such that

$$\int_l k_\nu(l)\,\rho(l)dl = k_\nu\,\rho l \tag{5.76}$$

Let's consider an isolated line, such as the spectral integration can be made over the whole line without superposition with any other line. In this case, $\Delta\nu$ can be extended to the interval $[-\infty, +\infty]$. The total absorptance of a line with gas concentration ρ is computed as

$$\int_{-\infty}^{\infty}[1-\tau_\nu]d\nu = \int_{-\infty}^{\infty}[1-\exp\{-k_\nu\rho l\}]d\nu \tag{5.77}$$

Since transmittance and absorptance are dimensionless quantities, the total absorptance has thus dimension of wavelength. The equivalent width $\mathbf{W}$, defined as

$$\mathbf{W} = \int_{-\infty}^{\infty}[1-\exp\{-k_\nu\rho l\}]d\nu \tag{5.78}$$

is the width (in cm^{-1}) of a rectangular line which would absorb completely radiation ($1-\tau_\nu{=}1$) and give the same total absorptance of the line ($\mathbf{W}$).

Weak lines

If absorption is weak, $k_\nu\rho\,l$ is small, the exponential can be developed and according to the normalization of the line shape, we obtain

$$\mathbf{W} = \int_{-\infty}^{\infty} k_\nu\rho l\,d\nu = \int_{-\infty}^{\infty} S\,g(\nu-\nu_{oi})\,\rho l\,d\nu = Su \tag{5.79}$$

Since the medium is homogeneous, $u = \rho\,l$ represents the total absorber amount encountered by radiation. Hence, absorption increases linearly with the total amount of absorber independently of pressure. It varies with temperature through the dependence of S with T.

Strong lines

For strong absorption, the center of the line is saturated, $(1-\exp() \approx 1)$, the computation of $\tau_{\Delta\nu}$ relies above all on the description of line wings. In this case $((\nu-\nu_o) \gg \alpha)$, α at the denominator may be omitted, with not change to the absorption, neither at the line center where the absorption is total, nor in the wings.

$$\mathbf{W} \approx 2\sqrt{S\alpha u} \tag{5.80}$$

The absorption depends on pressure through the Lorentz half width $\alpha(p,T) = \alpha_o\frac{p}{p_o}\sqrt{\frac{T_o}{T}}$.

b) Inhomogeneous paths

In the atmosphere, pressure and temperature generally vary along the path. In these conditions,

$$\mathbf{W} = \int_{\Delta\nu}[1-\exp\{-\int_z \rho(z)S(T(z))g_\nu(p(z),T(z))dz\}]\,d\nu \tag{5.81}$$

where $g_\nu(z)$ is the line shape at altitude z.

Then, we define an equivalent homogeneous atmosphere of constant pressure p_e, constant temperature T_e and amount of absorber u_e, that gives the same equivalent width.

$$\int_{\Delta\nu} [1 - \exp\{-\int_z n(z) S(T_e) g_\nu(p_e, T_e) dz\}]\, d\nu = \mathbf{W} \tag{5.82}$$

With the very common two-parameter Curtis-Godson approximation, the equivalent temperature T_e is set arbitrarily and p_e and u_e are tuned so as to satisfy the equivalence in the two limit regimes of weak and strong absorption.

For weak absorption, if we develop the exponential and keeping in mind the normalization of g_ν, a first condition is obtained as

$$\int_z n(z) S(T(z)) dz = \int_z n(z) S(T_e) dz \tag{5.83}$$

For strong absorption, if we consider again that the absorption at the center of the lines is saturated and that 1 - exp() only differs from 1 far from the center, we can write

$$\mathbf{W} = \int_{\Delta\nu} [1 - \exp\{-\int_z \frac{n(z)\, S(T(z))\, \alpha_L(p(z), T(z)}{\pi(\nu - \nu_o)^2} dz\}]\, d\nu \tag{5.84}$$

and similarly for W_o. Since the denominator is independent of altitude, it can be put out of the integration over z, the condition of equivalence in strong regime is

$$\int_z n(z)\, S(T(z))\, \alpha_L(p(z), T(z)) dz = \int_z n(z)\, S(T_e)\, \alpha_L(p_e, T_e) dz \tag{5.85}$$

From the first condition (5.83), the reference temperature T_e is arbitrarily fixed and the equivalent absorber amount u_e is

$$u_e = \int_z n(z) \frac{S(T(z))}{S(T_e)} dz \tag{5.86}$$

and the strong absorption condition gives

$$u_e \frac{\alpha_e}{\alpha_{Lo}} = \int_z n(z) \frac{\alpha_L(p(z), T(z))}{\alpha_{Lo}} \frac{S(T(z))}{S(T_e)} dz \tag{5.87}$$

A further simplification consists in omitting the influence of temperature variations, the first condition is then always satisfied and, if we write the variation of α_{Lo} with altitude, the second condition is

$$\int_z n(z)\, p(z)\, dz = p_e \int_z n(z)\, dz \tag{5.88}$$

c) Random band models

So far, we have introduced the concept of band model for a single isolated line. The absorption bands (rotation or vibration-rotation bands) correspond to the superposition of many spectral lines. The averaged transmittance over the spectral interval $\Delta\nu$ is then expressed by

$$\tau_{\Delta\nu}(z, z', \mu) = \frac{1}{\Delta\nu} \int_{\Delta\nu} \tau_\nu(z, z', \mu) d\nu = \frac{1}{\Delta\nu} \int_{\Delta\nu} \exp\{-\int_{z'}^{z} k_\nu(z'') \rho(z'') dz''\} \tag{5.89}$$

with

$$k_\nu(z") = \sum_i S_i \frac{\alpha_i(z'')}{\pi[(\nu - \nu_{oi})^2 + \alpha_i^2(z'')]} \tag{5.90}$$

The band model conceptually the simplest consists of describing the spectrum step by step in changing ν and in recalculating k_ν at each step. Since many lines can have, at a given wavenumber, a significant contribution to k_ν and since the line shape changes very quickly near the center, this type of model is of course time consuming. In practice, the strongest lines still have an influence as far as many hundreds of cm^{-1} from their center and the sharp variation of $g(\nu - \nu_o)$ requires integration steps as small as a few hundredth of cm^{-1}. The increasing computing power makes however their use more and more frequent. Band models which are made of simple physical approximations still remain very useful for a fast approach of the problems or when we consider very large spectral intervals and repeated calculations as it is the case in all problems where radiation is considered from the point of view of energy transfers (e.g. the computation of radiative heating rates in a GCM).

Here again, let's consider first the simplest case of a homogeneous path, with u the total amount of the absorbing gas. The most common band model is the statistic model. Its main hypothesis is that the lines are randomly distributed in the spectral interval, with a given distribution of intensities. Depending on this distribution, we distinguish two main models (see for example Paltridge and Platt, 1976, for more details):

- Goody's model (1952)

$$\tau_{\Delta\nu}(u) = \exp\{-\frac{\bar{S}u}{\bar{d}\sqrt{1+\frac{\bar{S}u}{\pi\bar{\alpha}}}}\} \tag{5.91}$$

- Malkmus' model (1967)

$$\tau_{\Delta\nu}(u) = \exp\{-\frac{\pi\bar{\alpha}}{2\bar{d}}[\sqrt{1+\frac{4\bar{S}u}{\pi\bar{\alpha}}}-1]\} \tag{5.92}$$

In these expressions, $\bar{S}$ is the averaged intensity, $\bar{d}$ is the averaged distance between two consecutive lines and $\bar{\alpha}$ is the averaged half-width. These parameters are related to the spectroscopic parameters of the interval:

$$\frac{\bar{S}}{\bar{d}} = \sum_j \frac{S_j}{\Delta\nu} \tag{5.93}$$

$$\frac{\pi\bar{\alpha}}{\bar{d}} = \frac{4}{\Delta\nu}\frac{[\sum_j \sqrt{S_j\alpha_j}]^2}{\sum_j S_j} \tag{5.94}$$

Provided that $\Delta\nu$ is large enough to contain many lines but still small enough so that the lines are randomly distributed over the whole interval, the statistical models describe quite well the absorption by atmospheric gases. However, these models have a major drawback whose influence increases when $\Delta\nu$ decreases: if the center of a given line is within the interval, the whole line, including its far wings is supposed to be in the same interval. To correct this drawback, Aoki (1978) suggested a method which takes into consideration only the part of that line present in $\Delta\nu$, the rest is treated as a continuum. Indeed, the tendency, now is to fit the band model parameters directly to line by line calculations. This is not always possible however: if the lines are distributed according to particular conditions, the main hypothesis of random distribution is not fulfilled and the fit is generally poor.

5.5.5 The problem of water vapour continuum

As seen on 5.4 the IR window (8-14 μm) is of major importance in the LW radiation transfer since it corresponds to the maximum of the Planck function for terrestrial temperatures. Unfortunately, the spectral absorption in that domain is not very well known. In addition to the 9.6 μm O_3 bands and the weak CO_2 bands, there is an important water vapour continuum, the origin of which is not well established. Although water vapour presents such a continuum absorption over the whole IR spectral interval, its major impacts is the IR window since it is there the only active absorption. Several parameterizations of the water vapour continuum absorption have been made (Bignell, 1970; Roberts et al., 1976; Clough et al., 1989) but questions remain about their validity over the whole spectrum.

5.6 Application to Remote Sensing of Water Vapour

In the previous pages, the general form of radiative transfer equation resulting from contributions between the boundary (l_o) and a given position x has been derived

$$L_\nu(x,\vec{s}) = L_\nu(l_o,\vec{s})\tau_\nu(l_o,x,\vec{s}) + \int_{l_o}^{x} J_\nu(l,\vec{s})\ \tau_\nu(l,x,\vec{s})\ dl \tag{5.95}$$

with the source function accounting for emission and scattering at position l

$$J_\nu(l,\vec{s}) = \underbrace{\sigma_\nu^{abs}(l)B_\nu(T(l))}_{emission} + \underbrace{\frac{\sigma_\nu^{sc}(l)}{4\pi}[\iint_{4\pi} p(\Theta)L_\nu(l,\vec{s}')d\omega']}_{scattering}$$

Emission is well described by the Planck function. Shortwave are only emitted at very high temperatures thus by the sun whereas the earth and atmosphere emit radiation in the longwave and microwave.

Scattering is caused by atmospheric constituents such as molecules and aerosols and cloud particles. Its magnitude depends mostly on the ratio between wavelength and the size of scattering particles. Problems in computing the scattering term of radiative transfer are twofold:

- it involves very expensive computation to account for multiple scattering, and interactions between scattering and absorption.
- the actual distribution of cloud and aerosols particles is unknown, due to their high variability in space and time.

It is thus difficult to calculate accurately how much radiation is scattered so that scattering is often a limitation to the retrieval of atmospheric parameters. In the visible and infra-red, where wavelength is small with respect to the size of scattering particles, the retrieval of atmospheric parameters is only possible in clear sky conditions. In the microwave, scattering is significant for very large particles only, those found in convective clouds and precipitation. Therefore, microwave instruments are considered to a certain extent as cloud penetrating sensors. In the following, we consider only cases in which scattering can be neglected, and the radiative transfer is computed for non scattering atmospheres.

5.6.1 Passive radiometry in the longwave and microwave

Molecular scattering is negligible at these wavelengths. Aerosols and cloud particles affect strongly infrared radiation and microwave in the case of large drops. These cases are not considered hereafter.

For a non scattering atmosphere, the RTE applies with $\sigma_\nu^{sc} = 0$ and consequently $\sigma_\nu^{ext} = \sigma_\nu^{abs}$. The radiance at a given position x is

$$L_\nu(x, \vec{s}) = L_\nu(l_o, \vec{s})\tau_\nu(l_o, x, \vec{s}) + \int_{l_o}^{x} \sigma_\nu^{abs}(l)\, B_\nu(T(l))\, \tau_\nu(l, x, \vec{s})\, dl \tag{5.96}$$

l_o being the position of the boundary.

Radiation at satellite level

For remote sensing applications, we are interested in the radiance reaching a satellite, located at the top of the atmosphere (TOA), in a direction $\vec{s}$ defined by its angle θ with respect to nadir. Considering an atmosphere with horizontal homogeneity (plane parallel hypothesis), optical depth can be expressed as a function of altitude z. The distance l to the satellite is related to z by $z = l\cos\theta$, so that $dl = dz/\cos\theta = dz/\mu$ (with the convention $\mu > 0$ for upwarding path and $\mu < 0$ downwards).

The radiance reaching the satellite has two origins:

Upward radiation between the surface and the satellite

With the above conventions, the upward radiation at TOA ($z = z_T$) can be rewritten:

$$L_\nu^{\uparrow 1}(z_T, \theta) = L_\nu^{\uparrow}(0, \theta)\, \tau_\nu(0, z_T, \theta) + \int_0^{z_T} B_\nu(T(z))\, \frac{d\tau_\nu(z, z_T, \theta)}{dz}\, dz \tag{5.97}$$

with the transmittance from z to z_T

$$\tau_\nu(z, z_T, \theta) = e^{-\frac{1}{\mu}\int_z^{z_T} \sigma_\nu^{abs}(z')dz'} = e^{-\frac{1}{\mu}\int_z^{z_T} k_\nu^{abs}(z')\rho(z')dz'}$$

and

$$\frac{d\tau_\nu(z, z_T, \theta)}{dz} = \frac{1}{\mu} k_\nu^{abs}(z)\rho(z) e^{-\frac{1}{\mu}\int_z^{z_T} k_\nu^{abs}(z')\rho(z')dz'}.$$

Where $k_\nu^{abs}(z')$ and $\rho(z')$ are respectively the absorption coefficient (per mass unit) and the density of each gas radiatively active at the wavelength.

$L_\nu^{\uparrow}(0, \theta)$ is the radiation emitted by the surface and corresponds to emission at surface skin temperature T_s with surface emittance ε_ν.

$$L_\nu^{\uparrow 1}(z_T, \theta) = \varepsilon_\nu\, B_\nu(T_s)\, \tau_\nu(0, z_T, \theta) + \int_0^{z_T} B_\nu(T(z))\, \frac{d\tau_\nu(z, z_T, \theta)}{dz}\, dz \tag{5.98}$$

Downward atmospheric radiation reflected by the surface and upwarding to the satellite

All the radiation reaching the surface is either reflected or absorbed (then reemitted). The surface is Lambertian (with isotropic reflectance) in longwave so that the reflected fraction can be expressed as $(1 - \varepsilon_\nu)$, ε_ν being the emittance. The radiation coming from space (boundary for the downward path) is negligible at these wavelengths. The downward radiance reaching the surface in a direction θ' is

$$L_\nu^{\downarrow}(0, \theta') = \int_{z_T}^{0} B_\nu(T(z)) \frac{d\tau_\nu(z, 0, \theta')}{dz}\, dz \tag{5.99}$$

Since the radiance reflected by the Lambertian surface results from contributions of all directions, the reflected term reaching the satellite is thus

$$L_\nu^{\uparrow 2}(z_T,\theta) = [\int_{-\frac{\pi}{2}}^{\frac{\pi}{2}} L_\nu^{\downarrow}(0,\theta')\,d\theta'](1-\varepsilon_\nu)\,\tau_\nu(0,z_T,\theta) \tag{5.100}$$

In the infrared, emittance ε_ν ranges between 1 for ocean and 0.85 for very dry surface like desert. In the microwave, smaller values are seen over ocean (between 0.4 and 0.7), with a strong dependence on surface wind speed. The value of emittance will decide whether introduce the second term or not. To make the comprehension easier, we will not consider the reflected term any further.

Weighting function

Introducing the weighting function

$$W_\nu(z,z_T,\theta) = \frac{d\tau_\nu(z,z_T,\theta)}{dz} = \tfrac{1}{\mu}\,k_\nu^{abs}(z)\rho(z)\,\tau_\nu(z,z_T,\theta)$$

the RTE becomes

$$L_\nu(z_T,\theta) = \underbrace{\varepsilon_\nu\,B_\nu(T_s)\,\tau_\nu(0,z_T,\theta)}_{surface} + \underbrace{\int_0^{z_T} B_\nu(T(z))\,W_\nu(z,z_T,\theta)\,dz}_{atmosphere} \tag{5.101}$$

the atmospheric term of the latter equation may be interpreted as the blackbody emission of several layers $[z, z+dz]$ at temperature $T(z)$ each contributing with a proper weighting factor $W_\nu(z,z_T,\theta)$ to the total radiation reaching the satellite.

When considering satellite remote sensing, emission of radiation decreases with altitude via the temperature dependence, but the transmittance from z to z_T compensates for it through (i) lower absorber concentration (ii) shorter the optical path. Consequently, the total radiation reaching the satellite is the contribution of high emitting layers (lower troposphere) which is strongly absorbed (small transmittance), of moderate emission from the mid troposphere with moderate absorption, and of small emission from the upper atmosphere which is weakly absorbed. The weighting function specifies the level from where the radiation originates, therefore it determines the layer which can be sounded from satellite at a particular wavelength.
The altitude at which a given weighting function is maximum depends on how active is molecular absorption at that particular wavelength. A strong absorption will make the weighting function peak higher in the atmosphere and on the contrary, channels located in the atmospheric window have their maximum of $W(z)$ close to the surface.

A set of channels with different absorptions can be used to sound different layers of the atmosphere. As an example, Figure 5.8 shows weighting functions associated to the 19 channels of TOVS instrument [2], that have been computed for a standard tropical atmosphere.

Remark: In the case of ground based measurements, it is the contribution from upper atmosphere (by itself small) that is mostly absorbed along its way to the surface so that ground-based passive radiometry is only sensitive to the low troposphere.

The width of weighting functions is limited by technological constraints. For a given central wavelength, the basic width of the weighting function is given by its monochromatic limit. This is nearly the case for microwave sensors where the current technology is able to provide virtually monochromatic measurements.

[2]TOVS for TIROS Operational Vertical Sounder

In the infrared, classical sensors are using filters whose spectral width $\Delta\nu$ is very large with respect to absorption lines. In this case, the gradient $\frac{dT_\nu}{dz}$ is not as sharp as for a single line since for each of the lines contained in $\Delta\nu$, the variation in absorption (through pressure and temperature dependence) does not occur at the same level z. Consequently, the weighting function is broader. New generation sensors (spectrometers or interferometers) provide measurements in much narrower spectral bands and the related weighting function are much sharper.

Inversion of radiance measurements

The computation of radiance knowing the profiles of atmospheric temperature, humidity and other gases, constitutes the direct problem, whereas inversion consists in extracting information on the atmosphere from the observed radiance. An ideal system for vertical profile inversion would have narrow weighting functions which do not overlap each other, so that radiance could be interpreted as a function of temperature in a single layer. As seen previously, the actual weighting functions are quite broad and overlap each other and this is a limitation to the vertical resolution of the set of sensors.

a) Information content

The expression for the weighting function

$$W_\nu(z, z_T, \theta) = \frac{1}{\mu} k_\nu^{abs}(z)\,\rho(z)\,e^{-\frac{1}{\mu}\int_z^{z_T} k_\nu^{abs}(z')\rho(z')dz'} \tag{5.102}$$

shows that the weighting function is dependent of

1. $k_\nu^{abs}(z)$: the spectroscopic properties of each gas which has significant absorption at this wavelength. These are functions of pressure (hence altitude) and to a minor extent of temperature.
2. $\rho(z)$ vertical profile (density) of these gases

Among the main absorbers, gases such as CO_2 and N_2O are uniformly distributed along the vertical and with no significant variations with time. They provide quite steady weighting functions for temperature sounding. For example the TOVS-HIRS (High-resolution Infrared Radiation Sounder) channels 1 to 5 are located on the same edge of the CO_2 absorption band between 14 and 15 μm , but with different absorption efficiencies which yield $W_\nu(z)$ peaking in different layers of the atmosphere. In H_2O absorption bands (e.g. channels HIRS 10, 11, and 12 of TOVS), the weighting function is much more variable since it changes with water vapour content so with meteorological situation. Note that when H_2O content increases, absorption is stronger so that the weighting function peaks higher in the atmosphere.

b) Temperature inversion

From equation (5.101), we understand that inversion of temperature profiles is easier than that of atmospheric constituents, since the atmospheric term of the RTE is the product of temperature profile (through $B_\nu(T(z))$) and of a weighting profile $W_\nu(z)$ that should strictly be called the "temperature weighting function".

With a set of measurements (subscript n) made at different wavelengths, whose $W_\nu(z)$ peak in different layers (subscript p), each measurement L_n can be expressed as a linear combination

$$L_n = \varepsilon_n B_{n,o} + \sum_p B_n(T_p) W_{n,p}(T_p, p, \rho) \tag{5.103}$$

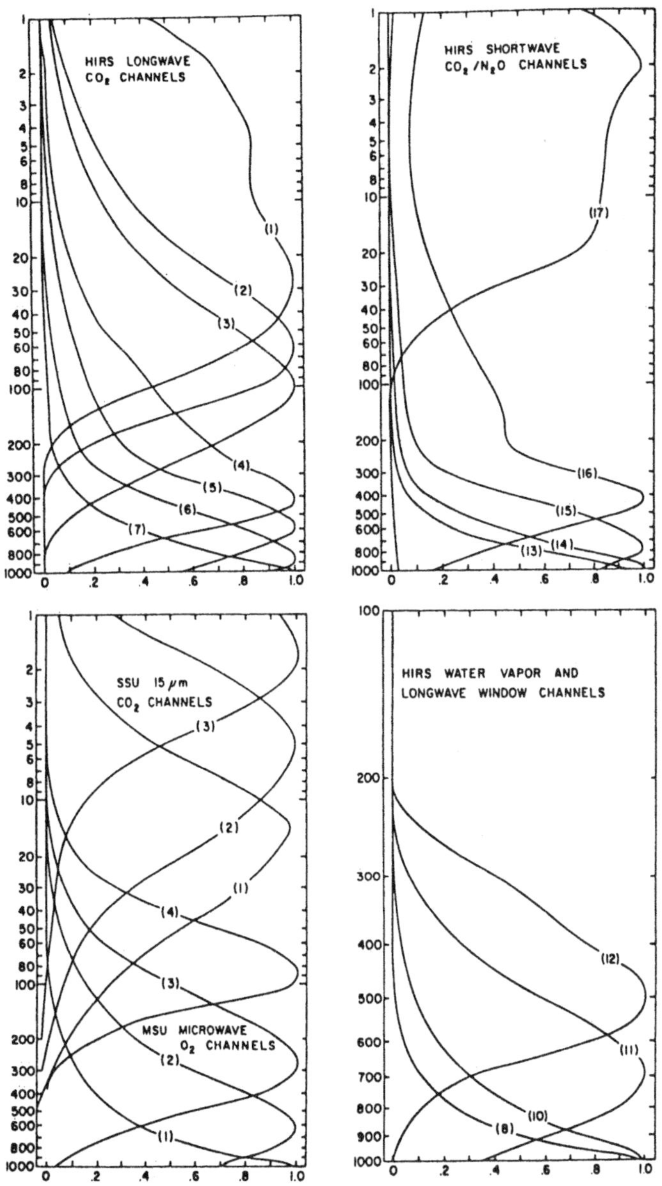

Figure 5.8: *TOVS weighting functions calculated for a given atmosphere (from Smith et al., 1979).*

where $B_n(T_p)$ is linked to the discrete temperature profile (through the Planck function) and $W_{n,p}$ is the discrete weighting factor associated to spectral channel n and layer p.

The following problems are solved by using suitable approximations and efficient algorithms.

- $B_n(T_p)$ varies with wavelength. This problem can be cleared by defining a reference function $B(T_p)$, and correcting the radiance measurement accordingly;
- Since absorption lines are only weakly dependent on temperature, $W_{n,p}(T_p)$ can be computed for a mean temperature profile so that an independent set of linear equations is obtained;
- the surface term can be included in the summation;
- the computation of the solution is subject to the conditioning of the system of equations, namely the number of independent radiance measurements (n) and of vertical layer (p), making the problem over-constrained or under-constrained. More sophisticated methods involve a-priori information (a guessed profile) together with its corresponding error and uses the radiance measurements to correct (accordingly to their respective errors) this profile. This approach is very close to the methods used in data analysis for numerical weather prediction, and often inversion has become a part of the data assimilation process (cf. Courtier, this volume, chapter 4).

c) Inversion of atmospheric constituents

Estimation of atmospheric constituents like water vapour or ozone is much more non linear since it is not possible to express the RTE as a product of the constituent profile and an independent "constituent weighting function".

From the atmospheric contribution of the RTE,

$$\int_0^{z_T} B_\nu(T(z))\, k_\nu^{abs}(z)\rho(z) \exp\{-\tfrac{1}{\mu}\textstyle\int_z^{z_T} k_\nu^{abs}(z')\rho(z')dz'\}\, dz \qquad (5.104)$$

a water vapour weighting function would be function of temperature profile and of the constituent profile itself $\rho(z)$ in a complicated way. In this case, it is no longer possible to linearize the equation. Again, although information on constituent profile exists in a set of radiance measurements, the inversion of such information requires specific methods. Data assimilation at ECMWF consider the problem as a whole, by computing the best estimation of a vector containing both temperature and humidity profiles considering an a-priori profile and its error (Eyre et al., 1993; McNally and Vesperini, 1995).

5.6.2 Passive radiometry in the shortwave

The shortwave band is commonly associated to the solar spectrum hence is usualy bounded by 0.2 μm and 4.3 μm. Atmospheric and surface emission can be considered as negligible at these wavelengths.

The dimension of atmospheric constituents (molecules and particles) relative to wavelength does not allow scattering to be neglected. Not only cloud, but also scattering by aerosols and Rayleigh scattering (mainly by H_2O molecules acting at very short wavelengths, up to 0.7 μm) have a significant impact on shortwave radiation.

RTE in the shortwave

The general form of radiative transfer equation resulting from contributions between the boundary (l_o) and a given position x is

$$L_\nu(x,\theta) = L_\nu(l_o,\theta)\tau_\nu(l_o,x,\theta) + \int_{l_o}^{x} J_\nu(l,\theta)\, \tau_\nu(l,x,\theta)\, dl \qquad (5.105)$$

The atmospheric source function, due to scattering only, becomes

$$J_\nu(l,\theta) = \frac{\sigma_\nu^{sc}}{4\pi}[\iint_{4\pi} p(\Theta)L_\nu(l,\vec{s}')d\omega']$$

In the shortwave, radiation at the boundary, $L_\nu(l_o,\theta)$, is the incoming solar radiation at the top of the atmosphere (TOA) in direction θ defined by θ_o and φ_o, respectively nadir and azimuth angles. This can be computed as $B_\nu(T_{sun})$ (cf. section 5.2.2) and will be hereafter referred to as E_ν.

Non-scattering atmosphere

Like for the longwave, the main difficulty is in estimating $J_\nu(l,\theta)$ because this involves very expensive computation and requires a knowledge of the structure of scattering particles. (cf. section 5.5.1). Retrieval of vertically integrated water vapour content from shortwave measurements is however possible in a non scattering atmosphere, by virtue of its total absorption effect. The wavelength must be choosen long enough to avoid Rayleigh scattering.

In a non scattering atmosphere, the shortwave radiance transmitted downwards from TOA in direction θ_o, φ_o at the surface level 0 is

$$L_\nu^\downarrow(0,\theta_o) = E_\nu(\theta_o,\varphi_o)\tau_\nu(z_T,0,\theta_o)$$

A fraction is reflected by the surface depending on its reflectivity, $r_\nu(\theta_o,\theta,\varphi-\varphi_o)$, and then transmitted upwards

$$L_\nu^\uparrow(z_T,0) = E_\nu(\theta_o,\varphi_o)\tau_\nu(z_T,0,\theta_o)r_\nu(\theta_o,\theta,\varphi-\varphi_o)\tau_\nu(0,z_T,\theta)$$

The main uncertainty is on the surface reflectance $r_\nu(\theta_o,\theta,\varphi-\varphi_o)$ due to its angular and spectral dependence.

Remark: The same method can be used to infer the water vapour content above a cloud, the latter being considered as a reflector. If the cloud top height z_c is known, the inferred quantity is the integrated water vapour content between z_c and the TOA.

Differential absorption method

The differential absorption method (Frouin et al., 1990; Gao and Goetz, 1990) is based on measuring radiances generated by the same source at two (or more) different wavelengths.

The ratio between these two measurements $L_{\Delta\nu_2}$ and $L_{\Delta\nu_1}$ made at satellite level is

$$\frac{L_{\Delta\nu_2}}{L_{\Delta\nu_1}} = \frac{E_{\Delta\nu_2}\,\tau_{\Delta\nu_2}(z_T,0,\theta_o)\,r_{\nu_2}(\theta_o,\theta,\varphi-\varphi_o)\,\tau_{\Delta\nu_2}(0,z_T,\theta)}{E_{\Delta\nu_1}\,\tau_{\Delta\nu_1}(z_T,0,\theta_o)\,r_{\nu_1}(\theta_o,\theta,\varphi-\varphi_o)\,\tau_{\Delta\nu_1}(0,z_T,\theta)} \quad (5.106)$$

A way to get free from the knowledge of r_ν consists in measuring radiation at two close wavelengths and assuming $r_{\nu_1}(\theta_o,\theta,\varphi-\varphi_o) = r_{\nu_2}(\theta_o,\theta,\varphi-\varphi_o)$. In that case, it is possible to express the ratio independently of the reflectance.

$$\frac{L_{\Delta\nu_2}}{L_{\Delta\nu_1}} = \frac{E_{\Delta\nu_2}\,\tau_{\Delta\nu_2}(z_T,0,\theta_o)\,\tau_{\Delta\nu_2}(0,z_T,\theta)}{E_{\Delta\nu_1}\,\tau_{\Delta\nu_1}(z_T,0,\theta_o)\,\tau_{\Delta\nu_1}(0,z_T,\theta)} \quad (5.107)$$

Actually, any sensor has a given spectral width $\Delta\nu$ so that the average transmittance over $\Delta\nu_1$ would be for channel 1:

$$\tau_{\Delta\nu_1} = \int_{\Delta\nu_1} \tau(\nu)d\nu$$

and accounting for absorbers, the total transmittance (for downwarding and upwarding paths) over bandwidth $\Delta\nu_1$

$$\tau_{\Delta\nu_1}(z_T, z_T, \theta_o, \theta) = \int_{\Delta\nu_1} \exp\{-\tfrac{1}{\mu_o}\int_0^{z_T} k_\nu^{abs}(z')\rho(z')dz'\} \exp\{-\tfrac{1}{\mu}\int_0^{z_T} k_\nu^{abs}(z')\rho(z')dz'\}d\nu \tag{5.108}$$

$$\tau_{\Delta\nu_1}(z_T, z_T, \theta_o, \theta) = \int_{\Delta\nu_1} \exp\{-(\tfrac{1}{\mu_o}+\tfrac{1}{\mu})\int_0^{z_T} k_\nu^{abs}(z')\rho(z')dz'\}d\nu$$

Note that $\tau_{\Delta\nu_1}(z_T, z_T, \theta_o, \theta) \neq \tau_{\Delta\nu_1}(z_T, 0, \theta_o)\,\tau_{\Delta\nu_1}(0, z_T, \theta)$.

The retrieval of water vapour content in the shortwave can be understood through the concept of equivalent absorber amount explained in section 5.5.4. The equivalent absorber amount (here water vapour) accounts for pressure and temperature dependence of the absorption line

$$u_e = \int_0^{z_T} \frac{p(z)}{p_e}\frac{T_e}{T(z)}\rho(z)\,dz$$

In the atmosphere, temperature, pressure and absorbers have a near exponential profile, with their associated scale heights H_p, H_T, and H_ρ. As mentioned by Frouin et al. (1990), the equivalent absorber amount can be expressed statistically as function of the absorber amount itself

$$u_e \simeq \int_0^{z_T} e^{-z/H_p}\, e^{z/H_T} \rho_o e^{-z/H_\rho} \simeq 0.8u$$

Using a band model to parameterize the spectral integration of the transmittance over band width $\Delta\nu$, for example the one of Goody (1952)

$$\tau_{\Delta\nu_1} = \exp\{-\frac{\bar{S}u}{d\sqrt{1+\bar{S}u/\pi\alpha}}\}$$

which in case of strong absorption $\bar{S}u \gg \pi\alpha$, the average transmittance reduces to

$$\tau_{\Delta\nu_1} \simeq e^{-\frac{\sqrt{\bar{S}u\pi\alpha}}{d}} = e^{-\beta\sqrt{\frac{1}{\mu_o}+\frac{1}{\mu}}\sqrt{u}}$$

that can be simplified defining β as a spectral constant which can be calculated for the corresponding channel.

The ratio between the two channels measurements is

$$\frac{L_{\Delta\nu_2}}{L_{\Delta\nu_1}} = e^{-(\beta_1-\beta_2)\sqrt{\frac{1}{\mu_o}+\frac{1}{\mu}}\sqrt{u}}$$

from which u can be inferred directly. An example showing the implementation of this method is given by Frouin et al. (1990).

5.7 Active Remote Sensing

Active remote sensing methods are also based on radiation measurements but in that case, the radiation is generated by the instrument itself, which means that its intensity is known and its wavelength is chosen. Active remote sensing involves the same radiative transfer theory as passive radiometry. Retrieval methods are based on extinction of radiation along the optical path or backscattering by atmospheric constituents. Radar (RAdiation Detection And Ranging) and LIDAR (LIght Detection And Ranging) are systems which contain transmitter and receiver.

Radar uses microwave transmitter whereas LIDAR uses a Laser. Most of them ("pulsed" radar or lidar) emit a pulse of radiation, and the time lag Δt between emission and measurement of backscattered radiation specifies the depth Δr of the sounded region since $\Delta r = c\Delta t/2$, with c speed of electromagnetic radiation. With several measurements at successive time lags after emission of the pulse, a profile can be built.

In the following, the radar and lidar equation are introduced, as well as the underlying physical concepts. Dedicated lectures on these topics are given by Ackerman (cloud radar), Joe (radar for precipitation), and Weitkamp (lidar measurements), this volume.

5.7.1 Radar equation

In a radar system, the transmitter generates a power of radiation E_o [W], in the microwave. With an isotropic antenna, this power would be spread over a sphere so that at distance r from the transmitter, the flux (power density crossing a surface unit) would be $F_{iso} = P/4\pi r^2$. The directional gain G of an antenna is thus defined with respect to F_{iso} following $G(\theta,\varphi) = F(\theta,\varphi)/F_{iso}$. It represents the gain brought by a directional antenna with respect to an isotropic one. The radiation crossing a surface unity at distance r from the transmitter is thus

$$F(r,\theta,\varphi) = \frac{E_o\, G(\theta,\varphi)}{4\pi r^2}\tau_\nu(0,r,\theta) \quad [W/m^2] \tag{5.109}$$

The transmittance $\tau_\nu(0,r,\theta)$ accounting for the extinction at this wavelength, as for passive remote sensing.

The volume illuminated acts as a scatterer and is described by its scattering cross section σ $[m^2]$ which accounts for (i) the effective target area which traps incident radiation, (ii) the loss by absorption and (iii) the fraction scattered backwards (phase function).

Radiation reflected backward is also spread over the sphere's surface and crosses the surface of the antenna A_a so that the radiation received is

$$E_{rec} = \frac{E_o\, G(\theta,\varphi)}{4\pi r^2}\tau_\nu^2(0,r,\theta)\frac{\sigma}{4\pi r^2}A_a \tag{5.110}$$

As mentioned above, information is extracted from σ, according to the size and shape of scattering particles, mainly cloud droplets and/or ice crystals.

5.7.2 LIDAR equation

The equation for LIDAR is very similar the radar except that by virtue of the coherence of the Laser source, the transmitted radiation can be considered as non diverging. There is thus only one $(4\pi r^2)$ term.

$$E_{rec} = E_o\, \tau_\nu^2(0,r,\theta)\frac{A}{4\pi r^2}\beta \tag{5.111}$$

with β backscattering cross-section, and A surface of the receiver.

Different types of Lidar are associated to the different types of scattering processes described in section 5.2.1. The Raman Lidar is based on incoherent scattering, in which visible radiation is transmitted by the Laser, interacts with matter and reemits radiation through cascades in the infrared range, giving information on many distinct gases. Associated energies are very small and signal processing represents a major part of the retrieval process. The Dual Lidar is based on differential absorption of radiation along the optical path, at two close wavelengths.

Acknowledgments

This lecture is based on Y. Fouquart's post-graduate lectures at the University of Lille, it also beneficiates from his lecture given at the NATO ASI summer school "Physically Based Modelling and Simulation of Climate and Climatic Change" (Erice, Sicily, 1986) and from a lecture given by M. Herman at a summer school organized by CNES (Marseille, August 1990). We also would like to thank Pedro Viterbo for his helpful comments on the manuscript.

5.8 References

Aoki T (1978) Semi-direct band models for the transmittance calculation. *J Meteorol Soc of Japan* 56: 112–120

Bignell KJ (1970) The water-vapour infrared continuum. *Quart J Roy Met Soc* 96: 390–404

Clough SA, Kneizys FX, Davies RW (1989) Line shape and the water vapour continuum. *Atm Res* 23: 229–241

Eyre JR, Kelly G, McNally AP, Andersson E, Persson A (1993) Assimilation of TOVS radiances through one-dimensional variational analysis. *Quart J Roy Met Soc* 119: 1427–1463

Fels SB (1979) Simple strategies for inclusion of Voigt effects in IR cooling rate calculations. *Appl Opt* 18: 2634–2637

Foot JS (1988) Near infrared radiances observed by the UK C130 multi-channel radiometer during the marine stratocumulus IFO and preliminary comparison with model calculations. *MRF Internal Note* 44: 1–10

Fouquart Y (1986) *Radiative transfer in climate modeling.* NATO Advanced Study Institute on Physically Based Modeling and Simulation of Climate and Climatic Change., Erice, Sicily, 11-23 May 1986

Frouin R, Deschamps P-Y, Lecomte P (1990) Determination from space of atmospheric total water vapor amounts by differential absorption near 940 nm: theory and airborne verification. *J Appl Meteor* 29: 448–460

Gao B-C, Goetz AFH (1990) Column atmospheric water vapor and vegetation liquid water retrievals from airborne imaging spectrometer data. *J Geophys Res* 95D: 3549–3564

Goody RM (1952) A statistical model for water vapour absorption. *Quart J Roy Met Soc* 78: 165–169

Goody RM, Yung YL (1989) *Atmospheric radiation. Theoretical basis.* Oxford University Press. 2nd edition, 1989

Herman M (1990) *Eléments de transfert radiatif.* CNES Summer School on The Middle Atmosphere and Space Observations, Marseille. Cépaduès-Editions, Toulouse

IAMAP radiation commission (1978) *Terminology and units of radiation quantities and measurements.* Ed. E. Raschke, University of Köln

Malkmus W (1967) Random Lorentz band models with exponential tailed S^{-1} line intensity distribution function. *J Opt Soc Amer* 57: 323–329

McNally AP, Vesperini M (1995) Variational analysis of humidity information from TOVS radiances at ECMWF. *Quart J Roy Met Soc*, submitted

Paltridge GW, Platt CMR (1976) *Radiative processes in meteorology and climatology.* Elsevier, Holland. 341 pp

Roberts RE, Selby JE, Biberman LM (1976) Infrared continuum absorption by atmospheric water vapour in the 8–12 μm window. *Appl Opt* 15: 2085–2090

Smith WL, Woolf HM, Hayden CM, Wark DQ, McMillin LM (1979) The TIROS-N Operational Vertical Sounder. *Bull Am Meteorol Soc* 60: 1177–1187

Ulaby FT, Moore RK, Fung AK (1986) Retrieval of Integrated *Precipitable Water Vapor and Cloud Liquid from Satellite Observations.* Microwave Remote Sensing, Vol 3: From Theory to Applications. Addison-Wesley

Further Reading

Liou K-N *An Introduction to Atmospheric Radiation.* Academic Press, Inc, 1980

Liou K-N *Radiation and cloud processes in the atmosphere. Theory, observation and modeling.* Oxford University Press, New York, 487 pp.,1992

Goody RM, Yung YL *Atmospheric radiation. Theoretical basis.* Oxford University Press, 2nd edition, 1989

Stephens GL *Remote sensing of the lower atmosphere. An introduction.* Oxford University Press, New York, 544 pp., 1994

Van de Hulst HC*Light scattering by small particles.* Dover Publ, Inc, 1957

Ulaby FT, Moore RK, Fung AK *Microwave Remote Sensing*, Vol 1: Fundamentals and radiometry. Addison-Wesley, 1986

Chapter 6

Extratropical Cyclones: Their Mesoscale Structure, Precipitation and Role in the Transport of Water

Ronald E. Stewart
Climate Processes and Earth Observation Division
Atmospheric Environment Service
Downsview, Ontario
Canada

Abstract

The focus of this article is the nature of extratropical cyclones and in particular their effects on the water cycle. These cloud systems produce much of the precipitation found in mid-latitudes. Embedded features within the overall storm, such as precipitation bands, account for the greatest precipitation and these features are often, but not always, linked with frontal surfaces. Depending on the background large scale forcing, stability and moisture availability, the degree to which these systems produce precipitation from the available vapour (the precipitation efficiency) varies widely. Other roles played by these systems are, for example, the transport of moisture to high regions of the troposphere and the moistening of low levels through the evaporation or sublimation of falling precipitation, but little quantitative work on such aspects has been conducted. Extratropical cyclones therefore directly affect the large scale water cycle through their impact on precipitation and the re-distribution of vapour throughout the troposphere. Remote sensing plays a critical role in advancing the study of extratropical cyclones and their impacts by providing quantitative observations of critical features that often cannot be measured by any other means.

6.1 Introduction

Extratropical cyclones with their wide expanse of layered cloud contribute substantially to the global water and energy cycles. They account for a major fraction of the precipitation in mid-latitudes and they are associated with some of the main avenues for the latitudinal transport of moisture through the atmosphere (see for example Lambert, 1988).

In order to fully appreciate their role on the water cycles globally and regionally, attention has to as well be paid to their internal structure. Organized embedded circulations, mesoscale

NATO ASI Series, Vol. I 45
Radiation and Water in the Climate System:
Remote Measurements
Edited by Ehrhard Raschke

features, with their enhanced vertical motions redistribute and alter the vertical profiles of momentum and moisture. There are a wide variety of such mesoscale features that need to be accounted for.

In addition, these cloud systems are responsible for much of the production of the precipitation that falls over vast regions of the world. This precipitation is produced under a wide variety of conditions, it occurs with a variety of rates and even forms. The exact processes contributing to this precipitation, including their often mesoscale origin, need to be addressed in order for this critical parameter, precipitation, to be properly simulated within climate models of the water cycle.

As well, the interaction between these layer cloud systems and orography often leads to regions of enhanced precipitation. Because mountains are the source regions for many of the world's rivers, the role of orography on the small scale distributions of precipitation from these clouds is a critical problem.

For many reasons then, the nature of cloud systems affects water cycles in the atmosphere. The purpose of this article is to summarize some of the key attributes of these cloud systems, with the emphasis being on their role on water cycles, and to point out some of the ways in which remote sensing contributes to the study of these systems and their impacts.

6.2 Mesoscale Structures

6.2.1 Occurrence and general nature

The overall attributes of extratropical cyclones are well appreciated and have been the focus of numerous studies. The founding of the Norwegian School of Meteorology some 75 years ago represented a major step in this regard (Bjerknes, 1919; Bjerknes and Solberg, 1922). Much subsequent work has been done and has for example clarified their thermal (Palmen and Newton, 1948) and kinematic "conveyor belt" structure (Harrold, 1973; Carlson, 1980; Browning, 1985). Figure 6.1 shows a typical example of such a situation; most of these studies have relied heavily on satellite information to provide an appropriate perspective.

A great deal of mesoscale organization occurs within the overall cloud regions (see for example Hobbs, 1978). These features are typically organized into quasi-two-dimensional structures (precipitation bands) near the surface fronts and these are often well-observed with radar systems (Browning and Harrold, 1970; Hobbs, 1978; Heymsfield, 1979; Saarikivi and Pukakka, 1990). A typical example of such a band within a winter storm is shown in Figure 6.2. Since these sub-structures interact with the larger scales, good physical understanding of these mesoscale features is essential for the accurate representation of the larger scale layer cloud field in global models.

As recently pointed out by Ryan (1995), the current classification of mesoscale structures within extratropical cyclones is largely based on the observational studies in the United Kingdom and the west coast of the USA. Regions of enhanced precipitation tend to occur within banded structures (see for example Houze and Hobbs, 1982; Browning, 1985). The bands have furthermore been classified by width into narrow or wide bands, although it should be noted that these studies will be biased to situations of maritime weather systems at mid-latitudes striking the western portion of continents.

Wide bands range in width up to about 150 km and are found embedded within and parallel to the frontal zone. These bands have been further classified by the storm sector in which they are located. An important characteristic of the warm frontal and warm sector clouds is the

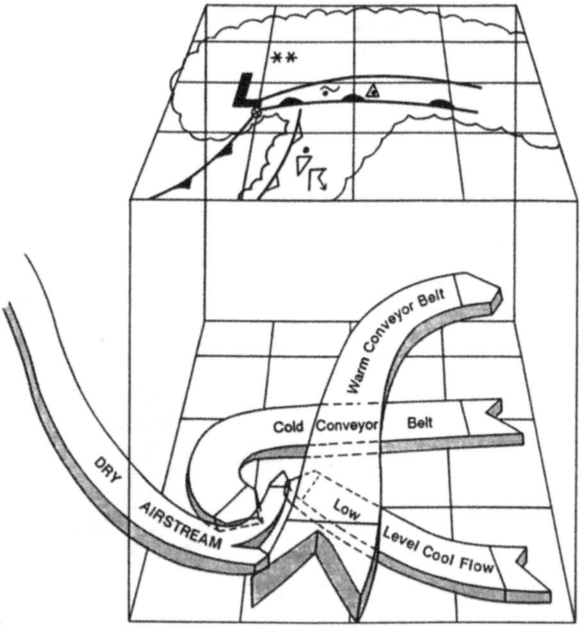

Figure 6.1: *The conveyor belt structure of an extra-tropical cyclone observed over Canada in 1986, as studied by Stewart and Macpherson (1989). CCB refers to cold conveyor belt, and WCB to warm conveyor belt. Standard symbols for precipitation indicate snow to the north of the warm front, rain and thunderstorms in the warm sector, and freezing rain and ice pellets adjacent to the warm front.*

snow "generating cells" described by, for example, Douglas et al. (1957). It was shown by, for example Hobbs (1978) that the enhanced precipitation at the surface was directly linked with the presence of such generating cells aloft, at least in many instances.

Narrow cold-frontal bands are 5-25 km across and tend to be generated by forced convection. These are typically associated with a narrow more intense updraft region, although a weaker downdraft is typically linked with the heavy precipitation.

Although there have not been adequate measurements in all regions of the world to verify the universal application of such results, Ryan (1995) does find that wide and narrow bands, as discussed above, appear to occur within regions of the world in which adequate measurements and analyses have been made. This includes studies from North America, Russia, Western Europe, Australia, Israel and China.

It should be noted that the mesoscale structure of extratropical cyclones can include precipitation bands that mainly reflect "radar reflectivity enhancements" rather than a physically-driven mesoscale circulation. The simple schematic of such a situation is shown in Figure 6.3. Within storms that produce both rain and snow at the surface, it is possible that the classic radar bright band reaches down towards the surface. With a horizontal depiction of the radar information from such a situation, there is an apparent precipitation band present that occurs right at the junction between the different forms of precipitation. One has to be very careful

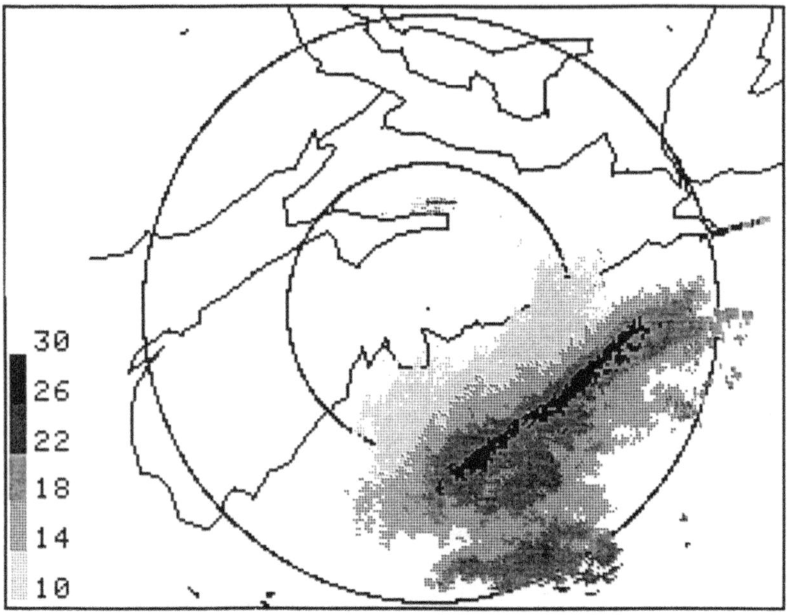

Figure 6.2: *Radar reflectivity (in dBZe) within a winter storm near Nova Scotia in 1986. The range rings are at 100 and 200 km. Adapted from Donaldson and Stewart (1989).*

in applying remote sensing techniques in extratropical cyclones; unusual situations such as this effect can lead to misinterpretations.

The discussion so far has concentrated on mesoscale features that are fundamentally linked with the systems. There are many examples of internal structures that are coupled to interactions between the cloud system and the underlying topography. Bergeron (1949) studied one particular feature, the persistence of precipitation along a Scandinavian coastline during the winter. He pointed out that a persistent circulation developed so that snow fell over the cold land and rain fell over the warm ocean. Stewart et al. (1990) examined an apparently similar feature along the coastline of Atlantic Canada. They suggested that convergence induced by changes in surface roughness was not the only process leading to this situation. They felt that the variation in diabatic heating (cooling) associated with the precipitation type change induced a secondary circulation that favoured the prolonged production of snow near the coastline. Similar features have subsequently been observed near sea ice edges.

6.3 Flow Fields and Moisture Patterns

Many cyclonic scale weather systems undergo a substantial evolution during their lifecycle. For example, newly generated systems over the Gulf Stream are characterized by substantial convection as cold continental air sweeps over the warm ocean currents in western ocean basins (Neiman and Shapiro, 1993). Later on, this situation evolves into a more stratiform cloud type in which convection can be almost completely absent, as revealed from sferics data for example,

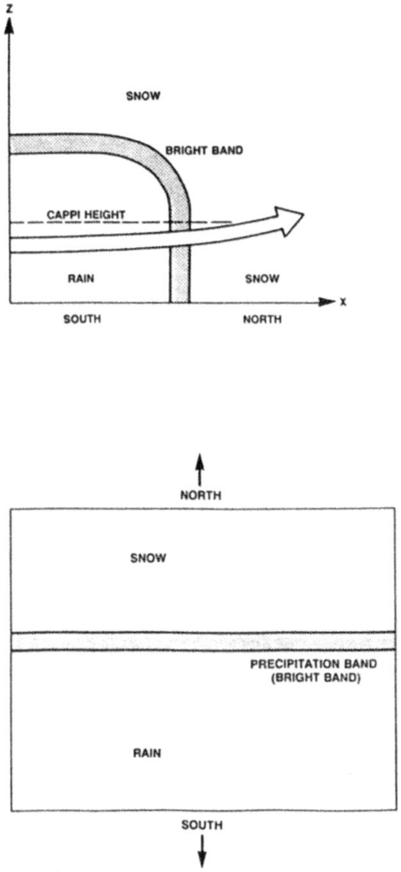

Figure 6.3: *The "production" of an apparent precipitation band by the downwards tilting of the radar bright band within storm producing rain and snow at the surface. Adapted from Stewart and King (1990).*

when they move over open ocean which in the case of the north Atlantic, can be quite cold. Satellite-borne lightning detection systems will allow for a much better quantitative assessment of such an evolution.

The overall cloud fields are produced by broad slantwise ascent within "conveyor belts" that redistribute heat, momentum and moisture over great distances (see for example Carlson, 1980; Browning and Monk, 1982; Saarikivi and Puhakka, 1990), as shown in Figure 6.1. These "conveyor belts" are synoptic scale in origin but mesoscale ascent may significantly influence their character. Such mesoscale motions certainly alter the local properties of the layered clouds and precipitation through modification of the profiles of moisture and temperature. The measurement of the pattern of mesoscale vertical motion is particularly important to quantitative studies of these processes.

Recently, Marecal and Lemaitre (1995) and Szeto and Stewart (1995) found that small scale precipitation and cloud processes can significantly modulate the local characteristics of the conveyor belts. Condensational heating can enhance the strength of the low level jet associated with the warm conveyor belt. In addition, they found that precipitation-induced diabatic effects, including melting snow as first studied by Findeisen (1940), can alter the baroclinicity near the surface front which in turn affects the structure and intensity of the (geostrophic) along-front flow (the cold conveyor belt) provided the forcing exists for a sufficient time.

The types of clouds linked with fronts vary greatly, particularly because different stabilities occur within different storm sectors and the supply of moisture is strongly influenced by vertical motions (Figure 6.4). There is still a strong organization to the types of clouds however with, for example, widespread cloud regions tending to occur ahead of the warm front, more convective in the warm sector, and low clouds often occurring near the low centre.

Figure 6.4: *NOAA satellite infrared image of the cloud structure of an extratropical cyclone over the eastern US at 18.00 UTC on March 12, 1994. The horizontal width of the image is approximately 2000 km.*

In addition, there is often a layered structure present in the cloud fields with thicknesses sometimes as small as a few hundred meters (Figure 6.5). The development of a number of ground-based radar systems that are sensitive enough to detect non-precipitating clouds should allow for a major advancement in the deciphering of the layered structure of extratropical cyclones.

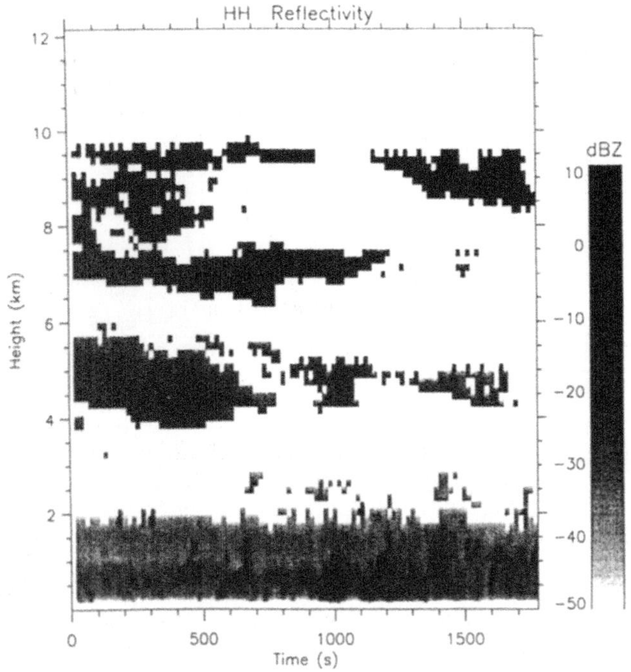

Figure 6.5: *An example of cloud layering within an extratropical cyclone over southern Ontario, Canada on September 12, 1995 as observed with a 3 cm wavelength, vertically-pointing radar. The sampling begins at 23.04 UTC.*

The production of clouds at different levels within the overall cloud system is also an indication of the role of these features on the transport of moisture throughout the troposphere. The warm conveyor belt in particular plays a critical role in lifting low level (and low latitude) moisture up to high levels of the troposphere. The extensive cirrus canopies linked with the systems is an excellent indication of the result of such a flow field. The common occurrence of high clouds over mid-latitude oceanic regions, as found by an analysis of satellite information by Wylie and Menzel (1989), is undoubtedly due to the production of cirrus by extratropical systems.

It is also apparent that the phase of the cloud particles varies within the overall system (Azarov et al., 1988; Koldaev et al., 1992; Mazin, 1995). It is expected that many of the clouds occurring within the systems are glaciated but regions of mixed phase and/or all liquid cloud certainly exist, as shown in Figure 6.6 from Cunningham (1951). It may be that some of these traits are an inherent characteristic of the cloud system itself and are a consequence of the processes controlling the overall organization. An example of this latter situation is the observed upper lid of the low level cold conveyor belt being at at 0°C within some extratropical storms (see for example Carlson, 1980; Stewart and Macpherson, 1989). Initially where the southern edge of the cold conveyor belt passes to the west of the warm conveyor belt, the cloud top will be at 0°C and so the cloud will be liquid. Remote sensing techniques such as polarized radar and radiometers are critical to properly observing such areas.

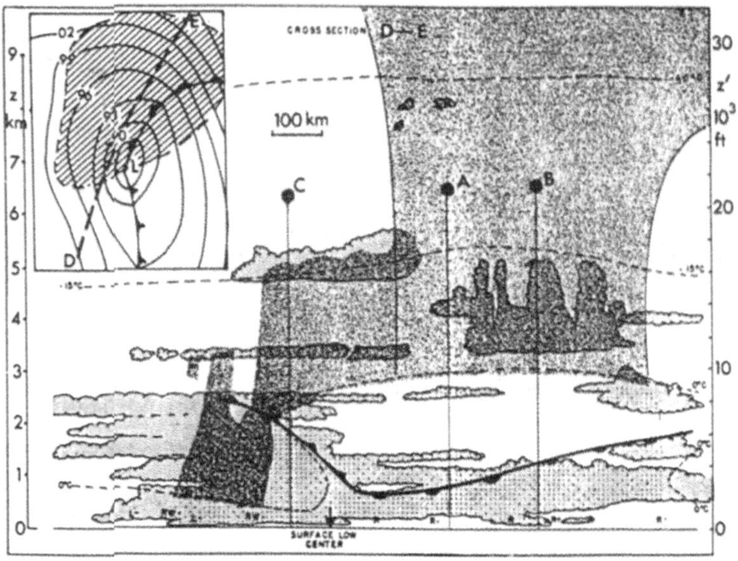

Figure 6.6: *Variations in the structure of an extratropical cyclone, as shown by Cunningham (1951). The darkest shading refers to clouds with mixed phase, the shaded area extending to the upper levels refers to all-ice clouds, and the remaining shaded areas are all-liquid regions.*

6.4 Precipitation Mechanisms

Precipitation represents a major transport mechanism for water. In addition, the phase changes linked with this process produce dramatic impacts on the thermal structure of the atmosphere. Consequently, this feature needs to be well-handled in models depicting these cloud systems.

Cunningham (1951) wrote one of the first articles stressing vertical profiles. He found that the vertical structure varied greatly within the storms. Quite different profiles of microphysical features occurred within different storm regions. He felt that a critical question affecting the actual vertical profiles was the role played by the initiation of precipitation. He found evidence that such initiation could either be internal to the general cloud field or else external due to the seeding of the cloud field through the settling of particles from above.

Precipitation formation within precipitating layer clouds can therefore occur through several mechanisms. Much of the precipitation is produced through ice-based processes requiring ice nuclei. Subsequent growth is through diffusional ice growth and through aggregation until the particles reach the melting layer. Drop interactions and condensational growth then dominate the subsequent development of precipitation.

In terms of actual processes producing precipitation within the different mesoscale structures, there are many articles that have discussed such processes. In the wide bands, it is quite common for only very low levels of liquid water to be present at temperatures below 0°C so that particle growth is dominated by sublimational growth onto ice crystals which can simultaneously be undergoing aggregational growth to produce snowflakes. Within narrow banded regions, it is not uncommon in contrast for significant liquid water to be present so that some degree of accretional growth occurs that can lead to rimed crystals or perhaps graupel. Over most regions and over much of the year, the final form of precipitation is rain that results from the melting of these ice forms of precipitation.

It should be added though that sometimes winter forms of precipitation have to be considered, above that typically associated with storms. Many of the winter forms of precipitation have formed at least partially through melting and refreezing (such as freezing rain and ice pellets), but it is not uncommon for precipitation to form through all liquid processes even when the temperatures are all below freezing (Kajikawa et al., 1988; Stewart, 1992). In addition, many of the forms of precipitation that can occur do so in close proximity to each other, or they can even occur simultaneously.

In terms of the various forms of precipitation, it is also the case that some of the heaviest precipitation linked with, for example, the warm front in such winter storms occurs in close proximity to the wide variety of precipitation forms (Figure 6.7). This fact certainly poses difficulties for properly measuring the actual rates under the harsh conditions that occur then, including the use of remote sensing tools such as radar.

6.5 Estimates of Water Transports from Observations

6.5.1 Complete cloud systems

One of the most fundamental aspects of these systems is their water budget. Although the assessment of the water budget appears to be rather straightforward, a major problem is the lack of appropriate measurements. In general, sounding information is the basic measurement that can be used but one has to be concerned with the locations of the soundings with respect to the system and there are ongoing concerns with the accuracy of the humidity sensing devices. In light of such concerns, it is essentially a requirement that satellite-based remote sensing is needed to achieve the objectives of such studies.

It should be noted that the flux of moisture that is derived from soundings within the clouds ignores contributions from the condensate itself. Especially when the condensate is snow (which has a low terminal velocity) and the temperature is well below at 0°C, the mixing ratio of the condensate can easily exceed 20% of the water vapour saturation mixing ratio, even for low precipitation rates. Moisture fluxes derived from the sounding information could therefore underestimate the actual fluxes under such conditions (Figure 6.8).

There are consequently few articles addressing the water budgets associated with individual weather systems, and, as indicated above, there are some uncertainties associated with their findings. However, Houze et al. (1976) and McBean and Stewart (1991) estimated these budgets within storms over the west coast of North America and over the north Pacific Ocean respectively (Figure 6.9). They both found that the main source of moisture was from the low levels east of the storms with net inflow rates less than or of the order of 100 $kg^{-1}s^{-1}m^{-1}$. At levels above 50 kPa, there was net divergence.

There are also few articles that directly discuss the precipitation efficiency of the systems. Here, precipitation efficiency is generally considered to be the ratio of surface precipitation to the available water vapour. However, Wexler and Atlas (1958) pointed out that the precipitation

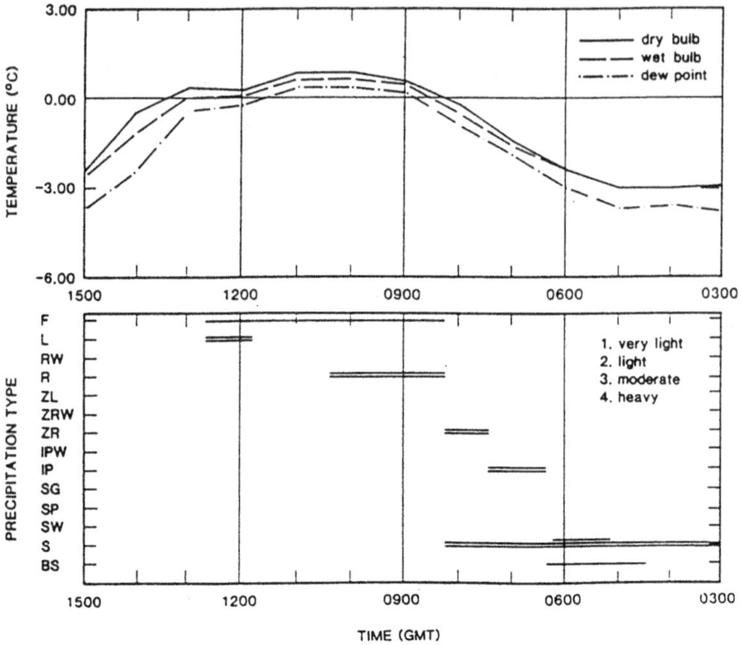

Figure 6.7: *An illustrative plot of surface temperature and precipitation conditions linked with the passage of warm fronts in many winter storms over Canada. The forms of weather or precipitation are: fog (F), drizzle (L), rain (R), freezing drizzle (ZL), freezing rain (ZR), ice pellets (IP), snow grains (SG), snow pellets (SP), snow (S) and blowing snow (BS). W indicates a shower.*

efficiency of a storm that they studied in the northeast U.S. was quite high. The results of McBean and Stewart (1991) implied that the precipitation efficiency in their case was close to one, but they acknowledged that their dataset was not definitive.

Precipitation efficiency depends on many aspects of the cloud structure. It is particularly sensitive to the processes that initiate precipitation and to those occurring below cloud base. One can certainly anticipate in a qualitative manner some general variations of the precipitation efficiency with sub-cloud conditions. In the high Arctic for example where dry sub-cloud conditions commonly prevail during the winter, there has to be a great loss due to sub-cloud sublimation and the precipitation efficiency must be low (Figure 6.10). In contrast, within storms along the East Coast of Canada, near-saturated conditions prevail down to near the surface (Stewart et al., 1995), little loss due to sublimation or evaporation is expected, and the precipitation efficiency should be higher.

6.5.2 Mesoscale water transport

Since mesoscale vertical velocities can greatly exceed those on the synoptic scale the transport properties of weather systems must be substantially influenced by these embedded circulations

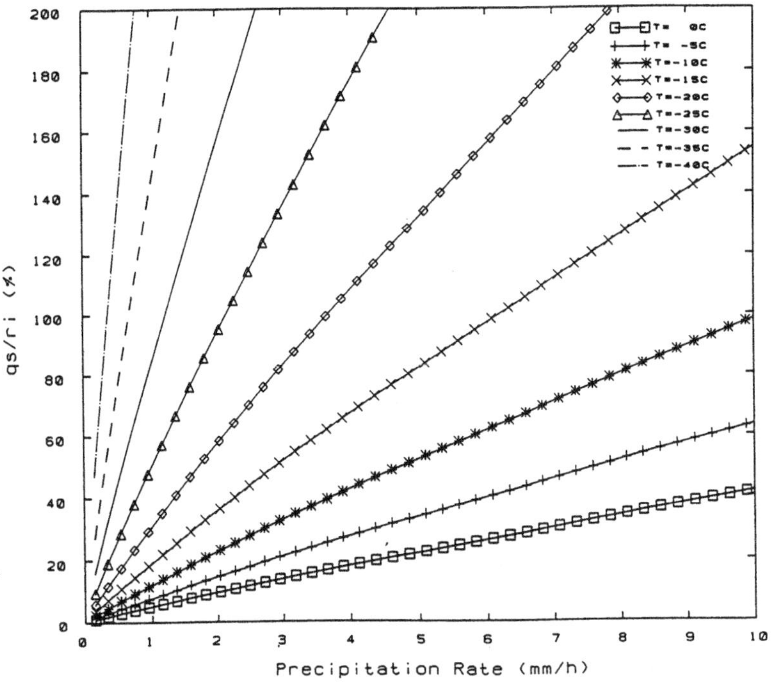

Figure 6.8: *The ratio of snow mixing ratio to total water substance mixing ratio as a function of precipitation rate for different atmospheric temperatures. It is assumed that the air is saturated with respect to ice.*

which, although they are mesoscale in extent across a frontal discontinuity, may extend for hundreds of kilometers along the front.

Some studies of mesoscale water budgets have in fact been conducted. On the west coast of the USA, both warm sector and wide cold frontal rainbands have cloud base at about 1 km when the low-level air beneath the warm conveyor belt is cool and moist (see for example Hobbs et al., 1980). Consequently, stratiform clouds are deep. The melting layer is typically 1–2 km above cloud base. When a cloud forms beneath the warm conveyor belt, the layer clouds have high precipitation efficiencies and subcloud evaporation is minimized. For example, Houze et al. (1981) estimated the efficiency of a warm frontal rainband over Washington state to be between 60–90%. However, earlier Hobbs et al. (1980) found that two wide cold frontal bands characterized by the "seeded-feeder" process had efficiencies of 80% and 20%. In the latter case, the efficiency was strongly affected by low-level evaporation.

An important characteristic of some flow fields is that the warm conveyor belt is ascending over a cold conveyor belt that is both cold and dry (for example, Browning and Monk, 1982). The warm frontal precipitation falling from the warm conveyor belt evaporates as it falls into the dry subcloud air. During summer in south-eastern Australia the sub-cloud layer beneath the warm conveyor belt is cloud free. As shown by Ryan at al. (1989), this hot airstream in the sub-cloud layer with relative humidities of less than 30% lead to very low efficiencies (Ryan and Wilson, 1985). In contrast over southern Australia during winter and spring, the

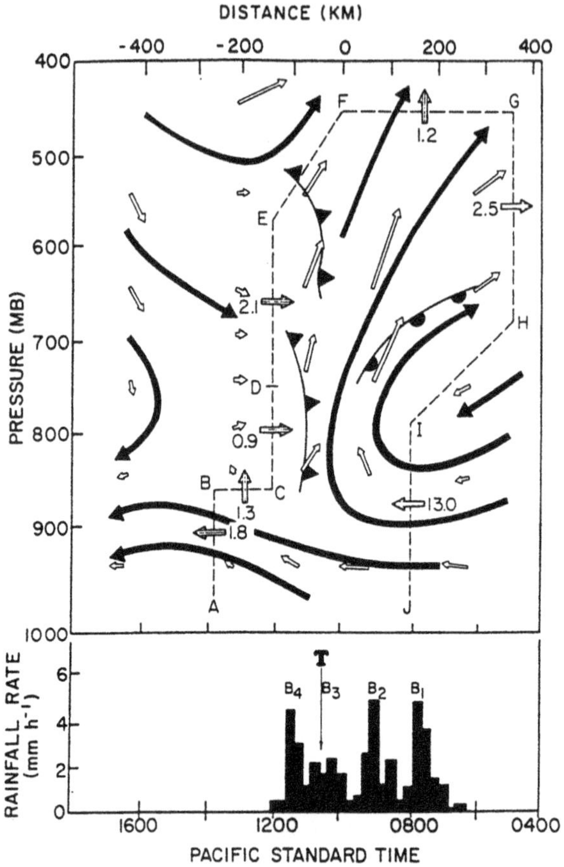

Figure 6.9: *The moisture budget of an extratropical cyclone over the north-east Pacific Ocean. The fluxes are in units of 10^4 g s^{-1} m^{-1}. Adapted from Houze et al. (1976). Letters on the diagram refer to band and frontal features described in that article.*

pre-frontal airstream associated with cutoff lows is both cooler and moister and cloud base is more typically 1–2 km. Under these conditions, it is not uncommon for a cloud system to be comprised of several layers each about 1 km thick and to be separated by a few hundred meters (Ryan, 1995).

A detailed examination of microphysical controls on extratropical cyclone budgets is certainly needed. It is certainly well-known that precipitation production often requires the seeder-feeder process in which the initial production of ice at cold, high levels is needed (see for example Rutledge and Hobbs, 1983). Much of the initial precipitation growth is furthermore initiated preferentially within generating cells induced by cloud top instability; microphysical conditions within these small convective elements is particularly crucial.

Smith (1992) has also asserted that the moistening of the upper atmosphere within these systems is assisted substantially through the direct lofting of ice crystals, as opposed to just

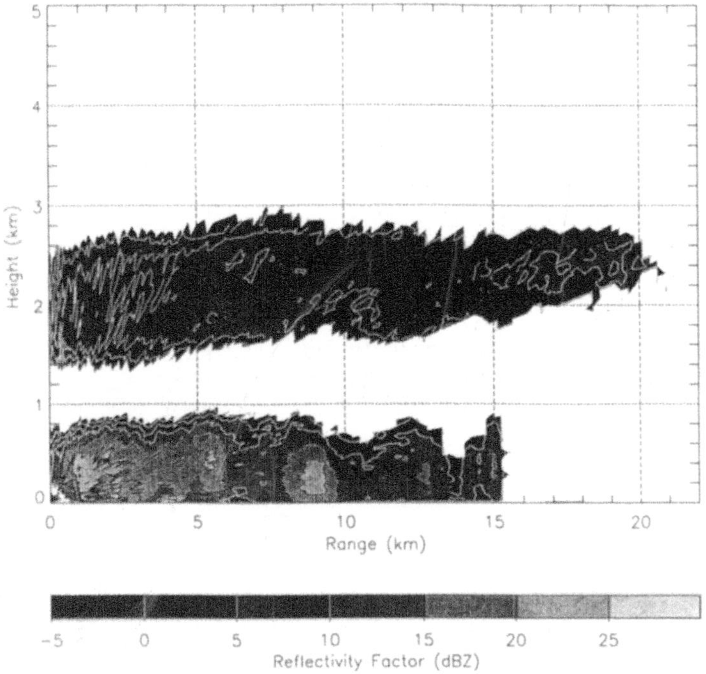

Figure 6.10: *Vertical radar profile within a light snow event on September 8, 1994 at Inuvik, near the Arctic coastline of northern Canada.*

the vertical ascent of water vapour. Moreover, this lofted ice can be so tenuous that its presence can be missed by infrared satellite measurements; deduced cloud top heights can be lower (by up to a few kilometers) than the actual situation (Figure 6.11). Similar observations have recently been made within warm fronts over northern Canada.

In addition, Lau et al. (1993) pointed out how critical the melting level was to the overall water budget of tropical convective systems; within extratropical cyclones this level can be present at all locations (in summer cases), it can occur in some sectors (in many winter cases), or it can be completely absent (in higher latitudes in the winter).

6.6 Numerical Simulations and Water Transports

6.6.1 Model description and the control case

To better understand the role of extratropical cyclones on the transport of water, some simple numerical simulations have been carried out. The model used in the simulations is a non-hydrostatic cloud model and is based on the one described by Szeto and Cho (1994). A detailed bulk water type cloud microphysical scheme is utilized in this study. Six types of water substance (water vapour, cloud water, cloud ice, rain, snow and ice pellets) are included

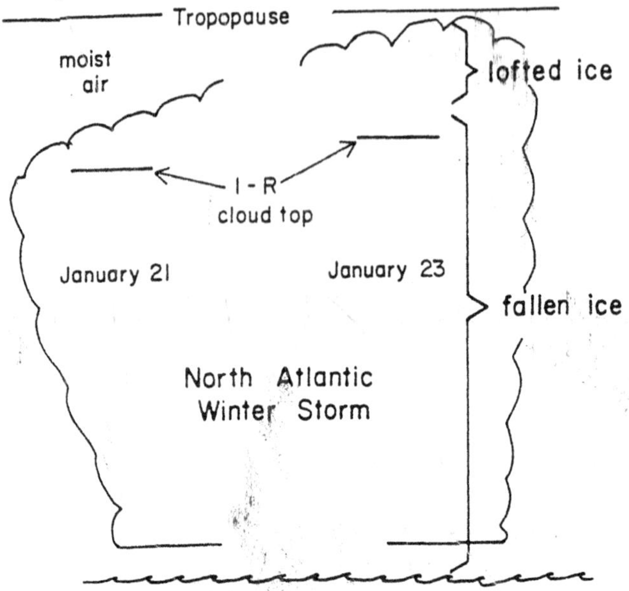

Figure 6.11: *Comparison of the cloud tops deduced by in-situ observations and by satellite infrared techniques (IR cloud top) within two extratropical cyclones in 1989. Adapted from Smith (1992).*

in the scheme. The horizontal and vertical grid lengths used in the simulations are 7.5 km and 330 m, respectively. This resolution should be sufficient for the simulation of the non-convective frontal precipitation features that are of interest here. It is assumed that there is weak convergence (10^{-5} s^{-1}) across the domain, there is a stable lapse rate ($\partial\theta/\partial z = 2.5$ K km^{-1}), there is a constant wind shear, and there is weak baroclinicity with maximum values of about 1.5°C (100 km)$^{-1}$. It should be added that a simple vertical profile of moisture is also assumed with relative humidity with respect to water being 85% up to 2 km, decreasing to 70% at 6 km, and finally to 35% at 13 km. The moisture profile, as indicated by relative humidity, was also assumed to initially and at all later times be the same as in the 2D plane being considered so that advection along the front does not change relative humidity.

The cloud and precipitation fields at steady state are shown in Figure 6.12. With the cold air on the right hand side of the figure, it is evident that the model simulated the classical organization near a warm front of a large overhang above a sub-saturated area. This overhang region containing cloud particles extends some 600 km and some 200 km of this region lies over areas with relative humidity with respect to ice less than 50%. The maximum precipitation rate at the surface was approximately 1.8 mm h^{-1} and this peak was located about 80 km to the cold side of the surface front and was in the form of snow with rain, a common situation within winter storms as discussed in Section 6.4. The region experiencing precipitation at the surface extended to about 350 km to the cold side of the front.

Approximately 40% of the vapour entering the system finally ends up as precipitation on the surface, which is much less than the inferred results of, for example, McBean and Stewart (1991). (Here, the sub-domain being considered in the calculations ranged from 600 km to the

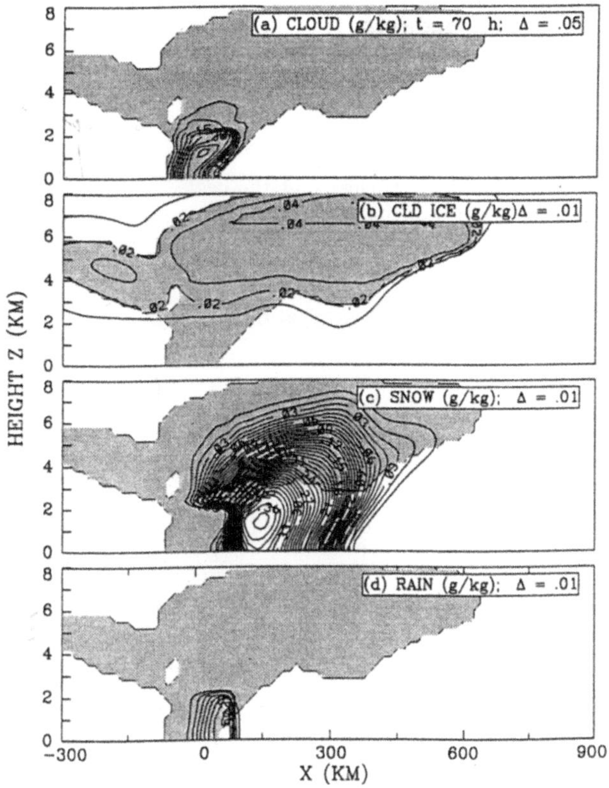

Figure 6.12: *A diagram illustrating the cloud and precipitation fields after 70 h of simulation when a steady state has been reached. The shading refers to the presence of any cloud liquid or cloud ice. The CLOUD panel refers to cloud liquid water, all condensate mixing ratios are in units of g* kg^{-1}, *and the delta entry refers to the contour interval.*

warm side of the surface front to some 1500 km to the cold side. The integrated precipitation rate at the surface within this domain was compared against the net moisture influx to compute the precipitation efficiency.) Interestingly, the maximum precipitation rates near the surface occur in regions with relative humidity near saturation; relatively little of this precipitation is then lost.

In terms of which processes are largely responsible for the loss of moisture, the sublimation of ice in the form of snow on the cold side of the surface front was the main factor, as inferred by radar by for example Harris (1977). There was little loss from the evaporation of rain but there was significant loss of cloud water; the regions where rain occurred were small and they also tended to occur within saturated conditions anyway.

By varying the initial temperature and relative humidity fields in the calculations, a better appreciation of their effect on precipitation efficiency was obtained. There is a general decrease in precipitation efficiency with decreasing low level moisture supply. The major reason for this is the enhanced relative role of sublimation in the drier situations. For the same initial relative humidity, precipitation efficiency decreased as temperature increased. There was more moisture available to the cloud system in the warmer cases, but the cloud base was higher, and consequently there was greater loss of the precipitation below cloud. The highest precipitation efficiency occurred under the coldest conditions. These situations have the lowest bases and so there is less opportunity for the precipitation to be lost. The low cloud bases were largely the result of the efficient manner in which sublimation led to the saturation of initially dry sub-cloud conditions.

Regardless of the actual situation, it is clear that the degree of sub-saturation below cloud makes a major impact on the results. As an indication of the sensitivity of the environment to degree of sub-saturation, Table 6.1 shows the rate of mass loss due to sublimation and

Relative Humidity (%)		**Evaporation**	**Sublimation**	
			100 kg m^{-3}	10 kg m^{-3}
T = 20°C	90	20		
	80	40		
	60	80		
	60 (5)	34		
T = 10°C	90	15		
	80	30		
	60	60		
	60 (5)	25		
T = 0°C	90	10	16	51
	80	20	30	84
	60	40	54	100
	60 (5)	17	28	82
T = -10°C	90		9	32
	80		18	57
	60		34	89
	60 (5)		18	54
T = -20°C	90		5	16
	80		10	30
	60		17	54
	60 (5)		9	28

Table 6.1: *Precipitation rate loss (in %) due to sub-saturation below cloud base for rain and snow using bulk formulations. Losses are calculated over a 1 km depth, including an initial precipitation rate of 1 mm h^{-1} except for cases of 5 mm h^{-1} which are indicated by brackets. Two snow densities (100 and 10 kg m^{-3}) are used.*

evaporation of precipitation at several temperatures under simple relative humidity values and assuming bulk formulations, as described by for example by Szeto and Cho (1994). One can see that even small degrees of sub-saturation can lead to considerable loss. In addition, it is clear that the form of precipitation makes a large impact on the degree of loss. Precipitation in the form of snow leads to losses about twice that experienced for rain under the same conditions, even if 100 kg m^{-3} snowflake densities are considered. The rate of precipitation also causes a substantial effect with lower rates being more susceptible to losses. In the case of snow, the density also makes a major impact on the loss rate with low density, slowly falling snow being most readily lost. It should be added that detailed calculations of spectral evolution in such cases reveal significant differences from the bulk approach; the proper evaluation of these processes for suitable inclusion within models remains elusive. Remote sensing techniques such as multi-parameter radar will be needed to better illustrate the nature of precipitation over the needed wide areas to properly observe these critical portions of the water cycle.

It needs to be emphasized though that interpretations of the precipitation efficiency results have to account for the fundamental changes that occur in the frontal cloud structure as a function of initial conditions. Even though initial large scale forcing may be the same, varying initial moisture and temperature conditions lead to differences in the latent heat exchange within the cloud which in turn affect, for example, the strength of the frontal zones and vertical velocity magnitudes. Such changes in the internal dynamic structure lead to changes in the degree of saturation, abundance of cloud water and distribution of precipitation. A variable such as precipitation efficiency is the net result of the complete system in which the moisture itself can affect the nature of the system in highly non-linear ways. The proper observation of key aspects of the water budget, including vertical velocity fields, is therefore quite critical in order to properly validate such model predictions; remote sensing offers the best hope of achieving this.

6.7 Concluding Remarks

Extratropical systems are very common features that play a role in the transport of water in the atmosphere. These cloud systems produce much of the precipitation that falls on the Earth's surface. Therefore, understanding regional water budgets must address these features. The detailed nature of these systems needs to be well-understood however if their contribution to the water budgets can be quantified under the wide variety of conditions that actually occur in the atmosphere. Although this article showed that detailed calculations reveal some of the sensitivities that need to be considered, far more work is needed to develop an acceptable manner of properly handling the crucial moisture re-distribution issues faced by climate modellers.

Programs being mounted to better simulate regional water budgets in different regions of the world will be a major driving force for the continuing assessment of the impacts of these weather systems. Efforts focussing on high latitude basins such as the Mackenzie Basin of Canada where snow is prevalent should be linked with low precipitation efficiencies due to the dominance of sublimation in low moisture conditions, although some instances in this environment may have a high efficiency if the initial snowfall is able to saturate the sub-cloud region. Warmer areas in which rain is more common, including the Baltic Sea region, should in general have higher efficiencies due in part to the availability of surface moisture.

Remote sensing tools play a critical role within studies aimed at better understanding the effects of extratropical cyclones on water cycles. Satellite-borne systems are very often used, radar are needed for clouds and precipitation studies, and other instruments are being or need to be used for observing other critical dynamic and moisture parameters.

Although the original focus of studying extratropical systems was their short-term weather prediction, a separate thrust is now being driven by the requirements of climate studies to better appreciate their role on regional and global water and energy cycles.

Acknowledgments

The author would like to thank Kit Szeto for his assistance in carrying out the simulations used in this article. He would also like to thank Bob Crawford, John Hanesiak and Jason Burford for their ongoing contributions and David Hudak, Vytas Kezys, Simon Haykin, and Hao Le (all of the Atmospheric Environmental Science Service of Canada, Toronto) for providing some of the figures used in this article.

6.8 References

Azarov AS, Buranbaev NF, Koldaev AV, Chernikov AA (1988) Determination of water reserves of winter clouds by radar radiometery. *Soviet Meteor Hydrology* 7: 15–23

Bergeron T (1949) The problem of artificial control of rainfall on the globe. Part III: The coastal maxima of precipitation in autumn and winter. *Tellus* 1:15–32

Bjerknes J (1919) On the structure of moving cyclones. *Geofys Publ* 1: 3–8

Bjerknes J, Solberg H (1922) Life cycles of cyclones and polar front theory of atmospheric circulation. *Geofys Publ* 3: 3–18

Browning KA, Harrold TW (1970) Air motion and precipitation growth within a cold front. *Quart J Roy Meteor Soc* 96: 369–389

Browning KA (1985) Conceptual models of precipitation systems. *ESA Journal* 9: 157–180

Browning KA, Monk GA (1982) A simple model for the synoptic analysis of cold fronts. *Quart J Roy Meteor Soc* 108: 435–452

Carlson TN (1980) Airflow through midlatitude cyclones and the comma cloud pattern. *Mon Wea Rev* 108: 1498–1509

Cunningham RM (1951) Some observations of natural precipitation processes. *Bull Amer Meteor Soc* 32: 334–343

Donaldson NR, Stewart RE (1989) On the precipitation regions within two storms affecting Atlantic Canada. *Atmos-Ocean* 27: 108–129

Douglas RH, Gunn KLS, Marshall JS (1957) Pattern in the vertical of snow generation. *J Meteor* 14: 95–114

Findeisen W (1940) The formation of the 0°C isothermal layer and fractocumulus under nimbostratus. *Meteor Z* 57: 49–54

Harrold TW (1973) Mechanisms influencing the distribution of precipitation within baroclinic disturbances. *Quart J Roy Meteor Soc* 99: 232–251

Harris FI (1977) The effects of evaporation at the base of ice precipitation layers: theory and radar observations. *J Atmos Sci* 34: 651–672

Heymsfield GM (1979) Doppler radar study of a warm front. *J Atmos Sci* 36: 2093–2107

Hobbs PV (1978) Organization and structure of clouds and precipitation on the mesoscale and microscale in cyclonic storms. *Rev Geoph Space Physics* 16: 741–755

Hobbs PV, Matejka TJ, Herzegh PH, Locatelli JD, Houze RA (1980) The mesoscale and microscale structure and organization of clouds and precipitation in midlatitude cyclones. I: A case study of a cold front. *J Atmos Sci* 37: 568–596

Houze RA, Locatelli JD, Hobbs PV (1976) Dynamics and cloud microphysics of the rainbands in an occluded frontal system. *J Atmos Sci* 33: 1921–1936

Houze RA, Rutledge SA, Matejka TJ, Hobbs PV (1981) The mesoscale and microscale structure and organization of clouds and precipitation in midlatitude cyclones. III: air motions and precipitation growth in a warm frontal rainband. *J Atmos Sci* 38: 639–649

Houze RA, Hobbs PV (1982) Organization and structure of precipitating cloud systems. *Adv Geoph* 24: 225–315

Kajikawa M, Sakurai K, Kikuchi K (1988) Characteristic features of supercooled raindrops of freezing precipitation. *J Meteor Soc Japan* 66: 393–398

Koldaev, AV, Melnichuk YV, Mironov AF (1992) Remote sensing of cloud liquid water space distribution. *Proc. 11th Intl Conf Clouds and Precip*, Montreal 573–577

Lambert SJ (1988) A cyclone climatology of the Canadian Climate Centre general circulation model. *J Climate* 1: 110–115

Lau KM, Sui CH, Tao WK (1993) A preliminary study of the tropical water cycle and its sensitivity to surface warming. *Bull Amer Meteor Soc* 74: 1313–1321

Marecal V, Lemaitre Y (1995) Importance of microphysical processes in the dynamics of a CSI mesoscale frontal band. *Quart J Roy Meteor Soc* 121: 301–318

Mazin IP (1995) Cloud water content in continental clouds of middle latitudes. *Atmos Res* 35: 283–297

McBean GA, Stewart RE (1991) Structure of a frontal system over the northeast Pacific Ocean. *Mon Wea Rev* 119: 997–1013

Neiman PJ, Shapiro MA (1993) The life cycle of an extratropical marine cyclone. Part II: mesoscale structure and diagnostics. *Mon Wea Rev* 121: 2177–2199

Palmen E, Newton CW (1948) A study of the mean wind and temperature distribution in the vicinity of the polar front in winter. *J. Meteor* 5: 220–226

Rutledge SA, Hobbs PV (1983) The mesoscale and microscale structure and organization of clouds and precipitation in midlatitude cyclones. VIII: a model for the seeder-feeder process in warm frontal rainbands. *J Atmos Sci* 40: 1185–1206

Ryan BF, Wilson KJ (1985) The Australian summertime cool change: III subsynoptic and mesoscale model. *Mon Wea Rev* 113: 224–240

Ryan BF, Wilson KJ, Zipser EJ (1989) Modification of the thermodynamic structure of the lower troposphere by the evaporation of precipitation ahead of a cold front. *Mon Wea Rev* 117: 138–153

Ryan BF (1995) On the global variation of precipitating layer clouds. *Bull Amer Meteor Soc*, in press

Saarikivi P, Pukakka T (1990) The structure and evolution of a wintertime occluded front. *Tellus* 42A: 122–139

Smith RB (1992) Deuterium in north Atlantic storm tops. *J Atmos Sci* 49: 2041–2057

Stewart RE (1992) Precipitation types in the transition regions of winter storms. *Bull Amer Meteor Soc* 73: 287–296

Stewart RE, King P (1990) Precipitation type transition regions in winter storms over southern Ontario. *J Geoph Res* 95: 22355–22368

Stewart RE, Lin CA, Macpherson SR (1990) The structure of a winter storm producing heavy precipitation over Nova Scotia. *Mon Wea Rev* 118: 411–426

Stewart RE, Macpherson SR (1989) Winter storm structure and melting-induced circulations. *Atmos-Ocean* 27: 87–107

Stewart RE, Yiu DT, Chung KK, Hudak DR, Lozowski EP, Oleskiw M, Sheppard BE, Szeto KK (1995) Weather condition associated with the passage of precipitation type transition regions over eastern Newfoundland. *Atmos-Ocean* 33: 25–53

Szeto KK, Cho HR (1994) A numerical simulation of squall lines. Part III: sensitivities to precipitation processes and the Coriolis force. *J Atmos Sci* 51: 1341–1351

Szeto KK, Stewart RE (1995) Effects of melting on frontogenesis. *J Atmos Sci* , submitted

Wexler R, Atlas D (1958) Moisture supply and growth of stratiform precipitation. *J Meteor* 35: 531–538

Wylie DP, Menzel WP (1989) Two years of cloud cover statistics using VAS. *J Clim Appl Meteor* 2: 380–392

PART B

REMOTE SENSING OF CLOUDS, RADIATION AND PRECIPITATION

Chapter 7

Clouds and the Radiative Heating of the Earth Surface-Atmosphere System

Rolf Stuhlmann
GKSS Research Center
21502 Geesthacht
Germany

Abstract

The Earth Radiation Budget Experiment (ERBE) and the Scanner for Radiation Budget (ScaRaB), for instance, provide excellent data sets to study the diurnal, seasonal and regional variations of the radiation balance of the climate system and how it is affected by the radiative forcing by clouds. These type of investigations are supplemented by use of imager data taken by the geostationary satellites (and also the polar orbiters for the more polward regions of the globe) which, with their high spatial and temporal resolution, describe the variability of the atmosphere and the embedded clouds. The ERB data sets are used to look at the energy balance at the top of the atmosphere (TOA) to determine the sinks and sources of the Earth surface-atmosphere circulation system and to extract the TOA cloud radiative forcing, which represents the net effect of clouds on the total system. But, even in case of a negligible TOA cloud forcing, there might be a strong effect of clouds on the weather and climate processes, which is felt through the changes in rediative interaction at the surface and in the atmosphere, separately. Therefore, it is essential to use the data also to study how the TOA energy balance is divided up between the surface and the atmosphere. Here, for instance, the Global Energy and Water Cycle Experiment (GEWEX) Surface Radiation (SRB) Project is retrieving surface radiation fluxes over the globe from the International Satellite Cloud Climatology Project (ISCCP) data. Doing that, the uncertainty in deriving the cloud base altitudes from satellite data leads to the largest errors in determining the net longwave fluxes at the surface. But, despite the problem of retrieving the cloud bottom heights from currently available global satellite data sets, first attempts are also made to study the effects of clouds on the radiative heating of the atmosphere itself. First results sustain that the effect of clouds in the shortwave spectral domain is primarily a cooling of the surface. In the longwave spectral domain the effect of clouds is primarily on the atmosphere, where the distribution of cloud atmospheric forcing strenghtens the meridional temperature gradient. To improve the efforts to determine the atmospheric heating profiles and the net longwave radiation budget at the surface, a global measurement of cloud layering and vertical cloud thickness is urgently needed. A space borne millimeter wave radar will provide the information about the vertical structure of clouds and

NATO ASI Series, Vol. I 45
Radiation and Water in the Climate System:
Remote Measurements
Edited by Ehrhard Raschke

cloudbase altitude, and, with suitable complementary measurements, is a promising method to provide estimates of water and ice mass.

7.1 Introduction

Within the next century, the regional distribution and magnitude of global environmental change will have a profound societal impact. This makes it urgently necessary to provide unequivocal answers to questions about theses effects. Up to now, there is still a lack of knowledge of the independent processes that affect our climate system. One problem, which still has to be worked on, is how good are the projections of future climate change caused by anthropogenic alterations of atmospheric compositions. Increasing greenhouse gases in the Earth's atmosphere will perturb the Earth's radiation balance, forcing future climate change. Current general circulation models predict a warming of the atmosphere in the range form 1.7° to 5.4°C for a doubling of carbon dioxide (Houghton et al., 1990). The large variaton in predicted temperature changes comes from differences in treatment of climate feedback mechanisms, which can either amplify of moderate the warming of the atmosphere depending on the model (Cess et al., 1989; 1990). A critical factor here is the strength and sign of the cloud-radiation feedback, since it plays a important role in determining the magnitude and geographical distribution of climate changes that result from natural or human forcing of climate change.

First investigations by a one-dimensional model which predict the temperature profile from a thermodynamic balance (Manabe and Strickler, 1964) showed that the global mean surface temperature is very close to the observed value of 288 K when clouds were included in the model calculations. Without clouds the computed temperature was about 13°C higher. About 20 years ago, several researchers started to investigate the role of cloud radiative effects within the climate system by using satellite observations to estimate the effect of clouds on the energy balance at the top of the atmosphere (e.g.: Cess, 1976; Cess et al., 1982; Ellis, 1978; Hartmann and Short, 1980; Ohring and Clapp, 1980; Ohring et al., 1981). Since then, the quality and detail of the observational data sets for cloud cover and Earth Radiation Budget (ERB) have greatly improved. Within this lecture the effect of clouds on the radiative heating of the Earth surface-atmosphere system will be discussed.

7.2 Radiation Budget TOA

Starting with the NASA Nimbus/ERB mission (Jacobowitz et al., 1984) and the more recent Earth Radiation Budget Experiment (ERBE) (Barkstrom and Smith, 1986) a climatology of the ERB with continually improving measurement accuracy was established. To obtain the radiation fluxes at the top of the atmosphere (TOA) from the ERBE satellites a sophisticated inversion method was applied (Smith et al., 1986) to account for the spectral sensitivity of the instruments, the angular variation of the radiances depending on surface type and amount of clouds present in the observed scene. In addition, a procedure was applied to obtain proper averages of the fluxes from temporally and spatially sparse samples (Brooks and Minnis, 1984). These data sets are used by the scientific community to determine the energy exchange between space and the Earth surface-atmosphere system.

The radiative heating of the Earth surface-atmosphere system is determined by its absorption of shortwave (0.2–4.0 μm) radiation emerging from the sun, which partly penetrates the system and partly is reflected back to space. This heating is compensated by a cooling determined by the exchange of longwave radiation (4–100 μm) with space, which is emitted by the Earth surface-atmosphere system dependent on its temperature and moisture profiles and cloud top height distributions.

$$Q_{RAD} = S_0\mu_0(1-\alpha) - OLR = S_0\mu_0 - RSR - OLR \qquad (7.1)$$

Equation 7.1 describes the Earth Radiation Budget at the top of the atmosphere, where, Q_{RAD}, is the absorbed net-radiation eithin the system, $S_0\mu_0$, the incoming solar radiation, α, the planetary albedo and, RSR and OLR, the reflected shortwave and outgoing longwave radiation to space, respectively. Provided that there is no climatic change in temperature of the Earth surface-atmosphere system, the annual and global average of the absorbed net-radiation, Q_{RAD}, has to be zero. Table 7.1 presents results of the global average absorbed shortwave ($S_0\mu_0(1-\alpha)$) and outgoing longwave (OLR) radiation as determined by Harrison et al. (1990) from data of the ERBE experiment.

Date	Outgoing Longwave Radiation (Wm^{-2})	Absorbed Shortwave Radiation (Wm^{-2})	Absorbed Netto Radiation (Wm^{-2})
April 1985	234.5	236.5	2
July 1985	237.5	234.4	−3.1
October 1985	234.1	243.0	8.9
January 1986	231.9	243.3	11.4
Annual	234.5	239.3	4.8

Table 7.1: *Summary of the Radiation Fluxes at the Top of the Atmosphere* (Wm^{-2}).

Table 7.1 shows that the global average ERB has an annual cycle, where energy is gained by the system during the northern winter, while energy is lost by the system during summer-time. The annual and global average absorbed shortwave radiation is 239.3 Wm^{-2} which, for a global annual mean solar irradiance of $S_0/4 = 342$ Wm^{-2}, corresponds to a shortwave reflected flux of $RSR = 102$ Wm^{-2} or an average albedo of $\alpha = 29.9\%$. The annual average outgoing longwave radiation is calculated as $OLR = 234.5$ Wm^{-2}, giving an imbalance in the annual average absorbed net radiation of $Q_{RAD} = 4.8$ Wm^{-2}. This imbalance of 4.8 Wm^{-2} is about the same size as the uncertainty of the measurement itself and thus, agrees that within this uncertainty limit our planet on average is in radiative balance.

On the contrary, on regional and zonal scales the ERB at the TOA has to be in imbalance with a positive net radiation near the equator and negative near the poles. Thus, the atmosphere and the oceans must transport the energy excess away from the tropics. Figures 7.1–7.3 present as example the regional distribution of monthly means of the radiation fluxes, OLR, RSR, and Q_{RAD} as derived for the top of the atmosphere.

The Intertropical Convergence Zone (ITCZ) is clearly visible in Fig. 7.1 showing the OLR. The high cloud tops, in particular above land areas, produce an OLR of below 240 Wm^{-2} with maxima up to 300 Wm^{-2}. North and south of 35° latitude the OLR shows a more zonal structure decreasing from about 240 Wm^{-2} to minima of about 180 Wm^{-2} at the North Pole and about 130 Wm^{-2} at the South Pole. The latitudinal structure of the RSR, presented in Fig. 7.2, reflects the typical conditions of solar illumination. Within April, the sun is overhead between 5°–15° north resulting in a larger RSR for the northern than the southern hemisphere. The longitudinal distribution of RSR reflects the underlying geographic distribution of land

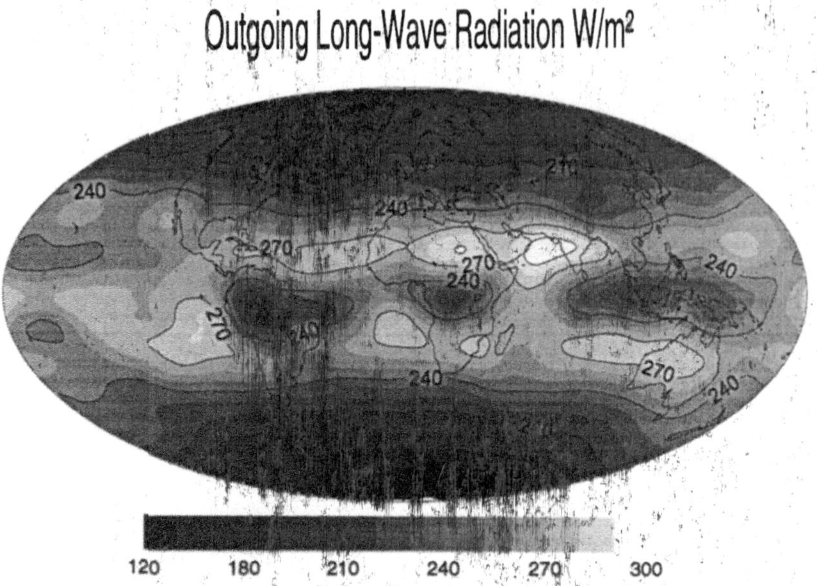

Figure 7.1: *Monthly mean outgoing longwave radiation (OLR) for April 1985.*

and snow and the presence and absence of clouds. The net radiation, presented in Fig. 7.3, has a more zonal character compared to the two individual components discussed above. The region with a maximum energy input of larger than 80 Wm^{-2} is located between the equator and 30° north, where above the Indian Ocean and West Pacific energy fluxes of larger than 100 Wm^{-2} are measured. The most striking feature is the strong anomaly above the Sahara desert. Here, the high surface temperature, causing a strong longwave energy loss to space and the bright surface albedo, causing a strong reflection of shortwave energy back to space, force the Earth surface-atmosphere system above parts of the Sahara even to a radiative energy loss of up to -10 Wm^{-2}. Since the Sahara desert is not cooled on longer time scales, the atmospheric circulation has to transport energy into this tropical region to compensate the radiation loss. The only mechanism by which such a compensation can be done is subsidence of air, causing a diabatic heating, which in turn evaporates the developing clouds above these regions.

Figure 7.4 presents the seasonal variations of the zonal average OLR and RSR, respectively. The seasonal variations in OLR reflect two major phenomena. The first is the summertime warming of the pole, followed by its wintertime cooling. The second major feature is the latitudinal wandering of the ITCZ between 10° north in July and 10° south in January. The RSR reflects similar meteorology as the OLR, but here the variations of the solar irradiance with latitude dominates the zonal profiles. Profiles of April and October 1985 are fairly symmetrical, while those of July 1985 and January 1986 are not. The reason for the nonsymmetrical behavior at the polar region is, that the Antarctic continent is much more reflective than the Arctic ocean. This difference in surface reflectivity is, in addition, amplified by the increase in solar irradiance of 6% from July to January.

The monitoring of tzhe ERB is now continued by the French-Russian-German Scanner for Radiation Budget (ScaRaB) on board the Meteor-3 satellite (Kandel et al., 1994) and will also

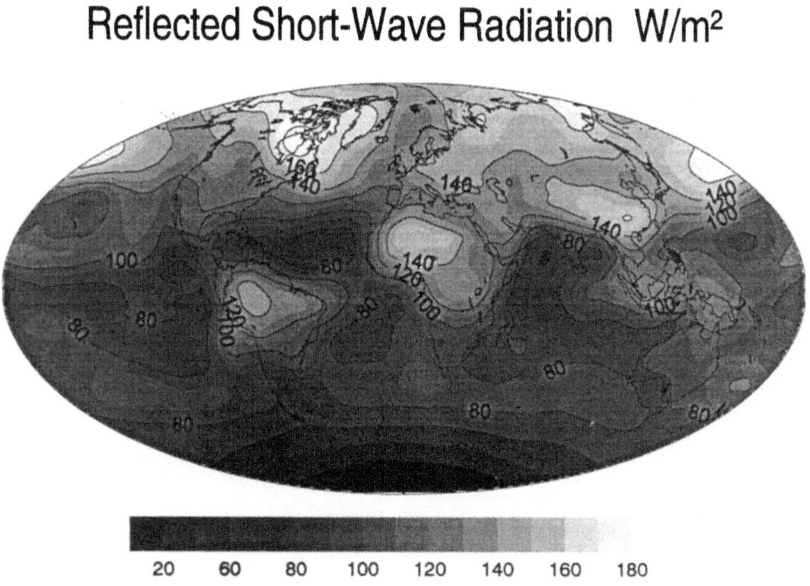

Figure 7.2: *Monthly mean reflected shortwave radiation (RSR) for April 1985.*

be continued by the NASA Clouds and the Earth's Energy System (CERES) scanner to be launched in 1997 (Dozier, 1994). Beside the establishment of an ERB climatology, a great effort has been undertaken by the International Satellite Cloud Climatology Project (ISCCP) to produce a global cloud climatology. The ISCCP uses narrowband radiances in the visible and thermal infrared obtained from geostationary and polar-orbiting operational meteorological satellites to determine several cloud parameters including cloud cover, cloud top pressure and cloud optical thickness (Rossow and Schiffer, 1991; Schiffer and Rossow, 1985).

7.3 Concept of 'Cloud Forcing'

The most common approach to quantify the interaction between clouds, radiation and climate is to look at the Earth radiation budget at the top of the atmosphere. But the determination of Q_{RAD} does not directly allow a separation of its dependence on clouds (type, amount) from that of others as solar zenith angle, surface albedo or absorption characteristics of the clear-sky atmosphere. To estimate the contribution of clouds, commonly the concept of 'cloud (radiative) forcing', an approach first suggested by Charlock and Ramanathan (1985), is applied to the data. This approach is based on the assumption that the net radiation, Q_{RAD}, under all-sky situations is a linear combination of a contribution of the clear-sky, Q_{RAD}^{clear}, and that of an overcast-sky, $Q_{RAD}^{overcast}$, weighthed by the fractional cloud cover, N.

$$Q_{RAD} = S_0\mu_0(1-\alpha) - OLR = (1-N)Q_{RAD}^{clear} + NQ_{RAD}^{overcast} = Q_{RAD}^{clear} + N(Q_{RAD}^{overcast} - Q_{RAD}^{clear}) \quad (7.2)$$

The last term on the right hand side of Equation 7.2 represents the cloud effect, the socalled 'cloud forcing' (CF), on the radiation budget at the top of the atmosphere.

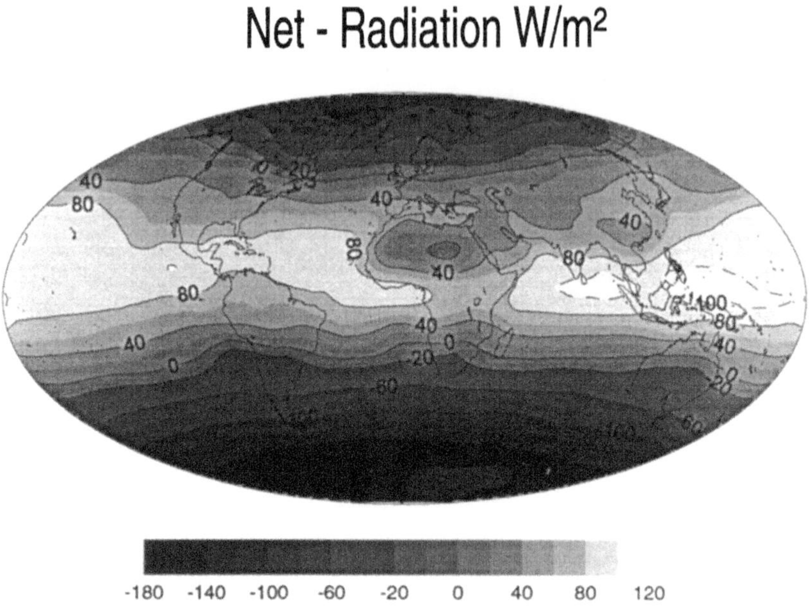

Figure 7.3: *Monthly mean net radiation (Q_{RAD}) for April 1985.*

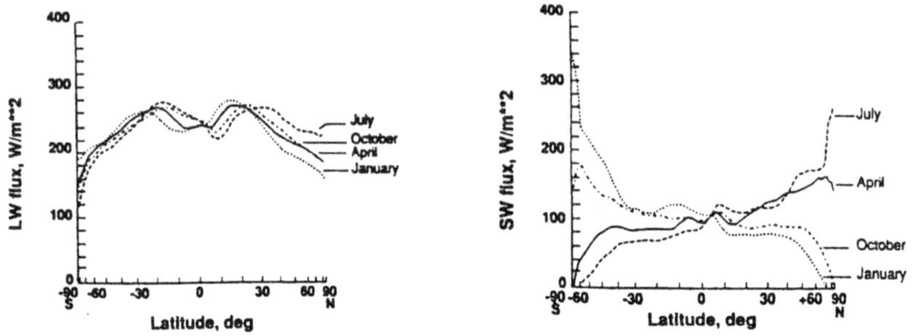

Figure 7.4: *Zonal average outgoing longwave and reflected shortwave radiation for the months April, July, October 1985 and January 1986 derived from ERBE.*

There is one problem in calculating the last term of Equation 7.2 directly from the satellite data. The foot prints of the satellite instruments (PIXELs) on Earth, determined by the instruments' instantaneous field of views (IFOVs), have a nonnegligible spatial extend compared to the spatial scale of broken cloudiness. The ERBE instruments, for instance, have a pixel size of about 60 km^2 at the sub satellite point. Thus, pixels partly filled by clouds and clear-sky is a very common situation, which does not allow to unequivocally determine N and $Q_{RAD}^{overcast}$ from one measurement. Re-arranging Equation 7.2, the cloud forcing (CF) can also be determined according to Equation 7.3.

$$CF = Q_{RAD} - Q_{RAD}^{clear} = CF_{SW} + CF_{LW} \tag{7.3}$$

By use of sophisticated scene identification algorithms, taking into account the spectral dependence of differing targets within different wavelength bands, it is possible to identify clear-sky pixels with a very high accuracy. Thus, the application of such a scene identification allows to estimate on a regional scale, 2.5° × 2.5° regions for ERBE, over a time period of one month the average clear-sky net radiation budget $\overline{Q}_{RAD}^{clear}$. Then, in general, the cloud forcing can be approximated according to Equation 7.3 by simply taking differences between these monthly mean clear-sky fluxes $\overline{Q}_{RAD}^{clear}$ and the monthly regional mean of the all sky radiation fluxes $\overline{Q}_{RAD}$ at the top of the atmosphere. In addition, it is common, as also shown in Equation 7.3, to separate the cloud forcing into a short-wave CF_{SW} and a long-wave CF_{LW} component, respectively.

In the shortwave range clouds usually reflect more incoming solar radiation than the clear-sky surfaces. This albedo effect serves to cool the Earth surface-atmosphere system as indicated in Equation 7.4. In the longwave range, clouds act like a greenhouse gas. They absorb a large fraction of radiation emitted from the surface at high termperature, but reemit less of it to space due to the low temperatures at the cloud tops, thus, heating the Earth surface-atmosphere system as shown in Equation 7.5.

$$CF_{SW} = -S_0\mu_0(\alpha - \alpha^{clear}) < 0 \tag{7.4}$$

$$CF_{LW} = -(OLR - OLR^{clear}) > 0 \tag{7.5}$$

Date	Clear-Sky Longwave	Clear-Sky Shortwave Absorbed	Longwave Cloud Forcing	Shortwave Cloud Forcing	Netto Cloud Forcing
April 1985	265.8	281.6	31.3	-45.1	-13.8
July 1985	267.6	281.1	30.1	-46.7	-16.6
October 1985	266.3	293.1	32.2	-50.1	-17.9
January 1986	262.5	295.0	30.6	-51.7	-21.1
Annual	265.6	287.7	31.1	-48.4	-17.3

Table 7.2: *Summary of Cloud Forcing at the Top of the Atmosphere (Wm^{-2}).*

Table 7.2 presents results of the global average absorbed clear-sky shortwave and clear-sky outgoing longwave radiation, as well as the components of the cloud forcing as determined by Harrison et al. (1990). Generally, there is a small seasonal effect on the global values of these parameters. The global annual shortwave cloud forcing is CF_{SW} = 48 Wm^{-2}. Mean global longwave cloud forcing CF_{LW} = 31 Wm^{-2}, resulting in a net cloud forcing of $CF = -17$ Wm^{-2}. April has a minimum net cloud forcing magnitude of $CF = -14$ Wm^{-2}, and January has the maximum value of $CF = -21$ Wm^{-2}.

As shown in Table 7.2, the quality of clouds, to almost perfectly absorb longwave radiation and at the same time to excellently reflect shortwave radiation, produce opposite effects on the radiation balance TOA. This effect of cancellation can be illstrated by a very simplified model taking into account the decrease in temperature with altitude in the atmosphere and the increase in reflection as a function of cloud albedo. On the condition of global annual average solar insolation and assuming an unrealistic atmosphere without longwave atmospheric absorption above the cloud tops, CF may be approximated as follows:

$$CF = -\frac{S_0}{4}\Delta\alpha + OLR^{clear} - \sigma(T_S - \Gamma Z_{CT})^4 \tag{7.6}$$

Here, $\Delta\alpha$ is the difference in albedo between the overcast- and the clear-sky and $\sigma(T_S - \Gamma Z_{CT})^4$ the approximated outgoing longwave radiation for an overcast cloud situation with its top at a height Z_{CT}, where T_S and Γ are the surface temperature and the atmospheric lapse rate, respectively. For numeric calculations a value of 265 Wm^{-2} can be taken from Table 7.2 for OLR^{clear}, where the surface temperature and lapse rate can be set to climatic values of 288 K and 6.5 K/km, respectively.

Figure 7.5 presents results calculated from Equation 7.6. As can be seen, the longwave effect is dominating above regions where the change in albedo is small, as for instance above deserts. The opposite is true for dark surfaces like oceans, where the change in albedo between an overcast and a clear-sky is largest. In general, Figure 7.5 shows that the effect of an increase in cloud albedo can be compensated by an increase in cloud top height. Since the assumption of a non absorbing atmosphere above the cloud top is unrealistic, in particular for low cloud tops, the range of validity of Figure 7.5 starts at cloud tops which are approximately higher than 5 km.

The regional distributions of the longwave, shortwave and net cloud forcing, determined for the month April 1985 from ERBE data, are shown in Figures 7.6 to 7.8. The longwave cloud forcing is largest with heatings of $CF_{LW} > 60$ Wm^{-2} above the continents within the ITCZ. Secondary maxima of $CF_{LW} > 40$ Wm^{-2} are found in the regions of the mid-latitude storm tracks. These all are areas where upper-level clouds are present, which strongly modulate the longwave part of the energy balance TOA as discussed with Figure 7.5. Since these clouds are also optical thick and, thus, bright, the general pattern of the shortwave cloud forcing is similar to the longwave one, but with the opposite sign. Areas of strongest cooling with values of $CF_{SW} < -70$ Wm^{-2} are found for the mid-latitude storm tracks within the northern hemisphere and for the ICTZ, while for the southern storm track a somewhat smaller cooling of $CF_{SW} < -50$ Wm^{-2} is found. Since both spectral components have opposite signs and, thus, tend to cancel each other the general pattern ot the net cloud forcing is different from the two former ones. We still see that the largest net cloud forcing with a cooling of $CF < -60$ Wm^{-2} is connected to the northern storm tracks. Other striking areas of negative net cloud forcing are those with the prevailing low-level stratus clouds located off the west coast of South America and Africa with $CF < -30$ Wm^{-2}, caused by the strong albedo but small temperature contrast between clouds and ocean. On the contrary, above the Sahara desert CF is positive with values larger than 5 Wm^{-2}, which can be attributed to the small albedo contrast above deserts.

Figure 7.9, Hartmann (1993), presents the zonal and seasonal averages of CF_{LW}, CF_{SW} and CF. The curves of the zonal average net cloud forcing show the importance of mid-latitude cloudiness, which especially during the summer months produce a strong shortwave cooling much larger than the corresponding longwave heating. As a result a strong seasonal dependence of CF is found polewards of about $\pm$ 30° with peak values at about $\pm$ 60° forced by the seasonal variation in solar insolation. In the convective regions of the tropics, the large shortwave and longwave components more of less cancel each other. The resulting net cloud forcing is negative

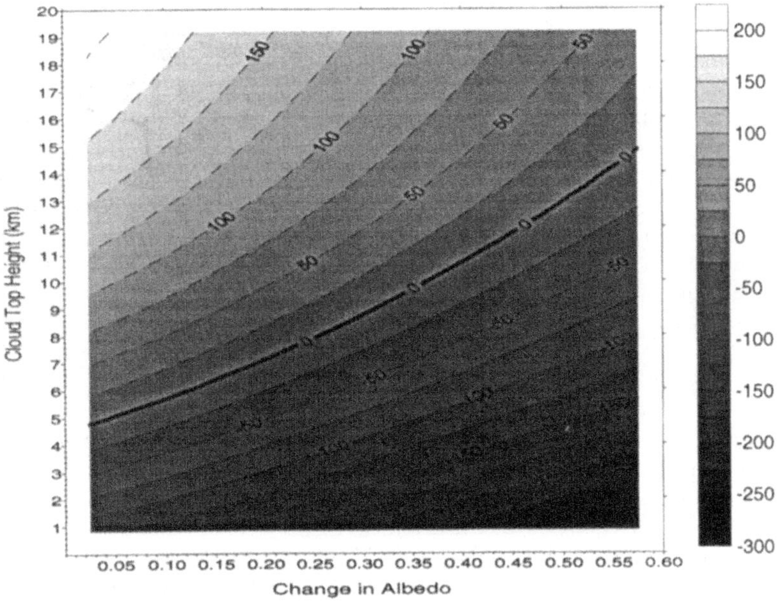

Figure 7.5: *Approximated net cloud forcing (Wm^{-2}), according to Equation 7.6, as a function of albedo change and change in cloud top height.*

but very small and almost independent of season and latitude. As presented in Table 7.2, the global annual CF is a cooling.

It is worthwhile to note that it is wrong to argue, that because clouds cool the present climate, they will likewise act to moderate global warming. It is, however, the change in CF, associated with a climatic change, that governs the cloud feedback and, thus, the strength of the warming. It is also important to note that even in those regions where CF is small, there might be a strong effect of clouds on the weather and climate processes. This effect is felt through the changes in radiative interaction at the surface and in the atmosphere, separately, as discussed below.

7.4 Effects of Clouds on the Surface Radiation Budget

The surface radiation budget, $Q_{RAD}(S)$, presented in Equation 7.7, is determined by the downward solar flux at the surface, DSR, the corresponding surface albedo, α_S, the downward longwave radiation at the surface, DLR, and the corresponding upwelling longwave radiation, depending on surface emittance, ϵ_S, and surface temperature, T_S. The components of the surface radiation balance, unfortunately, cannot be directly measured from satellites. To derive a global estimate of the surface radiation budget, about 20 years ago, methods were developed to link the incoming solar radiation at the surface as well as the surface albedo with satellite observations by means of models which account for atmospheric and cloud extinction (e.g.: Gautier et al., 1980; Möser and Raschke, 1983; Tarpley, 1979). An extended discussion of how

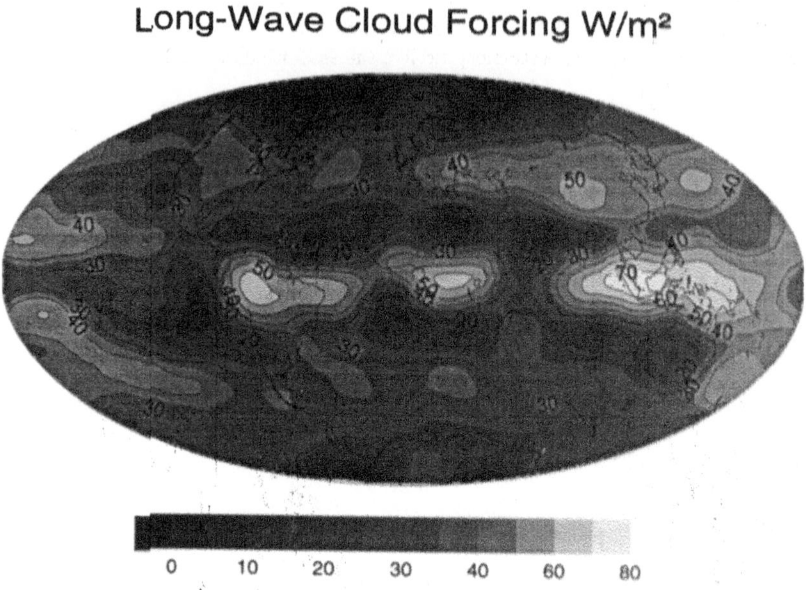

Figure 7.6: *Monthly mean longwave cloud forcing (CF_{LW}) for April 1985.*

applicable these models are, for instance, to derive reliable estimates of DSR from the satellites data is given by Schmetz (1991).

$$Q_{RAD} = DSR \cdot (1 - \alpha_S) + DLR - \epsilon_S \sigma T_S^4 \qquad (7.7)$$

Meanwhile, these techniques have been improved in a way to firstly produce more accurate results on the incoming solar radiation at the surface, and secondly to also estimate the long- wave radiation budget at the surface (e.g.: Bishop and Rossow, 1991; Darnell et al., 1992; Pinker and Laszlo, 1992; Stuhlmann et al., 1990; Zhang et al., 1994). For the later, so far, only preliminary results are published, since no information about the cloud base height distribution is available from operational satellite data. Lately, the World Climate Research Program (WCRP) Radiation Projects Office formed the Global Energy and Water Cycle Experiment (GEWEX) Surface Radiation Budget Project to meet a community need for a global, long-term record of the surface radiation budget. The SRB Project is now retrieving fluxes of short-wave surface radiation over the globe from the ISCCP data with a spatial resolution of 2.5° × 2.5° latitude and longitude (Alberta et al., 1994; Charlock et al., 1993; Whitlock et al., 1993). The SRB project will shortly extend the retrievals to the long-wave.

To look solely at the surface radiation budget, does not directly allow a separation of the effect of clouds (type, amount, optical thickness) from that of others like solar zenith angle, surface albedo, temperature profile or absorption characteristics of the clear-sky atmosphere. To estimate the effect of clouds on $Q_{RAD}(S)$ the concept of cloud forcing, developed for the radiation budget at the top of the atmosphere, can also be applied to the surface data (Gupta et al., 1993; Laszlo and Pinker, 1993; Rieland and Stuhlmann, 1993; Rossow and Zhang, 1995).

$$CF(S) = Q_{RAD}(S) - Q_{RAD}^{clear}(S) = CF_{SW}(S) + CF_{LW}(S) \qquad (7.8)$$

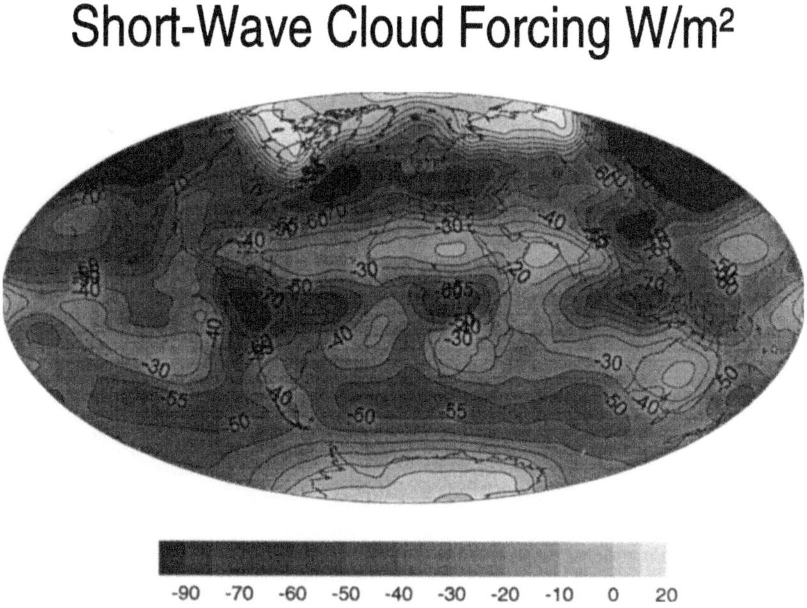

Figure 7.7: *Monthly mean shortwave cloud forcing (CF_{SW}) for April 1985.*

Similar as for the top of the atmosphere the cloud forcing at the surface, $CF(S)$, is calculated as the difference between the monthly averaged net radiation at the surface for all-sky conditions, $\overline{Q}_{RAD}(S)$, and that for clear-skies, $\overline{Q}_{RAD}^{clear}(S)$. The net cloud forcing at the surface again can be separated into its two components, $CF_{SW}(S)$, for the shortwave and, $CF_{LW}(S)$, for the longwave. A discussion on the estimate of the effect of clouds on solar surface absorption derived from different satellite observations is given in the paper by Keevallik and Rannik (1995). Most results published for the longwave cloud forcing at the surface are derived from climate model runs (Ramanathan, 1986; 1987; Slingo and Slingo, 1988; Randall, 1990). The few results based on satellite data strongly depend on the assumption of the cloud base altitude distribution and thus need further validation in the future.

Rieland and Stuhlmann (1993), for instance, present the solar cloud forcing at the surface derived from Meteosat data for a one year period (February 1985 to January 1986). Their annual average is presented in Figure 7.10. Compared to calculations for the top of the atmosphere it is found that both, $CF_{SW}(S)$ and CF_{SW} are of the same order of magnitude. The maximum cooling due to the albedo effect of clouds, $CF_{SW}(S)$, is about 80–100 Wm^{-2}. The regions of maximum cloud forcing are located above the north and south midlatitude oceans and correspond very well to those regions with a maximum in cloudiness. Secondary maxima with a cooling of about 60–70 Wm^{-2} appear above the regions of tropical convective clouds in the ITCZ and low stratus clouds off the western African coast. Minimum values of a cooling less than 30 Wm^{-2} are located above several near cloud-free subtropical regions. A very interesting area is that of the Sahara Desert. Here, cloud forcing at the top of the atmosphere is found to be a very small cooling of $CF_{SW} \approx 10$ Wm^{-2}, while at the surface the cooling, $CF_{SW}(S) >$ 30 Wm^{-2}, is significantly larger. The different behavior of these components indicates that

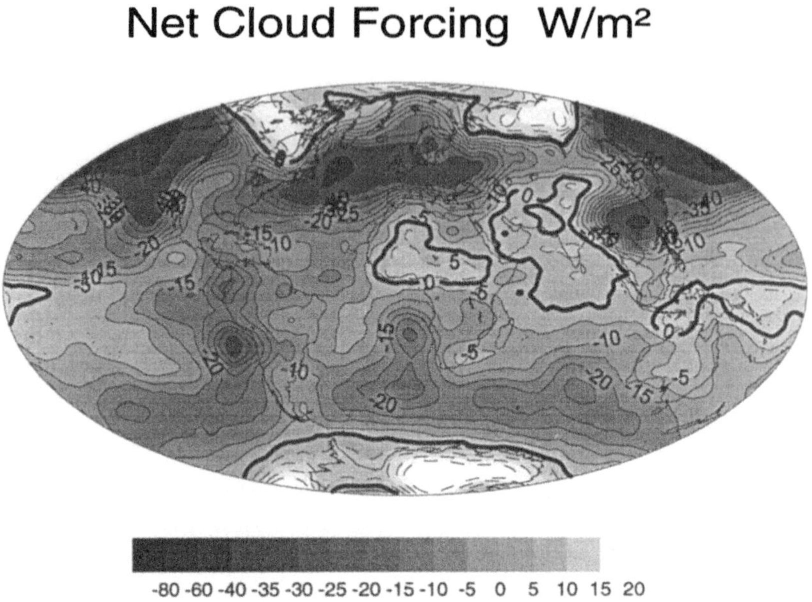

Figure 7.8: *Monthly mean net cloud forcing (CF) for April 1985.*

with the occurence of clouds above these regions, the absorption within the Earth surface-atmosphere system is almost unchanged, while at the surface it is drastically reduced. As a consequence, there must be a vertical redistribution of absorbed solar energy from the surface to the atmosphere.

The average value of $CF_{SW}(S)$ for the Meteosat region is calculated to be a cooling of 55 Wm^{-2} (Rieland and Stuhlmann, 1993), which compares well with the annual global average cooling of 52.5 Wm^{-2} calculated by Gupta et al. (1993) (see Table 7.3).

The regional distribution of longwave component $CF_{LW}(S)$ is not only determined by cloud cover and cloud emittance as measured by the satellites, but also by cloud base height, water vapor amount below the clouds, and the temperature gradient between surface and cloud base. The cloud base height as well as the water vapor and temperature below the clouds cannot be retrieved directly from operational satellite data, and, thus, have to be derived from crude assumptions which may cause systematic errors in the results.

Gupta et al. (1993) present the zonally averaged cloud forcing components at the surface for the months of January 1985 and July 1985. Their results for all three components, $CF_{SW}(S)$, $CF_{LW}(S)$ and $CF(S)$, are shown in Figure 7.11 from the top to the bottom, respectively, for both months.

The long-wave term, $CF_{LW}(S)$, mid panels in Figure 7.11, in general, is small but always positive. Largest values are found for the southern hemisphere in regions with a large amount of low level clouds and a low atmospheric water vapor content. Gupta et al. derive for the annual global average $CF_{LW}(S)$ a heating of the surface of about 34.6 Wm^{-2}.

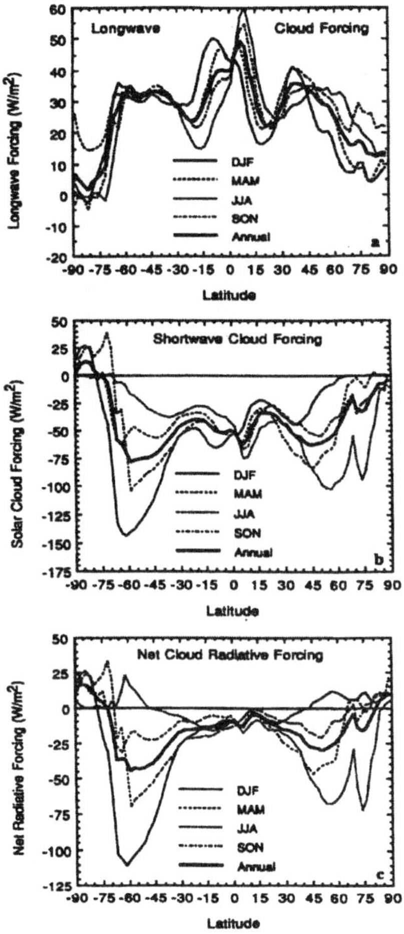

Figure 7.9: *Zonal and seasonal averages of a) CF_{LW}, b) CF_{SW} and c) CF (Hartmann, 1993).*

Comparing the seasonal zonal curves of CF_{SW} in Figure 7.9 with those of CF_{SW} for January and July in Figure 7.11, it is seen that both have a quite similar dependence on latitude and with season. In addition, Figure 7.11 shows that the net surface cloud forcing, $CF(S)$, is dominated by the shortwave cooling term, especially within the summer hemisphere. In contrast, the latitudinal dependence of the zonal components of CF_{LW} and $CF_{LW}(S)$ is quite different. For instance, in the ITCZ, the areas of high-level clouds, CF_{LW} is found to have a strong warming while $CF_{LW}(S)$ has only a very weak warming. The opposite effect is found for those regions with low-level clouds. In addition, the water vapor content has a strong effect on $CF_{LW}(S)$, as discussed above, but almost no effect on CF_{LW}. From these estimates it can be already concluded that the effect of clouds in the short-wave spectral domain for all areas of the globe is primarily a cooling of the surface, while in the longwave spectral domain the effect of clouds can be a significant warming or cooling of the atmosphere, depending on latitude. The effect of clouds on the atmospheric heating will be discussed in more detail below.

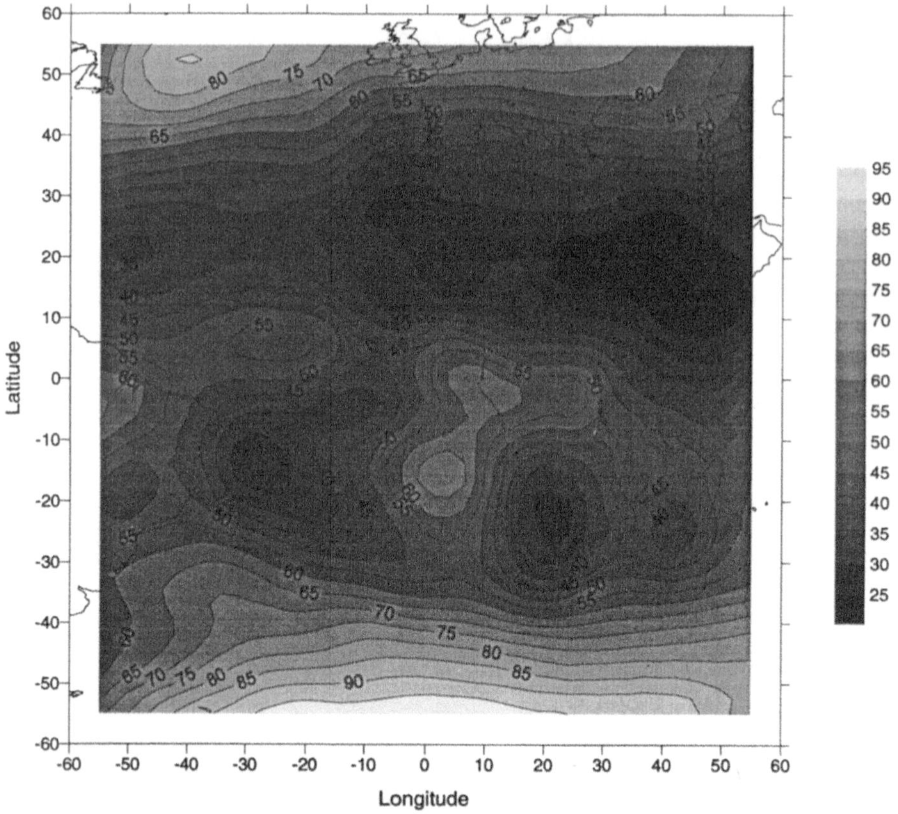

Figure 7.10: *Annual average solar cloud forcing at the surface, $CF_{SW}(S)$ in Wm^{-2}. Note that the positive values in this figure correspond to a cooling of the surface.*

Date	$\mathbf{CF_{LW}(S)}$ ($\mathbf{Wm^{-2}}$)	$\mathbf{CF_{SW}(S)}$ ($\mathbf{Wm^{-2}}$)	**CF(S)** ($\mathbf{Wm^{-2}}$)
April 1985	35.5	-48.3	-12.8
July 1985	33.6	-45.6	-12.0
October 1985	34.3	-55.9	-21.5
January 1986	34.6	-60.0	-25.1
Annual	34.6	-52.5	-17.9

Table 7.3: *Summary of global average surface cloud forcing components after Gupta et al. (1993).*

Table 7.3 presents the seasonal variation and the annual mean of the global average surface cloud forcing components, $CF_{LW}(S)$, $CF_{SW}(S)$, and $CF(S)$, as they were derived from satellite

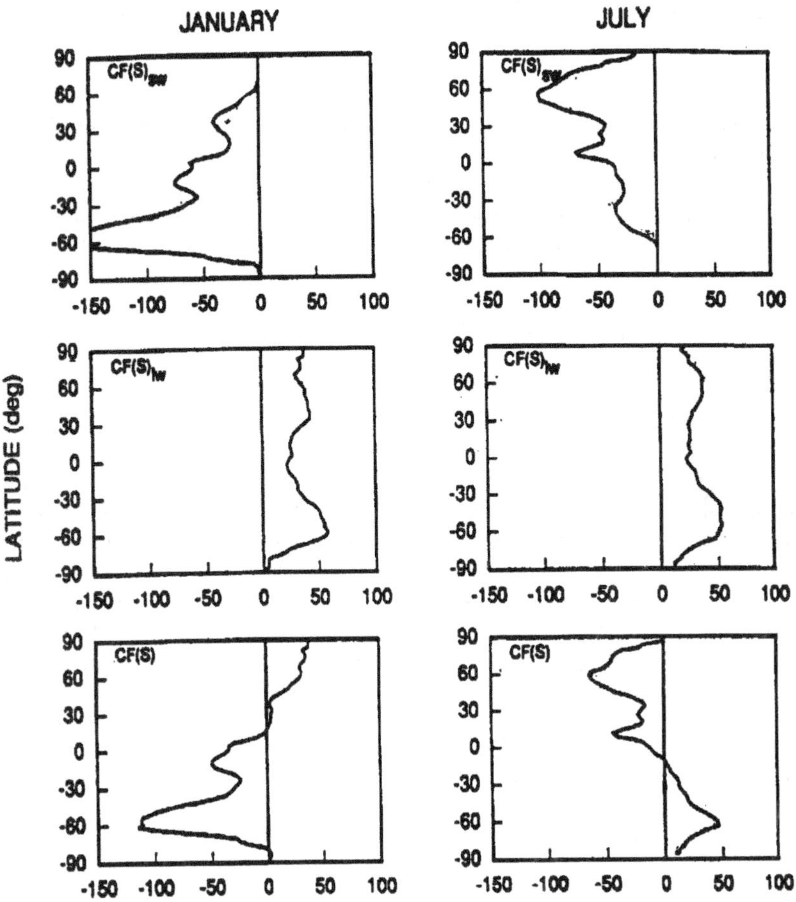

Figure 7.11: *Zonally averaged cloud forcing components, $CF_{SW}(S)$, $CF_{LW}(S)$ and $CF(S)$ in Wm^{-2} (after Gupta et al., 1993).*

measurements by Gupta et al. (1993). The global average longwave surface cloud forcing shows no seasonal variation with an annual mean heating of $CF_{LW}(S) = 34.6$ Wm^{-2}. On the contrary, the shortwave surface cloud forcing shows a strong dependence on average solar insolation, which is largest in January. The annual global average $CF_{SW}(S)$ is determined as a cooling of -52.5 Wm^{-2}. As expected, the seasonal dependence of the net surface cloud forcing is similar to the shortwave component, with an annual average cooling of $CF(S) = -17.9$ Wm^{-2}, which is almost identical to that determined for the top of the atmosphere as presented in Table 7.2.

7.5 Effects of Clouds on the Atmospheric Radiation Budget

As shown in many general circulation studies (e.g.: Johnson et al., 1985; Oort and Peixoto, 1983), it is the vertical and horizontal distribution of the total diabatic heating of the atmosphere which determines the energy conversion and the dynamic structure of the atmosphere. The full vertical profile of radiative flux divergence is also needed to determine the effect of radiation on the generation of Available Potential Energy (Lorenz, 1995; Stuhlmann and Smith, 1988). Thus, to understand how cloudiness will affect the general circulation and therefore weather and climate, it is necessary to estimate the change in the three-dimensional structure of the diabatic heating field as forced by a change in cloudiness.

The radiative heating of the total atmosphere and also the effect of clouds on that heating can be derived from a combination of the results for the top of the atmosphere and the surface. Results on the shortwave cloud forcing of the atmosphere, $CF_{SW}(A)$, are given by Laszlo and Pinker (1993), Rieland and Stuhlmann (1993) and Rossow and Zhang (1995). Figure 7.12 presents the annual average shortwave atmospheric cloud forcing, $CF_{SW}(A)$, determined for the Meteosat field of view by Rieland and Stuhlmann (1993). The figure shows that the shortwave atmospheric cloud forcing turns out to be about one order of magnitude less than that at the top and falls inside a range between a warming of $CF_{SW}(A) > 20$ Wm^{-2} and a cooling of $CF_{SW}(A) < -5$ Wm^{-2}. A short-wave cooling of the total atmospheric column is found for clouds within the tropics having high cloud tops. Here, the increase in backscattering of solar radiation by these clouds, which takes place in the upper tropospheric levels, strongly reduces the atmospheric short-wave absorption in comparison to a clear-sky situation. Contrary to this fact, clouds with tops at lower levels cause a small increase in shortwave absorption.

Because the interaction of clouds with shortwave radiation can cool or heat the total atmosperic column, $CF_{SW}(A)$ is not highly correlated with the mean cloud cover ($R = 0.37$). The largest atmospheric shortwave heating of $CF_{SW}(A) > 20$ Wm^{-2}, for example, is found for the Sahara, an area with only a very low amount of cloudiness, but a very highly reflective surface. In contrast, for the Atlantic ocean within the same latitudinal belt, a region with a high amount of cloudiness, but a very small surface albedo, only a small shortwave heating of $CF_{SW}(A) <$ 10 Wm^{-2} is found. As a result, it is concluded that the influence of clouds on $CF_{SW}(A)$ is not determined by the cloud properties alone but in combination with the spectral surface reflectance properties. An annual and regional average for the Meteosat field of view, $\pm$ 60° longitude and latitude, of a warming of $CF_{SW}(A) = 5$ Wm^{-2} is given by Rieland and Stuhlmann (1993), which corresponds to a 1.4% increase in atmospheric solar absorption due to cloudiness.

Similar to the above, the longwave cloud forcing of the atmosphere, $CF_{LW}(A)$, depends on the vertical distribution of the clouds within the atmosphere. The zonally averaged $CF_{LW}(A)$, as calculated by Gupta et al. (1993) for January 1986 and July 1985, and presented in Figure 7.13, show a strong warming of the atmosphere in the tropics, the regions where deep convective clouds are present. On the contrary, for the subtropics and midlatitudes of both hemispheres, the regions with prevailing middle and low level clouds, a moderate cooling is derived. The annual global average of $CF_{LW}(A)$ is calculated to be a cooling of -3.4 Wm^{-2} (Gupta et al., 1993). For hemispheric averages Gupta et al. found a strong seasonal change, which could change, for the northern hemisphere example, from a warming of 4.7 Wm^{-2} in July to a cooling of -7.3 Wm^{-2} in January. These changes are associated with the shift of the warm band in the tropics across the equator. The cooling, which is found outside the tropics, undergoes only minor seasonal variations.

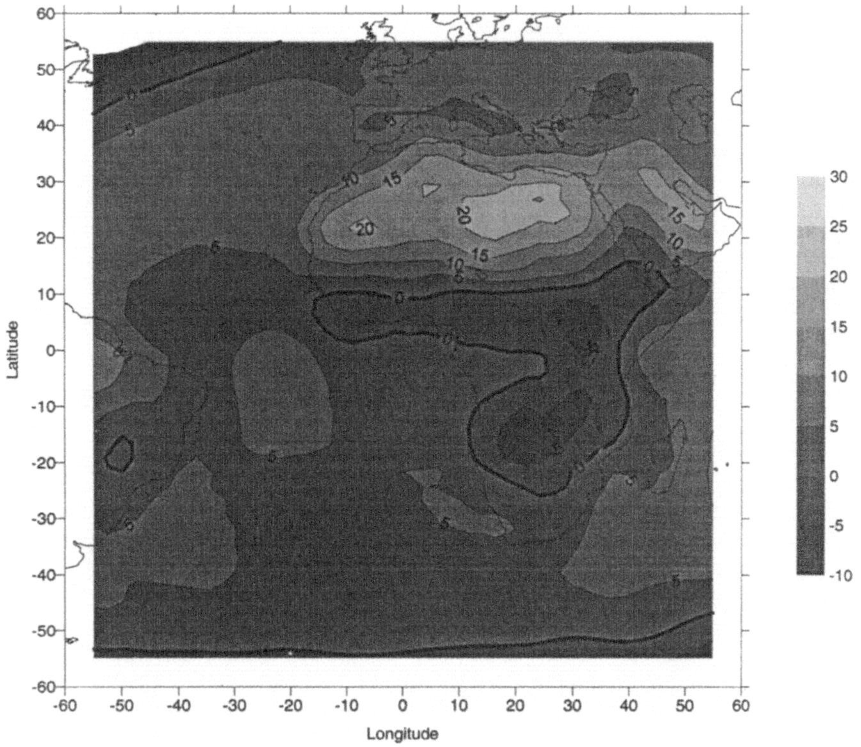

Figure 7.12: *Annual average solar cloud forcing of the atmosphere $CF_{SW}(A)$ in Wm^{-2}.*

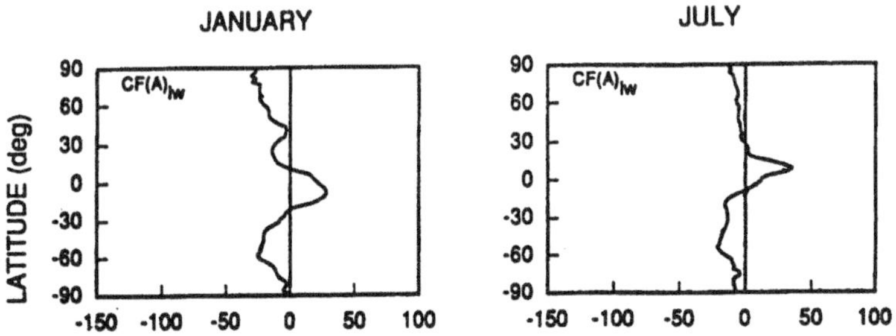

Figure 7.13: *Zonal average longwave cloud forcing of the atmosphere, $CF_{LW}(A)$, in Wm^{-2} for the months of January 1986 and July 1985 from Gupta et al. (1993).*

First sensitivity studies to retrieve the full radiative flux divergence profiles within the atmosphere have been made (Charlock et al., 1993; 1994; Stuhlmann et al., 1993). Satellite based retrievals of cloud parameters, such as those from the ISCCP for instance, are used as inputs to radiative transfer calculations to derive the vertical distribution of radiative heating $\frac{\partial T}{\partial t}(p)$ (K/day).

$$\frac{\partial T}{\partial t}(p) = -\frac{g}{c_p} \cdot \frac{\partial M}{\partial p} \tag{7.9}$$

$$M(p) = M_{SW}^{+}(p) - M_{SW}^{-}(p) + M_{LW}^{+}(p) - M_{LW}^{-}(p) \tag{7.10}$$

Here, $M_{SW/LW}^{+}(p)$ and $M_{SW/LW}^{-}(p)$ are the upward and downward shortwave and longwave exitances at the corresponding pressure levels as retrieved from satellite and auxilliary data by means of radiative transfer calculations. Similar to the discussion above, to describe the effect of clouds on the radiative heating profiles, a socalled "cloud generated radiative heating" ($CGRH(p)$) can be defined as the difference between the all-sky and clear-sky situation.

$$CGRH(p) = \frac{\partial T}{\partial t}(p) - \frac{\partial T}{\partial t}(p)^{clear} = CGRH(p)_{SW} + CGRH(p)_{LW} \tag{7.11}$$

Some preliminary results of the $CGRH(p)$ are presented for three tropospheric and two stratospheric layers for the month of April 1985 by Stuhlmann et al. (1993). These results are derived from a combination of Meteosat cloud- and ECMWF atmospheric state analyses, which are used as inputs for a delta-two stream radiative transfer calculation (Schmetz, 1984). The vertical extent of the clouds and, thus, the cloud base height, was assumed to depend on a combination of cloud optical thickness and cloud top temperature. The accuracy of this parameterization could not be tested. It was only shown that the profiles of the $CGRH(p)$ strongly depend on this parameterization, while if integrated over the total atmospheric column $CGRH$ is less sensitive to the type of parameterization used to reckon the cloud base heights.

Figure 7.14, after Stuhlmann et al. (1993), shows that the profiles of the $CGRH(p)$, averaged over the Meteosat field of view and split into three cloud classes, give an atmospheric shortwave heating, $CGRH(p)_{SW} > 0$, for those layers in which the clouds are embedded. The atmosphere beneath the clouds is cooled, $CGRH(p)_{SW} < 0$, because of the strong decrease in solar insolation. In contrast, a longwave atmospheric cooling, $CGRH(p)_{LW} < 0$, shows up for the layers in which the cloud tops are embedded and a heating, $CGRH(p)_{LW} > 0$, for the layers beneath the clouds. The cooling is caused by the additional radiation emitted to space which depends on the cloud top temperature and emittance. The heating is related to the longwave radiation emitted from the surface and trapped by the clouds. The net result is that low clouds cool the tropospheric layers between the surface and the tropopause region. The presence of mid-level clouds cause a heating in the lowest tropospheric layer, while the upper tropospheric layers are still cooled. Upper level cirrus and convective clouds cause a heating of all tropospheric layers.

Figure 7.15 presents the column integrated zonal net atmospheric cloud forcing, $CF(A)$, for two months and the annual mean, as calculated by Rossow and Zhang (1995), from the ISCCP data. Their results clearly show that, as net effect, the atmospheric cloud forcing will strengthen the meridional temperature gradient. This finding compares well with the longitudinal and latitudinal distribution of the tropospheric column integrated $CGRH$ of Stuhlmann et al. (1993). They find a heating for the region of the ITCZ and the Africal continent, and a cooling of the ocean regions north and south of the ITCZ. They also show that there is a strong land ocean contrast for the column integrated $CGRH$, which compares well with the results for the longwave atmospheric cloud forcing presented by Gupta et al. (1993).

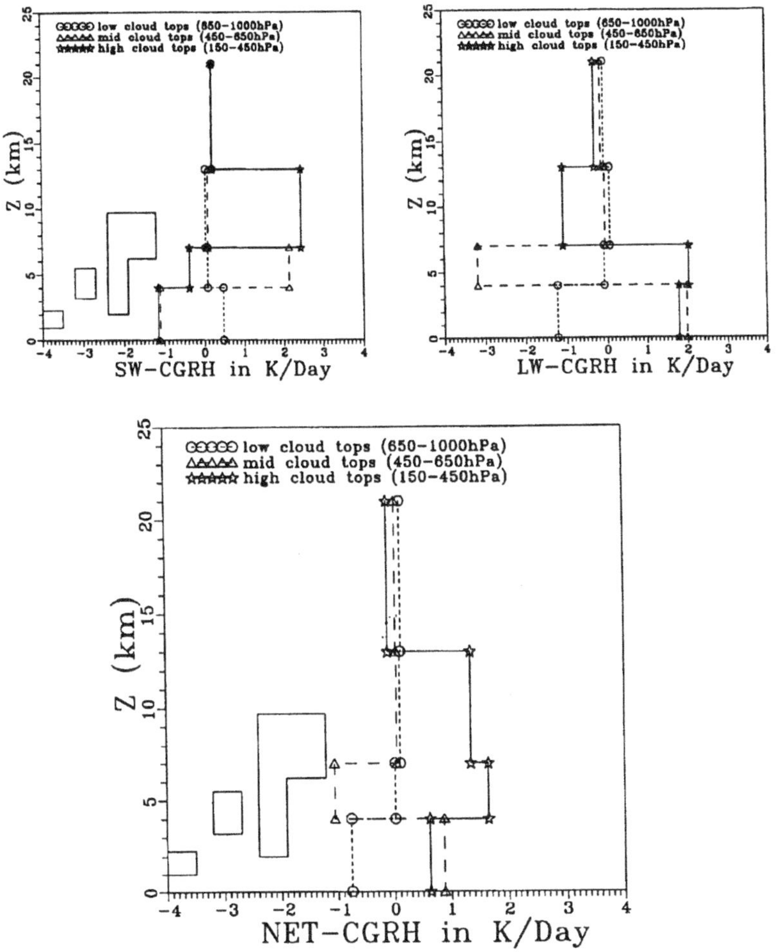

Figure 7.14: *Profiles of shortwave, $CGRH(p)_{SW}$, longwave $CGRH(p)_{LW}$ and net $CGRH(p)$ for three cloud classes averaged over a month and the net Meteosat field of view determined from data of April 1985 (Stuhlmann et al., 1993).*

7.6 Summary and Outlook

The effect of clouds on the radiative heating of the Earth surface-atmosphere system was discussed. The global annual averaged cloud forcing at the top of the atmosphere, CF, is a cooling of about -17 Wm^{-2}. This cooling does not imply that cloud forcing will reduce the atmosperic warming, which is predicted for a doubling of carbon dioxide. It will be the change in cloud forcing with climate change, which depends on the change in cloud pattern, that determines the cloud radiative feedback.

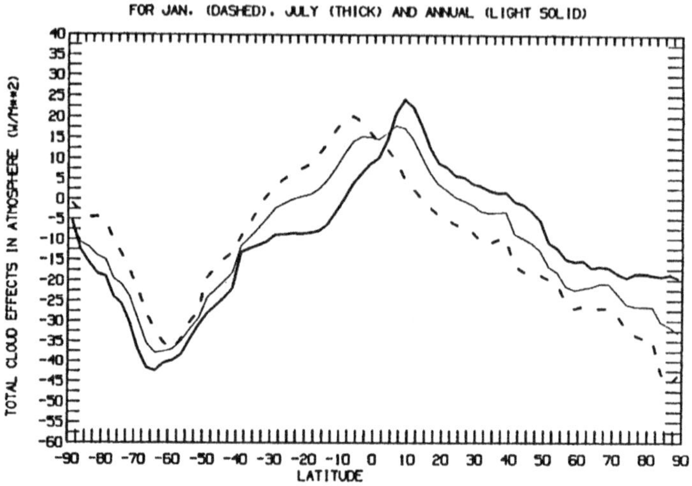

Figure 7.15: *Zonal average of the net atmospheric cloud forcing, $CF(A)$, for January (dashed line), July (thick solid line) and the annual mean (thin solid line) as calculated by Rossow and Zhang (1995) from ISCCP data.*

It was also discussed, that even for those regions with CF close to zero there is a radiative effect of clouds, which affects the surface and atmosphere separately. Within the short-wave spectral domain there is mainly a cooling of the surface by clouds. For the long-wave spectral domain the cloud effect is primarily on the atmosphere, where it tends to strengthen the meridional temperature gradient.

Beside looking at the radiative heating of the total atmospheric column it is also desirable to have a vertical profile of atmospheric heating. First attempts to derive the cloud effect on the radiative heating profile tend to show that low clouds will cool the troposphere within all layers. The presence of mid level clouds will cause a destabilization of the atmosphere by heating the lowest atmospheric layer and cooling the layers above. As cloud tops further grow in the atmosphere, the destabilization of the troposphere will change to a stabilization again, since the $CGRH(p)$ profile for high clouds shows a heating through all tropospheric layers with the largest values in the highest and the smallest values in the lowest layers. Thus, with increasing cloud top height, the atmospheric column integrated $CGRH$ will be an additional increasing heating source for the troposphere.

To derive the downwelling longwave radiation at the surface and the long-wave atmospheric heating rates, assumptions about cloud base altitudes have to be made, since they cannot be derived from the satellite measurements. These uncertainties will lead to systematic errors within the quantities discussed. The magnitude of uncertainty will depend on the types of data sets used for the retrievals.

If inputs for the radiative transfer calculations are taken from averaged cloud parameters within a 2.5° × 2.5° grid box, the main problem will be how to handle cloud overlap in the calculations (Charlock et al., 1994). The way cloud overlap is handled will have a major impact on the cloud generated radiative heating profiles. The uncertainties in the column integrated atmospheric heating and the downwelling long-wave radiation at the surface will be smaller, but still large.

If the cloud properties, which have to be used within the radiative transfer calculations to determine radiation fluxes $M^{+}_{SW/LW}(p)$ and $M^{-}_{SW/LW}(p)$ at the corresponding pressure levels, are directly retrieved from spatially high resolution (1–5 km) satellite data, there still exists the problem of deriving the correct cloud parameters for scattered inhomogeneous cloudiness. This problem of beam filling leads to an uncertainty in optical thickness and corresponding cloud cover of the detected cloud within a pixel and thus will lead to a systematic error in the parameterization of the vertical cloud extension (Stuhlmann, 1993). These uncertainties again will cause the largest errors for the $CGRH(p)$ profiles and somewhat smaller errors if the profiles are integrated over the troposphere.

The problem how to parameterize the clouds' vertical extension or the corresponding cloud bottom altitudes for the radiative flux divergence calculations requires more work in the future. Nevertheless, the uncertainties related to this problem are small enough for a preliminary and a more general discussion of cloud radiative effects on the Earth surface-atmosphere system as it was done here. In the future, it will be urgently needed to have a global measurement of cloud layering, vertical cloud thickness, and an understanding how much water exists in the cloud in the form of liquid and solid. This information coupled with knowledge about the cloud microphysics strongly dictates how much radiation is absorbed and reflected as well as governing the interchange of substances between different phases. Here, the science community together with the space agencies develop plans to study the feasibility of an international multi-sensor cloud-radiation satellite mission. Such a mission has to fill the gap of data necessary for determining the distribution of clouds, understanding of this distribution in relation to the governing processes, and quantifying the links between clouds, radiation and the water budget. Such a mission may include a millimeter wave cloud profiling radar (IGPO, 1994), a backscatter lidar, a sub-millimeter-wave radiometer, and a ERB-type instrument. A selection of such instruments would provide three-dimensional profiles of clouds fields, major aerosol layers together with a characterization of cirrus cloud extinction, ice mass and crystal size, and the related TOA ERB field.

7.7 References

Alberta TL, Charlock TP, Whitlock CH, Rose FG, DiPasquale R, Pinker R, Staylor WF, Gupta SK (1994) Climate observations with GEWEX Surface Radiation Budget Project data. *Proceedings of the 8th Conference on Atmospheric Radiation*, AMS, January 1994, Nashville, Tennessee, 22–24

Barkstrom BR, Smith GL (1986) The Earth Radiation Budget experiment: Science and implementation. *Rev Geophys* 24: 379–390

Bishop JKB, Rossow WB (1991) Spatial and temporal variability of global surface solar irradiance. *J Geophys Res* 96: 16.839–16.858

Brooks DR, Minnis P (1984) Simulation of the Earth's monthly average regional radiation balance derived from satellite measurements. *J Climate Appl Meteor* 23: 392–403

Cess RD (1976) Climate change, an appraisal of atmospheric feedback mechanisms employing zonal climatology. *J Atmos Sci* 33: 1831–1848

Cess RD, Briegleb BP, Lian MS (1982) Low-latitude cloudiness and climate feedback: comparative estimates from satellite data. *J Atmos Sci* 39: 53–59

Cess RD et al. (1989) Interpretation of cloud-climate feedback as produced by 14 atmospheric general circulation models. *Science* 245: 513–516

Cess RD et al. (1990) Intercomparison of climate feedback processes in 19 atmospheric general circulation models. *J Geophys Res* 95: 16.601–16.615

Charlock TP, Ramanathan V (1985) The albedo field and cloud radiative forcing produced by a general circulation model with internally generated cloud optics. *J Atmos Sci* 42: 1408–1429

Charlock TP, Whitlock CH, Alberta TL (1993) The GEWEX Surface Radiation Budget Project. *GEWEX News* 3, No. 1

Charlock TP, Rose FG, Yang S-K, Alberta T, Smith GL (1993) *An observational study of the interaction of clouds, radiation and the general circulation*, IRS 92: Current Problems in Atmospheric Radiation, eds. S. Keevallik and O. Kärner, A. DEEPAK Publishing, 151–154

Charlock TP, Rose FG, Alberta T, Smith GL, Rutan D, Manalo-Smith N, Bess TD, Minnis P (1994) Retrievals of the surface and atmospheric radiation budget: Tuning parameters with radiative transfer to balance pixel-scale ERBE data. *Proceedings of the 8th Conference on Atmospheric Radiation*, AMS, January 1994, Nashville, Tennessee, 435–437

Charlock TP, Rose FG, Alberta T, Smith GL, Rutan D, Manalo-Smith N, Minnis P, Wielicki B (1994) Cloud profiling radar requirements: Perspective from retrievals of the surface and atmospheric radiation budget and studies of atmospheric energetics, in *IGPO Publication Series* No. 10, Pasadena, California, B10–B21

Darnell WL, Staylor WF, Gupta SK, Ritchey NA, Wilber AC (1992) Seasonal variations of surface radiation budget derived from ISCCP C1 data. *J Geophys Res* 97: 15.714–15.760

Dozier J (1994) Planned EOS observations of the land, ocean and atmosphere. *Atmospheric Research* 31: 329–357

Ellis JS (1978) *Cloudiness, the planetary radiation budget and climate*, Colorado State University, Colorado, 129 pp

Gautier C, Diak G, Masse S (1980) A simple physical model to estimate incident solar radiation at the surface from GEOS satellite data. *J Appl Meteor* 19: 1005–1012

Gupta SK, Staylor WF, Darnell WL, Wibler AC, Ritchey NA (1993) Seasonal variations of surface and atmospheric columns as determined from ISCCP C1 data. *J Geophys Res* 98: 2703–2713

Harrison EF, Minnis P, Barkstrom BR, Ramanathan V, Cess RD, Gibson GG (1990) Seasonal variation of cloud radiative forcing derived from the Earth radiation budget experiment. *J Geophys Res* 95: 18.687–18.703

Hartmann DL, Short DA (1980) On the use of Earth radiation budget statistics for studies of clouds and climate. *J Atmos Sci* 37: 1233–1250

Hartmann DL (1993) *Radiative effects of clouds on Earth's climate, in Aerosol- Cloud-Climate Interaction*, ed. P.V. Hobbs, Academic Press, Inc., San Diego, 151–173

Houghton JT, Jenkins GJ, Ephraums JJ (1990) *Climate Change - The Scientific Assessment*, Cambridge University Press, 365pp

IGPO (1994) Utility and feasibility of a cloud profiling radar, Reprot of the GEWEX Tropical Workshop, *IGPO Publication Series* No. 10, Pasadena, California, 150 pp

Jacobowitz H, Soule HV, Kyle HL, House FB and the Nimbus-7 Experiment ERB Team (1984) The Earth Radiation Budget (ERB) experiment: An overview. *J Geophys Res* 89: 5021–5038

Johnson DR, Townsend RD, Wei M-Y (1985) The thermally coupled response of the planetary scale circulation to the global distribution of heat sources and sinks. *Tellus* 37: 106–125

Kandel RS, Monge J-L, Viollier M, Pakhormov LA, Adasko VI, Reitenbach RG, Raschke E, Stuhlmann R (1994) The ScaRaB project: Earth radiation budget observations from the meteor satellite. *Adv Space Res* 14: (1)47–(1)54

Keevallik S, Rannik U (1995) Solar surface absoption from different satellite observations. *Adv Space Res*, in press

Lorenz EN (1955) Available potential energy and the maintanance of the general circulation. *Tellus* 7: 157–167

Manabe S, Strickler RF (1964) *J Atmos Sci* 21: 361–385

Möser W, Raschke E (1983) Mapping of global radiation and of cloudiness from Meteosat image data. *Meteor Rundsch* 36: 33–41

Ohring G, Clapp PF (1980) The effect of changes in cloud amount on the net radiation at the top of the atmosphere. *J Atmos Sci* 37: 447–454

Ohring G, Clapp PF, Heddinghais TR, Krüger AF (1981) The quasi-global distribution of the sensitivity of Earth - atmosphere radiation budget to clouds. *J Atmos Sci* 38: 2539–2541

Oort AH, Peixoto JP (1983) Global angular momentum and energy balance requirements from observations. *Adv Geophys* 25: 355–490

Pinker RT, Laszlo I (1992) Modeling surface solar irradiance for satellite applications on a global scale. *J Appl Meteor* 31: 194–211

Ramanathan V (1986) Scientific use of surface radiation budget for climate studies, in *Surface Radiation Budget for Climate Applications*, ed. by J. T. Suttles and G. Ohring, NASA RP- 1169, Washington, D. C., 58–86

Ramanathan V, Cess RD, Harrisson EF, Minnis P, Barkstrom BR, Ahmad E, Hartmann DL (1989) Cloud-radiative forcing and climate: results from the Earth Radiation Budget Experiment. *Science* 243: 57–65

Randall DA (1990) Atmospheric and surface cloud radiative forcing, in *Proceedings of the Conference on Long-Term Monitoring of Earth's Radiation Budget*, ed. by B. R. Barkstrom, Orlando, Fla, 16–26

Rieland M, Stuhlmann R (1993) Towards the influence of clouds on the shortwave radiation budget of the Earth-atmosphere system estimated from satellite data. *J Appl Meteor* 32: 825–843

Rossow WB, Schiffer RA (1991) ISCCP cloud data products. *Bull Amer Meteor Soc* 72: 2–20

Rossow WB, Zhang Y-C (1995) Calculations of surface and top-of-atmosphere radiative fluxes from physical quantities based on ISCCP datasets, Part II: Validation and first results. *J Geophys Res* 100: 1167–1197

Schiffer RA, Rossow WB (1985) ISCCP global radiance data sets: A new resource for climate research. *Bull Amer Meteor Soc* 66: 1498–1505

Schmetz J (1984) On the parameterization of the radiative properties of broken clouds. *Tellus* 36: 417–432

Schmetz J, Mhita M, deBerg LV (1990) Meteosat observations of longwave cloud- radiative forcing for April 1985. *J Climate* 3: 784–791

Schmetz J (1991) Retrieval of surface radiation fluxes from satellite data: *Dyn Atmos Oceans* 16: 417–432

Slingo A, Slingo JM (1988) The response of a general circulation model to cloud longwave radiative forcing, I: Introduction and initial experiments. *Q J R Meteor Soc* 114: 1027–1062

Smith GL, Green RN, Raschke E, Avis LA, Suttles JT, Wielicki BA, Davies R (1986) Inversion methods for satellite studies of the Earth's radiation budget: Development of algorithms for the ERB mission. *Rev Geophys* 24: 407–421

Stuhlmann R, Smith GL (1988) A study of cloud-generated radiative heating and its generation of available potential energy Part II: Results for a climatological zonal mean January. *J Atmos Sci* 45: 3928–3943

Stuhlmann R, Rieland M, Raschke E (1990) An improvement of the IGMK model to derive total and diffuse solar radiation at the surface from satellite data. *J Appl Meteor* 29: 586–603

Stuhlmann R (1993) Feasibility study for calculating the cloud-generated radiative flux divergence from Meteosat imagery data. *GKSS P2 Report*, 85 pp

Stuhlmann R, Raschke E, Schmid U (1993) Cloud generated radiative heating from Meteosat data, IRS 92: *Current Problems in Atmospheric Radiation*, eds. S. Keevallik and O. Kärner, *A. DEEPAK Publishing*, 69–75

Tarpley JD (1979) Estimating incident solar radiation at the surface from geostationary satellite data. *J Appl Meteor* 18: 1172–1181

Whitlock CH, Charlock TP, Staylor WF, Pinker R, Laszlo I, DiPasquale R, Ritchey NA (1993) WCRP surface radiation budget shortwave data product description - Version 1.1, *NASA Technical Memorandum* 107747, 28 pp

Zhang Y-C, Rossow WB, Lacis AA (1994) Calculations of surface and top of atmosphere radiative fluxes from physical quantities based on ISCCP datasets, Part I: Method and sensitivity to input data uncertainties, in press

Chapter 8

Remote Sensing of Atmospheric Water Vapor

William B. Rossow
NASA Goddard Institute for Space Studies
New York, NY 10025
USA

8.1 Water Vapor in the Climate System

Many of the unique features of Earth's surface environment result from the presence of large amounts of water, mostly in liquid phase. However, key to many feedback processes in the climate system is the fact that some water is also present in both vapor and solid phase: exchanges between the different reservoirs of water in its different forms, mediated by the circulation of the atmosphere, make the behavior of the whole system much more complicated. Current concern with how sensitive the climate is to changes made by humans and how sensitive human societies are to changes in climate motivates interest in water in all its forms; however, the focus of this paper is on the roles played by water vapor in determining the energy content and circulation of the troposphere.

Earth's surface is divided into two very different regimes, solid land including glaciers, and liquid oceans; about 2.5% of all water resides on land, about 2% in the glaciers and about 0.5% as near-surface or surface water (Peixoto and Oort, 1992). The latter small amount of water is crucial to the survival of humans, directly for drinking and indirectly for food derived from the land biosphere. Moreover, "developed" society depends on many other uses of water that support an improved standard of living. This particularly important reservoir is maintained by a balance between supply by rain and snow fall over land, more than 60% of which re-evaporates, and loss by river runoff back into the ocean. Thus, our supply of water depends on the processes in the atmosphere by which water vapor is transported from the oceans to the land and converted into precipitation there.

The global atmosphere holds only 10^{-5} of the amount of water in the oceans and the net transport of water from the ocean surface, through the atmosphere, and onto the land surface is equivalent to about 1% of this water vapor transferred per day (Peixoto and Oort, 1992). Essentially, this transfer is a mixing process where the circulation of the atmosphere mixes water from areas of higher abundance over the ocean source to areas of lower abundance over

NATO ASI Series, Vol. I 45
Radiation and Water in the Climate System:
Remote Measurements
Edited by Ehrhard Raschke

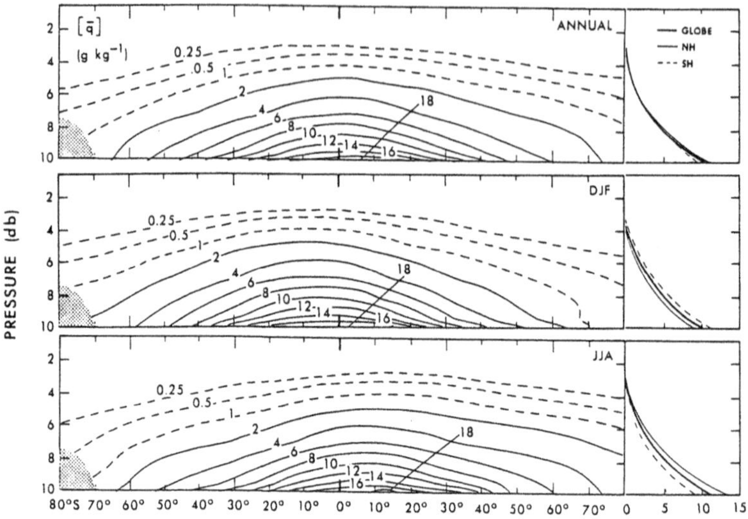

Figure 8.1: *Zonal-mean cross sections of the specific humidity in g kg^{-1} for annual, DJF, and JJA mean conditions. Vertical profiles of the hemispheric and global mean values are shown on the right (from Peixoto and Oort, 1992).*

the land sink, but the unusual physical properties of water add major complications to this process.

First, the large latent heat of water phase changes make the vapor pressure of water in equilibrium a very strong function of temperature. Hence, because of the variations of atmospheric temperature with altitude, latitude and season, abundance gradients of water vapor in local equilibrium are very large (Figure 8.1). There are two very important consequences of this fact: (1) significant water vapor abundance variations occur on spatial scales much smaller than the spatial scales of the most energetic motions of the atmospheric circulation and (2) the large scale atmospheric circulation produces significant deviations of the local water vapor abundance from equilibrium with temperature (Figure 8.2). Thus, the local abundance of water vapor and its variations on different time scales cannot be predicted from temperature alone, as if a thermodynamic equilibrium obtained (Gaffen et al., 1992; Bony and Duvel, 1994).

Second, the large latent heat also makes it more difficult to condense water vapor making possible a local mis-match between evaporation and precipitation (Figure 8.3). The former process occurs rapidly only if sufficient energy (sunlight) is available to maintain high surface temperatures and surface winds are sufficiently strong to carry off the evaporated vapor (maintaining the relative humidity below saturation). The latter process also occurs rapidly if vertical motions are strong even though the three-dimensional motions of the atmosphere usually produce weak surface winds where vertical motions are strong.

Third, the large latent heat of water phase changes significantly alters the energy budget that drives the atmospheric motions. Evaporation of water from the tropical ocean surface is the major cooling term that balances the heating by sunlight. Condensation of water vapor to form liquid precipitation constitutes the major source of heating for the tropical atmosphere. Thus, water transformations form the major heat exchange process between the tropical surface and atmosphere (Webster, 1994) and constitute a major source of energy for the atmospheric circulation.

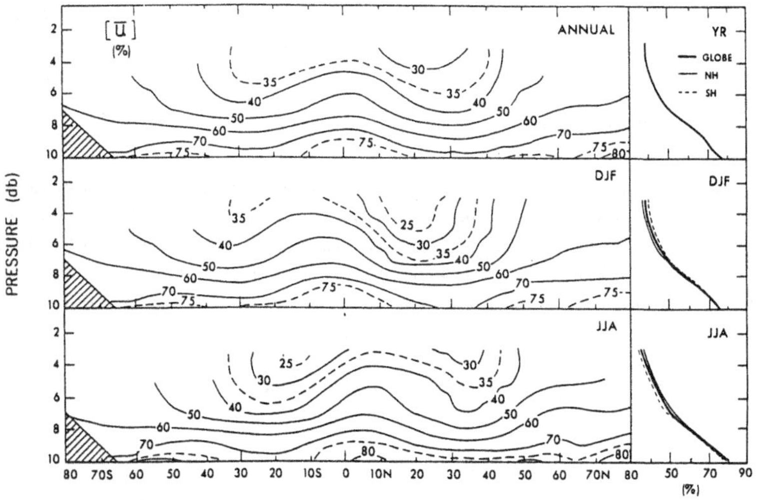

Figure 8.2: *Zonal-mean cross sections of the relative humidity in % for annual, DJF, and JJA mean conditions (from Peixoto and Oort, 1992).*

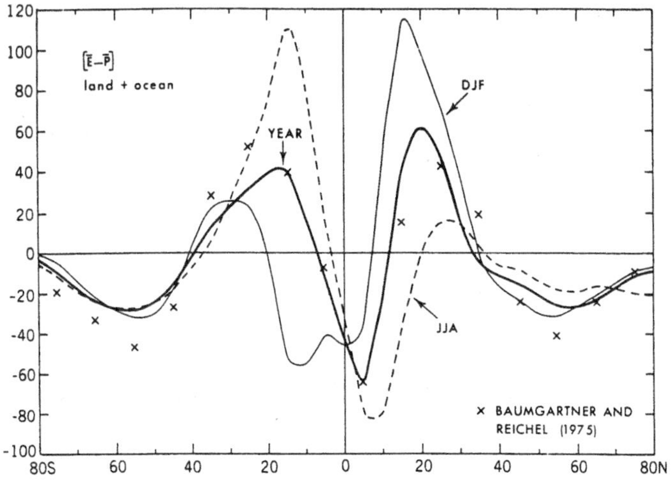

Figure 8.3: *Meridional profiles of the zonal-mean divergence of the total water vapor transport (div* **Q***) ≈ (E-P) in 0.01 m yr^{-1} for annual, DJF, and JJA mean conditions. Some annual-mean estimates of E-P by Baumgartner and Reichel (1975) are added for comparison (from Peixoto and Oort, 1992).*

Fourth, the deceptively simple water molecule produces one of the most complex interactions with radiation such that water vapor and clouds also dramatically alter the exchanges of radiation that control the energy budget. Water vapor is the major "greenhouse gas" in Earth's atmosphere producing a major negative feedback on the radiative cooling of the surface. The greenhouse effect arises because water vapor only weakly absorbs sunlight (Figure 8.4), but strongly absorbs infrared radiation at most wavelengths, except in the "window" from 8–12 μm (Figure 8.5). In the polar regions, the major sink of atmospheric energy is radiation to space which is moderated by clouds. Thus, water vapor transport into the polar regions determines the efficiency of this energy sink (cf., Serreze et al., 1995). When water vapor condenses to form clouds, two important changes occur: the absorption of infrared radiation becomes essentially uniform with wavelength, filling in the "window", and the water particles become strong reflectors of sunlight. Thus, on average, water vapor is the most significant influence on the transfer of infrared radiation and clouds are the most significant influence on the transfer of sunlight (e.g., Rossow and Zhang, 1995). Note that clouds play an equally important role in the radiation budget even though they represent only about 1% of the water in vapor form (e.g., Staelin et al., 1976; Lin and Rossow, 1996). These variations in the net radiation balance drive the atmospheric circulation.

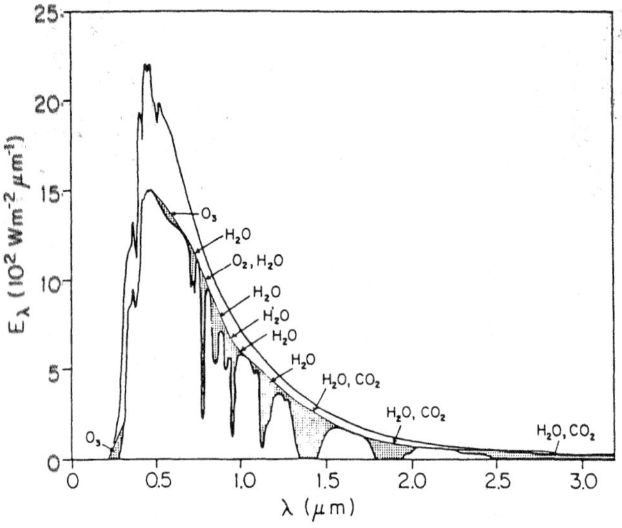

Figure 8.4: *Spectral distribution of solar irradiation at the top of the atmosphere and at sea level for average atmospheric conditions for the sun at zenith. The shaded areas represent absorption by various atmospheric gases. The unshaded area between the two curves represents the portion of the solar energy backscattered by the air, water vapor, dust, and aerosols and reflected by clouds. For the curve at the top of the atmosphere the integral* $\int E_\lambda \, d\lambda$ *1360* Wm^{-2} *represents the solar constant (adapted from Gast 1965), (from Peixoto and Oort, 1992).*

Fifth, the unusual variation of water density near the liquid-solid phase transition causes ice to float on liquid water. Because solid water reflects sunlight much more than liquid water, the surface heating is drastically reduced. However, heat diffusion in ice is also much less efficient than in liquid water, particularly when the latter is stirred by the ocean circulation. Hence the further loss of heat by the ocean surface is greatly reduced by the formation of sea ice. These

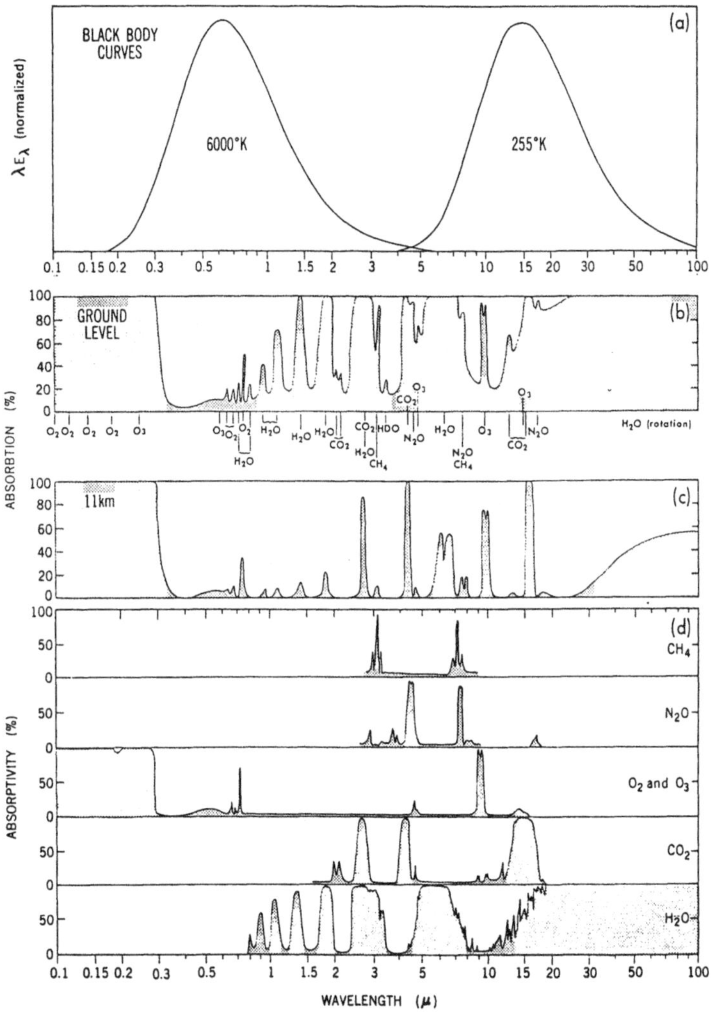

Figure 8.5: *Black body curves for the solar radiation (assumed to have a temperature of 600 K) and the terrestrial radiation (assumed to have a temperature of 255 K) (a); absorption spectra for the entire vertical extent of the atmosphere (b) and for the portion of the atmosphere above 11 km (c) after Goody (1964); and absorption spectra for the various atmospheric gases between the top of the atmosphere and the earth's surface (d) after Howard et al. (1955) (updated with data from Fels and Scharzkopf (1988), personal communication between 10 and 100 μm), (from Peixoto and Oort, 1992).*

effects make the energy balance of the polar atmosphere much more complicated (Curry and Ebert, 1992; Curry et al., 1995).

Water vapor is also involved in a number of crucial chemical cycles in our atmosphere that may control other factors important to climate: (1) sulfate chemistry is central to the formation of stratospheric aerosols (see references in Iraci et al., 1995) and about half of tropospheric aerosols (Charlson, 1993), (2) stratospheric aerosols, in turn, play a key role in stratospheric ozone loss processes (Solomon, 1988), and (3) the OH radical from the decomposition of water vapor plays key catalytic roles in the chemistry of several trace gases including ozone and methane.

8.2 Measurement Methods

Routine water vapor abundance measurements are made as part of weather observations by hygrometers at surface weather stations and in rawinsonde balloons (called RAOBS for short) that are launched a few times every day (Elliott and Gaffen, 1991). These are direct in situ measurements that are quite accurate at relatively high temperatures and vapor abundances. There are several ways to make this measurement; the most common are to measure the temperature difference between a dry and a moist, evaporating thermometer, to measure the change in size of a sensitive element caused by moisture absorption, to measure the temperature at which condensation begins, and to measure changes in electrical resistance caused by moisture absorption. The most common instruments today use the last method. These data, collected for more than 30 years at hundreds of locations, have provided most of what we know about the distribution and variations of water vapor in the atmosphere (Peixoto and Oort, 1992). However, there are three shortcomings of these observations. (1) Current instrument sensitivity significantly reduces the reliability of measurements of water vapor abundance in the upper troposphere (below temperatures of −40°C, cf. Elliott and Gaffen, 1991), particularly in the tropics (Rind et al., 1992), and in the whole polar troposphere, particularly in wintertime. (2) Geographic coverage of rawinsonde balloon launch sites is poor, particularly over oceans. (3) Most sites launch only two balloons per day at synoptic reporting times, which does not provide a longitudinally consistent nor complete representation of diurnal variations in water vapor.

The advent of weather satellites in the 1960's–1970's offered the possibility of making remote sensing water vapor measurements that are more complete than obtained from surface weather stations; however, this opportunity has not generally been exploited. Routine water vapor abundance profiles have been produced since 1979 from multi-wavelength infrared measurements by the High Resolution Infrared Radiation Sounder (HIRS) flown on the polar orbiting weather satellites operated by the National Oceanic and Atmospheric Administration (NOAA) in the United States (Smith and Woolf, 1976; Smith et al., 1979). Although providing global coverage (with somewhat reduced vertical resolution), these measurements are no more reliable in the upper tropical and polar troposphere than are the RAOBS: no results are reported at pressures below about 300 mb. Even though NOAA has generally operated two polar orbiters, the analyzed dataset does not provide measurements more frequently than about once per day. Hence, satellite water vapor measurements have only improved geographic coverage. A re-analysis of this dataset could provide more information about diurnal variations of temperature and water vapor.

Water vapor abundances can be determined from radiation measurements by determining the amount of absorption by water vapor at wavelengths selected to be relatively free of other effects. Sunphotometers on the surface, which point directly at the sun, can determine the total amount of water vapor in the column of atmosphere by comparing the strength of radiation in and out of absorption bands in the water vapor spectrum (Figure 8.4). This technique is not generally used from satellites because the surface absorbs most of the sunlight inside and outside the water vapor bands, greatly reducing the difference in radiation with wavelength. At infrared wavelengths, water vapor profiles can be inferred from the amount of emitted radiation observed at wavelengths where the absorption strength varies from weak to strong

(Figure 8.5), if the atmospheric temperature profile is known (e.g., Stephens, 1994). A similar technique can be applied at microwave wavelengths (e.g. Janssen, 1993). The infrared and microwave methods can be used both from the surface and from satellites. Infrared-based determinations of water vapor are limited by the occurrence of clouds that block the emissions from lower levels; hence, the available observations represent "clear" conditions (e.g. Wittmeyer and Vonder Haar, 1994), although the differences of total water vapor abundances between clear and cloudy conditions are not too large (Gaffen and Elliott, 1993). Note that, although the operational analysis of the infrared measurements from HIRS produces an estimate of water vapor abundance profiles, the analysis procedure does not actually make use of direct infrared radiation measurements as described here. Rather the method relates water vapor amounts to temperatures by an empirical relationship developed from comparisons to RAOBS (Smith and Woolf, 1976). Microwave observations available from current satellites limit retrievals to a single wavelength which may be affected by the presence of precipitating systems; hence only the total column amount of water vapor can be estimated (e.g. Tjemkes et al., 1991). Strong effects of land surface emissions have limited microwave studies to ocean areas outside of sea ice-covered regions. Within the past year a new multi-wavelength microwave instrument (called SSM/T2) has been flown, the first of an operational series that retrieve water vapor abundance profiles in the lower troposphere (Wilheit, 1990). Comparisons of monthly mean maps of total column water vapor amounts from RAOBS, HIRS and retrieved from the Special Sensor Microwave/Imager (SSM/I) flown on military weather satellites shows agreement to within about 20% (Stephens et al., 1994; Wittmeyer and Vonder Haar, 1994).

Several experimental water vapor measurement techniques have been studied though none has entered regular use. From the surface, use of a Raman lidar technique shows promise of providing very high vertical and time resolution observations of the vertical distribution of water vapor over individual surface sites (e.g. Melfi et al., 1989). To measure from space the very low water vapor amounts in the stratosphere (about 10^{-3} to 10^{-4} the amount of water near the surface), both the infrared and microwave techniques have been employed but with the instrument viewing Earth's atmosphere at the limb instead of pointed towards the surface: SAGE II used near-infrared measurements (Rind et al., 1992) and the Microwave Limb Scanner on the Upper Atmosphere Research Satellite uses microwave measurements (Barath et al., 1993). Because of a lack of accurate water vapor measurements in the upper troposphere, there have been several studies of the possible utility of radiation measurements at 6.7 μm wavelength from satellites to infer the total water vapor amount in the upper atmosphere (van de Berg et al., 1993; Wu et al., 1993,; Soden and Bretherton, 1994; Schmetz et al., 1995; Stephens et al., 1995; Bates et al., 1995).

8.3 Current Research Challenges

Accurate satellite determinations of sea surface temperatures and other quantities using radiation measured at infrared wavelengths are still hindered by uncertainties in the absorption by water vapor, particularly the so-called continuum absorption (cf. Figure 8.5). Retrieval of water vapor abundances at higher altitudes and latitudes are also affected by uncertainties in the absorption line spectrum of water vapor (cf. Stephens et al., 1995). Moreover, these uncertainties limit the accuracy of determining the precise effect of water vapor changes on the radiation balance in the atmosphere. There are some empirical models (e.g. Clough, 1993) that fit available laboratory observations reasonably well (though these laboratory measurements do not cover enough wavelength and temperature range and are particularly lacking at the temperatures encountered in the atmosphere) and, at least, two theories for the physical process that creates the continuum absorption. Recent work (Ma and Tipping, 1991; 1992; 1994) provides a revised theory of absorption line shapes that appears to explain the continuum absorption and its temperature dependence over both the infrared and microwave portions of the spectrum. Work remains to confirm the accuracy of this theory and to determine whether it can explain the continuum absorption at all wavelengths.

Quantifying the magnitude of vertical water vapor transport by different types of atmospheric motions is important for three reasons: (1) despite the much lower absolute abundance of water vapor in the upper troposphere, its effect on the surface radiation balance (the greenhouse effect) is almost as large as that of water vapor near the surface (Hansen et al., 1984; Arking, 1995), (2) the upper tropospheric humidity controls the occurrence of upper level cloudiness, particularly cirrus, that can have significant effects on the climate (e.g. Randall et al., 1984; Slingo and Slingo, 1988), and (3) transport of water out of the tropic oceans and into land areas and higher latitudes depends on the amount of water vapor injected into the upper tropical troposphere from the ocean surface (e.g. Del Genio et al., 1991). Uncertainties surrounding the first issue led to some controversy about the sign of water vapor feedback on climate change (Lindzen, 1980; Betts, 1990; Del Genio et al., 1994 and references therein). Although current observations suggest that seasonal warming of the upper troposphere in midlatitudes is accompanied by an increase in water vapor abundance as expected (e.g. Rind et al., 1991); nevertheless, the complexity of the water vapor budget even in models (Figure 8.6) suggests the need for much better observations of the space and time variations of water vapor in the upper troposphere to diagnose the processes affecting its distribution (Del Genio et al., 1994). In particular, observations are needed of water vapor vertical profiles extending into the lower stratosphere in both clear and cloudy conditions.

Analysis of RAOBS measurements of water vapor have already provided a lot of detail about the global and even vertical distribution of water vapor and its variations and horizontal transports (Peixoto and Oort, 1992). To motivate the need for better observations, estimates of the variations of water vapor abundance from RAOBS can be used to show the effects of current measurement uncertainties. The largest variations of column water vapor abundance are with geographic location and season: Figure 8.7 shows the estimated variations. From the warmer equator to the colder poles, the annual mean water vapor abundance decreases by more than a factor of ten; however, in many locations the amplitude of the seasonal variations is as large as the annual mean value. Estimated uncertainties in even the annual mean values (e.g. Elliott and Gaffen, 1991) are about 10-20% of the mean values, which is smaller than the average seasonal variations. The relationship of seasonal water vapor and temperature variations demonstrates that the water vapor abundance is significantly altered by the large scale circulation (Stephens, 1990; Gaffen et al., 1992), particularly in the tropics, and cannot be treated as a simple function of temperature, as is still done in many analyses.

Although the seasonal variations of water vapor are probably reliably measured, estimated variations of water vapor from day to day and with longitude (Figure 8.8) are estimated to be only about 20–25% and 10% of the mean value, respectively (Peixoto and Oort, 1992), both comparable to the possible measurement uncertainties. This, in turn, makes it difficult to determine the magnitude of water vapor transports by wave motions in the atmosphere. Since the mean winds and water vapor abundances are well determined, the average flux of water vapor from east to west and from south to north can be determined (Figures 8.9a and 8.10a). The significant features of the east-to-west water vapor transports by the mean winds highlight the general westerly flows at midlatitudes and easterly flows at low latitudes and show the largest fluxes occur over the central ocean basins from the warmer western ocean currents to the colder eastern currents (Figure 8.9a). The blocking action of the continents, especially by high mountain ranges is readily apparent. A significant intraseasonal variation of westward water vapor flux at low latitudes is associated with the seasonal monsoons of Asia (Chen et

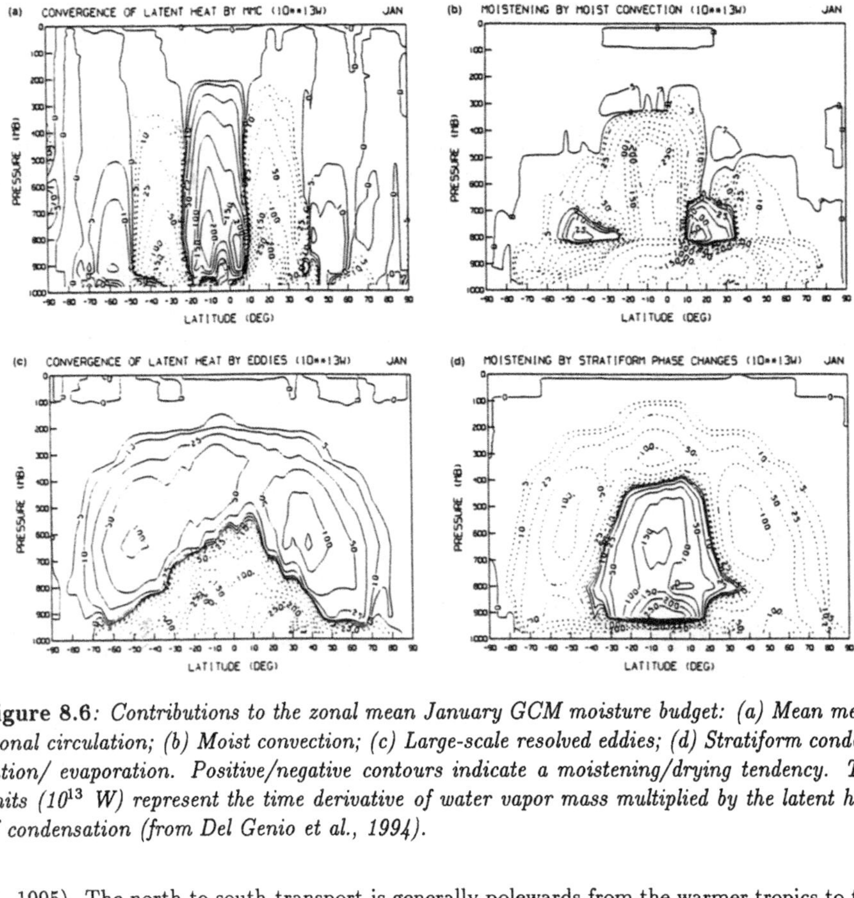

Figure 8.6: *Contributions to the zonal mean January GCM moisture budget: (a) Mean meridional circulation; (b) Moist convection; (c) Large-scale resolved eddies; (d) Stratiform condensation/ evaporation. Positive/negative contours indicate a moistening/drying tendency. The units (10^{13} W) represent the time derivative of water vapor mass multiplied by the latent heat of condensation (from Del Genio et al., 1994).*

al., 1995). The north-to-south transport is generally polewards from the warmer tropics to the colder polar regions with a notable exception in the Indian Ocean where the mean transport is from the ocean south of the equator onto the Asian continent (Figure 8.10a). Whereas the wave motions only produce east-to- west transports that are about 10% of that produced by the mean winds (Figure 8.9b), they produce comparable or larger north-to-south fluxes as the mean winds, particularly at midlatitudes (Figure 8.10b). An analysis by Chen et al. (1995) shows an annual cycle in water flux from the winter hemisphere to the summer hemisphere consistent with larger precipitation amounts in the summer hemisphere. However, the accuracy of the wave transports is probably very low, particularly over oceans where the major water fluxes appear and there are very few RAOBS available.

The need for better space-time resolution in water vapor measurements, particularly over oceans and polar regions is motivated by three subtle effects of these transports on the climate. (1) If there were no topography or large longitudinal extents covered by land, there would probably be no net flux of water vapor in the east-west direction on average. However, the mountain ranges force the formation of precipitation, thereby removing water vapor from the zonal flow and forming barriers to east-west water vapor transport. The consequent river runoff from these

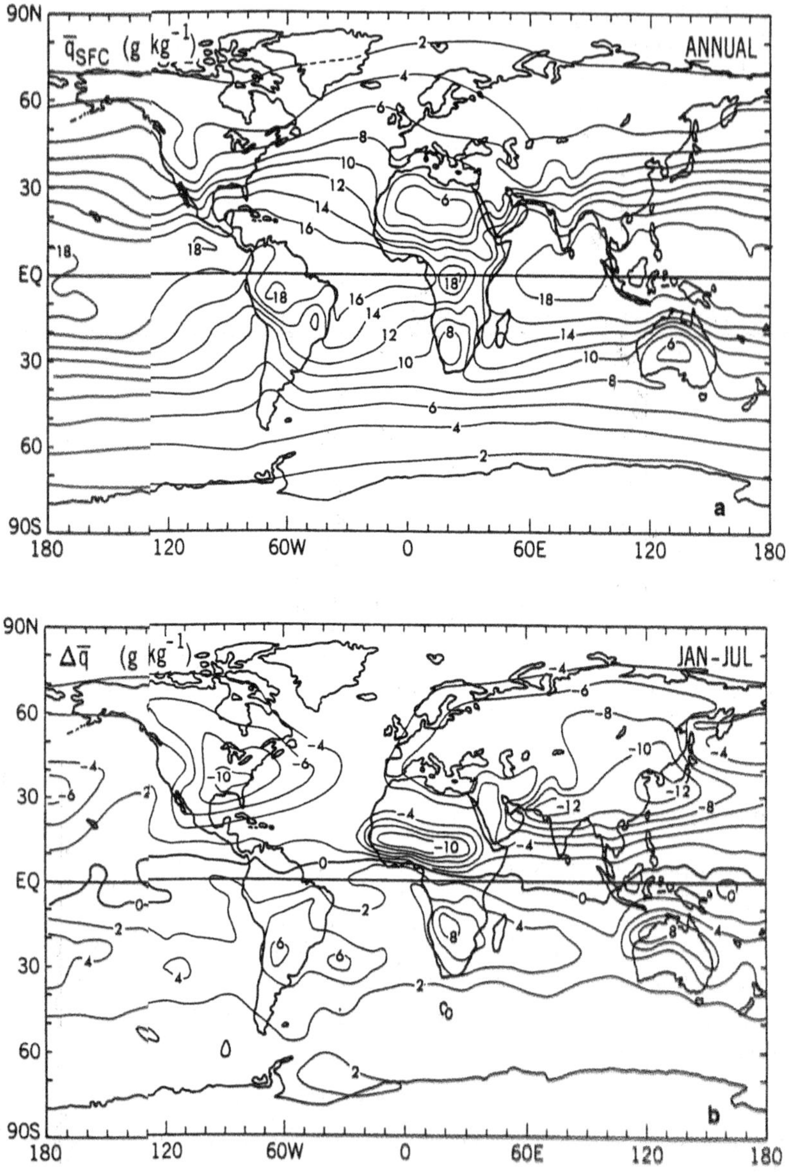

Figure 8.7: *Global distribution of the surface specific humidity for annual-mean (a) and January-minus-July conditions (b), of the vertical-mean specific humidity (from Peixoto and Oort, 1992).*

mountains may or may not return this water to the ocean basin where it was evaporated. One significant consequence of this process is that there are rapid net transports of water vapor from one ocean basin to another, though they are hard to estimate accurately with available data,

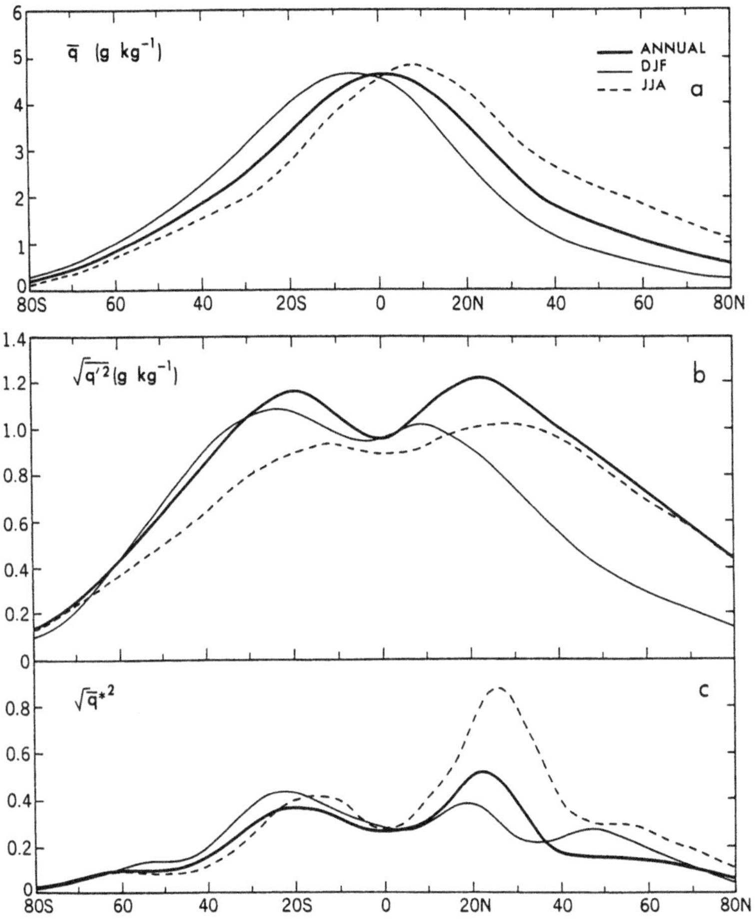

Figure 8.8: *Meridional profiles of the vertical- and zonal-mean values of the time-mean specific humidity (a), the day-to-day standard deviation of the specific humidity (b), and the east-west standard deviation of the time-mean specific humidity (c) in g kg*$^{-1}$ *for annual, DJF, and JJA mean conditions (from Peixoto and Oort, 1992).*

that produce differences in salinity that affect the organization of the global ocean circulation system (Zaucker et al., 1994). (2) The net vapor transport to polar ice sheets determines whether water is stored in them or released from them. This balance leads to significant changes in sea level and possibly climate, but current day data are not accurate enough to estimate this net flux. (3) Finally, the most important net flux of water vapor is to the land where some of it falls as precipitation to provide all of the water that is used by humans, directly and indirectly. This net flux represents only about 1% of the atmospheric water vapor per day, so an accurate estimate with available data is difficult. Because the RAOBS coverage of many continents is quite good, we have more accurate estimates of this flux for North America and Europe; but estimates for South America and Asia are very poor. Moreover, we lack sufficient

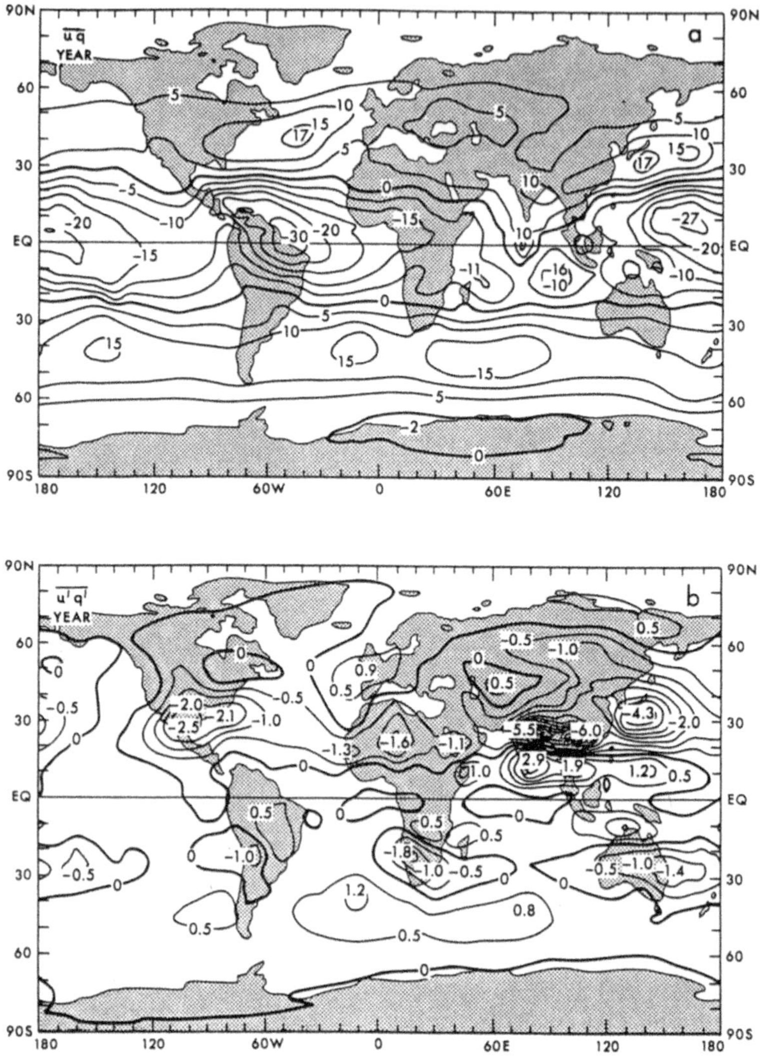

Figure 8.9: *Global distribution of the vertical-mean water vapor transport in the zonal direction by all motions (a) and transient eddies (b) in m* s^{-1} *g* kg^{-1} *for annual-mean conditions. Positive values indicate eastward transport of moisture. The vertically integrated zonal transports in units of 10 kg* m^{-1} s^{-1} *can be obtained by multiplying the fields by* p_0 / g *(≈ 1 over oceans and low-level land) (from Peixoto and Oort, 1992).*

data to separate the different dynamical processes contributing to this important net flux and, therefore, cannot test our climate prediction models adequately.

Although there are improvements in water vapor measurements becoming available, none of these satellite systems does a thorough job of measuring vertical profiles of water vapor from the

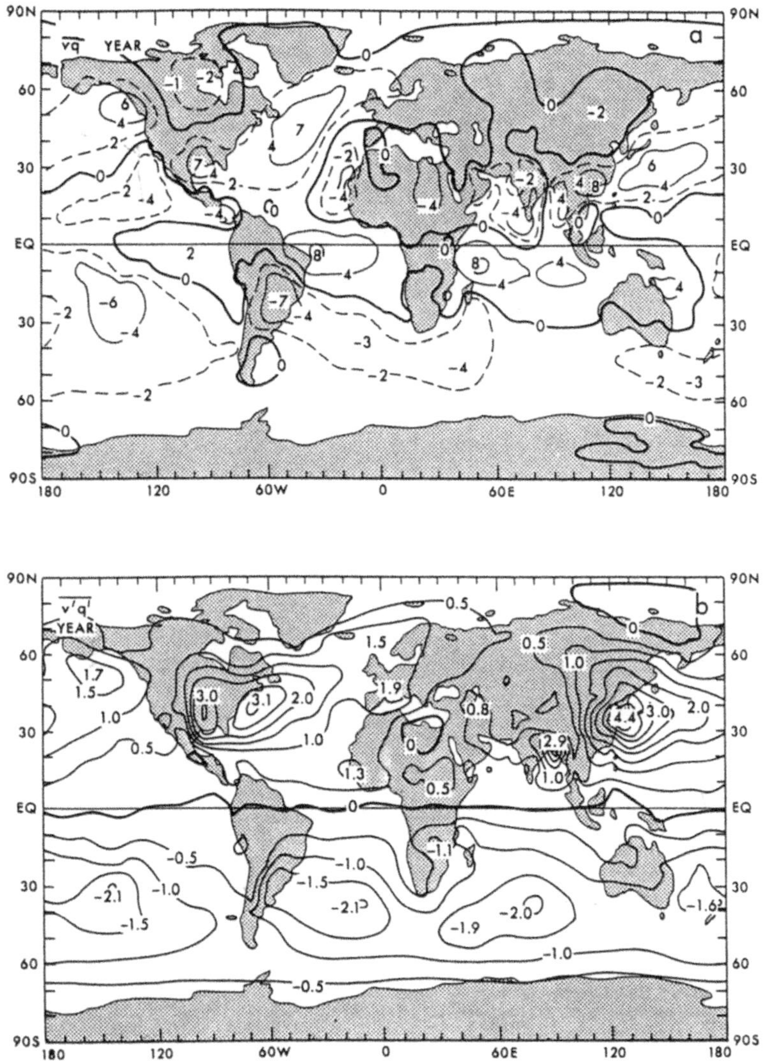

Figure 8.10: *Global distribution of the vertical-mean meridional (northward) transport of water vapor by all motions (a) and transient eddies (b) in m s^{-1} g kg^{-1} for annual-mean conditions. To convert to ϕ in units of 10 kg m^{-1} s^{-1}, multiply by p_0/g ($\approx$ 1 over oceans and low-level land) (from Peixoto and Oort, 1992).*

surface up into the lower stratosphere over both ocean and land in clear and cloudy conditions with enough time and space resolution to separate the various dynamic and cloud processes that affect the transport and abundance of atmospheric water vapor. Improvements could be made with instruments measuring more complete infrared and microwave spectra. The limitations of other instruments arise both because the emission spectrum from water vapor is complicated and temperature dependent and because other factors, such as the surface and clouds, must be removed from the satellite measurements to isolate the water vapor variations in detail. Figure 8.11 illustrates in some detail the water vapor spectrum in the infrared and microwave wavelength regions. The large number of absorption lines with widely varying strengths and the continuum absorption in these spectra, if known accurate from theoretical calculations and laboratory measurements and measured together, would provide a much more detailed description of the complex variations of water vapor in our atmosphere.

8.4 References

Arking A (1996) Water vapor and lapse rate feedback: Insight from a one-dimensional climate model. *J Climate*, in press

Barath FT, Chavez MC, Cofield RE, Flower DA, Frerking MA, Gram MB, Harris WM, Holden JR, Jarnot RF, Kloezeman WG, Klose GJ, Lau GK, Loo MS, Maddison BJ, Mattauch RJ, McKinney RP, Peckham GE, Pickett HM, Siebes G, Soltis FS, Suttie RA, Tarsala JA, Waters JW, Wilson WJ (1993) The Upper Atmosphere Research Satellite Microwave Limb Sounder instrument. *J Geophys Res* 98: 10.751–10.762

Bates JJ, Wu X, Jackson DL (1996) Interannual variability of upper troposphere water vapor band brightness temperature. *J Climate*, in press

Betts AK (1990) Greenhouse warming and the tropical water vapor budget. *Bull Amer Meteor Soc* 71: 1465–1467

Bony S, Duvel J-P (1994) Influence of the vertical structure of the atmosphere on the seasonal variation of precipitable water and greenhouse effect. *J Geophys Res* 99: 12.963–12.980

Charlson RJ (1993) Gas-to-particle conversion and CCN production. In *Dimethylsuphide, Oceans, Atmosphere and Climate.* G. Restelli and G. Angeletti, eds., Kluwer Academic, Norwell, Mass

Chen T-C, Chen J-M, Pfaendtner J (1995) Low-frequency variations in the atmospheric branch of the global hydrological cycle. *J Climate* 8: 92–107

Clough SA (1993) Radiative transfer model development in support of the atmospheric radiation measurement program. *Proc. Third ARM Science Team Meeting*, Norman, OK, 11–17

Curry JA, Ebert EE (1992) Annual cycle of radiation fluxes over the Arctic Ocean: Sensitivity to cloud optical properties. *J Climate,* 5: 1267–1280

Curry JA, Schramm JL, Ebert EE (1995) Sea-ice albedo climate feedback mechanism. *J Climate* 8: 240–247

Elliott WP, Gaffen DJ (1991) On the utility of radiosonde humidity archives for climate studies. *Bull Amer Meteor Soc* 72: 1507–1520

Del Genio AD, Lacis AA, Reudy RA (1991) Simulations of the effect of a warmer climate on atmospheric humidity. *Nature* 351: 382–385

Del Genio AD, Kovari W, Yao M-S (1994) Climatic implications of the seasonal variation of upper troposphere water vapor. *Geophys Res Lett* 21: 2701–2704

Gaffen DJ, Elliott WP (1993) Column water vapor content in clear and cloudy skies. *J Climate* 6: 2278–2287

Gaffen DJ, Elliott WP, Robock A (1992) Relationships between tropospheric water vapor and surface temperature as observed by radiosondes. *Geophys Res Lett* 19: 1839–1842

Gaffen DJ, Robock A, Elliott WP (199?) Annual cycles of tropospheric water vapor. *J Geophys Res* 97: 18.185–18.193

Hansen J, Lacis A, Rind D, Russell G, Stone P, Fung I, Ruedy R, Lerner J (1984) Analysis of feedback mechanisms. In *Climate Processes and Climate Sensitivity*, Maurice Ewing Ser., Vol. 5, Geophys. Monogr. 29, (J.E. Hansen and T. Takahashi, eds.), AGU, Washington, D.C., 130–163

Iraci LT, Middlebrook AM, Tolbert MA (1995) Laboratory studies of the formation of polar stratospheric clouds: Nitric acid condensation on thin sulfuric acid films. *J Geophys Res* 100: 20.969–20.977

Jackson DL, Stephens GL (1995) A study of SSM/I-derived columnar water vapor over the global oceans. *J Climate* 8: 2025–2038

Janssen MA (1993) *Atmospheric Remote Sensing by Microwave Radiometry.* Wiley, New York, 572 pp

Lin B, Rossow WB (1996) Cloud water path observations over oceans using ISCCP and SSM/I data. *J Climate,* in press

Lindzen RS (1990) Some coolness concerning global warming. *Bull Amer Meteor Soc* 71: 288–299

Ma Q, Tipping RH (1991) A far wing line shape theory and its application to the water continuum absorption in the infrared region, (I). *J Chem Phys* 95: 6290–6301

Ma Q, Tipping RH (1992) A far wing line shape theory and its application to the foreign-broadened water continuum absorption (III). *J Chem Phys* 97: 818–828

Ma Q, Tipping RH (1994) The detailed balance requirement and general empirical formalisms for continuum absorption. *J Quant Spectrosc Radiative Trans* 51: 751–757

Melfi SH, Whiteman D, Ferrare R (1989) Observation of atmospheric fronts using Raman lidar moisture measurements. *J Appl Meteor* 28: 789–806

Peixoto JP, Oort AH (1992) *Physics of Climate.* American Institute of Physics, New York, 520 pp

Randall D, Harshvardhan, Dazlich DA , Corsetti TC (1984) Interactions among radiation, convection and large-scale dynamics in a general circulation model. *J Atmos Sci* 46: 1943–1970

Rind D, Chiou EW, Chu W, Oltmans S, Lerner J, Larsen J, McCormick MP, McMaster L (1992) Overview of the Stratospheric Aerosol and Gas Experiment II water vapor observations: Method, validation and data characteristics. *J Geophys Res* 98: 4835–4856

Rossow WB, Zhang Y-C (1995) Calculation of surface and top-of-atmosphere radiative fluxes from physical quantities based on ISCCP datasets, Part II: Validation and first results. *J Geophys Res* 100: 1167–1197

Schmetz J, Menzel WP, Velden C, Wu X, van de Berg L, Nieman S, Hayden C, Holmlund K, Geijo C (1995) Monthly mean large-scale analyses of upper tropospheric humidity and wind field divergence from three geostationary satellites. *Bull Amer Meteor Soc* 76: 1578–1584

Serreze MC, Barry RG, Welsh JE (1995) Atmospheric water vapor characteristics of 70 N. *J Climate* 8: 719–731

Slingo A, Slingo JM (1988) The response of a general circulation model to cloud longwave radiative forcing, 1, Introduction and initial experiment. *Q J Roy Meteor Soc* 112: 1027–1062

Smith WL, Woolf HM (1976) The use of statistical covariance matrices for interpreting satellite sounding radiometer observations. *J Atmos Sci* 33: 1127–1140

Smith WL, Woolf HM, Hayden CM, Wark DQ, McMillin LM (1979) The TIROS-N Operational Vertical Sounder. *Bull Amer Meteor Soc* 60: 117–118

Soden BJ, Bretherton FP (1994) Evaluation of water vapor distribution in general circulation models using satellite observations. *J Geophys Res* 99: 1184–1210

Solomon S (1988) The mystery of the Antarctic ozone "hole." *Rev Geophys* 26: 131–148

Staelin DH, Kunzi KF, Pettyjohn RL, Poon RKL, Wilcox RW, Waters JW (1976) Remote sensing of atmospheric water vapor and liquid water with the Nimbus 5 microwave spectrometer. *J Appl Meteor* 15: 1204–1214

Stephens GL (1990) On the relationship between water vapor over oceans and sea surface temperature. *J Climate* 3: 634–645

Stephens GL (1994) *Remote Sensing of the Lower Atmosphere.* Oxford University, New York, 523 pp

Stephens GL, Jackson DL, Bates JJ (1994) A comparison of SSM/I and TOVS column water vapor data over the global oceans. *Meteorol Atmos Phys* 54: 183–201

Stephens GL, Jackson DL, Wittmeyer I (1996) Global observations of upper- tropospheric water vapor derived from TOVS radiance data. *J Climate* 9, in press

Tjemkes SA, Stephens GL, Jackson DL (1991) Space-borne observations of columnar water vapor: SSMI observations and algorithm. *J Geophys Res* 96: 10.941–10.954

van de Berg L, Pyomjamsri A, Schmetz J (1993) Monthly mean upper tropospheric humidities in cloud-free areas from Meteosat observations. *Int J Climatol* 11: 819–826

Wilheit TT (1990) An algorithm for retrieving water vapor profiles in clear and cloudy atmospheres from 183 GHz radiometric measurements: Simulation studies. *J Appl Meteor* 29: 508–515

Wittmeyer IL, Vonder Haar TH (1994) Analysis of the global ISCCP TOVS water vapor climatology. *J Climate* 7: 325–333

Wu X, Bates JJ, Khalsa SJS (1993) A climatology of the water vapor band brightness temperatures from NOAA operational satellites. *J Climate* 6: 1282–1300

Zaucker F, Stocker TF, Broecker WS (1994) Atmospheric freshwater fluxes and their effect on the global thermohaline circulation. *J Geophys Res* 99: 12.443–12.457

General Reading

Goody RM, Yung YL (1989) *Atmospheric Radiation, Theoretical Basis.* Oxford, New York, 519 p

Janssen MA (1993) *Atmospheric Remote Sensing by Microwave Radiometry.* Wiley, New York, 572 pp

Peixoto JP, Oort AH (1992) *Physics of Climate.* American Institute of Physics, New York, 520 pp

Stephens GL (1994) *Remote Sensing of the Lower Atmosphere.* Oxford University, New York, 523 pp

Webster PJ (1994) The role of hydrological processes in ocean-atmosphere interactions. *Rev Geophys* 32: 427–476

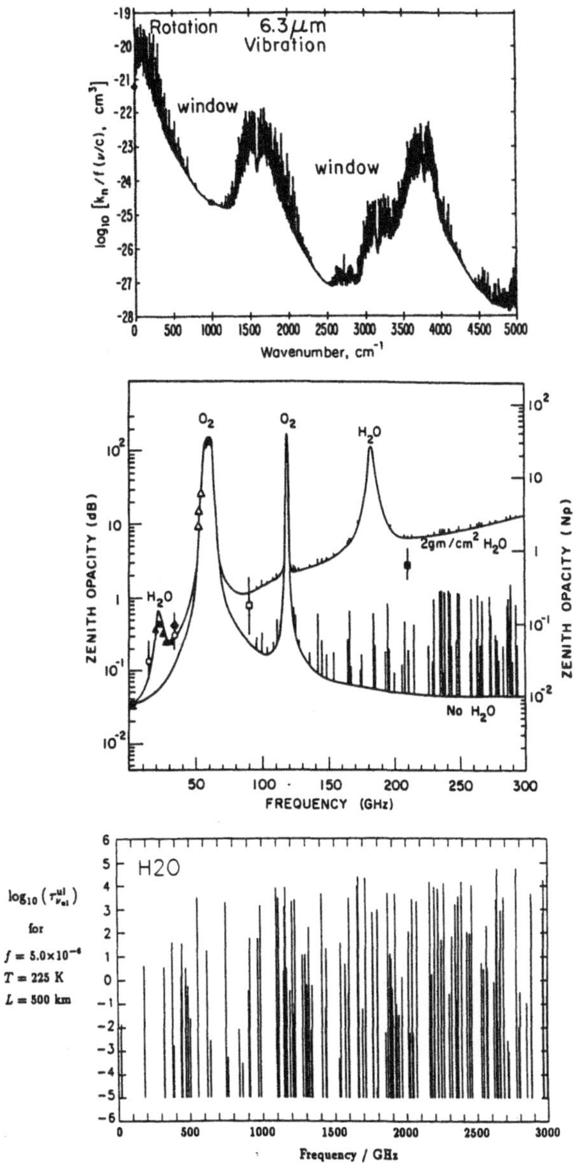

Figure 8.11: *(a) Theoretical absorption coefficients for pure water vapor at 1 bar and at 296 K. The vertical axis is proportional to the absorption coefficient divided by a factor that is approximately $\tilde{\nu}/c$ (modified from Clough et al., 1980), (from Stephens, 1994). (b) Background microwave opacities and for a wet (H_2O = 2 gm/cm^3) and completely dry atmosphere. The background transmission of the atmosphere is defined by microwave absorption by O_2, H_2O, and stratospheric O_3 (reprinted from Waters 22), (from Janssen, 1993). (c) Microwave spectrum of H_2O, (from Janssen, 1993).*

Chapter 9

Properties of Clouds and Cloud Systems

William B. Rossow
NASA Goddard Institute for Space Studies
New York, NY 10025
USA

9.1 Introduction

Observations of clouds from the surface, aircraft and spacecraft give the impression of great complexity and variability. Quantitative descriptions of the variety of clouds and determination of its causes have been limited by lack of comprehensive and systematic observations that cover the full range of space and time scales over which clouds vary. Now, there are two cloud datasets that cover scales from 30 km and 3 hr up to the whole globe for more than a decade, giving us, for the first time, a truly global view of clouds and their variability. One dataset provides information on cloud cover, cloud base heights and cloud morphology from several decades of surface weather observations (Warren et al., 1986, 1988; Hahn et al., 1995). These observations can be supplemented by balloon measurements of the profiles of atmospheric temperature and humidity that can provide information about cloud layer structure (Poore et al., 1995; Wang and Rossow, 1995). Much more detailed cloud observations of global cloudiness are possible using satellite remote sensing. A comprehensive survey of cloud cover, cloud top temperature/pressure and visible optical thickness has been produced from more than a decade of observations from the international constellation of weather satellites (Schiffer and Rossow, 1983; Rossow and Schiffer, 1991). Newer satellite instruments and analysis techniques extend the list of cloud properties that can be measured by satellites. Combining all of these results provides the best available survey of cloud properties and also indicates what satellite instruments should be included in a future cloud observing system.

9.2 Remote Sensing of Clouds

Satellites measure radiation, so a physical model of the interaction of radiation in cloudy atmospheres is needed to retrieve cloud properties from the satellite measurements (e.g. Goody

NATO ASI Series, Vol. I 45
Radiation and Water in the Climate System:
Remote Measurements
Edited by Ehrhard Raschke

and Yung, 1989; Stephens, 1994). The type of radiation measured determines the kind of radiative transfer model. The intensity of reflected sunlight (0.30–4.0 μm wavelengths) can be measured, as well as its variation with wavelength or, since the source is uni-directional, the amount and direction of polarization can be determined. The intensity of thermal emission in the infrared (4.0–200 μm wavelengths) or microwave (1 mm–10 cm wavelengths) can be measured as well as its variations with wavelength; microwave polarization is also a useful indication of scattering processes. All of these approaches depend on natural sources of radiation, but satellite instruments can also emit coherent forms of radiation and measure their interaction with clouds: lidars emit radiation in the wavelength ranges of sunlight and infrared and radars emit microwave radiation. These types of radiation are much easier to separate from natural radiation because they are emitted at sharply defined wavelengths and polarizations. Table 9.1 lists various properties of clouds that will be discussed in later sections and remote sensing techniques that have been developed to measure them.

Optical thickness	Intensity of reflected sunlight, Spectrum of emitted IR, Intensity of reflected lidar or radar, Intensity of emitted microwave
Particle size	Spectrum and/or Polarization of reflected sunlight, Spectrum of emitted IR, Spectrum and/or Polarization of scattered microwave
Particle shape	Angular distribution of Intensity and/or Polarization of reflected sunlight or lidar
Particle size variance	Polarization of reflected sunlight
Number density	Derived from optical thickness and particle size
Liquid water path	Derived from optical thickness (and particle size)
Ice water path	Derived from optical thickness (and particle size)
Precipitation	Intensity and/or polarization of emitted microwave, Intensity of reflected radar
Cloud top location	Spectrum and/or Polarization of reflected sunlight, Solar extinction, Intensity and/or spectrum of emitted IR, Timing of returned lidar or radar pulse
Cloud base location	Spectrum of emitted IR, Timing of returned lidar or radar pulse

Table 9.1: *Satellite remote sensing techniques used to determine cloud properties.*

The most extensively used method to measure cloud optical thickness from satellites employs measurements of the intensity of reflected sunlight (Rossow et al., 1989; Rossow and Schiffer, 1991). This method is less sensitive at low optical thicknesses (< 1), but still effective at very large values (> 60). Although the spectrum of emitted IR can be used to determine lower optical thicknesses (< 10) (Carlson et al., 1993), this technique has not been used for Earth observations. A simpler approach uses IR intensities measured from satellites at a few discrete wavelengths to estimate lower optical thicknesses from transmitted thermal emission (Wylie and Menzel, 1989; Wylie et al., 1994). The results from all of these techniques depend on particle size, whereas measurements of thermal emission in the microwave can be used to measure optical thickness (or water path) for liquid water clouds over oceans independently of particle size. Microwave techniques are insensitive to low cloud optical thicknesses (< 7) but are more sensitive at very large values (> 100). Several limited satellite microwave analyses have been completed (Greenwald et al., 1993; Liu and Curry, 1992, 1993; Lin and Rossow, 1994). Ice crystals in clouds are too small to affect microwave emissions; but combining reflected sunlight (or IR spectra) with microwave can be used to estimate ice water path from satellites (Lin and Rossow, 1994). Lidar and millimeter radar scattering can also be used to estimate ice water path (Sassen et al., 1989; Kropfli, 1995): lidar is most sensitive to very low optical thicknesses (< 0.5), whereas radar remains useful up to very large values (> 60). These techniques are more sensitive to particle sizes, however. Radar has not been used on satellites yet; one lidar has flown on the Shuttle (McCormick et al., 1995).

Cloud particle sizes (dimensions in the range from 6–60 μm) have been determined for liquid water clouds on Earth from satellite measurements of the wavelength dependence of reflected sunlight (Han et al., 1994). IR spectra can also be used, especially for the larger size particles occurring in ice clouds (Carlson et al., 1993), but this has not been done for Earth observations. Size information for precipitation-sized particles (dimensions > 500 μm) can be determined from centimeter radar scattering, but this has not been done from satellites yet (an experimental mission is planned for 1997: the Tropical Rainfall Measuring Mission).

Retrieval of particle sizes is especially sensitive to particle shape when using observations of reflected sunlight; hence, measurements of the angular distribution of reflected intensity and polarization can be used, in principle, to infer particle shape but this technique has not been tried before.

There are many different methods for determining cloud top location from satellite measurements. Two that have not been used for Earth observations but have been used on other planets are measuring the strength of gas absorption features, giving the amount of gas over the cloud, and measuring the amount of Rayleigh scattering from the polarization of reflected sunlight, giving the pressure at cloud top (Travis et al., 1978). Timing the return pulses from lidar and radar gives a more direct estimate of cloud top height, but only the former has recently been attempted from space (LITE mission on Shuttle). Two more extensively used methods are determining heights at which sunlight is extinguished when the Earth's limb occults the sun (Woodbury and McCormick, 1986; Kent et al., 1993; Wang et al., 1995) and determining cloud top temperature from IR emissions (Rossow et al., 1989; Wylie and Menzel, 1989; Wylie et al., 1994).

Cloud base location can sometimes be determined from the spectrum of IR emission, if cloud optical thickness is low enough (Minnis et al., 1995). Also, lidar can determine cloud base if it can penetrate the whole cloud layer (Sassen et al., 1989). Millimeter radar can penetrate much thicker clouds and would provide a more general survey of cloud vertical structure, but such an instrument has never been flown on a satellite.

9.3 Summary of Observed Cloud Properties

Clouds exhibit a very wide range of properties from the thinnest wisps of ice cirrus clouds at 15–20 km altitude or small clumps of liquid cumulus clouds near the surface to the violent storm clouds that extend throughout the whole depth of the troposphere and produce heavy rainfall. Cirrus have very little water content, occupy thin layers, and may have patchy horizontal coverage. Storm cloud systems have nearly the maximum possible water content, occupy multiple or very deep layers, and can cover areas more than 1000 km across. Table 9.2 lists cloud properties in order of decreasing significance in determining microphysical processes in clouds, including precipitation, and in determining radiative transfer in clouds. Figures 9.1–9.4 summarize the distributions of some of these properties obtained from extensive surveys of weather observations and satellite remote sensing.

	Microphysics Rank	Radiation Rank
Cloud cover	1	1
Liquid/Ice water path	2	2
Layer thickness	2	4
Particle size	3	5
Vertical structure	4	4
Particle size variance	5	6
Top/base height/temperature	6	3
Variance of water content	7	5
Particle shape	8	6

Table 9.2: *Cloud properties in order of decreasing significance for microphysical (first number) and radiative transfer (second number) processes.*

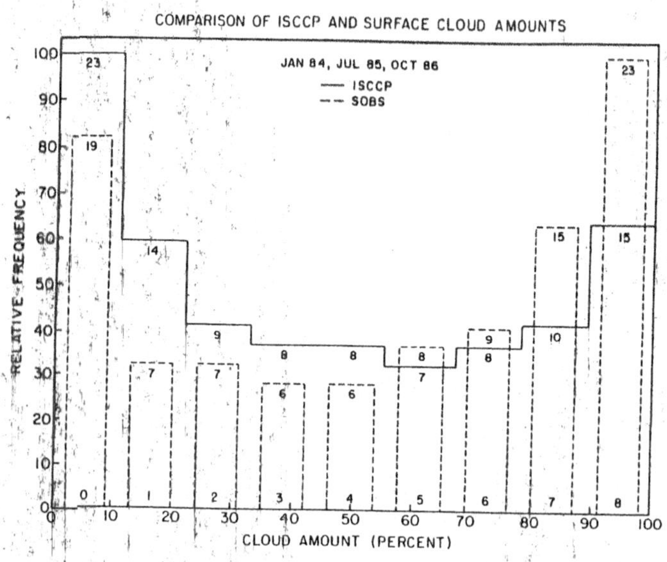

Figure 9.1: *Frequency distributions of cloud amount observed from satellites (solid line) and from the surface (dashed line) (from Rossow et al., 1993).*

Cloud amount, even at the relatively large spatial scale of the satellite results (280 km) is still dominated by either complete absence or complete cover (Rossow et al., 1993), producing a bi-modal distribution in Figure 9.1. Most of the other cloud property distributions have a characteristic shape: most clouds have properties in a relatively narrow range near the most frequent value, but about a quarter to one third of the population has values spread over a very much larger range. Cloud layer thicknesses are concentrated below 1 km, but some clouds have thicknesses greater than 8 km (Figure 9.2). Even the height distributions for two-layer clouds show a concentration of the lower layers in the boundary layer near 1 km height, but the upper layers are spread throughout the troposphere (Figure 9.3). This explains the broad distribution of cloud top pressures (Figure 9.4a); however, the corresponding optical thicknesses are highly concentrated near values of about 5 with a long tail extending beyond 100 (Figure 9.4b). Droplet radii in liquid water clouds show two populations, one near 7–10 μm over land and a second near 12–15 μm over oceans; however, there are some clouds that exhibit values greater than 20 μm that usually occur with very large optical thicknesses (Han et al., 1994). Liquid water path distributions grow progressively broader going from non-precipitating warm clouds, through non-precipitating cold clouds and precipitating warm clouds, to precipitating cold clouds (Lin and Rossow, 1996); but all distributions show some much larger values. Thus, a typical cloud can be defined by the most frequently occurring characteristics, but there are a significant number of clouds that have properties far from typical. These two different kinds of clouds, typical and atypical, can play different roles in the climate, as discussed later.

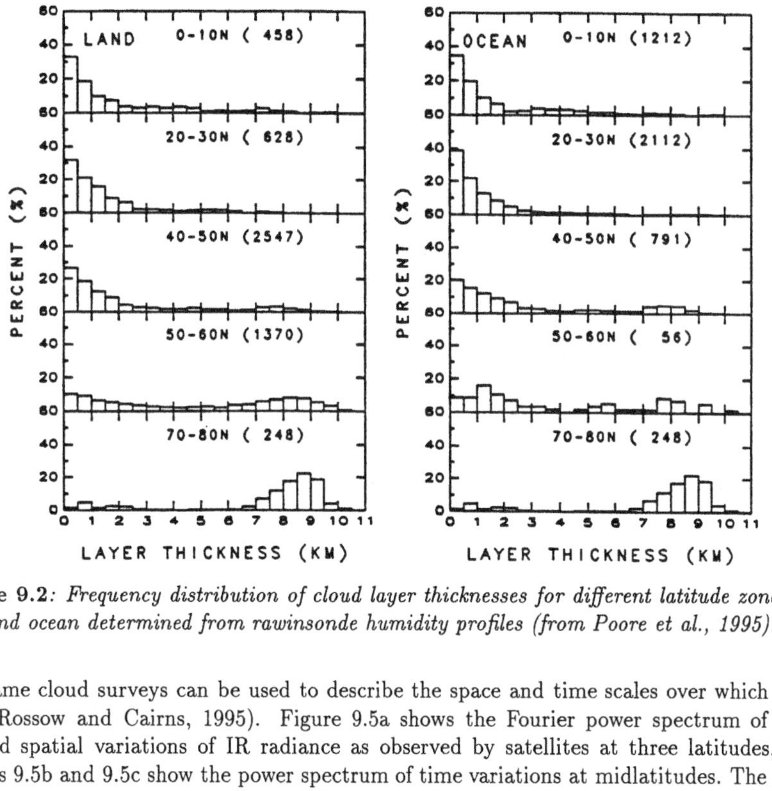

Figure 9.2: *Frequency distribution of cloud layer thicknesses for different latitude zones over land and ocean determined from rawinsonde humidity profiles (from Poore et al., 1995).*

The same cloud surveys can be used to describe the space and time scales over which clouds vary (Rossow and Cairns, 1995). Figure 9.5a shows the Fourier power spectrum of cloud-induced spatial variations of IR radiance as observed by satellites at three latitudes, while Figures 9.5b and 9.5c show the power spectrum of time variations at midlatitudes. The shapes of these spectra show that the largest variations occur at large space and time scales, of order $\sim$ 10,000 km and $\sim$ 10 days. The most notable exception is significant variations over a diurnal time scale, particularly at the larger spatial scale of a whole latitude zone (Figure 9.5b). The other important feature of cloud variations is that the space and time scales are coupled: very rapid variations occur primarily at the smaller spatial scales and planetary scale cloud features change significantly only over long time periods. Figure 9.6 shows that the Fourier power spectrum for tropical cloudiness becomes steeper (less variability at smaller scales) as the satellite observations are averaged over time; only for an average over 10 days is the power reduced at the largest spatial scales. There is little coherent spatial structure in the daily variability of cloud cover exhibited in an Empirical Orthogonal Function (EOF)

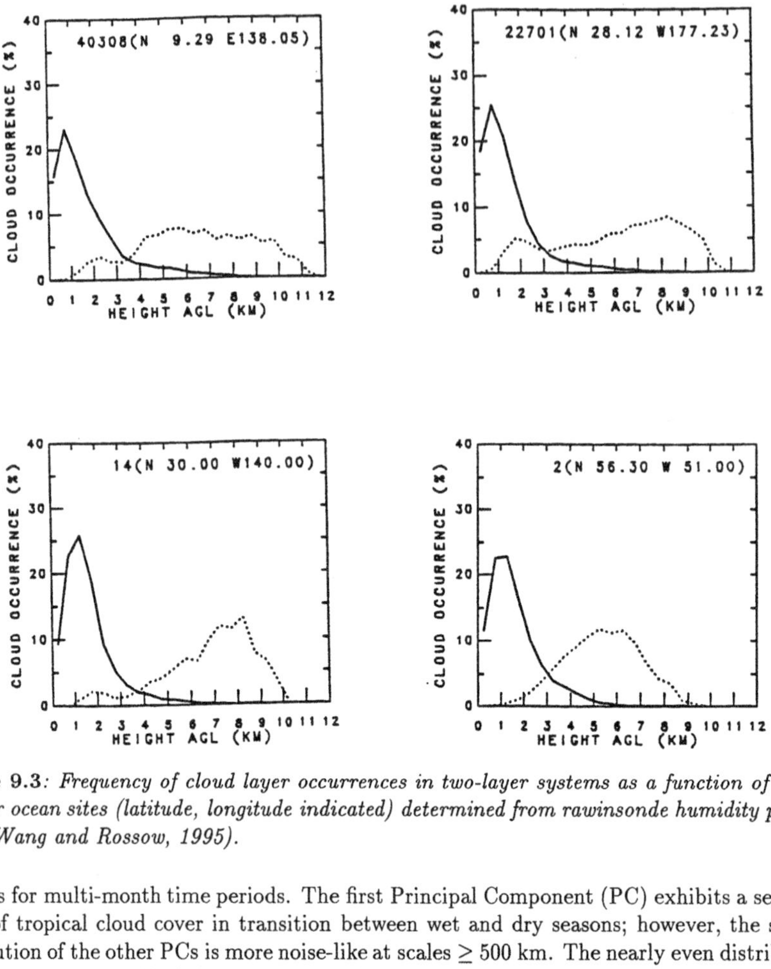

Figure 9.3: *Frequency of cloud layer occurrences in two-layer systems as a function of height for four ocean sites (latitude, longitude indicated) determined from rawinsonde humidity profiles (from Wang and Rossow, 1995).*

analysis for multi-month time periods. The first Principal Component (PC) exhibits a seasonal trend of tropical cloud cover in transition between wet and dry seasons; however, the spatial distribution of the other PCs is more noise-like at scales $\geq$ 500 km. The nearly even distribution of the variance over all of the eigenvectors (only the first one is significant) also shows that the daily variations have little coherent structure (except for diurnal variations which have been removed), although there is a hint of the subtropical jet streams in the optical thickness distributions. Thus, on time scales of $\sim$ 30 days, there is little large scale structure in the cloud variations.

9.4 Global Patterns of Cloud Variations

There is much more coherent structure in the large scale patterns of cloud variation on longer time scales. This is shown by an EOF analysis performed on global monthly mean maps of cloud properties that eliminate the smaller scales of variation. Figure 9.7 shows the first four and only significant PCs and their eigenvectors (EV) for cloud cover (similar results obtain for cloud optical thickness and top pressure). The first notable feature of the cloud cover variation is that the largest changes occur only with location: more than 70% of the variation is contained

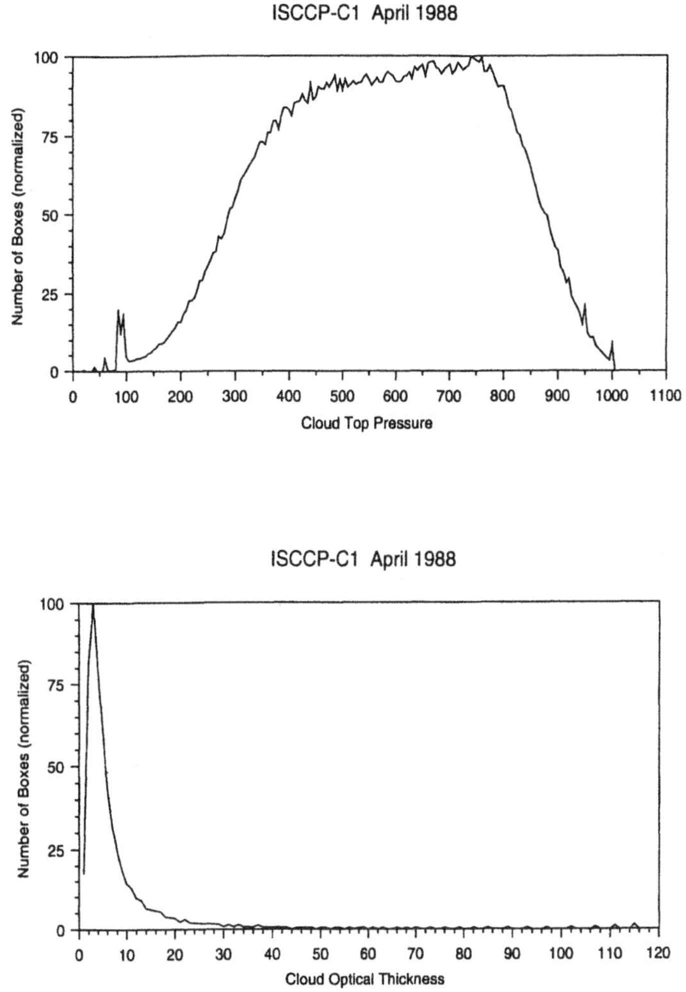

Figure 9.4: *Frequency distributions of (a) cloud top pressure and (b) visible optical thickness observed from satellites (from Rossow and Cairns, 1995).*

in the first PC which is essentially constant with time (Rossow et al., 1993; Rossow and Cairns, 1995). For cloud optical thickness and top pressure the first PC explains 60–65% of the total variability. The characteristic global pattern of this geographic variability marks the classical climate regimes produced by the large scale atmospheric circulation.

At lower latitudes, regimes dominated by mean upward motions and intense convective activity are associated with large cloud cover (positive contours in the first PC), low optical thicknesses and water contents, and a predominance of high level and low level cloudiness. Embedded in these regions is the highest concentration of dense precipitating systems. The low latitude regimes over oceans, dominated by mean downward motions with convection confined to the boundary layer, are associated with two types of low-level cloud regimes: one with very small

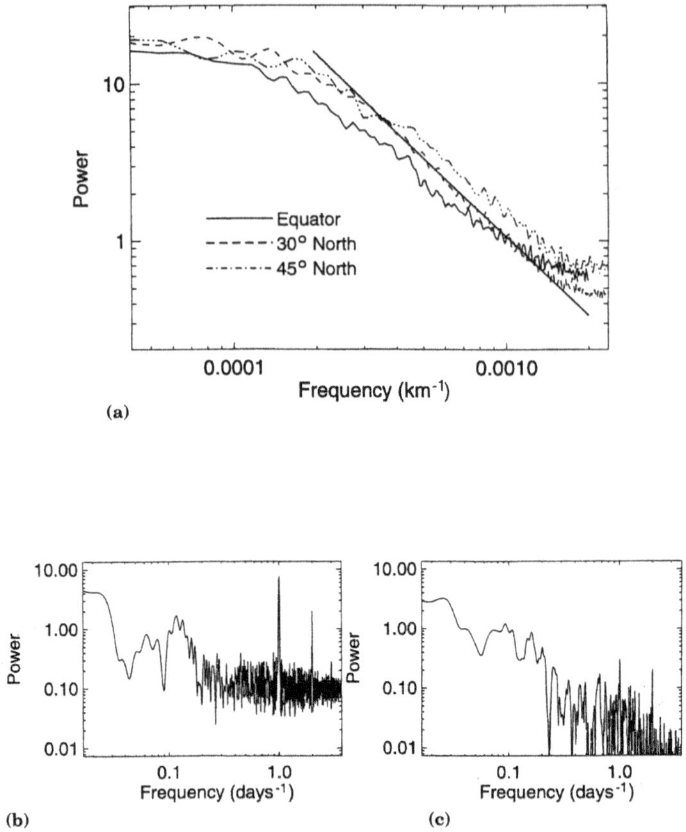

Figure 9.5: *Fourier power spectra of the (a) space and (b,c) time variations of cloudiness (from Rossow and Cairns, 1995). Analysis for spatial variations is performed at the equator (solid line), at 30° N (dashed line) and at 45° N (dot-dash line). A line representing a (-2)-power law is shown for reference. The time variations are analyzed for a whole latitude zone (b, 45° N) and for one location (c, Europe) in the same zone.*

cloud cover and optical thicknesses and one with very large cloud cover and moderate optical thicknesses. Over land, deserts, the downward motion regime exhibits very low cloud cover and optical thickness and a predominance of high-level cloudiness. Midlatitudes, where large scale cyclonic storms dominate the general circulation over oceans, are characterized by large cloud cover, large optical thicknesses, cloud tops in the middle troposphere and cloud bases near the surface. Although the frequency of dense precipitating clouds is less at midlatitudes, the area-average cloud optical thicknesses are larger than in the tropics. Polar regions present a more complicated and poorly measured situation ranging from fairly dense, low-level cloudiness in summertime to fairly tenuous, middle and high level cloudiness in wintertime (e.g., Curry and Ebert, 1992).

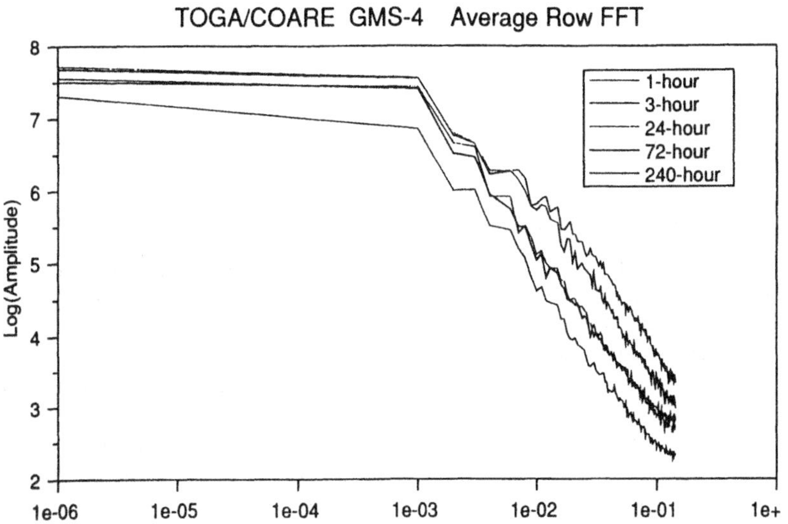

Figure 9.6: *Fourier power spectra of space variations in infrared satellite images of the western equatorial Pacific (the TOGA-COARE region), progressing downward from a single image to images averaged over the indicated time intervals (from Rossow and Cairns, 1995).*

The second notable feature of the cloud cover variation is that the largest time variation is the annual cycle (second and third PC in Figure 9.7) which is dramatically different at lower, middle, and higher latitudes (Rossow et al., 1993). The largest annual variations in cloud properties occur in the tropics with the seasonal shifts of the Inter Tropical Convergence Zone (ITCZ), although polar cloudiness may undergo relatively large seasonal variations in cloud water content. The tropical annual cycle exhibits increases in cloud cover, optical thickness, and cloud top height from wintertime (dry season) to summertime (wet season). In midlatitudes, the magnitude of the annual cycle is much smaller, particularly over oceans, and exhibits decreases in cloud cover, optical thickness and cloud top height from wintertime (cold season) to summertime (warm season). In some land areas, however, the summertime cloud top heights and cloud cover might be slightly larger than wintertime values. Polar region cloudiness may not exhibit much seasonal variation in cloud cover, but current poor observations suggest some changes in cloud heights and optical thicknesses may occur (Curry and Ebert, 1992).

The fact that the annual cycle is represented by two PCs, one with a phase coinciding with the solstices and one with the equinoxes, indicates that the amplitude and phase of the annual cycle vary with location, so that the description given above is only an approximation for a much more complicated variation. Significant differences between the hemispheres appear in the vicinity of the Asian monsoons, which are more extensive in boreal summer than in austral summer, in the eastern Pacific ITCZ, which remains in the northern hemisphere year round, over Brazil, which exhibits much more cloudiness in summer than winter, and over the two eastern Pacific marine stratus regions, which vary in phase over the year in both hemispheres. Although the southern hemisphere exhibits much less annual variation in temperature than the northern hemisphere, it exhibits much more annual variation in total cloud amount.

The third notable feature of the cloud variations shown in Figure 9.7 is that variations on time scales longer than one year are very small: the fourth PC (2% of the variation) includes

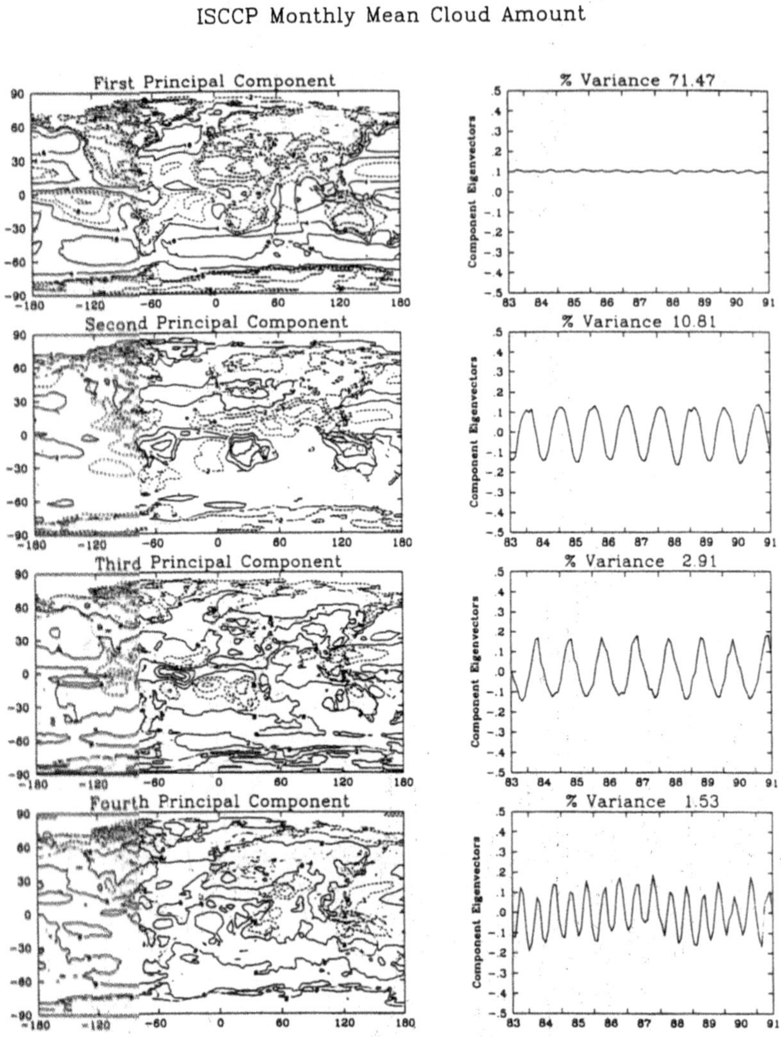

Figure 9.7: *Empirical Orthogonal Function analysis of monthly mean, total cloud cover variations from ISCCP over the whole globe for eight years (1983–1991) (from Rossow and Cairns, 1995).*

a semi-annual variation and a slow variation over the whole data record. Thus, despite the common impression that clouds vary a lot, most of this variation is confined to time scales less than one month; on longer time scales the cloudiness of Earth is remarkably constant (Rossow et al., 1993; Rossow and Cairns, 1995). However, this does not mean that small, but systematic variations of cloud do not occur and that they are not important. Rather, given the estimated sensitivity of the climate to small changes in the radiative balance (e.g. Hansen et al., 1984), even cloud changes of a few percent can be important. Figure 9.8 illustrates one of the better known systematic changes in cloudiness that occurs on a time scale of 3–10 years, namely the El Niño-Southern Oscillation (ENSO) events. The figure shows the EV for global high-level cloudiness, with the mean annual cycle removed, from two satellite datasets (NIMBUS-7: Stowe et al., 1988; ISCCP) compared with the variations from the mean of tropical sea level pressure differences. This figure shows that, while there is some similarity in the changes associated with these two ENSO events, there are also large differences in detail, so that a proper characterization of the average ENSO variations requires a long enough data record to sample many (at least 3–10) such events. The currently available datasets cover only about four events.

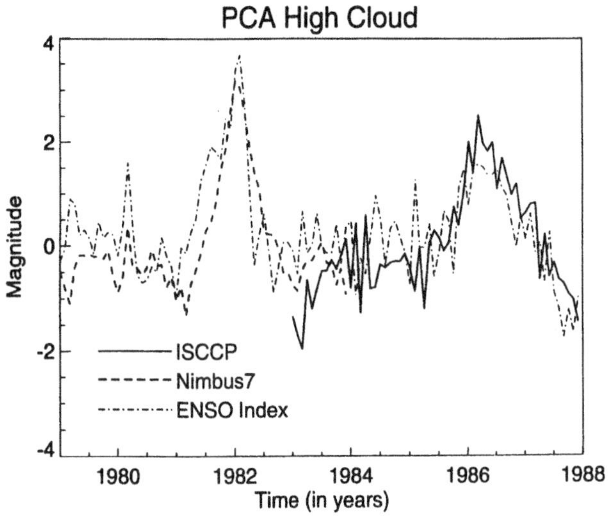

Figure 9.8: *Empirical Orthogonal Function analysis of the NIMBUS-7 and ISCCP high-level cloud amounts compared with the ENSO index for the El Niños in 1982–83 and 1986–87 (from Rossow and Cairns, 1995).*

The one systematic variation of clouds at time scales less than one month is the diurnal cycle (Figure 9.5b). Figure 9.9 illustrates some of the variety of cloud diurnal cycles exhibited in different climate regimes (Carlson et al., 1996). In general the diurnal variation amplitudes are largest at low latitudes. A notable feature is that the phase of the diurnal cycle of low-level cloudiness over oceans is very different than over land (not shown). Low-level clouds over land increase to maximum cover and height in early afternoon, coincident with or slightly later than the maximum surface temperatures resulting from the peak solar heating of the day, whereas low-level clouds over ocean increase to maximum cover and height near or just before dawn, even though maximum surface temperatures still occur in early afternoon. The other notable feature is that the diurnal variations of middle and high-level clouds over both ocean and land have a different phase than that of low-level clouds and that the diurnal variation of high-level clouds over land varies significantly with season, being largest in local summertime (Figure 9.9).

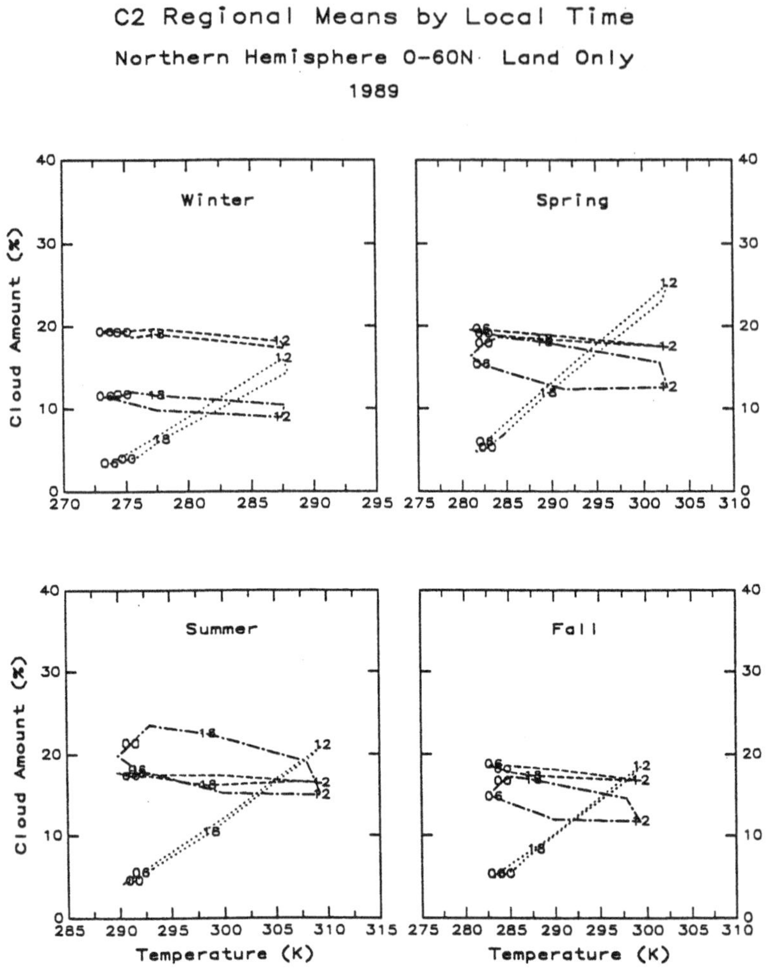

Figure 9.9: *Diurnal phase diagrams of cloudiness for different climate regimes.*

9.5 Classification of Clouds

Clouds are classified in surface observations on the basis of the estimated height of their bases above the local surface topography and on their shape (e.g. Warren et al., 1986). While cloud base heights of low-level clouds can be estimated with good accuracy, the discrimination between middle and high clouds may depend more on morphology (cf. Wang and Rossow, 1995). There are three main height classes: low (0–2 km), middle (2–4 km at higher latitudes and 2–6 km at lower latitudes), and high. There are three main cloud types or shapes: cumuloform, stratoform and cirroform. Cumuloform clouds form under the influence of small scale convective turbulence, whereas stratoform clouds are associated with larger scale synoptic motions. Cirroform clouds are indicative of a change of phase from liquid to ice. Different cloud types are associated with different kinds of weather: most notably fair weather is associated with frequent occurrence of cumulus clouds (low-level cumuloform) and/or cirrus (high-level cirroform) clouds, whereas foul weather clouds producing precipitation are associated with cumulonimbus (vertically extensive cumuloform) clouds and nimbostratus (vertically extensive and thick stratoform) clouds. Figure 9.10 compares the zonal average amounts of cirrus and cumulus as determined from both surface and satellite observations. Both these types of clouds have relatively uniform latitudinal distributions. Cumulonimbus and nimbostratus clouds, together with all high clouds, have noticeable concentrations in the tropics and midlatitudes, respectively, associated with the regions that have the most frequent and active storms in our atmosphere. These locations are also associated with higher total cloud amounts (Figure 9.7).

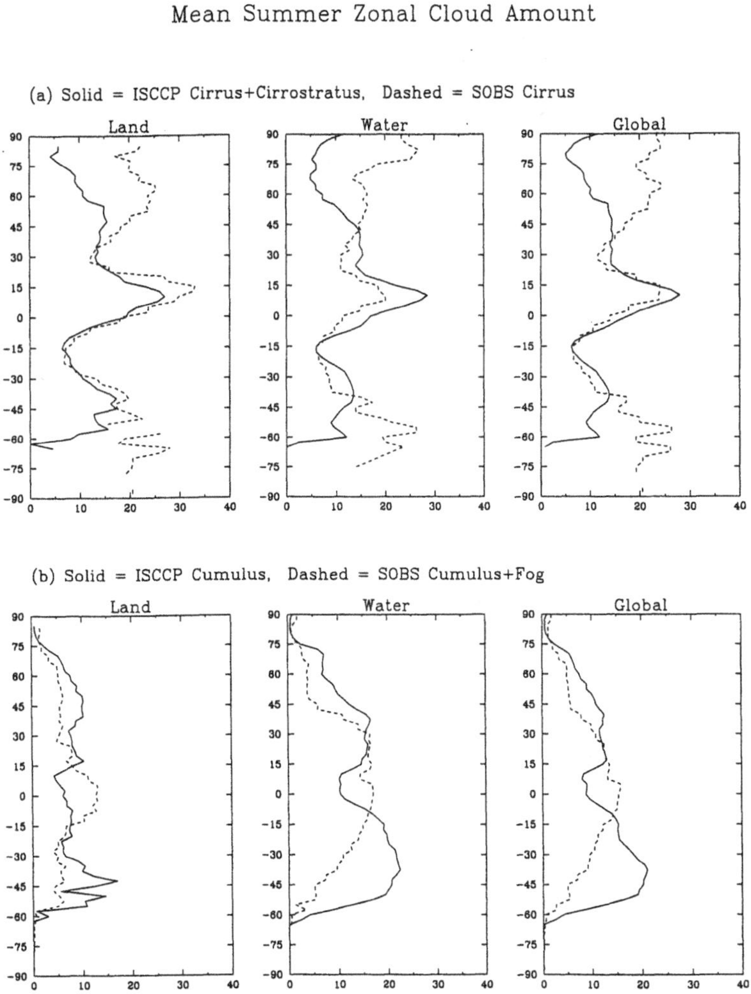

Figure 9.10: *Zonal, seasonal mean (a) cirrus and (b) cumulus cloud amounts from surface weather observations (SOBS - dashed lines) and from satellite (ISCCP - solid lines).*

Clouds are classified in satellite observations by the location of their tops and their optical thicknesses (Rossow and Schiffer, 1991; see also Liu et al., 1995). Three height classes are defined by cloud top pressures (equivalent to height above mean sea level): low-level clouds have top pressures > 680 hPa (height 3 km), middle clouds have top pressures < 680 hPa and > 440 hPa (height 6 km), and high clouds have top pressures < 440 hPa. The optical thickness categories are more complicated as shown in Figure 9.11. The distributions of cloud properties in three different climate regimes are shown in Figure 9.12. We note that in the subtropical

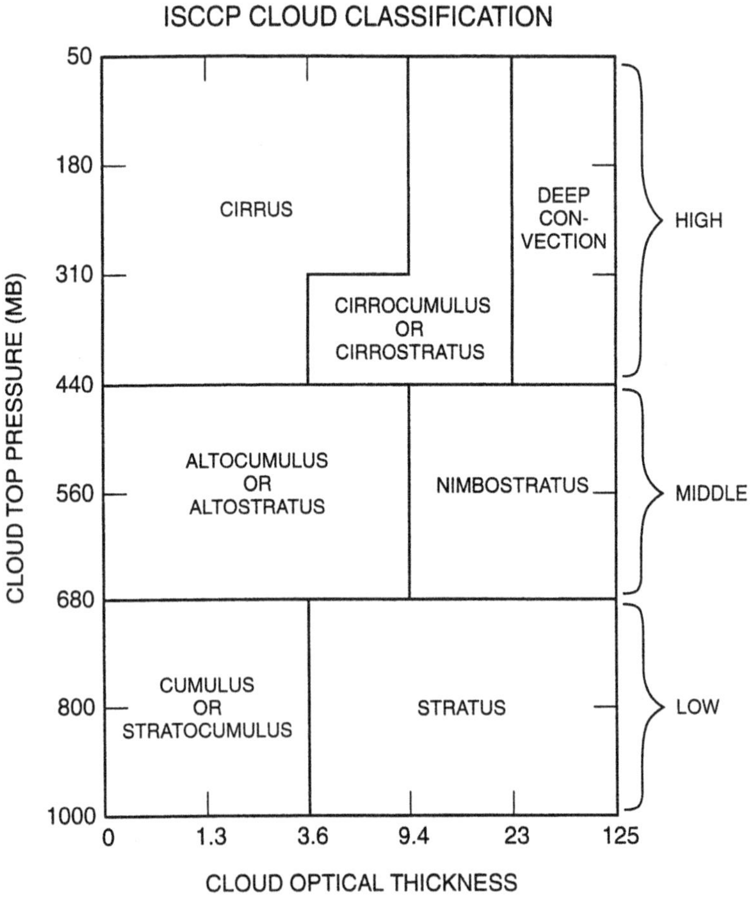

Figure 9.11: *Cloud classification scheme using satellite-determined cloud top pressures and optical thicknesses (from Rossow and Schiffer, 1991). Cloud type names are meant to be qualitative only.*

regimes the most frequent type is low-level optically thin clouds. In midlatitudes there is a mixture of low and high-level optically thin and middle-to-high level and optically thick clouds. In the tropics, in addition to a shift of the optically thicker clouds to a much higher level, a third type of cloud is apparent with moderately high tops and moderate optical thicknesses.

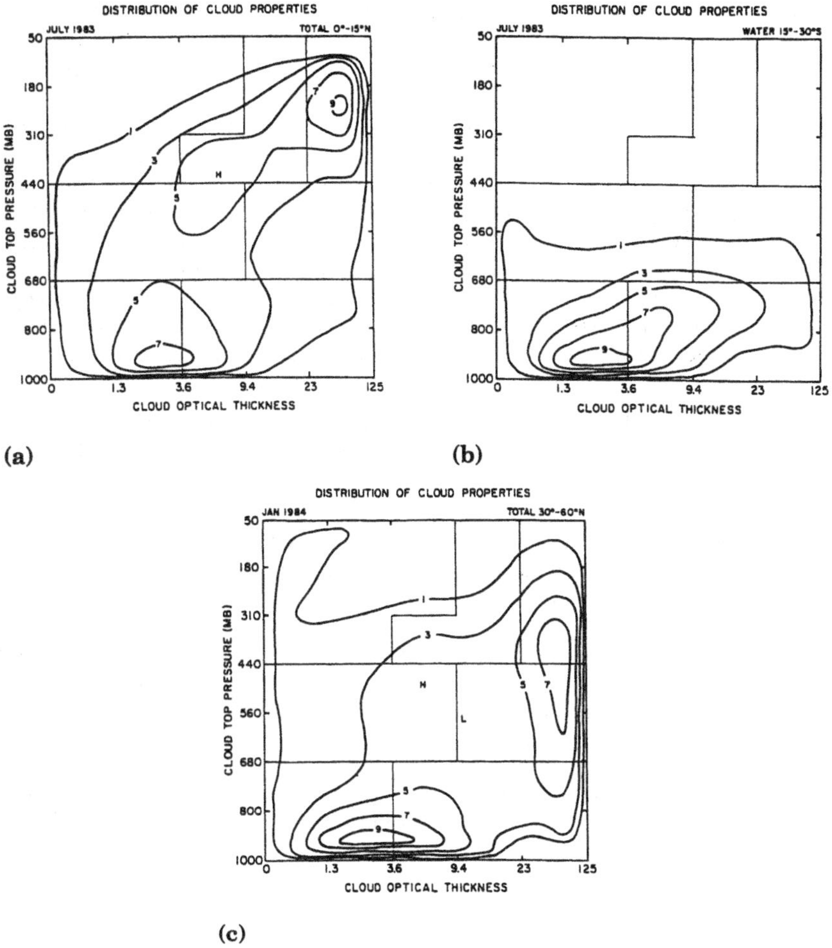

Figure 9.12: *Frequency distributions of clouds classified in satellite observations for (a) the tropics, (b) the subtropical oceans and (c) midlatitude land areas.*

The latter is identified as the stratoform anvil cloud in large convective complexes (Machado and Rossow, 1993).

Satellite measurements of microwave wavelength radiation can indicate the presence of rainfall and the combination of microwave and visible/infrared measurements can be used to estimate the approximate mixture of ice and liquid water (Lin and Rossow, 1994; 1996), as shown in Figure 9.13.

Coincident observations indicate a close link between the amounts of ice and liquid water in the clouds and the presence of rainfall: rainfall is much more likely in clouds with optical thicknesses ≥ 60 (equivalent to water path values ≥ 40 mg/cm^2), but even more likely if cloud tops are high enough that significant ice is present (cf., Liu et al., 1995). Hence, we identify the cloud in the upper right hand corner of Figure 9.12 to be associated with precipitation produced in storms. Plotting the geographic distribution of these cloud types (Figure 9.14) shows a striking resemblance to the distribution determined by surface observers. The similarity of the distributions observed from surface and satellite of cirrus and cumulus cloud amounts and cumulonimbus and nimbostratus suggests a primary link between the weather producing clouds and their observed radiative properties.

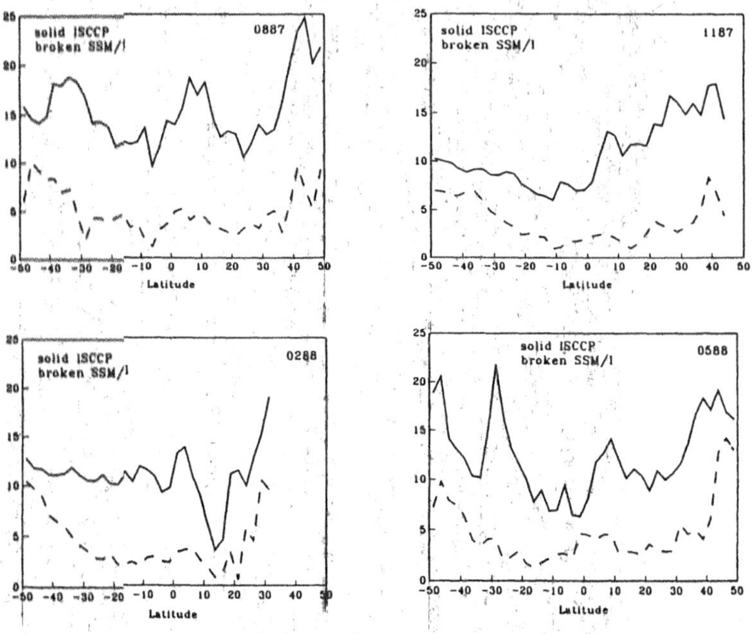

Figure 9.13: *Zonal, monthly mean liquid (dashed line) and total (solid line) water path for non-precipitating clouds in four seasonal months: the difference between total and liquid water path is approximately the ice water path (from Lin and Rossow, 1996).*

9.6 Cloud Systems

Comparison of the large scale pattern of the geographic distribution of cloud properties to some attributes of the mean atmospheric circulation shows that the main dynamic process that controls cloudiness in the atmosphere is vertical motions in the general atmospheric circulation. This association of clouds and vertical motions has long been studied at much smaller scales (Cotton and Anthes, 1989). Although the mean vertical motion pattern resembles the mean pattern of cloud properties, it is the vertical motions in individual storm systems that determine the largest values of cloud water content and top height. The shapes of the cloud property distributions suggest a two-population model: one large background population composed of optically thinner clouds with small vertical extents and one relatively sparse population composed of optically thicker clouds with much larger vertical extents. Averaging the observations produces values more similar to the more frequent (more typical) clouds in the background population. The background population dominates the radiation balance (Rossow and Zhang, 1995), but the smaller storm population produces the precipitation (Lin and Rossow, 1996). The object of further research is to determine what atmospheric conditions produce these two cloud populations and what relation exists between them.

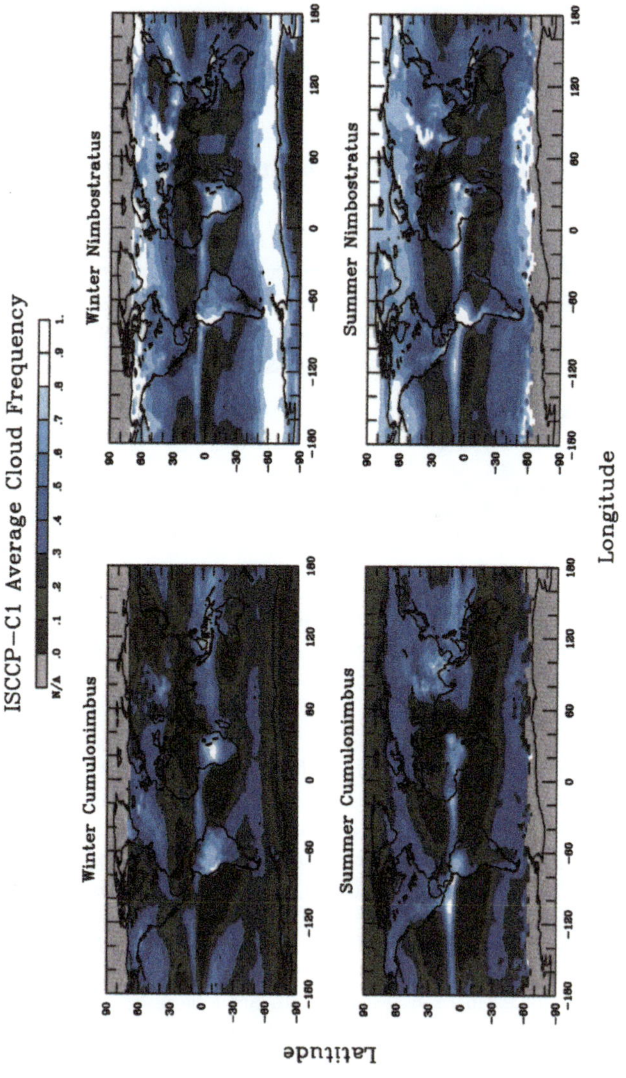

Figure 9.14: *Geographic distributions of seasonal mean frequencies of occurrence of cumulonimbus and nimbostratus (see Figure 9.11) from the ISCCP satellite dataset.*

What atmospheric conditions produce deep convection in the tropics and what kinds of clouds are produced? The simplest answer to the first question is that surface heating by sunlight produces buoyant parcels of moisture-laden air near the surface that, once condensation begins, can rise all the way to the tropopause. This simple model predicts a concentration of convection over regions with high surface temperatures; but the actual distribution of such convective systems contradicts this simple prediction in several ways. The hottest surface temperatures appear in North Africa and other nearly cloud-free deserts where there is no convection. The hottest ocean water is in the western Pacific and eastern Indian ocean, yet only the former is persistently covered by convection. A closer look at the western Pacific shows that convection does not occur everywhere, suggesting other factors that influence the occurrence of convection (Figure 9.15). Combining satellite observations of cloud properties with surface weather observations shows that it is the large scale circulation and its influence on the boundary layer humidity that sometimes inhibits convection over warm oceans and that the ocean temperature is important for triggering convection because the tropical atmosphere is more nearly isothermal than the ocean surface (Fu et al., 1990, 1994).

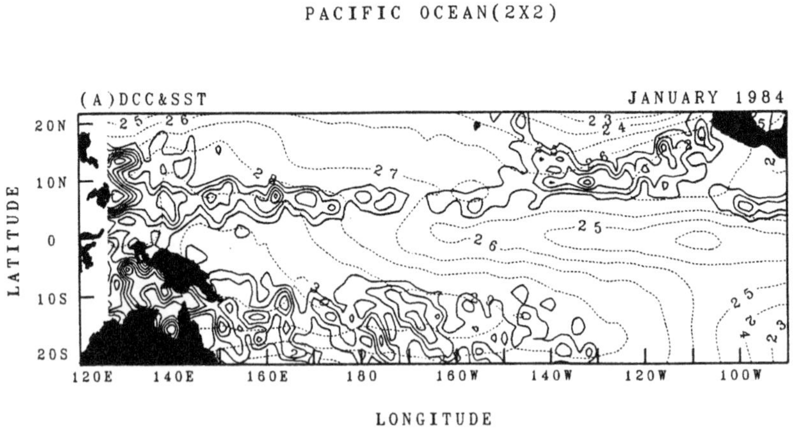

Figure 9.15: *Average distribution of sea surface temperature in °C (dashed contours) and frequency of occurrence of deep convective cloudiness (solid contours) for January 1984 over the tropical Pacific (from Fu et al., 1994).*

The first survey of the cloud properties associated with deep convection shows two notable features: strong concentration of convection over land (but not the hottest land areas) and the strong variations of cloud properties with the size of the system (Machado and Rossow, 1993). There is a particularly suggestive pattern to the size dependence: as the system size grows, so does the area covered by convective towers. As they grow larger in horizontal dimension, they extend higher into the atmosphere (lower cloud top temperatures in Figure 9.16). Surface and aircraft observations have also correlated the size and vertical extent of convective towers with the strength of the updrafts (e.g. Betts, 1973; Zipser and LeMone, 1980). The main change in the stratoform cloud component of these systems as their size increases is that their optical thicknesses or albedos increase (Figure 9.16). Since these stratoform clouds tend to have nearly constant base and top locations (e.g., Gamache and Houze, 1982; Tollerud and Esbensen, 1985), the albedo increase is equivalent to a water content increase, suggesting that precipitation also becomes more frequent (Lin and Rossow, 1996).

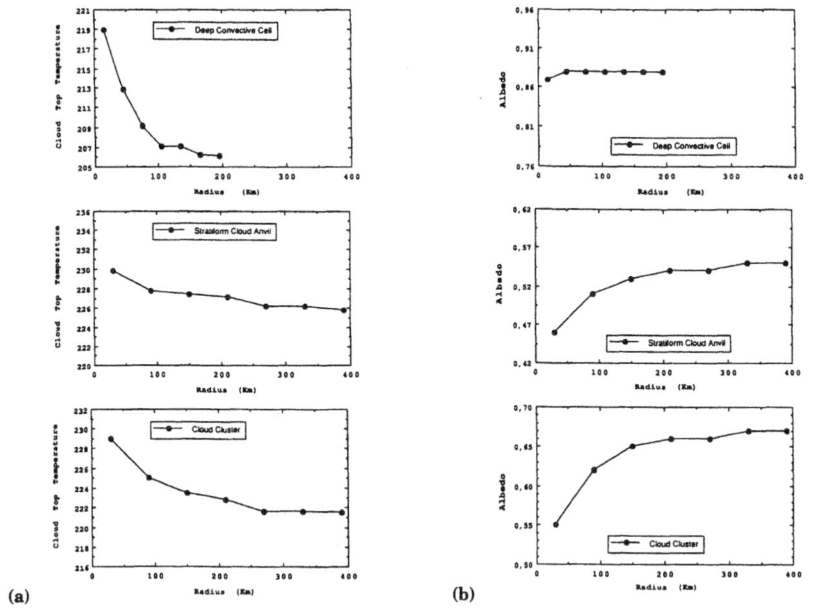

Figure 9.16: *Variation of average (a) cloud top temperature and (b) cloud visible albedo with the radius of tropical convective complexes for the convective towers (upper panel), the stratoform anvil clouds (middle panel) and the whole cloud cluster (lower panel) from satellite observations (from Machado and Rossow, 1993).*

What kinds of clouds are produced by midlatitude cyclones? This type of investigation is only just beginning. Early results correlating cloud types determined from satellite data with meteorological information on winds and temperatures (Lau and Crane, 1995) show that the optically thicker and generally higher clouds occur in association with midlatitude cyclones, particularly in those parts of these systems where vertical motions are strongest (Figure 9.17). Moreover, the high-topped and optically thicker clouds concentrate in the warm sector where rising motions predominate, while the thin cirrus clouds congregate in the cold sector where sinking motions predominate (Figure 9.18). These early results encourage studies to determine quantitative relationships between the strength of these storm systems and the properties of the clouds they produce. Two kinds of non-storm clouds are receiving a lot of attention: cirrus clouds far from either convective complexes or cyclones (e.g. Sassen et al., 1989) and marine boundary layer cloudiness in the subtropics, primarily in the higher cloud cover regimes (e.g. Betts and Boers, 1990). Since this research is in its early stages, explanations connecting the larger scale atmospheric motions to the properties of these clouds are not available as yet.

9.7 Future Studies

The last few studies described above point the way to research that has just begun and has the potential to determine how atmospheric motions affect cloud properties and precipitation. By combining the satellite and surface cloud measurements with observations of atmospheric conditions (temperature, humidity, winds) for many individual storms, the relationships between

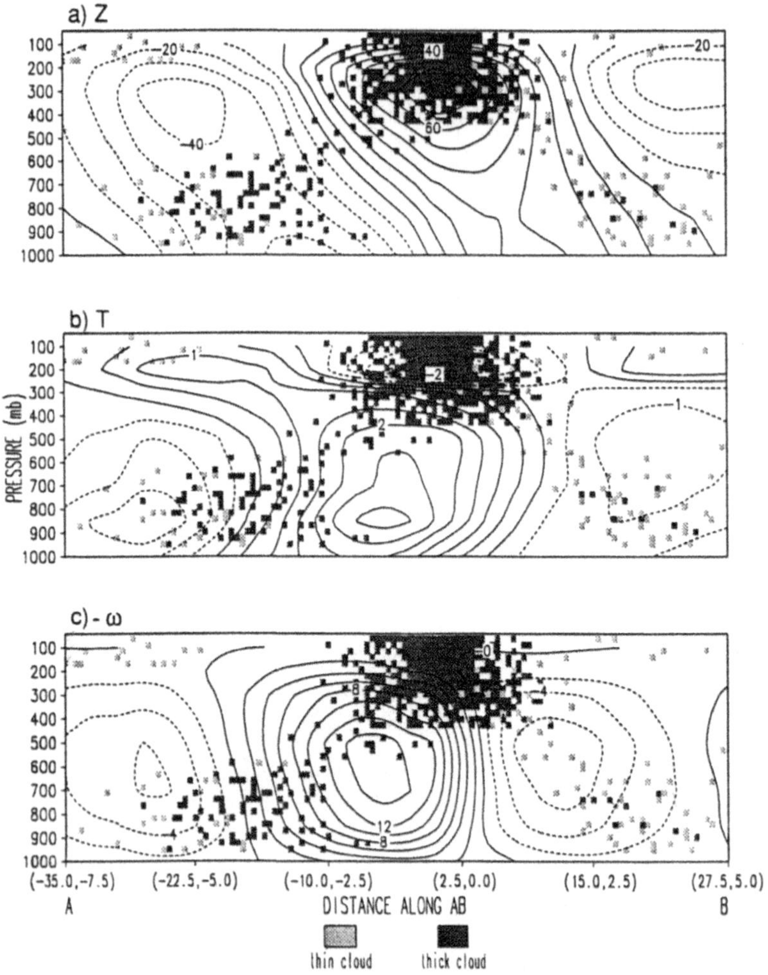

Figure 9.17: *Composite distribution of clouds with different optical thicknesses, indicated by the shading in small squares, as function of top pressure and position along the motion track of midlatitude cyclones compared with the distribution of anomalies in geopotential height (Z), temperature (T), and vertical motion (ω) (from Lau and Crane, 1995).*

the strength of these storms and the cloud properties may be discerned. Similarly conditions in fair weather may linked to the properties of boundary layer and cirrus clouds. The satellite cloud properties can be used to determine their effects on the radiation balance (e.g. Rossow and Zhang, 1995). Other satellite measurements can be combined to determine the relation of cloud properties and precipitation (e.g. Lin and Rossow, 1996). For the first time, we have global, multi-year collections of all of these quantities organized together that would allow this type of study to be conducted. However, a key missing element is the capability to determine these cloud properties at the same time at scales 5 km and 30 min, as would be possible if

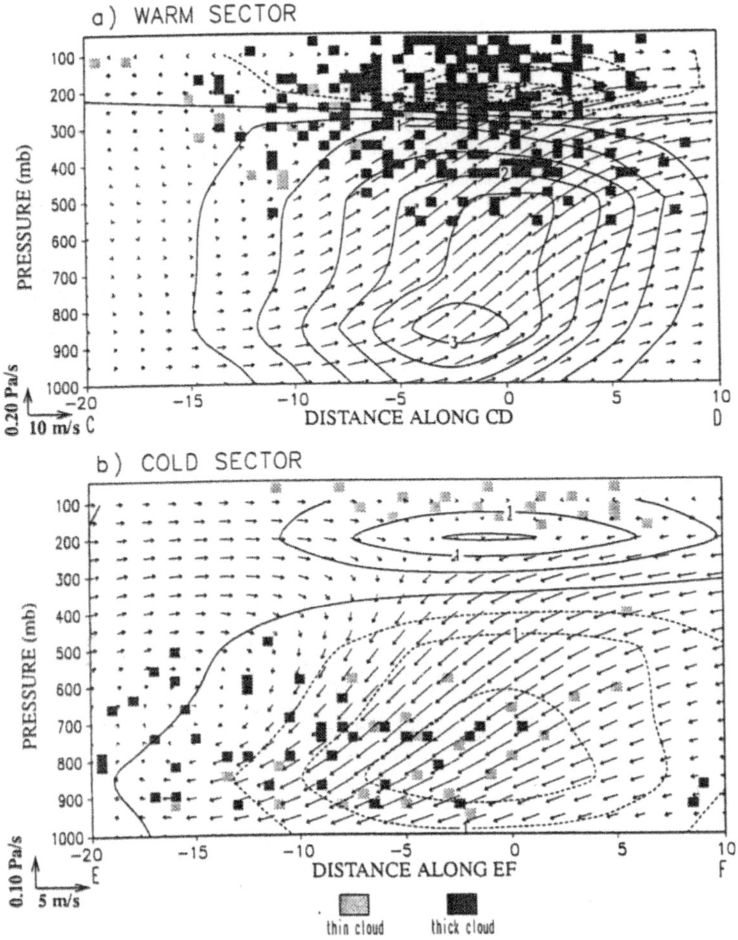

Figure 9.18: *Composite distribution of clouds with different optical thicknesses, indicated by the shading in small squares, as function of top pressure and across the motion track of midlatitude cyclones compared with the distribution of anomalies in temperature (contours) and winds (arrows) in the (a) warm sector and (b) cold sector of the cyclone (from Lau and Crane, 1995).*

several instruments were on a single spacecraft in geostationary orbit. Such a cloud-physics satellite mission would still have to include a cloud profiling radar to determine cloud vertical structures. Such a mission would also be valuable in lower orbit where measurements would be limited to larger scales.

For more information, see the ISCCP Web Home Page at URL = http://isccp.giss.Nasa.gov

9.8 References

Betts AK, Boers R (1990) A cloudiness transition in a marine boundary layer. *J Atmos Sci* 47: 1480–1497

Carlson BE, Lacis AA, Rossow WB (1993) Tropospheric gas composition and cloud structure of the Jovian North Equatorial Belt. *J Geophys Res* 98: 5251–5290

Carlson BE, Cairns B, Rossow WB (1996) Spatial and temporal characterization of diurnal cloud variability using ISCCP. *J Climate*, in press

Curry JA, Ebert EE (1992) Annual cycle of radiation fluxes over the Arctic Ocean: Sensitivity to cloud optical properties. *J Climate* 5: 1267–1280

Fu R, Del Genio AD, Rossow WB (1990) Behavior of deep convective clouds in the tropical Pacific deduced from ISCCP radiance data. *J Climate* 3: 1129–1152

Fu R, Del Genio AD, Rossow WB (1994) Influence of ocean surface conditions on atmospheric vertical thermodynamic structure and deep convection. *J Climate* 7: 1092–1108

Gamache JF, Houze RA (1982) Mesoscale air motions associated with a tropical squall line. *Mon Wea Rev* 110: 118–135

Goody RM, Yung YL (1989) *Atmospheric Radiation, Theoretical Basis.* Oxford, New York, 519 pp

Greenwald TJ, Stephens GL, Vonder Haar TH, Jackson DL (1993) A physical retrieval of cloud liquid water over global oceans using Special Sensor Microwave/Image (SSM/I) observations. *J Geophys Res* 98: 18.471–18.488

Hahn CJ, Warren SG, London J (1995) The effect of moonlight on observation of cloud cover at night and application to cloud climatology. *J Climate* 8: 1429–1446

Han Q-Y, Rossow WB, Lacis AA (1994) Nearglobal survey of effective cloud droplet radii in liquid water clouds using ISCCP data. *J Climate* 7: 465–497

Hansen JE, Travis LD (1979) Light scattering in planetary atmospheres. *Space Sci Rev* 16: 527–610

Hansen J, Lacis A, Rind D, Russell G, Stone P, Fung I, Ruedy R, Lerner J (1984) Climate sensitivity: analysis of feedback mechanisms. Climate Processes and Climate Sensitivity, *Geophys. Monogr. Ser.* Vol. 29, AGU. J.E. Hansen and T. Takahashi (eds.), Washington, D.C., 130–163

Kent GS, Winker DM, Osborn MT, Skeens KM (1993) A model for the separation of cloud and aerosol in SAGE II occultation data. *J Geophys Res* 98: 20.725–20.735

Kropfli RA, Matsarov SY, Uttal T, Oss BW, Frisch AS, Clark KA, Partran BW, Snider JB (1995) Studies of cloud microphysics with millimeter radar. *Atmos Res* 35: 299–313

Lau N-C, Crane MW (1995) A satellite view of the synoptic-scale organization of clouds in midlatitude and tropical circulation systems. *Mon Wea Rev* 123: 1984–2006

Liao X, Rossow WB, Rind D (1995) Comparison between SAGE II and ISCCP high-level clouds, Part I: Global and zonal mean cloud amounts. *J Geophys Res* 100: 1121–1135

Liao X, Rossow WB, Rind D (1995) Comparison between SAGE II and ISCCP high-level clouds, Part II: Locating cloud tops. *J Geophys Res* 100: 1137–1147

Lin B, Rossow WB (1994) Observations of cloud liquid water path over oceans: Optical and microwave remote sensing methods. *J Geophys Res* 99: 20.907–20.927

Lin B, Rossow WB (1996) Cloud water path observations over oceans using ISCCP and SSM/I data. *J Climate,* in press

Liu G, Curry JA (1992) Retrieval of precipitation from satellite microwave measurements using both emission and scattering. *J Geophys Res* 97: 9959–9974

Liu G, Curry JA (1992) Determination of characteristic features of cloud liquid water from satellite microwave measurements. *J Geophys Res* 98: 5069–5092

Liu G, Curry JA, Sheu R-S (1995) Classification of clouds over the western equatorial Pacific Ocean using combined infrared and microwave satellite data. *J Geophys Res* 100: 13.811–13.826

Poore K, Wang J-H, Rossow WB (1995) Cloud layer thicknesses from a combination of surface and upper air observations. *J Climate* 8: 550–568

Rossow WB, Garder LC, Lacis AA (1989) Global, seasonal cloud variations from satellite radiance measurements. Part I: Sensitivity of analysis. *J Climate* 2: 419–462

Rossow WB, Schiffer RA (1991) ISCCP cloud data products. *Bull Amer Meteor Soc* 72: 2–20

Rossow WB, Walker AW, Garder LC (1993) Comparison of ISCCP and other cloud amounts. *J Climate* 6: 2394–2418

Rossow WB, Cairns B (1995) Monitoring changes of clouds. *Climatic Change,* 31: 305–341

Rossow WB, Zhang Y-C (1995) Calculation of surface and topofatmosphere radiative fluxes from physical quantities based on ISCCP datasets, Part II: Validation and first results. *J Geophys Res* 100: 1167–1197

Sassen K, Starr DO, Uttal T (1989) Mesoscale and microscale structure of cirrus clouds: Three case studies. *J Atmos Sci* 46: 371–396

Schiffer RA, Rossow WB (1983) The International Satellite Cloud Climatology Project (ISCCP): The first project of the World Climate Research Programme. *Bull Amer Meteor Soc* 64: 779–784

Stephens GL (1994) *Remote Sensing of the Lower Atmosphere. An Introduction.* Oxford, New York, 523 pp

Stowe LL, Wellemeyer CG, Eck TF, Yeh HYM and the NIMBUS-7 Cloud Data Processing Team (1988) NIMBUS-7 global cloud climatology. Part I: Algorithms and validation. *J Climate* 1: 445–470

Tollerud EI, Esbensen SK (1985) A composite life cycle of nonsquall mesoscale convective systems over the tropical ocean. Part I: Kinematic fields. *J Atmos Sci* 42: 823–837

Wang J, Rossow WB (1995) Determination of cloud vertical structure from upper air observations. *J Appl Meteor* 34: 2243–2258

Wang P-H, McCormick MP, Minnis P, Kent GS, Yue GK, Skeens KM (1995) A method for estimating vertical distribution of the SAGE II opaque cloud frequency. *Geophys Res Lett* 22: 243–246

Warren SG, Hahn CJ, London J, Chervin RM, Jenne RL (1986) Global distribution of total cloud and cloud type amounts over land. 29 pp. + 200 maps, (NTIS number DE87-00-6903)

Warren SG, Hahn CJ, London J, Chervin RM, Jenne RL (1988) Global distribution of total cloud and cloud type amounts over the ocean. 42 pp. + 170 maps, (NTIS number DE90-00-3187)

Woodbury GE, McCormick MP (1986) Zonal and geographical distribution of cirrus clouds determined from SAGE II data. *J Geophys Res* 91: 2775–2785

Wylie DP, Menzel WP (1989) Two years of cloud cover statistics using VAS. *J Climate* 2: 380–392

Wylie DP, Menzel WP, Woolf HM, Strabala KI (1994) Four years of global cirrus cloud statistics using HIRS. *J Climate* 7: 1972–1986

Zhang Y-C, Rossow WB Lacis AA (1995) Calculation of surface and topofatmosphere radiative fluxes from physical quantities based on ISCCP datasets, Part I: Method and sensitivity to input data uncertainties. *J Geophys Res* 100: 1149–1165

Zipser EJ, LeMone MA (1980) Cumulonimbus vertical velocity events in GATE. Part II: Synthesis and model core structure. *J Atmos Sci* 37: 2458–2469

General Reading

Cotton WR, Anthes RA (1989) *Storm and Cloud Dynamics.* Academic, San Diego, 880 pp

Goody RM, Yung YL (1989) *Atmospheric Radiation, Theoretical Basis.* Oxford, New York, 519 pp

Houze RA (1993) *Cloud Dynamics.* Academic, San Diego, 573 pp

Ludlam FH (1980) *Clouds and Storms.* Pennsylvania State University Press, University Park, 405 pp

Stephens GL (1994) *Remote Sensing of the Lower Atmosphere, An Introduction.* Oxford, New York, 523 pp

Chapter 10

Lidar Measurements: Atmospheric Constituents, Clouds, and Ground Reflectance

Claus Weitkamp
Institut für Physikalische und Chemische Analytik
GKSS-Forschungszentrum Geesthacht GmbH
D-21494 Geesthacht
Germany

10.1 Introduction

Vision is a capability developed to a high degree of perfection in many zoological species including man. In addition to shape, color, texture and movement, distance can also be inferred. Distance measurement schemes, however, tend to get increasingly inaccurate as soon as objects are farther away than 10^2 to 10^4 m, depending on circumstances. As, with very few exceptions, vision is a *passive* process that relies on external sources of illumination, the straightforward and accurate time-of-flight determination of distance is not available to us naturally.

To *actively* illuminate an object with a short pulse of light and determine its distance from a measurement of the arrival time of its image, an instrument is required. We call instruments of this kind *lidars*. The time-of-flight scheme is not limited to the use of light, but works equally well with other kinds of radiation (cf. Table 10.1). Also lidar is not just one technique; lidar instruments vary greatly according to which physical process of interaction of light with the atmospheric constituents is used. These processes are chiefly absorption and scattering.

Although in principle all gaseous constituents of the atmosphere that are present in concentrations $\geq 10^{-9}$ parts by volume and all physical parameters can be measured, lidar techniques are still in a state of rapid development. In this contribution lidar techniques are presented that serve the measurement of gaseous and particulate constituents of the atmosphere; techniques for pressure, temperature, wind, gravity and radiation fields are not considered.

Section 10.2 is a survey of the processes that lead to a lidar return signal and of the algorithms used for the retrieval of the interesting atmospheric properties from the lidar returns. It also summarizes what problems must be solved to extract information on the earth's surface from an optical return signal to an airborne or spaceborne remote sensor; these ground-reflectance measurements are not depth-resolving as are lidars. Other non-depth resolving techniques such

NATO ASI Series, Vol. I 45
Radiation and Water in the Climate System:
Remote Measurements
Edited by Ehrhard Raschke

Acronym	Signification	Depth-resolving
SONAR	SOund NAvigation Ranging	yes
RADAR	RAdiowave Detection And Ranging	yes
SODAR	SOund Detection And Ranging	yes
LIDAR	LIght Detection And Ranging, or: Light Identification, Detection, And Ranging	yes
DAS LIDAR	Differential Absorption and Scattering Lidar	yes
DIAL	DIfferential Absorption Lidar	varying
DOAS	Differential Optical Absorption Spectrometry	no

Table 10.1: *Acronyms*

as long-path differential optical absorption spectroscopy (DOAS) that provides *average* concentration data only are not dealt with in this chapter. Section 10.3 shows examples of each type of lidar and presents a few selected results. In Section 10.4 past and present techniques to measure atmospheric water vapor are briefly reviewed. In Section 10.5 some recent developments in the field are reported.

10.2 Theory

If a short pulse of radiation is transmitted into the atmosphere, part of it will be scattered in the backward direction and reach the transmitter at time t. From the speed of light c the distance x is easily determined as

$$x = \frac{ct}{2}. \tag{10.1}$$

Let us assume that the receiver and transmitter are located in the same place, that the light source is a laser, and that its pulse duration Γ and average power P_0 during the pulse characterize well the transmitted pulse of energy

$$E_0 = \Gamma P_0. \tag{10.2}$$

The atmospheric return signal, unlike, say, the radar return from a well-localized object such as a ship or mountain, will then show a broad time distribution given by

$$P(x) = \frac{c\Gamma}{2} P_0 \frac{A\eta O(x)}{x^2} \beta(\pi, \lambda, x) \tau^2(\lambda, x). \tag{10.3}$$

A and η are the area and efficiency of the detector, O is the overlap integral between the transmitted beam and the receiver field of view, $\beta(\pi)$ is the backscatter coefficient and τ the transmission of the atmosphere. β and τ and thus P will generally be different for different wavelengths λ. β and τ also depend on distance x, O is equal to unity for large distances for which the laser beam is entirely inside the field of view, and decreases for closer distances (Fig. 10.1).

Lidar is not just one technique, but a family of techniques which make use of the different kinds of interactions that light undergoes in the atmosphere and which greatly vary in size and cost of the apparatus, complexity of the data reduction algorithms, dependence on ancillary information, and significance of the results.

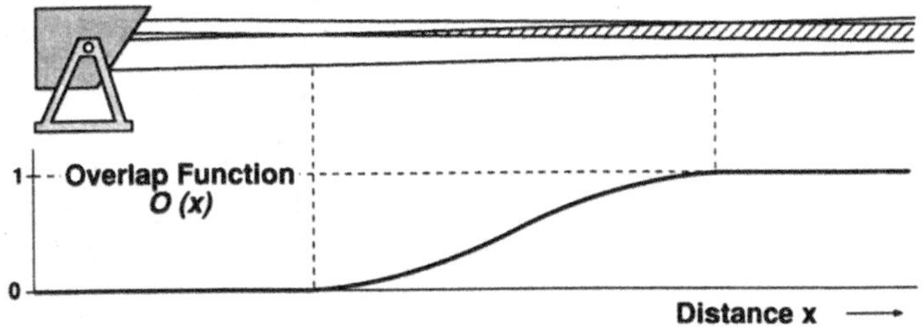

Figure 10.1: *Schematic representation of the overlap integral $O(x)$ of the laser beam with the receiver field of view, as a function of distance x.*

10.2.1 Mie lidar

In the lower troposphere where aerosols such as dust, haze and fog are the dominant scatterer, most of the lidar backscatter radiation comes from particles that are comparable in size with the wavelength of the scattered light. The theory of this process was first described by Gustav Mie (1868–1957). Although Mie's theory rigorously applies to spherical and homogeneous particles only, it has become customary to call devices Mie lidars that register elastically backscattered laser radiation and analyze the return signals for the distribution and properties of the scattering aerosol.

In Eq. (10.3) the transmittance τ decreases monotonically with distance x according to

$$\tau(\lambda, x) = \exp\left[-\int_0^x \alpha(\lambda, \xi) d\xi\right] \tag{10.4}$$

where α is the extinction coefficient: clearly the local extinction coefficient at point ξ must be integrated out to distance x to obtain the transmission between the lidar and the point x where the backscattering occurs; the power 2 to which τ is raised reflects extinction of the primary as well as the backscattered light.

Introducing Eq. (10.4) into Eq. (10.3) leads to an equation with one measured quantity (P) and two unknowns (α and β). A way out of this dilemma is what is today known as Klett's inversion method (Klett, 1981; Fernald, 1984). It consists in the postulate of a relation

$$\beta = F\alpha^{\kappa} \tag{10.5}$$

between the unknowns, F and κ being two constants. The lidar equation then turns into an ordinary differential equation of the Riccati type, the solution of which is

$$\alpha = \frac{e^{\frac{S-S_0}{\kappa}}}{\frac{1}{\alpha_0} - \frac{2}{\kappa}\int_{x_0}^{x} e^{\frac{S-S_0}{\kappa}} d\xi} \tag{10.6}$$

with the abbreviation

$$S(x) = \ln[x^2 P(x)] \tag{10.7}$$

and a set of boundary values

$$\alpha_0 = \alpha(x_0), \quad S_0 = S(x_0) \tag{10.8}$$

at some given distance x_0 for which the extinction coefficient α is sufficiently well known. This can, e.g., be the case at the beginning of the lidar range at which an in-situ device is available for a measurement of the local extinction coefficient α_0. Unfortunately, extinction coefficients determined from the regression formula (10.6) depend critically on the choice of α_0; in fact, the omnipresent noise on the measured signals P is sufficient to get the denominator on the right-hand side of Eq. (10.6) so close to zero that absurd or (meaningless) negative absorption coefficients occur, unless unrealistically large values of the constants F or κ are used. This procedure usually called forward integration because Eq. (10.6) is solved by beginning at the near end of the range is therefore applicable to special cases only. Now the boundary values (10.8) can be defined at any point within the lidar range, including the most remote point x_m with

$$\alpha_0 = \alpha(x_m), \quad S_0 = S(x_m) \tag{10.9}$$

then entering into Eq. (10.6). This small modification turns the minus sign into a plus sign and thus removes the singularity of Eq. (10.6). This backward integration is today the procedure of choice. It is carried out in such a way that the lidar range is chosen long enough to reach into a region where aerosol extinction is negligible relative to molecular extinction. Molecular extinction is easily calculated from Rayleigh theory and is given by

$$\alpha_{mol}(x) = \frac{8\pi \mathrm{sr}}{3}\beta_{mol}(\pi, \lambda), \tag{10.10}$$

with the wavelength-dependent 180° molecular scattering coefficient (Ref. 3)

$$\beta_{mol}(\pi, \lambda) = 1,47\left(\frac{550\mathrm{nm}}{\lambda}\right)^4 10^{-6}\mathrm{m}^{-1}\mathrm{sr}^{-1}. \tag{10.11}$$

The parameter κ in Eq. (10.5) is for all practical purposes set equal to 1, and from empirical data a value for the ratio α/β is chosen so that backward integration can be carried out. In doing so, care must be taken to avoid numerous pitfalls.

For one thing, backward integration has a tendency to flatten the data and to "iron out" structures clearly visible in forward integration and perceptible in the atmosphere with the naked eye (Kunz, 1987). Second, independent measurements of α and β (Ansmann et al., 1991) show that the ratio is not spatially constant even in one type of cloud. As can be seen from Fig. 10.2, lidar ratios of about 2 sr prevail in the upper part of a cirrus above 8 km, whereas values around 12 sr are measured in the lower parts between 7 and 8 km. Thirdly, lidar ratios in clouds such as cirrus change with time. In six consecutive measurements of 2.5 minutes duration each, only the second, third, and fourth measurement of a cirrus between 9.3 and 11.7 km height were correctly represented by an average lidar ratio of 13 sr, whereas for measurement no. 1 a value of 16 and for measurement no. 5 a value of 20 sr is more appropriate; for measurement no. 6 the lidar ratio is even larger and was not determined (Fig. 10.3).

A convenient way for the determination of the average lidar ratio of a cloud or other mass of aerosol is the consecutive use of forward *and* backward integration and variation of the lidar ratio until both results agree. To avoid the problems associated with forward integration the procedure should be started with low α/β ratios. Figure 10.4 shows a sequence of α profiles generated in this way. However, and fourthly, the results obtained must not be mistaken as the real profile of the atmospheric values of α. Figure 10.5a shows a case in which the Klett forward and backward profiles agree quite well with the result of an independent measurement of α and β. In Fig. 10.5b this is not the case: the independently measured profile (dotted line) differs markedly from the Klett result, although the forward (solid line) and backward result (dashed line) agree so well that they can hardly be distinguished. It must be kept in mind that the same measurement data were used for the generation of the three curves of Fig. 10.5b, with an additional measurement of β available for the dotted profile. In summary, forward-backward integration appear to be a viable means of determination of the average lidar ratio provided he following conditions are met:

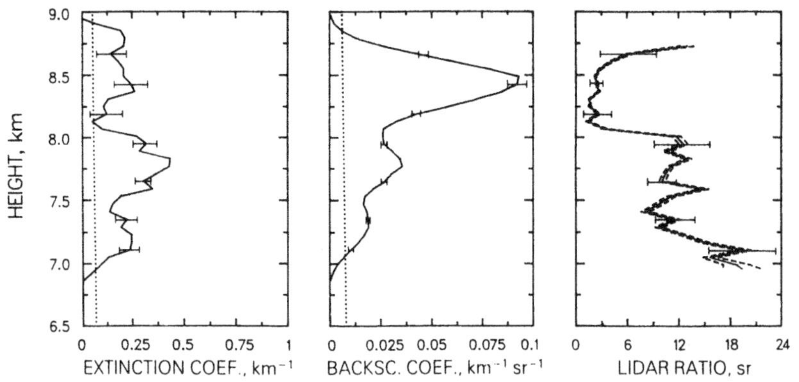

Figure 10.2: *Aerosol extinction and backscatter coefficient and lidar ratio measured with a Raman lidar on 20 September 1989 above Norderney. Error bars indicate statistical uncertainty. Dotted lines represent the Rayleigh contributions to extinction and scattering. The dashed curves on either side of the lidar ratio trace indicate the effect of ozone absorption (low lidar ratio - no ozone, high lidar ratio - twice the standard atmospheric ozone concentration). Note the variability of the lidar ratio within the cloud. From Ansmann et al. (1991).*

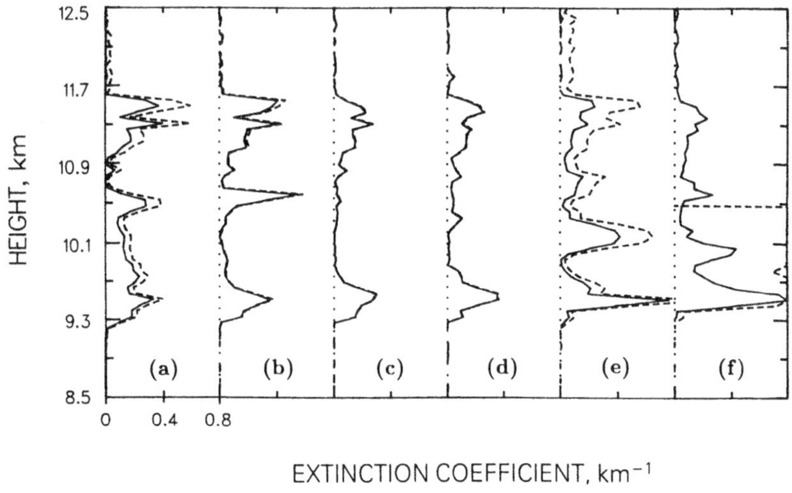

Figure 10.3: *Solution of the Klett method with forward (dashed lines) and backward integration (solid lines) with a constant lidar ratio of 13 sr. Reference heights z_0 = 6 km, z_m = 12 km, boundary values $\alpha(z_0) = \alpha(z_m) = 0$. Data taken on 18 October 1989 between 11:29 and 11:44 h local time at Norderney. Sampling time for each of the six profiles is 150 s. The lidar ratio chosen is appropriate for cases c and d only. From Ansmann et al. (1991).*

- Reliable near-end and remote-end boundary values of the aerosol extinction coefficient must be available.

- The average aerosol lidar ratio must be constant during the measurement.
- The aerosol lidar ratio must be spatially constant over the region of interest.
- An appropriate gas absorption correction must have been carried out prior to the application of the Klett inversion algorithm.

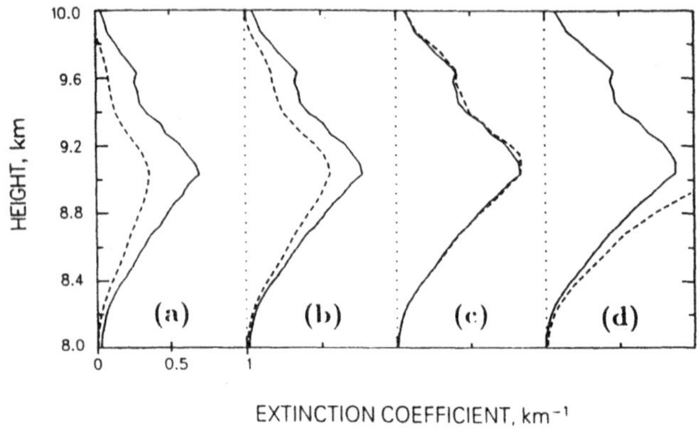

Figure 10.4: *Klett forward (dashed lines) and backward integration (solid lines) with average lidar ratio of 10, 13, 16, and 19 sr (a through d). Reference heights and boundary values as in Fig. 10.3. Data taken on 24 October 1989 at Norderney. From Ansmann et al. (1991).*

10.2.2 Differential absorption and scattering lidar

Gas concentrations cannot be determined from one elastic lidar return signal alone. Scattering by gases ist not characteristic, whereas optical absorption by many gases does show distinct features that vary rapidly with wavelength in some spectral regions. The idea is to use two wavelengths instead of one. If these wavelengths are chosen such that one is absorbed by the gas of interest whereas the other is not, and if the "on-resonance" or signal wavelength and the "off-resonance" or reference wavelength are so closely spaced that the scattering and absorption properties of the remaining atmosphere are the same or nearly the same, then the concentration distribution of the gas of interest can be inferred from the ratio of the on-resonance and off-resonance return signals or, more precisely, from the *change* of this ratio with distance.

For the mathematical treatment we have to remember that the (macroscopic) extinction coefficient α is the product of particle number density N and extinction cross section which for all practical purposes is equal to the absorption cross section σ. Denoting now with α the absorption coefficent of the atmosphere without the additional gas of interest, we can write

$$\tau = \int_0^x [\alpha(\lambda, \xi) + N(\xi)\sigma(\lambda)]d\xi. \tag{10.12}$$

Equation (10.3) then takes the form

$$P(x, \lambda_i) = \frac{c\Delta t}{2} P_0(\lambda_i) \frac{A\eta O(x)}{x^2} \beta(x) \exp\left\{-2\int_0^x [\alpha(\xi) + N(\xi)\sigma(\lambda_i)]d\xi\right\}. \tag{10.13}$$

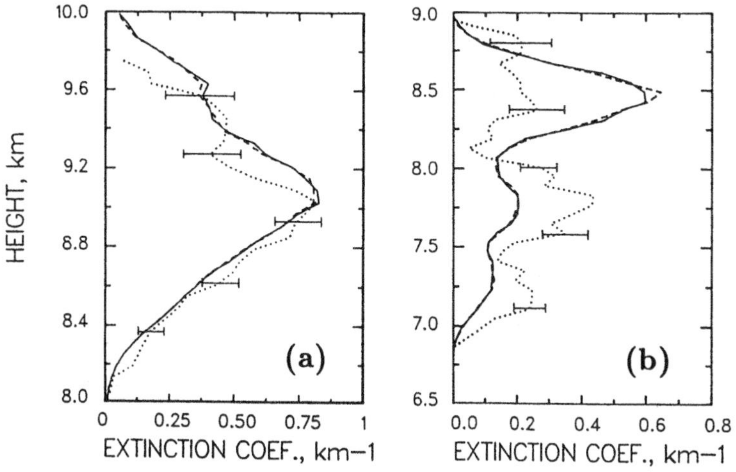

Figure 10.5: *Aerosol extinction coefficient from a Raman lidar (dotted lines) and from the same instrument used as an elastic-backscatter lidar only (solid lines: Klett forward integration, dashed lines: Klett backward integration). Optimum Klett (average) lidar ratios are 15.7 (left) and 7.3 sr (right). Data taken in Norderney on 24 October (left) and 20 September 1989 (right). From Ansmann et al. (1991).*

Let i be 0 and 1, $\lambda_{0,1}$ denote the reference and signal wavelengths, and $P(x, \lambda_{0,1})$ be the reference and signal lidar returns. If the two Equations (10.13) are then devided by each other, wavelength-independent factors such as c, A, O, x^2 cancel, depth-independent factors like Γ and β can be reduced to a constant, and if λ_0 and λ_1 are so close together that for a moment the wavelength dependence of β and α are neglected, we directly get the profile

$$N(x) = \frac{1}{2[\sigma(\lambda_1) - \sigma(\lambda_0)]} \frac{d}{dx} [\ln P(x, \lambda_0) - \ln P(x, \lambda_1)] \tag{10.14}$$

of the unknown gas molecule density. For practical purposes the mass density

$$C(x) = MN(x) \tag{10.15}$$

is often the preferred figure, M being the mass of one molecule of the gas of interest. The effects of calculating the ratio of the two signals, the logarithm, and its spatial derivative are illustrated in Fig. 10.6. Note that Eqs. (10.14) and (10.15) allow the calculation of the concentration profile in *closed form* and from the measured signals alone, provided the cross sections σ at the on-resonance and off-resonance wavelengths are known. These can be determined to any desired accuracy in a laboratory experiment. DAS lidar is therefore sometimes considered as an absolute method of concentration determinations that does not depend on a calibration, the role of the calibration having been taken by the previous measurement of the absorption cross sections.

Looking at Eq. (10.13), we notice that the simplifications made to arrive at Eq. (10.14) need more detailed consideration. Formally three correction terms must be added to the expressions (10.14) and (10.15), leading to

$$C(x) = \frac{M}{2\Delta\sigma} \frac{d}{dR} \left[\ln \frac{P(\lambda_0, x)}{P(\lambda_1, x)} \right] + A + B + G \tag{10.16}$$

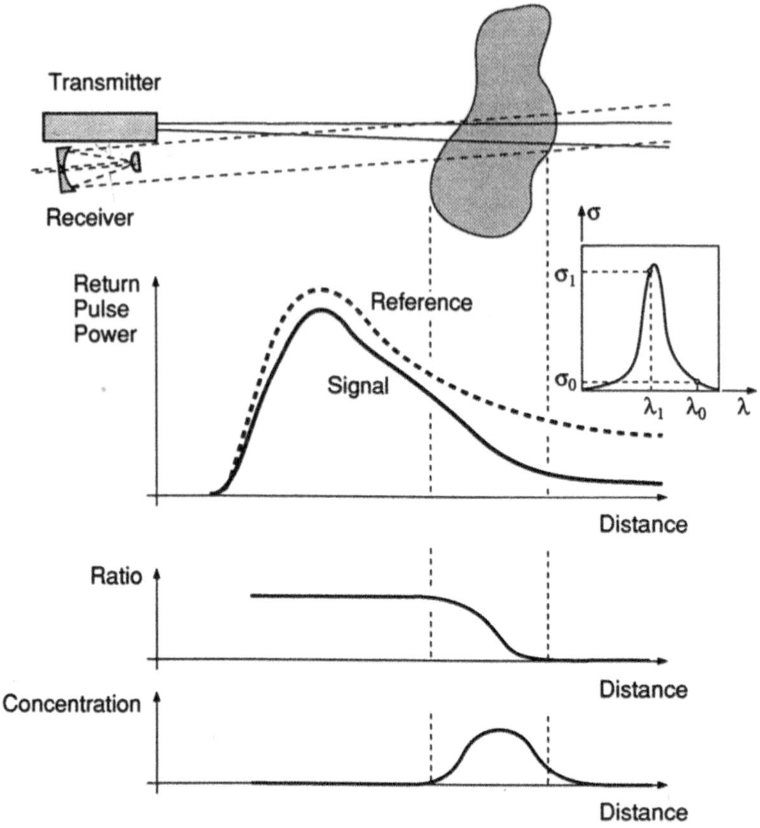

Figure 10.6: *Principle of differential absorption and scattering lidar. Return signals at measurement and reference wavelengths* λ_1, λ_0 *are attenuated differently if scattered light originates in or beyond a gas cloud. The derivative of the logarithm of the signal ratio is proportional to the gas concentration.*

with

$$A = -\frac{M}{2\Delta\sigma}[\alpha_{mol}(\lambda_1, x) - \alpha_{mol}(\lambda_0, x) + \alpha_{aer}(\lambda_1, x) - \alpha_{aer}(\lambda_0, x)], \tag{10.17}$$

$$B = -\frac{M}{2\Delta\sigma}\frac{d}{dR}\left[\ln\frac{\beta_{mol}(\lambda_0, x) + \beta_{aer}(\lambda_0, x)}{\beta_{mol}(\lambda_0, x) + \beta_{aer}(\lambda_0, x)}\right], \tag{10.18}$$

$$G = -\frac{M}{\Delta\sigma}\sum_i \frac{C_i(x)}{M_i}[\sigma_i(\lambda_1) - \sigma_i(\lambda_0)]. \tag{10.19}$$

A takes care of the fact that the average atmosphere absorbs approximately equally at the on- and off-resonance wavelength, but residual differences remain that are so much the larger the farther apart λ_1 and λ_0 are chosen. The molecular contribution (α_{mol}), although strongly wavelength-dependent, can very accurately be determined from the theory of Rayleigh scattering and its λ^{-4} law. Aerosol absorption is harder to evaluate, but depends much less on wavelength and is normally on the same order of magnitude as the molecular contribution for $\lambda_1 - \lambda_0$ wavelength differences of a few nanometers.

B corrects for differences in backscatter. Here gradients, not the magnitude of atmospheric backscatter itself cause noticeable effects. Again the molecular part is not critical, but the aerosol contribution can be so important that no reliable concentration data can be deduced from the lidar signals at the onset of a thick cloud with sharp boundary.

G is the gas interference term. It is obvious that if other gases i than the one to be measured show sizeable differential absorption at wavelengths λ_1 and λ_0 and can be present in concentrations high enough so the product of particle density and differential cross section gets comparable to that of the gas of interest, the effect must be taken into account. The term G can usually be neglected in measurements of nitrogen dioxide, water vapor and a few other gases in the visible and near infrared, but plays a role in ultraviolet DAS lidar and in middle-infrared DAS lidar around 10 μm. If and only if $\lambda_1 < \lambda_0$ then A and B are negative. G is negative for those interfering gases for which $\sigma(\lambda_1) > \sigma(\lambda_0)$, as is the case by definition for the gas of interest.

10.2.3 Raman lidar

Prompt inelastic scattering of light which is characterized by a shift of the scattered radiation with respect to the primary radiation is called Raman scattering. The shift originates from the capability of the scattering molecule to absorb part of the energy of the incoming photon

$$E_P = \frac{hc}{\lambda_P} \tag{10.20}$$

leaving behind a scattered photon of energy

$$E_R = \frac{hc}{\lambda_P} - \Delta E. \tag{10.21}$$

Here E_P and E_R are the energies of the primary and Raman-scattered photons, λ_P and λ_R the corresponding wavelengths, and h and c Planck's constant and the speed of light. ΔE is the amount of energy absorbed by the molecule. Clearly, ΔE cannot take any value but is determined by the vibrational and rotational properties of the molecule.

In this context only vibration-rotation transitions and only so-called Stokes scattering, in which ΔE is positive, are considered. It may be mentioned, however, that purely rotational Raman scattering and anti-Stokes transitions, in which the molecule *adds* energy to the photon, can both be used in other types of lidar which aim, e.g., at the measurement of atmospheric temperature profiles (Vaughan et al., 1993; Zeyn et al., 1994).

For Raman lidar only one laser is needed. The radiation collected by the receiver optics contains a mixture of wavelengths (cf. Fig. 10.7) that must be separated by a spectrometer or monochromator. Usually a number of narrow wavelength regions are "watched" simultaneously, the corresponding optics being called a "channel" and the whole instrument "polychromator".

The signal from distance x that is observed at the Raman wavelength λ_R if the primary (laser) wavelength is λ_P is given by

$$P(\lambda_P, \lambda_R, x) = \frac{c\Delta t}{2} P_0 \frac{A\eta O(x)}{x^2} N(x) \frac{d\sigma(\pi, \lambda_P, \lambda_R)}{d\Omega} \exp\left\{ - \int_0^x [\alpha_P(\xi) + \alpha_R(\xi)] d\xi \right\}. \tag{10.22}$$

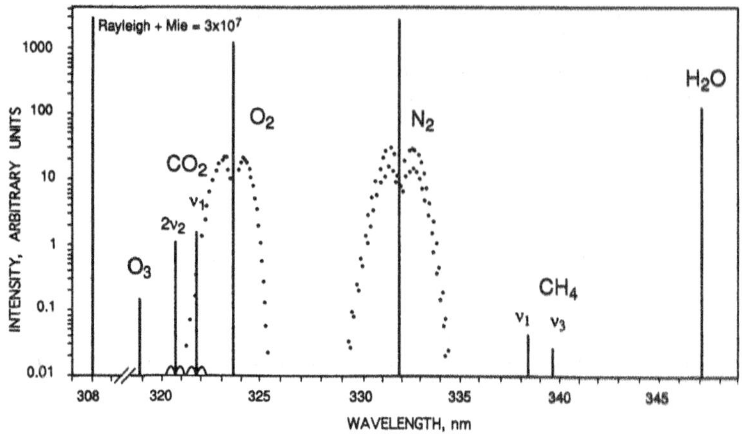

Figure 10.7: *Calculated Raman spectrum of atmospheric air for a primary wavelength of 308 nm with 100% relative humidity at 300 K, normalized to the intensity of the CO_2 $2\nu_2$ Q branch. The most intense Raman line (Q branch of N_2) is more than 10 000 times weaker than the elastic (Mie and Rayleigh) peak.*

N is again the spatial density of scattering molecules, $d\sigma(\pi, \lambda_P, \lambda_R)/d\Omega$ is the differential Raman backscatter cross section, and

$$\tau_{P,R} = \exp \int_0^x -\alpha_{P,R}(\xi) d\xi \tag{10.23}$$

the extinction of the primary and Raman radiation on their way between the place of scattering (x) and the lidar.

Similarly to the case of DAS lidar, Eq. (10.22) can now be written down for different gases, say, nitrogen and water vapor. If the equations are divided by each other, wavelength-independent parameters including the attenuation of the primary radiation cancel, and depth-independent parameters can be reduced to a constant. The result can be written as

$$N_{H_2O}(x) = H \frac{P(\lambda_P, \lambda_{R,H_2O}, x)}{P(\lambda_P, \lambda_{R,N_2}, x)} \frac{\tau_{R,N_2}(x)}{\tau_{R,H_2O}(x)} N_{N_2}(x). \tag{10.24}$$

This means that the water molecule density is easily determined from the density of nitrogen the mixing ratio of which has been known to vary very little.

The constant H contains the optical efficiencies of the two channels of the polychromator and also the Raman cross sections of the two gases. The latter are available in the literature and could be measured to even better accuracy if required, but the former are tedious to determine and subject to variation if small changes are made on the geometry of the receiver or polychromator. It is therefore customary to adjust the constant H experimentally by comparison of the lidar profile at one point with the result of a carefully calibrated local measurement.

At short distance the difference in atmospheric extinction of the Raman scattered radiation from H_2O and N_2 is small, but cannot be neglected for measurements beyond the planetary boundary layer or when aerosols are present. The molecular contribution to the third factor in Eq. (10.24) is easily determined from Rayleigh scattering theory with a simple atmospheric density model. For the determination of differences due to aerosols another channel can be useful that senses oxygen, another gas with a height-independent mixing ratio. From the deviation of the ratio of the signals measured in the O_2 and N_2 channels from the expected ratio the attenuation behavior of the aerosol can be deduced and corrected for.

Raman lidar offers yet another interesting aspect as to the determination of aerosol properties. Whereas the elastic backscatter signal is affected by both the backscattering and the extinction coefficient of an aerosol, the Raman signal undergoes extinction by the aerosols, but remains unaffected from its backscattering behavior. With an elastic and one (N_2 or O_2) Raman channel it is thus possible to measure the aerosol contributions to atmospheric backscattering and extinction, β and α, separately and independently of assumptions about relations between the two.

The mathematical treatment of the procedure is developed in Ansmann et al. (1990), a detailed error analysis can be found in Ansmann et al. (1992).

10.2.4 Fluorescence lidar

Fluorescence lidar is based on the direct absorption of radiation into a molecular or atomic transition. Differently from the Raman effect the irradiating photons have to satisfy a resonance condition. Accordingly, fluorescence is orders of magnitude more intense than Raman scattering. Although fluorescence always relies on some resonance effect, the return of the atom or molecule to its ground state with emission of the primary wavelength is often but inconsistently called resonance fluorescence, whereas this term is not used when deexcitation occurs to another level and the return radiation is shifted in wavelength.

In the troposphere and lower stratosphere applications of fluorescence lidar have so far remained unimportant. Naturally occuring atmospheric gases do not fluoresce, and fluorescence of pollutants such as aromatic hydrocarbons is too weak for quantitative measurements. The technique has, however, been used for the pursuit of plumes or puff-like emissions to which a fluorescent tracer like fire orange was purposely added (Kyle et al., 1982); in one case the material could be traced up to several hundred kilometers downwind from the emission, although 50 kg of dye had to be used for the experiment (Uthe et al., 1985).

Fluorescence lidar has been very successfully used for stratospheric studies of atomic and ionic species present in heights between 75 and 110 km. Sodium, potassium, lithium and other elements have been observed (Sandford and Gibson 1970; Hake et al., 1972; Felix et al., 1973; Mégie et al., 1978; Pelon et al., 1980; Beatty et al., 1988; Tilgner and von Zahn, 1988). In the maximum of the distribution around 90 km the concentration of Na atoms is only 5 particles per mm^3, and thus almost 16 orders of magnitude less than the density of air molecules at standard conditions. Concentrations of K and Li are even lower. Measurements have been made down to 10^{-3} of the maximum concentration. This proves the sensitivity of the method which lends itself to the study of the movements of the upper part of our atmosphere known as gravity waves (Hauchecorne and Chanin, 1980).

10.2.5 DIALEX and ground reflection

If the measurement system is mounted on an airplane or satellite and the ground is used as a (topographic) reflector, the differential absorption technique with two wavelengths can be used even with a non-depth-resolving system (and, thus, cw lasers) to determine the *total amount* or *average concentration* of a gas in the air column. Such a system called DIALEX was proposed by Wiesemann et al. (1978) and Boscher et al. (1980). Although not a lidar, a device of this kind could be useful for rapid mapping of average loads of pollutants over extended areas.

Clearly, the reflection coefficient of the surface would have to be the same for the signal and reference wavelength, or at least show a spatially independent ratio. It turned out, however, that, at least in the middle infrared where CO_2 lasers represent a convenient source for airborne measurements, reflexion coefficients vary greatly and constitute the major source of uncertainty in the method.

Attempts have therefore been made to conversely use reflection properties of the ground for information about the nature of the soil, rock, or vegetation responsible for the optical echo from the planet's surface. An in-depth analysis (Schlüssel et al., 1994) shows that the intensity of the return signal is determined by no less than 10 ground-related parameters. These parameters include the single-scattering albedos and scattering phase function parameters of soil and vegetation and the leaf area index. Even if up to seven of them have only marginal influence on the magnitude of the reflection coefficient or can be treated in some analytical or empirical way, not much relevant information can so far be extracted from the measured backscatter signal except in a very few special cases such as remote prospection of the surface of the moon for deposits of ilmenite, a titanium mineral rich in oxygen and a candidate for a supplier of O_2 for a permanently-manned lunar station (Melendrez et al., 1994).

10.3 Instruments and Applications - Case Studies

Of the numerous practical implementations of the lidar principle, only a very few will be presented here; the following paragraphs are meant to illustrate the preceding section, rather than comprehensively cover the subject. Examples are all from the author's laboratory, primarily because of easy access to the data and the pictorial material, published or unpublished.

10.3.1 Shipborne IR DAS lidar: HCl distribution, transport, degradation, and lifetime in the marine atmosphere

One of the earliest shipborne DAS lidars was used between 1979 and 1982 for the detection of hydrogen chloride in plumes of incineration ships. These ships were used to destroy large quantities of toxic, highly chlorinated chemical waste that was otherwise hard to dispose of. Incineration was measured to be a clean (no Cl_2, very little NO_x) and very efficient ($>$ 99.98% destruction efficiency, $<10^{-14}$ total dioxins and furanes) process, only the hydrogen chloride to which the chlorine was converted was considered a potentially harmful emission. With up to 83% of Cl contents and a waste throughput of 10 t/h for each incinerator, considerable amounts of the gas were released into the atmosphere. To investigate the concentration distributions in the plume, to elucidate degradation processes, to determine the lifetime, to provide basic data for the application of numerical transport models, to facilitate the choice of incineration sites and to assess the overall environmental effects of the technique, a lidar system was built and mounted on a small ocean-going vessel, the R/V TABASIS.

The system (cf. Fig. 10.8) was equipped with only one laser, a pulsed deuterium fluoride laser that can be made to emit on several lines between 3.5 and 4.0 μm. With a minimum pulse-to-pulse delay of 200 ms and a maximum pulse-pair repetition frequency of 1 Hz, the system did not meet the requirement of $<$ 1-ms pulse delay; this was one of the reasons for the relatively poor sensitivity which, however, was fully adequate for measurements in the plumes of the incineration ships.

Table 10.2 gives the technical data of the lidar. The reference line is sufficiently far from the center of an HCl absorption line so the absorption cross section is only 0.056 x 10^{-24}

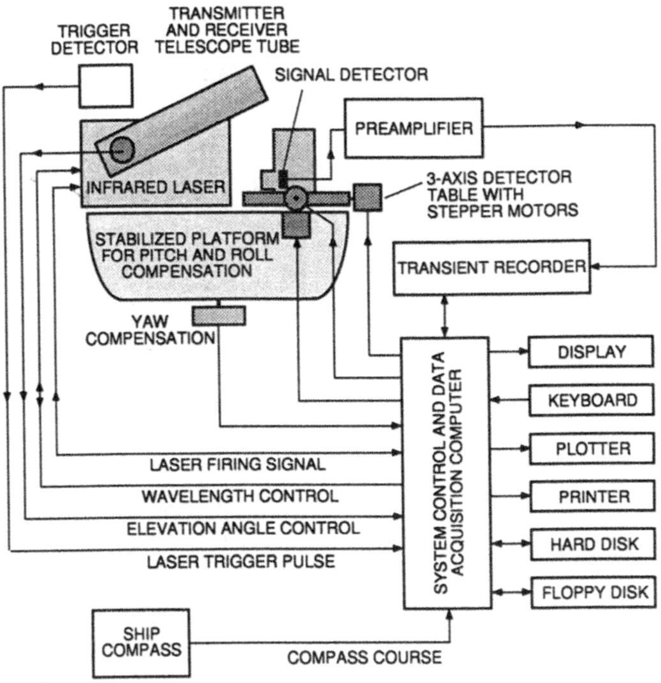

Figure 10.8: *Block diagram of shipborne HCl DAS lidar.*

m^2, as compared to a value of 20.4 x 10^{-24} m^2 for the signal line. Great care was taken to determine cross sensitivities. Plusminus data in Table 10.2 indicate that only an upper limit of the respective cross sections could be found so the sign of the effect could not be determined. However, even if all cross sensitivities have the same sign, the combined effect amounts to little more than 5% of the sensitivity limit and thus remains negligible.

The measurements yield a number of unexpected results. 98% of the emitted HCl was on average present in gaseous form and thus amenable to the lidar measurements, as opposed to only 2% as predicted by some early critics of the proposal. Plumes were usually not visible or visible to the eye from large distance only as one or several faint yellowish stripes, but could clearly be located with the lidar. When water vapor condensation occurred the distribution of (aerosol) backscatter coefficients did not always coincide with the distribution of HCl. Cross sections through plumes were patchy rather than smooth and Gaussian-shaped (Fig. 10.9). Plume dispersion in both height and width was much slower than at land, due to greatly reduced surface roughness and frequently occurring stable stratification, with little vertical exchange. Average 1/e lifetime of the HCl in the marine atmosphere was 30 minutes, with variations of a factor of 3 in both directions, depending on weather conditions (Heinrich et al., 1986; Weitkamp 1988); for dispersion over land an HCl lifetime between 10 and 20 h had previously been reported. A simple unidirectional wind model yielded a ratio of incineration-related to natural HCl crossing the shoreline of the North Sea of 1 part in 10^9, with more refined trajectory models indicating that air masses often go in circles over the sea before reaching the coast, thus giving the HCl even more time for degradation over the water. All data considered,

Laser		DF, pulsed
Emission lines:	signal	DF P_2 (3), 3.636 24 μm
	reference	DF P_2 (5), 3.698 23 μm
Absorption lines:	signal	$H^{37}Cl$ P(6), 3.636 19 μm
	reference	$H^{35}Cl$ P(8), 3.699 59 μm
Pulse energy:	signal	10 mJ
	reference	30 mJ
Pulse-pair repetition rate		1 Hz
Wavelength selection		Grating
Receiver optics		0.50-m afocal Cassegrain
Detector		InSb, liquid-nitrogen cooled
Amplifier bandwidth		500 kHz to 2 MHz, adjustable
Digitizer		10 bit, 10 MHz
Sensitivity		300 ppb
Depth resolution		>5 m
Pulse pairs per profile		15 to 150
Minimum/Maximum distance		100/500 to 300/1500 m, adjustable
Cross sensitivity:	CO_2	$\pm$ 6 ppb @ 300 ppm CO_2
	CH_4	+ 5 ppb @ 2 ppm CH_4
	H_2O	+ 4 ppb @ 2.4% (100% r.h. at 20°C)
	N_2	$\pm$ 2 ppb @ 78%

Table 10.2: *Technical data of shipborne HCl lidar.*

the results indicate that HCl from at-sea incineration of organochlorine waste does in no way threaten marine or terrestrial ecosystems. From a scientific point of view there is no reason for phasing out the technique as was done several years later upon pressure by a number of groups that claim to fight for the protection of the environment.

10.3.2 ARGOS: O_3, SO_2, aerosols

The gases that contribute most to urban air pollution are sulfur dioxide, nitrogen oxides, and ozone; other gases, even if important from a photochemical, toxicologic or olfactory point of view, occur in much smaller concentrations. A system that is well suited for the measurement of SO_2, NO_2, O_3 and a number of other gases is ARGOS, the acronym standing for "Advanced Remote Gaseous Oxides Sensor". ARGOS, although also a DAS lidar, differs from the HCl lidar in a number of important features. For one thing, ARGOS contains two lasers that are fired with 50 μs of delay to produce a signal and a reference pulse well within the critical 1-ms interval. Next, wavelengths of ARGOS are freely tunable so optimum on and off-resonance wavelengths can be chosen for the measurement of each gas. While delivering pulse energies of the same order of magnitude, the repetition rate is 10 times higher, allowing shorter measurement times. Finally ARGOS is equipped with an acoustic wind profiler that provides the data necessary for the transformation of measured concentration distributions into emission and flux or transport values. For full mobility on land ARGOS is built into an 8.6-m standard truck. Figure 10.10 shows a block diagram of the system. Technical data are given in Table 10.3. Table 10.4 gives the performance data of ARGOS. Although measured for SO_2, they differ quite little for O_3 and NO_2; it must be noted, however, that not all of these data are obtained simultaneously

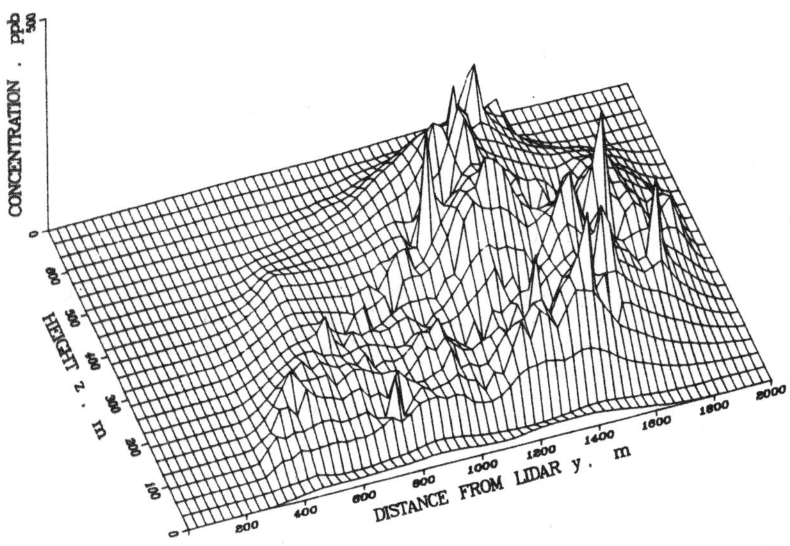

Figure 10.9: *Distribution of HCl in a cross section through the plume of incineration ship MATTHIAS II. Measurement on 24 June 1982, 15:51 to 16:23 h CET, at a distance of 5.7 km from the incinerator.*

and that, e.g., the best possible sensitivity and depth resolution require longer measurements and cannot be obtained throughout the whole range.

Lasers	2 Nd:YAG-pumped dye lasers
Output energies	>15 mJ
Pulse-pair repetition rate	10 Hz
Pulse emission delay	50 μs
Wavelength tuning	Grating
Wavelength measurement and surveillance	Echelle grating spectrometry of dye laser fundamentals
Receiver optics	0.30-m Newtonian
Compression of lidar signal dynamics	Geometrical, diameter of diaphragms 0.2/0.5/0.8/1.0
Detector	PMT Hamamatsu R 212
Digitizer	10 bit, 20 MHz

Table 10.3: *Technical data of mobile DAS lidar ARGOS.*

Measurements of ozone differ from NO_2 and SO_2 measurements in that the absorption cross section of ozone does not show well-defined peaks and valleys, but an almost continuous decrease from a maximum around 255 nm to 0.5, 0.2 and 0.1 times the maximum around 275, 285, and 292 nm (Molina and Molina, 1986). On the other hand, SO_2 has a marked structure in this wavelength range, with cross sections comparable to those of ozone.

Depth resolution	15 m
Range	3 km
Measurement time per profile	1 minute
Sensitivity/precision (1σ)	3 ppb

Table 10.4: *ARGOS performance data (for SO_2 under conditions of ≥ 20 km visibility).*

Care must therefore be taken to use only those pairs of wavelengths for which SO_2 cross sections are minimum and equal. Of the 13 pairs of suitable wavelengths determined from high-resolution SO_2 cross-section measurements (Weitkamp et al., 1992) the optimum pair was determined in an optimization procedure based on the analysis of Eq. (10.16) and taking into account parameters such as range, sensitivity, and systematic error due to aerosols and aerosol gradients (Goers 1994). The results shown in Fig. 10.11 clearly demonstrate that the on/off-resonance pair of 280.91/282.72 nm is a good choice as for longer wavelengths the (negative) error increases, whereas shorter wavelengths (not seen from Fig. 10.11) result in range limitations as soon as medium or high O_3 concentrations are encountered.

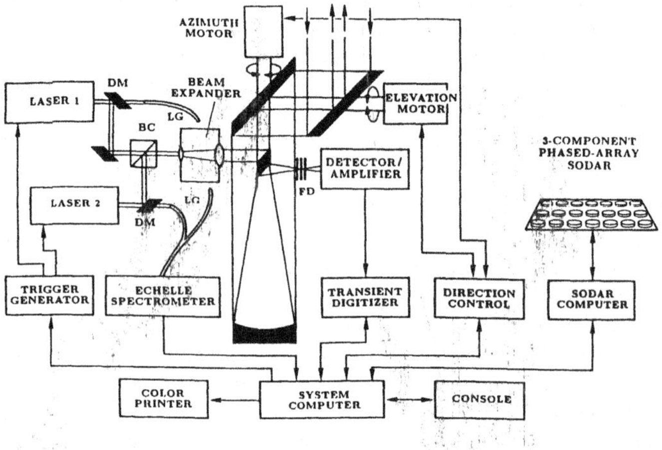

Figure 10.10: *Block diagram of the ARGOS lidar system.*

In several applications ARGOS proved to be an excellent system for the measurement of tropospheric ozone concentration distributions. Time series of concentration profiles from the ground to the top of the mixing layer taken in a rural and a nearby urban environment during a fair-weather summer smog episode occasionally showed O_3 concentrations in excess of 250 $\mu g/m^3$, both at the rural and the urban site. The maxima usually occurred between 600 and 800 m above ground. Surface concentrations did not exceed 150 $\mu g/m^3$. An interesting feature of the time series was a nocturnal maximum in addition to the early-afternoon maximum easily explained by photochemical processes. The nocturnal maximum only appeared aloft, not at ground level, and is attributed to the advection of high-ozone air masses that formed during the day and contain too little oxidizable material (such as NO) to reduce the ozone. Sulfur

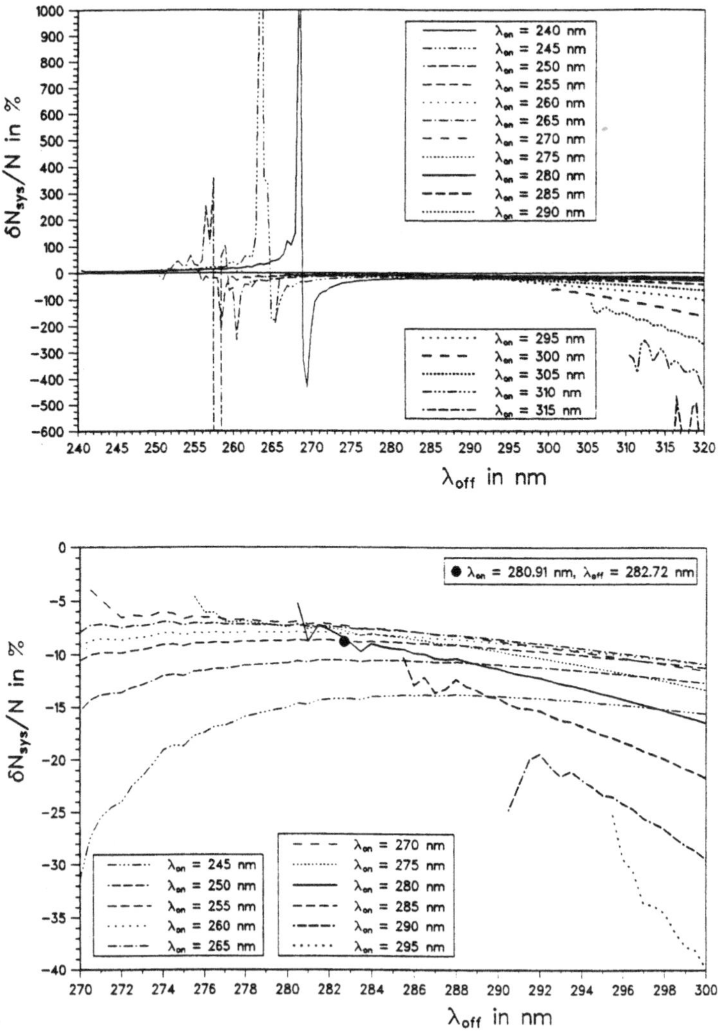

Figure 10.11: *Systematic uncertainty of the ozone concentration, $\delta N_{sys}/N$, measured in a layer with sharp aerosol decrease at the top of the planetary boundary layer for a number of signal wavelengths λ_{on} as a function of the reference wavelengths λ_{off}. The lower part is an extended-scale representation of the region of minimum errors around 285 nm (from Goers, 1994).*

dioxide, although on a slow decrease in most countries of Central Europe, continue to be an important air pollutant that causes millions of dollars of damage every year. ARGOS was used to show that the DIAL technique is well suited for an automatic detection of uncontrolled leaks in the scrubber and exhaust system of a zinc smelter; large quantities of SO_2 can so far escape undetected, particularly if the kind of damage described occurs at night or during weekends.

Another example in which SO_2 distribution measurements with ARGOS played an important role is the determination of the minimum safe distance between inhabited areas and natural-gas production facilities. In case of a blowout caused by a breaking pipe or valve people living downwind of such a facility can be affected both by fire and by the toxic effects of gaseous sulfur compounds contained in so-called sour-gas deposits. As results of model calculations differed by a factor of 100 for relatively small variations in boundary conditions, a blowout was simulated by the release and ignition of a large quantity of sour gas, and the resulting distribution of SO_2 and aerosols was measured with ARGOS.

A third example for SO_2 measurements with ARGOS is a campaign in the Erzgebirge, a mountain range a little more than 1000 m high extending from southwest to northeast and separating Germany from the Czech Republic. Once densely forested, the area suffered from heavy damage when mining and roasting of sulfidic ore began in the middle ages. In the early 19th century silver and zinc mining was gradually given up, but after a short phase of recovery forest damage took over again and has gotten so severe today that bare rock appears in many places, leaving authorities helpless as to how reforestation efforts should be started. Although it had been clear that SO_2 was still the culprit, and that large quantities of the gas are released from coal-fired power plants is the Czech Basin, the exact mechanism responsible for the magnitude of the effect had been unclear. Measurements made with ARGOS from several hilltops down into the adjacent valleys and up into the troposphere revealed that exceptionally poor vertical mixing occurs during long period of time and that the gas, instead of being mixed and diluted, "creeps" through the valleys and up the slopes, strongly damaging the trees' leaves and needles. Local sources, however, also play a role as was measured at Bad Schandau where the River Elbe passes through a gorge in the mountains. Lidar measurements up the river (and upwind) showed comparatively little SO_2, whereas massive SO_2 concentrations were observed downwind of the town. An unexplained phenomenon was the occurrence of SO_2 in the valley on one side, yet not in the one on the other side of a mountain top, indicating that such "microorographic" effects deserve further investigation.

10.3.3 Combined Raman lidar: moisture, cirrus, Pinatubo dust

Because of the small Raman scattering cross sections, the use of the Raman lidar technique is limited to gases with abundances of 10^{-4} or more. Even so, high-power lasers, large receiver telescopes and sensitive detectors along with efficient background-blocking devices are required. Unlike DAS lidar, on the other hand, Raman lidar is not a doubly-differential technique and thus much less sensitive to optical misalignment, small wavelength shifts, and all sources of noise except photon statistics.

A schematic diagram of the GKSS combined Raman lidar setup is shown in Fig. 10.12. For wavelength separation a grating or a filter polychromator (Fig. 10.13) can be used. In either case sufficient suppression of the strong elastic backscatter signal in the Raman channels must be guaranteed. Suppression factors required are 10^7 for the O_2 and N_2 channels and 10^{10} or better for the H_2O and CO_2 channels. Table 10.5 gives the most important technical data of the system (Ansmann et al., 1992).

The combined Raman lidar was used for many determinations of moisture profiles and profiles of aerosol properties. Its appeal in cloud studies is the capability to simultaneously measure water vapor, backscattering, and extinction. In the early days of the system numerous comparisons were made with results of radiosonde ascents. Deviations at low and medium altitudes between the lidar and radiosonde data could be explained by the fact that the lidar is Eulerian in nature, whereas the radiosonde can be considered as "semi-Langrangian" as it travels with the air mass that it measures - horizontally, but not vertically. At greater altitudes, however,

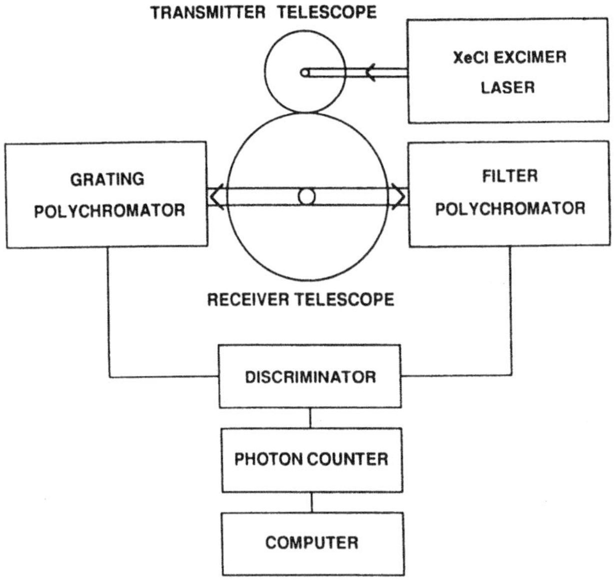

Figure 10.12: *Schematic diagram of the combined Raman lidar.*

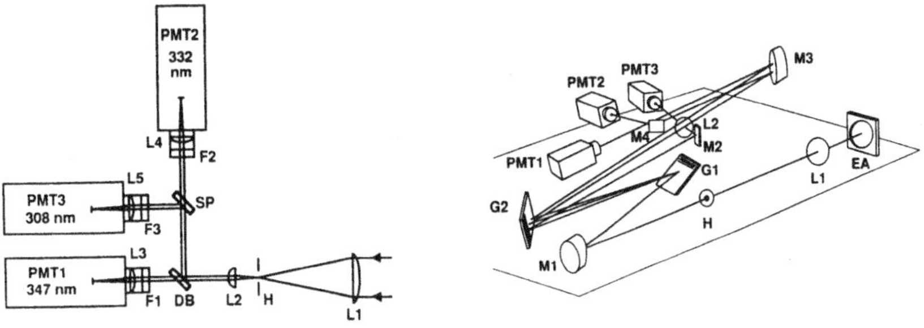

Figure 10.13: *Filter polychromator (left) and grating polychromator (right), schematic. L lenses, EA and H entrance aperture, DB dictroic beam splitter, SP uncoated fused silica plate, F interference filters, M mirrors, G1 echelle grating, G2 holographic grating, PMT photomultiplier tubes.*

systematic deviations occurred that could not be explained from this difference. In Fig. 10.14 the maximum values of these radiosoundings are plotted *vs* temperature. It later turned out that the process used for the measurement of moisture with the radiosondes is not suited in temperatures below −40°C, and routine users of radiosonde data had to take this shortcoming into account for their practical applications.

Measurements of cirrus is another example in which the combined Raman lidar can provide valuable results. The separate measurement of the height profiles of particle backscatter and extinction coefficients, β and α, not only allows a direct assessment of geometric and radiative

Laser	XeCl excimer
Wavelength	308 nm
Output energy	270 mJ
Repetition rate	250 Hz
Receiver optics	0.80-m afocal Cassegrain
Detectors	PMT Thorn EMI 9893 ClB 350
Data acquisition system	3-input-channel multichannel scaler
Maximum count rate	300 MHz
Minimum time-bin width	100 ns
Number of time bins	3 x 1024

Table 10.5: *Technical data of combined Raman elastic-backscatter lidar.*

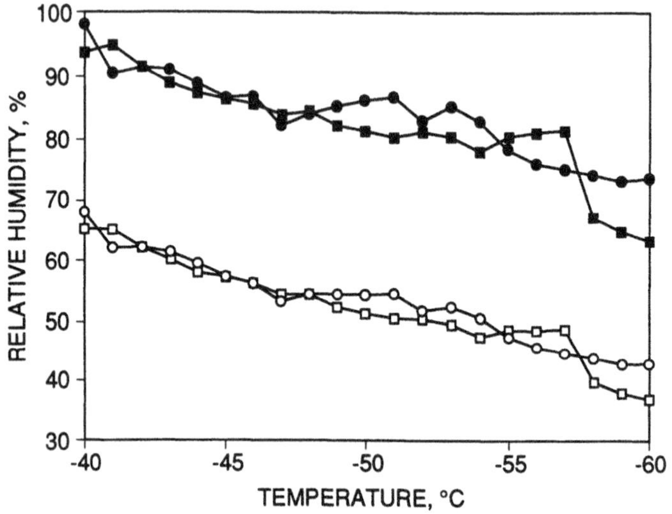

Figure 10.14: *Maximum humidity values ever measured by radiosonde during ICE 89 at the Sylt (circles) and Norderney site (squares), as a function of temperature. 93 and 38 radiosonde ascents were performed. The relative humidity related to water and related to ice is indicated by open and closed symbols, respectively. Clearly these data are extremely unlikely to represent the correct moisture values, and contradict the observed lidar data.*

properties of the cirrus, it also yields microphysical properties such as the mean square (or "effective") particle radius which is related with the lidar ratio α/β and was seen to vary greatly between different parts of the cirrus. In the framework of the International Cirrus Experiment ICE '89 (Raschke et al., 1990) systematic studies of cirrus fields were made from three North-Sea islands in the German Bight, Norderney, Heligoland, and Sylt (Ansmann et al., 1993). Cirrus data above Norderney were taken with α and β measured independently with the combined Raman lidar, whereas the Heligoland and Sylt data were obtained using Klett's inversion method. As is shown in Fig. 10.15 optical and geometric thickness appear to correlate quite well above Norderney, and moderately well above Heligoland, while no such correlation between thickness and transparency of the clouds is seen above Sylt.

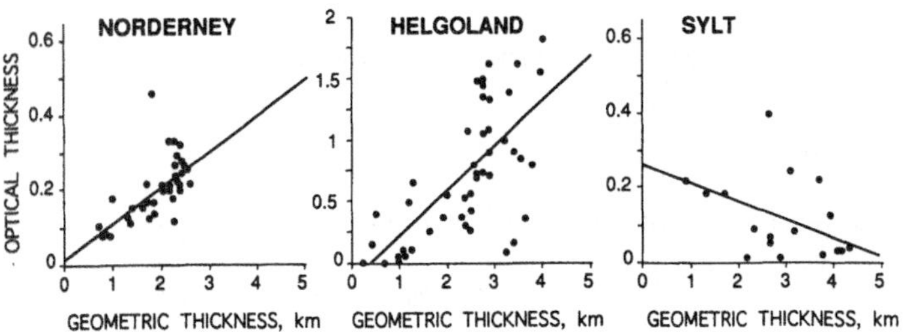

Figure 10.15: *Optical thickness vs geometric thickness of a cirrus deck measured on 18 October 1989 on three islands of the German Bight.*

The determination of aerosol properties is even more important when large-scale catastrophes such as the eruption of a volcano emit substances that subsequently turn into aerosols and get distributed over large parts of the globe. The outbreak of Mount Pinatubo on the Philippines in June 1991 was such an event. In August 1991 irregularities of the lidar return signals were first observed over Geesthacht. From then on until the end of 1992 measurements were made twice a week and continued with somewhat smaller frequency of observation later. Geometric, optical, and microphysical properties were determined both height-resolved and column-integrated (cf. Fig. 10.16 and Table 10.6). An interesting capability of the lidar measurements is to extend from the stratosphere through the tropopause into the upper troposphere. The volcanic aerosol often reaches down well below the tropopause and is easily distinguished here from cirrus and other aerosol by its microphysical properties.

Lidar ratio		25 sr
Effective particle radius		0.30 to 0.35 μm
Geometric thickness of layer		12 km
Optical thickness:	stratosphere only	0.15
	total	0.20
Column mass		0.05 g/m^2
Column-integrated particle surface		2.5 x 10^{11} μm^2/m^2
Mass density:		4 μg/m^3
	maximum	25 μg/m^3
Surface density:		20 x 10^6 μm^2/m^3
	maximum	80 x 10^6 μm^2/m^3
	background 1978/79	0.7 x 10^6 μm^2/m^3
Total mass, earth surface equivalent:	1992, stratosphere only	25 Mt
	1992, total layer	> 30 Mt
Average sink velocity, 1992		3.6 km/a
Mass half life		300 d

Table 10.6: *Pinatubo stratospheric aerosol average data (August 1991 to December 1992).*

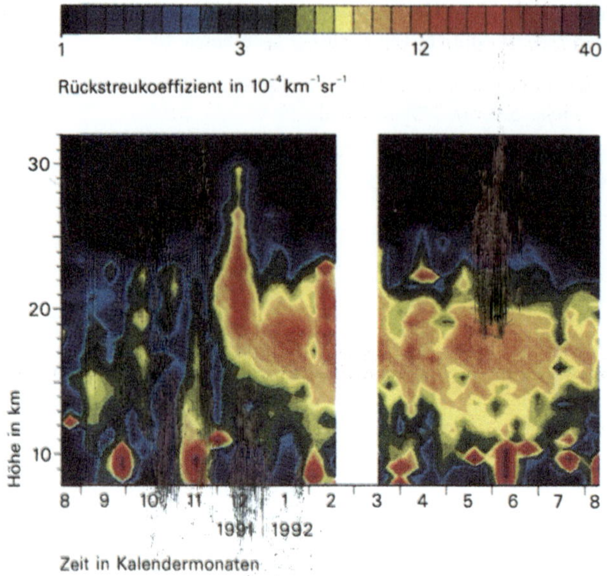

Figure 10.16: *Aerosol content of the lower stratosphere as a function of height and time, for a 12-month period after the eruption of Mount Pinatubo on 15 June 1991, over Geesthacht. No data could be taken between 15 February and 15 March 1992. Text on graph must read (code bar): BACKSCATTER COEFFICIENT, 10^{-4} km^{-1} sr^{-1} (vertical axis): HEIGHT, km (horizontal axis): TIME, month and year.*

The measured data are important for the global effects of volcanic stratospheric aerosol on the attenuation of the primary solar radiation, on the reduction of the surface radiative cooling, on the modification of the global circulation pattern, and on ozone reduction through heterogeneous chemical reactions in the stratosphere. In spring 1992, the measurements showed the first polar stratospheric clouds ever observed as far south as 54°N as a kind of byproduct.

10.4 Water Vapor Lidar Overview

10.4.1 DAS lidars

Historically, the first lidars to measure moisture were infrared differential absorption and scattering lidars operating in the 10-μm wavelength range. Following the pioneering work of Murray et al. (1976) and Murray et al. (1979), the technique has seen many improvements and refinements up to our days (Baker, 1983; Hardesty, 1984; Grant et al., 1987; Barbini et al., 1990).

10-μm lidars are inherently eyesafe. As the CO_2 lasers used are normally operated in a line-tunable regime, no wavelength or power stabilization is required, and wavelength selection is easy. CO_2 lasers have excellent wall-plug efficiency.

On the other hand, a number of drawbacks must be faced. Rayleigh backscatter is weak at 10 μm, so the technique relies more on Mie backscatter than other lidar variants. It is therefore better suited for horizontal than for vertical profiling. As aerosol has a tendency to change with moisture conditions, backscatter coefficients also change with moisture, and usually do so differently for different laser lines. CO_2 laser pulses show a long (1 μs) tail that is hard to correct for if good depth resolution is required. Finally line strengths are temperature-dependent, making a simultaneous measurement of the temperature profile desirable.

Therefore the visible and near-visible infrared wavelength range was also investigated. Browell et al. (1979) used a ruby-laser-pumped dye laser along with the ruby fundamental. Later Nd:YAG-laser-pumped dye lasers (Browell et al., 1981; Cahen et al., 1982; Ehret and Renger, 1988; Ehret et al., 1993), excimer-pumped dye lasers (Bösenberg et al., 1990), as well as alexandrite (Higdon et al., 1993) and titanium-sapphire lasers (Wulfmeyer et al., 1995) were all used for the purpose.

The temperature and pressure dependence, i.e., the height dependence of parameters such as the wavelength and cross section of the on-resonance line, the Rayleigh Doppler broadening of the backscatter signal, possible shift and broadening of laser emission lines as well as laser spectral impurity all had to be investigated (Schotland, 1974; Measures, 1977; Zuev et al., 1983; Ansmann, 1984; Brown, 1985; Zuev et al., 1985; Bösenberg, 1985; Ansmann and Bösenberg, 1987; Ansmann, 1989) and have made DAS lidar moisture measurements a challenging venture up to our days.

10.4.2 Raman lidars

It was therefore templing to try the Raman lidar concept for moisture determination. First measurements were carried out with frequency-doubled ruby (Melfi et al., 1969) and frequency-quadrupled Nd:YAG lasers (Renaut et al., 1980). Frequency-doubled and frequency-tripled Nd:YAG lasers were then used because their emission wavelengths provide better sensitivity or range (Bukin et al., 1985; Melfi and Whiteman, 1985; Vaughan et al., 1988). A big step forward was made when excimer lasers came into existence; KrF (Cooney et al., 1985; Eichinger et al., 1994) and XeCl (Weitkamp et al., 1992; Bisson and Goldsmith, 1993) have remained the most popular lasers until now.

Today double-laser Raman lidar systems (McGee et al., 1993; Reichardt et al., 1995) provide the best results because of their capability to measure ozone, temperature, extinction and backscatter in addition to the water vapor mixing ratio and because they can apply the novel principle of Raman DIAL, a differential-absorption technique not based on Rayleigh-Mie, but on Raman scattering in the atmosphere.

Clearly, Raman cross sections which are $\leq 10^{-33}$ m^2 sr^{-1} require powerful lasers, high pulse repetition rates, large receiver telescopes and sensitive detectors. Skylight interference makes the use of narrow fields of view and of appropriate filters or other wavelength-selective and wavelength-blocking devices mandatory. The geometric adjustment is time-consuming, particularly for the low-intensity water vapor channel. The best choice of the wavelength (cf. Table 10.7) is difficult and depends somewhat on the purpose of the measurement: At long wavelengths the λ^{-4} dependence of Raman scattering results in small signals superimposed by intense skylight, at short wavelengths tropospheric ozone reduces the range so drastically that no reasonable range can be obtained. It appears to be today generally accepted that except for studies of the lowest part of the troposphere, wavelengths between 308 and 355 nm are best suited for moisture profiling with Raman lidar.

Apart from these difficulties, Raman moisture lidar offers a number of advantages over other lidar schemes. For Raman lidar only one laser is needed. Its wavelength, width, and stability are uncritical. The geometric alignment is not critical once the polychromator is properly adjusted.

Laser	Wavelength λ_0 (nm)	λ_{N_2} (nm)	λ_{H_2O} (nm)	Atmosphere absorption @λ_0	λ_{H_2O} in solar blind region	In practical use
KrF	248	263	273	Very strong	Yes	Yes
KrF + CO_2S1	257	273	284	Very strong	Yes	No
Nd:YLF x 4	263	280	291	Strong	Yes	No
Nd:YAG x 4	266	284	295	Moderate	Barely	Little
XeCl	308	332	357	Weak	No	Yes
Ruby x 2	347	378	398	Absent	No	In early days
XeF	353	383	403	Absent	No	Yes
Nd:YAG x 3	355	388	408	Absent	No	Yes

Table 10.7: *Water-Vapor Raman Lidars: Optimum Choice of Radiation Sources.*

Data analysis is straightforward, easy, fast, and simple. And, perhaps most important, aerosols may affect the range, but bear no influence of the accuracy of the result.

10.5 Recent Developments

Although moisture lidar and lidar for the determination of atmospheric constituents in general is available in several variants that operate on a routine basis, the field is still in rapid development both as to the improvement of components and to the proposal of entirely new concepts.

10.5.1 Broadband-emission lidar with narrowband detection of absorption (BELINDA)

The DAS lidar scheme has so far suffered from the necessity to emit into the atmosphere two pulses of different, yet carefully controlled wavelengths. For most practical purposes this meant the use of two lasers. The BELINDA concept is based on the emission of *one* laser pulse comparable in width with the width of the absorption line of interest and the analysis of the backscattered radiation in *two* pairs of wavelength windows, one pair close to (but not at the maximum of) the absorption line, playing the part of the on-resonance, and the other pair in the wings of the line, playing the role of the off-resonance radiation in a classical DAS lidar. The concept is now being investigated for its practicability (Theopold et al., 1993; Linow et al., 1994).

10.5.2 Pulse lasers for the 1–4 μm wavelength range

Water vapor as well as a few other gases show distinct absorption in the wavelength range between 1 and 4 μm. This wavelength range depends less on the presence of aerosols for backscattering than does the 10-μm range, and radiation above 1.4 μm is totally eyesafe. Promising laser materials for the generation of tunable radiation in this wavelength range are TmHo:YAG (Ghibaudo and Krawczyk, 1992) and TmCrHo:YAG (Lasarev et al., 1994; Killinger et al., 1994); frequency mixing in nonlinear crystals is also a possible concept for the exploitation of near-ir wavelengths for differential absorption and scattering lidar.

10.5.3 Near-IR statistical (cw) DAS lidar

Instead of using repetitively-pulsed lasers and deriving the depth information from the r eturn-pulse arrival time according to Eq. (10.1), we can also obtain depth profiles from the cross-correlation function of a statistically modulated continuous-wave laser beam transmitted into the atmosphere with the observed temporal distribution of the return signal (Takeuchi et al., 1983). In view of a revival of this technique, tunable multi-wavelength external-cavity diode lasers (Papen et al., 1994) and also Ar^+/Ti:Al_2O_3-pumped single-frequency HoTm:YLF lasers (Killinger et al., 1994) that are presently being developed may be interesting sources of radiation.

10.5.4 New components

Of the numerous innovations that occur in optics only a few shall be mentioned here because in the author's mind they bear the greatest potential for bringing down the size, weight, cost and operation effort associated with present-day lidars and at the same time improving the performance of these devices.

Werner et al. (1992) proposed the use of an all-optical amplifier instead of the currently used electronic devices. Avalanche photodiodes (APDs) of both silicon (Dautel et al., 1993) and germanium (Tanaka et al., 1994) are likely to soon replace conventional photodetectors. A few gaps remain to be closed in our understanding of the efficient protection of photomultiplier tubes by electric (Bristow et al., 1994) and optical switching (Lehmann et al., 1994). Finally Langford et al. (1994) propose a signal dithering technique suited for extending the effective range of conventional transient recorders.

What ground-based lidar can measure	What ground-based lidar cannot measure
Atmospheric constituents	Atmospheric constituents
$\leq 10^{-8}$ parts by volume	$< 10^{-8}$ parts by volume
Thin clouds and other aerosols:	Thin clouds and other aerosols:
geometry	exact shape of particles
cloud coverage	full particle size distribution
scattering ratio	
multiple/single scattering ratio	
lidar ratio	
depolarization	
effective particle radius	
mass density	
surface density	
Thick clouds:	Thick clouds:
cloud base	properties beyond OD ~ 2
properties @ OD ~ 2	

Table 10.8: *Lidar Measurements: Atmospheric Constituents, Clouds, and Ground Reflectance Summary.*

10.6 Conclusion

In the present paper an attempt is made to describe the lidar principle of operation and to show a few examples of lidar applications in the measurement of atmospheric constituents including aerosols. The concentration limit for lidars can be stated to be around 10 parts in 10^9, or 10 ppb, although a few substances can be measured in concentrations many orders of magnitude below that limit. Measurement capabilities in clouds are summarized in Table 10.8.

For gaseous pollutants immission, emission, and transport data can be obtained, lifetimes determined, and sources localized. Horizontal, vertical and slant profiles as well as fans of such profiles provide three-dimensional distributions. Snapshots and time series can be generated; the potential of measuring rare events by automatically triggered systems is only beginning to be investigated.

What is still difficult to achieve with lidar is all-weather operation, although measurements in light rain can be carried out. Also fully automated long-term unattended service and, thus, worldwide target coverage by lidars on board satellites is not available today except for the very simplest implementations of the technique.

10.7 References

For abbreviations (ILRC, ISTP, ORSA, SPIE) see Section "Additional Reading"

Ansmann A (1984) Diplomarbeit, Universität Hamburg

Ansmann A (1989) Bodengebundene DIAL-Wasserdampfmessung: Berücksichtigung der Dopplerverbreiterung der Laserlinie durch Rayleighrückstreuung. Dissertation, Universität Hamburg. *Hamburger Geophysikalische Einzelschriften*, Reihe A, Heft 89: 69 p

Ansmann A, Bösenberg J (1987) Correction scheme for spectral broadening by Rayleigh scattering in differential absorption lidar measurements of water vapor in the troposphere. *Applied Optics* 26: 3026–3032

Ansmann A, Riebesell M, Weitkamp C (1990) Measurement of atmospheric aerosol extinction profiles with a Raman lidar. *Optics Letters* 15: 746–748

Ansmann A, Bösenberg J, Brogniez G, Elouragini S, Flamant PH, Klapheck K, Linn H, Menenger L, Michaelis W, Riebesell M, Senff Ch, Thro P-Y, Wandinger U, Weitkamp C (1993) Lidar Network Observations of Cirrus Morphological and Scattering Properties during the International Cirrus Experiment 1989: The 18 October 1989 Case Study and Statistical Analysis. *Journal of Applied Meteorology* 32: 1608–1622

Ansmann A, Riebesell M, Wandinger U, Weitkamp C, Michaelis W (1991a) Combined Raman elastic-backscatter lidar for the independent measurement of aerosol backscatter and extinction profiles. Report GKSS 91/E/42: 8 p

Ansmann A, Riebesell M, Wandinger U, Weitkamp C, Michaelis W (1991b) Klett forward-backward integration for model-independent determination of the aerosol extinction-to-backscatter ratio. GKSS 91/E/43: 8p

Ansmann A, Riebesell M, Wandinger U, Weitkamp C, Voss E, Lahmann W, Michaelis W (1992) Combined Raman Elastic-Backscatter LIDAR for Vertical Profiling of Moisture, Aerosol Extinction, Backscatter, and LIDAR Ratio. *Applied Physics* B 55: 18–28

Ansmann A, Wandinger U, Riebesell M, Weitkamp C, Michaelis W (1992) Independent measurement of extinction and backscatter profiles in cirrus clouds by using a combined Raman elastic-backscatter lidar. *Applied Optics* 31: 7113–7131

Ansmann A, Wandinger U, Weitkamp C (1993) One-Year Observations of Mount-Pinatubo Aerosol with an Advanced Raman Lidar over Germany at 53.5°N. *Geophysical Research Letters* 20: 711–714

Baker PW (1983) Atmospheric water vapor differential absorption measurements on vertical paths with a CO_2 lidar. *Applied Optics* 22: 2257–2264

Barbini R, Colao F, Palucci A, Ribezzo S, Orlando S (1990) Remote Sounding of Atmospheric Water Vapour from the ENEA DIAL Station. 15ILRC II: 156–159

Bisson SE, Goldsmith JEM (1993) Daytime Tropospheric Water Vapor Profile Measurements with a Raman Lidar. 5ORSA: 19–22

Boscher J, Englisch W, Wiesemann W (1980) Differentielle Absorptions-Spektroskopie mit dem Fernanalysesystem Dialex. *Laser + Elektro Optik* 12 (3): 17–22

Bösenberg J (1985) Measurements of the pressure shift of water vapor absorption lines by simultaneous photoacoustic spectroscopy. *Applied Optics* 24: 3531–3534

Bösenberg J, Senff C, Thro PY (1990) DIAL Measurements of Water Vapor in the Troposphere: Assessment of Accuracy. 15ILRC II:170–172

Braun WC (1985) Simplified calculations for accuracy of a lidar dial system to measure atmospheric H_2O vapor and temperature. *Applied Optics* 24: 109–117

Bristow M, Bundy D, Wright A (1994) Photomultipliers and Gating Circuits Suitable for Differential Absorption Lidars. 17ILRC: 191–193

Browell EV, Wilkerson TD, McIllrath TJ (1979) Water vapor differential absorption lidar development and evaluation. *Applied Optics* 18: 3474–3483

Browell EV, Carter AF, Wilkerson TD (1981) Airborne differential absorption lidar system for water vapor investigations. *Optical Engineering* 20 (1): 084–090

Bukin OA, Stolyarchuk YuS, Tyapkin VA (1985) Measurement of Moisture-Content Profiles in the Bottom Layer of the Atmosphere by the Method of Spontaneous Light-Scattering Spectroscopy. *JAS Bulg* 43: 631–636

Cahen C, Mégie G, Flamant P (1982) Lidar Monitoring of the Water Vapor Cycle in the Troposphere. *Journal of Applied Meteorology* 21: 1506–1515

Cooney J, Petri K, Salik A (1985) Measurements of high resolution atmospheric water-vapor profiles by use of a solar blind Raman lidar. *Applied Optics* 24: 104–108

Dautet H, Deschamps P, Dion B, McGregor AD, McSween D, McIntyre RJ, Trottier C, Webb PP (1993) Photon counting technique with silicon avalanche photodiodes. *Applied Optics* 32 : 3894–3900

Ehret G, Renger W (1988) Airborne Water Vapor DIAL. 14ILRC: 190–191

Ehret G, Kiemle C, Renger W, Simmet G (1993) Airborne remote sensing of tropospheric water vapor with a near-infrared differential absorption lidar system. *Applied Optics* 32: 4534–4551

Eichinger WE, Cooper DI, Archuletta FL, Hof D, Holtkamp DB, Karl RR, Quick CR, Tiee J (1994) Development of a scanning, solar-blind, water Raman lidar. *Applied Optics* 33: 3923–3932

Fernald FG (1984) Analysis of atmospheric lidar observations. Some comments. *Applied Optics* 23: 652–653

Ghibaudo J-B, Krawczyk R (1992) Water vapor, temperature and wind velocity measurements from space using 2 μm Tm:Ho;YAG. SPIE 1714: 258–269

Goers U-B (1994) Laserfernmessung von Schwefeldioxid und Ozon in der unteren Troposphäre mit Hilfe der differentiellen Absorption und Streuung unter den Bedingungen des mobilen Einsatzes und der besonderen Berücksichtigung des Einflusses von Grenzschichtaerosolen. Dissertation, Universität Hamburg, 1994 . Report GKSS 94/E/52: 147p

Grant WB, Margolis JS, Brothers AM, Tratt DM (1987) CO2 DIAL measurements of water vapor. *Applied Optics* 26: 3033–3042

Hardesty RM (1984) Coherent DIAL measurement of range-resolved water vapor concentration. *Applied Optics* 23: 2545–2553

Hauchecorne A, Chanin M-L (1980) Planetary waves-mean flow interaction in the middle atmosphere: Numerical modeling and lidar observations. *Annales Geophysicae* 6: 409–416

Heinrich H-J, Eck I, Weitkamp C (1986) The distribution of hydrogen chloride in plumes of incineration vessels: remote measurement of concentration distributions and determination of dilution and degradation parameters. Report GKSS 86/E/44: 162p

Killinger DK, Vaidyanathan M, He C, Taczak T (1994) High-Resolution Spectral Studies of Ho Lasers for Lidar/DIAL Applications. 17ILRC: 298–300

Killinger DK, Mooradian A, eds. (1983) Optical and Laser Remote Sensing. *Springer*, Berlin Heidelberg New York, 383p

Klein V, Werner C (1993) *Fernmessung von Luftverunreinigungen.* Springer, Berlin Heidelberg New York London Paris Tokyo Hongkong Barcelona Budapest, 254p

Klett JD (1981) Stable analytical inversion solution for processing lidar returns. *Applied Optics* 20: 211–220

Kobayashi T (1987) Techniques for Laser Remote Sensing of the Environment. *Remote Sensing Reviews* 3: 1–56

Kunz GJ (1987) Lidar and missing clouds. *Applied Optics* 26: 1161

Kyle TG, Barr S, Clements WE (1982) Fluorescent particle lidar. *Applied Optics* 21:14–15

Langford AO, O'Leary TJ, Proffitt MH (1994) Extending the Dynamic Range of Differential Absorption Lidar Measurements through Large-Scale Dithering. 17ILRC: 173–174

Lasarev VV, Matvienko GG , Ponomarev YN, Rybalko VS, Tyryshkin IS (1994) The Design of Eye Safety YAG: TmCrHo Pulsed Laser and Study the Energy Losses of its Radiation in Air and Gas-Aerosol Media. 17ILRC: 275–276

Lehmann S, Wulfmeyer V, Bösenberg J (1994) A Time Dependent Attenuation for Dynamic Range Reduction of Lidar-Signals. 17ILRC: 289–290

Linow S, Theopold F, Weitkamp C, Michaelis W (1994) Properties of a Double-Cavity Etalon. 3ISTP: 366–369

McGee TJ, Gross M, Ferrare R, Heaps W, Singh U (1993) Raman Dial measurements of stratospheric ozone in the presence of volcanic aerosols. *Geophysical Research Letters* 20: 955–958

Measures RM (1977) Lidar equation analysis allowing for target lifetime, laser pulse duration, and detector integration period. *Applied Optics* 16:1092–1103

Measures RM (1984) *Laser Remote Sensing.* Wiley-Interscience, New York Chichester Brisbane Toronto Singapore: 510 p

Melendrez DE, Johnson JR, Larson SM, Singer RB (1994) Remote sensing of potential lunar resources 2. High spatial resolution mapping of spectral reflectance ratios and implications for nearside mare TiO_2 content. *Journal of Geophysical Research* 99 (E3): 5601- 5619

Melfi SH, Lawrence JD, McCormick MP (1969) Observation of Raman Scattering by Water in the Atmosphere. *Applied Physics Letters* 15: 295–297

Melfi SH, Whitemann D (1985) Observation of Lower-Atmospheric Moisture Structure and Its Evolution using a Raman Lidar. *Bulletin of the American Meteorological Society* 66: 1288–1292

Molina LT, Molina MJ (1986) Absolute Absorption Cross Sections of Ozone in the 185- to 350-nm Wavelength Range. *Journal of Geophysical Research* 91: 14501–14508

Murray ER, Hake RD, van der Laan JE, Hawley JG (1976) Atmospheric water vapor measurements with an infrared (10-μm) differential-absorption lidar system. *Applied Physics Letters* 28: 542–543

Murray ER, Powell DD, van der Laan JE (1980) Measurement of average atmospheric temperature using a CO_2 laser radar. *Applied Optics* 19: 1794–1797

Papen GC, Murphy GM, Koch GJ, Dejule RY, Kaliski RW (1994) Tunable Multiple Wavelength External Cavity Diode Lasers for Remote Sensing Applications. 17ILRC: 194–195

Raschke E, Schmetz J, Heintzenberg J, Kandel R, Saunders R (1990) The International Cirrus Experiment (ICE) - A joint European Effort. *ESA Journal* 14: 192–199

Renaut D, Pourny JC, Capitini R (1980) Daytime Raman-lidar measurements of water vapor. *Optics Letters* 5: 233–235

Schlüssel G, Dickinson RE, Privette JL, Emery WJ, Kokaly R (1994) Modeling the bidirectional reflectance distribution function of mixed finite plant canopies and soil. *Journal of Geophysical Research* 99: 10577–10600

Schotland RM (1974) Errors in the Lidar Measurement of Atmospheric Gases by Differential Absorption. *Journal of Applied Meteorology* 13: 71–77

Takeuchi N, Sugimoto N, Baba H, Sakurai K (1983) Random modulation cw lidar. *Applied Optics* 22: 1382–1386

Tanaka M, Sakurai S, Kobayashi F, Saito Y, Kano T, Nomura A (1994) Possibility of Photon Counting in Near-Infrared (0.8 - 1.5 μm) Region by Ge-APD. 17ILRC: 291–294

Theopold F, Weitkamp C, Michaelis W (1993) Double-cavity étalon in the near infrared. *Optics Letters* 18: 253–254. Report GKSS 93/E/15

Uthe EE, Viezee W, Morley BM, Ching JKS (1985) Airborne Lidar Tracking of Fluorescent Tracers for Atmospheric Transport and Diffusion Studies. *Bulletin of the American Meteorological Society* 66: 1255–1262

Vaughan G, Wareing D P, Thomas L, Mitev V (1988) Humidity measurements in the free troposphere using Raman backscatter. *Quarterly Journal of the Royal Meteorological Society* 114: 1471–1484

Vaughan G, Wareing DP, Peper SJ, Thomas L, Mitev V (1993) Atmospheric temperature measurements made by Rotational Raman scattering. *Applied Optics* 32: 2758–2764

Weitkamp C (1988) Infrared lidar measurement of the diffusion of hydrogen chloride from seaborne waste incineration. In: R Kesselring, FK Kneubühl eds: *Fourth International Conference on Infrared Physics*, ETH Zürich, Switzerland, 22-26 August 1988. Proceedings, Zürich 1988: 218–226

Weitkamp C, Thomsen O, Bisling P (1992) Mess- und Vergleichswellenlängen zur Elimination von SO_2-Querempfindlichkeiten bei der Lidar-Fernmessung troposphärischen Ozons. *Laser und Optoelektronik* 24 (2): 46–52

Werner Ch, Murphy E, Schwiesow R (1992) Analysis of Optical Amplifiers applied to Short-Wavelength Doppler Lidars using Direct Detection. SPIE 1714: 284–290

Wiesemann W, Beck R, Englisch W, Gürs K (1978) In-Flight Test of a Continuous Laser Remote Sensing System. *Applied Physics* 15: 257–260

Wulfmeyer V, Bösenberg J, Lehmann S, Senff C, Schmitz St (1994) Injection-seeded alexandrite ring laser: performance and application in a water-vapor differential absorption lidar. *Optics Letters* 20: 638–640

Zeyn J, Voss E, Lahmann W, Weitkamp C, Michaelis W (1994) Daytime temperature lidar based on rotational Raman scattering. 3ISTP 2: 262–265

Zuev VV (1983) Lidar differential absorption and scattering technique: theory. *Applied Optics* 22: 3733–3741

Zuev VV, Ponomarev Yu N, Solodow AM, Tikhomirov BA, Romanovsky OA (1985) Influence of the shift of H_2O absorption lines with air pressure on the accuracy of the atmospheric humidity profiles measured by the differential-absorption method. *Optics Letters* 10: 318–320

Additional Reading

Becherer RJ, Werner C editors (1992). *Lidar for Remote Sensing.* SPIE Volume 1714. Bellingham, WA, USA: SPIE - The International Society for Optical Engineering, 336 p

Killinger DK, Mooradian A, eds. (1983) *Optical and Laser Remote Sensing.* Springer, Berlin Heidelberg New York, 383 p

Kobayashi T (1987) Techniques for Laser Remote Sensing of the Environment. *Remote Sensing Reviews* 3: 1–56

Klein V, Werner C (1993) *Fernmessung von Luftverunreinigungen.* Springer, Berlin Heidelberg New York London Paris Tokyo Hongkong Barcelona Budapest, 254 p

Measures RM (1984) *Laser Remote Sensing.* Wiley-Interscience, New York Chichester Brisbane Toronto Singapore: 510 p

Weitkamp C (1990) Lidar. In Ruck B ed. (1990) *Lasermethoden in der Strömungsmeßtechnik*, Stuttgart AT-Fachverlag, 151–208

A prolific source of information is also the proceedings of several series of conferences such as the Topical Meetings on Optical Remote Sensing of the Atmosphere (ORSA), the International Symposia on Tropospheric Profiling: Needs and Technologies (ISTP) and the International Laser Radar Conferences (ILRC):

3ORSA 1990. 12–15 February 1990, Incline Village, NV, USA, Optical Society of America, 1990 Technical Digest Series Volume 4, 650 p

4ORSA 1991, 18–21 Novemberr 1991, Williamsburg, VA, USA, OSA, 1991 Technical Digest Series Volume 18, 332 p

5ORSA 1993, 8–12 March 1993, Salt Lake City, UT, USA, OSA, 1993 Technical Digest Series Volume 5, 468 p

6ORSA 1995, 6–10 February 1995, Salt Lake City, UT, USA, OSA, 1995 Technical Digest Series Volume 2, 236 p

1ISTP, 31 May to 3 June 1988, Boulder, CO, USA, National Center for Atmospheric Research, 260, 10 p

2ISTP, 10 to 13 September 1991, Boulder, CO, USA, National Center for Atmospheric Research, 214 p

3ISTP, 30 August to 2 September 1994, Hamburg, Germany, Max-Planck-Gesellschaft zur Förderung der Wissenschaften, Volume 1 p. 1–200, Volume 2 p. 201–462

12ILRC, 13 to 17 August 1984, Aix-en-Provence, France, 451 p

13ILRC, 11 to 16 August 1986, Toronto, ON, Canada, National Aeronautics and Space Administration, NASA Conference Publication 2431, 321 p

14ILRC, 20–23 June 1988, Innichen-San Candido, Italy, 512 p

15ILRC, 23–27 July 1990, Tomsk, USSR, Institute of Atmospheric Optics, Volume 1, 404 p., Volume 2, 430 p

16ILRC, 20–24 July 1992, Cambridge, MA, USA, National Aeronautics and Space Administration, NASA Conference Publication 3158, Volume 1 p. 1–380, Volume 2 p. 381–732

17ILRC, 25–29 July 1994, Sendai, Japan, Laser Radar Society of Japan, 592 p

18ILRC, Summer 1996, to be held in Berlin, Germany

Chapter 11

Retrieval of Precipitation from Satellites

C. Simmer
Institut für Meereskunde
Düsternbrooker Weg 20, 24105 Kiel
Germany

11.1 Introduction

For land areas, precipitation is one of the most important processes for the biosphere, including mankind. For the oceans, precipitation is a source for freshwater and thus an important forcing term for the thermohaline circulation. In the atmosphere, precipitation liberates latent energy, which is a main forcing term for atmospheric motions, especially in the tropics. The global average precipitation is equivalent to a water column of about 1 m height. This amounts to a continuous energy supply to the atmosphere of about 80 Wm^{-2}, which is roughly 30% of the incoming solar energy. Mainly caused by its strong coupling to the vertical velocity field, precipitation is extremely variable in space and time.

Over most areas of the globe, especially over sea, there are not enough direct measurements for the needs of an adequate understanding of the hydrological cycle. In addition, measurements over land are hampered by orographic effects. Over sea, errors are caused by the air-flow obstruction caused by the ship. Remote sensing from the surface, e.g. by radar, enable high temporal sampling and larger areal coverage, but cannot be applied globally due to high costs for equipment and personnel. In addition, the inversion of radar measurements is not straightforward and has problems of its own like uncertainy of dropsize spectra and the increased elevation of the radar beam from the earth surface with increasing distance. Remote sensing from satellites is left as the only practible method to achieve global coverage of precipitation estimates.

11.2 General problems related to remote sensing of rain

Remote sensing of any parameter is always based on the interpretation of the intensity field of electromagnetic radiation (radiance) emanating from the raining cloud and its environment.

NATO ASI Series, Vol. I 45
Radiation and Water in the Climate System:
Remote Measurements
Edited by Ehrhard Raschke

This radiation originates either from the earth-atmosphere system itself or from the sun (passive remote sensing) or has been produced by an artificial radiation source (active remote sensing). We will see, that in the case of precipitation the interpretation process always consists of the following two steps:

1. Parameters describing the activity of the raining process are determined from the radiance field.

2. These parameters are interpreted in terms of the rain rate at the surface.

In the following, we abbreviate the rainrate at the surface with RR. We must accept that there is no direct signal of RR in the measured radiances. The closest we can get to RR with passive methods is a rough measure of liquid water or ice in the atmospheric column. Active methods like the rain radar on TRMM (Tropical Rainfall Measurements Mission, e.g. Simpson et al., 1988) will in the future provide profiling of thee quantities and also estimates of rain-drop fall velocities, but there will never be a direct equivalent to RR.

A prerequisite to allow remote sensing of rain from satellites is transparency of the nonraining atmosphere for electromagnetic radiation. We can see from Fig. 11.1, that except in the far infrared between 15 μm and 1 mm there are regions (intersected by gas absorption line and bands) in every part of the spectrum, where rain retrieval is in general possible.

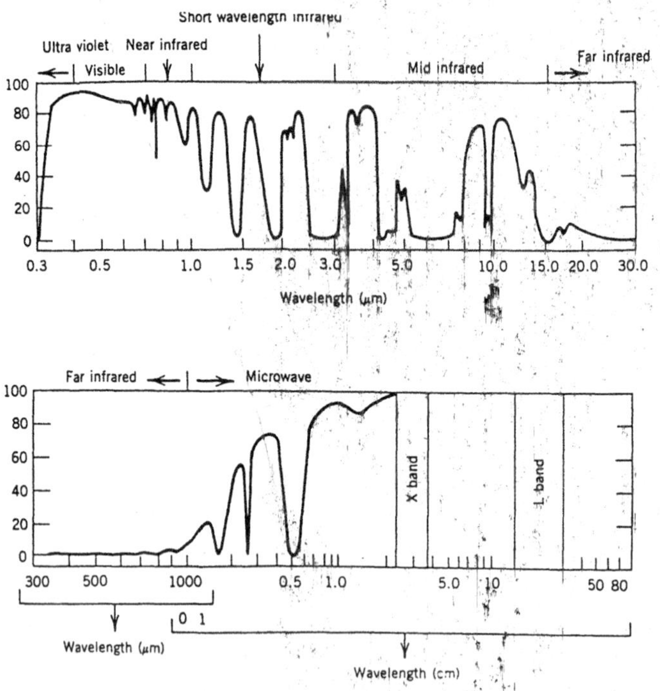

Figure 11.1: *Nadir transmittance of a cloud-free atmosphere (schematic).*

11.3 Interaction between atmospheric hydrometeors and radiation

The earth atmosphere is roughly a mixture of gases, aerosols, and hydrometeors, which differ by size, shape, and substance. The naturally occuring interaction processes between atmospheric constituents and electromagnetic radiation are absorption due to resonance effects caused by changes of the electronic, vibrational, and rotational states of molecules and emission, absorption, and scattering caused by inhomogeneities of the complex refractive index m (or dielectric constant ϵ) of the atmosphere. For scattering to be efficient, these inhomogeneities must occur in a spatial scale comparable to the wavelength. Thus, the non-dimensional Mie-parameter χ, defined as being proportional to the ratio between the spatial dimension of the inhomogeneity, i.e. the radius of the particle r, and the wavelength λ of the electromagnetic radiation

$$\chi = \frac{2\pi r}{\lambda} \tag{11.1}$$

serves as one measure of the importance and indicates the type of scattering effects (Fig. 11.2). At very low values of χ scattering is negligible. With increasing χ scattering becomes increasingly important and can be described by Rayleigh scattering. Maximum scattering effects are observed for χ around 3 to 10 (i.e. particle size is in the range of the wavelength) and for homogeneous spherical particles Mie-theory serves as an appropiate quantitative description. Far more complicated treatment is necessary and is still not at hand, when the non-spherical irregular particles are considered. For large χ (particle size much larger than the wavelength) scattering decreases again approaching the geometrical optics limit, which can be adequately modelled by ray-tracing techniques.

It follows that scattering at molecules and aerosols is an important process in the solar spectral range but at longer wavelengths scattering is mainly caused by atmospheric hydrometeors, namely cloud and rain particles. Cloud and rain particles, however, interact with electromagnetic radiation of almost the whole spectrum significantly via scattering processes. This leads to multiple scattering effects, which renders the relation between cloud and rain particles and radiances outside the cloud highly non-linear. This non-linear relation causes additional problems when — typical for satellite measurements — retrieval must be performed on spatial averages of radiances (beam-filling problem).

Beside the spatial dimensions of the inhomogeneities in the atmosphere caused by the hydrometeors, the value of the complex index of refraction $m = m' + im''$ compared to its value for the gaseous atmosphere must be considered. The larger the difference, the larger the scattering effects will be. The value of the real part m' determines the scattering efficiency and the value of the imaginary part m'' is responsible for absorption within the hydrometeors. Due to multiple scattering both effects cannot be estimated separately, because every scattering event can give rise to another absorption event which increases the overall absorption within an ensemble of hydrometeors.

For water and ice both parts of the refractive index are quite complicated and irregular functions of wavelength and temperature. Some values for water are given in Tab. 11.1. The real part of m starts at values around 1.3 in the visible range, goes through a minimum of about 1.1 in the thermal window around 10 μm, and increases towards the microwave frequencies to almost 10. The imaginary part starts at very low values in the visible, increases steeply to about 0.4 in the thermal infrared, then more slowly up to around 3 in the mid-microwaves from where it decreases again. Except in the very long microwaves, where m decreases with temperature, both parts of the refractive index increase somewhat with temperature. Values for ice are of

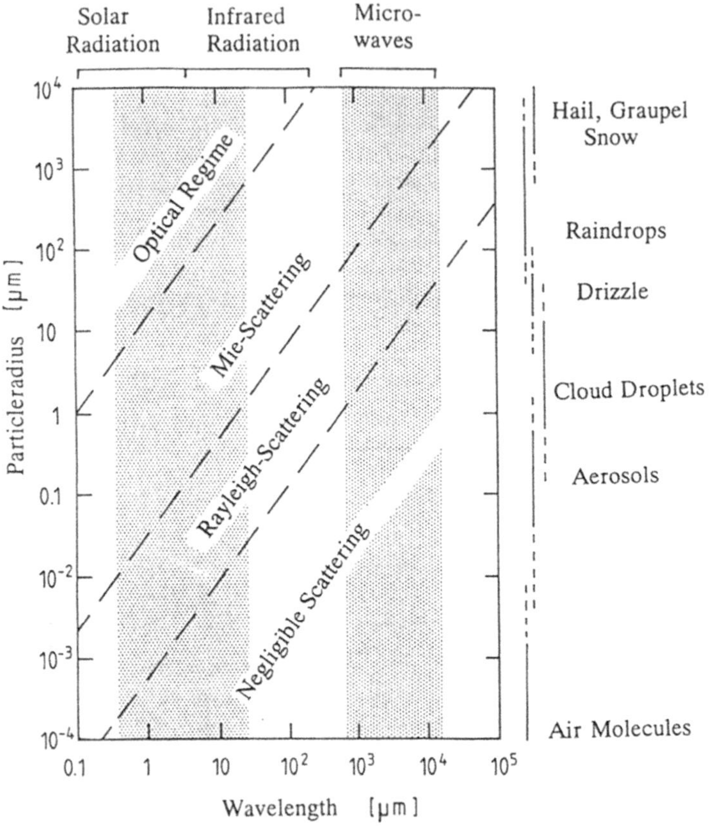

Figure 11.2: *Schematic description of scattering regimes for atmospheric particles as a function of particle size and and wavelength. The dashed lines do not represent constant values of the Mie-parameter (adopted from Simmer, 1994).*

λ	m	λ	m
0.55 μm	1.333–1.96 × 10^{-9}i	500 μm	2.22–0.74i^*
0.75 μm	1.330–1.56 × 10^{-7}i	1 mm	2.50–1.09i^*
0.5 μm	1.371–0.272i	2 mm	2.560–0.894i^*
11μm	1.153 - 0.0986i	5 mm	3.192–1.766i^*
60μm	1.703–0.587i^*	1 cm	4.221–2.526i^*
100 μm	2.06–0.551i^*	2 cm	5.837–3.005i^*
200μm	2.08–0.509i^*	10 cm	9.035–1.387i

Table 11.1: *Complex refractive index of pure water at 273 K (with * for 297 K).*

the same magnitude and show similar spectral behaviour as for water for small wavelengths up to the thermal infrared window. In the microwaves the real part approaches a constant value of about 1.78 and the imaginary part is very small.

We discuss the extinction and absorption coefficients for cloud and rain, which follow from the spectral characteristics of the refractive index and the typical droplet size distributions, on the basis of Fig. 11.3 based on calculations by Deirmendjian (1975), which shows very clearly

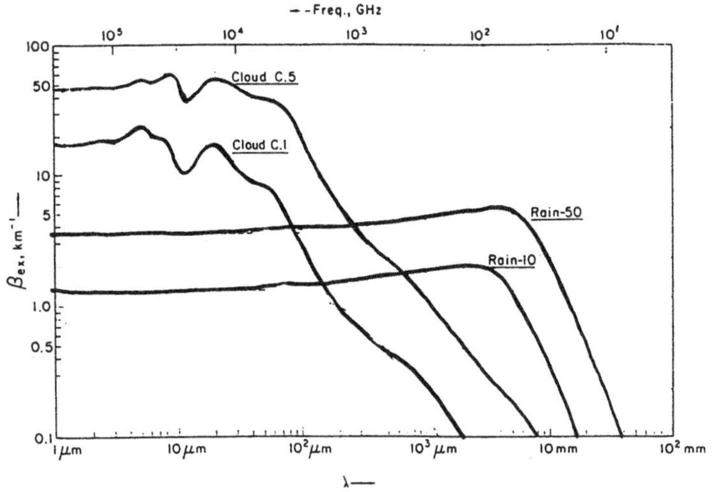

Figure 11.3: *Theoretical cloud and rain extinction coefficients for two cloud and two rain types based on data from Deirmendjian (1975).*

the role both types of hydrometeors play in radiation interaction processes within the earth atmosphere. Compared are the volume extinction coefficients computed from Mie theory for two cloud droplet size distributions and from two rain droplet size distributions. The size distributions are assumed to be of the so-called modified Gamma distribution

$$n(r) = ar^{\alpha}e^{-br^{\gamma}} \tag{11.2}$$

with r the particle radius and a, c, b, and γ constants. This function vanishes for very small and very large radii and has a single maximum. The cloud labeled C.1 is a typical fair weather cumulus (mode radius $r_c = 4$ μm), C.5 corresponds to a nimbostratus cloud ($r_c = 6$ μm). The rain type labeled R.10 corresponds to rain spectra just below a raining cloud when the rain intensity at ground is about 10 mm/h ($r_c = 0.333$ mm) while R.50 represent raindrop spectra at ground level during rainfall rates of 50 mm/h ($r_c = 0.6$ mm).

In the solar and terrestrial infrared spectrum up to a wavelength of 100 to 200 μm clouds are close to opaque. Their optical depth is generally above 10 at least for clouds with vertical dimensions over 500 m. The rain, on the other hand, exhibits optical depths of about one order of magnitude less, which explains why we can easily see through rain if it is not too intensive. At 1 mm wavelength—this corresponds about to the beginning of the microwave region at 300 GHz where the atmosphere becomes transparent again—the extinction by rain has slowly increased above the values for clouds attaining rather flat maxima in the centimeter wavelength range. These maxima, however, are far lower than the values reached by clouds in the solar and thermal infrared spectral range.

11.3.1 Solar and thermal infrared

From Fig. 11.2 we see that scattering processes at all atmospheric hydrometeors is important in the solar and thermal IR, and that we have to apply Mie-theory in any case. Although the extinction coefficients of clouds in the visible are not very different from those in the thermal infrared (Fig. 11.3), clouds are more opaque in the infrared. In the visible the imaginary part of the index of refraction m'' is very low (Tab. 11.1) giving rise to a single scattering albedo ω_0 (fraction of the extinction coefficient caused by scattering) close to unity. In the IR m'' is much higher causing ω_0 to decrease down to values of 0.5. Thus, after only a few scattering events any infrared photon will be absorbed by the cloud particles while there are still high chances for a visible photon to escape the cloud. The opacity of clouds in the IR enables the quite accurate determination of its top temperature in the thermal infrared window.

But from the above it is obvious that it is impossible to sense rain from satellites directly in the visible or thermal infrared regions. The rain generating cloud will almost always shield very effectively the rain it produces from the satellite view. But the rain producing cloud will be detectable against the cloud-free atmosphere due to its much higher optical depth. Thus rain intensity must always be completely indirectly inferred from the cloud parameters we are able to sense, like its reflectivity in the solar spectral range, which is dependent on the cloud liquid water and on the cloud drop radius (clouds with smaller droplets are brighter because small droplets scatter visible radiation more effectively), and its top temperature, which can be measured in the thermal infrared because of their high opacity.

11.3.2 Microwaves

In the microwaves both clouds and rain have optical depths much lower than clouds in the visible and thermal infrared (Fig. 11.3). It follows that except for very high rain intensities cloudy and rain-bearing areas will not be deliniated in satellite images as sharply as clouds are in the visible and thermal infrared. The signal reaching the satellite sensor will be influenced also by the interior of the raining cloud, which is a desirable effect if we want to retrieve precipiation.

Scattering in clouds is small in most of the microwave region (Fig. 11.2) and can often be neglected without large errors. For frequencies up to about 50 GHz scattering at raindrops can be adequately described by Rayleigh scattering characterized by a phase function which is symmetric in the forward and backward directions (Fig. 11.4a,b). At higher frequencies forward scattering increases (Fig. 11.4c,d), and Mie-theory must be applied. For ice particles, but also for large raindrops, the particles differ from ideal spheres and become complicated computationally. Ray-tracing methods which are used in the visible where the hydrometeors are much larger than the wavelength cannot be used in the Mie-regime.

Over large parts of the microwave region the refractive index of water increases with decreasing temperature. This leads to an increasing mass absorption coefficient with cloud height (Fig. 11.5). This behaviour counteracts the effect of decreasing thermal emission caused by the temperature decrease with height. Thus radiances at frequencies below 50 GHz measured above non-raining clouds are largely independent of the vertical distribution of cloud water and enable the retrieval of integrated cloud water content in this spectral region (e.g. Karstens et al., 1994). Emission by cirrus clouds is very small (Fig. 11.5) and can be neglected in most cases.

Compared to the non-raining clouds rain droplet spectra lead to much higher mass extinction coefficients than non-raining clouds and a drastic increase of the scattering coefficient (Fig.

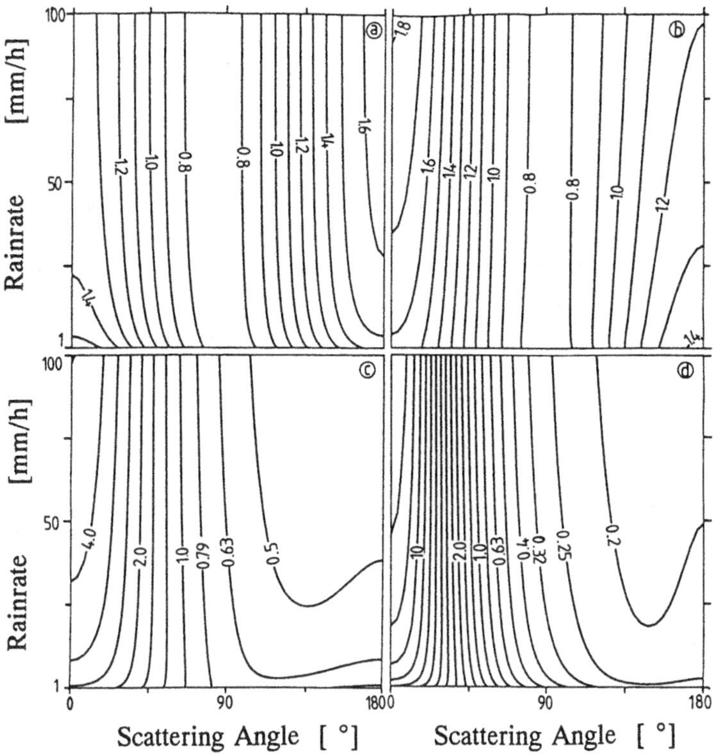

Figure 11.4: *Scalar phase function as a function of rainrate and scattering angle for (a) 18 GHz, (b) 37 GHz, (c) 85 GHz, and (d) 183 GHz.*

11.6). It is important to note that the coefficients change by up to 10% if different types of rain droplet spectra are used (Fig. 11.7). Because changes in the emerging radiances at the top of the rain layer must be expected additional information is needed to invert the satellite measured radiances to rain rates.

Rain generation does often involve precipitation-size ice particles. Different from cirrus cloud particles precipitation-size ice interacts very effectively with microwave radiation via scattering (Fig. 11.8), especially at the higher frequencies. For all frequencies the volume absorption coefficient is much larger for water than for ice. The volume scattering coefficient σ_s of ice is about half of the one for water for 19.35 and 37 GHz but at 85 GHz σ_s for ice even exceeds the value for water. We can summarize and conclude:

- Scattering is negligible for cloud droplets up to about 100 GHz. Rayleigh scattering is applicable for higher frequencies.
- Cirrus shows negligible interaction with microwaves at almost all frequencies.
- Up to 100 GHz extinction by rain is much higher than by clouds.
- Scattering is important for rain particles at all frequencies: Rayleigh scattering can be applied below 50 GHz, Mie scattering must be applied above 50 GHz.

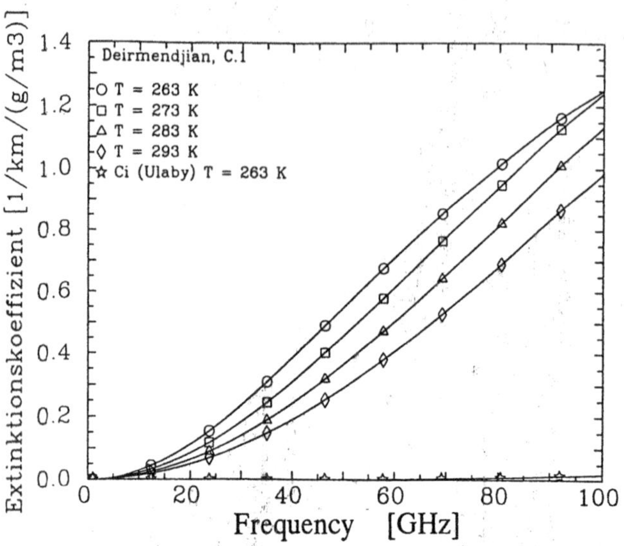

Figure 11.5: *Mass extinction coefficient for a C.1 cloud at different temperatures. The mass extinction coefficient of a cirrus cloud is shown for comparison.*

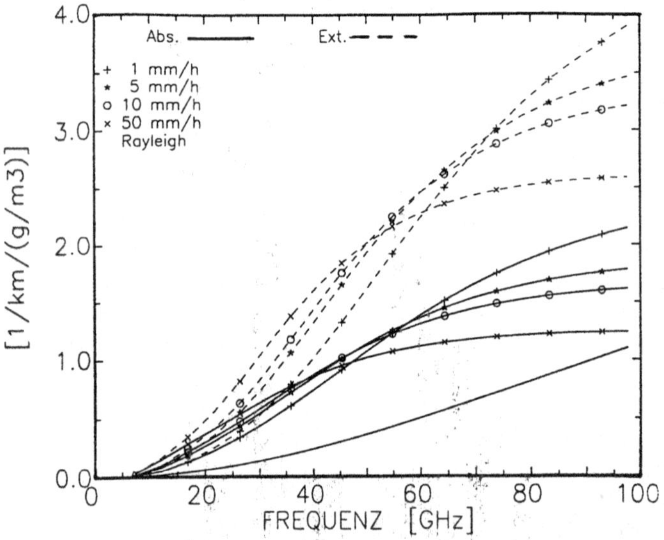

Figure 11.6: *Mass extinction and absorption coefficient for different rain rates using the dropsize distribution by Willis (1984) compared to the mass extinction coefficient of a cloud.*

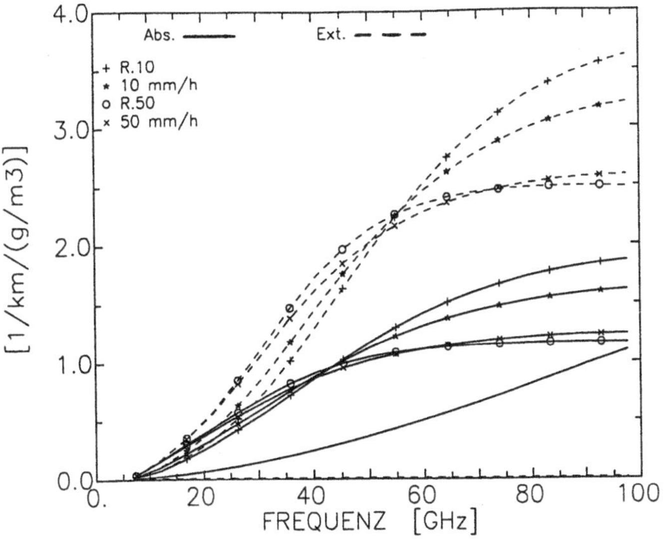

Figure 11.7: *Comparison of mass extinction and absorption coefficients computed on the basis of two different types of raindrop spectra for rain rates of 10 and 50 mm/h. The curves labeled R.10 and R.50 refer to the dropsize distributions by Deirmendjian (1975), the others to Willis (1984).*

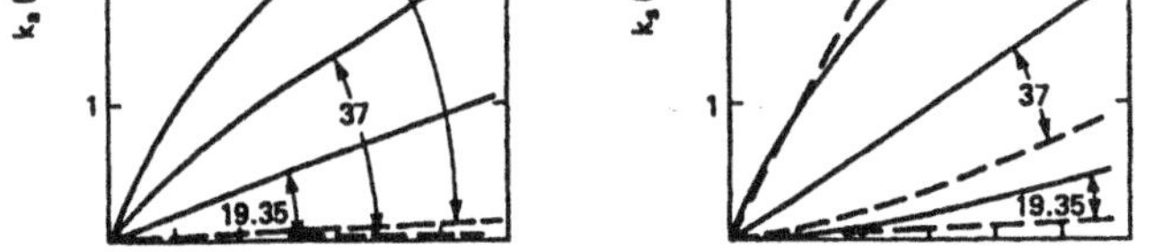

Figure 11.8: *Volume absorption and scattering coefficient of precipitation size water and ice particles as a function of the rainrate computed for Marshall-Palmer dropsize distributions (Marshall and Palmer, 1948).*

- Interaction parameters of rain may vary by 10% depending on droplet spectrum.
- Absorption/emission by precipitation-size ice is much smaller than by liquid partides, but the importance of scattering by precipitation-size ice increases with frequency.

11.4 Signal of precipitation in remote sensing measurements

According to the dominant interaction processes between electromagnetic radiation and the raining cloud, we can distinguish four types of signals predominantly observed in four frequency ranges of the electromagnetic spectrum. The first three signal types are the basis of passive techniques in the three spectral ranges of the electromagnetic spectrum with atmospheric windows, namely the visible/infrared, the high-frequency microwaves ($\nu > 50$ GHz), and the middle-frequency microwaves (10 GHz $< \nu <$ 50 GHz). A schematic description of the rain signal for the three passive techniques is given in Fig. 11.9 on which the discussion of the three following paragraphs is based. The fourth type is related to active methods in the low-frequency microwaves ($\nu <$ 10 GHz).

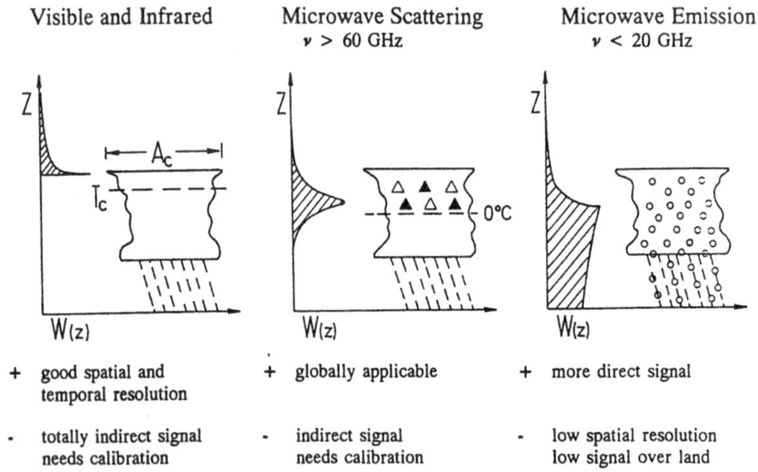

Figure 11.9: *The signal of precipitation clouds in passive remote sensing techniques (for discussion see text).*

11.4.1 Passive visible/infrared

Raindrops are practically invisible from space in these frequencies. The interaction of electromagnetic radiation with precipitation-size particles is low, and scattering and emission by the rain-producing cloud particles above hide the rain from the satellite view. In the thermal infrared the sensors can observe only the very upper parts of the clouds allowing the detection of the cloud-top temperature. No direct rain signal can be received at the satellite. In the visible the situation is even worse. Since raining clouds will contain large amounts of liquid water and ice making them optically thick no direct signal of the rain below can be detected from the satellite. The only quantitative signal is the cloud albedo which is a function of the cloud water content and the size distribution of the cloud droplets, thus also no direct rain signal here. Usually the radiances emanating from the cloud tops are interpreted in terms of top temperature, areal extent, and their temporal evolution (Griffith et al., 1978). From these parameters RR is inferred by tatistical relations between the convective activity of the cloud (described by the above parameters) and RR derived from in-situ data or ground-based radar.

11.4.2 Passive high-frequency microwaves

At frequencies above 60 GHz the interaction intensity between radiation and precipitation has its maximum (Fig. 11.3). Raining clouds are almost opaque and in addition to absorption scattering by both liquid and frozen precipitation-size particles is important. Accordingly raining clouds appear cold compared to the environment. The radiance depression is dependent on the amount of scatterers in the upper part of the cloud. Since ice particles play a key role in precipitation generation we can estimate RR from the observed radiance depression. The microwave reflection and emission of water surfaces is highly polarized in contrast to emission and scattering by atmospheric constituents. So, in addition to the radiance depression the polarization reduction serves as another piece of information over ocean areas. The main advantage of the high-frequency method is its relative independence of the signal of the surface below the cloud. Thus the signal can be observed over both land and ocean surfaces.

11.4.3 Passive mid- to low-frequency microwaves

Below 40 GHz raining clouds become increasingly transparent for microwave radiation, below 10 GHz emission dominates and the emitted radiation is almost proportional to the rain water content in the atmospheric column. If the height of the water column and its profile structure is known RR can be retrieved. As above, additional information is provided by the polarization reduction. Over land the polarization of the background radiation is low and the signal of the rain cloud can barely be distinguished from its surroundings. This is the main disadvantage of the low-frequency approach. At higher frequencies absorption/emission and scattering by precipitation particles increase. This leads to a strongly non-linear signal dependence on RR and to saturation at decreasing RR with increasing frequency. This non-linear relation is a serious problem, because rain cells are typically smaller than the resolution of satellite radiometers (e.g. 30 to 50 km for SSM/I). Compared to the methods described above the low-frequency approach is based on a more direct relation between signal and RR. This method is, however, restricted to oceanic areas.

11.4.4 Rain radar

Rain radar measure the backscattering of microwave radiation at even lower frequeucies. Atmosphere and cloud drops interfere only weakly with the signal, and under certain assumptions about the drop-size distribution RR can be inferred from the backscatter intensity. This method is based on the most direct relation between signal and RR of all the methods discussed. The ranging capability will also allow to determine rain water profiles. There have been no rain radars in satellite orbits up to now. On the TRMM-satellite a rain radar is planned to be used together with passive instruments in the microwave, infrared, and visible spectrum. Thus TRMM will allow for the testing and/or calibration of the passive techniques. Rain radars are the topic in other parts of the book and will not be discussed further here.

11.5 Satellite radiometers with precipitation-sensing capabilities

The VIS/IR techniques are applicable to dual-channel data of the common geostationary and polar-orbiting satellites, which are in orbit since the early days of satellite remote sensing beginning in the sixties of this century. Today five operational geostationary satellites (METEOSAT, GOES-E, GOES-W, GMS, and INDSAT) are almost currently in orbit and provide visible and infrared window measurements at least once an hour up to about 50° latitude. The data gaps in the high latitudes are filled by at least two operational NOAA (National Oceanographic and Atmospheric Administration) polar orbiting satellites. Additional measurements are available by similar russian polar orbiters of the METEOR-type and a range of experimental satellites.

Longer term microwave observations of precipitation began 1972 with the single channel (19.35 GHz, horizontally polarized) cross-track scanning ESMR (Electronically Scanned Micowave Radiometer) on Nimbus-5 (Wilheit et al., 1976). Three years later another ESMR with 37 GHz, dual-polarization, was launched on Nimbus-7. For the first time a conical scan geometry was used to accomplish a constant nadir view angle of 50°. Application for rainrate retrieval were shown e.g. by Weinman and Guetter (1977) and Rodgers et al. (1979). Preceeded by the SMMR (Scanning Multichannel Microwave Radiometer) instrument on Seasat, which functioned only for three months (see Gloersen and Barath (1977) for applications to rainrate retrieval), SMMR on Nimbus-7 provided five years of dual-polarized microwave measurements at 6.6, 10, 18, 21, and 37 GHz, which have been extensively used and is still used for rainrate retrieval over ocean (e.g. Spencer et al., 1983a; Petty and Katsaros,1990) and land surfaces (e.g. Spencer et al., 1983b).

Without any gap SMMR was followed by the SSM/I (Special Sensor Microwave/Imager) on the operational DMSP (Defense Meteorological Satellite Program) series of the U.S. DoD (Departement of Defense). SSM/I comprises 7 channels (19 GHz, 37 GHz, and 85 GHz dual-polarized, 22 GHz only vertically polarized) and measures at a constant view angle of 53° over a swath width of 1400 km. Since the beginning of the nineties always at least two DMSP satellites are in operational orbit (currently DMSP-F10, DMSP-F12, DMSP-F13). Most of the algorithms discussed in the following sections have been derived for SSM/I. In addition to the SSM/I instrument the DMSP satellites carry also the microwave sounders SSM/T (for temperature profiling) and SSM/T2 (for water vapour profiling), and the VIS/IR OLS (Operational Line Scanner) radiometer. Attempts to use these radiometers together with the SSM/I for precipitation retrieval are currently studied.

It should be mentioned that other microwave radiometers suitable for precipitation retrieval (e.g. MSR (Microwave Scanning Radiometer) on the Japanese MOS (Marine Observing Satellite) satellites, SAMIR (Satellite Microwave Radiometer) on the Indian Bhashkara satellites, ATSR (Along-Track Scanning Radiometer) on the European ERS (European Research Satellite) satellites) have been or still are in orbit, but due to the restricted availability of the data these instruments have not been much studied.

In the near future additional suitable measurements will be available from the AMSU (Advanced Microwave Sounding Unit) radiometer on the NOAA operational satellites, TMI (TRMM Microwave Imager) on the TRMM satellite, and MIMR (Multichannel Imaging Microwave Radiometer) on the planned European polar orbiting METOP satellite series.

11.6 Remote sensing methods

11.6.1 Visible and infrared

Several methods have been developed to use visible and infrared radiation measurements to retrieve rainrate estimates:

- GPI-Type Methods
- Outgoing Longwave Radiation (OLR)
- Cloud droplet size in the upper cloud
- Lightning

Since lightning is produced by charge separation processes which involve precipitation-size hydrometeors lightning has some relation to the convective activity within a cloud. Thus a connection between the frequency of lightning detectable from satellites and rainrate at the surface can be expected. But not every raining cloud produces lightning, which limits the applicability of these methods to only one class of precipitation.

Raining clouds must have a fraction of large droplets in the upper part of the cloud. Since Mie-scattering is important in the visible and infrared spectral range raining clouds can in principal be distinguished from non-raining clouds by these effects. This methods are still in the development stage.

The outgoing longwave radiation is dominated by the cloud-top temperature. The colder the cloud the lower the OLR. Methods based on OLR measurements are in principle very similar to the GPI-type methods, which are the most frequently used techniques. They are based on the assumption that the rain produced in convective clouds is proportional to the area A_C with cloud-top temperatures below a predefined threshold T_C:

$$RR \propto A_C \quad \text{with} \quad A_C \quad \text{area with} \quad T < T_C \tag{11.3}$$

By using the infrared observations from geostationary satellites and radar-observed rainfall data for the GARP (Global Atmospheric Research Program) Atlantic Tropical Experiment (GATE), Arkin (1979) found that the correlation between 6-hour rainfall and the fractional coverage of cold clouds is above 0.8 when the fractional coverage is defined by T_C between 230 and 240 K. These findings and further analysis (Arkin and Meisner, 1987) gave rise to the foundation of the GOES (Geostationary Operational Environmental Satellite) Precipitation Index (GPI)

$$RR \quad \text{mm/h} = 3.0 \quad \text{mm/h} \quad \times \quad \text{fractional area of } 2.5 \times 2.5 \text{ with} \quad T < 235K \tag{11.4}$$

The method works in the tropics over land and ocean, but becomes increasingly inaccurate at higher latitudes. The main advantage of the method is the high temporal resolution because visible/infrared sensors can be used on satellites in geostationary orbits. Comparison with radar and raingage estimates have shown that (11.3) is indeed a valid assumption, but T_C can differ from case to case. The constant of proportionality has a regional dependence and may vary by more than 100% from month to month (Arkin and Xie, 1994). Thus large efforts are made to determine T_C and the constant of proportionality a from additional information. Candidates are the time evolution, the spatial inhomogeneity of the cloud, and information from visible sensors or microwave measurements, if available.

Three techniques have been decribed to increase the quality and flexibility of the GPI:

Microwave Adjusted GPI: The constant of proportionality a is assumed to be a function of longitude and latitude and computed by $a = RR(SSM/I)/RR(GPI)$ from monthly coincidences of the DMSP and GMS-satellites (Adler et al., 1993).

NAWT (Negri-Adler Wetzel-Technique): In a first step the area of raining clouds is defined by $T_C < 235$K. For the coldest 10% of the cloud the rain rate is assumed to be 8 mm/h; for the next warmest 40% of the cloud area 2 mm/h is assumed, and the rest of the area is assumed to be covered with non-raining clouds (Negri et al., 1984; Negri and Adler, 1993).

CST (Convective-Stratiform Technique): The area of raining clouds is defined by $T_C <$ 235K. For this area the relative minima in cloud-top temperature distribution and the slopes of the cloud-top temperature in the vicinity of the minima are determined. It follows a division into stratiform (2 mm/h assumed rainrate) and convective rain (8 mm/h assumed rainrate) using both qualifiers (Adler and Negri, 1988; 1993).

11.6.2 Microwaves

In the microwave spectral region the radiances are more directly related to the hydrometeors in the satellite field of view. Contrary to the visible and infrared spectral region the whole state of the atmosphere decribed by the vertical profiles of temperature, water vapour, cloud and rain liquid and ice profiles, and the shape of the particles determine the outgoing radiances at the top of the atmosphere. So any parameter retrieval is at least partially an inversion of the radiative transfer equation. Concerning rain basically three paths have been followed to construct rain retrieval algorithms:

Complete Inversion: A complete description of the state of the atmosphere is sought which, by feeding its physical description into the equation of radiative transfer, results into radiances sufficiently close to the measured radiances to be inverted.

Statistical Inversion: A functional relation is sought between the radiances and the desired parameter. This function may be linear or non-linear.

Indexing: By physical reasoning indices are derived from the radiances which can be related to the effect of precipitation, like the transmission of the atmosphere or the effect scattering by upper ice particles has on the radiances. The relation between these indices and the rain rate is determined by assumptions about the structure of the hydrometeor profiles.

Published rain rate retrieval algorithms can seldomely be associated with only one of these paths. The number of variables influencing the radiances is far greater than the number of spectral channels available for inversion. In complete inversion algorithms the reduction of unknowns is achieved by making assumptions about the structure of the profiles of temperature, humidity, and hydrometeors. These assumptions are mostly extracted from statistical analysis. Indexing methods are subject to similar restrictions when the relations between the indices and the rain rate is determined. In the following paragraphs examples for each class of will be given.

11.6.3 Complete inversion

A prerequisite for a complete inversion algorithm is a radiative transfer model. The algorithm searches for a state of the atmosphere which reproduces up to the assumed error of the radiometer the measured radiances. Two principles can be followed to achieve this goal:

- Newton iteration

- Library search

The Newton iteration is a straight-forward method which is extensively described e.g. in Houghton et al. (1984). If x_n is a vector containing the parameters which describe as a first guess the state of the atmosphere an improved guess x_{n+1} can be computed by

$$x_{n+1} = x_n + \left(S_x^{-1} + K_n^t S_\epsilon^{-1} K_n\right)^{-1} \left(K_n^t S_\epsilon^{-1}(r - r_n) + S_x^{-1}(x_0 - x_n)\right) \tag{11.5}$$

with S_x the known covariance matrix of the states of the atmosphere, S_ϵ the error covariance matrix of the radiometer measurements r, $K_n = \partial r_n/\partial x$ the matrix of partial derivatives of the modeled radiances r_n with respect to the parameters of the state of the atmosphere x_n and x_0 any state of the atmosphere which serves together with S_x as a constraint to the solution. The inversion of the matrix K_n requires that K_n is quadratic implying that the state of the atmosphere must be described by a number of parameters less or equal to the number of measurements available. Applications of this type of methods in the passive microwaves have been shown by Francis et al., (1983), Wentz (1983), and Olson (1989), and in combination with an indexing method by Petty (1994a; 1994b). To avoid the numerous integrations of the radiative transfer equation the matrices of the partial derivatives for the whole expected range can be computed in advance. The direct inversion produces a state of the atmosphere described by a largely reduced set of parameters. The constraints serve to avoid the iteration into a possible (from the viewpoint of radiative transfer) but meteorologically impossible state of the atmosphere.

The library approach starts with a large set of possible states of the atmosphere for which the corresponding radiances are computed using a radiative transfer model. Inversion is performed basically by finding one state of the atmosphere which has modelled radiances equal or very close to the measured radiances. The states of the atmosphere which fill the library can be based on the reduced parameter sets used for direct inversion; or it can be based on measured or modelled states of the atmosphere. Partial applications of this type of algorithm have been shown by Kummerow and Giglio (1994) and Simmer and Liu (1995). The advantage of the library method is the relatively simple extension to other radiometric channels or even other types of measurements.

11.6.4 Statistical inversion

The basis of statistical inversion procedures is, similar to the library methods, a data set containg physical descriptions of states of the atmosphere together with the corresponding radiances. The states of the atmosphere may origin from measurements or dynamic models, and the radiances may be measured or modeled with radiative transfer codes. Using the concept of minimizing a cost function, e.g. in the least squares sense, a functional relation is sought between the radiances and an arbitrary parameter describing the state of the atmosphere like rainfall at the surface. At least three methods can be distinguished:

- linear regression
- non-linear regression
- neural networks

Linear regression methods are based on the assumption, that a linear relation exists between the parameters describing the state of the atmosphere x and the radiances r, which can be written in the form of a matrix equation:

$$x = Dr \tag{11.6}$$

The elements of the matrix D are then determined by minimizing the quadratic difference between the estimated parameters and the exact ones for the data set leading to the equation (e.g. Houghton et al., 1984)

$$D = S_{r,x} S_{r,r}^{-1} \tag{11.7}$$

where $S_{r,x}$, is the covariance matrix of the radiances with the parameters and $S_{r,r}$ the covariance matrix of the radiances. Instead of using the radiances and the sought parameters itself in (11.6) transformations of both can be used to linearize the relation between both. This is especially necessary for rain rate retrieval. A typical example of this method is the algorithm by Bauer and Schlüssel (1993):

$$\begin{aligned} \log_{10} RR &= 14.66 - 0.7488 \times 10^{11} T_{19V}^{-4} - 0.04503 T_{22V} \\ &+ 0.5064 \times 10^5 T_{19H}^{-2} - 0.5990 \times 10^5 T_{37H}^{-2} \\ &- 0.1172 \times 10^{-3} (T_{37V} - T_{19H})^2 \end{aligned} \tag{11.8}$$

which has been derived for measurements of the Special Sensor Microwave/Imager (SSM/I) on the polar orbiting DMSP (Defense Meteorological Satellite Program) satellites. The rainrate RR is given in mm/h and the T_{nnC} are the equivalent blackbody temperatures at nn GHz for the linear polarization C with V for vertical and H for horizontal polarization.

The inversion procedure described by (11.7) is only applicable for linear relations as described by (11.6). If nonlinear relations are assumed mostly iterative procedures quite similar to the Newton technique, described in connection with the direct inversion methods, are used. The minimum of the cost function is found by using its derivatives with respect to the coefficients of the proposed relations (e.g. Press et al., 1986).

Neural networks (Hertz et al., 1991) in connection with learning by error backpropagation (Rumelhart et al., 1986) constitute in a way a generalization of nonlinear regression methods by avoiding any assumptions about the functional form of the relation between radiances and the parameters to be retrieved. Neural networks are increasingly used in remote sensing of atmospheric and surface parameters (e.g. Lee et al., 1990; Davis et al., 1993; Churnside et al., 1994; Stogryn et al., 1994). Applications to rain retrieval from microwaves have been recently shown by Hsu et al. (1994) and Zhang and Scofield (1994).

11.6.5 Indexing

The idea of indexing is to transform the measured radiation temperatures into an index which is physically related to rainfall and a monotonic function of the rainfall intensity. Based on the signal of rainfall as discussed in the previous sections three radiometric quantities fulfill these requirements, and give rise to the following techniques:

Emission Methods: Over the oceans low-frequency microwave radiation increases with the content of hydrometeors.

Attenuation/Polarization Methods: The transmission of the atmospheric column decreases with the content of hydrometeors.

Scattering Methods: The brightness temperature depression due to precipitation-size ice in the upper part of the cloud increases with the amount and size of scattering particles.

Indexing methods in some way or another try to estimate these parameters from the radiation temperatures and use simple cloud models or statistical means to infer rainrate from these. The first two methods can only be applied over the oceans because the cold (emission methods) and polarized (attenuation/polarization methods) ocean surface background is a prerequisite to cause these kind of signals. The scattering methods can be applied over both ocean and land surfaces.

Emission methods

A typical emission algorithm has been derived by Prabhakara et al. (1992):

$$RR = \left[\exp\{\beta(W)(T_{37H} - T^*)\}^{1.7} - 1\right]\gamma(W) \tag{11.9}$$

with β, γ and T^* empirical functions of the total water vapour content W. The latter can be derived quite accurately by microwave measurements. According to (11.9) the rain rate increases monotonically but in a nonlinear fashion with increasing radiation temperature of the 37 GHz channel measurements. The threshold for the occurrance of rain T^* on the 37 GHz radiation temperature is assumed to be dependent on the state of the atmosphere itself, which is parameterized by the amount of total water vapour in the atmospheric column.

Attenuation/polarization methods

The microwave radiation emanating from the ocean surface is radiatively cold and highly polarized. When the atmosphere becomes optically thick due to a raining cloud the signal becomes less an less polarized. Thus the polarization difference at a fixed frequency,

$$P = T_V - T_H \tag{11.10}$$

often simply called polarization, is an index for the transmittance of the atmosphere and thus for the amount of hydrometeor extinction. It has the following properties:

- For the usual observation angle around 50° P is at least over water surface maximum for an optically thin i.e. cloud-free atmosphere.
- P decreases both with increasing optical thickness of the atmosphere and with the area covered by an optically thick cloud.
- P does not depend on scattering, because both polarizations are affected in the same way.
- The effect of near-surface wind and of the gas atmosphere on P can be reduced by normalizing with the value of P_{clear} for the cloud-free environment obtained from nearby observations: $P' = P/P_{clear}$.
- It can be shown that (e.g. Petty and Katsaros, 1990a; Petty, 1994a) that $P' \simeq \tau^{\alpha}$ with α between 1.5 and 2.

The higher the frequency the more the relation between the normalized polarization and the rainrate will depend on the areal extend of the raining cell and not so much on the rainrate itself. Thus the transformation of R' into a rainrate will also depend on the type of precipitation.

Scattering methods

Scattering algorithms are based on the assumption that precipitation is always produced with the help of the ice phase via the Bergeron-Findeisen process. The precipitation-size ice particles in the upper part of the raining cloud cause scattering of microwave radiation which leads to a radiation temperature depression (Fig. 11.8), which is assumed to be a monotonic function of the rain itensity. To derive the scattering-based depression the radiation temperatures without precipitation size ice particles must be estimated. This idea has been put forward first by Spencer (1986) and refined by Spencer et al. (1989) who derived the so-called Polarization

Corrected Temperature (PCT) which can be approximated independently of the observing frequency by:

$$PCT = 1.818T_V - 0.818T_H \tag{11.11}$$

PCT can be interpreted as the radiation temperature of the lower troposphere if it were a blackbody. Thus in the case of no scatterers PCT has an almost constant value close to the thermodynamic temperature of the top of the lower troposphere, which can be estimated by climatological means. Scattering reduces PCT below these climatological values. The difference between actual PCT and the climatological value can then be used as an index for precipitation.

Petty and Katsaros (1990b) derived from similar principles a scattering index S, which they define with the help of the normalized polarization P' by:

$$S = P'T_{V,clear} + (1 - P')T_C - T_V \tag{11.12}$$

with T_V the measured vertically polarized radiation temperature, $T_{V,clear}$ the clear sky temperature derived from surrounding measurements, and T_C the hypothetical (unpolarized) radiation temperature of an opaque boundary layer without rain induced scattering. S can be interpreted approximately as the PCT depression caused by scattering (Petty, 1994a) and has to be related to the rainrate by assumptions about the vertical structure of the hydrometeor profile. Petty and Katsaros (1990b) used the simple relation $RR = 0.25(S - 10)$.

Finally, Grody (1991) has derived a scattering index for 85 GHz of similar interpretation by first predicting the radiation temperature without scatterers using a statistical relation based on the 19 GHz and 22 GHz channels of SSM/I which are only weakly influenced by scattering. The relation used by Adler et al. (1993):

$$RR = \frac{(251 - T_{85H})}{4.91} \tag{11.13}$$

can be interpreted as a simplification of this concept by assuming a fixed non-scatter radiation temperature for 85 GHz. A prerequisite for this algorithm is a screening algorithm which separates raining from non-raining conditions (see below).

11.6.6 Special problems of microwave-based rainfall determination over land surfaces

Over land surfaces the surface emittance is much higher (0.80–0.95) than over water surfaces. Thus the effective radiation temperature of the surface is very similar to the effective temperature of the atmosphere. Also the polarization of rain-free areas is already low. From this follows that emission and attenuation/polarization methods are difficult to apply, and the less direct scattering methods must be used.

Since the surface emittance depends on surface structure, vegetation, soil moisture, the surface signal is very variable and screening algorithms, which differentiate between rain and no-rain cases, are very important. At present these screening algorithms have decision-tree structure with partially surface and season dependent thresholds applied to individual radiation temperatures, polarization, spectral gradients, and scattering indices (e.g. Grody, 1991).

11.7 Comparison of algorithm performance

When comparing the performance of satellite-based rain retrieval algorithms commonly three problems are encountered:

- The satellite data of the comparison data base are never free from errors ranging from spurious complete radiometer failure (which must be detected) over calibration problems (these especially affect algorithms based on radiative transfer simulations) to navigation errors when earth location of the data is crucial (e.g. when data sets are blended with each other or results from different data sources are compared).
- There are by now no accepted large scale validation data sets. Ground-truth data obtained from rain-gauges are desirable but the network densities are much too low, especially over the oceans. Monthly means can be compared at best because of the different sampling characteristics of satellite (instant area averages) and rain-gauge measurements (time-integrated point measurements). Radar measurements provide areal averages, which can be better compared with satellite measurements, but the results of radar-derived rainfall estimates have their own problems, which must be carefully evaluated before accepting the data for validation purposes. There are strong indications that radar data are biased as a function of the distance from radar caused by the increasing beam distance rom the surface (Petty, personal communication). One of the most recent of the larger satellite retrieval programs (PIP-1, see below) decided not to use radar data at all for their purpose.
- Lacking adequate validation data sets evaluation of algorithm performance is mostly based on algorithm intercomparison. While it can be agreed on the type of results to compare (monthly means, standard deviations, range etc.) and the means how to compare (simple differences, correlation coefficients, etc.) there is no commonly agreed basis for ranking the different algorithms.

11.7.1 Programs

There have been many attempts to compare and validate satellite rain retrieval algorithms. Two of the most recently completed comparison projects involving a range of algorithms and research groups from many laboratories are the First Algorithm Intercomparison Project (AIP/1) within the Global Precipitation Climatology Project (GPCP) (Arkin and Xie, 1994) and the First Precipitation Intercomparison Project (PIP-1) of the NASA-led WetNet Project (Barrett et al., 1994; 1995). The former concentrates on the evaluation of IR-based algorithms while the latter is directed more to passive microwave methods. In the following we will shortly describe these projects and summarize their mayor results.

AIP/1 (GPCP)

GPCP was established by the World Climate Research Programme (WCRP) to produce global analyses of area- and time-averaged precipitation for use in climate research. To achieve this goal satellite-based VIS/IR-retrieval algorithms are used primarily to be able to make use of the high time and space sampling frequency obtained from geostationary satellites. Within AIP/1 rainfall estimates derived from the visible and IR measurements of the Geostationary Meteorological Satellite (GMS) and from DMSP passive microwave observations are compared with rainfall derived from a combination of data from 15 precipitation radars and over 1300 automated raingauges over the Japanese islands and the adjacent ocean regions during the June and Mid-July to Mid-August periods of 1989. Of the 17 compared rainfall estimates 7 are based on IR-estimates alone, four combine IR and visible data, three combine SSM/I and IR data two use model forcasts and one uses climatological information together with IR data.

We have learned in earlier chapters that IR-methods are based on the statistical relation between rainfall intensity at the surface and the areal and time coverage of clouds with cloud-top temperatures (determined from the IR-signal) below a certain threshold. The results for the region under investigation can be summarized as follows (Arkin and Xie, 1994):

- The threshold temperature of $-28°$ (used in (11.3)) is applicable.
- Maximum correlations found were between 0.7 and 0.8, somewhat lower than obtained from the GARP results.
- The optimal value of the constant of proportionality a and of the threshold temperature T_C depend on area size and the time interval used.
- a and T_C depend on the region.
- For the same region a can vary from month to month by 100%.
- No improvement was obtained using additional predictors derived from IR, like spatial inhomogeneity.

PIP-1 (WetNet)

The NASA-led WetNet Project is primarily focussed on algorithms based on data from the U.S. military DMSP (Defence Meteorological Satellite Program) satellite SSM/I (Special Sensor Microwave Imager) microwave radiometer. A central part of the WetNet work program is a series of Precipitation Intercomparison Projects (PIPs). The first of these projects, PIP-1, was intended primarily to evaluate existing passive microwave algorithms both in relation to each other and also against conventional (rain gauge) data sets. Within the PIP groups involved in the development of rain retrieval algorithms agree on a common data base, which is first analyzed independently by the groups. In PIP-1, intercomparisons of global rainfall estimates for August, September, October and November of the first year of SSM/I data, 1987, have been undertaken for 15 algorithms based on passive microwave DMSP-SSM/I image data, one algorithm based on passive microwave NOAA (National Oceanographic and Atmospheric Administration) MSU (Microwave Sounding Unit) sounder data, one infrared image data-based algorithm, one combined passive microwave/infrared image data based algorithm, one numerical weather prediction model and rainfall observations from rain-gauges.

The contributors to PIP-1 agreed to use the following procedure to evaluate the results of the different algorithms and to attempt a ranking, which could not be final having all the problems mentioned above in mind.

1. Subjective ("eye-ball") inspections of the graphical displays was used to judge, whether the obtained monthly distributions of rain fall were reasonable and in broad agreement with what is known both on the global and regional scale.
2. Quantitative comparisons were made with the results of the Global Precipitation Climatology Project (GPCP) prepared by the Global Precipitation Climatology Center in Offenbach, Germany (Rudolf et al., 1992), which is generally recognized as the best available rain-gauge based data set so far.
3. The authors of the rain-retrieval algorithms were asked to make an assessment of their own algorithms based on the knowledge of their strengths and shortcomings in view of the results obtained in 1. and 2.

Finally, the PIP-1 Steering Commitee made an attempt to conclude on a broad ranking of the algorithms based on the results of 1. to 3. As to be expected all compared algorithms did reproduce the known structures of precipitation, namely high values over the ITCZ and the storm tracks connected to the cyclones of the midlatitudes, and the low values related to the subtropical highs. But there is large quantitative disagreement. Even for the monthly and zonally averaged precipitation differences of up to 100% for the ITCZ are not unusual. Some algorithms predict substantial rainfall almost everywhere while others display at the same time large areas with no rainfall at all. Key results of PIP-1 were (Barrett et al., 1994; 1995):

- No single satellite method of global rainfall estimation was found to be always better than all others, but several methods were each generally better over some major regions of the world.

- Scattering/differencing microwave techniques perform better over land and emission based algorithms perform better over ocean areas.

- Algorithms seem to perform better in regions and climatic conditions for which they have been calibrated, which suggests that re-calibration and especially better ground-truth data would lead to further improvements.

- Differences between different algorithms were particularly large in centres of trade-wind anticyclones, over land but especially over hot and cold deserts, and in oceanic zones peripheral to the Arctic and Antarctic ice sheets.

- Solely SSM/I-based techniques achieved results at least comparable to any other method tested including IR-based techniques (despite their much higher temporal sampling) and weather forecast models.

11.7.2 A remark

Interestingly, both comparison programmes concluded that precipitation estimated by methods, which combine microwave with infrared data, do not in general score better, or did even worse than algorithms only based on SSM/I measurements. It is certainly premature and wrong to conclude from these results, that the combination of different spectral ranges will not increase the quality of precipitation estimates from satellites. It must be kept in mind, that the data sets used in combination originated from different sensors with different scanning geometries, different resolutions, or even from different satellites. To my opinion, these results merely give the indication, that much more work is necessary to develop schemes to combine measurements of such diverse characteristics, or to build satellites with sensor specifications, which make these technically difficult problems obsolete.

11.8 Remaining problems

There are a number of reasons which are responsible for the large observed scatter of the results of the various algorithms discussed above. The most important ones will be dicussed in the following.

11.8.1 Different footprint resolution of microwave radiometers

The angular beam width θ of microwave radiometer depends on frequency and is roughly given by:

$$\theta = \frac{\lambda}{d} \tag{11.14}$$

with λ the wavelength of the measured radiation and d the diameter of the measuring antenna. Due to technical restriction (one common antenna for several frequencies) the footprints of the different channels of a microwave radiometers differ from each other substantially. For SSM/I the footprint size of the 19 GHz channels is in the range of 50 km while the high-frequency 85 GHz channels have resolutions of 15 km (Fig. 11.10). Farrar et al. (1994) have quantified

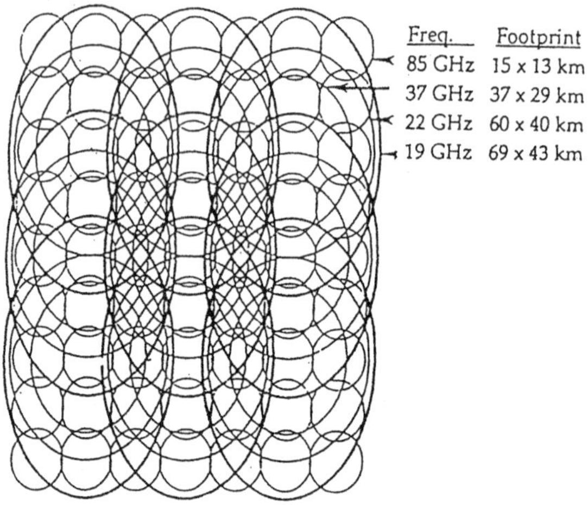

Figure 11.10: *Schematic description of the SSM/I footprint size for the different channels.*

this effect by applying a range of algorithms to original SSM/I data and to the same data set but with unified resolution. The latter was obtained by removing the overlap between the low-frequency channels with the Backus-Gilbert approach (e.g. Robinson et al., 1992). Farrar et al. (1994) found that due to the non-linear relationship between rainrate and brightness temperatures single channel emission-based algorithms showed an increase by 5–6% and single-channel scattering based algorithm showed decreases between 10% and 20% when resolution unification was applied and the results space-time averaged. The differences were lower for multichannel algorithms. These results are consistent with the findings by Simmer and Liu (1995) who tested the response of the regression algorithm by Bauer and Schlüssel (1993) and a library based approach if subjected to resolution homogenization.

11.8.2 Insufficient temporal sampling by microwave radiometers

To allow sufficient spatial resolution by microwave radiometers from satellites these instruments must be flown at present and in the near future on polar platforms only. If two satellites are in operation only up to four measurements per day are available over most regions (Fig. 11.11). Since rainfall is over most regions an event of relative low probability and short duration this time sampling is insufficient for climatological applications.

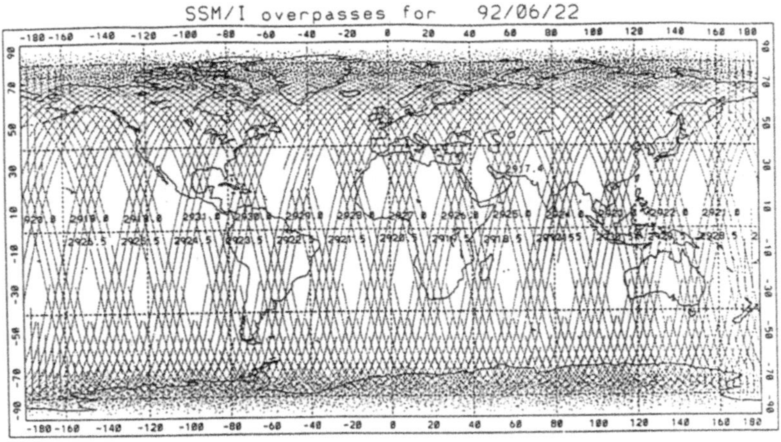

Figure 11.11: *Data coverage by SSM/I during one day.*

Wilheit et al. (1991) proposed an interesting procedure to overcome this deficiency by assuming that the number of occurrances of monthly rain totals within an area should exhibit a log-normal frequency distribution. Preliminar rain rates are retrieved for $5° \times 5°$ boxes over one month by an emission algorithm based on a simplified cloud/precipitation model using the 19 GHz and 22 GHz channels of SSM/I. The height of the freezing level, which is a necessary piece of information for these types of algorithms, is estimated as a constant for the whole grid-area and the month from the frequency distributions of the 19 GHz and 22 GHz channel measurements. The observed rain rate frequency distribution is then used to fit an log-normal distribution overlapping a normal distribution representative for the non-rain cases. Fitting parameters are besides a constant the logarithmic mean rainrate, the logarithmic standard deviation of rainfall, and the distiribution parameters of the normal distribution for the rain-free areas. Basically, this method predicts the seldomly observed, and thus uncertain, but very important high rainrate occurrances from the more frequently observed low rainrate cases. The method is, however, restricted to the tropical oceanic atmosphere due to the emission algorithm used and the assumptions of constant freezing level over a large area and time.

11.8.3 Three-dimensional geometry

Most algorithms for the microwave region are based on results obtained from one-dimensional radiative transfer simulations. Microwave radiometers, however, measure at a view zenith angle around 50 degrees looking mostly at the side walls of the clouds. Especially over the oceans three-dimensional effects caused by the reflection of radiation off the water surface can amount to several tens of degrees in radiation temperature (Liu and Simmer, 1994). The effect on the retrieval algorithms depends on the signal which mainly used by the algorithm. A study by Petty (1994a) gives an indication that emission based algorithms over the oceans will suffer more heavily from three-dimensional effects than the scattering-based algorithms.

11.8.4 Aspherical hydrometeors

The shape of large raindrops differs from spheres and the particles exhibit predominant orientations due to the air flow (e.g. Pruppacher and Klett, 1978), and hail particles are also inhomogeneous inside. Usually these particles are treated as homogeneous spheres in the simulations, causing errors which have not been quantified so far. Some authors take into account aspherical particles but only in an empirical approach to correct for polarization differences observed by SSM/I above optically thick rain clouds (e.g. Kummerow and Giglio, 1994). Theories to take into accout non-spherical particles are known (e.g. Wauben and Hovenier, 1992; Jin, 1992; Wauben et al., 1993; Mishchenko and Travis, 1994) but adequate information about the distribution of these particles in precipitating clouds is still not at hand.

11.8.5 Variability of the state of the atmosphere

The main reason for the observed discrepancies in quantitative rain retrieval by different algorithms is that the precipitation process has many more degrees of freedom than can be fixed by only a few number of radiation measurements. Any algorithm inherently assumes its own cloud/precipitation model and most of the differences can be retraced to differences between those. To reduce the number of degrees of freedom, measurements from different spectral ranges must be combined and sound assumptions must be made about the vertical and horizontal structure of the observed raining system. One possibility is the assimilation of the satellite measurements into dynamic cloud models which has been pursued recently by several authors (e.g. Mugnai et al., 1992; Smith et al., 1992; Kummerow and Giglio, 1994).

11.9 Outlook

From the vast amount of published simple 1-line either statistically-based or indexing-based rainrate retrieval algorithms, and from the large scatter when the results are compared with each other or with independent estimates it is clear, that the determination of rainrate at the surface from satellite measurements is still an unsolved problem. But we must also admit, that rain retrieval from satellites is at the present time also the most accurate way to estimate rainfall at least from the global to the regional scale.

It can be observed from still on-going studies within PIP-2 and PIP-3, that the estimates from the more complicated microwave-based algorithms, which take into account explicitly the different footprint sizes and antenna patterns of the radiometers and head at consistency between observed and modelled radiances, do converge on similar results. This makes clear that rainfall can only be accurately determined when a complete description of the physical parameters of atmosphere and surface, both vertically and horizontally, is part of the retrieval. If this is accepted it is obvious, that apt consideration of other spectral channels like the visible and infrared window channels and sounding channels in both infrared and microwave will further enhance the quality of rainrate retrieval and thus our knowledge of a most important variable in the climate system.

11.10 References

Adler RF, Negri AJ (1988) A satellite infrared technique to estimate tropical convective and stratiform rainfall. *Journal of Applied Meteorology* 27: 30–51

Adler RF, Negri AJ, Keen PR, Hakkarinen IM (1993) Estimation of monthly rainfall over Japan and surrounding waters from a combination of low-orbit microwave and geosynchronous IR data. *Journal of Applied Meteorology* 32: 335–356

Arkin PA (1979) The relationship between fractional coverage of high cloud and rainfall accumulations during GATE over the B-scale array. *Monthly Weather Review* 107: 1382–1387

Arkin PA, Meiner BN (1987) The relationship between large scale convective rainfall and cold cloud over the western hemisphere during 1982–84. *Monthly Weather Review* 115: 51–74

Arkin PA, Xie P (1994) The Global Precipitation Climatology Project: First Algorithm Intercomparison Project. *Bulletin of the American Meteorological Society* 75: 401–419

Barrett EC, Adler RF, Bauer P, Berg W, Chang A, Ferraro R, Ferryday J, Goodman S, Hong Y, Janowiak J, Kidd C, Kniveton D, Morrissey M, Olson W, Petty G, Rudolf B, Shibata A, Smith E, Spencer R (1994) The First WetNet Precipitation Intercomparison Project (PIP-1): Interpretation of Results. *Remote Sensing Reviews* 11: 303–373

Barrett EC, Kidd C, Kniveton D (1995) The First WetNet Precipitation Intercomparison Project (PIP-l): Reflections on the results. In *Proceeding of IGARSS'95*, Firenze 10–14 July 1995, IEEE Catalog Number 95CH35770. 649–651

Bauer P, Schlüssel P (1993) Rainfall, total water, ice water, and water vapor over sea from polarized microwave simulations and Special Sensor Microwave/Imager data. *Journal of Geophyical Research* 8: 20737–20759

Churnside JH, Sternitz TA, Schröder JA (1994) Temperature Profiling with Neural Network Inversion of Microwave Radiometer Data. *Journal of Atmosphere and Ocean Technology* 11: 105–109

Davis DT, Chen Z, Tcang L, Hwang J-N, Chang ACT (1993) Retrieval of snow parameters by iterative inversion of a neural network. *IEEE Transactions on Geocience and Remote Sensing* 31: 842–852

Deirmendjian D (1975) Far-infrared and submillimeter wave attenuation by clouds and rain. *Journal of Applied Meteorology* 14: 1584–1593

Farrar MR, Smith EA, Xiang X (1994) The impact of spatial resolution enhancement of SSM/I microwave brightness temperatures on rainfall retrieval algorithms. *Journal of Applied Meteorology* 33: 313–333

Francis CR, Thomas DP, Windsor PL (1983) The evaluation of SMMR retrieval algorithms. In Allan, T. D., editor, *Satellite Remote Sensing*, Ellis Horwood Limited, pp 481–498

Gloersen P, Barath FT (1977) A scanning multichannel microwave radiometer for Nimbus-G and Seasat-A. *IEEE Journal of Oceanic Engineering* OE-2, 172–178

Griffith CG, Woodley WL, Grube PG, Martin DW, Stout J, Skidar DN (1978) Rain estimation from geosynchronous satellite imagery - visible and infrared studies. *Monthly Weather Review* 106: 1153–1171

Grody NC (1991) Classification of snow cover and precipitation using the Special Sensor Microwave Imager. *Journal of Geophysical Research* 96: 7423–7435

Hertz J, Krogh A, Palmer RG (1991) *Introduction to the theory of neural computation.* Addison-Wesley

Houghton JT, Taylor FW, Rodgers CD (1984) *Remote sounding of atmospheres.* Cambridge Planetary Science Series. 343p

Hsu K, Sorooshian S, Gupta VK, Gao X (1994) Application of artificial neural network models in rain rate retrieval from satellite passive microwave imagery. *EOS* 75: 31

Jin Y-Q (1992) A Mueller matrix approach to complete poarimetric scattering from a layer of non-uniformly oriented, non-spherical scatterers. *Journal of Quantitative Spectroscopy and Radiative Transfer* 48(3): 295–306

Karstens U, Simmer C, Ruprecht E (1994) Remote sensing of cloud liquid water content. *Meteorology and Atmospheric Physics* 54: 157–171

Kummerow C, Giglio L (1994) A passive microwave technique for estimating rainfall and vertical structure information from space. Part I: Algorithm Description. *Journal of Applied Meteorology* 33: 3–18

Lee J, Weger RC, Sengupta SK, Welch RM (1990) A Neural Network Approach to Cloud Classification. *IEEE lansactions on Geoscience and Remote Sensing* 38(5): 846–855

Liu Q, Simmer C (1994) 3-D radiation effects of a raining cloud. In *Proceedings of the Specialist Meeting on Microwave Radiometry and Remote Sensing of the Environment*, Rome, page 7p

Marhall JS, Palmer Mck (1948) The distribution of raindrops with size. *J Meteorol* 5: 165–166

Mishchenko MI, Travis LD (1994) Light scattering by polydisperse, rotationally symmetric nonspherical particles: linear polarization. *Journal of Quantitative Spectroscopy and Radiative Transfer* 51: 579–778

Mugnai A, Smith EA, Tripoli GJ (1992) Foundations for statistical physical precipitation retrieval from passive microwave measurements. Part I: Brightness temperature properties of a time dependent cloud radiation model. *Journal of Applied Meteorology* 31: 506–531

Negri AJ, Adler RF (1993) An intercomparison of three satellite infrared rainfall techniques over Japan and surrounding waters. *Journal of Applied Meteorology* 32: 357–373

Negri AJ, Adler RF, Wetzel PJ (1984) Satellite rain estimation: An analysis of the Griffith-Woodley technique. *Journal of Climate and Applied Meteorology* 2: 1565–1576

Olson WS (1989) Physical retrieval of rainfall rates over the ocean by multispectral microwave radiometry: Application to tropical cyclones. *Journal of Geophysical Research* 94: 2267–2280

Petty G (1994a) Physical Retrievals of Over-Ocean Rain Rate from Multichannel Microwave Imagery. Part I: Theoretical Characteristics of Normalized Polarization and Scattering Indices. *Meteorology and Atmospheric Physics* 54: 79–99

Petty G (1994b) Physical Retrievals of Over-Ocean Rain Rate from Multichannel Microwave Imagery. Part II: Algorithm Implementation. *Meteorology and Atmospheric Physics* 54: 101–121

Petty GW, Katsaros KB (1990a) Precipitation observed over the South China Sea by the Nimbus-7 scanning multichannel microwave radiometer during winter MONEX. *Journal of Applied Meteorology* 29: 273–287

Petty GW, Katsaros KB (1990b) New geophysical algorithms for the Special Sensor Microwave Imager. In *Proceedings 5th Conference on Satellite Meteorology and Oceanography*, London UK. 247–251

Prabhakara C, Dalu G, Sushanisi R, Nucciarone JJ, Liberti GL (1992) Rainfall over oceans: Remote sensing from satellite microwave radiometers. *Metorology and Atmospheric Physics* 47: 177–199

Press WH, Flannery BP, Teukolsky SA, Vetterling WT (1986) *Numerical Recipes, the art of scientific computing.* Cambridge University Press, 818p

Pruppacher HR, Klett JD (1978) *Microphysics of clouds and precipitation.* D. Reidel Publishing Company, 714p

Robinson WD, Kummerow C, Olson WS (1992) A technique for enhancing and matching the resolution of microwave measurements from the SSM/I instrument. *IEEE Transactions on Geoscience and Remote Sensing* 30: 419–429

Rodgers E, Siddalingaiah H, Chang ATC, Wilheit TT (1979) A statistical technique for determining rainfall over land employing Nimbus-6 ESMR measurements. *Journal of Applied Meteorology* 18: 978–991

Rudolf B, Hauschild H, Reiss U (1992) Beiträge zum Weltzentrum für Niederschlags-klimatologie. Contributions to the Global Precipitation Climatology Centre. *Meteorologische Zeitschrift* 1: 7–84

Rumelhart DE, Hinton GE, Williams RJ (1986) Learning representations by backropagating errors. *Nature* 323: 533–536

Simmer C (1994) *Satellitenfernerkundung hydrologischer Parameter der Atmosphäre mit Mikrowellen.* Verlag Dr. Kovac, 314p

Simmer C, Liu Q (1995) Determination of rainfall over the ocean from SSM/I measurements with a library approach. In *Proceedings of IGARSS'95*, IEEE Catalog No. 95CH35770 pages 1892–1894

Simpson J, Adler RF, North GR (1988) A proposed tropical rainfall measuring mission (TRMM) satellite. *Bulletin of the American Meteorological Society* 69: 278–295

Smith EA, Mugnai A, Cooper HJ, Tripoli GJ, Xiang X (1992) Foundations for statistical-physical precipitation retrieval from passive microwave satellite measurements: Part I: Brightness-temperature properties of a time-dependent cloud-radiation model. *Journal of Applied Meteorology* 31: 506–531

Spencer RW (1986) A satellite passive 37 GHz scattering-based method for measuring oceanic rain rates. *Journal of Climate and Applied Meteorology* 25: 754–766

Spencer RW, Hinton BB, Olson WS (1983a) Nimbus-7 37 GHz radiances correlated with radar rain rates over the Gulf of Mexico. *Journal of Climate and Applied Meteorology* 22: 2095–2099

Spencer RW, Olson WS, Wu R, Martin DW, Weinman JA, Santek DA (1983b) Heavy thunderstorms observed over land by the Nimbus-7 scanning multichannel microwave radiometer. *Journal of Climate and Applied Meteorology* 22: 1041–1046

Spencer RW, Goodman HM, Hood RE (1991) Precipitation retrieval over land and ocean with the SSM/I: Identification and characteristics of the scattering signal. *Journal of Atmosphere and Ocean Technology*6: 254–273

Stogryn AP, Butler CT, Bartolac TJ (1994) Ocean surface wind retrievals from Special Sensor Microwave imager data with neural networks. *Journal of Geophysical Reearch* 99: 981–984

Wauben WMF, Hovenier JW (1992) Polarized radiation of an atmosphere containing randomly-oriented spheroids. *Journal of Quantitative Spectroscopy and Radiative Transfer* 47: 491–504

Wauben WMF, de Haan JF, Hovenier JW (1993) Influence of particle shape on the polarized radiation in planetary atmospheres. *Journal of Quantitative Spectroscopy and Radiative Transfer* 50: 237–246

Weinman JA, Guetter PJ (1977) Determination of rainfall distributions from microwave radiation measured by the Nimbus s ESMR. *Journal of Applied Meteorology* 1: 437–442

Wentz FJ (1983) A model for ocean microwave brightness temperatures. *Journal of Geophysical Research* 88: 1892–1908

Wilheit TT, Theon JS, Shenk WE, Allison LJ, Rodgers EB (1976) Meteorological interpretations of the images from the Nimbus-5 Electrically Scanned Microwave Radiometer. *Journal of Applied Meteorology* 15: 166–172

Wilheit TT, Chang ATC, Chiu LS (1991) Retrieval of monthly rainfall indices from microwave radiometric Measurements using probability distribution functions. *Journal of Atmosphere and Ocean Technology* 8: 118–136

Willis PT (1984) Functional fits to some observed drop-size distributions and parametrizations of rain. *Journal of the Atmospheric Sciences* 41: 1648–1661

Zhang M, Scofield RA (1994) Artificial neural network techniques for estimating heavy rainfall and recognizing cloud mergers from satellite data. *International Journal of Remote Sensing* 15: 3241–3261

Chapter 12

Precipitation at the Ground: Radar Techniques

Paul Joe
Atmospheric Environment Service of Canada
Meteorological Research Branch
Cloud Physics Research Division
4905 Dufferin St., Downsview, Ontario
Canada, M3H 5T4

Radar can estimate precipitation over a large area (of order 100 km radius) with high spatial (of order 1 km) and temporal (of order 10 minute) resolution. The standard measurement of rainfall is by rain gauges which make essentially point measurements. A very dense network of rain gauges is required to match the radar precipitation fields due to the spatial and temporal variability. This is costly and not very practical. It is only recently that radar has been able to provide adequate estimates of precipitation for hydrological applications. There are many choices, issues and assumptions made when using radar for precipitation monitoring. These range from theoretical, empirical, technological and processing considerations. Basic radar theory and implementation considerations as it relates to precipitation measurements will be discussed. The basic radar equation will be presented with an emphasis on the assumptions made in its derivation. Hardware considerations such as beamwidth, wavelength and signal processing will be discussed. The intelligent use of Doppler signal processing in operational networks can help in the elimination of many radar echo artifacts resulting from ground targets or anomalous propagation. The signal processing for polarization has the potential to classify the precipitation targets and produce better estimates of heavy rainfalls. Advanced data processing can be used to extrapolate the radar echoes measured aloft to the ground, to account for local enhancement or shadowing effects and to adjust for bright band effects. Real-time adjustment procedures using telemetered rain gauges have been proposed to reduce the discrepancy between the gauge and radar measurements. While most of the work has been with rain, snowfall studies have been limited. This is in part due to the difficulty in making accurate snow fall measurements. There is no accepted standard for the high temporal measurement (of order 10 minute) of snow fall. The limited snowfall studies show a reasonable average relationship valid for daily accumulations but short term accumulations show that there are various relationships which are attributed to changes in crystal type. Radar has played a significant role in the study of processes leading to the evolution of precipitation type and distributions. This is a result of the sensitivity of the radar backscatter power to drop size. In addition, the Doppler velocity

NATO ASI Series, Vol. I 45
Radiation and Water in the Climate System:
Remote Measurements
Edited by Ehrhard Raschke

spectrum from a vertically pointing radar can be transformed into drop size distributions if the air motion is known. The technology for hydrometeor identification has only recently advanced to a promising stage. Polarization radar for the discrimination of hail, ice and water appear to be possible. Dual wavelength studies making use of either differential attenuation or diffential backscatter have had a resurgence for precipitation estimation.

12.1 TheWeather Radar

Meteorological radars are capable of detecting precipitation and variations of the refractive index in the atmosphere. The latter may be generated by local variations of temperature or humidity. Radar echoes may also be produced from airplanes, dust, birds or insects. The meteorological radars having characteristics best suited for atmospheric observation and investigation transmit electromagnetic pulses in the 3–10 GHz frequency range (10–3 cm wavelength, respectively), and are designed for detecting, mapping and measuring precipitation intensity, their motion and perhaps their type. Higher frequencies (35 and 95 GHz), are used to detect smaller hydrometeors, such as cloud or even fog droplets. Although this has valuable applications in cloud physics research, these frequencies generally are not used in operational weather forecasting because of the excessive attenuation of the radar signal by the intervening medium.

At lower frequencies, radars known as strato-tropospheric or wind profiling radars are capable of detecting variations of the refractive index of clear air. The radars detect precipitation as well. They are usually fixed beam systems.

The returned signal from the transmitted pulse encountering a weather target, called an echo, has an amplitude, a phase and a polarization. Most operational radars worldwide are still limited to analysis of the amplitude which is related to the size distribution and numbers of particles in the (pulse) volume illuminated by the radar beam. The amplitude is used to determine a parameter called the reflectance factor (Z) which is used to estimate the mass of precipitation per unit volume (M) or the intensity of precipitation (R) through the use of empirical relations. A primary application is thus to detect, map and estimate the precipitation at ground level instantaneously, nearly continuously, and over large areas.

Some research radars have used reflectance factors measured at two orthogonal polarizations of the transmitted and received waveform. Research continues to determine the value and potential of polarization systems for precipitation measurement and target state, but operational systems do not exist at present. Doppler radars have the capability of determining the phase difference between the transmitted and received pulse. The rate of change of the phase difference is a measure of the mean Doppler velocity of the particles. The mean Doppler velocity is a reflectance weighted average of the radial components of the displacement velocities of the hydrometeors in the pulse volume. The Doppler spectrum width is a measure of the spatial variability of the velocities and may provide some indication of the wind shear and turbulence.

Weather radar can serve many purposes. The focus of this paper will be on the ability of the radar to estimate precipitation. Its use for other applications will be briefly described. We proceed by discussing radar technology, basic radar theory, applications and then the various techniques for the measurement of precipitation. We briefly discuss the use of high frequency radars for the detection of clouds.

12.2 Radar Technology

12.2.1 Principles of radar measurement

The principles of radar and the observation of weather phenomena were established in the 1940's. Since that time great strides have been made in improving equipment, signal and data processing and its interpretation. The interested reader should consult some of the texts for greater detail. Good references are: Skolnik (1970) for engineering and equipment aspects; Sauvageot (1982), Battan (1981) and Collier (1989) for meteorological phenomena and applications; Atlas (1964, 1990) for general review; Rinehart (1991) for modern techniques; and Doviak and Zrnic (1993) for Doppler radar principles and applications. A brief summary of the principles follows.

Most meteorological radars are pulsed radars, that is, electromagnetic waves at fixed preferred frequencies are transmitted from a directional antenna into the atmosphere in a rapid succession of short pulses. Fig. 12.1 symbolically diagrams a directional radar antenna emitting a pulsed shaped beam of electromagnetic energy over the curved earth surface and illuminating a portion of a meteorological target. Many of the physical limitations and constraints on the observation technique are immediately apparent from the diagram. For example, there is a limit to the minimum altitude that can be observed at far ranges due to the curvature of the earth.

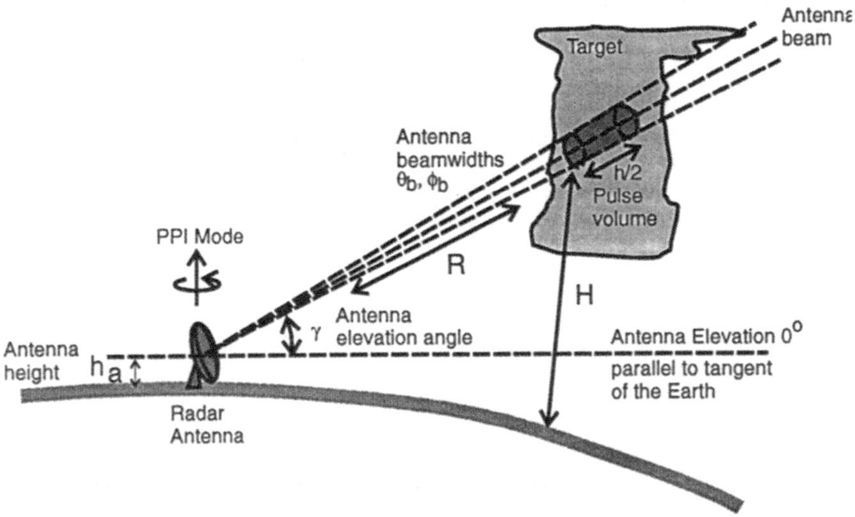

Figure 12.1: *Propagation of electromagnetic waves through the atmosphere for a pulse weather radar. Note: h_a is the height of the antenna above the Earth's surface, R is the range, $h/2$ is the length of the pulse and H is the height of the pulse above the Earth's surface. The figure is drawn with a radius of 4/3 times the radius of the Earth to account for the refraction effects of the atmosphere. In this representation, the radar beam travels in straight lines.*

A parabolic reflector in the antenna system concentrates the electromagnetic energy in a conical shaped beam which is highly directional. The width of the beam increases with range, for example, a nominal 1 degree beam spreads to 0.9, 1.7 and 3.5 km at ranges of 50, 100, and 200 km, respectively. The short bursts of electromagnetic energy are absorbed and scattered by any meteorological targets encountered. Some of the scattered energy is reflected back to the radar antenna and receiver. Between successive pulses, the receiver listens for any return of the wave. The return signal from the target is commonly referred to as the radar echo.

The strength of the signal reflected back to the radar receiver from the target is a function of the concentration, sizes and water phase of the precipitation particles comprising the target. The power return, P_r, therefore provides some measure of the characteristics of the meteorological target and is, but not uniquely, related to a precipitation rate for any given form of precipitation. The radar range equation relates the power-return from the target to the radar characteristics and parameters of the target.

The power measurements are determined by the total power backscattered by the target within a volume being sampled at any one instant - the pulse volume. The power may be integrated later in both space and time. The pulse volume dimensions are dependent on the radar pulse length in space (h) and the antenna beam widths in the vertical (ϕ_b) and the horizontal (θ_b). The antenna beam widths, and therefore the pulse volume, increase with range.

Since the power which arrives back at the radar at the same instant of time is involved in a two-way path, the pulse-volume-length is only one-half pulse length in space ($h/2$) and is invariant with range. The location of the pulse volume in space is determined by the orientation of the antenna in azimuth and elevation and the range to the target; the range (r) is determined by the time required for the pulse to travel to the target and be reflected back to the radar. The target is therefore uniquely defined in space by measurements of range and azimuth and elevation angles, with due considerations for the resolution determined by h, ϕ_b and θ_b.

Particles within the pulse volume are continuously shuffling relative to one another. This results in phase effects in the scattered signal and intensity fluctuations about the mean target intensity. Little significance can be attached to a single echo intensity measurement from a weather target, such as from a single pulse or hit. At least 25 to 30 pulses must be integrated to obtain a reasonable mean intensity. This is normally done electronically in an integrator circuit. Further averaging of pulses in range and azimuth and in time is frequently done to increase the sampling size and accuracy of estimate. It follows that the space resolution is coarser.

An alternative to the pulse weather radar is the frequency modulated continuous wave (FM-CW) weather radar (Richter, 1969). Two antennae are needed since the radar system is listening at the same time as transmission. Range is determined by frequency modulation. Often, a linear chirp waveform is used where the frequency increases with time over a given period. Therefore, echoes received at different frequencies are due to echoes at different ranges. These are usually range limited and used in special applications (Russchenberg, 1995; van Gorp, 1995).

12.2.2 The radar range equation for precipitation targets

The total energy backscattered is the sum of the energy backscattered by each of the scattering particles in the pulse volume. Using this target model and electromagnetic theory, Probert-Jones (1962) developed an equation relating the echo power received by the radar to the parameters of the radar and the targets' range and scattering characteristics. It is accepted as being a reliable relationship providing quantitative reflectance measurements with good accuracy.

$$\overline{P}_r = \frac{\pi^3}{1024 \ln 2} \frac{P_t h G^2 \theta_b \phi_b}{\lambda^2} \frac{|K|^2 10^{-18} Z}{r^2} \tag{12.1}$$

Symbol	Description	Units
$\overline{P}_r$	The power received back at the radar, averaged over several pulses	watts
P_t	The peak power of the pulse transmitted by the radar	watts
h	The pulse length in space	meter
G	The gain of the antenna over an isotropic	
θ_b, ϕ_b	The horizontal and vertical beamwidths respectively of the antenna radiation pattern at the -3 dB level of one way transmission	radians
λ	The wavelength of the transmitted wave	meter
$\|K\|^2$	The refractive index factor of the target	
r	The slant range from the radar to the target	meter
Z	The radar reflectance factor, usually taken as the equivalent reflectance factor Z_e when the target characteristics are not well known	mm^6m^{-3}

Table 12.1: *Radar equation symbols.*

where all the units are in MKS system except for Z which is in mm^6 m^{-3}. See Table 12.1 for description of symbols.

The parameters in the radar equation have been grouped into those depending on the characteristics of the radar and those depending on the range and characteristics of the target. The radar parameters, except for the transmitted power, are essentially fixed. If the transmitter is operated and maintained at a constant output (as it should be) then the equation can be simplified to:

$$\overline{P}_r = \frac{C|K|^2 Z}{r^2} \tag{12.2}$$

where C is the radar constant.

There are a number of basic assumptions inherent in development of the equation which have varying importance in the application and interpretation of the results. They are reasonably realistic but the conditions are not always met exactly. These assumptions are summarized as follows:

- The scattering precipitation particles in the target volume are homogeneous dielectric spheres whose diameters are small compared to the wavelength, that is, $D < 0.06\lambda$, for strict application of Rayleigh scattering approximations.
- The pulse volume is completely filled with randomly scattered precipitation particles.
- The reflectance factor Z is uniform throughout the sampled pulse volume and constant during the sampling interval.
- The particles are all water drops or all ice particles, i.e., all particles have the same refractive index factor $|K|^2$.
- Multiple scattering (among particles) is negligible.
- There is no attenuation in the intervening medium between the radar and measurement point (target).

- The incident and backscattered waves are linearly polarized.
- The main lobe of the antenna radiation pattern is of Gaussian shape.
- The antenna is a parabolic reflector type of circular cross section.
- The gain of the antenna is known or can be calculated with sufficient accuracy.
- The contribution of the side lobes to the received power is negligible.
- Absorption of the transmitted signal by ground clutter in the beam is negligible.
- The peak power transmitted (P_t) is the actual power transmitted at the antenna, that is, all waveguide losses, etc. are considered.
- The average power measured (P_r) is averaged over a sufficient number of pulses or independent samples to be representative of the target pulse volume.

This simplified expression relates the echo power ($\overline{P}_r$) measured by the radar to the radar reflectance factor, Z, which in turn is related to the rainfall rate, R. These factors and their relationship are important in interpreting the intensity of the target and in estimating precipitation amounts from radar measurements. In spite of the many assumptions of varying importance the expression provides a reasonable estimate of the target mass. This estimate can be improved by further consideration of factors in the assumptions.

12.2.3 Non-coherent radar

The power backscattered from weather targets is of the order of 10^{-8} to 10^{-15} watts, covering a range of about 70 dB from the strongest to weakest targets detectable. To adequately cover this range of signal, a logarithmic receiver is normally used. Linear receivers with 90 dB dynamic range are just being introduced in research radars (Keeler et al., 1995). Many pulses must be averaged in the processor to provide a significant measurement; the pulses can be integrated in different ways but usually in a digital form, and must account for the logarithmic scale. In practice, for a typical system, the signal at the antenna is received, amplified logarithmically, averaged over many pulses, corrected for (log) averaging and converted to a reflectance factor using the radar range equation.

The reflectance factor is the most important parameter for radar interpretation. The factor derives from the Rayleigh scattering model and theoretically is defined as the sum of particle (or drop) diameters to the sixth power per unit volume.

$$Z = \sum_{vol} D^6 \tag{12.3}$$

In many cases, the numbers of particles, composition and shape are not known and an equivalent or effective reflectance factor Z_e is defined. Snow and ice particles must refer to an equivalent Z and represents the Z assuming the backscattering particles were all spherical drops.

A common practice is to work in a logarithmic scale or dBZ units which are numerically defined as:

$$dBZ = 10 \log_{10} Z_e \tag{12.4}$$

Volumetric observations of the atmosphere are normally made by scanning the antenna pseudo-horizontally and elevating by steps in the vertical at each revolution. An important consideration is the resolution of the targets. Parabolic reflector antennas are used to focus the waves into a conical shaped beam. Larger reflectors create narrower beams, greater resolution and sensitivity at increasing costs. The beamwidth (θ_b), the points where the power is one-half that of the axis, is dependent also on the wavelength and can be approximated by:

$$\theta_b = \frac{70\lambda}{D_{antenna}} \tag{12.5}$$

The units of θ_b are degrees. $D_{antenna}$ is the diameter of the parabolic antenna dish. Good weather-radars have beamwidths of 0.5 to 1 degree.

The useful range of weather radars, except for long range detection of thunderstorms, is of the order of 200 km since at that range, the 1° beam at an elevation angle of 0.5° is already at a height of 4 km above the curved earth surface with a beamwidth of the order of 1.5 km or greater. For good quantitative precipitation measurements, the maximum effective range is less than 130 km, since the beam is too high for good ground estimates, beam spreading reduces resolution and the measurement can be affected by underfilling with target. Technically, there is a maximum unambiguous range determined by the pulse repetition frequency since the range must be measured during the listening period between pulses. At usual pulse repetition frequencies (PRF), this is not a problem. For example, for a PRF equal to 250 pulses per second, the maximum range is 600 km. At higher PRFs required for good Doppler systems, the range will be greatly reduced, although new developments in progress will ameliorate this situation (Joe et al., 1994).

12.2.4 Doppler radar

The development and introduction of Doppler weather radars to weather surveillance provides a new dimension to the observations. Doppler radar provides a measure of the target(s) velocity along a radial from the radar in a direction either towards or away from the radar. A further advantage of the Doppler technique is the greater effective sensitivity to low reflectance targets close to the radar noise level when velocity fields can still be distinguished in a noisy Z field.

At the normal speeds of meteorological targets, the frequency shift is relatively small compared to the radar frequency and very difficult to measure. An easier task is to retain the phase of the transmitted pulse, compare it with the phase of the received pulse and then determine the change in phase between successive pulses. The time rate of change of the phase is directly related to the frequency shift which in turn is directly related to the target velocity - the Doppler effect. If the phase changes by greater than ± 180° the velocity is not determined unambiguously but will be biased by some multiple of the Nyquist interval (the highest unambiguous velocity that can be measured by a Doppler radar). An additional processing step is required to retrieve the correct velocity.

The maximum unambiguous Doppler velocity depends on the radar wavelength (λ) and the pulse repetition frequency (PRF) and can be expressed as:

$$V_{max} = \pm\frac{PRF\lambda}{4} \tag{12.6}$$

The maximum unambiguous range can be expressed as:

$$r_{max} = \frac{c}{PRF \times 2} \tag{12.7}$$

Thus V_{max} and r_{max} are related by the equation:

$$V_{max} r_{max} = \pm \frac{\lambda c}{8} \tag{12.8}$$

These relationships show the limits imposed by the selection of the PRF. A high PRF is desirable to increase the unambiguous velocity; a low PRF is desirable to increase the radar range. A compromise is required until better technology is available to retrieve the information unambiguously outside these limits. The relationship also shows that the longer wavelengths have higher limits. In numerical terms, for a typical S-band radar with a PRF of 1000 Hz, $V_{max} = \pm 25$ m s^{-1}, while for an X-band radar $V_{max} = \pm 8$ m s^{-1}.

Some Doppler radars are fully coherent radars; that is, their transmitters employ very stable microwave sources, usually Klystrons, which maintain the same phase from pulse to pulse. A semicoherent radar is more common; the microwave source is usually a magnetron, where the phase of successive pulses is random. The Doppler information is extracted in the processing by retaining the phase of each transmitted pulse for comparison. The advantage of the magnetron is their lower cost. The advantages of the Klystron are the greater stability and ability to produce higher power. In general, non-coherent radars can be converted relatively easily to a semicoherent Doppler system. The conversion should also include the more stable coaxial type magnetron.

Both reflectance factor and velocity data are extracted from the Doppler radar system. The target is typically a large number of hydrometeors (rain drops, snow flakes, ice pellets, hail, etc.) of all shapes and sizes moving at different speeds due to the turbulent motion within the volume and their falling speeds. The velocity field is therefore a spectrum of velocities - the Doppler spectrum (See Fig. 12.2).

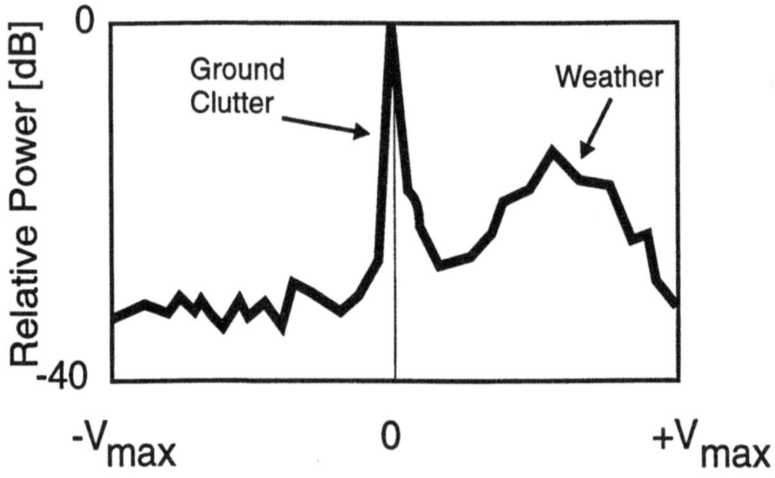

Figure 12.2: *The Doppler spectrum of a weather echo and a ground target. The ground target contribution is centered around zero and is much narrower than the weather echo.*

Two systems of varying complexity are used to process the Doppler parameters. The simpler pulse pair processing (PP) system uses the comparison of successive pulses in the time domain to extract a mean velocity and spectrum width. The second and more complex system requiring very fast processing capability uses a Fast Fourier Transform (FFT) algorithm on a time series of data to produce a full spectrum of the velocities in each sample volume or unit.

Ground clutter appearing in the Doppler spectrum will have zero velocity while weather echoes have a broad velocity interval (see Fig. 12.2). The ground echo power can then be almost eliminated, even within precipitation echoes, while retaining the precipitation echo power (Passarelli et al., 1981). This is a major advantage of spectral processing. Digital filtering in the time domain to notch out the ground clutter will also eliminate weather echo near the zero velocity band. Stripped of clutter echo, the significant meteorological parameters can be computed.

12.2.5 Polarization diversity radars

Experiments with polarization diversity radars have been performed for many years to determine their potential for enhanced radar observations of the weather. Some promising studies point towards the possibility of determining better drop size distribution information and subsequently a better measure of rainfall amounts and of differentiating between hydrometeor types. The basic premise of polarization is to detect the microdifferences amongst the hyrometeors in a radar volume.

There are two basic radar techniques in current usage. In one case, a circularly polarized wave is transmitted and the amount of power in the co-polar and cross-polar sense are measured as well as the correlation between the orthogonal polarizations. In the other case, pulses are transmitted alternately with horizontal then vertical polarization and the returned signals at each polarization are measured either simultaneously or in sequence. Most conventional weather radars are horizontally polarized in order to maximize the backscatter from large raindrops. Conversion of these radars lead to the popularity of the latter type of polarization system. Matching the beams, switching polarizations and measurement of small differences in signals are formidable tasks requiring great care in applying the techniques.

12.2.6 Radar accuracy requirements

The accuracy requirements depend on the most important application(s) of the radar observations. Modern radars appropriately installed, calibrated and maintained are relatively stable and do not produce significant measurement errors. External factors such as ground clutter effects, anomalous propagation, attenuation and propagation effects, beam effects, target composition particularly with variations and changes in the vertical, and rain rate-reflectance relationship inadequacies contribute most to the inaccuracy. Considering only errors attributable to the radar system, the measurable radar parameters can be determined with an acceptable accuracy (see Table 12.2).

12.3 Propagation and Scattering of Radar Signals

12.3.1 Scattering and propagation effects

Electromagnetic waves propagate in straight lines in a homogeneous medium with the speed of light. The Earth's atmosphere is not homogeneous and microwaves undergo refraction, absorption and scattering along their propagation path. The atmosphere is usually vertically stratified and the rays change direction depending on the changes in height of the refractive

Parameter	Definition	Acceptable Accuracy
ϕ	Azimuth angle	.1 deg
γ	Elevation angle	.1 deg
V_r	Mean Doppler velocity	.25 m/s
Z	Reflectance factor	1 dBZ
σ_v	Doppler spectrum width	1 m/s

Table 12.2: *Accuracy requirements. These figures are relative to a standard Gaussian spectrum with a variance smaller than* 4 m^2s^{-2}. *Velocity accuracy deteriorates when the spectrum width grows, while reflectance accuracy improves.*

index which is a function of temperature and moisture. When the waves encounter precipitation and clouds, part of the energy is absorbed and part is scattered in all directions, part back to the radar site.

12.3.2 Refraction in the atmosphere

The amount of bending of electromagnetic waves can be predicted using the vertical profile of the temperature and moisture (see Bean and Dutton, 1966). Under normal atmospheric conditions the waves travel in a curve bending earthward slightly. The ray path can bend either upwards (subrefraction) or more earthward (superrefraction). In either case the altitude of the beam will be in error using the standard atmosphere assumption.

From a precipitation measurement standpoint the greatest problem occurs under superrefractive or ducting conditions. The ray can bend sufficiently to strike the earth and cause ground echoes not normally encountered. The phenomenon occurs when the index of refraction decreases rapidly with height, for example, this can occur with an increase in temperature and a decrease in moisture with height. This condition is referred to as anomalous propagation, ANAPROP or AP.

12.3.3 Attenuation in the atmosphere

Microwaves are subject to attenuation due to atmospheric gases, clouds and precipitation by absorption and scattering.

Attenuation by gases

Gases attenuate microwaves in the 3–10 cm bands. Absorption by atmospheric gases is due mainly to water vapor and oxygen molecules. Attenuation by water vapor is directly proportional to the pressure and absolute humidity and increases almost linearly with decreasing temperature. The concentration of oxygen over the world to altitudes of 20 km is relatively uniform and attenuation is proportional to the square of the pressure and also a function of temperature. Attenuation by gases is dependent on the climate and the season but this variation is small. Attenuation is significant at weather radar wavelengths and can amount to 3 to 4 dB over a range of 200 km. Attenuation can be computed as a function of range and applied as a correction to the precipitation field.

Wavelength (cm)	Relation (dB km^{-1})
10	$0.000343R^{0.97}$
5	$0.0018R^{1.05}$
3.2	$0.01R^{1.21}$

Table 12.3: *One Way Attenuation Relationships for Hydrometeors. After Burrows and Attwood (1949), one way specific attenuations at 18°C, R is in units of mm/hr.*

Attenuation by hydrometeors

Attenuation by hydrometeors can result from both absorption and scattering. It is the most significant source of attenuation and can be substantial. It is dependent on the shape, sizes, numbers and composition of the particles. This dependence has made it very difficult to overcome in any quantitative way using radar observations alone.

Attenuation is dependent on wavelength. At 10 cm wavelengths the attenuation is rather small while at 3 cm it is quite significant. At 5 cm the attenuation may be acceptable for many climates particularly in the high mid-latitudes (see Table 12.3). Wavelengths below 5 cm are not recommended for good precipitation measurement.

For precipitation estimates by radar, some general statements can be made with regard to the magnitude of attenuation. The attenuation is dependent on water mass of the target, thus heavier rains attenuate more; clouds with much smaller mass attenuate less. Ice particles attenuate much less than liquid particles.

Clouds and ice clouds cause little attenuation and, except where extreme precision is required, can be ignored. Snow or ice particles or a hailstone which can grow to a size much larger than a rain drop, and then become wet due to partial melting, experience a large increase in reflectance and in attenuation properties. Melting of falling precipitation reaching the melting level can produce very high reflectance and attenuation values that distort precipitation estimates.

12.3.4 Scattering by clouds and precipitation

The signal power detected and processed by the radar is power backscattered by the target, or hydrometeors. The backscattering cross section (σ_b) is defined as the area of an isotropic scatterer which would return to the emitting source the same amount of power as the actual target. The backscattering cross section of spherical particles was first determined by Mie (1908). Rayleigh found that if the ratio of the particle diameter to the wavelength was equal or less than 0.06 a simpler expression could be used to determine the backscatter cross section (see Fig. 12.3):

$$\sigma_b = \frac{\pi^5 |K|^2 D^6}{\lambda^4} \tag{12.9}$$

where $|K|^2$ is the refractive index factor and is equal to 0.93 for liquid water and 0.197 for ice, D is the drop diameter and λ is the wavelength. This relationship is used in the development of the radar range equation. The basic equation for determining the target intensity from the radar power measurements is:

$$Z = \frac{C\overline{P}_r r^2}{|K|^2} \tag{12.10}$$

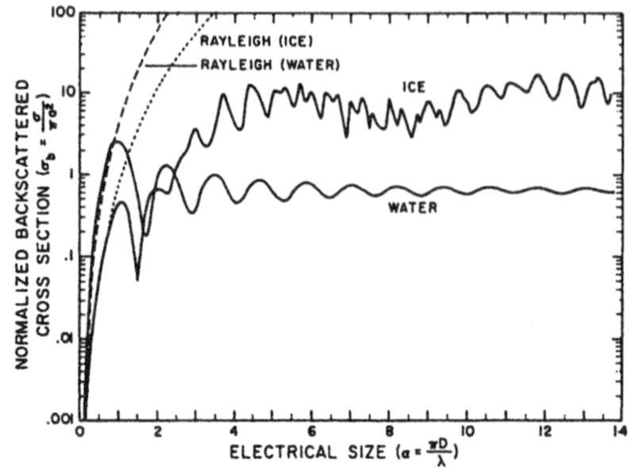

Figure 12.3: *Rayleigh and Mie Scattering. Rayleigh scattering is an approximation to Mie scattering for small (relative to wavelength) particles.*

where C is the radar constant for the system and other symbols are defined previously.

At 5 and 10 cm wavelengths, the Rayleigh approximation is valid for most practical purposes unless hailstones are present. Large concentrations of ice mixed with liquid can cause anomalies, particularly near the melting level. By taking account of the refractive index factor for ice (i.e., $|K|^2 = 0.208$) and choosing an appropriate relation for the reflective factor and precipitation rate (Z_e vs. R), precipitation amounts can be estimated reasonably well in snow conditions. The value of 0.208, instead of 0.197 for ice, accounts for the change in particle diameter for water and ice particles of equal mass.

12.3.5 Scattering in clear air

In regions without precipitating clouds, i.e. clear air, it has been found that echoes are mostly due to insects or to strong gradients of refractive index in the atmosphere. The echoes are of very low intensity, and detected only by very sensitive radars. Equivalent Z_e values for clear air phenomena appear in the range of -5 to -55 dBZ although these are not true Z parameters, the physical process generating the echoes being entirely different. For precipitation measurement, these echoes are a minor noise in the signal. Echoes due to refractive index fluctuations can usually be associated with some meteorological phenomenon such as a sea breeze, thunderstorm outflows or mixing in the boundary layer. Clear air echoes also can be associated with birds and insects in surprisingly low concentrations. Echo strengths of 5 to 35 dBZ are not unusual especially during migrations (see Table 12.4).

Although normal radar processing would interpret the signal in terms of Z or R, the scattering properties of the clear atmosphere are quite different from that of hydrometeors and most often expressed in terms of the structure parameter of refractive index, C_n^2, a measure of the mean-square fluctuations of the refractive index as a function of distance (see Gossard and Strauch, 1983).

Object	σ_b (m^2)
Aircraft	10 to 1000
Man	0.14 to 1.05
Weather Balloon	0.01
Birds	0.001 to 0.01
Bees, dragonflies, moths	3×10^{-6} to 10^{-5}
2 mm water drop	1.8×10^{-10}

Table 12.4: *Typical backscatter cross-sections for various targets.*

12.3.6 Sources of error

Fig. 12.4 schematically illustrates problems in interpreting radar measurements (after Browning, 1987). Other sources of errors are described below.

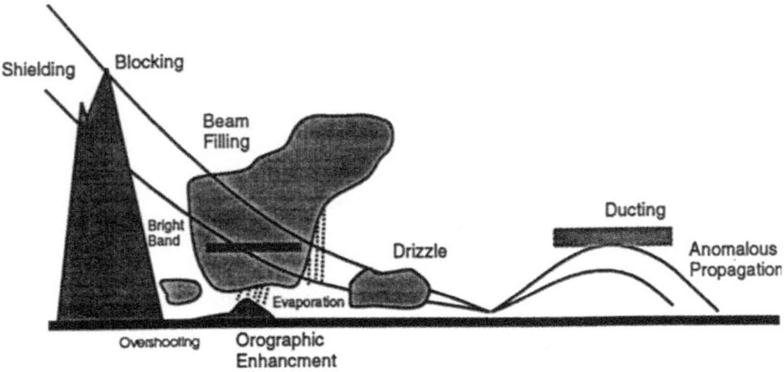

Figure 12.4: *The figure illustrates some of the possible sources of errors for the radar detection of precipitation. The figure is adapted after one by Browning (1987).*

- **Radar Beam Filling:** In many cases, and especially at large ranges from the radar, the pulse volume is not completely filled with homogeneous precipitation. Precipitation intensities often vary widely on small scales, and at large distances from the radar the pulse volume increases in size which averages out the variations. At the same time, the effects of earth curvature become significant and the radar beam overshoots the top of the echo. In general, the measurements may not be considered as quantitative beyond a range of 100 km and are underestimated.

- **Non Uniformity Of Vertical Distribution Of Precipitation:** The first parameter of interest when making radar measurements is usually the precipitation at ground level. Because of the effects of beam width, beam tilting and earth curvature, measurements are made by averaging over a large volume located at some altitude. The variations of reflectance with height may contribute to significant errors. Over and underestimation of the precipitation rate can occur. Of particular annoyance is the presence of the bright band where enhanced reflectivities due to large wet (melting) snowflakes lead to an overestimate of the amount of falling precipitation (see Fig. 12.5).

- **Variations in $Z - R$ Relationship:** The variations in types and size distribution of hydrometeors cannot be estimated in operational measurements and are sometimes an important source of error. The basic assumption is that the target is rain but this is not always the case.

- **Attenuation By Intervening Precipitation:** Attenuation by rain may be important, especially at the shorter radar wavelengths (5 cm and 3 cm). Attenuation by snow, although less than for rain, may be important over long path lengths. Precipitation intensities are underestimated due to this effect.

- **Beam Blocking:** Depending on the radar installation, the radar beam may be partly or completely occulted by topography or obstacles located between the radar and the target. This results in underestimations of reflectance, and hence of rainfall rate.

- **Attenuation Due To A Wet Radome:** Most radars antennas are protected from wind and rain by a radome, usually made of fiberglass. Under intense rainfall, the surface of the radome may become coated with a thin film of water or ice, resulting in a strong azimuth dependent attenuation (Donaldson, 1990).

- **Electromagnetic Interference:** Electromagnetic interference from other radars or devices such as microwave links or the sun may be an important factor of error in some cases. This type of problem is easily recognized by observation.

- **Ground and Other Clutter:** Contamination of rain echoes by ground clutter may bring very large errors in precipitation estimation. The ground clutter should first be minimized by good antenna engineering and a good choice of the radar location. This effect may be greatly reduced by a combination of hardware ground clutter suppression devices (Aoyagi, 1983), signal and data processing.

  Chaff, birds, refractive inhomogenieties and insects may also produce echoes detectable by the radar.

- **Anomalous Propagation:** Anomalous propagation distorts the radar beam path, and has the effect of increasing ground clutter by refracting the beam towards the ground. Anomalous propagation is frequent in some regions, when the atmosphere is subject to strong decreases in humidity and/or increases in temperature with height. Clutter returns due to anomalous propagation may be very misleading to untrained human observers.

- **Antenna Accuracy:** The antenna position may be known within .1 degrees with a well engineered system. Errors may also be produced by the excessive width of the radar beam or by the presence of sidelobes, in the presence of clutter or of strong precipitation echoes.

- **Electronics Stability:** Modern electronic systems are subject to small variations with time. This may be controlled by using a well engineered monitoring system, which will keep the variations of the electronics within less than 1 dB, or activate an alarm when a fault is detected.

- **Processing Accuracy:** The signal processing must be designed to take the best advantage of the sampling capacities of the rest of the system. It must keep to a minimum, the variances of the estimation of reflectance, Doppler velocity and spectrum width. Range and velocity aliasing may be important sources of error.

- **Radar Range Equation:** In interpreting radar received power measurements in terms of the meteorological parameter Z by the radar range equation, there are many assumptions. Non-conformity with the assumptions can cause error.

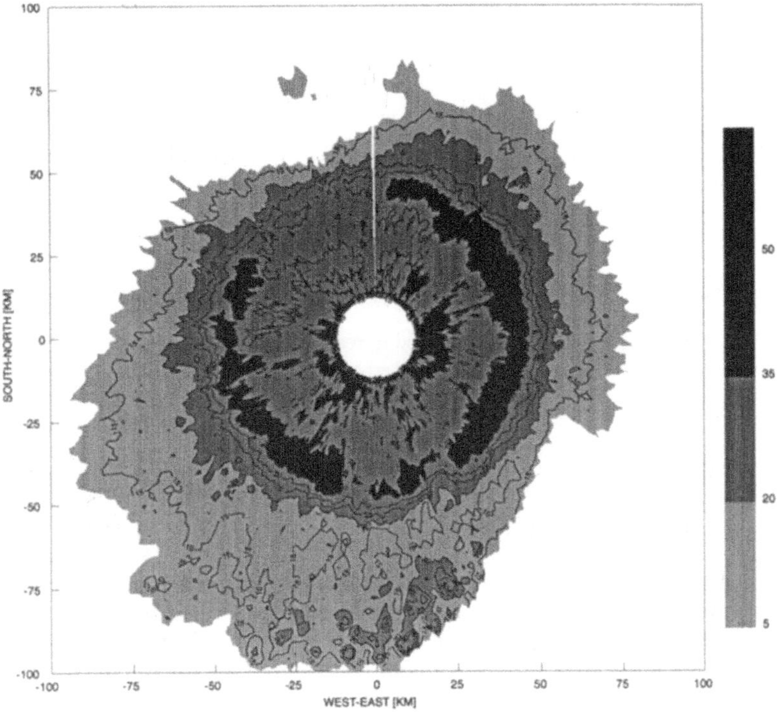

Figure 12.5: *A plan position indicator (PPI) display of a bright band. The dark ring is due to large wet snowflakes that generate high reflectivities. The ring effect is due to the conical scanning pattern of the radar where increasing range corresponds also to increasing height. See Fig. 12.23 for a vertical cross-section of the same precipitation system.*

12.4 Velocity Measurements

A fundamental problem in operational use of any pulse Doppler radar is the removal of ambiguity in Doppler mean velocity estimates, that is, velocity aliasing. Discrete equi-spaced samples of a time-varying function result in a maximum unambiguous frequency equal to one-half the sampling frequency (f_S). Frequencies greater than $f_S/2$ are aliased (folded) into the Nyquist co-interval ($\pm f_S/2$) and are interpreted as velocities within $V_{max} = \pm\lambda f_S/4$, where λ is the wavelength of transmitted energy.

Techniques to de-alias the velocity include dual PRF techniques or continuity techniques. In the former, radial velocity estimates are collected at two different PRFs with different maximum unambiguous velocities and combined to yield a new estimate of the radial velocity with an extended unambiguous velocity (Crozier et al., 1991; Sauvageot, 1982). For example, a C-Band radar using PRFs of 1200 and 900 Hz have nominal unambiguous velocities of 16 and 12 m/s, respectively. The amount of aliasing can be deduced from the difference between the two velocity estimates to de-alias the velocity to an extended velocity range of ± 48 m s^{-1}.

Continuity techniques rely on having sufficient echo to discern that there are aliased velocities and correcting them by assuming velocity continuity (i.e., assuming no discontinuous jumps of $2 \times V_{max}$). See Doviak and Zrnic (1993) for a detailed discussion.

There is also a range limitation imposed by the use of high PRFs ($\sim$1000 Hz). Echoes beyond the maximum range will be aliased back into the primary range. For radars with coherent transmitters (for example, Klystron systems), the echoes will appear within the primary range. For coherent-on-receive systems, the second trip echoes will have very noisy velocity characteristics due to the randomly transmitted phases and can be eliminated to some extent (Joe et al., 1995).

12.4.1 Measurement of velocity fields

A great deal of information can be determined in real-time from a single Doppler radar. Color displays of single-Doppler radial velocity patterns aid in the real-time interpretation of the associated reflectance fields, and can reveal important features not evident in the reflectance structures alone. Such a capability is of particular importance in the identification and tracking of severe storms. On typical color displays, velocities between $\pm V_{max}$ are assigned one of 8–15 colors or more. Velocities extending beyond the Nyquist interval enter the scale of colors at the opposite end. This process may be repeated if the velocities are aliased more than one Nyquist interval.

Doppler radar can also be used to derive vertical profiles of horizontal winds. When the radar's antenna is tilted above horizontal, increasing range implies increasing height, and a profile of wind with height can be obtained by sinusoidal curve-fitting to the observed data (termed velocity-azimuth display or VAD technique after Lhermitte and Atlas, 1961). The winds along the zero radial velocity contour are perpendicular to the radar beam axis. Typical elevated conical scan patterns in widespread warm frontal precipitation reveal an S-shaped zero radial velocity contour as the mean wind veers with height (Wood and Brown, 1986). On other occasions, closed contours representing jets are evident.

Since the measurement accuracy is good, divergence estimates can also be obtained employing the VAD technique. This technique cannot be accurately applied during periods of convective precipitation around the radar. The VAD technique seems well suited for winds from precipitation systems associated with extratropical and tropical cyclones.

Moderately powerful, sensitive Doppler radars have successfully obtained VAD wind profiles and divergence estimates in the optically clear boundary layer during all but the coldest months, up to heights of 3–5 km AGL. In the radar's clear-air mode, a time-series of measurements of divergence and derived vertical velocity is particularly useful in nowcasting the probability of deep convection.

Since the mid 1970's, experiments have been made for measuring three dimensional wind fields using multiple Doppler arrays. Measurements made at a given location inside a precipitation area may be combined, by using a proper geometrical transformation, in order to obtain the three wind components. Many significant advances in the knowledge of cloud dynamics have been obtained from multiple Doppler radar systems (Ray et al., 1980).

12.5 Meteorological Application

12.5.1 Meteorological requirements

The radar characteristics of any one radar will not be ideal for all applications. Radars observations have been found most useful for:

- severe weather detection, tracking and warning,
- surveillance of synoptic and mesoscale weather systems, and
- estimation of precipitation amounts.

The selection criteria of a radar system is usually optimized to meet several applications but can also be specified to best meet a specific application of major importance. The choice of wavelength, beamwidth, pulse length and PRF have particular consequences.

12.5.2 Operational severe weather detection and warning

Radar is the only realistic means of monitoring severe weather over a wide area. Radar echo intensities, area and patterns can be used to identify areas of severe weather. These storms include thunderstorms with probable hail and damaging winds. Doppler radar which can identify and provide a measure of intense winds associated with gust fronts, downbursts and tornadoes adds a new dimension. Nominal range of coverage is about 200 km, which is sufficient for local short range forecasting and warning. Radar networks are used to extend the coverage. Effective interpretation requires alert and well trained personnel to provide effective warnings at present and until automated algorithms and storm models have been developed for the local areas.

Reflectance signatures for some severe storms are well known. The most commonly used criterion for conventional radar detection of potentially-severe thunderstorms today is reflectance intensity. Operational forecasters are advised to look for regions of high reflectivities (50 dBZ or greater), hook echoes, overhangs and other echo shapes to warn of tornadoes or of severe thunderstorms, but the false alarm rate is high (see Fig. 12.6; Lemon et al., 1978). These signatures extend to the spiral-bands and eyewall structure for hurricanes and hook or finger-like echoes for tornadoes and hail (Donaldson, 1970).

In Doppler data, the existence of the mesocyclone has led to a useful severe storm detection technique. A mesocyclone is a vertical column of rising cyclonically rotating air typically 10–20 km across. The mesocyclone signature (or velocity couplet) is observed to form in mid-levels of a storm and descend to cloud base, coincident with tornado development (Burgess, 1976; Burgess and Lemon, 1990). This behavior has led to improved tornado-warning lead times of 20 min or longer, during quasi-operational experiments in Oklahoma (JDOP, 1979). Most of the Doppler observations on this relationship have been made in the U.S. and it is not known if it can be generalized yet (Houze et al., 1993; Crozier et al., 1991). During experiments in Oklahoma, roughly 50% of all mesocyclones produced verified tornadoes; as well, all storms with violent tornadoes formed in environments with strong shear and possessed strong mesocyclones (Burgess and Lemon, 1990).

The second signature — Tornado Vortex Signature — (TVS) is produced by the tornado itself and is the location of a very tight circulation in a region much smaller than that of the

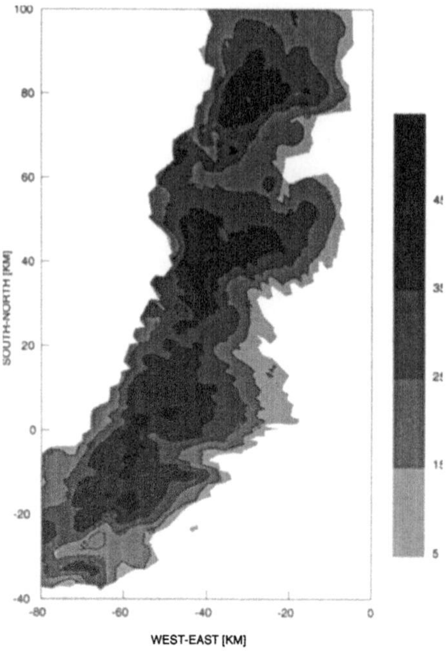

Figure 12.6: *A Constant Altitude Plan Position Indicator reflectance map of a line of tornadic storms. The storm located at approximate coordinates −40km West-East and 30km South-North shows the hook echo feature associated with a mesocyclone which is a parent circulation to the tornado. At this time an intense tornado was on the ground.*

mesocyclone in which it is imbedded. In some cases, the TVS has been detected aloft nearly half an hour or more before a tornado touched the ground (Brown and Lemon, 1976).

Strong, gusty winds are often observed near convective activity. Those produced by small, intense downdrafts of less than 4 km diameter impinging on the ground have been termed "microbursts". The more common intense downdrafts larger than 4 km across are called "downbursts." Low level divergence signatures of downbursts have been routinely made with Terminal Doppler Weather Radars (TDWR) for the protection of aircraft during takeoff and landing. These radars are specially built for limited area surveillance and repeated rapid scanning of the air space around the airport terminals. The microburst has a life cycle between 10–20 minutes which require specialized radar systems for effective detection. In this application, the radar-computer system automatically provides warnings to the air traffic control tower (Michelson, Shrader and Wilson, 1990).

Doppler radar studies of the role of boundary-layer convergence lines in new thunderstorm formations support earlier satellite cloud-arc studies. There are indications that mesoscale boundary-layer convergence lines (including intersecting gust fronts from prior convection) play a major role in determining where and when storms will form. Wilson and Schreiber (1986) have documented and explained several cases of tornadogenesis by nonprecipitation induced wind shear lines, as observed by Doppler radar (Mueller and Carbone, 1987).

12.5.3 Surveillance of synoptic and mesoscale systems

Radars can provide a nearly continuous surveillance of weather related to synoptic and mesoscale storms over a large area (say a range of 220 km, area 125,000 km^2) if unimpeded by hills (see Fig. 12.7). Due to ground clutter at short ranges and the earth curvature which places the

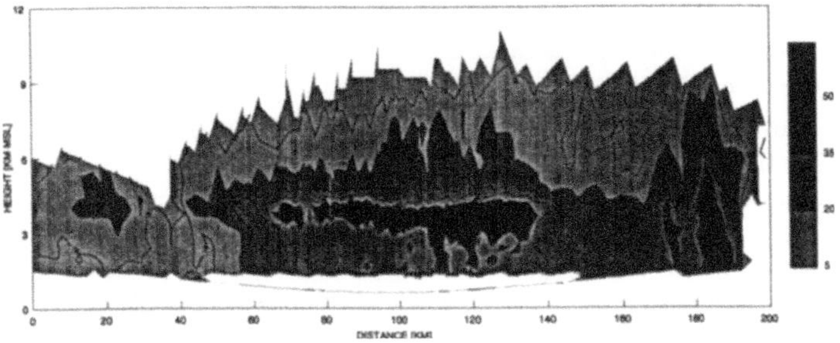

Figure 12.7: *An example of surveillance of wide spread weather. The echo to the south of the radar (center of image) is the remnants of Hurricane Hugo in its extratropical stage. The echo to the northwest of the radar is the precipitation associated with a cold front. The two weather systems are 'colliding', eventually the precipitation associated with Hugo accelerated ahead of the cold front (Joe et al., 1992; Abraham et al., 1991).*

radar beam height at increasing height above ground level with increasing range, the maximum practical range for weather observation is about 200 km. Over large water areas, other means of observation are often not available or possible. Networks of radars can extend the coverage and may be cost effective (Collier, 1991).

In regions where very heavy and extensive precipitation is a common problem, selection of a 10 cm radar with lesser attenuation in precipitation, may be warranted. In other areas such as the mid-latitudes 5 cm radars which are subject to greater attenuation may be quite effective at much less cost. The 3 cm wavelength suffers from too much attenuation in precipitation to be very effective except for very light rain or snow situations. Doppler radar helps to reduce ground clutter effects. Narrower beamwidths provide better resolution of patterns and greater effectiveness at longer range.

Recent improvements in digital radar data processing and display techniques have led to the development of new quantitative, radar-based products for hydrometeorological applications. A number of countries are using such radar products with numerical models for operational flood forecasting and control (e.g., see Cluckie and Owens, 1989).

Since the late 1970's, advanced color displays and mini-computers have been used to provide time lapse and zoom capabilities for radar data. The British Frontiers system (Browning and Collier, 1982; Collier, 1989), the Japanese AmeDAS system, the French ARAMIS system (Cheze, 1989) and the U.S. PROFS system allow the user to interact and produce composite color displays from several remote radars at once, as well as to blend the radar data with other types of information.

12.6 Precipitation Measurements

12.6.1 Precipitation estimation

Radar has a long history of use in estimating the intensity and thereby the amount and distribution of precipitation falling within the radar viewing area with a good resolution in time and space. Most studies have been associated with rainfall but snow measurements can also be made with appropriate allowances for target composition. Readers should consult reviews by Joss and Waldvogel (1990) and Smith (1990) for a comprehensive discussion on the state of the art, the techniques, the problems and pitfalls, and the effectiveness and accuracy. Ground level precipitation estimates from typical radar systems are made for areas of 2 km^2, successively for 5–10 minute periods using low elevation PPI scans with beamwidths of 1°.

When the radar beam or its sidelobes encounter ground targets, persistent echoes occur which can add up in time to appear as large rain amounts if no precautions are taken in the data analysis. A clutter map in the computer memory may be used to eliminate clutter. But as clutter, and especially that caused by anomalous propagation, is variable in time and space, clutter maps are difficult to use in a general context. Doppler techniques hold promise for reducing these difficulties but some problems are likely to remain (Passarelli et al., 1982).

These and other questions, such as the choice of the wavelength, errors caused by attenuation, considerations when choosing a radar site for hydrological applications, hardware calibration of radar systems, sampling and averaging and the meteorological adjustment of radar data, are discussed in Joss and Waldvogel (1990) and Smith (1990). The present brief treatment considers only rainfall measurements; little operational experience is available about radar measurements of snow and even less about measurements of hail.

12.6.2 Precipitation characteristics which affect radar measurements

Drop size distributions

Returned power or radar reflectance factor is the basic precipitation parameter because of its commonality. Basic to the interpretation of Z is its relationship to the drop size distribution (DSD) since Z is equal to the sum of the diameter of drops to the sixth power per unit volume. The reader is referred to textbooks on cloud physics (Pruppacher and Klett, 1978; Rogers and Yau, 1989; Mason, 1971) which discuss the formation of clouds and precipitation.

Marshall-Palmer (1948) (hereafter referred to as M-P) observed that rain DSD's are exponentially distributed (see Fig. 12.8) and can be empirically expressed as:

$$N(D) = N_0 \exp(-\Lambda D) \tag{12.11}$$

$$\Lambda = 4.1 R^{-0.21} \quad \text{mm}^{-1} \tag{12.12}$$

$$N_0 = 8 \times 10^3 \quad \text{m}^{-3}\text{mm}^{-1} \tag{12.13}$$

Higher order gamma formulations have been proposed (Ulbrich, 1983; Cataneo and Stout, 1968; Austin and Geotis, 1979) but the formulation above remains the most popular.

For hail, exponential distributions have also been observed (Douglas, 1964; Hitschfeld and Stauder, 1965; Federer and Waldvogel, 1975; Spahn and Smith, 1976; Cheng and English, 1983). Various results have been expressed as

$$N_0 = 115\Lambda^{3.63} \tag{12.14}$$

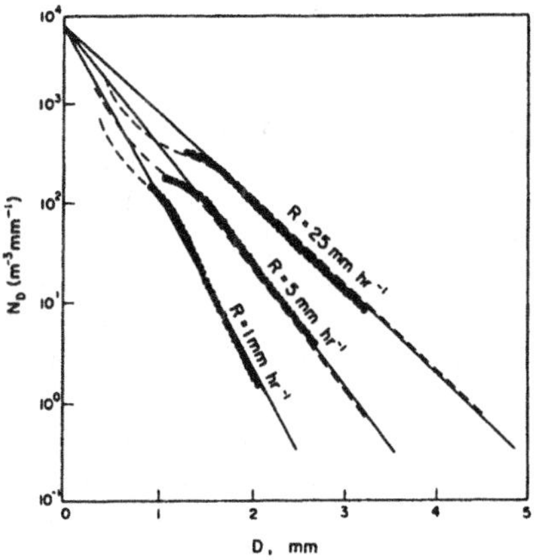

Figure 12.8: *Marshall and Palmer (1948) observed exponential drop size distributions. They observed that the slope of the distribution was a function of the rainrate but the functions had a common intercept (after Battan, 1981).*

For snow, exponential distributions of the melted equivalent diameters have also been found. There are many uncertainties in the measurement of the dimension, mass and type of snow crystals and aggregates and larger uncertainties in the equivalent precipitation rate are expected. Table 12.5 shows two parameterizations of the exponential distribution (Gunn and Marshall, 1958; Sekhon and Srivastava, 1970).

Parameter	Gunn/Marshall	Sekhon/Srivastava
Λ	$25.5R^{-0.48}$	$22.9R^{-0.45}$
N_0	$3.8 \times 10^{-2} R^{-0.87}$	$2.5 \times 10^{-2} R^{-0.94}$

Table 12.5: *Exponential Snow Parameters.*

Terminal velocity

To convert DSDs to rainrates, a terminal velocity measurements is needed. Gunn and Kinzer (1949) made measurements at sea level and can be expressed as (Atlas et al., 1973):

$$w_t(D) = 9.65 - 10.3\exp(-600D) \quad \text{(MKS units)} \tag{12.15}$$

The formulation was extended by Foote and duToit (1969) to lower pressures.

For snow, the situation is much more complex and dependent on the crystal type. Fig. 12.9 shows the classical textbook data on fall crystal fallspeeds. Langleben (1954) found that $w_t = kD_m^n$, where D_m is the melted diameter in cm, and $k = 160,\ 234$ and $n = 0.3$ for dendrites and columns/plates, respectively. For snow aggregates, Gunn and Marshall (1958) found

$$w_t = 0.98D_e^{0.31} \tag{12.16}$$

where D_e is the equivalent diameter (in mm) of the melted snow aggregate.

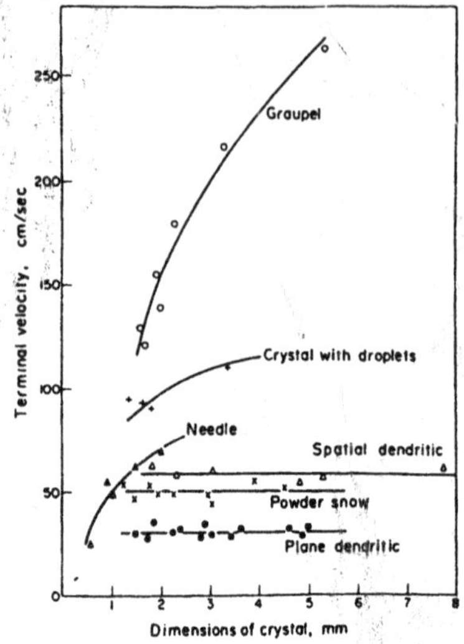

Figure 12.9: *Terminal velocities for various crystal types.*

For hail, Matson and Huggins (1980) found:

$$w_t = 3.62D^{0.5} \tag{12.17}$$

for D in mm. At 1.5 km, R is about 10% larger than at the surface due to velocity effects.

Derived relationships

Under the assumption of an exponential distribution, relationships between Z, M and R can be derived.

$$M = \frac{\pi\rho_w}{6}\int_0^\infty D^3N(D)dD = \pi\rho_w N_0/\Lambda^4 = \pi\rho_w N_0 D_0^4/3.67^4 \tag{12.18}$$

and the median volume diameter is $D_0 = 3.67/\Lambda$.

$$Z = \int_0^\infty D^6 dD = N_0 6!\Lambda^{-7} = N_0 6! D_0^7/3.67^7 \tag{12.19}$$

and

$$R = \frac{\pi\rho_w}{6}\int_0^\infty D^3N(D)w_t(D)dD = \frac{\pi N_0}{\Lambda^4}[9.65 - \frac{10.3}{(1+600/\Lambda)^4}] = \frac{\pi N_0 D_0^4}{3.67^4}[9.65 - \frac{10.3}{(1+600\times D_0/3.67)^4}] \tag{12.20}$$

in SI units.

12.6.3 $Z - R$ Techniques

Empirical $Z - R$ relations and the variations from storm to storm and within individual storms have been the subject of many studies over the past forty years. The rainfall rate (R), a product of the mass content (M) and the fall velocity (w_t) in a radar volume. This is roughly proportional to the 4th power of the particle diameters whereas the radar reflectance (Z) is proportional to the 6th power of particle diameters. Thus, the natural variability in drop-size distributions is an important source of uncertainty in radar measurements of precipitation.

An empirical $Z-R$ relation can be obtained from measured drop-size distributions (Marshall-Palmer, 1948). An alternative is to compare Z measured aloft by the radar with R measured at the ground. The latter approach also reflects any differences between the radar detected precipitation aloft and that which reaches the ground and has the advantage that it also takes into account errors in the radar calibration, but the result is not strictly a $Z-R$ relationship and it may not apply to other radars. Differences in radar calibration could lead to different $Z-R$ relationship for the same DSD.

The possibility of accounting for part of the variability of $Z-R$ relation by stratifying storms according to rain type (such as convective, stratiform, orographic) has received a good deal of attention (Battan, 1981). The improvements achieved are not substantial and questions remain as to the practicality of applying this technique on an real-time operational basis. After averaging over time and/or space, the errors associated with these variations, will rarely exceed a factor of two in rain rate. At longer ranges (>130 km), errors caused by the inability to observe the precipitation close to the ground and beam filling are usually dominant.

Fig. 12.10 shows the variation in empirically determined $Z-R$ relationships. In spite of all the

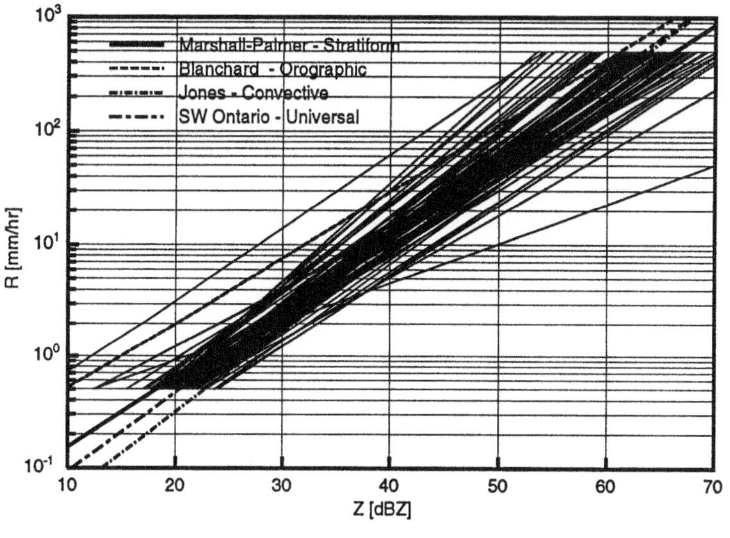

Figure 12.10: *A graphical presentation of over 69 $Z-R$ relationships (Battan, 1981) to illustrate the volume of research and the variation in $Z-R$ results. Four specific relationships for different rain types are highlighted.*

work, it has been difficult to displace the Marshall-Palmer relationship given by $Z = 200R^{1.6}$. From Fig. 12.10, the MP relationship roughly falls in the middle of all the measurements.

Snow and hail are not spherical and do not have the same dielectric properties as rain. Marshall and Gunn (1952) suggested that for a weak dielectric like ice, the backscattering is the same as that for a sphere with equivalent mass and that the shape is immaterial. They found exponential distributions similar to the rain results except the intercept was not a constant (see Fig. 12.11).

Accounting for the change is particle density, $Z_i = 0.224 Z_w$ and Sekhon and Srivastava (1970) found that $Z_w = 1780 R_s^{2.21}$.

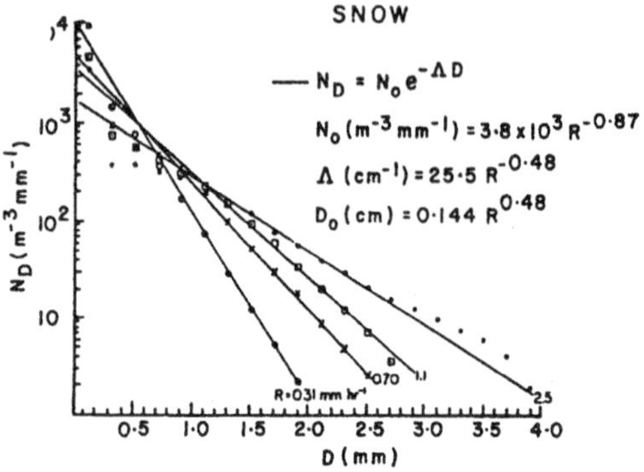

Figure 12.11: *Classical $Z - S$ relationships found by Marshall and Gunn (1952).*

Comparisons with 1 minute snowfall measurements challenge the notion of the shape being unimportant for short term applications. Fig. 12.12 shows a graph of $Z - S$ for different crystal types. On the other hand, Fig. 12.13 shows the close relationship after a month of snow accumulation. The conclusion that one reaches is that a climatological (daily or longer) $Z - S$ relationship exists for snow but instantaneous (1 minute) relationships can be quite different.

For hail, there is a lack of simultaneous measurements of hail and radar. The limited Cheng and English (1983) data lead Torlaschi et al. (1984) to produce $Z_H = 5.38 \times 10^6 [\ln(88/R_H)]^{-3.37}$. But there is not a widely accepted single parameter technique to identify hail. Thresholding at a high reflectance level is the most popular technique. Waldvogel et al. (1979) suggested a storm with 45 dBZ at a height of 1.4 km above 0°C will produce hail.

12.6.4 Area-Time integral

The $Z - R$ approach attempts to measure rainfall rate over short time periods and in real-time. Byers (1948) noticed a close relationship between the total amount of rainfall and the area and duration of a rain shower. It is essentially independent of the rain intensity distribution within the storm. From maps of radar rainfall accumulation maps (using the M-P $Z - R$), good agreement with rain gauges have been found (Leber et al, 1961) which supported this idea. Without using a $Z - R$ relationship, Doneaud et al. (1984, 1987) found a relationship between the Area Time Integral (ATI) with a very small dispersion (see Fig. 12.14):

$$\text{Rain volume} \quad (\text{km}^2\text{mm}) = 3.68 ATI^{1.01} \tag{12.21}$$

where ATI is in units of km^2 h. This relationship is dependent on the threshold selected to define the rain area. While this has limited use in real-time short term forecasting applications, its real value should be in climatological studies and applications.

12.6.5 Streamflow techniques

Recently, hydrologists interested in flooding situations have used radar data as precipitation fields to drive hydrologic run-off models (Kouwen et al., 1993; Kouwen and Garland, 1989). In

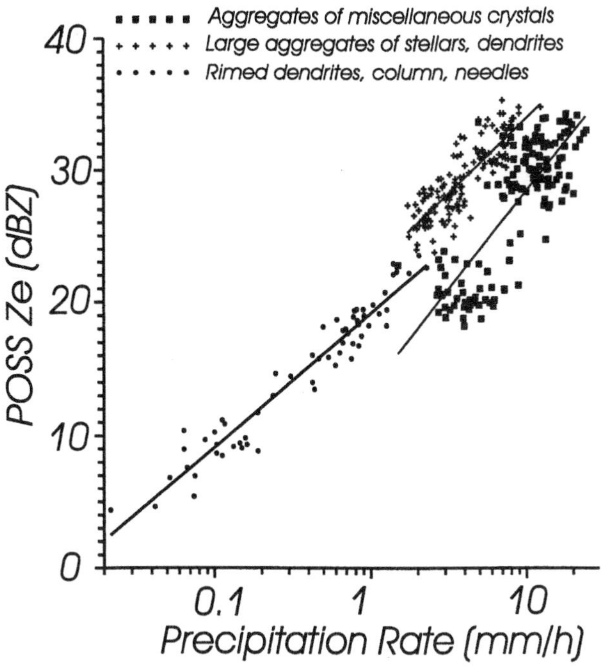

Figure 12.12: *$Z-S$ for different crystal types from a vertically pointing X-band Doppler radar designed for present weather classification (Sheppard, 1990). The crystal type classification was manually determined (Figure courtesy of Brian Sheppard).*

this approach, the area, duration, spatial distribution of the rain fall pattern and underlying terrain characteristics are used to compute the flow of excess river water. Similar to the ATI method, accumulation over space and time averages the radar data while the rainfall pattern and terrain characteristics adds another level of sophistication. Fig. 12.15 shows hydrographs of measured river water flow and predicted water flow from radar. The hydrographs are matched at the start of the sequence to account for lags due to pre-existing soil moisture conditions. Then the data are independently generated. The drawback of the technique is the reliability of the terrain run-off model. In the example of Fig. 12.15, the hydrological model is objectively determined from satellite imagery. The favorable comparison demonstrates the utility of radar data for the spatial representation of rainfall patterns and the possibility of calibrating radars using river flow data.

12.6.6 Attenuation techniques

Since heavy rains can attenuate the intensity of the transmitted radar signal, therefore, the estimation of rainfall rate may be determined through the estimation of attenuation. The attenuation or extinction cross section σ_e is equal to the total absorption and scatter cross-section (see Fig. 12.16). The specific attenuation (K) is given by:

$$K = 4.34 \times 10^3 \int_0^\infty N(D)\sigma_e(D)dD \tag{12.22}$$

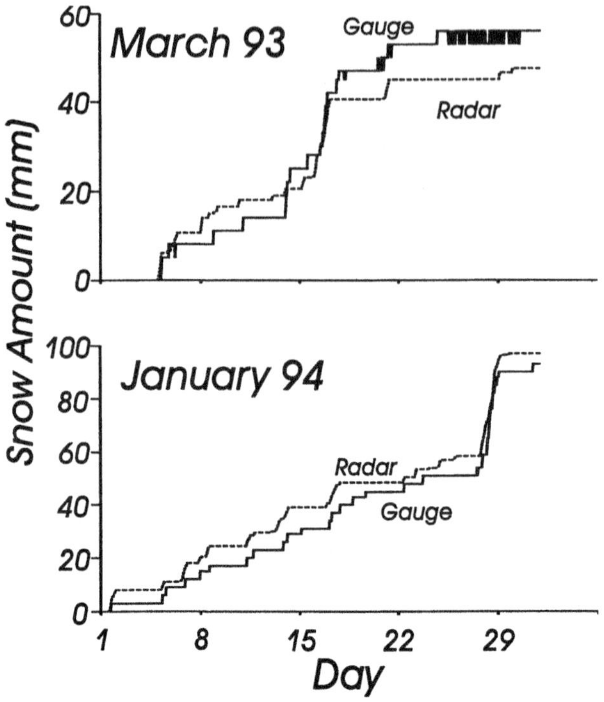

Figure 12.13: *Snowfall accumulation from radar and a Nipher shielded snow gauge for two months using the Sekhon and Srivastava (1970) $Z-S$ relationship. The monthly accumulations in March 93/January 94 are much closer than would be implied by the previous figure showing that deviations in $Z-S$ relationship average to zero (Figure courtesy of Nick Kouwen).*

At the top of Fig. 12.16, drop sizes are presented in millimeters. A typical raindrop is of 1–2 mm in diameter. So the normalized extinction or attenuation for wavelengths smaller than 1 cm, now commonly employed for cloud sensing, is quite severe - more than an order of magnitude greater than a 3 cm radar. (Another problem of millimeter wave radar data is that large hydrometeors will scatter in the Mie regime and interpretation of the information is problematic).

If the terminal velocity and the specific attenuation can be approximated by power law relations, that is, $w_t(D) = 386.6D^{0.67}$ and $\sigma_e(D) = CD^n$ and if $n = 3.67$ then rainrate and specific attenuation are linearly related and therefore independent of drop size distribution. This is true at $\lambda = 0.86$ cm (Atlas and Ulbrich, 1974). Fig. 12.17 shows the dispersion of the relationship for various wavelengths.

Not suprisingly, there are major practical problems due to the strong attenuation which limit the rainrate magnitude and range of coverage — typically, 20 mmh^{-1} and 30 km for two way measurements. The results are also path averaged. Another problem is that you also need a known target at a distance or a bistatic system to estimate the amount of attenuation. The edge of the mainlobe and side lobes can reflect off the ground for quasi-horizontal systems leading to poor power estimations due to these multipath reflections.

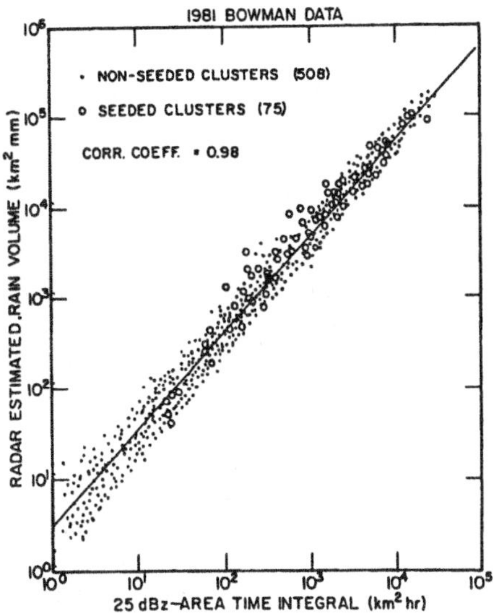

Figure 12.14: *Radar estimated and ATI estimated rainfall amounts showing that total rainfall amounts are dependent on the area and duration of the rain event and independent of the spatial intensity distribution (after Doneaud et al., 1987).*

12.6.7 Differential phase techniques

Electromagnetic waves experience phase shifts as they propagate through precipitation since it is an anisotropic media. Large raindrops are oblate and fall with a minor vertical axis (see Fig. 12.18). This results in horizontally polarized electromagnetic fields experiencing larger phase shifts than vertically polarized fields. The specific differential phase K_{DP}, which is a measure of this difference, requires alternately or simultaneously transmitted polarized waves and is given by:

$$K_{DP} = \frac{180\lambda}{\pi} Re[\int_0^{D_m} [f_h(D_e) - f_v(D_e)]N(D_e)dD_e] \qquad (12.23)$$

where f_h and f_v are the horizontal and vertical forward scatter coefficients.

The quantity in the square brackets is of the form aD_e^b where $b = 4.24$ at $\lambda = 10$ cm. Since the exponent is close to 3.67 or 3, this quantity has a weak dependence on DSD since M and R are proportional to D^3 and $D^{3.67}$, respectively. Sachidananda and Zrnic (1987) found that $R = 5.1(K_{DP}\lambda)^{0.866}$ and English et al. (1991) found that $R = 37K_{DP}^0.866$ (see Fig. 12.19).

At high rain rates (> 70 mmh^{-1}), the K_{DP} approach is more accurate than a $Z - R$ approach (Chandrasekar et al., 1990). It is insensitive to hail in a rain-hail mix, independent of system calibration, independent of rain attenuation, beam blockage and beam filling since these factors do not affect the differential phase shifts.

12.6.8 Dual wavelength

In this technique, the difference in backscatter at an attenuating and non-attenuating wavelength is used to deduce the DSD parameters. This could be classified as an attenuation

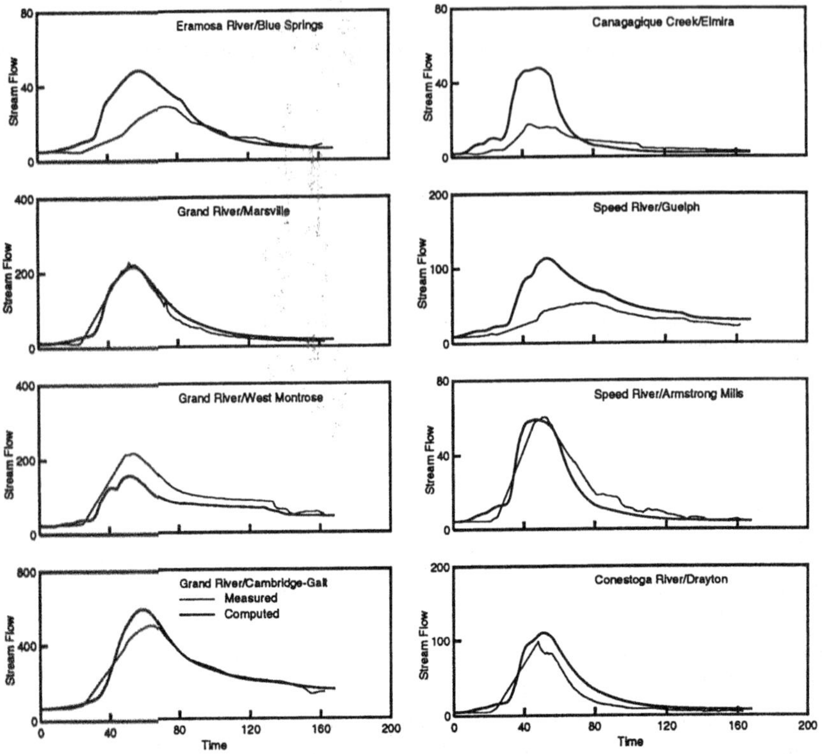

Figure 12.15: *Radar and measured hydrographs. The radar data is processed in collaboration with an objectively determined hydrological model of the underlying terrain (Figure courtesy of Nick Kouwen).*

technique where attenuation is estimated using the returned signal from the non-attenuating wavelength rather than the backscatter from a fixed target. It can be shown that:

$$\frac{\overline{K_s}}{\overline{Z}} = [4.4 \times 10^3 \Gamma(n+1) C'/6!] \Lambda^{6-n} \tag{12.24}$$

where n is the exponent in the power law approximation for the normalized extinction cross-section and is equal to 1.67 for $\lambda = 0.86$ cm. K_s, the specific attenuation, and Z are measured quantities and determine Λ.

Small differences are determined from two large numbers and therefore difficulties arise from the statistical fluctuations of the two signals (Eccles and Mueller, 1973). This requires accurate calibration and matched beams. Severe or total attenuation of the shorter wavelength limits its effective range.

If the target is small with respect to both wavelengths and therefore within the Rayleigh scattering regime, the ratio of reflectivities is independent of drop size distributions. When the target is large with respect to one or both wavelengths, there are Mie scattering effects. There is differential backscatter and the reflectance factor ratio changes and this can be used to identify hail (Eccles and Atlas, 1973; Tuttle and Rinehart, 1983).

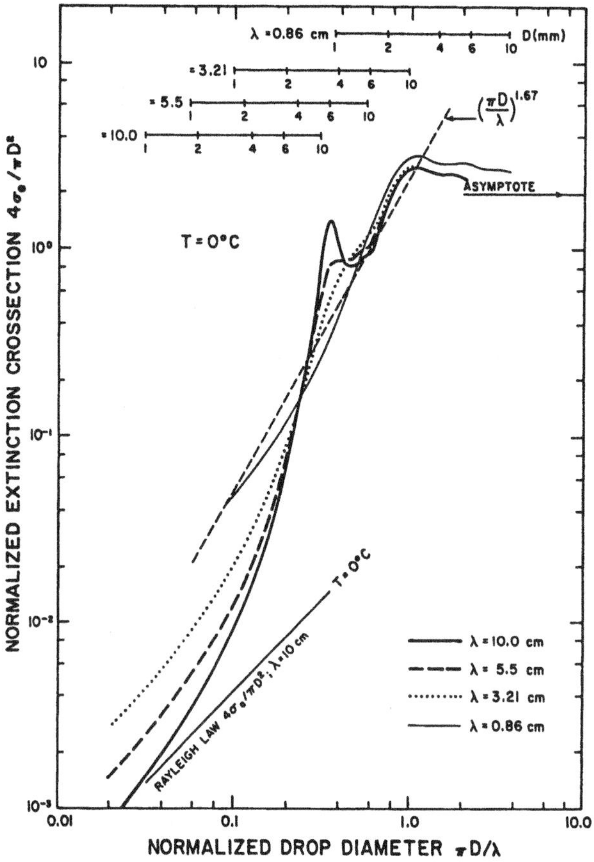

Figure 12.16: *Normalized extinction coefficient (after Doviak and Zrnic, 1993).*

12.6.9 Dual polarization

Z_{DR} is the ratio of the horizontal to the vertical polarized return powers. This is a parameter produced by radars that have dual linear polarizations. It is indicative of of the mean oblateness of the precipitation particles in the resolution volume. Assuming a negative exponential distribution, Z_{DR} is independent of N_0 and therefore uniquely determines Λ. The latter is related to the median drop size which in turn can be combined with Z_H to retreive N_0 (see Fig. 12.20). The use of Z_{DR} and Z_H as a dual parameter technique to retrieve the exponential DSD parameters has been evaluated (Seliga and Bringi, 1976; Al-Khatib et al., 1979). The technique is limited by the form of the DSD model. That is, the technique is sensitive to the largest and most oblate of drops and so the maximum diameter, D_{max}, is another independent variable. However, Plank (1977) found that the maximum diameter was a function of rainrate, that is, $D_{max}\Lambda = 7.5$.

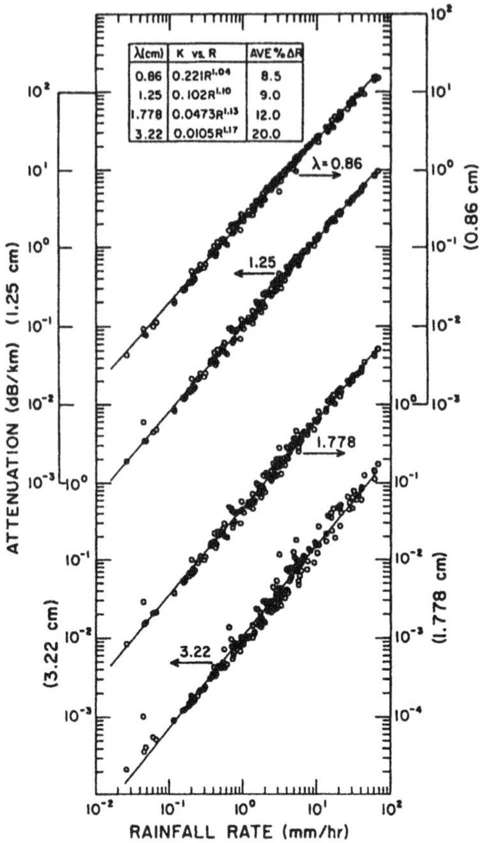

Figure 12.17: *Specific attenuation and rainrate for various wavelengths showing the small variance in the attenuation-rainrate relationship for* $\lambda = 0.86$ *cm. (After Atlas and Ulbrich, 1974).*

Another way of formulating the dual polarization estimate of rain was provided by Sachidananda and Zrnic (1987) who expressed rain rate as a function of Z_H and Z_V, rather than Z_{DR} and Z_H. That is,

$$R = 6.84 \times 10^{-3} Z_H^{-3.86} Z_V^{4.86} \tag{12.25}$$

There is still considerable controversy, however, as to whether or not this technique has promise for operational use in the measurement of precipitation (English et al., 1991). Small differences in large quantities require accurate calibration and non-interfering radomes are two of the operational problems. At close ranges (and therefore with high spatial resolution) polarization diversity radars may give valuable information about precipitation particle distributions and other parameters pertinent to cloud physics. At longer ranges one cannot be sure that the radar beam is filled with a homogeneous distribution of hydrometeors, so the empirical relationship of the polarimetric signature to the drop size distribution has increased uncertainty.

If multiparameter techniques worked perfectly one could reduce the error caused by $Z - R$ from 33% to 17% (Ulbrich and Atlas, 1984; Balakrishnan et al., 1989; Chandrasekar et al., 1990; see

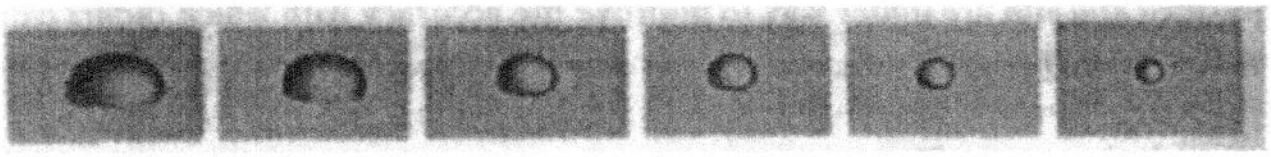

Figure 12.18: *Drop shape as a function of size. From left to right, the drop sizes are: 8.00, 7.35, 5.80, 5.30, 3.45 and 2.70 mm (from Pruppacher and Beard, 1970). These observations form the basis for the use of polarization techniques.*

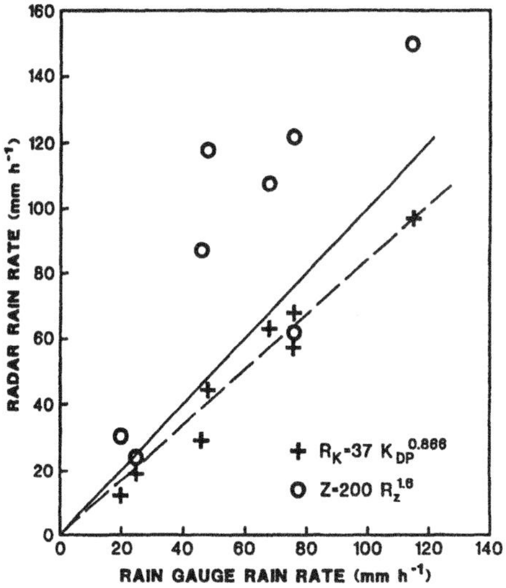

Figure 12.19: *Radar and raingauge intercomparison using a Z and a K_{DP} technique showing the superiority of the latter technique for high rainrates (after English et al., 1991).*

Table 12.6). For the ranges required (area covered) for hydrological applications, correction for other biases already discussed are usually much greater, perhaps by an order of magnitude or more.

12.6.10 Precipitation typing

The use of polarization radar affords the possibility to classify the precipitation by type since the hydrometeor shape may be measured. Large hail are oblate in shape and fall with a minor horizontal axis. Therefore, a vertically polarized beam will have a stronger backscatter signal and Z_{DR} should be negative whereas large rain falls with its minor axis vertically and therefore

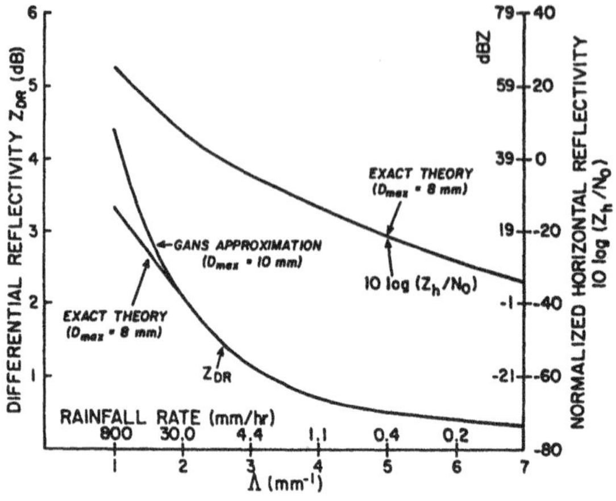

Figure 12.20: *Z_{DR} and Z approach to determine the parameters of an exponential DSD. Z_{DR} is used to determine Λ, the Z is used to determine N_0 (after Seliga and Bringi, 1976).*

should exhibit positive Z_{DR}. This simple formulation is made more complex when particle aerodynamics are considered. Large oblate hail also wobbles as it falls effectively reducing the Z_{DR} on a statistical basis (Stewart and List, 1982).

Since K_{DP} is also sensitive to shape, it could also provide particle shape information. One such relationship is: $Z = 8\log(2K_{DP}) + 49$ where K_{DP} smaller than this would indicate the presence of hail and larger values would indicate rain (Doviak and Zrnic, 1993). If one assumes that small hail is isotropic, then it will not contribute to K_{DP} and thus K_{DP} is due only to rain. Using K_{DP} and R, and the M-P $Z - R$ relationship, we can therefore calculate the relative contributions of rain and hail to the radar reflectance factor; that is, $Z_R = 65,800K_{DP}^{1.386}$ and $Z_H = Z - Z_R$ where the units are mm^6 m^{-3}.

There are other parameters - correlation (ρ) and L_{DR} - which may have some value but their interpretation is unclear at this point.

12.6.11 Raingauge-radar techniques

A conceptually attractive approach is to use a network of gauges as a reference against which to compare and adjust the radar estimates. This approach has an intuitive appeal, but suffers from a fundamental limitation. Rainfall patterns are not homogeneous and therefore there is no reference to establish the accuracy of the gauge network.

Gauges are the accepted standard for rainfall measurements at a point even though automated systems suffer significant wind effects (Neff, 1977; Zawadzki, 1975). The accuracy for area estimates of rainfall is dependent on the density of the gauges and variability of the rainfall pattern. In concept, adjustment of the radar fields with a few point measurements by raingauges should improve the areal estimation of rainfall. This has the advantages of compensating for calibration errors and for inappropriate $Z - R$ relationships (for whatever reason).

Brandes (1975) suggested integrating over time the radar rainfall estimates for points near a raingauge, calculating the radar adjustment factor necessary to match the gauge estimates and then adjusting the entire radar field. The procedure is amenable to a iterative process to reduce the difference between the radar and raingauge estimates of the precipitation pattern. Improvements of the radar estimates of 24% to 14% have been demonstrated (Brandes, 1975) and were better than raingauges alone for large areas, low gauge density and long duration rainfalls (see Fig. 12.21).

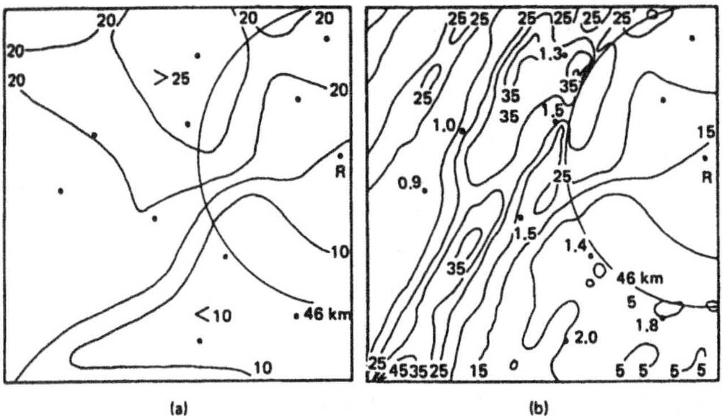

Figure 12.21: *Rainfall fields derived from raingauges and radar derived fields adjusted by raingauges. The inherent detail with radar is obvious (after Brandes, 1975).*

There is general agreement that intercomparisons with gauges should be made routinely, as a check on the radar and gauge performance, and that appropriate adjustments should be made if a radar bias is clearly indicated. In situations where the radar estimates are far from the mark due to radar calibration or other problems, such adjustments can bring about significant improvements.

However, the adjustments do not automatically assure improvements in the radar estimates, and sometimes the adjusted estimates are poorer than the original ones. This is especially true for convective rainfall where the vertical extent of echo mitigates the difficulties associated with the vertical profile, and the gauge data are suspect because of unrepresentative sampling.

The general guideline is that the adjustments will produce consistent improvements only when the systematic differences (i.e. the bias) between the gauge and radar rainfall estimates are larger than the standard deviation of the random scatter of the gauge versus radar comparisons. That guideline allows one to judge whether gauge data should be used to make adjustments. This leads to the idea that the available data should be tested before any adjustment is actually applied. Various methods for accomplishing this have been explored, but at this time there is no widely accepted approach.

12.6.12 Vertically pointing measurements

A Doppler radar in vertically pointing mode can measure the velocity spectrum to determine the DSD. That is, $S_n(w - w_t)dw_t = D^6 N(D)dD/Z$. In the absence of air motion, each component of the Doppler velocity spectrum is the same as the terminal velocity and hence drop size. The amplitude of each frequency component is related to the number of drops of that drop size. The problem is that, except near the ground, there is usually an updraft or downdraft velocity (w) that must be determined. Three potential methods to determine w, include: using the measured reflectance, assume a Marshall-Palmer relationship to estimate the mean terminal velocity and then retrieve the air velocity by the difference from the mean Doppler velocity (Rogers, 1964); using a minimum detectable size assumption which is assumed to move with the air (Battan, 1981); or exploiting the clear air velocity component to the spectrum for large wavelength systems (Wakasugi et al., 1987). However, errors of only ± 0.25 m s^{-1} in vertical velocity can cause errors of 100% in drop number concentrations (Atlas et. al., 1973).

12.6.13 Vertical profile effects

Highly variable vertical reflectance profiles are observed, both within a given storm and from storm to storm because of growth or evaporation of precipitation, air motion and change of phase (ice and water in the melting layer, or bright band). This will lead to a non-uniform distribution of reflectance with height. As the beamwidth increases with range, the profile will be averaged resulting in a smoother vertical distribution of reflectance (see Fig. 12.22).

In convective rainfall, the advantages of radar are most pronounced and experience shows less difficulty with the vertical profile problem. But in stratiform rain or snow, the vertical profile becomes more important (see Fig. 12.23). As the beam becomes wider with increasing distance, the vertical distance between the sample volume of the radar and the ground usually increases too. Therefore, the differences between estimates of rainfall by radar and the rain measured at the ground also increases. Reflectance generally decreases with height. Therefore, radar estimations underestimate the rain measured at the surface.

No one method of compensating for the effects of the vertical reflectance profile in real time is widely accepted. A climatological mean range-dependent factor can be applied to obtain a first-order correction. Different factors may be appropriate for different storm types (for example, convective versus stratiform).

The vertical profiles in storms vary with location and time, and the lowest level visible to the radar usually varies because of irregularities in the radar horizon. Consequently a point-by-point correction process using a representative vertical profile for each zone of concern may be needed to obtain the best results.

Representative profiles can be obtained from the radar volume scan data themselves, climatological summaries, or storm models (Joss and Lee, 1993). At long ranges, for low level storms, and especially when low antenna elevations are blocked by obstacles such as mountains, the underestimate may be severe (the spreading of the beam also becomes a problem in convective rain).

The basic procedure of deducing rainfall rates from measured radar reflectivities for hydrological applications requires the following steps:

- Making sure that the hardware is stable by calibration and maintenance.
- Correcting for errors using the vertical reflectance profile.
- Taking into account all the information about the $Z_e - R$ relationship and deducing the rainfall.
- Adjustment with raingauges.

The first three parts are based on known physical laws and only the last one uses a statistical approach to compensate for residual errors. This allows the statistical methods to work most efficiently. After making the profile corrections, one should use a reflectance - rainrate relationship appropriate to the situation, geography as well as season, to deduce the value of R (see previous discussion).

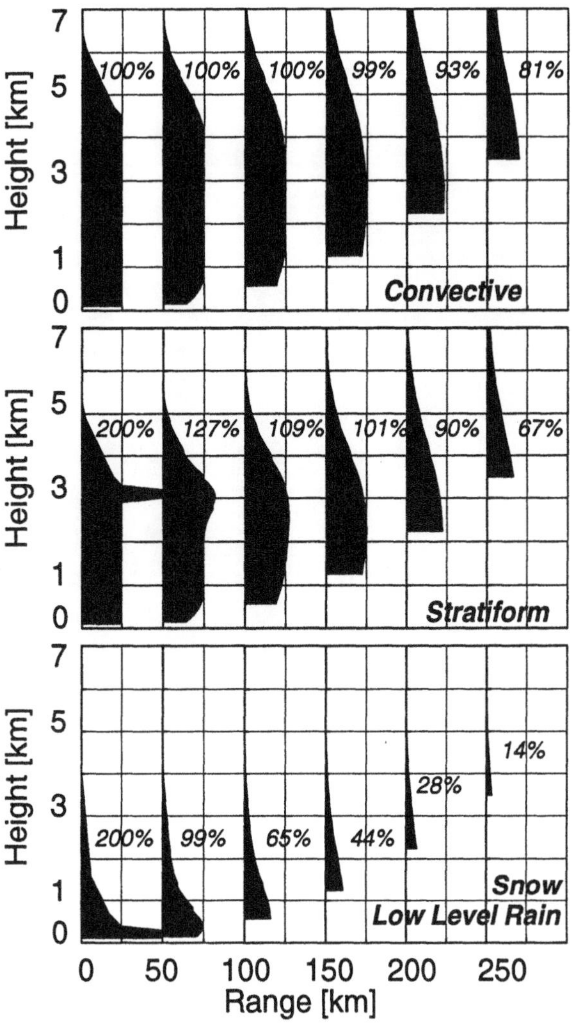

Figure 12.22: *Vertical reflectance profiles as a function of range. The figure is adapted from Joss and Waldvogel (1990). Averaging in range and beam filling are the dominant reasons for underestimation of precipitation. The figure illustrates that the least effect is experienced in convective situations.*

12.6.14 Frontiers

The United Kingdom Meteorological Office (UKMO) has developed an interactive graphical approach to the correction of radar data to adjust for the various sources of errors. The system is called FRONTIERS (Conway, 1987) and has been in continuous use in the CFO/UKMO since 1986. The objective is to produce the best possible estimate of the intensity and areal extent of the rainfall using all available data sources in the forecast office.

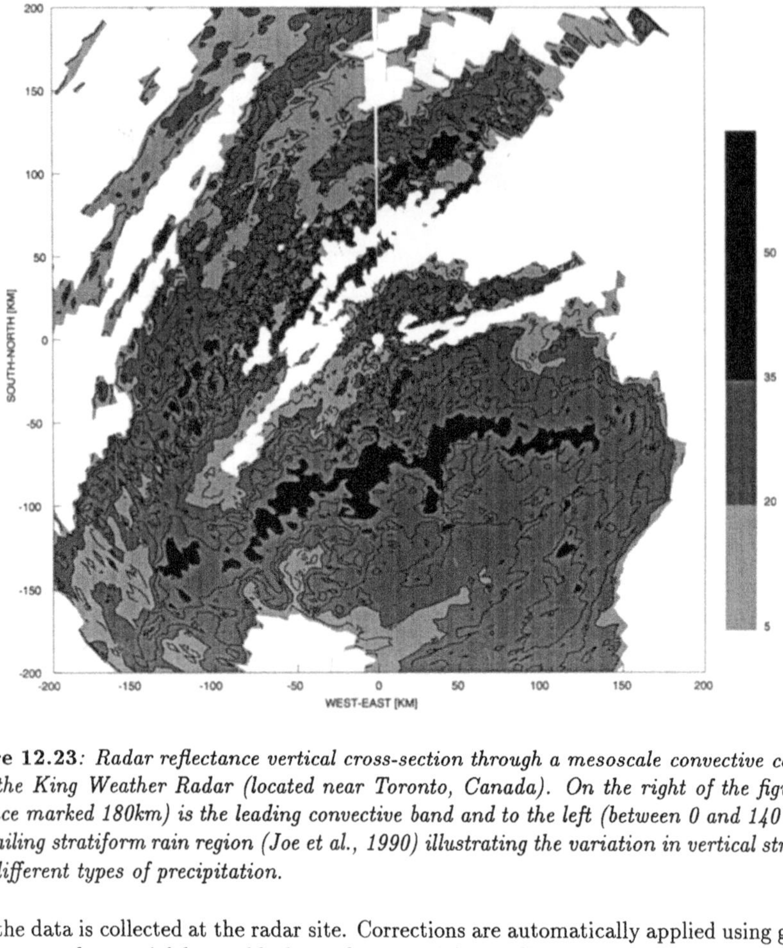

Figure 12.23: *Radar reflectance vertical cross-section through a mesoscale convective complex from the King Weather Radar (located near Toronto, Canada). On the right of the figure (at distance marked 180km) is the leading convective band and to the left (between 0 and 140 km) is the trailing stratiform rain region (Joe et al., 1990) illustrating the variation in vertical structure with different types of precipitation.*

First the data is collected at the radar site. Corrections are automatically applied using ground clutter map, for partial beam blockage, for partial beam filling. The data is converted to rainfall using the M-P relationship and corrected for attenuation. The data are adjusted using telemetered raingauges depending on topography and precipitation type. In spite of all these adjustments, they are not enough to compensate for all of the problems. In fact, the range and raingauge adjustments are removed by the operator. Spurious echoes due to ground clutter and anomalous propagation echoare interactively removed, adjustments for beam filling, orographic enhancement and then the raingauge adjustments are reapplied. The system is human intensive though efforts are underway to automate the system as much as possible (Kitchen et al., 1994; Hardaker et al., 1995).

12.7 Summary

Over the years much research has been directed towards exploring the use of weather radar for measuring rain. In general, radar measurements of rain, deduced from measured reflectance by using an empirical $Z - R$ relation, agree well with gauge measurements close to the radar. Variations in the DSDs, the inability to measure precipitation close enough to the ground, beam averaging and filling are the main shortcomings of weather radar.

To make the best possible use of radar, the radar site should be chosen such that precipitation can be seen by the radar as close as possible to the ground. Seen means that there is no shielding or clutter echoes, or that the influence of clutter can be eliminated, for instance by Doppler processing. Wavelength and antenna size should be chosen such that a suitable compromise between attenuation caused by precipitation and good spatial resolution is achieved. Systems should be rigorously maintained and quality controlled. Measurements of reflectance should be corrected for errors originating from the averaging effect of a broadened beamwidth and/or for the vertical profile of reflectance. The three dimensional echo pattern, together with knowledge about the position of the radar and the orography around it, allows for the correction in real time for a large fraction of the vertical profile problem, though the optimal technique remains to be found.

The sample size must be adequate for the application. For hydrological applications, and especially when adjusting radar estimates with gauges, it is desirable to integrate the data over a number of hours and/or square kilometers. Polarimetric measurements may provide some further improvement, but it has yet to be demonstrated that the limited operating conditions, additional cost, complexity and risk of misinterpreting polarization measurements can be justified for operational applications in hydrology.

The principle advantages of radar are its high spatial and temporal resolution, wide area coverage and immediacy (real-time data). Radar also has the capability of measuring over inaccessible areas such as lakes and to follow a floating target or a convective complex in a real-time sequence, for instance to make a short term forecast. It is unlikely that radar will ever completely replace the raingauge, since gauges can provide additional information and are essential for adjusting and/or checking the radar indications. Good quantitative information is already obtained from radar networks in many places.

12.8 Acknowledgments

Portions of this document were extracted from the WMO CIMO Guide, Chapter 9, P. Joe and C.L. Crozier (Ed.), 1995. Other contributors to that document were: J. Aoyagi, M. Gilet, J. Golden, J. Joss, L.P. Merritt, T.M. Puhakka and P.L. Smith.

12.9 References

Abraham J, Macdonald K, Joe P (1991) The interaction of hurricane Hugo with the Mid-Latitude Westerlies, 19th Conference on Hurricanes and Tropical Meteorology, May 6–10, Miami, Florida

Al-Khatib HN, Seliga TA, Bringi VN (1979) Differential reflectance and its use in the radar measurement of rainfall. *Atmos Sci Prog Rep* AS-S-106, Ohio State University, Columbus

Aoyagi J (1983) A study of the MTI weather radar system for rejecting ground clutter. *Papers in Meteor and Geophys* 33 (4), 187–243

Atlas D (1964) *Advances in Radar Meteorology. Advances in Geophysics*, Vol. 10, Editors H.E. Landsberg and J. Van Meighen. New York, Academic Press 317–479

Atlas D (Editor) (1990) Radar in Meteorology: Battan Memorial and 40th Anniversary Radar Meteorology Conference. Boston, *Amer Meteor Soc*, 806 pp

Atlas D, Scrivastava RC, Sekhon RS (1973) Doppler radar characteristics of precipitation at vertical incidence. *Re Geophy and Space Phys* 11 (1), 1–35

Atlas D, Ulbrich CW (1974) The physical basis for attenuation-rainfall relationships and the measurement rainfall parameters by combined attenuation and radar methods. *J Rech Atmos* 8(1-2), 275–298

Austin PM, Geotis SG (1979) Raindrop sizes and related parameters for GATE. *J Appl Met* 18: 569–575

Balakrishnan N, Zrnic DS, Goldhirsh J, Rowland J (1989) Comparison of simulated rain rates from disdrometer data employing polarimetric radar algorithms. *J Atmos Oceanic Technol* 6: 476–486

Battan LJ (1981) *Radar Observation of the Atmosphere.* Chicago, Univ. of Chicago Press, 324 pp

Bean BR, Dutton EJ (1966) *Radio Meteorology.* Washington, D.C., U.S. Govt. Printing Office, 435 pp

Brandes EA (1975) Optimizing rainfall estimates with the aid of radar. *J Appl Met* 14: 1339–1345

Brown RA, Lemon LR (1976) Single Doppler radar vortex recognition, Part II: Tornadic vortex signatures. Preprints 17th Radar Meteor. Conf. Boston, *Amer Meteor Soc* 104–109

Browning KA (1987) Towards the more effective use of radar and satellite imagery in weather forecasting, In *Weather Radar and Flood Forecasting*, Ed. by V.K. Collinge and C. Kirby, John Wiley and Sons, 239–269, pp269

Browning KA, Collier CG (1982) *An integrated radar-satellite nowcasting system in the U.K.* (Editor: K.A. Browning). London, Academic Press, 47–61

Burgess DW (1976) Single Doppler radar vortex recognition, Part I: Mesocyclone signatures. Preprints 17th Radar Meteor. Conf. Boston, *Amer Meteor Soc*, 97–103

Burgess DW, Lemon LR (1990) Severe thunderstorm detection by radar, Chapter 30a, Radar in Meteorology: Battan Memorial and 40th Anniversary Radar Meteorology Conference (D. Atlas, Editor). Boston, *Amer Meteor Soc*, 619–647

Burrows CR, Attwood SS (1949) *Radio wave propagation*, Academic Press, N.Y.

Byers HR (1948) The use of radar in determining the amount of rain falling over a small area. *Trans Amer Geophys Union*, 187–196

Cataneo R, Stout GE (1968) A method for estimating rainfall rate-radar reflectance relationships. *J Met* 8: 815–819

Chandrasekar V, Bringi VN, Balakrishnan N, Zrnic D (1990) Error structure of multiparameter radar and surface measurements of rainfall. Part III: Specifi differntial phase. *J Atmos and Oceanic Tech* 7: 621–629

Cheng L, English M (1983) A relationship between hailstone concentration and size. *J Atmos Sci* 40: 203–213

Cheze JL (1989) Une revue du programme ARAMIS. International Seminar on "Weather Radar Networking", Brussels, Belgium, Sept. 5-8. (COST-73 - ECC/ Cooperation in Science and Technology). Geneva, WMO, 75–80

Collier CG (1991) *International Weather Radar Networking*, Kluwer Academic Press, 332pp

Collier CG (1989) *Applications of Weather Radar Systems, A Guide to Uses of Radar Data in Meteorology and Hydrology.* Chichester, England, 294 pp

Conway BJ (1987) FRONTIERS: An operation system for nowcasting precipitation, Proc. Symp. Mesoscale Analysis and Forecasting, Vanvouver, ESA SP-282, 233–239

Crozier CL, Joe P, Scott JW, Herscovitch HN, Nichols TR (1991) The King City operational Doppler radar: development, all season applications and forecasting. *Atmosphere-Ocean* 29: 479–516

Cluckie ID, Owens MD (1989) Real-Time Rainfall Run-Off Models and Use of Weather Radar Information. New York, John Wiley and Sons, Dennis, A.S., C.A. Schock and A. Koscielski, 1970: Characteristics of hail storms of western South Dakota. *J Appl Meteor* 9: 127–135

Donaldson NR (1990) An example of attenuation by wet snow on a radar dome. *Atmosphere-Ocean* 29: 699–711

Donaldson RV Jr. (1970) Vortex signature recognition by a Doppler radar. *J Appl Meteor* 9: 661–670

Doneaud AA, Ionescu-Niscov S, Priegnitz DL, Smith PL (1984) The area-time integral as an inicator for convective rain volumes. *J Climate Appl Met* 23: 555–561

Doneaud AA, Miller JR Jr., Johnson LR, Vonder Haar TH, Laybe P (1987) The area-time integral technique to estimate convective rain volumes over areas applied to satellite data - preliminary investigation. *J Climate Appl Met* 26: 156–169

Douglas RH (1964) Hail size distributions, 11th Radar Met. Conf., Boulder, CO, AMS, 146–149

Doviak RJ, Zrnic DS (1993) *Doppler Radar and Weather Observations* (2nd Edition). San Diego, Calif., Academic Press, 562 pp

Eccles PJ, Atlas D (1973) A dual-wavelength radar hail detector. *J Appl Meteorol* 12: 847–854

Eccles PJ, Mueller EA (1973) X-band attenuation and liquid water content estimation by a dual-wavelength radar. *J Appl Meteorol* 10: 1252–1259

English M, Kochtubajda B, Barlow F, Holt A, McGuinness R (1991) Radar measurement of rainfall by differential propagation phase: a pilot experiment. *Atmos Ocean* 29: 357–380

Federer N, Waldvogel A (1975) Hail and raindrop size distributions from a Swiss multicell storm. *J Appl Meteorol* 14: 91–97

Foote GB, duToit PS (1969) Terminal velocity of raindrops aloft. *J Appl Meteorol* 8: 249–253

Gossard EE, Strauch RG (1983) *Radar Observations of Clear Air and Clouds.* Amsterdam, Elsevier Sci Publ, 280 pp

Gunn KLS, Kinzer GD (1949) The terminal velocity of fall for water droplets in stagnant air. *J Met* 6: 243–248

Gunn KLS, Marshall JS (1958) The distribution of size of aggregate snowflakes. *J Meteorol* 15: 452–466

Hardaker PJ, Holt AR, Collier CG (1995) A melting-layer model and its use in correcting for the bright band in sing-polarization radar echoes. *Quart Journ of the Royal Met Soc* 495–525

Hitchfeld W, Stauder M (1965) *The temperature of hailstones*, Alberta Hail Studies, 1964, Sci. Rep. MW-42, Stormy Weather Group, McGill Univ., Montreal, Canada

Houze RA Jr., Schmid W, Fovell RG, Schiesser H-H (1993) Hailstorms in Switzerland:left movers, right movers and false hooks. *Mon Wea Rev* 121: 3345–3370

JDOP Staff (1979) Final Report of the Joint Doppler Operational Project. *NOAA Tech. Memo. ERL NSSL-86.* Norman, Okla., Nat Severe Storms Lab, 84 pp

Joe PI, Nichols TR, Crozier CL (1990) Airflow patterns in two mesoscale convective systems. *Atmos Res* 25: 167–194

Joe PI, Abraham and J, Macdonald K (1991) Doppler Radar Study of Post-tropical HUGO Interacting with a Cold Front, 25th Radar Conference, Paris, France, 678–682

Joe P, Crozier CL (Ed.) (1995) CIMO Guide, Radar Measurements, Chapter 9, WMO, submitted

Joe P, Scott J, Siggia A, Passarelli R (1994) Random phase processing for the recovery of second trip echoes, COST-75 International Seminar 20–23 September

Joe P, Passarelli R, Siggia A, Scott J (1995) Second trip unfolding by phase diversity processing, Preprints 27th Conf. on Radar Met., Vail, CO, *Amer Met Soc* 770–772

Joss J, Waldvogel A (1990) Precipitation Measurement and Hydrology. Chapter 29a, Radar in Meteorology: Battan Memorial and 40th Anniversary Radar Meteorology Conference (D. Atlas, Editor). Boston, *Amer Meteor Soc* 577–606

Joss J, Lee RW (1993) Scan strategy, clutter suppression calibration and vertical profile corrections, 26th Intern. Conf. on Radar Met., *Amer Met Soc Norman,* OK, 390–392

Langleben MP (1954) The terminal velocity of snowflakes. *Quart J Roy Met Soc* 80: 174–181

Leber GW, Merrit CJ, Robertson JP (1961) WSR-57 analysis of heavy rains, 9th Weather Radar Conf 102–105

Lemon LR, Burgess DW, Brown RA (1978) Tornadic storm air flow and morphology derived from single Doppler radar measurements. *Mon Wea Rev* 106: 48–61

Lhermitte R, Atlas D (1961) Precipitation motion by pulse Doppler radar. Preprints 9th Wea. Radar Conf., Boston, *Amer Meteor Soc,* 218–233

Keeler RJ, Hwang CA, Loew E (1995) Pulse compression weather radar, 27th Conf. on Radar Met., Vail, CO, *Amer Met Soc,* 767–769

Kitchen M, Brown R, Davies AG (1994) Real-time correction of weather radar data for the effects of bright band, range and orographic growth in widespread convection. *Quart Journ of the Royal Met Soc* 120: 1231–1254

Kouwen N, Garland G (1989) Resolution considerations in using radar rainfall data for flood forecasting. *Can Journ of Civil Engineering* 16: 279–289

Kouwen N, Soulis ED, Pietroniro A, Donald JD, Harrington RA (1993) Group response units for distributed hydrologic modeling. *Journ of Water Resources Planning and Management* ASCE, 119(3)

Marshall JS, Gunn KLS (1952) Measurement of snow parameters by radar. *J Met* 9: 322–327

Marshall JS, McK. Palmer W (1948) The distribution of raindrops with size. *J Meteor* 5: 165–166

Mason BJ (1971) *The physics of clouds,* Oxford University Press, pp 671

Matson RJ, Huggins AW (1980) The direction measurement of sizes, shapes and kinematics of falling hailstones. *J Atmos Sci* 37: 1107–1125

Michelson M, Shrader WW, Wilson JG (1990) Terminal Doppler weather radar. *Microwave J* 32(2): 139–148

Mie G (1908) Beiträge zur Optik trüber Medien, speziell kolloidaler Metallösungen. *Ann Phys* 25: 377–445

Mueller EA, Carbone RE (1987) Dynamics of thunderstorm outflow. *J Atmos Sci* 44: 1879–1898

Neff EL (1977) How much rain does a rain gage gauge? *J Hydrol* 35: 213–220

Passarelli R, Romanik P, Geotis SG, Siggia AD (1981) Ground clutter rejection in the frequency domain, 20th Conf. Radar Me., *AMS*, 308–313

Plank VG (1977) Hydrometeor data and analytical-theoretical investigations pertaining to the SAMS rain erosion program of the 1972-73 season at Wallops Island, Virginia, AFGL/SAMS Rep. No. 5, AFGL-TR-77-0149, AFGL, Hanscom AFB, Mass.

Probert-Jones JR (1962) The radar equation in meteorology. *Quart J Roy Met Soc* 88: 485–495

Pruppacher HR, Beard K (1970) A wind tunnel investigation of the internal circulations and shape of water drops falling at terminal velocity in air. *Quart J Roy Met Soc* 96: 247–256

Pruppacher HR, Klett JD (1978) *The microphysics of clouds and precipitation*, D. Reidel Publishing Co, 714 pp

Ray PS, Ziegler CL, Serafin RJ, Baumgartner W (1980) Single and multiple Doppler radar observations of tornadic storms. *Mon Wea Rev* 108: 1607–1625

Richter JH (1969) High resolution tropospheric radar sounding. *Radio Science* 4: 1261–68

Rinehart RE (1991) *Radar for Meteorologists.* Grand Forks, N. Dakota, U.S.A., Univ. of N. Dakota, Dept Atmos Sci, 334 pp

Rogers RR (1964) An extension of the $Z - R$ relation for Doppler radar, *Proc. World Conf. Radar Met.*, 158–161

Rogers RR, Yau MK (1989) *A short course in cloud physics*, 3rd edition, Pergamon Press, pp 293

Russchenberg J (1995) 27th Conference on Radar Meteorology, Vail, CO, *Amer Met Soc*

Sachidananda M, Zrnic DS (1987) Rain rate estimates from differential polarization measurements. *J Atmos Oceanic Technol* 4: 588–598

Sauvageot H (1982) *Radarmeteorologie.* Paris, Eyrolles, 296 pp

Sekhon RS, Srivastava RC (1970) Snow-size spectra and radar reflectance. *J Atmos Sci* 27: 299–307

Seliga TA, Bringi VN (1976) Potential use of radar differential reflectance measurements at orthogonal polarizations for measuring precipitation. *J Appl Meteor* 15: 69–76

Sheppard BE (1990) The measurement of raindrop size distributions using a small Doppler radar. *J Atmos Oceanic Technol* 31: 689–699

Skolnik M, (Editor) (1970) *Radar Handbook.* New York, McGraw-Hill, 39 Chapters

Smith PL (1990) Precipitation Measurement and Hydrology: Panel Report, Chapter 29b, Radar in Meteorology: Battan Memorial and 40th Anniversary Radar Meteorology Conference (D. Atlas, Editor). Boston, *Amer Meteor Soc*, 607–618

Stewart RE, List R (1982) Gryational motion of disks during free fall. *Phys Fluids* 26: 920–927

Spahn JF, Smith PL Jr. (1976) Some charcteristics of hailstone size distributions inside hailstorms, 17th Radar Met. Conf., 534–537

Torlachi E, Humphries RG, Barge BL (1984) Circular polarization for precipitation measurements. *Radio Sci* 19: 193–200

Tuttle JD, Rinehart RE (1983) Attenuation correction in dual-wavelength analysis. *J Climate Appl Met* 22: 1914–1921

Ulbrich CW (1983) Natural variations in the analytical form of the raindrop-size distribution. *J Clim Appl Met* 22: 1764–1775

Ulbrich CW, Atlas D (1984) Assessment of the contribution of differential polarization to improve rainfall measurement. *Radio Science* 19 (1): 49–57

van Gorp JJ (1995) To be or not to be a microburst, 27th Conference on Radar Met., Vail, CO, Amer Met Soc

Wakasugi K, Mizutani A, Matsuo M, Fukao S, Kato S (1987) Further discussion on deriving drop-size distribution and vertical air velocities directly from VHF Doppler radar spectra. *J Atmos Oceanic Technol* 4: 170–179

Waldvogel A, Federer B, Grimm P (1979) Criteria for the detection of hail cells. *J Appl Met* 18: 1521–1525

Wilson JW, Schreiber WE (1986) Initiation of convective storms at radar- observed boundary-layer convergence lines. *Mon Wea Rev* 114: 2516–2536

Wood VT, Brown RA (1986) Single Doppler velocity signature interpretation of nondivergent environmental winds. *J Atmos and Oceanic Tech* 3: 114–128

Zawadzki II (1975) On radar-raingauge comparison. *Journ Applied Met* 14: 909–928

12.10 Appendix A: Radar Terminology, Characteristics and Units

Tables 12.6–12.9 summarize the terminology, characteristics and units commonly used in weather radar.

The common weather radars use a magnetron transmitter and are called non-coherent, that is the phase of each pulse is random. True Doppler radars use a Klystron transmitter where the phase of every pulse is the same. The phase measurements are used to determine the radial velocity of any target encountered. Magnetrons can also be used to measure velocities in non-coherent Dopplers and in this case electronics must be incorporated to save the phase of each transmitted pulse to compare with the received pulse. Generally, due to the higher costs of Klystron systems, the use of magnetron systems have been favored.

Spherical rain drops become elliptically shaped with the major axis in the horizontal plane when falling freely in the atmosphere. The oblateness of the drop is related to the drop size. The oblateness then becomes a measure of the diameter. The power backscattered from an oblate spheroid is larger for a horizontally polarized wave than for the vertically polarized wave provided the drops satisfy Rayleigh scattering. The difference is some measure of the oblateness from which it is postulated, also a measure of the drop diameter. Using suitable assumptions a drop size distribution can be derived and thus a rainfall rate.

The differential reflectance, called Z_{DR}, is defined as ten times the logarithm of the ratio of the horizontally polarized reflectance Z_H, and the vertically polarized reflectance Z_V. Comparisons of the equivalent reflectance factor Z_e and the differential reflectance Z_{DR} suggests the target may be separated as being hail, rain, drizzle or snow.

Relationships have generally been found to take the form:

$$Z = AR^b \tag{12.26}$$

where A and b are constants. The relationship is not unique and very many empirical relations have been developed for various climates or localities and storm types. Nominal and typical values for the index and exponent are: $A = 200$, $b = 1.60$ (Marshall and Palmer, 1948).

The equation is developed under a number of assumptions that may not always be totally valid. Nevertheless, history and experience has shown the relationship in most instances provides a good estimate of precipitation at the ground unless there are obvious anomalies.

The three dimensional echo pattern, together with knowledge about the position of the radar and the orography around it, allows for the correction in real time for a large fraction of the vertical profile problem. This correction allows us to extend the region in which an accuracy acceptable for many hydrological applications is obtained.

Technique	Magnitude of Errors (%) (Balakrishnan et al., 1989)	Condition (mm hr^{-1}) (Chandrasekar et al., 1990)
Z	30-35	< 20
Z_{DR}	20-25	20 - 70
K_{DP}	10- 15	> 70

Table 12.6: *Error Estimates and Optimum Rainfall Conditions.*

Radar Band	Frequency	Wavelength	Nominal	
UHF	300–1000 MHz	1–0.3m	70 cm	
L	1000–2000 MHz	0.3–0.15 m	20 cm	*
S	2000–4000 MHz	15–7.5 cm	10 cm	*
C	4000–8000MHz	7.5–3.75 cm	5 cm	*
X	8000–12500 MHz	3.75–2.4 cm	3 cm	
K_u	12.5–18 GHz	2.4–1.66 cm	1.5 cm	
K	18–26.5 GHz	1.66–1.13 cm	1.25 cm	
K_a	26.5–40 GHz	1.13–0.75 cm	0.86 cm	

Table 12.7: *Radar frequency bands.*

Symbol	Parameter	Units
Z_e	equivalent or effective radar reflectance	mm^6 m^{-3} or dBZ
V_r	Mean Radial Velocity	m s^{-1}
σ_v	spectrum width	m s^{-1}
Z_{DR}	differential reflectance	dB
CDR	circular depolarization ratio	
L_{DR}	linear depolarization ratio	
K_{DP}	propagation phase	degree / km
ρ	correlation	

Table 12.8: *Meteorological radar parameters and units.*

Symbol	Parameter	Units
c	speed of light	m s^{-1}
f	transmitted frequency	Hz
f_d	Doppler frequency shift	Hz
P_r	received power	mW or dBm
P_t	transmitted power	kW
PRF	pulse repetition frequency	Hz
T	pulse repetition time (=1/PRF)	ms
ω	antenna rotation rate	deg s^{-1} or rpm
λ	transmitted wavelength	cm
ϕ	azimuth angle	deg
θ_b	beamwidth between half power points	deg
τ	pulse width	ms
γ	elevation angle	deg

Table 12.9: *Physical radar parameters and units.*

Chapter 13

Ground-Based Remote Sensing of Cloud Properties Using Millimeter-Wave Radar

Eugene E. Clothiaux, T.P. Ackerman and David M. Babb
Department of Meteorology, Pennsylvania State University
502 Walker Building, University Park, PA 16802
United States of America

Abstract

The elements of radar signal processing are described in the context of climate applications of radar. To this end descriptions of radar hardware are minimized and emphasis is placed on describing the transformation of the radar signal as it propagates from the transmitter up into the atmosphere and then back to the radar receiver after a scattering event. For the final radar observables to be of use to the climate community, they must be turned into quantitative measures of the location and properties of clouds. Descriptions of recent algorithm development efforts are described to illustrate approaches and techniques that are being developed to accomplish the goal of producing quantitative descriptions of cloud properties from the radar observables.

13.1 Introduction

The importance of clouds to the present climate of the Earth is unquestioned. As the Earth system undergoes change, both natural and human-induced, the physical properties of clouds can be expected to change. To understand the current role of clouds in the climate system, as well as possible roles in future climate systems, observation of the current cloud fields and their evolution in time is necessary. Unfortunately, even some of the most basic features of clouds, such as their frequency of occurrence or geometrical thickness, are difficult to determine. Recent satellite observational studies, such as the International Satellite Cloud Climatology Project, are leading to a significant improvement in understanding the horizontal distribution of clouds; however, these kinds of satellite analyses are only beginning to address the issue of the vertical distribution of the observed cloud fields. With the technologies currently available, the most

NATO ASI Series, Vol. I 45
Radiation and Water in the Climate System:
Remote Measurements
Edited by Ehrhard Raschke

useful instruments for studying the vertical distribution of clouds are active remote sensors, such as radar and lidar. Radar is the only device that can map out the vertical distribution of most clouds under all weather conditions, whereas lidar is sensitive to tenuous upper level clouds in otherwise clear sky conditions. In this article we will discuss the role of radar in the remote sensing of clouds and their properties.

Radar remote sensing of cloud properties can be viewed as a two step process. First, the actual power levels that are eventually recorded in one form or another within the radar unit must be interpreted; that is, they must be related to physical quantities that are of interest and use in inferring the properties of clouds. Second, the methods by which the radar-derived quantities are used to infer actual cloud properties must be developed. The first step is, in many ways, a solved problem. There are standard, accurate signal processing techniques for calculating the reflectivity, mean Doppler velocity and mean Doppler width from the receiver signal voltages that result when a meteorological target scatters transmitted electromagnetic radiation back to the radar receiver. The second step is by no means a solved problem and much current research is being devoted to the development of cloud property retrieval algorithms using radar-derived quantities as input. The discussion that follows is organized around these two steps. We first discuss the issues of radar signal processing and then we describe the results of some of the research that attempts to infer various cloud properties using the radar-derived quantities as inputs.

Fortunately, the signal processing techniques applied to cloud-sensing millimeter-wave radars are identical to the techniques developed since the late 1940s for the radar remote sensing of rain and other interesting weather conditions (Atlas, 1990). There are several published books that describe to various levels of detail the methods of radar-signal processing when observing meteorological targets. As an introduction to radar-remote sensing, Battan (1973) and Sauvageot (1992) are good starting points. Doviak and Zrnic (1984; 1993) and Ulaby et al. (1982) provide a much more thorough description about each aspect of a radar and its use in observing meteorological targets; however, the increased levels of rigor make the discussions more difficult to understand without a prior knowledge of the subject. The work of Doviak and Zrnic (1993) is especially relevant to the information that we are presenting here. Therefore, we have adopted the approach of trying to tie our discussion as closely as possible to their work, while attempting to provide enough discussion and figures to clarify issues that we initially found difficult to understand.

Retrieval of cloud properties using radar and related observations is a relatively new subject. Our treatment here summarizes several recently published methods by outlining the basic physics and essential mathematical relationships of each. Because this field is still in its infancy, the methods have not been extensively compared or verified. Thus we are unable to present an evaluation of the relative accuracy of the methods.

13.2 The Radar Signal

To keep the following discussion of radar-signal processing steps as concrete as possible, we will describe each step in the context of the 94-GHz cloud radar developed at the Pennsylvania State University (Figures 13.1 and 13.2). Our examples are largely drawn from this system because we are intimately familiar with its operation (Clothiaux et al., 1995). As Figure 13.2 illustrates, there are essentially two distinct components to the radar hardware: the transmitter and the receiver. Therefore, our discussion is divided into two parts. First, we describe the transmission of the electromagnetic wave from the extended interactive oscillator out to the meteorological target and finally back to the receiver antenna. Since most descriptions of radar remote sensing

Figure 13.1: *The Penn State 94-GHz radar.*

are couched in terms of power densities, we then describe this same transmission process using power densities as the physical quantity of interest. The second part of the discussion follows the return signal from the output port of the receiver antenna through the receiver chain. We describe the effects of the meteorological target motions on the return signal in terms of the voltages that eventually emerge from the I and Q demodulators (Figure 13.2).

Before launching into a description of the transmission process we first mention one property of complex numbers that we use in our discussion of electromagnetic fields. Let

$$A = A_0 e^{i\phi} \tag{13.1}$$

represent an arbitrary complex number. The operator $\mathcal{R}$ means take the real part of A; that is,

$$A_{part}^{real} = \mathcal{R}(A) = \mathcal{R}(A_0 e^{i\phi}) = \mathcal{R}(A_0 \cos\phi + iA_0 \sin\phi) = A_0 \cos\phi. \tag{13.2}$$

Even though electromagnetic waves are physical entities, they are often efficiently represented by complex value expressions during mathematical manipulation (Bohren and Huffman, 1983, Section 2.2; hereafter BH, 2.2). For example, the plane, monochromatic electric field

$$E(r,t) = E_0 \cos(kr - \omega t) \tag{13.3a}$$

can be represented by

$$E(r,t) = E_0 e^{i(kr-\omega t)} \tag{13.3b}$$

with the understanding that the true field is described by the real part of the above expression.

13.2.1 The Electromagnetic field description of the transmitted signal

The Electromagnetic Field in the Transmitter Waveguide

For pulsed radars such as the one illustrated in Figure 13.1, the primary function of the transmitter is to send up into the atmosphere a pulsed beam of electromagnetic radiation. To generate pulses of electromagnetic radiation at the carrier frequency (93.95 GHz, or 3.19 mm, for the

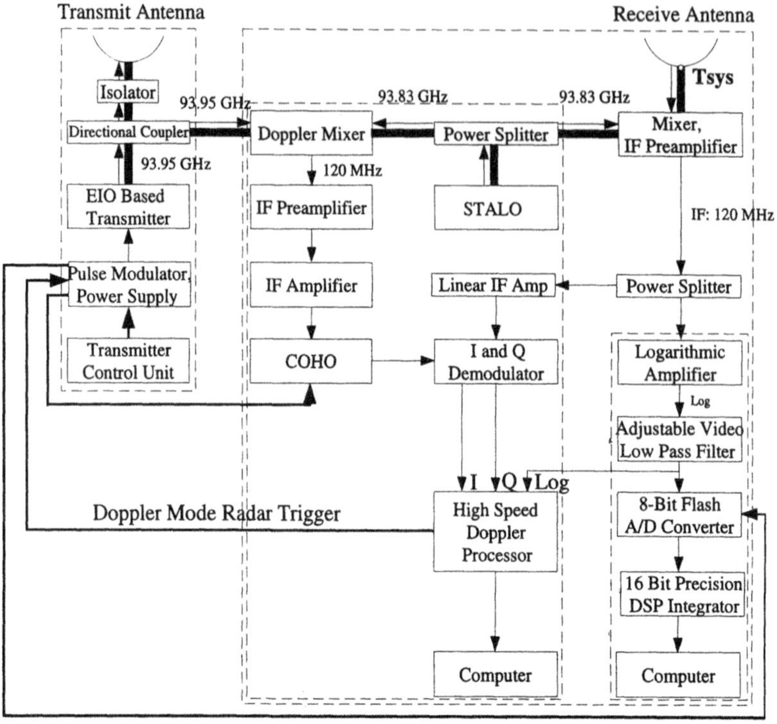

Figure 13.2: *Block diagram of the radar hardware. The two outer dashed blocks enclose the transmitter and receiver components of the radar. Within the receiver box, there are two more dashed blocks: one encloses the incoherent path processing hardware and the other the coherent, or Doppler, processing hardware.*

Penn State radar), an extended interactive oscillator, or EIO, that generates electromagnetic radiation at the carrier frequency is gated by a pulse modulator. A sketch of the resultant, pulsed electromagnetic wave that travels from the oscillator out to the transmitter antenna is illustrated in Figure 13.3. The theory of propagating waves in rectangular shaped waveguides is well documented (e.g., Reitz et al., 1979); the result for the electric field components of a transverse wave propagating parallel to the interior space of the waveguide in the z direction is

$$E_x = E_{0x} \cos(\frac{m\pi x}{a}) \sin(\frac{n\pi y}{b}) \sin(\frac{2\pi z}{\lambda_g} - \omega t), \tag{13.4a}$$

$$E_y = E_{0y} \sin(\frac{m\pi x}{a}) \sin(\frac{n\pi y}{b}) \sin(\frac{2\pi z}{\lambda_g} - \omega t) \tag{13.4b}$$

and $E_z = 0$. The wavelength λ_g of the radiation within the waveguide and along the direction of propagation is given by

$$\frac{2\pi}{\lambda_g}^2 = \frac{2\pi}{\lambda_c}^2 - \frac{m\pi}{a}^2 - \frac{n\pi}{b}^2, \tag{13.4c}$$

where a and b are waveguide dimensions, $\lambda_c = 2\pi/\omega_c$ and ω_c is the carrier frequency (Figure 13.4a). For the Penn State radar, the waveguide dimensions are $a = 2.5$ mm and $b = 1.3$ mm.

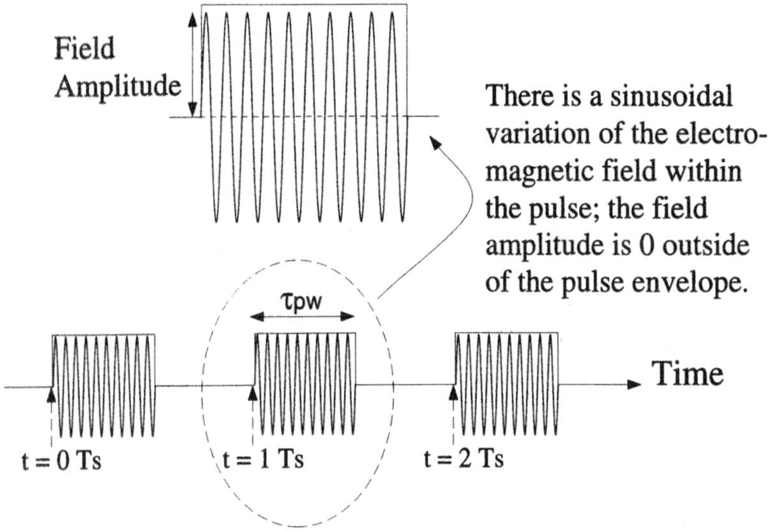

Figure 13.3: *Cartoon illustration of the electromagnetic wave leaving the transmitter oscillator. The pulse modulator generates the rectangular waveform which allows the oscillator output to be pulsed.*

Since $\lambda_c = 3.19$ mm and λ_g must be larger than zero for a physically meaningful solution, the only viable choices of m and n are one and zero, respectively. Therefore, $E_x = 0$ and E_y reduces to

$$E_y = E_{0y} \sin(\frac{\pi x}{a}) \sin(\frac{2\pi z}{\lambda_g} - \omega t). \qquad (13.4d)$$

The electromagnetic wave within the waveguide is transverse, sinusoidal in time and polarized with the plane of polarization being defined by the direction of propagation of the wave and the direction of the electric field (Figure 13.4b).

The Transmitted Electromagnetic Field Far from the Radar

After suffering losses in power on its journey to the transmitter antenna, the electromagnetic wave emerging from the waveguide is sent up into the atmosphere by the transmitter antenna (Figure 13.5). At a large distance r from the antenna, i.e., in the far field of the antenna ($2\pi r/\lambda_c \gg 1$), the electromagnetic wave is polarized with both its electric and magnetic field components perpendicular to its direction of propagation, which is radially away from the radar. The electric field component of the wave is parallel to the plane of polarization illustrated in Figure 13.4b. The mathematical description of the electric field at a point far away from the antenna is (BH, 3.2)

$$E_t(r, \theta, \phi, t) = \frac{A(\theta, \phi)}{r} \cos(k_c r - \omega_c t + \phi_t), \qquad (13.5a)$$

where ω_c is the carrier frequency, t is the time, r is the distance from the antenna, c is the speed of light and ϕ_t sets the initial phase of the transmitted wave. The amplitude $A(\theta, \phi)$ depends upon the power delivered to the transmitter antenna and the gain and beam pattern of the transmitter antenna (Figure 13.6). In complex language notation

$$E_t(r, \theta, \phi, t) = \mathcal{R}(\frac{A(\theta, \phi) e^{i\phi_t}}{r} e^{i(k_c r - \omega_c t)}). \qquad (13.5b)$$

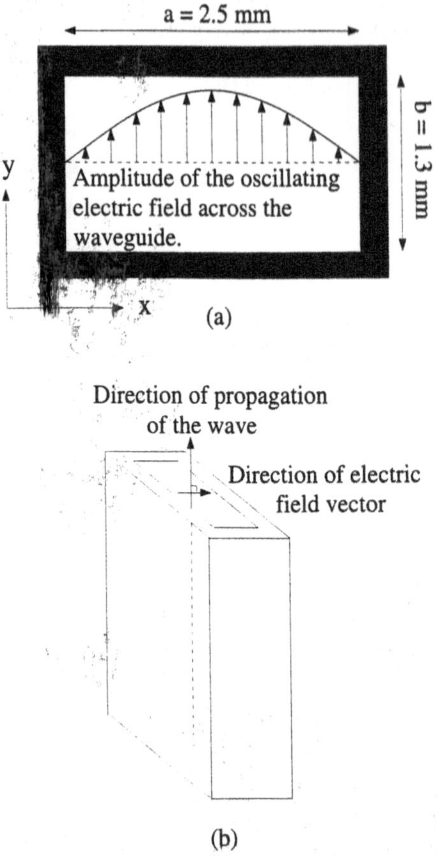

Figure 13.4: *a) Cross section of the waveguide connecting the oscillator to the transmitter antenna. The electromagnetic field that propagates through the waveguide is parallel to the y-axis. b) The plane of polarization of the transmitted wave is defined by the direction of propagation of the wave and the direction of the electric field.*

The Scattered Electromagnetic Field at the Receiver Antenna

Now suppose that at r the pulse encounters an object of some sort. The electric field scattered by the object back to the radar is

$$\begin{aligned} \mathbf{E}_s(r,\theta,\phi,t) &= \frac{S_{2,i}(180^\circ)e^{-i(\omega_m t-\phi_s)}}{k_c r} e^{ik_c r} \frac{A(\theta,\phi)e^{i\phi_t}}{r} e^{i(k_c r-\omega_c t)} \hat{\mathbf{e}}_{same} \\ &+ \frac{S_{4,i}(180^\circ)e^{-i(\omega_m t-\phi_s)}}{k_c r} e^{ik_c r} \frac{A(\theta,\phi)e^{i\phi_t}}{r} e^{i(k_c r-\omega_c t)} \hat{\mathbf{e}}_{cross} \end{aligned} \tag{13.6}$$

where $S_{2,i}(180^\circ)$ and $S_{4,i}(180^\circ)$ are the magnitudes of two of the amplitude scattering matrix elements (BH, 3.2). The phases associated with these two matrix elements have two parts: one part ϕ_s that represents phase changes upon scattering and a second part $\omega_m t$ due to the

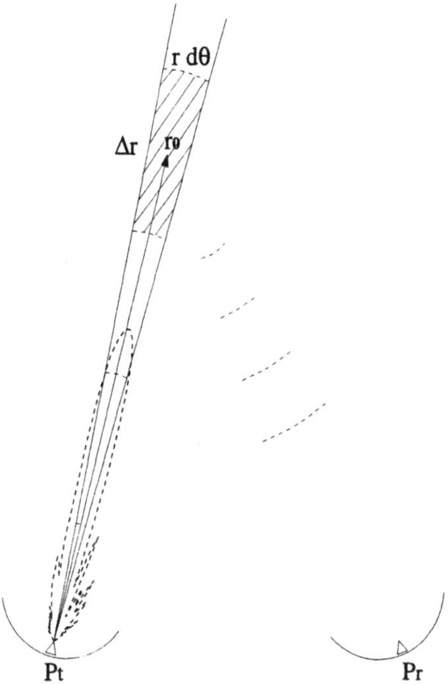

Figure 13.5: *Cartoon illustration of the propagation of the electromagnetic field from the transmitter antenna, to the atmosphere and then back to the receiver antenna.*

motion of the particle toward or away from the radar. Although the magnitude of the cross polarization matrix element $S_{4,i}(180°)$ is generally small, it may contain useful information about the properties of the scatterer. (Note that the Penn State radar is insensitive to the cross polarization electric field component in its present configuration because this component of the field does not propagate through the circular-to-rectangular waveguide hardware that connects the receiver antenna to the mixer/preamplifier assembly.) Because of the generally small amplitude of $S_{4,i}(180°)$ and the fact that polarization studies are beyond the scope of this article, we neglect the cross polarization component at this point and write the scattered electric field at the radar receiver antenna as

$$E_s(r,\theta,\phi,t) = \frac{S_{2,i}(180°)e^{-i(\omega_m t-\phi_s)}}{k_c r}e^{ik_c r}\frac{A(\theta,\phi)e^{i\phi_t}}{r}e^{i(k_c r-\omega_c t)}. \tag{13.7}$$

The scattered electric field is now polarized and we drop the bold face on E_s to explicitly state that there is only one remaining component. Rearranging (13.7) leads to

$$E_s(r,\theta,\phi,t) = A^s e^{i[2k_c r-(\omega_c+\omega_m)t+\phi_s+\phi_t]} \tag{13.8a}$$

where

$$A^s = A^s(r,\theta,\phi) = \frac{S_{2,i}(180°)A(\theta,\phi)}{k_c r^2}. \tag{13.8b}$$

At this point we assume that both the transmitter phase ϕ_t and the phase shift ϕ_s upon scattering are constants, even though ϕ_s may be a function of time for such meteorological

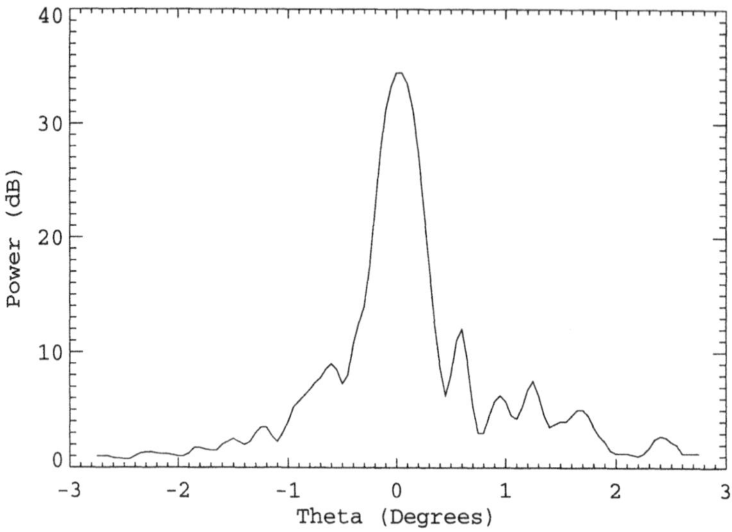

Figure 13.6: *Beam pattern of the Penn State 94-GHz radar transmitter antenna.*

targets as vibrating water drops or tumbling ice particles (Doviak and Zrnic, 1993, Section 3.4.3; hereafter DZ, 3.4.3). The term $\omega_m t$ represents the shift in phase of the scattered wave in the direction of the radar due to the relative radial velocity of the target with respect to the radar and $2k_c r = 2\pi(2r/\lambda_c)$ is the number of wavelengths along the round trip path between the radar and the target in units of radians.

From Fields to Power Densities

In the radar literature the electromagnetic wave propagating through the atmosphere is usually described in terms of the time-averaged power density over one cycle of the carrier wave. The instantaneous power density of the field incident on the scatterer is

$$P_{scat}(r,\theta,\phi,t) \propto E_t^2(r,\theta,\phi,t). \tag{13.9a}$$

The proportionality factor η_0 takes into account the electric and magnetic nature of the radiation:

$$P_{scat}(r,\theta,\phi,t) = \frac{1}{\eta_0} E_t^2(r,\theta,\phi,t). \tag{13.9b}$$

Averaging $P_{scat}(r,\theta,\phi,t)$ over one cycle of the wave yields

$$P_{scat}(r,\theta,\phi) = \frac{1}{\eta_0}\left(\frac{A(\theta,\phi)}{r}\right)^2 \left[\frac{1}{2\pi/\omega_c}\int_0^{2\pi/\omega_c} \cos^2(k_c r - \omega_c t + \phi_t)dt\right]. \tag{13.10a}$$

The factor in brackets equals 1/2 so that the power density at the scatterer becomes (DZ, 2.1)

$$P_{scat}(r,\theta,\phi) = \frac{1}{2\eta_0}\left(\frac{A(\theta,\phi)}{r}\right)^2. \tag{13.10b}$$

In a similar fashion one can demonstrate that the average return power density at the receiver antenna is

$$P_{ant}(r,\theta,\phi) = \frac{(A^s)^2}{2\eta_0} = \frac{1}{2\eta_0}\left(\frac{S_{2,i}(180^\circ)}{k_c r}\right)^2 \left(\frac{A(\theta,\phi)}{r}\right)^2. \tag{13.11a}$$

The backscatter cross section σ_b is defined by (BH, 4.6)

$$\sigma_b = 4\pi \left(\frac{S_{2,i}(180^\circ)}{k_c} \right)^2 , \tag{13.11b}$$

so that the power density at the receiver antenna becomes

$$P_{ant}(r, \theta, \phi) = \frac{1}{8\pi\eta_0} \frac{\sigma_b}{r^2} \left(\frac{A(\theta, \phi)}{r} \right)^2 . \tag{13.11c}$$

We are now in a position to compare this description of the transmission process with the more widely used description based on power densities.

13.2.2 The power description of the transmitted signal

Scattering from a Single Isolated Particle

In terms of the properties of the radar, let P_t represent the average power density over a single cycle of the electromagnetic radiation that is delivered to the transmitter antenna input port. If the wave is stable over the duration of the pulse (i.e., the amplitude and frequency of the sinusoid in Figure 13.3 do not change over the pulse.), P_t also represents the average power over the entire pulse. If the antenna were to radiate isotropically, the power density a distance r away would be $P_t/4\pi r^2$; however, the transmitter antenna provides a gain G_t along the main beam axis with a falloff in power away from the axis given by the beam pattern $f^2(\theta, \phi)$. Therefore, the power density a distance r from the antenna in the direction $(\theta,\ \phi)$ is (DZ, 3.1)

$$\left(\frac{P_t}{4\pi r^2} \right) \left(f_t^2(\theta, \phi) \right) (G_t) \tag{13.12}$$

As the beam travels a distance r, it is attenuated by atmospheric constituents, such as oxygen, water vapor and hydrometeors. Letting $k_e(r)$ represent the extinction coefficient of the atmosphere along the beam path (units of m^{-1}), the attenuation is (DZ, 3.3)

$$l_t^{-1} = e^{-\int_0^r k_e(r) dr} \tag{13.13}$$

where, in keeping with Doviak and Zrnic (DZ, 3.3), we have written the loss as a quantity greater than one. Therefore, (13.12) becomes

$$P_{scat}(r, \theta, \phi) = \left(\frac{P_t}{4\pi r^2} \right) \left(f_t^2(\theta, \phi) \right) (G_t) \left(l_t^{-1} \right) . \tag{13.14}$$

Comparing (13.10b) and (13.14), the electric field amplitude $A(\theta, \phi)$ in terms of the radar's parameters is

$$A(\theta, \phi) = \sqrt{\frac{P_t G_t \eta_0}{2\pi l_t}} f_t(\theta, \phi). \tag{13.15}$$

Now suppose that at r the pulse encounters an object of some sort. The amount of energy scattered by the object in any direction is characterized by the object's scattering cross section $\sigma(\theta', \phi')$. The cross section is defined so that the power density at the radar receiver antenna is (DZ, 3.2)

$$P_{ant}(r, \theta, \phi) = \left(\frac{P_t}{4\pi r^2} \right) \left(f_t^2(\theta, \phi) \right) (G_t) \left(l^{t-1} \right) \left(\frac{\sigma(\theta', \phi')}{4\pi r^2} \right) \left(l_r^{-1} \right) , \tag{13.16}$$

where we have incorporated attenuation from the object to the radar receiver. The primes in the definition of σ indicate that the receiver antenna may not be collocated with the transmitter antenna. However, for most radar applications the transmitter and receiver antennas are collocated. In this case, we are interested in only the backscatter cross section and we set $\sigma(\theta', \phi') = \sigma_b$. Note that if we set (13.11a) equal to (13.16), we have

$$A^s(r,\theta,\phi) = \sqrt{\frac{P_t G_t \eta_0 \sigma_b}{8\pi^2 l_t l_r}} \frac{f_t(\theta,\phi)}{r} \tag{13.17}$$

for the scattered field amplitude at the receiver antenna.

The receiver antenna has an effective aperture $A_e(\theta,\phi)$, so the power delivered by the antenna to the receiver is

$$P_{rec}(r,\theta,\phi) = \left(\frac{P_t}{4\pi r^2}\right)\left(f_t^2(\theta,\phi)\right)(G_t)\left(l_t^{-1}\right)\left(\frac{\sigma_b}{4\pi r^2}\right)\left(l_r^{-1}\right)\left(A_e(\theta,\phi)\right). \tag{13.18}$$

Doviak and Zrnic (DZ, 3.4) demonstrate that

$$A_e(\theta,\phi) = (\frac{G_r \lambda^2}{4\pi}) f_r^2(\theta,\phi), \tag{13.19}$$

so the average return power at the receiver antenna output port becomes

$$P_{rec}(r,\theta,\phi) = \left(\frac{P_t}{4\pi r^2}\right)\left(f_t^2(\theta,\phi)\right)(G_t)\left(l_t^{-1}\right)\left(\frac{\sigma_b}{4\pi r^2}\right)\left(l_r^{-1}\right)\left(\frac{G_r \lambda^2}{4\pi} f_r^2(\theta,\phi)\right), \tag{13.20}$$

which is known as the radar equation for a single scatterer. For most radars, including the Penn State system, $G_t = G_r = G$, $f_t = f_r = f$ and $l_t = l_r = l$. Therefore, we write the radar equation for scattering by an isolated target as (DZ, 3.4)

$$P_{rec}(r,\theta,\phi) = \frac{P_t f^4(\theta,\phi) G^2 \sigma_b \lambda_c^2}{(4\pi)^3 r^4 l^2}. \tag{13.21}$$

Scattering from Many Particles

Meteorological targets, unfortunately, are not point targets. They are usually collections of particles described by a distribution function $N(\mathbf{D}, \mathbf{r})$, where N represents the number of particles per unit volume with characteristic size $\mathbf{D}$ at location $\mathbf{r}$. For most applications in the literature, the targets are generally treated as spheres with $\mathbf{D}$ simply representing the diameter D of the particles. If $\sigma(\mathbf{D})$ describes the backscattering cross section of particles with characteristic size $\mathbf{D}$, then the total backscattering cross section per unit volume is (DZ, 4.4)

$$\eta(\mathbf{r}) = \int_{\mathbf{D}} \sigma(\mathbf{D}) N(\mathbf{D}, \mathbf{r}) d\mathbf{D}. \tag{13.22}$$

Substituting $\eta(\mathbf{r})dV$ for σ_b in (13.21), where dV is the volume element in spherical coordinates, i.e., $dV = r^2 \sin\theta d\theta d\phi dr$, the return power from the resolution volume becomes (DZ, 4.4)

$$P_{rec}(r,\theta,\phi) = \int_{r_b}^{r_t} \int_0^{\pi} \int_0^{2\pi} \frac{P_t f^4(\theta,\phi) G^2 [\eta(\mathbf{r}) dV] \lambda_c^2}{(4\pi)^3 r^4 l^2}. \tag{13.23}$$

The limits r_b and r_t are set by the pulse width and the distance to the target (Figures 13.7 and 13.8). Even though the integration over θ is from 0 to π, most of the contribution to the integral comes from the primary forward lobe of the antenna pattern. For the Penn State cloud radar the beamwidth at the half-maximum of the primary lobe is 0.24°. If the resolution volume is sufficiently small, $\eta(r, \theta, \phi) \approx \eta(\mathbf{r_0})$ and (13.23) becomes

$$P_{rec}(\mathbf{r_0}) = \int_{r_b}^{r_t} \int_0^{\pi} \int_0^{2\pi} \frac{P_t G^2 \lambda^2 \eta(\mathbf{r_0})}{(4\pi)^3 r_0^2 l^2(\mathbf{r_0})} f^4(\theta, \phi) \sin\theta d\theta d\phi dr. \tag{13.24}$$

For a circularly symmetric, Gaussian-shaped beam pattern (DZ, 4.4)

$$\int_0^{\pi} \int_0^{2\pi} f^4(\theta, \phi) \sin\theta d\theta d\phi = \frac{\pi \theta_1^2}{8 \ln 2}, \tag{13.25}$$

and the received power becomes (DZ, 4.4)

$$P_{rec}(\mathbf{r_0}) = \frac{P_t G^2 \lambda^2 \eta(\mathbf{r_0}) c \tau_{pw} \pi \theta_1^2}{(4\pi)^3 r_0^2 l^2(\mathbf{r_0}) 16 \ln 2}. \tag{13.26}$$

At first glance the derivation of P_{rec} seems innocuous enough, but there are some implicit assumptions in the derivation that hide a fair bit of the complexity that is at the heart of weather-radar signal processing. We will discuss these issues in the context of the signals that emerge from the I and Q demodulators, but before doing so we want to discuss the limits on the integral in (13.23) in a bit more detail.

13.3 The Radar Resolution Volume

13.3.1 The radar resolution volume along the beam direction

At this point we want to emphasize the difference between a resolution volume containing a single target and one containing many targets. Consider a single pulse travelling through space along a given direction (Figure 13.7). Now suppose a single scatterer is located at a distance $r_0 = ct_0$ from the radar. The leading edge of the pulse encounters the object at time t_0. Since there are no other scatterers involved in this example, the back edge of the pulse undergoes no scattering until it, too, encounters the object at r_0. Scatter from the leading edge of the pulse reaches the radar at time $2t_0$, whereas scatter from the back edge of the pulse does not reach the receiver antenna until time $2t_0 + \tau_{pw}$ where τ_{pw} is the transmitted pulse width in units of time. At any time t such that $2t_0 \leq t \leq 2t_0 + \tau_{pw}$, the receiver antenna is illuminated only from radiation originating from the target located at $r_0 = ct_0$. Therefore, analysis of the radar signal over any time interval during this time period leads to information on the scatterer at r_0.

The scenario is quite different when many scatterers are located in and around the scatterer at r_0 (Figure 13.8). As before, consider the case when the leading edge of the pulse encounters the object at r_0 at time $t = t_0$ (Figure 13.8b). The backscatter from this encounter reaches position $r_0 - c\tau_{pw}/2$ at time $t_0 + \tau_{pw}/2$. At this time, however, the back edge of the pulse is also at location $r_0 - c\tau_{pw}/2$ (Figure 13.8c). If the back edge of the pulse undergoes scattering at this point, then the scatter from the leading edge of the pulse at position r_0 reaches the receiver antenna at exactly the same time as the scatter from the trailing edge of the pulse at $r_0 - c\tau_{pw}/2$ (Figure 13.8d). In fact, at time $2t_0$ the radar receives scattered waves from all points located between $r_0 - c\tau_{pw}/2$ and r_0; this region of space is the resolution volume of the radar at the time $2t_0$ (Figure 13.8d).

As in the single scatterer case, the radar receives a scattered field from position r_0 for all times t such that $2t_0 \leq t \leq 2t_0 + \tau_{pw}$; however, at the time $2t_0$ the radar also receives energy from the

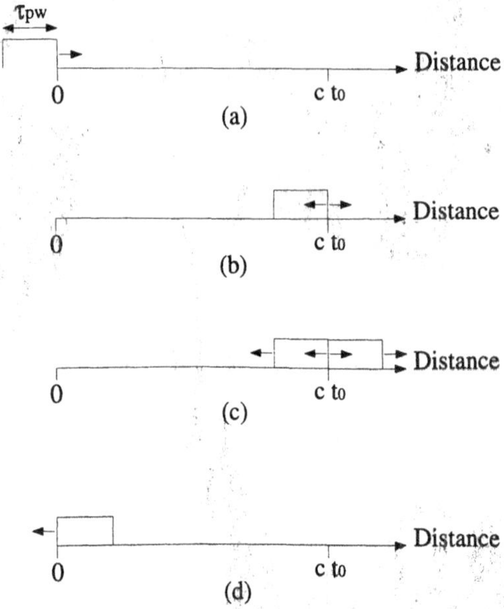

Figure 13.7: *Cartoon illustration of a transmitted pulse scattering off a single, isolated scatterer. The pulse width is τ_{pw}, and the distance from the radar to the target is ct_0.*

region between $r_0 - c\tau_{pw}/2$ and r_0, whereas at time $2t_0 + \tau_{pw}$ it receives energy from all points between r_0 and $r_0 + c\tau_{pw}/2$. It no longer makes sense to analyze the properties of the pulse from $2t_0 \leq t \leq 2t_0 + \tau_{pw}$ to infer properties about the scatterer located at position r_0 since the radar is receiving energy from different points in space during this interval. To study the properties of scatterers in a fixed region of space, say from $r_0 - c\tau_{pw}/2$ to r_0, only the radiation that reaches the receiver antenna at the precise time of $2t_0$ is of value.

Let T_s represent the time between radar pulses and let τ_s represent the time between power samples that are collected by the radar receiver (Figure 13.9). In this setup the first pulse is sent out at time $0T_s$ and samples are collected at times $0T_s + \tau_s, 0T_s + 2\tau_s$, etc. The second pulse leaves the transmitter antenna at time $1T_s$ and samples are then collected at times $1T_s + \tau_s, 1T_s + 2\tau_s$, etc. The number of distinct resolution volumes in space from which the radar collects information is equal to the number of samples N_g that are collected by the radar between any two radar pulses. The samples collected at the times $0T_s + j\tau_s, 1T_s + j\tau_s$, etc. are from the same resolution volume whose center is located at

$$r_j = c\frac{j\tau_s}{2} - \frac{c\tau_{pw}}{4} = \frac{c}{2}\left(j\tau_s - \frac{\tau_{pw}}{2}\right). \tag{13.27}$$

Therefore, the radar receiver generates a series of snap shots of the I and Q voltages that correspond to specific regions in space. The region in space is easily indentified from the delay $t = j\tau_s$ between the initiation of the pulse and the time the signal is sampled, as well as the pulse width τ_{pw} itself. The time interval between each sample is simply τ_s. With this notation in hand we write the limits r_b and r_t in (13.23) as $(c/2)(j\tau_s - \tau_{pw})$ and $(c/2)(j\tau_s)$, respectively.

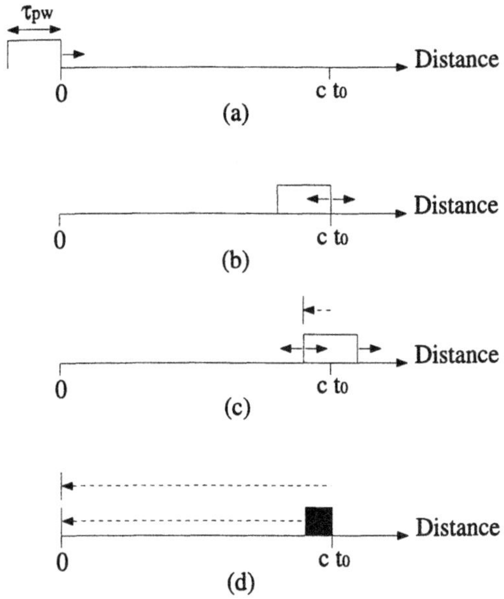

Figure 13.8: *Cartoon illustration of a transmitted pulse scattering off a collection of particles indicated by the filled square in (d). The pulse width is τ_{pw} and the distance from the radar to the center of the target is $ct_0 - c\tau_{pw}/4$.*

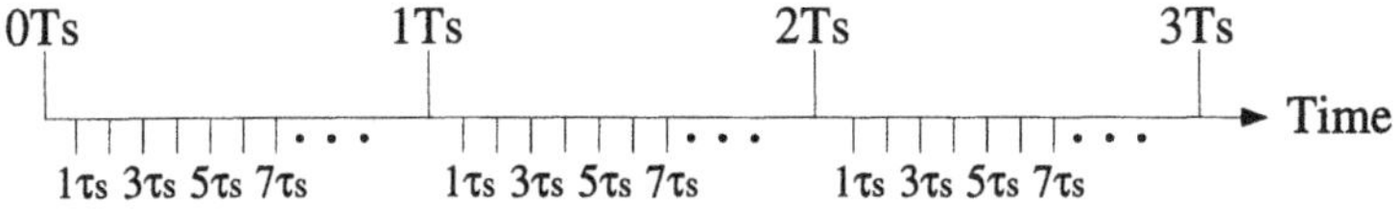

Figure 13.9: *The time between radar pulses is T_s and the time between receiver power samples is τ_s.*

13.3.2 The radar resolution volume perpendicular to the beam direction

The electromagnetic wave propagating from the transmitter antenna to the atmosphere is not contained within a single, well-defined direction. Rather, the field leaving the antenna (actually scattered by the antenna) to the atmosphere sets up interference patterns so that the strength of the field is significantly different from 0 in a number of different directions (Figure 13.6). Fortunately, the power in the forward lobe is 177.8 times, or 22.5 dB above, the power in the next most significant side lobe. For all practical purposes, therefore, the side lobes can be neglected in power calculations without introducing significant error. For example, the derivation of (13.25) is based on a single, Gaussian-shaped antenna beam in the forward direction. The side lobes can lead to returned power from objects that are located to the side of the radar, which can be problematic; however, these ground clutter issues are insignificant for the Penn State radar since the beam is narrow. The nonzero width of the beam within the main lobe leads to a resolution volume that increases in size with distance from the radar.

13.4 The Radar Receiver

13.4.1 Signal transformations in the radar receiver

From the receiver antenna output port the electromagnetic radiation propagates along a short stretch of waveguide to the mixer/preamplifier assembly. In the mixer/preamplifier unit the physical nature of the signal is transformed from a time-dependent, transverse electromagnetic wave at the input to a time-dependent voltage at the output which is subsequently passed through the rest of the signal processing chain by coaxial cable. Furthermore, the high carrier frequency ω_c is transformed to a much lower intermediate frequency ω_i. We write the signal voltage at the ouput of the mixer/preamplifier unit as

$$V(t) = V_s e^{i[2k_c r - (\omega_i + \omega_m)t + \phi_s + \phi_t]}, \tag{13.28}$$

where the amplitude V_s depends upon gains and losses in the mixer/preamplifier assembly. At this point the signal is split into two branches: one branch is input to the incoherent receiver path and the other branch passes to the coherent, or Doppler, receiver unit. We describe the coherent receiver here and refer the reader to the article of Clothiaux et al. (1995) for a description of the incoherent path power estimates produced by the Penn State radar.

The signal passes to the I and Q demodulators with only power gains and losses affecting it. In the I and Q demodulators an important transformation of the signal occurs. To remove ω_i from the signal, the physical voltage, i.e., the real part of (13.28),

$$V(t) = V_s \cos \alpha(t), \tag{13.29a}$$

where

$$\alpha(t) = 2k_c r - (\omega_i + \omega_m)t + \phi_s + \phi_t, \tag{13.29b}$$

is mulitplied by the two reference voltages

$$V_{ref}^{I}(t) = V_{ref,0} \cos(\beta(t)) \tag{13.30a}$$

and

$$V_{ref}^{Q}(t) = V_{ref,0} \cos(\beta(t) - \frac{\pi}{2}), \tag{13.30b}$$

where $\beta(t) = -w_i t$. The reference voltages have their origin in a sample of the transmitted signal that is eventually input to the I and Q demodulators through the coherent oscillator, or COHO (Figure 13.2). The multiplication process in the demodulators performs a transformation of the signal that is comparable to the trigonometric identity

$$\cos(\alpha)\cos(\beta) = \frac{1}{2}[\cos(\alpha + \beta) + \cos(\alpha - \beta)]. \tag{13.31}$$

A low-pass filter at the output of the demodulators removes the high frequency term containing $\cos(\alpha+\beta)$, together with any other frequecies much greater than $\alpha-\beta$ due to nonlinearities in the mixer, with the result that

$$I(t) = \text{Low Frequency Part}[V(t)V^I_{ref}(t)] = V_0\cos[2k_c r - \omega_m t + \phi_s + \phi_t], \tag{13.32a}$$

and

$$Q(t) = \text{Low Frequency Part}[V(t)V^Q_{ref}(t)] = V_0\sin[2k_c r - \omega_m t + \phi_s + \phi_t]. \tag{13.32b}$$

We emphasize that no matter what the nature of the electomagnetic field is at the receiver antenna output port, the result of the coherent path signal processing is to strip off the carrier frequency and create the I and Q signal voltages. The voltage amplitudes of I and Q are always a faithful representation of the electromagnetic field strength at the receiver antenna output port. If the gains G_{sys} and losses L_{sys} from the receiver antenna to the output of the demodulators are known, then the amplitudes of I and Q can be used to produce estimates of the electromagnetic field strength at the receiver antenna. Furthermore, the variations in time of the I and Q signals enable one to produce power density spectra of the signal from which estimates of the velocities of the scatterers relative to the radar are derived.

13.4.2 Estimating received power from an isolated scatterer

The power in the electromagnetic field at the receiver antenna due to scattering from a single, isolated particle is proportional to the real part of the electric field squared averaged over a period $T_{ant} = 2\pi/(\omega_c+\omega_m)$ of the Doppler shifted wave:

$$P_{ant} = \frac{(A^s)^2}{\eta_0}\left[\frac{1}{T_{ant}}\int_0^{T_{ant}} \cos^2[2k_c r - (\omega_c+\omega_m)t + \phi_s + \phi_t]dt\right]. \tag{13.33}$$

As before, the factor in brackets is 1/2 so that

$$P_{ant} = \frac{(A^s)^2}{2\eta_0}. \tag{13.34}$$

Therefore, apart from a constant of proportionality, the power averaged over a cycle of the wave is simply the wave amplitude squared and divided by two.

This same power can be estimated from either the I or Q signal voltages. Consider $I(t)$ from (13.32a). The instantaneous power in $I(t)$ is

$$P_{iq}(t) = \frac{I^2(t)}{R_{ref}}, \tag{13.35}$$

where R_{ref} is the appropriate impedance. Therefore, the power averaged over the period $T_{iq} = 2\pi/\omega_m$ is

$$P_{iq} = \frac{V_0^2}{R_{ref}}\frac{1}{T_{iq}}\int_0^{T_{iq}} \cos^2[2k_c r - \omega_m t + \phi_s + \phi_t]dt = \frac{V_0^2}{2R_{ref}}. \tag{13.36}$$

Assuming that we know G_{sys} and L_{sys}, as well as the receiver antenna aperture $A_e(\theta,\phi)$, we have

$$P_{ant} = \frac{P_{iq}}{A_e(\theta,\phi)G_{sys}L_{sys}^{-1}}. \tag{13.37}$$

Since $I(t)$ and $Q(t)$ are sinusoidal in time, there is a much simpler way to determine P_{iq}; however, we must use a complex value representation for the voltage:

$$P_{iq} = \frac{[I(t)][I^*(t)]}{2R_{ref}} = \frac{1}{2R_{ref}}[V_0 e^{-i(2k_c r - \omega_m t + \phi_s + \phi_t)}][V_0 e^{i(2k_c r - \omega_m t + \phi_s + \phi_t)}] = \frac{V_0^2}{2R_{ref}}, \tag{13.38}$$

where I^* is the complex conjugate of I. At this point it is instructive to expand the exponentials in (13.38) using Euler's equation to get

$$P_{iq} = \frac{V_0^2}{2R_{ref}}\left(\cos^2[2k_c r - \omega_m t + \phi_s + \phi_t] + \sin^2[2k_c r - \omega_m t + \phi_s + \phi_t]\right). \tag{13.39}$$

Comparing (13.32) and (13.39) we can write

$$P_{iq} = \frac{1}{R_{ref}}\left(\frac{I^2}{2} + \frac{Q^2}{2}\right). \tag{13.40}$$

Therefore, given instantaneous measurements of $I(t)$ and $Q(t)$, (13.40) can be used as a simple recipe to calculate the power over a cycle of the wave. This method of obtaining the average power in the I and Q signals is true only if they are sinusoids over the cycle.

13.4.3 Estimating received power from many scatterers

Now that we have calculated the backscattered power due to a single particle, let us see what happens when we introduce a second scatterer in the resolution volume. The electromagnetic field scattered to the radar becomes

$$E_s(t) = A_m^s e^{i[2k_c r_m - (\omega_c + \omega_{d,m})t + \phi_{s,m} + \phi_t]} + A_n^s e^{i[2k_c r_n - (\omega_c + \omega_{d,n})t + \phi_{s,n} + \phi_t]}. \tag{13.41}$$

The power averaged over a time interval large compared to $2\pi/\omega_c$ but small compared to the time intervals $2\pi/\omega_{d,m}$ and $2\pi/\omega_{d,n}$ is

$$P_{ant}(t) = \frac{E_s(t)E_s^*(t)}{2\eta_0} \tag{13.42a}$$

or

$$\begin{aligned} P_{ant}(t) &= \frac{1}{2\eta_0}((A_m^s)^2 + (A_n^s)^2 + A_m^s A_n^s e^{i[2k_c(r_m - r_n) - (\omega_{d,m} - \omega_{d,n})t + (\phi_{s,m} - \phi_{s,n})]} \\ &\quad + A_n^s A_m^s e^{-i[2k_c(r_m - r_n) - (\omega_{d,m} - \omega_{d,n})t + (\phi_{s,m} - \phi_{s,n})]}). \end{aligned} \tag{13.42b}$$

Expanding the exponentials yields

$$P_{ant}(t) = \frac{1}{2\eta_0}\left((A_m^s)^2 + (A_n^s)^2 + 2A_m^s A_n^s \cos[2k_c(r_m - r_n) - (\omega_{d,m} - \omega_{d,n})t + (\phi_{s,m} - \phi_{s,n})]\right). \tag{13.43}$$

If one considers all of the scatterers N_s per unit volume, the power becomes

$$P_{ant}(t) = \frac{1}{2\eta_0}\sum_{m=1}^{N_s}(A_m^s)^2 + \frac{1}{2\eta_0}\sum_{m \neq n}^{N_s} A_m^s A_n^s e^{-i[2k_c(r_m - r_n) - (\omega_{d,m} - \omega_{d,n})t + (\phi_{s,m} - \phi_{s,n})]}. \tag{13.44}$$

The information dependent on the total cross section of the particle is contained in the first sum. Therefore, if we can adopt a processing strategy that drives the contribution of the second term to zero, we can recover the total cross section in the resolution volume. Fortunately, if P_{ant} is measured repeatedly over a time interval long compared with the time intervals $2\pi/(\omega_{d,m} - \omega_{d,n}), m \neq n$, and averaged, the second sum does indeed go to zero. Therefore, we can write

$$P_{ant} = \lim_{N_n \to \infty} \left[\frac{1}{N_n} \sum_{j=1}^{N_n} P_{ant,j} \right] = \frac{1}{2\eta_0} \sum_{m=1}^{N_s} (A_m^s)^2, \tag{13.45}$$

where N_n is the number of samples in the average. Therefore, (13.26) for $P_{rec}(\mathbf{r_0})$ holds true as long as many power samples are collected over a sufficiently long period of time and then averaged.

When many scatterers are in the resolution volume, the scattered electric field at the receiver antenna is

$$E_s(t) = \sum_{m=1}^{N_s} A_m^s e^{i[2k_c r_m - (\omega_c - \omega_{d,m})t + \phi_{s,m} + \phi_t]}. \tag{13.46}$$

In terms of I and Q we have (DZ, 4.3)

$$I = \mathcal{R}\left(\sum_{m=1}^{N_s} V_m e^{i[2k_c r_m - \omega_{d,m} t + \phi_{s,m} + \phi_t]} \right) \tag{13.47a}$$

and

$$Q = \mathcal{I}\left(\sum_{m=1}^{N_s} V_m e^{i[2k_c r_m - \omega_{d,m} t + \phi_{s,m} + \phi_t]} \right), \tag{13.47b}$$

or

$$I = \sum_{m=1}^{N_s} V_m \cos[2k_c r_m - \omega_{d,m} t + \phi_{s,m} + \phi_t] \tag{13.48a}$$

and

$$Q = \sum_{m=1}^{N_s} V_m \sin[2k_c r_m - \omega_{d,m} t + \phi_{s,m} + \phi_t]. \tag{13.48b}$$

Using (13.8b), (13.34), (13.37) and (13.38), we can write the voltage amplitudes V_m as

$$V_m = \sqrt{2 R_{ref} G_{sys} L_{sys}^{-1} A_e(\theta,\phi) P_{ant}} \tag{13.49a}$$

or

$$V_m = \sqrt{\frac{R_{ref} G_{sys} L_{sys}^{-1} A_e(\theta,\phi)}{\eta_0 l} \frac{S_{2,m}(180°) A(\theta,\phi)}{k_c r^2}}, \tag{13.49b}$$

where we have included the loss l that we neglected in (13.8b).

Note that we can write the sums for I and Q as

$$I = |V_0(\tau_s, T_s)| \cos[\theta(\tau_s, T_s)] \tag{13.50a}$$

and

$$Q = |V_0(\tau_s, T_s)| \sin[\theta(\tau_s, T_s)], \tag{13.50b}$$

where τ_s indicates the time interval between the radar sending out a pulse and receiving the backscatter from the pulse (i.e., τ_s represents range) and T_s is the time interval between pulses (DZ, 4.3). Since the summed I and Q voltages can be represented as sinusoids, we can estimate the power over a cycle using (13.40).

13.5 An Example

Before discussing the properties of the I and Q voltages, as well as the Doppler moments, we thought that it would be instructive to illustrate all of the previous discussion with a realistic, but computer simulated, "meteorological target". To this end we illustrate the results of a forward computer simulation in which we explicity calculate the I and Q voltages for a specified drop size distribution located within a specified resolution volume. *In situ* measurements by particle counting probes on the University of Wyoming King Air aircraft during a field experiment in the Central Pennsylvania area produced the drop size distribution illustrated in Figure 13.10, which we use as input to the simulation. Multiplying the number densities measured in

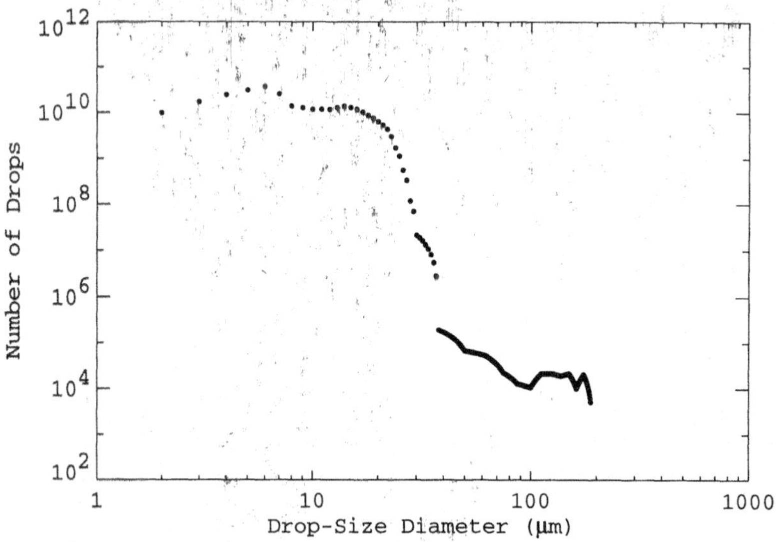

Figure 13.10: *Drop size distribution measured by FSSP, 2D-C and 2D-P probes on board the University of Wyoming King Air aircraft during an experiment at Rock Springs, Pennsylvania, in the Spring of 1995. The measurements were obtained in a stratus deck when the aircraft alititude was approximately 1 km above the surface.*

each bin by the radar resolution volume yields the total number of drops $N_{inbin}^{orig}[i]$ in each bin i that we must take into account in the simulation. Unfortunately, as Figure 13.10 illustrates, the number of drops in the radar resolution volume (3.08×10^{11} in this example) is too large to simulate on modern day computers, so we devised the following scheme to make the computer memory requirements manageable.

The exact expression for I, with a similar expression for Q, is

$$I = \sum_{m=1}^{3.08\times 10^{11}} V_m \cos[2k_c r_m - \omega_{d,m} t + \phi_{s,m} + \phi_t]. \tag{13.51}$$

Instead of using (13.51) in the simulations, we replace it with

$$I = \sum_{i=1}^{N_{bins}} \sum_{j=1}^{N_{inbin[i]}} S_{(i,j)} V_{(i,j)} \cos[2k_c r_{(i,j)} - \omega_{d,(i,j)} t + \phi_{s,i} + \phi_t] \tag{13.52a}$$

where

$$N_{inbin[i]} = \begin{cases} N_{binmax}, & \text{if } N_{binmax} \le N^{orig}_{inbin}[i], \\ N^{orig}_{inbin}[i] & \text{if } N^{orig}_{inbin}[i] \le N_{binmax}, \end{cases} \tag{13.52b}$$

$$S_{(i,j)} = \begin{cases} \sqrt{N^{orig}_{inbin}[i]\ \mathrm{DIV} N_{binmax} + 1} & \text{if } j < (N^{orig}_{inbin}[i] \% N_{binmax}), \\ \sqrt{N^{orig}_{inbin}[i]\ \mathrm{DIV} N_{binmax}} & \text{if } j \ge (N^{orig}_{inbin}[i] \% N_{binmax}), \end{cases} \tag{13.52c}$$

$V_{(i,j)}$, $r_{(i,j)}$ and $\omega_{d,(i,j)}$ are properties attributed to the jth particle in the ith bin, and $\phi_{s,i}$ is the phase shift upon scattering from particle sizes corresponding to the ith bin. The operation DIV implies integer division with no remainder and % stands for the modulo operation. For the results we present here $N_{binmax} = 1000$. The set of equations (13.52) attempts to produce the same fluctuations and magnitudes in the I voltages as the exact expression (13.51). The motivation for this scheme is the observation that nearly identical distributions of voltages over many samples, say 100 000, can be produced by expressions such as

$$I_a = \sum_{i=1}^{5000} \cos(2\pi r_i) \tag{13.53a}$$

and

$$I_b = \sum_{i=1}^{1000} \sqrt{5} \cos(2\pi r_i), \tag{13.53b}$$

where r_i is a pseudorandom number between 0 and 1 and the factor $\sqrt{5}$ in (13.53b) scales the amplitude of I_b to match that of I_a in (13.53a) (Figure 13.11). The square root factor in

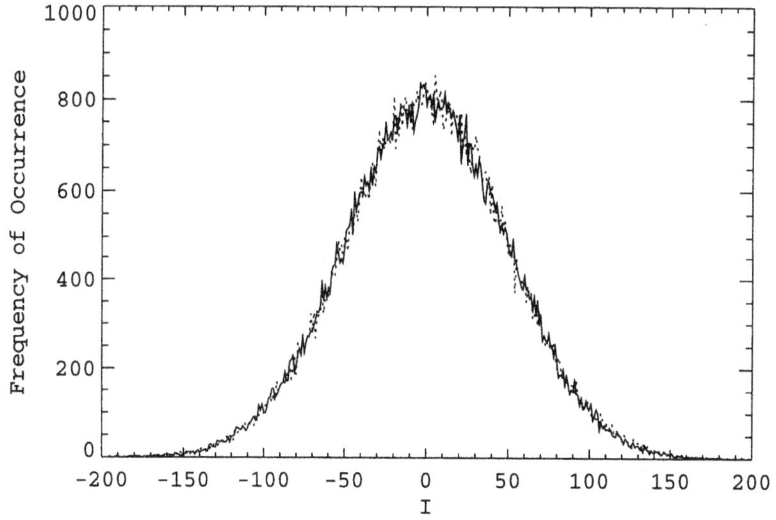

Figure 13.11: *Frequency of occurrence of I_a (solid line) and I_b (dashed line) defined in (13.53a) and (13.53b), respectively. We used 100 000 samples to generate the histograms.*

(13.53b) corresponds to the scale factor $S_{(i,j)}$ in (13.52) and $V_{(i,j)}$ is implicitly set to one both in (13.53a) and (13.53b).

In the simulation we use a pseudorandom number generator to randomly distribute the particles within the model radar resolution volume. Since the distribution in Figure 13.10 contained only liquid water, we assume that the drops are liquid water spheres. Therefore, the theory for scattering from spherical particles, i.e., Mie theory, is an appropriate tool to calculate the scattering amplitude matrix element $S_2(180°)$ of the drops as a function of size. The equations for computing $S_2(180°)$ come from Bohren and Huffman (1983); in fact, Bohren and Huffman (BH, 4.6) provide the equation for $S_2(180°)$ and they also provide code that computes $S_2(180°)$ (BH, Appendix A). Once the location $r_{(i,j)}$ and scattering matrix element $S_2(180°)$ are computed for each particle, we compute the resulting voltage amplitude $V_{(i,j)}$ using (13.49b). We then calculate the I and Q voltages using (13.52) with time $t = 0$. To simulate the particles falling within the resolution volume, we simply must advance the time t in (13.52). Since we must take snapshots of the I and Q voltages at the pulse repitition frequency f_{prf} of the radar, we advance the time in increments of $T_s = l/f_{prf}$. The Doppler shifts in (13.52) are related to the particle fall velocities by

$$\omega_m = -\left(\frac{4\pi v_m}{\lambda_c}\right). \tag{13.54}$$

In the simulation we assume that the particle fall speeds are related to the particle drop diameter by

$$v = \begin{cases} 4D(1 - e^{-12D}), & \text{if } \; D < .75mm; \\ 9.65 - 10.43e^{-0.6D}, & \text{if } \; .75mm < D, \end{cases} \tag{13.55}$$

where D is the diameter of the particle in mm and the velocity v has units of m/s (Rogers et al., 1993). The resulting series of I and Q voltages are illustrated in Figure 13.12.

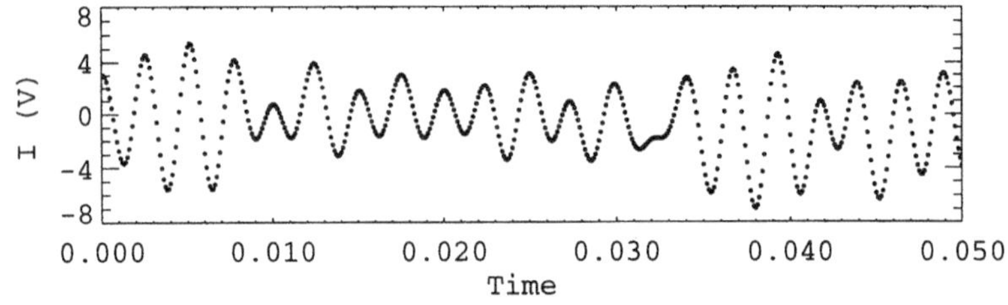

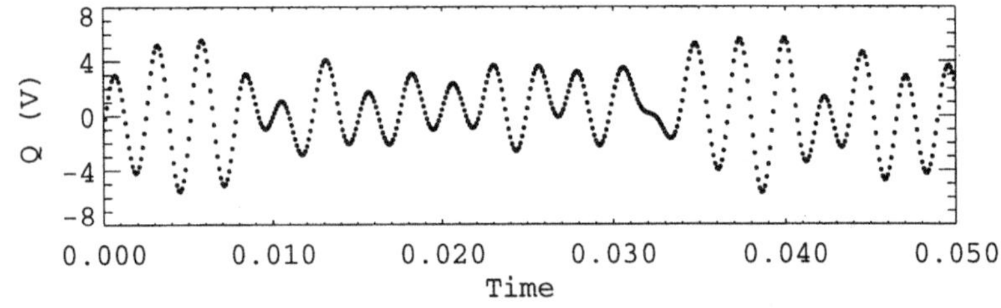

Figure 13.12: *The first 50 I and Q signal voltages generated by the forward simulation of the Penn State 94-GHz cloud radar using the particle drop size distribution in Figure 13.10 as input.*

13.6 Signal Statistics

13.6.1 I and Q Voltage statistics

Now that we have the I and Q voltages, and hence the received power P_{ant}, in hand, we can now obtain estimates of some of their statistical properties. Recall that both the I and Q voltages contain sums of sines and cosines with arguments that contain the term

$$2k_c r_m - \omega_{d,m} t + \phi_{s,m} + \phi_t. \tag{13.56}$$

We cannot emphasize the next two points strongly enough. The first point is that the remainder of $2k_c r_m \% 2\pi$ is a random variable with a uniform distribution over the interval from 0 to 2π. That is, since the exact locations of the scatterers are random across the resolution volume, there is no fixed relation between $2k_c r$ from one particle to the next and the phases $2k_c r_m \% 2\pi$ end up uniformly distributed over 0 to 2π. Therefore, the cosines and sines in the sums for I and Q are random variables distributed between -1 and 1 with a probability density function that is symmetric about 0. The second point is that a variable that is the sum of random variables of equal magnitudes tends to be Gaussian-distributed when a large number of terms are involved. Since the I and Q voltage amplitudes V_m are due to a well-behaved particle size distribution where no one magnitude is significantly larger than all of the other magnitudes, a series of measurements of either I and Q, if long enough, produces values that are Gaussian-distributed (Figure 13.13).

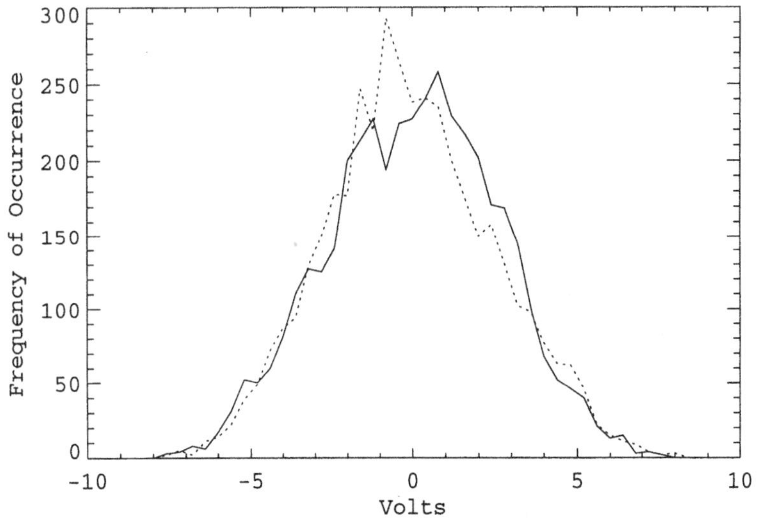

Figure 13.13: *Frequency of occurrence of I (solid line) and Q (dashed line) for the 4096 samples of the forward simulation.*

The I and Q voltages have a number of other interesting properties. At a single, isolated instant in time one cannot predict the exact value of I because one cannot know ahead of time the exact location of each of the particles in the resolution volume. This is not to say that one cannot predict the value of I an instant after a measurement of I; if two samples of

I are obtained in a vanishingly short period of time, their values converge. (In Figure 13.12 neighboring I voltages in time are always similar in magnitude.) Knowing I at one instant does provide some useful information in figuring out the values of I at later not to far in the distant future times. The autocorrelation function defined by

$$R(L) = \frac{\sum_{m=1}^{N_n-L}(I_m - \overline{I})(I_{m+L} - \overline{I})}{\sqrt{\sum_{m=1}^{N_n-L}(I_m - \overline{I})^2}\sqrt{\sum_{m=1}^{N_n-L}(I_{m+L} - \overline{I})}} \tag{13.57}$$

provides a quantitative assessment of how the I voltages are correlated when offset by a lag of L. As Figure 13.14 illustrates, the simulated I voltages are strongly correlated for small lag times and the correlation decreases as the lag time increases from zero. These same statements hold true for Q as well.

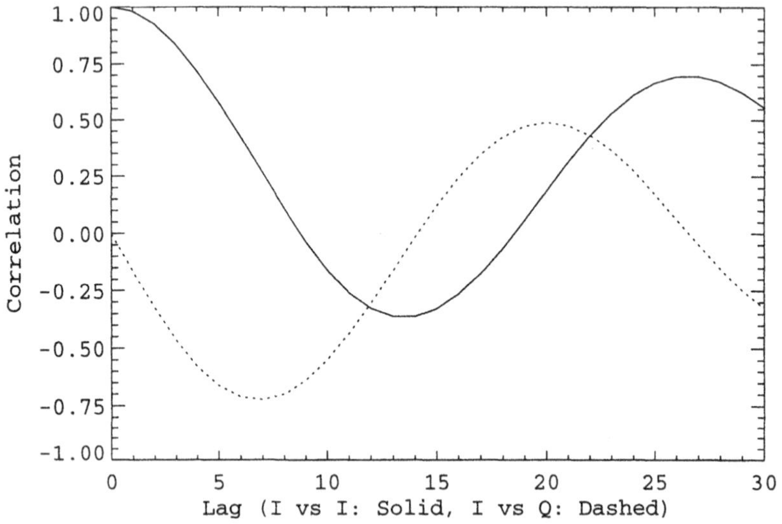

Figure 13.14: *Correlation of I with itself (solid line) and with Q (dashed line) for lag times from 0 to 30 T_s, where $T_s = 0.0001$ is the simulated pulse repetition period for the forward simulation.*

I and Q, however, are not correlated; that is, knowing the value of I at some time does not provide any useful information in figuring out the value of Q at the same time and vice versa. In this case the correlation coefficient defined by

$$R(L) = \frac{\sum_{m=1}^{N_n-L}(I_m - \overline{I})(Q_{m+L} - \overline{Q})}{\sqrt{\sum_{m=1}^{N_n-L}(I_m - \overline{I})^2}\sqrt{\sum_{m=1}^{N_n-L}(Q_{m+L} - \overline{Q})}} \tag{13.58}$$

with the lag $L = 0$ should be close to 0. For the time series in Figure 13.12, the correlation coefficient for lag $L = 0$ is indeed close to 0. Since the same random processes are controlling the evolution of I and Q, knowing I at one instant does provide some useful information in figuring out the values of Q at later not too far in the distant future times and vice versa. Consequently, the magnitude of the correlation between I and Q as defined by (13.58) increases as the lag L increases from zero (Figure 13.14).

13.6.2 Signal power statistics

Since I and Q are Gaussian-distributed independent processes, we have (DZ, 4.4)

$$p(I) = \frac{1}{\sqrt{2\pi}\sigma} e^{-I^2/2\sigma^2}, \tag{13.59a}$$

$$p(Q) = \frac{1}{\sqrt{2\pi}\sigma} e^{-Q^2/2\sigma^2}, \tag{13.59b}$$

and

$$p(I \cap Q) = p(I)p(Q) = \frac{1}{2\pi\sigma^2} e^{(-I^2-Q^2)/2\sigma^2} \tag{13.59c}$$

For the present discussion we simply note that the power P_{ant} goes as $I^2 + Q^2$. Therefore, it should come as no surprize that we can write (DZ, 4.4)

$$P_{ant} = \frac{1}{2\sigma^2} e^{-P_{ant}/2\sigma^2} \tag{13.60}$$

as the power probability density function for a large number of power returns from rain or perhaps a cloud. This distribution is an exponential distribution with a peak probability of zero and a mean value of $P_0 = 2\sigma^2$.

Since we already know that a single power measurement is, for all practical purposes, useless, knowing that individual power measurements are exponentially distributed is not really that helpful. What we really want to know is the probability density function for an average of N power samples P_{ant}, or P_N. After some mathematical manipulations, that include application of the residue theorem from the theory of complex numbers, the problem is solved. The resulting probability density function is (Marshall and Hitschfeld, 1953)

$$p(P_N) = \frac{N^N P_N^{N-1}}{(P_0)^N (N-1)} e^{-NP_N/P_0}, \tag{13.61}$$

which we graph in Figure 13.15 for a range of values of N and a given, fixed value of σ.

13.6.3 Mixer noise and the minimum detectable received signal

Now that we have characterized the scattered signal that comes back to the radar, we must now sort it out from the noise introduced by the detection process. Radars, unlike lidars, have one principle source of noise, namely, the mixer. The mixer noise process is governed by Johnson shot noise. The noise voltage is assumed to be a Gaussian-distributed random process with a mean voltage of zero and a standard deviation of

$$\sigma_v = \sqrt{P_n}, \tag{13.62}$$

where P_n is the average mixer noise power. ($P_n = 1.12 \times 10^{-9}$ mW for the Penn State radar receiver.) That is, the probability of a noise voltage is

$$p(V_n) = \frac{1}{\sqrt{2\pi}\sigma_v} e^{-V_n^2/2\sigma_v^2}. \tag{13.63}$$

Comparing (13.59) and (13.63), we find that both the mixer noise and the I and Q voltages are Gaussian-distributed. An important difference between these time series, however, is that the mixer noise process is not correlated in time. No matter how fast noise voltage samples

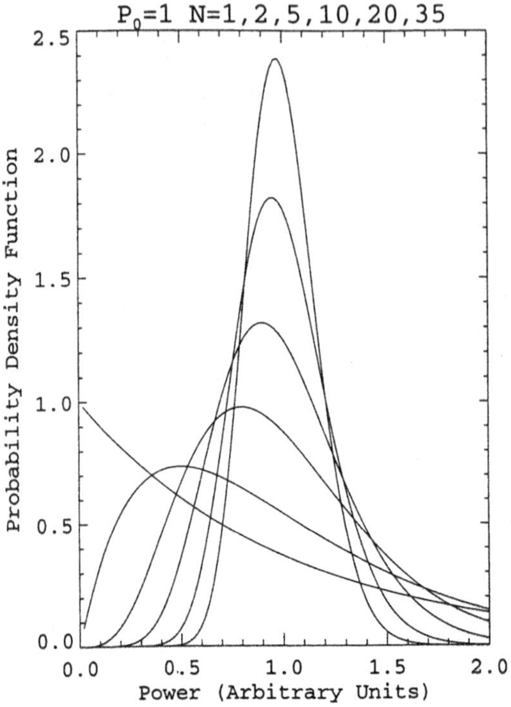

Figure 13.15: *Probability density function defined by (13.61) using a mean power of $P_0 = 1$ and averaging 1, 2, 5, 10, 20 and 35 samples.*

are collected from the mixer, there is no guarantee that the resulting voltages will be close in magnitude. For a time series of noise power samples the autocorrelation function is essentially zero for all lag times L. Another difference is the probability density function that describes the distribution of averaged noise voltages. The probability density function of an average of N noise power samples is essentially Gaussian-distributed with a standard deviation of $\sigma_v/\sqrt{N}$, as opposed to the more complicated (13.61) which describes the corresponding probability density function for averages of N signal power samples.

The goal now becomes finding the signal power P_{ant} in the noise power P_n. As an illustration, we have produced sets of simulated voltages that are representative of atmospheric return signals (i.e., exponentially distributed) and the mixer noise process and the question becomes can the former be found in the later when they are summed together. The series of illustrations that follow in Figure 13.16 address just this issue. The voltages illustrated in Figure 13.16 represent an extreme situation that allows us to demonstrate the power of averaging samples in order to a recover a weak atmospheric return signal.

Using a pseudorandom number generator, we generated 100 voltage samples with the correct statistical properties of noise. We squared each voltage sample to obtain the noise power and we then plotted the resulting points as the first 100 samples in Figure 13.16A1. We then generated another 100 noise power samples as before. This time, however, we also generated 100 atmospheric return, or signal, power samples with the correct statistical properties, again using

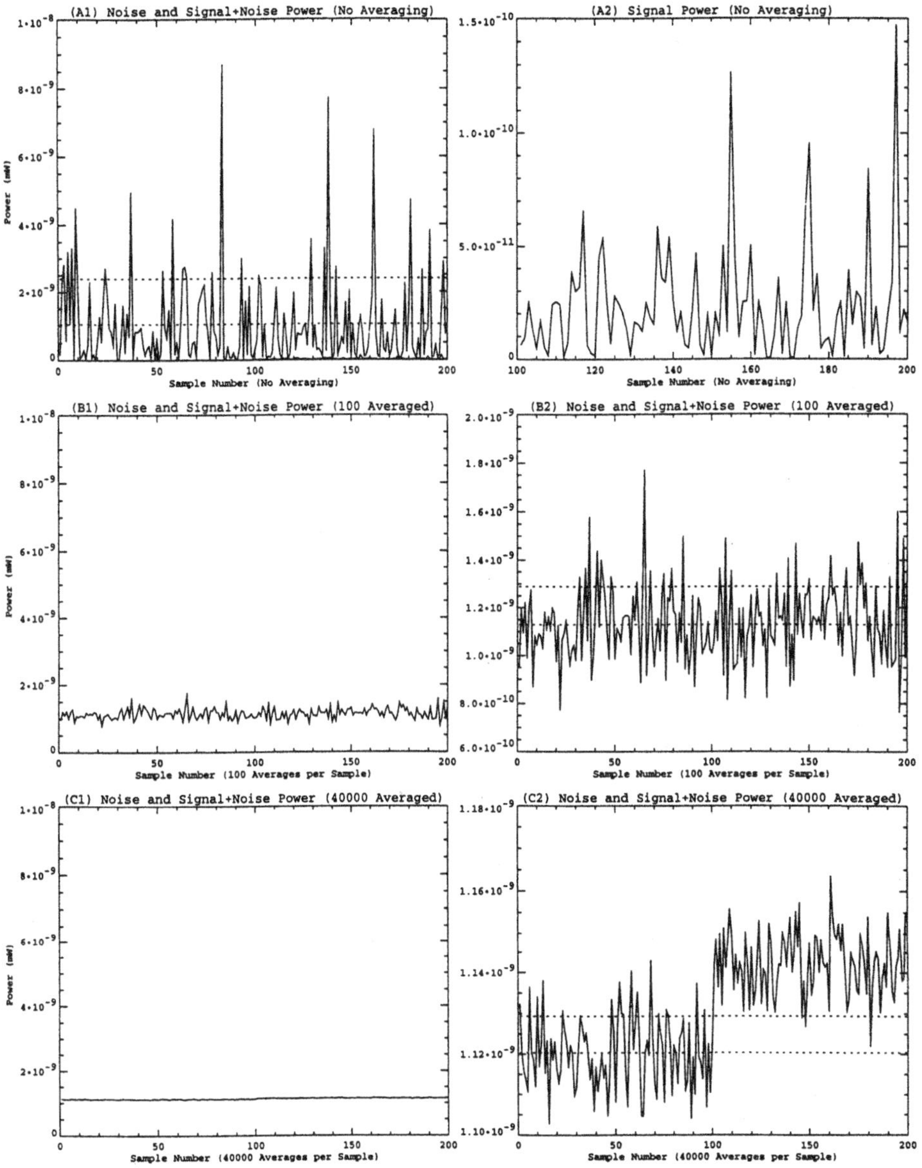

Figure 13.16: *An example of the average power at the output of a radar receiver when 1, 100 and 40 000 instantaneous noise and signal plus noise power samples are averaged. In (A1), (B2) and (C1), the first 100 averaged power samples contain only noise power, whereas the last 100 averaged power samples contain both noise and signal power contributes to the signal. (A2) is a magnification of the signal power samples that we added to the noise power in the last 100 samples of (A1). We plotted these same samples on (A1), but their values were so much lower than the noise power that they are hardly noticeable in (A1). (B2) and (C2) are identical to (B1) and (C1), respectively, except for the scale of the y-axis. The lower dashed line in (A1), (B2) and (C2) is at the level of the mean value of the first 100 points, i.e, at the level of the mean noise, for each figure. The upper dashed line equals the mean of the first 100 samples plus one standard deviation of these same samples.*

a pseudorandom generator. We added one signal power sample to each noise power sample and plotted the results as samples 101 to 200 in Figure 13.16A1. Since the signal power levels are small compared to the noise power levels, i.e., look for the second solid line in Figure 13.16A1, it looks as though nothing has changed from the first 100 samples. The top dashed line in the figure has a power level of 2.4×10^{-9} mW, which we obtained by adding the standard deviation of the first 100 noise power samples to the mean obtained from these same samples. We use this power level to differentiate signal plus noise from noise: everything above the dashed line is considered as signal plus noise, whereas everything below the dashed line is classified as pure noise. Applying this criterion to the series in Figure 13.16A1 leads to poor results since many of the first 100 power samples are classified as signal plus noise and almost all of the last 100 power samples are identified as noise. To detect the power associated with the signal we must reduce the standard deviation of both the noise and signal plus noise samples; to accomplish this we must average samples.

To this end we generated 10 000 noise power samples as before. We averaged the noise samples in blocks of 100 before plotting them in Figure 13.16B1. We then generated 10 000 noise power samples, 10 000 signal power samples, added a distinct signal power sample to each noise sample and averaged the summed samples in blocks of 100 to produce samples 101 to 200 in Figure 13.16B1. In Figure 13.16B2 we magnify the y-axis to show that we still get poor separation of the signal plus noise from the noise if we use the same criterion as in Figure 13.16A1.

Finally, we averaged blocks of 40 000 noise samples and signal plus noise samples to produce Figure 13.16C1. Notice that the standard deviations of both the noise and signal plus noise samples are much reduced compared either to Figure 13.16A1 or Figure 13.16B1. More importantly, the magnified plot in Figure 13.16C2 demonstrates that the criterion for separating noise from signal plus noise performs much better on this data; the number of signal plus noise samples that are identified as noise are greatly reduced from before. Therefore, by averaging more samples the average of the noise samples converges to the actual mean radar noise power and the standard deviation goes to zero. Furthermore, the signal plus noise samples converge on the mean noise power plus the mean sigal power also with a vanishing standard deviation. If enough averaging is performed on the samples, the standard deviations become sufficiently small that the atmospheric return process is unmasked. We use this standard technique to separate signal plus noise from pure noise when processing the Penn State radar data and we use a simple filter on the results in an attempt to remove the type of misclassifications that are illustrated in Figure 13.16C2 (Clothiaux et al., 1995). Once we have obtained an estimate of the signal power by subtracting the noise power from the signal plus noise power, we can use the radar equation (13.26) to solve for the reflectivity $\eta(\mathbf{r_0})$, which is one of the radar observables.

13.7 The Doppler Moments

From the I and Q voltages we obtained an estimate of the quantity $\eta(\mathbf{r_0})$ which provides information about cloud drop locations. $\eta(\mathbf{r_0})$ also provides information about the particle size distribution, but only through the integral of the backscatter cross section across all particles in the radar resolution volume. We are now at the point where we take the I and Q voltages and manipulate them in order to retrieve information about the motion of the scatterers in the resolution volume. As we illustrate, the power density spectra enables us to better estimate the particles sizes in the resolution volume.

The two techniques that we use are the fast fourier transform (FFT) and the pulse pair autocovariance algorithm. We apply these algorithms to the simulated I and Q time series in Section 13.5 and we then interpret the results in the context of the particle size distribution (Figure

13.10) and particle fall velocities (Equation 13.55) that we used to generate them in the first place. This discussion is intended to be illustrative and by no means describes which technique works best and under what conditions. For a discussion of these issues, Doviak and Zrnic (DZ, 6.4–6.5) is an excellent starting point.

13.7.1 Power density spectra

Extraction of the power density spectra from the I and Q voltages time series is relatively straightforward (DZ, 5.1). Both the I and Q time series from Section 13.5 have exactly 4096, or 2^{12}, samples. Since the number of samples is a power of 2, we can use any one of the standard complex FFT algorithms to generate the power density spectra from the time series (e.g., Press et al., 1988, Section 12.2). Using the I and Q time series in Figure 13.12 as the real and complex array inputs, respectively, to the FFT leads to the power density spectrum S illustrated in Figure 13.17, where we have transformed the abscissa from frequency to velocity using (13.54). The maximum and minimum velocities in the spectrum are set by the pulse repitition frequency $f_{prf} = 10000\ \mathrm{s}^{-1}$ of the radar. Substituting $\pm f_{prf}$ into (13.54) leads to

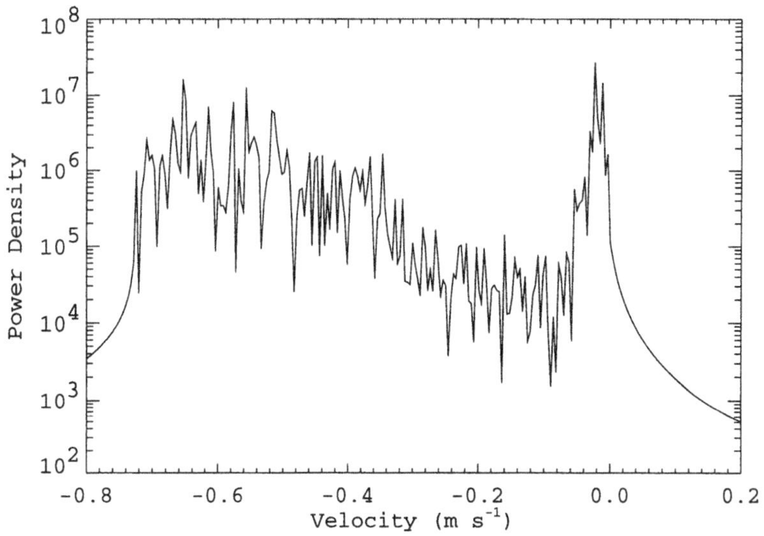

Figure 13.17: *Power density spectrum produced by applying a fast fourier transform, or FFT, to the 4096 I and Q voltages generated in the forward simulation. Negative velocities are downward.*

the range of velocities over which the spectrum is valid. The velocity interval between power density spectral points is set by the number of samples input to the FFT. Since the Penn State radar performs a FFT on 512 samples at a time, we broke the I and Q time series into 8 segments of 512, or 2^9, points each. We then performed a FFT on each 512 segment and averaged the resulting 8 power density spectra to produce Figure 13.18.

The power density $S(m)$ at each velocity $v(m)$ in Figures 13.17 and 13.18 is directly related to the total backscattering cross section of all the particles moving with a radial velocity $v(m)$ with respect to the radar. To further illustrate this point, in Figure 13.19 we plot

$$\sigma(\mathbf{D})N(\mathbf{D},\mathbf{r})d\mathbf{D}V_{res}, \tag{13.64}$$

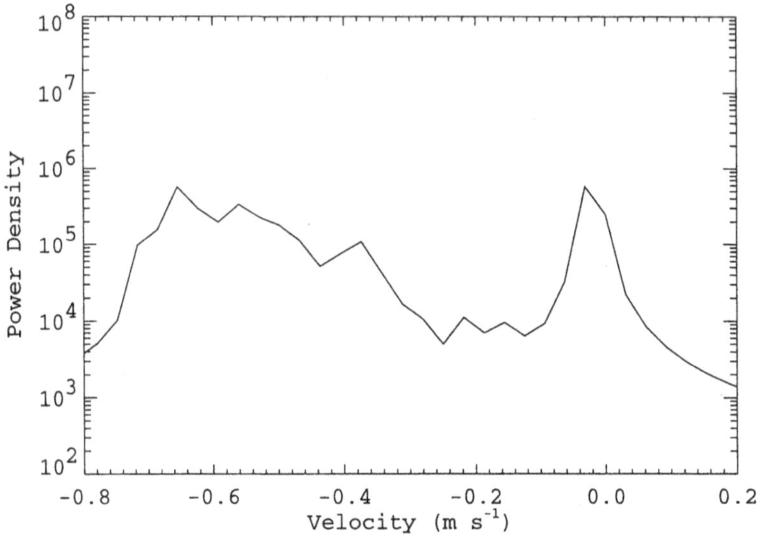

Figure 13.18: *Power density spectrum produced by applying a fast fourier transform, or FFT, to the eight 512 sample blocks of the forward simulation and then averaging the resulting spectra. Negative velocities are downward.*

where V_{res}, is the resolution volume, for the drop size distribution illustrated in Figure 13.10 as a function of the velocity that we obtained from the empirical relation in (13.55). Note that the power density spectra in Figure 13.18 is a faithful representation of Figure 13.19. In fact, for this highly idealized simulation, given an accurate radar calibration and the radar equation in (13.26), we can use the power density at each velocity $v(m)$ to directly infer the number density of the drops in the resolution volume whose sizes correspond to the velocity $v(m)$. This kind of argument is often used in current research aimed at retrieving particle size distributions from Doppler radar.

In many cases storing the power density spectra generated by the radar signal processors is beyond the capability of the computers that are running the radar. In these instances the power density spectra are used to calculate estimates of the first and second moments of the spectra, namely, the mean power-weighted velocity and the spectral width, i.e., the standard deviation of the power-weighted velocities about the mean. The mean power-weighted radial velocity within the resolution volume is defined by (DZ, 5.2)

$$\overline{v} = \sum_{m=-N_n/2}^{N_n/2} v(m) S_n(m) \tag{13.65}$$

and the spectral width σ_v is defined by (DZ, 5.2)

$$\sigma_v^2 = \sum_{m=-N_n/2}^{N_n/2} [v(m) - \overline{v}]^2 S_n(m), \tag{13.66}$$

where $S_n(m)$ is the normalized power density spectra at velocity $v(m)$ (DZ, 5.2), i.e.,

$$S_n(m) = \frac{S(m)}{\sum_{m=-N_n/2}^{N_n/2} S(m)}, \tag{13.67}$$

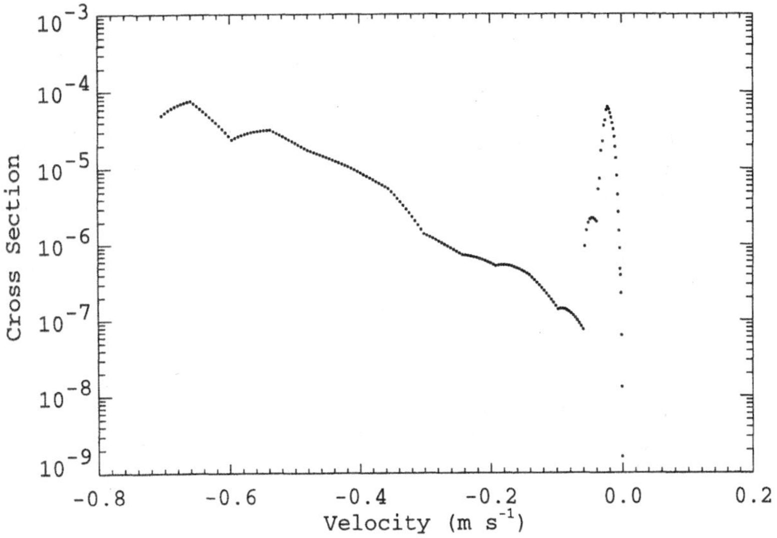

Figure 13.19: *The total backscattering cross section for all of the particles in the radar resolution volume of the forward simulation as a function of particle size plotted against the particle size fall velocity. Negative velocities are downward.*

and the velocity $v(m)$ is given by (DZ, 3.4.3 and 5.1)

$$v(m) = \left(\frac{m}{N_n/2}\right)\left(\frac{\lambda}{4T_s}\right). \tag{13.68}$$

Applying these equations to the spectrum in Figure 13.17 produces a velocity of -0.419 ms^{-1} and a spectral width of 0.261 ms^{-1}. These two quantities, together with the received power, or zero moment, are the radar observables generally reported at most Doppler radar weather sites.

13.7.2 Pulse pair

Performing a FFT on the I and Q time series in an operational setting requires a fast digital signal processing chip with sufficient memory to store and process the data, as well as a fast analog-to-digital convertor that can cover the dynamic range of the radar without introducing any nonlinearities. Since this kind of technology has become available only recently, early investigators in the field of weather-radar research developed methods of producing estimates of the three Doppler moments directly from the I and Q time series. One popular technique that is still in use today is the pulse pair autocovariance technique. A description of the pulse pair algorithm is provided by Doviak and Zrnic (DZ, 6.4-6.5). We simply apply their results to the I and Q time series illustrated in Figure 13.12 and compare the resulting moments with values derived from the power density spectra. If the complex voltage is

$$V(m) = I(m) + iQ(m) \quad (1 \leq m \leq N_n, N_n = 4096), \tag{13.69}$$

then the autocovariance of V at a lag corresponding to the interpulse period $T_s = 1/f_{prf}$ is (DZ, 6.4)

$$\mathbf{R} = \frac{1}{N_n - 1} \sum_{m=1}^{N_s - 1} V^*(m)V(m+1). \tag{13.70}$$

R is a complex number with a real part R_r given by

$$R_r = \frac{1}{N_n - 1} \sum_{m=1}^{N_s - 1} [I(m)I(m+1) + Q(m)Q(m+1)] \tag{13.71a}$$

and a complex part given by

$$R_c = \frac{1}{N_n - 1} \sum_{m=1}^{N_s - 1} [I(m)Q(m+1) - Q(m)I(m+1)]. \tag{13.71b}$$

The amplitude of **R** is

$$R = \sqrt{R_r^2 + R_c^2} \tag{13.72a}$$

and the phase is

$$\phi_R = \tan^{-1}(\frac{R_c}{R_r}). \tag{13.72b}$$

The pulse pair estimate of the mean power-weighted radial velocity of the particles in the resolution volume is (DZ, 6.4)

$$v = -\left(\frac{\lambda}{4\pi T_s}\right) \phi_R, \tag{13.73}$$

where ϕ_R is given in radians, and an estimate of the spectral width is (DZ, 6.5)

$$\sigma_v = \left(\frac{\lambda}{2\pi T_s \sqrt{2}}\right) \left|\ln(\frac{S}{R})\right|^{1/2} \quad (S > R), \tag{13.74a}$$

where (DZ, 6.5)

$$S = \frac{1}{N_n} \sum_{m=1}^{N_n} |V(m)|^2 - P_n \tag{13.74b}$$

and P_n is an estimate of the mean radar noise power. Applying (13.73) and (13.74) to the simulated I and Q time series produces an estimate of -0.423 ms^{-1} for the mean radial velocity and 0.255 ms^{-1} for the spectral width.

13.8 Cloud Properties from Ground-Based Remote Sensing

13.8.1 A climate model perspective

The preceding sections have described radar wave propagation and signal processing and shown how the measured power return can be related to the properties of the scatterers in the sample volume. In our previous articles (Clothiaux et al., 1995; Syrett et al., 1995) we have shown how millimeter-wave radar can be used to provide a unique and detailed look at clouds, including both their three-dimensional structure and microphysical structure. Quite recently, we and others interested in millimeter-wave radar technology have begun investigating how to translate the radar observations into quantitative data on cloud properties. To some extent, this can be done using the radar alone. However, much more powerful results can be achieved by using the radar observations in combination with observations from other instruments. In this section, we provide an overview of the various techniques being employed. Before discussing these in detail, however, it is useful to consider the data needs from the perspective of the climate community.

As we have demonstrated above, millimeter-wave radar can provide a detailed and unique look at cloud structure, including some aspects of cloud microphysics. It is important, however, that we take the next step of translating this information into quantitative data on cloud properties. Before discussing this process in detail, it is useful to consider from the perspective of the climate community what the data needs are.

One of the major uncertainties in current climate models is the treatment of clouds as embodied in the model parameterizations, and the impact of those clouds on the model radiation fields. Cloud parameterizations vary in their input requirements and output variables but, as a general rule, the output must specify cloud location (height and thickness) and amount of condensed phase (liquid or ice). The latter is then converted in some fashion to an optical depth that can be used as an input to the model radiative codes. More detailed parameterizations also produce information about cloud particle size and the vertical distribution of the condensed phase as outputs. Therefore, for a given set of parameterization inputs, verification of a parameterization relies on validating the following basic predicted quantities: number of cloud layers, base and top height of each layer, and either the amount of condensed phase or the particle drop size distributions within each layer. Note that if the particle drop size distribution within a layer is known, then the amount of condensed phase can be inferred. Because climate models at best simulate the statistical distribution of these quantities, it is important that we also look at our ability to provide a temporal distribution of these observed properties.

13.8.2 Models of cloud drop size distributions

Before describing several techniques proposed in the literature for retrieving water vapor path WVP, liquid water content LWC and path LWP and ice water content IWC and path IWP, as well as the cloud particle number density distribution, it is useful to set forth the notation that is commonly employed in descriptions of the retrievals. For cloud particles let $n(D)$ represent the number of particles per unit volume with a size between D and $D+dD$, where dD is a small increment in the pertinent dimension of the particles. For spheres D represents the particle diameter, whereas for nonspherical particles D may perhaps represent the maximum dimension or the diameter of a spherical particle that has the same volume as the particle in question. The total number of particles per unit volume becomes

$$N_t = \int_{D_{min}}^{D_{max}} n(D)dD, \tag{13.75}$$

where we have adopted the notation of Flatau et al. (1989). For analytic convenience D_{min} and D_{max} are generally set to 0 and ∞, respectively; the error this appoximation introduces is generally small.

The particle size density function $n(D)$ is usually modelled either by a lognormal distribution or the general modified gamma distribution. These two distributions are useful because they fit the observed distributions of cloud particles and the moments of the distributions can be defined in terms of known functions. The lognormal distribution is defined by

$$n_{log}(D) = \frac{N_t}{\sqrt{2\pi}\sigma D} \exp\left[-\left(\frac{\ln(D/D_n)}{\sqrt{2}\sigma}\right)^2\right], \tag{13.76a}$$

where σ and D_n are parameters. The moments $I_{log}(p)$ of n_{log} are defined by

$$I_{log}(p) = \frac{1}{N_t}\int_0^\infty D^p n_{log}(D)dD = D_n^p F_{log}(p), \tag{13.76b}$$

where

$$F_{log}(p) = \exp\left(\frac{\sigma^2 p^2}{2}\right). \tag{13.76c}$$

The general modified gamma distribution is defined by

$$n_{mg}(D) = \frac{N_t c}{\Gamma(\nu)}\left(\frac{D}{D_n}\right)^{c\nu-1}\frac{1}{D_n}\exp\left[-\left(\frac{D}{D_n}\right)^c\right], \tag{13.77a}$$

where $\Gamma(\nu)$ is the complete gamma function and c, ν and D_n are parameters. The moments $I_{mg}(p)$ of the distribution are

$$I_{mg}(p) = \frac{1}{N_t}\int_0^\infty D^p n_{mg}(D)dD = D_n^p F_{mg}(p), \tag{13.77b}$$

where

$$F_{mg}(p) = \frac{\Gamma(\nu + p/c)}{\Gamma(\nu)}. \tag{13.77c}$$

For a given choice of model, the goal of the retrievals is to determine the model free parameters, together with N_t, as a function of height. The liquid water content LWC and ice water content IWC follow from

$$Content(z) = \int_0^\infty \rho V(D)n(D)dD, \tag{13.78a}$$

where ρ is the density of the particles under consideration and $V(D)$ is the volume of the particles with dimension D. The liquid water path LWP and ice water path IWP follow from integrating the LWC and IWC in a vertical column:

$$Path = \int_0^\infty Content(z)dz, \tag{13.78b}$$

13.8.3 Cloud boundaries from radar reflectivity

Although detection of cloud location and thickness, hereafter simply referred to as cloud boundaries, from radar seems relatively straightforward, construction of an accurate, reliable algorithm to process large quantities of data is challenging. Since a fairly detailed description of our algorithm is given by Clothiaux et al. (1995), only the principal features are outlined here.

The starting point for any cloud boundary algorithm is a definition of the minimum detectable signal for the radar. This quantity is related to the physical characteristics of the radar, the number of pulses that are being averaged, and the noise power of the signal processing chain. Our current algorithm begins by defining all signals more than one standard deviation above the radar receiver noise as representing a possible significant cloud detection. This threshold includes, however, a significant number of noise events (Figure 13.20a). We then produce a cloud mask (i.e., a binary two-dimensional image of cloud occurrence in time and height) by smoothing this basic image using a box-car filter in time and space. The presence or absence of a possible significant detection at the 24 nearest-neighbors surrounding a particular pixel are used to estimate whether or not that particular pixel is a significant cloud detection. As Figure 13.20b demonstrates, our processing removes the noise and produces a contiguous cloud mask which can then be used to identify cloud boundaries in an automated fashion.

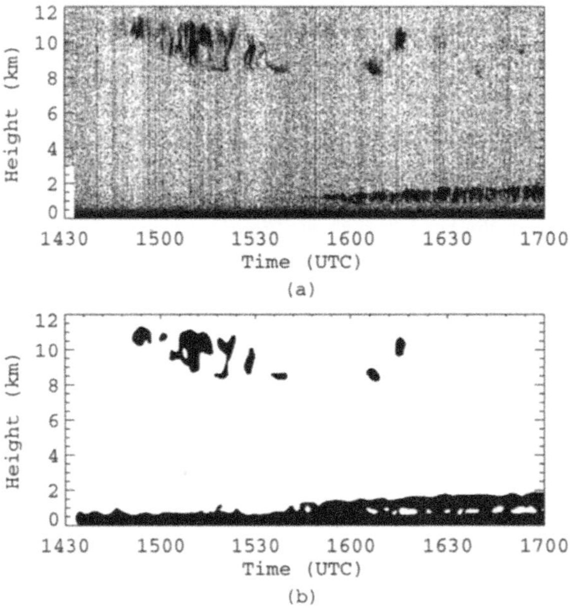

Figure 13.20: *(a) Receiver power output returns greater than one standard deviation above the noise are indicated by the black dots and all other power returns are indicated by white dots. (b) The binary cloud mask image produced by applying a two-dimensional boxcar filter to the image in (a).*

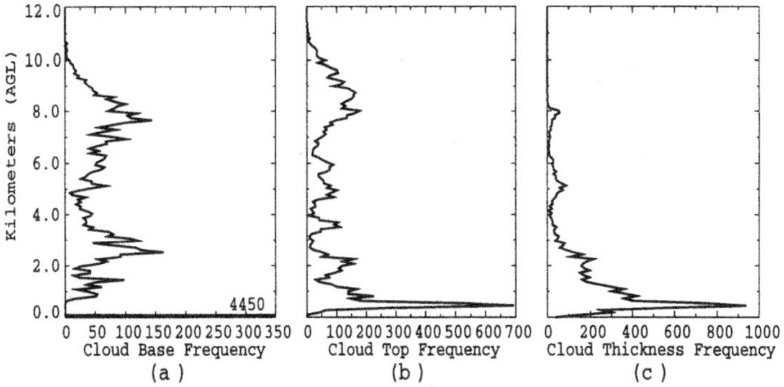

Figure 13.21: *Frequency of occurrence of (a) cloud base height, (b) cloud top height and (c) cloud thickness as recorded by the Penn State 94-GHz cloud radar during the FIRE II experiment in Coffeyville, Kansas.*

An example of cloud statistics generated in this fashion is shown in Figure 13.21. These data were acquired by the Penn State 94-GHz cloud radar during the First ISCCP Regional Experiment, Phase II, (FIRE II) held in Coffeyville, KS, in late 1991 (Uttal et al., 1995). The cloud base histogram (Figure 13.21a) shows distinct maxima in the upper troposphere around 8 km and in the lower troposphere around 3 km, with a pronounced minima in between. The former are, of course, cirrus clouds, while the latter is a combination of boundary layer and deep convective clouds. Due to the extreme sensitivity of millimeter-wave radar to small particles, they often obtain significant return from boundary layer aerosol, particularly under humid conditions. These radars are even more sensitive to the larger precipitation drops, making it difficult to devise an algorithm that can automatically differentiate between the visual cloud base and the constituents below cloud base. For radiative and microphysical purposes, we generally want to know the visual cloud base, i.e., the base of the cloud droplet distribution, rather than the precipitation base. In our ongoing research we are exploring the use of concurrent ceilometer or lidar cloud base data to identify periods of precipitation, thereby allowing us to correct the radar cloud base statistics so that they contain only information about clouds and not the precipitation emanating from them. Another possibility for inferring the cloud base height in the presence of drizzle is bimodality in the radar Doppler spectra, where one mode corresponds to the cloud particle returns and the other to the drizzle drop signals.

Cloud top heights are generally much easier to infer from radar reflectivity time-height cross sections. The cloud top distribution from FIRE II (Figure 13.21b) shows a distinct maximum around 9 km with a secondary maxima below about 6 km. The upper maximum is largely cirrus but includes a few occurrences of deep convective clouds.

The cloud base and top data can be differenced to produce frequency distributions of cloud thickness (Figure 13.21c). These distributions are dominated by cloud thicknesses on the order of 0.5 to 3 km. While we expect this result to be robust, there is some exaggeration of the effect in these histograms because the radar was not operated during periods of the most severe convection. There are cloud thicknesses that span the complete troposphere (9 to 12 km), indicating that some deep convective clouds were sampled.

A unique aspect of cloud radar is its ability to detect multiple cloud layers. Using the cloud mask as a starting point, we can count the number of cloud boundary bottoms and tops profile-by-profile to estimate the frequency of occurrence of multiple cloud layer systems. During the FIRE II experiment, clouds occurred approximately 60% of the time (Figure 13.22). (For these statistics, we did not include layers with cloud tops below 800 m in order to avoid confusion with aerosol layers, insects, etc.) When clouds did occur over the radar, they were multilayered 33% of the time. Using data from two radars, as well as a lidar, Uttal et al. (1995) concluded that during the FIRE II observational period when clouds occurred they had a better than 50% chance of being multilayered. We conclude that multiple layered cloud systems are not rare and to accurately estimate their vertical distribution cloud radar is essential.

13.8.4 Cloud properties in a vertical atmospheric column

Cloud Liquid Water Path from Passive Radiometry

Given that cloud boundaries can be determined, the most important cloud bulk property that we would like to obtain from our observations is the column amount of condensed water, either the liquid water path LWP or ice water path IWP. Measurement of the condensed liquid water column amount is perhaps most easily inferred from a determination of the atmospheric optical depth at some wavelength, where the optical depth is defined as

$$\tau_\lambda = \int_0^\infty \sum_{i=1}^{N_a} k_{\lambda,i}(z)\rho_i(z)dz. \tag{13.79}$$

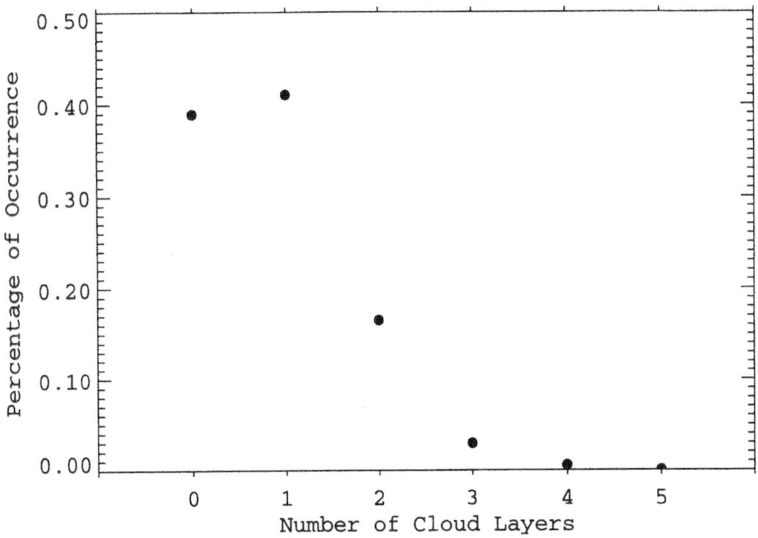

Figure 13.22: *Percentage of occurrence of multiple cloud layers as recorded by the Penn state 94-GHz cloud radar during the FIRE II experiment in Coffeyville, Kansas.*

In the definition of τ_λ, N_a is the number of atmospheric constituents radiatively active at the wavelength λ , $k_{\lambda,i}(z)$ is the volume-averaged, wavelength-dependent mass extinction coefficient of the ith species and $\rho_i(z)$ is the density of the ith species. Defining $\overline{k_{\lambda,i}}$ as

$$\overline{k_{\lambda,i}} = \frac{\int_0^\infty k_{\lambda,i}\rho_i(z)dz}{\int_0^\infty \rho_i(z)dz}, \tag{13.80}$$

we can write the equation for τ_λ as

$$\tau_\lambda = \sum_{i=1}^{N_a} \left[\overline{k_{\lambda,i}} \int_0^\infty \rho_i(z)dz \right]. \tag{13.81}$$

Now the integral

$$\int_0^\infty \rho_i(z)dz \tag{13.82}$$

is just the column amount of the ith species and is the quantity that we want to infer from measurements of τ_λ. For example, near 24 and 32 GHz, water vapor, cloud liquid water and oxygen are the dominant radiatively active atmospheric constituents. Since the column amount of oxygen is relatively constant in time and space, we can assume its atmospheric optical depth to be the constant $\tau_{\lambda,oxy}$. Therefore, (13.81) becomes

$$\tau_\lambda = \tau_{\lambda,oxy} + \overline{k_{\lambda,v}} \int_0^\infty \rho_v(z)dz + \overline{k_{\lambda,l}} \int_0^\infty \rho_l(z)dz, \tag{13.83}$$

where $\rho_v(z)$ and $\rho_l(z)$ are the vertical density profiles of water vapor and cloud liquid water, respectively. The two integrals on the right hand side of the above equation are simply the water vapor path WVP and the liquid water path LWP, respectively. Since we have measurements at two frequencies, we have two optical depth equations in two unknowns, assuming that $\overline{k_{\lambda,v}}$ and $\overline{k_{\lambda,l}}$ are known, which can be inverted to yield WVP and LWP. Importantly, the inversion is well-behaved because the total optical depths at the two frequencies have different dependencies on the water vapor and cloud liquid water column amounts. Furthermore, the retrieved WVP and LWP are not seriously affected by the presence of ice particles in the atmosphere since ice water is not radiatively active at these two wavelengths.

Cloud Liquid and Ice Water Content from Radar Reflectivity
In an alternative approach using only radar reflectivities Liao and Sassen (1994) theoretically compute the radar reflectivity Z and liquid water content LWC for liquid water drop distributions generated in an adiabatic cloud model. Performing a regression of the liquid water content versus the radar reflectivity for distributions generated with a number of different cloud model parameters, Liao and Sassen (1994) arrived at the relationship

$$LWC = 0.49(N_t Z)^{0.55}, \qquad (13.84a)$$

where N_t is the total number of drops per unit volume and depends upon the initial number of cloud condensation nuclei in the cloud model simulation. Liao and Sassen (1994) performed a similar regression analysis of the ice water content versus radar reflectivity using a number of different *in situ* observations by aircraft with the result that

$$IWC = 7.49 Z_e^{0.78}. \qquad (13.84b)$$

Liao and Sassen (1994) found their empirically derived relationships to be consistent with the results of a number of earlier investigations that adopted a similar approach.

Cloud Drop Size Distributions from Passive Radiometry and Radar Reflectivity
Matrosov et al. (1992) combine a measurement of brightness temperature between 10 and 12 μm and particle radar reflectivity at 9.3 GHz to infer the average particle concentration N_t and the average scaling diameter D_n across the depth of an ice cloud layer. (Actually, Matrosov et al. (1992) retrieve a diameter D_m that splits the cirrus cloud particle distribution into two equal volumes of cloud drops, as opposed to the scaling diameter D_n of a modified gamma distribution.) Their strategy consists of using the measured brightness temperature, or radiance, to infer the cloud optical depth. From the cloud optical depth they are in turn able to solve for the product $N_t D_n^2$. The reflectivity obtained from the radar then allows them to solve for N_t and D_n individually.

Consider the case of a vertically and horizontally homogeneous cirrus cloud layer with no underlying water clouds. Using a two-stream radiative transfer model in the cloud layer and ignoring any sources of radiation above the cloud layer, Matrosov et al. (1992) derive a simple parameterization for the nadir directed radiance $I_{cb}^{\downarrow}$ at the base of the cloud layer:

$$I_{cb}^{\downarrow} = B(T_{cb})\left[1 - \exp(-\alpha_0 \tau^0)\right], \qquad (13.85)$$

where $B(T_{cb})$ is the blackbody radiance for a physical temperature T_{cb} at the base of the cloud layer, τ^0 is the optical depth of the cloud layer and α_0 (= 0.7) is a free parameter used to achieve the best prediction of $I_{cb}^{\downarrow}$ from $B(T_{cb})$ and τ^0 for a number of forward two-stream model calculations using a range of values both for the cloud optical thickness and the particle size distribution parameters N_t and D_n. To infer $I_{cb}^{\downarrow}$ from measured radiances $I_m^{\downarrow}$ one must know the atmospheric transmittance $\mathcal{T}_a$ between cloud base and the surface:

$$I_{cb}^{\downarrow} = [I_m^{\downarrow} - B(T_{mr})(1 - \mathcal{T}_a)]/\mathcal{T}_a, \qquad (13.86)$$

where $B(T_{mr})$ is the blackbody radiance of the subcloud atmosphere with a mean temperature T_{mr} and $I_m^{\downarrow}$ is the radiance measured by the instrument. Therefore, with knowledge of $\mathcal{T}_a$, T_{mr}

and T_{cb}, which can be estimated from radiosonde profiles, the measured radiance $I_m^{\downarrow}$ can be inverted to optical thickness τ^0.

For a vertically and horizontally homogeneous cloud layer the optical depth of the cloud layer is

$$\tau_\lambda = \left[\int_0^\infty Q_{ext}(D)\pi\left(\frac{D}{2}\right)^2 n_{mg}(D)dD\right] H_c, \tag{13.87}$$

where $Q_{ext}(D)$ is the extinction efficiency of the cloud particles with size dimension D and H_c is the cloud layer thickness. Assuming that most cirrus ice particles are larger than 10 to 12 μm, $Q_{ext}(D) \approx 2$ and τ_λ becomes

$$\tau_\lambda = \left[\frac{\pi}{2}\int_0^\infty D^2 n_{mg}(D)dD\right] H_c. \tag{13.88}$$

Using the moments equation (13.77b) with $c = 1$ and $\nu = 2$, which Matrosov et al. (1992) argue is adequate for most cirrus cloud particle distributions, τ_λ becomes

$$\tau_\lambda = 3\pi N_t D_n^2 H_c. \tag{13.89}$$

Therefore, with a zenith directed radiance measurement one can infer the value of the product $N_t D_n^2$. Now the radar reflectivity Z is defined by

$$Z = \int_0^\infty D^6 n_{mg}(D)dD, \tag{13.90a}$$

or

$$Z = 2016 N_t D_n^6, \tag{13.90b}$$

where we have used (13.77b). Since there are now two equations, i.e., (13.89) and (13.90b), we can solve for the two unknowns N_t and D_n. If the cloud layer is not vertically homogeneous, Matrosov et al. (1992) replace Z in (13.90b) with the average value of Z from the top to the bottom of the cloud layer.

Cloud Drop Size Distributions from Passive Radiometry, Radar Reflectivity and Radar Velocity

In order to retrieve N_t and D_n as a function of height within a cirrus cloud layer, and thereby allow the IWC to vary with height within the cloud, Matrosov et al. (1994) developed a retrieval algorithm combining a measurement of the brightness temperature, or radiance, between 10 and 12 μm with radar measurements of the particle reflectivity and Doppler velocity as a function of height within the cloud layer. As before, Matrosov et al. (1994) assume that $c = 1$ and they demonstrate that the retrievals are not extremely sensitive to the value of ν so we set $\nu = 2$. For a vertically inhomogeneous cloud layer the optical depth defined in (13.87) becomes

$$\tau_\lambda = \sum_{j=1}^{M_g}\left\{\left[\int_0^\infty Q_{ext}(D)\pi\left(\frac{D}{2}\right)^2 n_{mg}(D)dD\right] H_{c,j}\right\}, \tag{13.91}$$

where $H_{c,j}$ is now the radar resolution volume spacing and N_g is the number of radar resolution volumes in the cloud layer. Note that the number of range gates with cloud return can be obtained from the location of the cloud boundaries. Again assuming that $Q_{ext}(D) \approx 2$ for all D the optical depth becomes

$$\tau_\lambda = \sum_{j=1}^{M_g}\left\{2016\pi N_{t,j} D_{n,j}^6 H_{c,j}\right\}. \tag{13.92}$$

The measured optical depth $\tau_{m,ir}$ is inferred from the infrared window measurements of sky brightness temperature using (13.85).

As before, the radar reflectivity is defined by (13.90b), where N_t and D_n are now a function of height, i.e., $N_{t,j}$ and $D_{n,j}$ respectively. The reflectivity-weighted mean Doppler velocity due to all of the particles in a radar resolution volume is

$$V_{f,j} = \frac{\int_0^\infty v_f(D) D^6 n_{mg}(D) dD}{\int_0^\infty D^6 n_{mg}(D) dD.} \tag{13.93}$$

Based on Pruppacher and Klett (1978), Matrosov et al. (1994) assume that the fall speed of the particles are related to the diameter by

$$v_f(D) = AD^B \tag{13.94}$$

where A and B are constants that depend upon the crystal shape. Substituting (13.94) into (13.93) and using (13.77b) yields

$$V_{f,j} = A \left[\frac{\Gamma(\nu + B + 6)}{\Gamma(\nu + 6)} \right] D_{n,j}^B, \tag{13.95}$$

where air density and viscosity effects have been neglected.

At this point $2N_g + 2$ unknowns have entered into the set of equations, i.e., $N_{t,j}$ and $D_{n,j}$ where $j = 1, \ldots, N_g, A$ and B. However, there are only $2N_g + 1$ measurements consisting of Z_j and $V_{f,j}$, where $j = 1, \ldots, N_g$, and $\tau_{m,ir}$. Since the variation of B is relatively small, i.e., 0.75 to 1.4, Matrosov et al. (1994) set $B = 1$. Picking an initial value of A_0 for A, Matrosov et al. (1994) solve (13.90b) and (13.95) for $N_{t,j}$ and $D_{n,j}$ at each range gate. Computing the infrared window optical depth τ_{ir} from (13.92) and comparing the result with the measured optical depth $\tau_{m,ir}$, they are able to derive a new value for A, i.e.,

$$A = A_0 \left(\frac{\tau_{m,ir}}{\tau_{ir}} \right)^{B/4}, \tag{13.96}$$

such that when (13.92) is recomputed with this value of A and the existing values of $N_{j,t}$ and $D_{j,t}$, the measured optical depth $\tau_{m,ir}$ results. Using A from (13.96) to solve (13.90b) and (13.95) for $N_{t,j}$ and $D_{n,j}$, Matrosov et al. (1994) arrive at their final estimates of the particle concentration and scaling diameter as a function of height within the cloud layer. The ice water contents IWC and the ice water path follow immediately from (13.78a) and (13.78b), respectively.

For the retrievals to work, the reflectivity-weighted mean Doppler velocity $V_{f,j}$ must represent the fall velocities of the particles in the air with no vertical motion. Since there are generally updrafts and downdrafts within a cloud, the raw radar derived value of $V_{f,j}$ cannot be used in (13.95). Based on work by Orr and Kropfli (1993), Matrosov et al. (1994) circumvent this problem by averaging over several hours all of the Doppler velocities $V_{f,j}$ at each resolution volume whose corresponding reflectivities fall within a narrow reflectivity bin. If the reflectivity value is indicative of a specific particle distribution within the radar resolution volume and if the average vertical motion of the air over the averaging interval is near zero, then the average velocity generated for each reflectivity bin of each resolution volume corresponds to the vertical fall speed of the particles in still air at the resolution volume location. Therefore, the reflectivity observed for each resolution volume is used as an index into the reflectivity bins from which an estimate of the average, still air particle fall speed is recovered.

Cloud Drop Size Distribtions from the First Three Radar Doppler Moments

Frisch et al. (1995) utilize the first three spectral moments, i.e., reflectivity, Doppler velocity and Doppler width, to retrieve the particle size distributions in light drizzle and stratus clouds with no drizzle. For both of these conditions, Frisch et al. (1995) assume that a lognormal distribution is an adequate model. Therefore, their retrieval amounts to inferring the particle number density N_t, scaling diameter D_n and distribution width σ, of a lognormal distribution from the three Doppler moments.

For a lognormal distribution the reflectivity Z_j is defined by

$$Z_j = \int_0^\infty D^6 n_{log}(D)dD = N_{t,j} D_{n,j}^6 e^{18\sigma_j^2}, \qquad (13.97)$$

where j represents a particular resolution volume. Based on Gossard et al. (1990), Frisch et al. (1995) assume a linear relationship between the drizzle-particle fall speed v_f and the particle diameter D:

$$v_f(D) = \frac{D-b}{a} \qquad (13.98)$$

where $a = 2.4 \times 10^{-4}$s and $b = 2.0 \times 10^{-5}$m. Therefore, the first Doppler moment $V_{f,j}$ is

$$V_{f,j} = \frac{\int_0^\infty [(D-b)/a] D^6 n_{log}(D)dD}{\int_0^\infty D^6 n_{log}(D)dD} = \frac{D_{n,j}}{a} e^{13\sigma_j^2/2} - \frac{b}{a}, \qquad (13.99a)$$

while the second Doppler moment W_j is defined by

$$W_j = \frac{\int_0^\infty [(D-b)/a - V_{f,j}] D^6 n_{log}(D)dD}{\int_0^\infty D^6 n_{log}(D)dD} = \left(\frac{D_{n,j}}{a}\right)^2 e^{13\sigma_j^2}(e^{\sigma_j^2} - 1). \qquad (13.99b)$$

Solving for $\exp(13\sigma_j^2)$ in (13.99a) and substituting the result into (13.99b), one has

$$\sigma_j = \{\ln[1 + W_j/(V_{f,j} + b/a)]\}^{1/2}. \qquad (13.100)$$

For drizzle either below cloud base or within a cloud of a negligible reflectivity, Z_j, $V_{f,j}$ and W_j are due to properties of the drizzle. Therefore, we have three unknowns, i.e., $N_{t,j}$, $D_{n,j}$ and σ_j, and three equations, i.e., (13.97), (13.99a) and (13.99b), with which to solve for them. From (13.100) one can solve for σ_j. Using the retrieved value of σ_j and (13.99b), one can then solve for $D_{n,j}$. Finally, one can solve for $N_{t,j}$ by substituting the retrieved values of σ_j and $D_{n,j}$ into (13.97).

For stratus clouds that consist of small drops and no drizzle, only the measurement of Z_j produces useful information about the cloud drops. In this case, $V_{f,j}$ and W_j are due to the mean vertical air motion and turbulence, respectively, since the small drops are presumably moving with the air. Therefore, we now have N_g useful measurements, i.e., Z_j for $j = 1, \ldots, N_g$, where N_g is the number of radar resolution volumes providing significant returns from the cloud drops. To reduce the number of unknowns, Frisch et al. (1995) assume that $\sigma_j = \sigma = 0.35$ for all j, $N_{t,j} = N_t$ for all j with N_t unknown, and $D_{n,j}$ is unknown and changes from one resolution volume to the next. All together we now have $N_g + 1$ unknowns. To have a well-constrained retrieval of $D_{n,j}$ and N_t, one additional measurement is needed; therefore, Frisch et al. (1995) assume that the liquid water path LWP is measured as well.

For this situation the reflectivity becomes

$$Z = N_t \, D_{n,j} e^{18\sigma^2}, \qquad (13.101)$$

where N_t is a constant and $\sigma = 0.35$. To solve for $D_{n,j}$ in (13.101) one must know N_t, which can be estimated from the measured LWP. The liquid water content LWC_j at the jth resolution volume is

$$LWC_j = \int_0^\infty \rho_w \left[\frac{4}{3}\pi \left(\frac{D}{2}\right)^3\right] n_{log}(D)dD = \left(\frac{\pi \rho_w N_t}{6}\right) D_{n,j}^3 e^{9\sigma^2/2}. \qquad (13.102)$$

Solving for $D_{n,j}$ in (13.101) and substituting the result into (13.102) yields

$$LWC_j = \left(\frac{\pi \rho_w N_t}{6}\right) [Z_j^{1/2} N_t^{-1/2} e^{-18\sigma^2/2}] e^{9\sigma^2/2}. \tag{13.103}$$

Integrating the liquid water content over the depth of the cloud layer, we have the liquid water path LWP:

$$LWP = \sum_{j=1}^{N_g} LWC_j H_{c,j} = \left(\frac{\pi \rho_w}{6}\right) N_t^{1/2} e^{-9\sigma^2/2} \left(\sum_{j=1}^{N_g} Z_j^{1/2}\right) H_{c,j}. \tag{13.104}$$

Substituting the measured LWP and radar reflectivities Z_j into (13.104), one can solve for N_t. Once N_t is in hand, $D_{n,j}$ can be obtained from (13.101) and the LWC_j from (13.102).

Cloud Drop Size Distributions from Radar Doppler Spectra

Gossard (1994) presents a technique that uses the entire Doppler spectrum in a retrieval of the cloud drop size distribution. The innovation in Gossard's retrieval is that the effects of turbulence and any mean vertical wind motion within the radar resolution volume on the Doppler spectrum are explicity accounted for in the retrieval. Assuming that the radar wavelength is much larger than the cloud drops, the density of the reflectivity Z as a function of diameter D i.e., the integrand of (13.90a), is

$$S_z^q(D) = n_{mg}(D) D^6, \tag{13.105a}$$

where the superscript q indicates that turbulence is not included in the definition of $S_z^q(D)$. In Figure 13.23a we plot $n_{mg}(D)$ and $S_z^q(D)$ for gamma distribution parameters $N_t = 6.54 \times 10^6$ m^{-3}, $\nu = 1$, and $D_n = 0.0233$ mm. Since ν is one, the drop size distribution in Figure 13.23a is exponential. The choice of $D_n = 0.0233$ mm forces the maximum value of $S_z^q(D)$ in Figure 13.23a to occur at $D_m = 0.14$ mm since the modal value of $S_z^q(D)$ occurs at $D_m = D_n(6+\nu-1)$. Changing variables in (13.105a) from the drop diameter D to the drop fall speed v, where we assume a known relationship $D = D(v)$, we have

$$S_z^q(v) = S_z^q(D(v)) \frac{dD}{dv}(v). \tag{13.105b}$$

For the spectrum $S_z^q(D)$ in Figure 13.23a, the corresponding fall velocity spectrum $S_z^q(v)$ is illustrated in Figure 13.23b, where we use (Rogers et al., 1993)

$$v(D) = 4 \left(\frac{\rho}{\rho_0}\right)^{-0.5} D(1 - e^{-12D}). \tag{13.106}$$

Again, the superscript q in $S_z^q(v)$ indicates a quiet air spectrum with particle fall velocities that are not affected by turbulence. If turbulence is present, then the spectrum $S_z^q(v)$ must be convolved with the turbulence spectrum to produce the final spectrum $S_z^t(v)$:

$$S_z^t(v) = \int_0^\infty S_z^q(v') \frac{1}{\sqrt{\pi v_\sigma^2}} e^{-[(v'-v)/v_\sigma]^2} dv', \tag{13.107}$$

where, like Gossard (1994), we have assumed Gaussian turbulence. The results of such a convolution are illustrated in Figure 13.23b for v_σ equal to 0.4 and 0.8. Apart from mean vertical wind motions within the radar resolution volume, which shifts the spectra in Figure 13.23b either to the left or right, $S_z^t(v)$ corresponds to the observed radar Doppler spectrum and is one of the outputs of a cloud radar. Gossard's technique attempts to retrieve the parameters N_t and D_n given spectra $S_z^t(v)$ that are potentially influenced by turbulence and mean vertical air motion within the radar resolution volume.

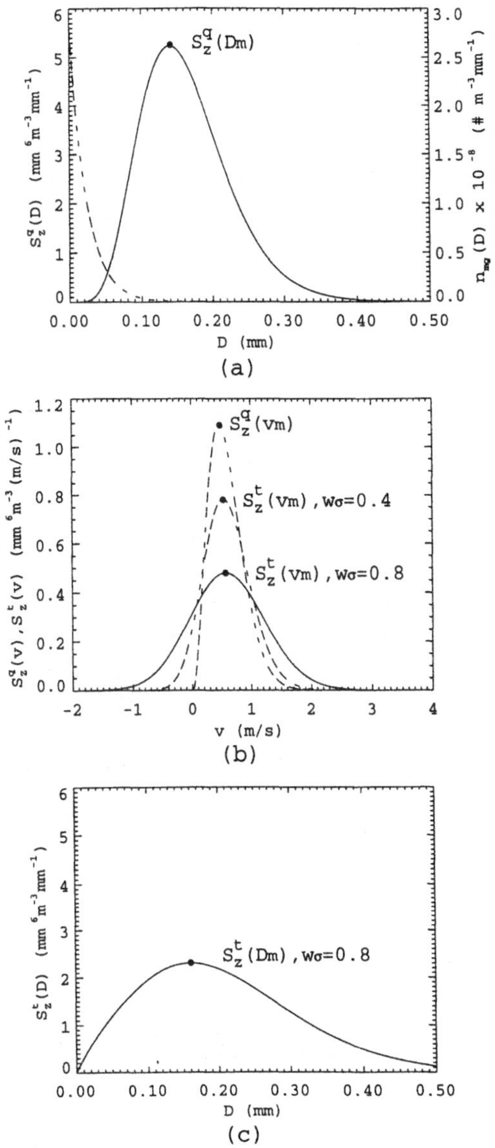

Figure 13.23: *(a) Cloud drop number density $n_{mg}(D)$ (dashed line) and reflectivity density $S_z^q(D)$ (solid line) for gamma distribution parameters $N_t = 6.5 \times 10^6 m^{-3}$, $D_n = 0.233$ mm and $\nu = 1$. (b) Reflectivity density versus drop fall velocity for no turbulence (upper dashed line), turbulence with $v_\sigma = w_\sigma = 0.4$ (lower dashed line) and turbulence with $v_\sigma = w_\sigma = 0.8$ (solid line). Positive velocities are downward. (c) Reflectivity density $S_z^t(D)$ that results from changing variables in $S_z^t(v)$ from drop fall velocity v to drop diameter D.*

If one simply changes variables in $S_z^t(v)$ from v to D, i.e.,

$$S_z^t(D) = S_z^t(v(D))\frac{dv}{dD}(D), \tag{13.108}$$

the resulting spectrum may look nothing like the original spectrum (Figure 13.23c) because of the effects of turbulence and perhaps a mean vertical wind within the radar resolution volume. Obviously, any drop size retrievals based solely on the spectrum $S_z^t(D)$ will contain serious errors. Note, however, that the ratio of the peak of $S_z^q(D)$ (e.g., Figure 13.23a) to the peak of $S_z^t(D)$ (e.g., Figure 13.23c) is a function only of the turbulent width v_σ and the modal diameter D_m and can be calculated for all possible combinations of these two parameters, just as we have done to generate the three panels in Figure 13.23. This fact motivates Gossard (1994) to define the deconvolution factor DF as the ratio

$$DF = DF(D_m, v_\sigma) = S_z^q(D_m) / \left[S_z^t(v_m) \frac{dv}{dD}(D_m) \right], \tag{13.109}$$

where v_m is the modal value of the spectrum $S_z^t(v)$ (e.g., the peak of the solid line (a) in Figure 13.23b) and the factor $S_z^t(v_m)[dv/dD(D_m)]$ is an estimate for the mode of $S_z^t(D_m)$ (e.g., the peak of $S_z^t(D)$ in Figure 13.23c). We are now in a position to illustrate the basic equation that Gossard uses to retrieve the modal diameter D_m of $S_z^q(D)$.

The reflectivity factor defined in (13.90a) can be integrated with the aid of (13.77) yielding

$$Z = N_t \left(\frac{D_m}{6+\alpha} \right)^6 \frac{(\alpha+6)!}{\alpha!} \tag{13.110}$$

where we let $\alpha = \nu - 1$ in accordance with Gossard (1994). Rearranging factors, Gossard writes Z as

$$Z = S_z^q(D_m) f(\alpha) D_m, \tag{13.111}$$

where the modal value of $S_z^q(D)$ is

$$S_z^q(D_m) = \frac{N_t(\alpha+6)^{1+\alpha} e^{-(6+\alpha)} D_m^6}{\alpha! D_m} \tag{13.111a}$$

and

$$f(\alpha) = \frac{(\alpha+6)! e^{(6+\alpha)}}{(\alpha+6)^{(7+\alpha)}}. \tag{13.111b}$$

Solving for $S_z^q(D_m)$ in (13.109) and substituting the result into (13.111) gives

$$D_m - Z / \left[f(\alpha) S_z^t(v_m) \frac{dv}{dD}(D_m) DF(D_m, v_\sigma) \right] = 0. \tag{13.112}$$

The reflectivity Z and mode $S_z^t(v_m)$ of the Doppler spectra are radar observables. If we assume a value for α, e.g., zero, one or two, then the only two unknowns in (13.112) are D_m and v_σ. As Gossard demonstrates, the part of the peak in $S_z^t(v)$ with negative Doppler velocities, i.e., particles moving upwards and away from the radar, is a function of the turbulent width v_σ; hence, the width v_σ can be recovered by fitting a Gaussian function to this region of the spectrum. Once an estimate of v_σ is in hand, one can numerically solve (13.111a) for D_m. Given D_m and Z, N_t follows from (13.111) and (13.112). Note that the value of v_m does not explicitly enter into the retrieval; consequently, updrafts and downdrafts within the radar resolution volume do not affect the outcome of the retrieval.

13.9 Concluding Comments

Millimeter-wave radar is proving to be a tremendously useful tool for the study of cloud properties. A number of papers have demonstrated the usefulness of radar reflectivity measurements for qualitative studies of cloud behavior, as well as quantitative determinations of cloud boundaries. However, the real power of radar lies in combining reflectivity and Doppler velocity measurements with lidar cloud base and passive radiometric measurements. Radiometric measurements typically provide integral measurements of total condensed phase. By using these as constraints on cloud particle retrievals from the active sensors, we can devise methods that allow us to define the vertical distribution of cloud particle size and condensed water content. This is a remarkable achievement. Further, the instruments currently being developed are much more robust than their earlier counterparts. Millimeter-wave radar and small lidar systems can be run continuously and unattended, as can passive radiometers. Various investigators have demonstrated this capability with individual instruments, and now the Atmospheric Radiation Measurement (ARM) program, sponsored by the United States Department of Energy, is putting instruments systems together for long term operations.

Given this progress, development of routine signal processing techniques for radars and retrieval algorithms for instrument systems must be viewed as high priority research. Much like satellites provide measurements of spatial variability, these suites of groundbased instruments can be used to continuously measure cloud statistics, provided that algorithms are available to process the raw data streams. These statistics will have many uses, including model verification and satellite retrieval validation. Most importantly, the unique perspective on cloud properties provided by these retrievals and statistics will improve our understanding of cloud physics and cloud processes.

Acknowledgments

We are indebted to many colleagues for their help as we have delved into radar signal processing and interpretation. We are especially indebted to Bruce Albrecht, whose efforts were instrumental in the development of a millimeter-wave cloud radar here in the Department of Meteorology at the Pennsylvania State University. Robert Peters engineered the radar system, with supporting help from J. Thomas Treaster, and he has patiently answered all of our questions about its operation. Gerald Mace has worked unceasingly to collect important data sets with the radar and, as we begin to apply cloud retrievals of the sort described in this document to these data sets, his contributions will become increasingly apparent. The core of this paper originated with a number of discussions with our colleague Hans Verlinde as he designed and taught a Radar Meteorology course for the first time here in the Department of Meteorology. The number of references to various scientists in the National Atmospheric and Oceanic Administration attests to their significant contributions to this field of research. This work would not have been possible without early support from the National Aeronautics and Space Administration (NAG-1-999) and sustained support from the Department of Energy Atmospheric Radiation Measurements Program (DE-FG02-9OER61071).

13.10 References

Atlas D, Ed (1990) *Radar in Meteorology.* American Meteorological Society, Boston, Massachusetts

Battan LJ (1973) *Radar Observation of the Atmosphere.* University of Chicago Press, Chicago, Illinois

Bohren CF, Huffman DR (1983) *Absorption and Scattering of Light by Small Particles.* John Wiley and Sons, New York, New York

Clothiaux EE, Miller MA, Albrecht BA, Ackerman TP, Verlinde J, Babb DM, Peters RM, Syrett WJ (1995) An evaluation of a 94-GHz radar for remote sensing of cloud properties. *J Atmos Oceanic Technol* 12: 201–229

Doviak RJ, Zrnic DS (1984) *Doppler Radar and Weather Observations.* Academic Press, San Diego, California

Doviak RJ, Zrnic DS (1993) *Doppler Radar and Weather Observations.* Second Edition. Academic Press, San Diego, California

Flatau PJ, Tripoli GJ, Verlinde J, Cotton WR (1989) *The CSU-RAMS Cloud Microphysics Module: General Theory and Documentation.* Paper No. 451, Colorado State University, Department of Atmospheric Science, Fort Collins, Colorado

Frisch AS, Fairall CW, Snider JB (1995) Measurement of stratus cloud and drizzle parameters in ASTEX with a Ka-band doppler radar and a microwave radiometer. *J Atmos Sci* 52(16): 2788–2799

Gossard EE, Strauch RG, Rogers RR (1990) Evolution of drop-size distributions in liquid precipitation observed by ground-based Doppler radar. *J Atmos Oceanic Technol* 7: 815–828

Gossard EE (1994) Measurement of cloud droplet size spectra by doppler radar. *J Atmos Oceanic Technol* 11: 712–726

Liao L, Sassen K (1994) Investigation of relationships between Ka-band radar reflectivity and ice and liquid water contents. *Atmos Res* 34: 231–248

Marshall JS, Hitschfeld W (1953) Interpretation of the fluctuating echo from randomly distributed scatterers. *Can J Phys* 31(1): 962–995

Matrosov SY, Uttal T, Snider JB, Kropfli RA (1992) Estimation of ice cloud parameters from ground-based infrared radiometer and radar measurements. *J Geophy Res* 97(D11): 11567–11574

Matrosov SY, Orr BW, Kropfli RA, Snider JB (1994) Retrieval of vertical profiles of cirrus cloud microphysical parameters from doppler radar and infrared radiometer measurements. *J Appl Meteor* 33: 617–626

Orr BW, Kropfli RA (1993) Estimation of cirrus particle fallspeeds from vertically pointing Doppler radar. Preprints, 26th *Int. Conf. on Radar Meteorology*, Norman, OK, Amer. Meteor. Soc., 588–590

Press WH, Flannery BP, Teukolsky SA, Vetterling WT (1988) *Numerical Recipes in C.* Cambridge University Press, Cambridge, England.

Pruppacher HR, Klett JD (1978) *Microphysics of Clouds and Precipitation.* D. Reidel, Boston, Massachusetts.

Reitz JR, Milford FJ, Christy RW (1979) *Foundations of Electromagnetic Theory.* Third Edition. Addison-Wesley, Reading, Massachusetts.

Rogers RR, Baumgardner D, Ethier SA, Carter DA, Ecklund WL (1993) Comparison of raindrop size distributions measured by radar wind profiler and by airplane. *J Appl Meteor* 32: 694–699

Sauvageot H (1992) *Radar Meteorology.* Artech House, Inc., Norwood, Massachussetts.

Syrett WJ, Albrecht BA, Clothiaux EE (1995) Vertical cloud structure in a midlatitude cyclone from a 94-GHz radar. *Monthly Weather Review*, accepted

Ulaby FT, Moore RK, Fung AK (1982) *Microwave Remote Sensing*, Vol. II, Radar Remote Sensing and Surface Scattering and Emission Theory. Addison-Wesley, Reading, Massachusetts.

Uttal T, Clothiaux EE, Ackerman TP, Intrieri JM, Eberhard WL (1995) Cloud boundary statistics during FIRE II. *J Atmos Sci*, December

PART C

PROPERTIES AND PROCESSES AT GROUND

Chapter 14

Remote Sensing of Soil Moisture

Simonetta Paloscia
Consiglio Nazionale delle Ricerche
Istituto di Ricerca sulle Onde Elettromagnetiche
via Panciatichi 64, 50127 Firenze
Italy

Abstract

Remote sensing methods are of fundamental importance for soil moisture monitoring both on a global and on a regional scale. Experimental research has demonstrated that, while observations in the visible and thermal infrared regions of the electromagnetic spectrum are able to detect the surface moisture of bare soils, microwave sensors are the most promising tools for obtaining quantitative estimates of this parameter. Valuable research on this subject has been performed worldwide, mainly in USA and Europe, and the results are reviewed in this paper on the basis of the most advanced international research.

14.1 Introduction

Water is one of the most widely spread substances on our planet and it is present basically in all substances, including rocks. The chemical characteristics of water, which exists in all three states (liquid, solid and gaseous), make this element very peculiar and of fundamental importance for life on Earth. Abundance or shortage of water has always affected the existence of plants and animals and consequently of mankind.

14.1.1 Hydrological cycle

These preliminary remarks are necessary for explaining the importance of the study of the hydrological cycle and of the laws which regulate the Soil-Plant-Atmosphere-Continuum (SPAC), especially on a global scale, in that the understanding of the mechanisms which regulate the distribution of water is essential for protecting water resources and promoting a rational use of them.

The physics of water transport in SPAC is described by the general flux equation which regulates water transport in soils (F): $F=K*(\Delta\psi/\Delta\ t)$, where the gradient of soil water potential $\Delta\psi$ is related to the hydraulic conductivity (K, in cm/sec) which is not constant with soil

NATO ASI Series, Vol. I 45
Radiation and Water in the Climate System:
Remote Measurements
Edited by Ehrhard Raschke

water content and also decreases quite rapidly as soil moisture decreases. This means that the gradients in soil water potential can change significantly in order to transport the same amount of water through the soil-root medium to the atmosphere. Changes in the water potentials of soil, roots and atmosphere influence the transpiration stream and therefore the amount of water released in the atmosphere through the evapotranspirative mechanisms (Denmead, 1976; Kramer, 1969).

The hydrological cycle is the recirculation of water from ocean to water vapour through the evaporation processes, the transformation of water vapour in precipitation and its return in the cycle through infiltration and evapotranspiration again. Although the process is longer and more complicated than it has been outlined here, we can focus on some simplified aspects. The water stored in soil and vegetation is an important fraction of the total amount of water involved in the hydrological cycle and represents a conspicuous quantity subtracted to the rapid loss due to runoff. The Soil Moisture Content (SMC), besides representing a water resource for plants, affects the separation of infiltration from runoff and it is therefore an important parameter for assessing erosion hazards and estimating evapotranspiration. A timely and precise information concerning water conditions of the soil-vegetation system can allow, therefore, a correct schedule of sowing and of irrigation, limiting wastes or lacks of water, and a reliable forecast of final yield (Kramer, 1969; Engman and Gurney, 1991).

There are basically two methods available for obtaining a reliable estimate of soil moisture content: in situ and remote sensing methods. In situ methods allow to retrieve a precise estimation of SMC in a very concentrated spot and a lot of samples must be collected in order to have a representative sampling of the examined area. Moreover, such a vast sampling is time-consuming and needs the work of several people if a reliable measurement of SMC has to be made in a relatively short time.

The recent development of remote sensing sensors has made it possible to begin the study of hydrological parameters in a quantitative way. The possibility to monitor a large area in a very short time and measure integrated values of some geophysical parameters instead of several sparse points is in fact very interesting. In particular, these methods can significantly contribute to the estimate of one of the key parameters involved in water balance which is the moisture content of soils on both a global and a regional scale, whose measurement is of primary importance in the process of modelling the hydrological cycle. Another advantage of remote sensing techniques is the possibility of checking the evolution in time of these parameters and thus improve hydrological models prediction by using repeated satellite passages.

In the following, after a short summary of theoretical statements, the state-of-art applications of remote sensing methods to soil moisture estimation will be reviewed.

14.2 Remote sensing methods

The electromagnetic spectrum covered by the various sensing methods ranges from very short wavelengths of gamma and x rays (mÅ) to very long wavelengths of radio bands (km) and it is shown in Fig. 14.1. Electromagnetic thermal radiation is described by the Planck law for which a blackbody at a temperature close to that of the Earth's surface ($\approx$ 300K) reaches a maximum in the thermal infrared region (10–14 μm) (Fig. 14.2). In the microwave range the spectral brightness (i.e. the radiative power per square meter, per unit of solid angle and per unit bandwidth of frequency) of a blackbody can be approximated by the Rayleigh-Jeans law, which directly relates the spectral radiance to the surface temperature (e.g. Ulaby et al., 1981).

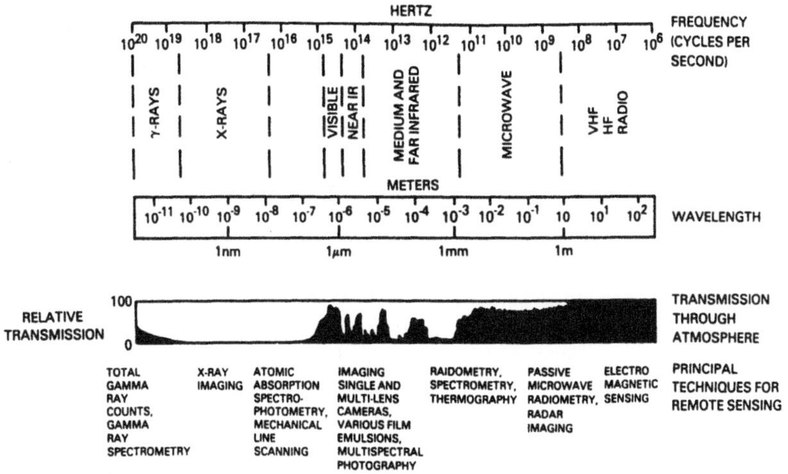

Figure 14.1: *An illustration of the electromagnetic spectrum showing the relationship between wavelength and frequency, the common names for spectral bands and the relative atmospheric transmission: The principal techniques used in remote sensing are also shown for their respective spectral region (After Engman and Gurney, 1991).*

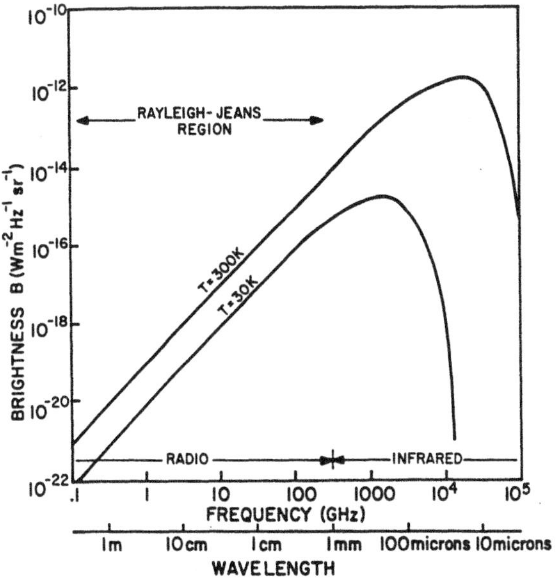

Figure 14.2: *Planck's law brightness curves for blackbodies at $T = 30K$ and $T = 300K$ (After Manual of Remote Sensing, 1983).*

Remote sensing techniques mainly use 3 regions of the electromagnetic spectrum: the visible band (0.40–0.70 μm), the infrared band (0.7–100 μm) separated into near, middle, far and thermal infrared, and the microwave band (1 mm–30 cm). In the visible (VIS) and near-infrared bands (NIR) (0.4–3.0 μm) solar radiation is partially reflected and partially absorbed by the earth's surface: whereas in water the visible radiation can penetrate even for several meters, in solids it is absorbed only for a very thin layer a few microns deep. In these regions of the spectrum remote sensing information is obtained by measuring the solar radiation and refers to the surface of observed bodies. In both thermal infrared (TIR) and microwave (MW) regions sensors are sensitive to the energy emitted from natural bodies due to a thermal effect and can measure their surface temperature; nevertheless in the microwave range the emitted and reflected radiation depends much more on the characteristics of the surfaces and therefore can give significant information concerning some geophysical parameters, namely soil moisture, biomass, ice depth, snow water equivalent, etc. Moreover, microwaves have a penetration power much higher than shorter wavelengths and thus they are the only way for investigating sub-surface characteristics. It is known that the atmosphere strongly attenuates all the γ, x and UV radiation, it is transparent in the VIS range and, except for some bands, has a still high absorptance in the infrared region. Moreover, energy propagation in the different bands depends on meteorological conditions in that clouds strongly absorb visible and infrared radiation up to millimetric waves ($\lambda < 1$ cm), while they influence much less the propagation of microwaves. This fact alllows the use of microwave sensors when it is necessary to operate independently from meteorological conditions.

Remote sensing sensors are substantially of two types: passive and active. To the first group belong the sensors (radiometers, spectrometers) able to measure the radiation coming from the observed surfaces: solar radiation (in the VIS-NIR range) and radiation directly emitted from the surface for thermal effect (in the TIR-MW range). To the second group belong the sensors (radar, laser) which have their own illuminating source able to emit in various bands of the spectrum. The most largely used active sensor is the Side Looking Radar (SLAR) in the two versions: Real Aperture Radar (RAR) and Synthetic Aperture Radar (SAR). Both sensors use very short pulses to obtain a good resolution across the flight line. In a RAR longitudinal resolution is determined by the physical dimensions of the antenna and the observation distance, whereas for SAR the resolution is independent from the distance in that the motion of the platform is used to obtain a long synthetic antenna. SAR has performances higher than RAR, however it is a highly complicated sensor both for the system and for the signal processing.

Electromagnetic parameters usually measured by remote sensing sensors are the following:

- *Radiance* (Wm^{-2} $ster^{-1}$):

It is the radiant flux per solid angle unit which passes through a surface coming from a direction, divided by the element of the surface itself projected on the plane perpendicular to this direction. Since the radiance varies with the wavelength a *spectral radiance* has been defined (Wm^{-2} $ster^{-1}$ m^{-1}). The concept of spectral radiance corresponds to that of *brightness* (Wm^{-2} $ster^{-1}$ H^{-1}) used in microwave radiometry where it more usually refers to spectral variations to frequency instead of wavelength.

- *Reflectance*:

It is the ratio between the radiant flux coming from reflection or scattering from an element of the surface and the incident flux on the element itself. In the microwave technique the characteristics of scattering of a target are expressed from the *radar cross section* (m^2), i.e. the area intercepting that quantity of power which, if it was isotropically scattered, would produce an echo equal to the one received from the object. When large objects are observed, the *normalized radar cross section* is preferably introduced, defined as the radar section per unit of surface area.

- *Emittance / Emissivity*:

Since natural bodies are not perfect emitters, their radiance is reduced by a factor, called emittance or emissivity[1] in the microwave range, defined as the ratio between the flux emitted by a surface element and the flux emitted by a black body at the same temperature. Its value ranges between 0 (perfect reflector) to 1 (black body).

14.2.1 Visible and infrared bands

Experimental research has demonstrated that, in theory, observations in the visible, thermal infrared and microwave regions of the electromagnetic spectrum, are able to give some information on the surface SMC of bare soils. In the optical range, soil spectral reflectance decreases as water is added to soil; however these variations of the reflectance are often in the same order or even lower than those caused by the physical or morphological differences existing among natural bare soils. Moreover, the observed variations in reflectance are in the order of about 10% for soil moisture values ranging between 10% and 35% and the optical response is influenced by a very thin soil layer (Idso et al., 1975). Considering that on canopy covered soil the situation is more complicated a widespread application of this technique is somewhat invalidated.

Thermal infrared sensors can be more useful than optical ones, because soil surface temperature is influenced more markedly than reflectance by the moisture conditions of a relatively deeper layer of soil. Observations in the thermal infrared band (8–14 μm) allow the measurement of surface temperature, which depends on repartition of net radiation flux between latent (evapotranspiration) and sensible heat. Infrared thermal emission depends on SMC in that wet regions are generally cooler than dry ones during the day and warmer at night. But even in this case the uncertainty in the relation between surface temperature and moisture is very high. Surface temperature, mainly if compared with air temperature, can give qualitative information on the evapotranspirative conditions of the surface. If quantitative information concerning the real amount of water contained in the soil is required, a method consists in computing the thermal inertia of the soil obtained by measuring the maximum and minimum daily surface temperature. As a matter of fact the amplitude of the diurnal range of soil surface temperature is a function of thermal conductivity and heat capacity (both dependent on SMC) as well as of meteorological factors which drive the diurnal variations of surface temperature. In most cases evaporation reduces the amplitude of diurnal surface temperature, so that the day/night temperature difference depends on some combination of SMC and evapotranspiration.

During the past decade many authors have reported on methods to derive the thermal inertia of bare and vegetated surfaces from infrared satellite data (e.g. Rosema et al., 1978; Becker et al., 1988), and most of the efforts in using these data have been addressed to the direct estimate of evapotranspiration. On this subject much research has been performed at the Institute National de la Recherche Agronomique in Versailles and Avignon (France) (e.g. Seguin et al., 1983), and by several groups in the Netherlands. The HCMM satellite, launched in April 1978, was the first satellite devoted to acquiring high-resolution thermal data at optimum period of the diurnal cycle. The HCMM carried out a two-channel radiometer (0.55–1.1 and 10.5–12.5 μm) in a sun-synchronous orbit at an altitude of 620 km and collected data at approximately 02.30 and 13.30 local standard time with a repeat cycle of 5 or 16 days. (Heilman and Moore, 1982). HCMM data collected at different dates over US test-sites were analyzed in order to evaluate the thermal data used in estimating near-surface soil moisture in a complex agricultural landscape. From this experiment a highly significant correlation (Fig. 14.3) was obtained between estimated soil temperature and soil moisture content.

During the TELL-US project which was set up as a collaboration agreement among several European institutes under the leadership of the JRC, in several European countries the volumetric

[1] instead of "emissivity" often the term "emittance" is used.

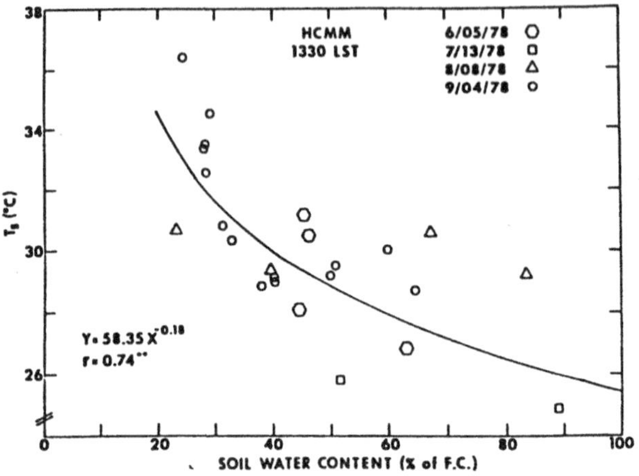

Figure 14.3: *Relationship between predicted soil surface temperature (T_s) and soil water content in the 0–4 cm layer of the soil profile, expressed as a percentage of field capacity (FC). Predictions were made using HCMM radiometric temperatures (after Heilman and Moore, 1982).*

moisture content was also estimated on the basis of National Aeronautics Space Administration's (NASA) HCMM infrared data, and using the relation between thermal inertia and SMC (Rosema et al., 1978). The results showed a reasonable agreement between measured and calculated average soil moisture content in the top soil layer with a tendency to underestimate the parameter.

Other intersting results were obtained by Carlson et al. (1984) using Synchronous Meteorogical Satellite (SMS)/GOES data. Thermal infrared (10.0–12.6 μm) measurements were used to determine the surface temperature from which the moisture available patterns of agricultural areas have been obtained using the Carlson Method (CM). The model is based on a surface energy balance at the ground-air interface, with fluxes of momentum, sensible and latent heat exchange through the surface interface layer, a turbulent surface layer and a convective mixing layer in the atmosphere and a diffusive heat flux in a substrate layer. The soil moisture retrieved from the model was then compared with the antecendent precipitation index (API), which is a parameter well correlated with soil moisture amounts over large regions, obtaining an average correlation coefficient of 0.7.

One of the main problems of these shortwave bands is that, as already stated, the measurement of soil moisture concerns a layer of only a few microns. From these data it is therefore extremely hard to retrieve the profiles of moisture in the root zone which is the most interesting for hydrological applications. Measurements of surface temperature and reflectance integrated with meteorological data sets can allow some hypotheses on the moisture of deeper layers but with significant errors. A method to bypass this hamper can be sometimes the use of plants as indicators of SMC, analyzing their water stress through reflectance variations and determining the water status of soil. Jackson (1982) set a plant water stress index (CWSI) based on the difference between air and canopy temperature and compared the temporal behaviours of this index with soil moisture. Although there is a substantial agreement between these two factors, in that as the soil moisture decreases the stress index increases, the crop water stress index does not drop to its lowest values immediately after irrigation, when soil moisture reaches its highest values, probabily because plants need some time to rehydrate. This fact confirms that a unique relationship between plant temperature and soil moisture does not exist and that the same alterations can be caused by both drought and diseases or temporary water stress due to, for instance, windy days. Moreover, other parameters such as net radiation and air vapour deficit are necessary to define the evapotranspirative conditions of vegetation mainly in a humid environment (Campbell and Norman, 1990).

14.3 Microwave Sensors

Extensive investigations have pointed out that, due to the large difference between the microwave dielectric properties of dry soil and liquid water, microwave sensors are by far the most efficient SMC detectors, although some limitations still exist on their use from spacecraft (Haefner and Pampaloni, 1992). Two kinds of sensors can be used: (i) radiometers able to measure thermal emission from soil, and (ii) radars which detect echoes of electromagnetic energy self-transmitted by them and reflected back to the sensor. In both cases the received radiation can be related to the moisture of bare or vegetated soil, according to the observation frequency, and for a layer of about 5 cm. As it will be discussed in the following the best frequency from an operative point of view for soil moisture measurements is around 1.5 GHz (approximately 21 cm wavelength).

14.3.1 Dielectric properties of soils

The two fundamental parameters measured by active and passive microwave sensors are respectively the backscattering coefficient (σ°), which describes the scattering characteristics of a surface in the direction of the sensor, and the brightness temperature (T_b), which measures microwave thermal emission. Both these parameters are strongly influenced by the dielectric properties of the observed media, which are characterized by their complex dielectric constant or permittivity. The dielectric constant (ϵ) is a measure of the response of the soil to an electromagnetic wave. This response is composed of two parts (real and imaginary) which determine the wave velocity and the energy losses in the medium, respectively. The square root of ϵ determines the index of refraction for the soil. The dielectric constant of most dielectric natural media is between 1 and 6 and increases up to 80 for free liquid water. This is the reason for the very high sensitivity of microwaves to the moisture contained in the observed bodies, depending on the molecule's ability to align its dipole moment along an applied field; anything that hinders the molecule's rotation (e.g. freezing, tight binding to a soil particle, etc.) therefore reduces the ϵ of water. Since the first water quantity added to dry soil is tightly bound to the surface of the particles, it will cause only a small increase of soil's ϵ. As more water is added, above a transition value of moisture (W_t), the soil's ϵ rapidly increases because the additional molecules are far from the soil particle surface and free to rotate (Schmugge, 1980). This explains the effects of soil texture in that the surface area depends on the particle size or texture; clay soils in fact, with a greater surface area per volume, hold more bound water molecules than sandy soils.

Soils are heterogeneous systems containing solid particles, bound and free water, air, voids, salts, etc. and their permittivity depends on that of their constituents but it is not a weighted average. Many empirical and theoretical models have been suggested to relate the dielectric constant of the mixture to that of its constituents, but the exact relationship is not yet well understood. An extensive investigation, which covers laboratory measurements as well as empirical and theoretical dielectric mixing models in the frequency range between 1.4 and 18 GHz, has been carried out at the University of Kansas (USA) by Hallikainen et al. (1985). The measurements

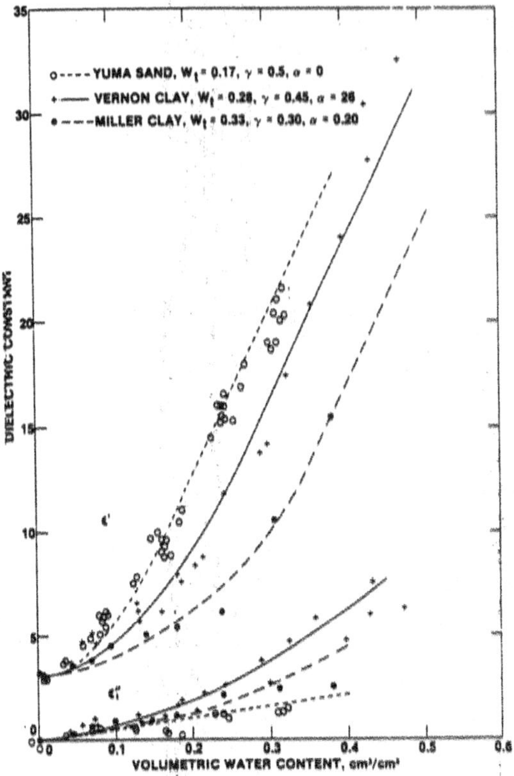

Figure 14.4: *A comparison between the calculated dielectric constant (real and imaginary parts) and the measured values for three different soil types as a function of volumetric soil moisture, λ=21 cm (after Wang and Schmugge, 1980).*

have been summarized by polynomial expressions dependent upon both the volumetric SMC and the percentage of sand and clay contained in the soil. In Fig. 14.4 a comparison between the calculated dielectric constants (both real and imaginary parts) from the empirical model by Wang and Schmugge (1980) and the laboratory measured values, at a wavelength of 21 cm, are shown for three texture classes of soils. This model uses the mixing of the dielectric constant of ice, water, rock and air and treats the transition moisture value as an adjustable parameter, according to the following expressions:

$$\epsilon_{soil} = W_c[\epsilon_i + (\epsilon_w - \epsilon_i)W_c/W_t * \gamma] + (P - W_c)\epsilon_a + (1 - P)\epsilon_r, \qquad (W_c \leq W_t) \qquad (14.1)$$

$$\epsilon_{soil} = W_t[\epsilon_i + (\epsilon_w - \epsilon_i) * \gamma] + (W_c - W_t)\epsilon_w + (P - W_c)\epsilon_a + (1 - P)\epsilon_r, \quad (W_c > W_t) \qquad (14.2)$$

where P is the porosity of dry soil; ϵ_i, ϵ_w, ϵ_a, ϵ_r are the dielectric constants of ice, water, air, and rock, respectively; W_c and W_t are the volumetric water content and the transition moisture.

To eliminate dependence upon adjustable parameters, Dobson et al. (1985) developed a physical soil model that is dependent on measurable soil characteristics only and requires no adjustable parameters to fit experimentally measured data. The model uses two components: the bound water fraction and the free water fraction, according to the pore-size distribution calculated from the particle-size distribution.

Theoretical studies on the permittivity of dielectric mixtures have been carried out in The Netherlands (de Loor, 1982) and at the Helsinki University of Technology as well (Shivola, 1989). Shutko and Reutov (1982) compared several current models with experimental data and concluded that, in the frequency range between 1 and 10 GHz, a two component refractive formula which considers only the volumetric fractions of dry matter and free water, is a good working model for most soil types. Dielectric measurements of soils in the 0.5–1.0 GHz range have been performed at the University of Bern (Stotzer et al., 1986) with a surface and a volume sensor.

14.3.2 Passive sensor measurements

Microwave radiation from soils can be expressed through the Rayleigh Jeans law, by means of an equivalent ("brightness") temperature T_b which, for homogeneous media, is related to the thermodynamic temperature T of the emitting surface through a factor e (emissivity) which takes into account the loss of energy at a surface and depends on the surface and dielectric characteristics of soils. Radiation collected by an antenna observing terrain (T_A) can be expressed as follows:

$$T_A = t(H) * [rT_{sky} + (1-r)T_{soil}] + T_{atm} \qquad (14.3)$$

where $t(H)$ is the atmospheric transmittance for a radiometer at a height H above the soil, r is the smooth surface reflectance (1-e), T_{soil} and T_{atm} are the thermometric temperatures of soil and atmosphere, respectively and T_{sky} is the contribution from the reflected sky brightness. The atmospheric transmission (for wavelengths higher than 5 cm which are typically used for soil moisture measurements) is around 99%, while T_{soil} and T_{atm} are usually both lower than 5K. Thus these terms can be neglected and T_b can be expressed as follows:

$$T_b = (1-r)T_{soil} = eT_{soil} \qquad (14.4)$$

Since e is dependent on the dielectric characterisitcs of soil, in the usual range of soil moisture a decrease of e from 0.95 to 0.60 can be expected with a consequent change of T_b of about 100K.

As a first approximation, the soil emissivity of smooth surfaces can be computed from the Fresnel relations (e.g. Ulaby et al., 1981). In Fig. 14.5 observed and predicted relationships between emissivity and volumetric soil moisture content for a bare smooth homogeneous soil at 2.25 cm and 18 cm of wavelength are shown. In the diagram variations of emissivity from $\approx$ 0.95 to $\approx$ 0.60 as the volumetric SMC changes from 0 to $\approx$ 40% are appreciable (Shutko, 1982). When soil temperature and absorption coefficient change with depth, microwave emission from soil is due to contributions from all depths with appropriate weights and can be predicted by means of theoretical models, which are mainly based on the radiative transfer theory (Schmugge and Choudhury, 1981). According to the models, the fraction of radiation upwelling from soil is determined by the dielectric properties of a surface layer whose thickness is in the order of a few tenths of a wavelength (Basharinov et al., 1979). Consequently, sensors operating at the longer wavelengths of the microwave region are the most suitable for collecting information on thick soil layers. An example of this sensitivity is shown in the diagram of Fig. 14.6, where a profile of T_b recorded using a 21 cm airborne radiometer is represented together with a profile of surface temperature measured with a thermal infrared radiometer. Whereas the infrared temperature does not change very much, the fluctuations of brightness temperature of

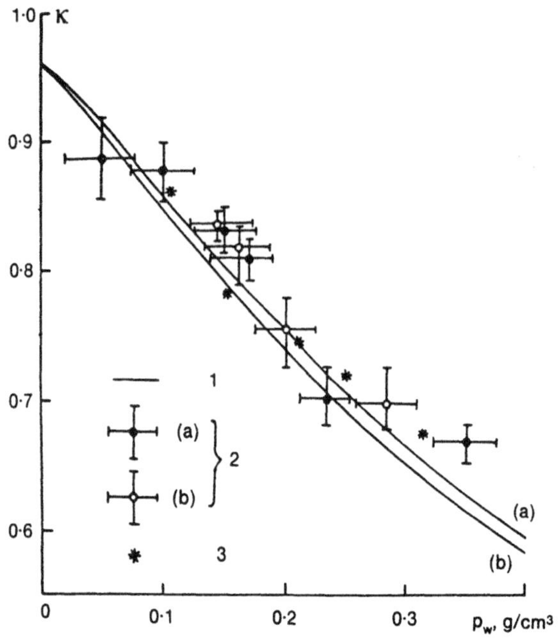

Figure 14.5: *Calculated (lines) and measured emissivity as a function of SMC_v at 2.25 cm (a) and 18 cm (b) of wavelength. (2) field and (3) lab measurements (after Shutko, 1982).*

the microwave radiometer are more enhanced and related to the variations of moisture of the surface.

Thermal microwave emission is measured by means of very sensitive and stable calibrated receivers connected to a low loss antenna. An outline of a microwave radiometer is given in Fig. 14.7. Much experimental research has been done all over the world to test the potential of microwave radiometers in measuring SMC and to determine the best observation parameters. Most activities have been performed in the US (e.g. Jackson et al., 1982; Schmugge et al., 1986), and by Russian scientists (e.g. Shutko, 1982) who were the first to conduct large scale aircraft investigations to test microwave radiometers in water management. Although vegetation cover and surface roughness effects reduce the sensitivity of MW to soil moisture, the results obtained have shown that soil moisture can be determined under a wide range of canopy conditions and roughnesses.

As an example, the first results obtained by US scientists after several aircraft campaigns have been here summarized. Jackson et al. (1984) evaluated aircraft data collected between 1978 and 1980 on some rangeland watershed near Chickasha (Oklahoma) using microwave radiometers operating at C- and L-bands (6 cm and 21 cm of wavelength, respectively). In Fig. 14.8 results obtained by comparing the normalized brightness temperature T_n, which is the ratio between the microwave brightness temperature and the surface temperature, at L-band (a) (21 cm wavelength) and at C- band (b) versus the 0–5 cm soil moisture are represented. It is evident that L-band data show a much better sensitivity to soil moisture than C-band data; this sensitivity can be expressed as the regression slope parameter that at L-band (H polarization and 0° incidence angle) is 0.0067 per percentage of soil moisture and it is an average value of the sensitivities found by Ulaby et al. (1983) both on bare soil and corn.

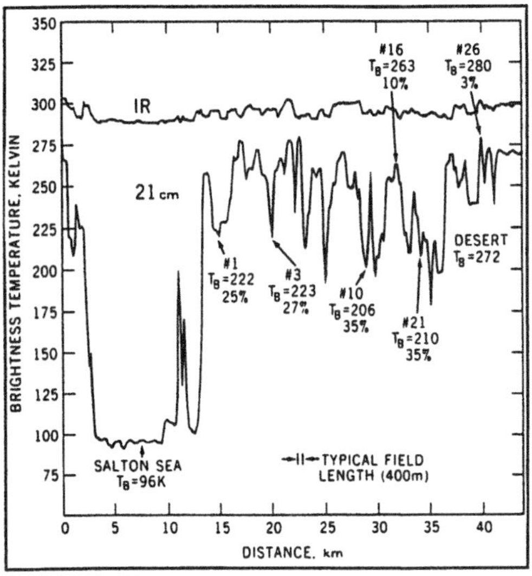

Figure 14.6: *Measurements of thermal infrafed (IR) and microwave brightness (T_b, at 21 cm wavelength) temperatures vs. distance for a flight path in California. Arrows indicate T_b and moisture values (in %) for some fields (after Schmugge and Becker, 1991).*

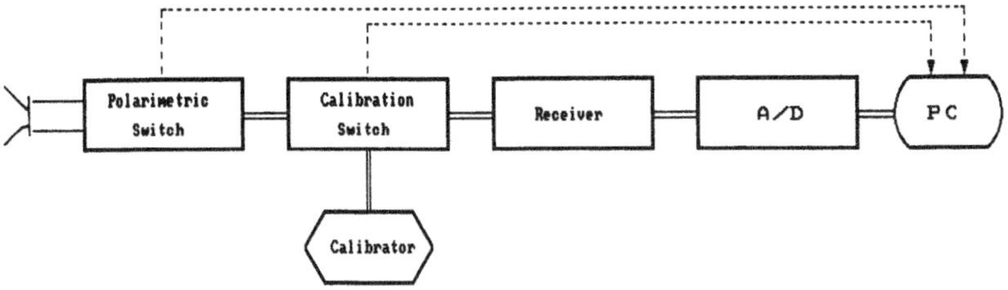

Figure 14.7: *Scheme of a microwave radiometer.*

Experimental research on microwave emission from agricultural fields was started at the Consiglio Nazionale delle Ricerche (CNR) in Florence (Italy) in 1980, using the Instrument for Radio Observations of the Earth (IROE). The latter was composed of X- (9.6 Gz), Ka- (37 GHz), L- (1.4 GHz) band dual polarized profiling radiometers combined with a thermal infrared sensor (8–14 μm) (Pampaloni, 1981; Bonsignori et al., 1987). The IROE in this version was operated in the spring/summer of 1988, during the AGRIRAD '88 campaign, both from a ground based platform and from a helicopter. In the first case samples of soils contained in a tank were observed at different incidence angles, while in the second case an agricultural area with different crops was surveyed during different periods of crop growth, so that several types of soils and crops were observed in a short time (Paloscia et al., 1989). The first approach gave

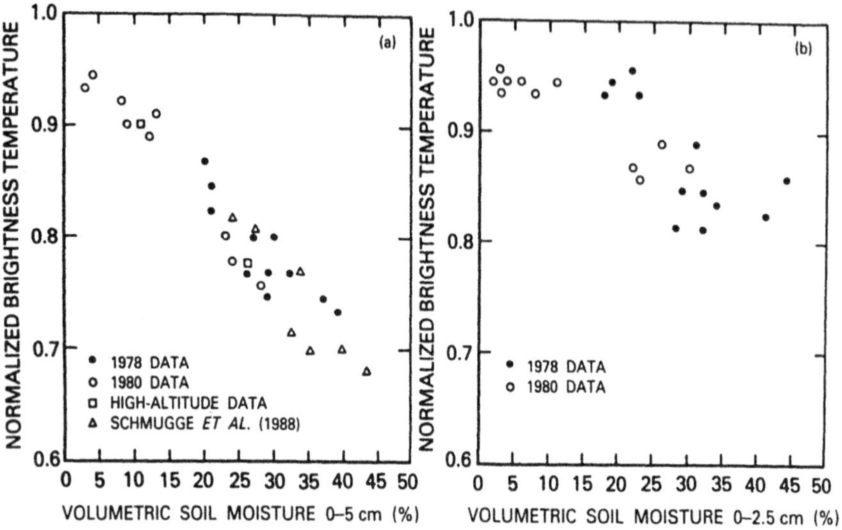

Figure 14.8*: Aircraft observations of emissivity and soil moisture over rangelands water shed (a) L-band (21 cm wavelength), $\theta = 0°$, H pol. (b) C-band (6 cm wavelength), $\theta = 0°$, H pol. (after Jackson and Schmugge, 1989).*

the possibility to control some geophysical parameters and guaranteed greater precision, whilst aircraft measurements allowed a much wider sampling under natural conditions.

On the basis of the previous results the sensitivity to SMC of microwave emission to the volumetric moisture content (SMC_v) of the most characteristic thickness for each wavelength was checked on smooth soil. As an example, T_n measured by ground-based sensors at 1.4 GHz, 10 GHz and 36 GHz at $\theta = 20°$ on a smooth sandy-clay soil sample is shown in Fig. 14.10 as a function of volumetric moisture of different soil thicknesses. The diagram points out that emission at both L and X bands is well correlated to SMC_v of the first 5 cm and 2.5 cm layers respectively, with a sensitivity in the order of about 3 K per % of SMC_v. On the contrary, the sensitivity of the K_a band emission is very low though correlated with the first 1 cm layer.

A summary of L band data collected using airborne sensors over natural agricultural surfaces is represented in Fig. 14.11, where the normalized temperature at H pol. and $\theta = 10°$ is shown as a function of gravimetric soil moisture (SMC_g) of two layers of soil: 0–5 cm (a) and 0–20 cm (b). Data were collected during the three flights from different types of bare soils and vegetated fields (sunflower and corn). In the diagrams the average spatial variability of moisture in the fields is represented by its mean standard deviation. Since many fields were covered with dense vegetation, and bare soil had a high surface roughness, the sensitivity of T_n to SMC was not very high and the points on the diagrams are fairly spread out. The slopes of the regression lines and the correlation coefficients, computed for each layer, are quite similar, i.e. sensitivity and correlation to SMC are fairly high even for a soil layer down to about 20 cm. However, the sensitivity to the SMC of so deep a layer of soil, even under vegetation, may be due to some correlation between SMC at 5 cm and 20 cm depth. For most fields the SMC profile was indeed rather regular, generally increasing with depth, as it is rather usual in summer time, when there is a strong evaporative demand of the atmosphere.

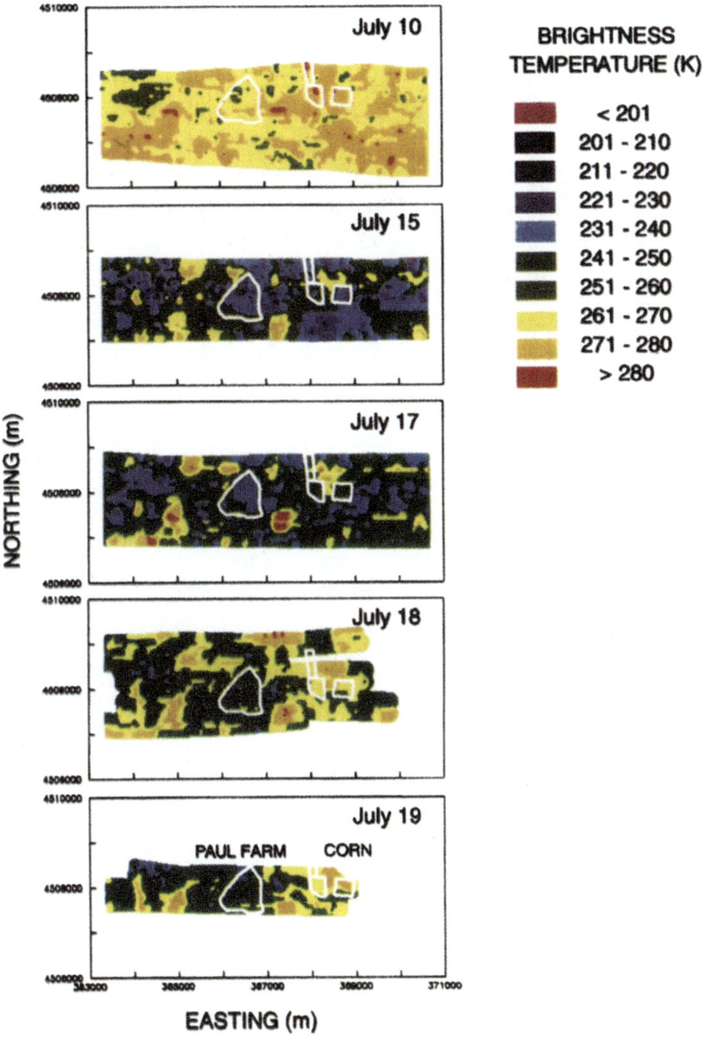

Figure 14.9: *Brightness temperature maps obtained using the PBMR (Push Broom Microwave Radiometer) at L-band (H pol.) for five days of July 1990 (before and after a rainfall) (after Jackson et al., 1994).*

From these measurements, worked out by means of ground based and airborne sensors, it is confirmed that microwave emission from soil is very well correlated to its moisture content and that, at frequencies around 1.5 GHz or lower, the sensitivity to SMC of the first 5 cm layer is fairly high even for vegetated fields and ranges between ≈ 3.5 K/0.01 g cm^{-3} for smooth bare surfaces to ≈ 1K/0.01 g cm^{-3} due to the effect of surface roughness and presence of vegetation. A good correlation was also found with SMC of layers down to 20 cm; however, due a to fairly

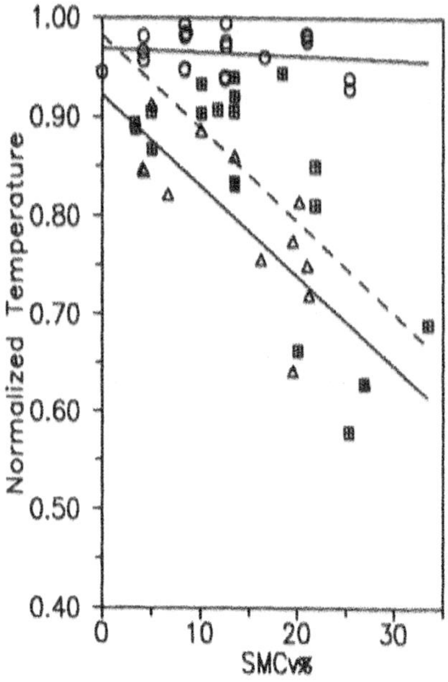

Figure 14.10: *T_n (H pol., $\theta = 20°$) at 1.4 GHz (triangles), 10 GHz (squares) and 36 GHz (circles) as a function of SMC_v of smooth soil. T_n data at 1.4, 10 and 36 GHz have been correlated to the SMC of the 0–5cm, 0–2.5cm and 0–1cm layer, respectively (after Paloscia et al., 1993).*

uniform soil moisture profile it has not been possible to state in a definitive way which is the thickness of soil layer sampled by the radiometer (Pampaloni et al., 1990). Moreover, these investigations pointed out that, on bare soils with a relatively smooth surface, the emission at 10 GHz is quite sensitive to the SMC of the first cm soil layer while the radiation at 36 GHz was found to be mostly influenced by roughness.

One of the more recent extensive experiments on this subject has been the Machydro'90 campaign conducted in Pennsylvania in 1990 to study the synergistic use of remote sensing sensors in multitemporal hydrological studies (Jackson et al., 1994). A new type of sensor, the 'pushbroom' microwave radiometer (PBMR), was flown to produce brightness temperature maps. The PBMR operate at L-band in H polarization, at an incidence angle between 8° and 24°. The ground resolution obtained at a flight altitude of 300 m is 90 m. In Fig. 14.9 five brightness temperature maps obtained at different dates on July 1990 are shown. During dry conditions (July 10) dry surface soils and forests have similar high values of T_b with very little indication of geographic features which becomes instead quite apparent on July 15 following a rainfall event, together with the effects of the various vegetation covers. As expected, the forested areas maintain their high initial values of T_b. There are a few areas that have very low values of T_b that are for the most part associated with stream channels and small water bodies. It was interesting to observe that the variability in the PBMR image was not as high as suggested by the field structure, which means that some of the variations in vegetation cover do not highly affect microwave data. In the sequence of the images the drying trend after the rainfall is evident. In Table 14.1 a comparison between the T_b measured on a selected area of the image

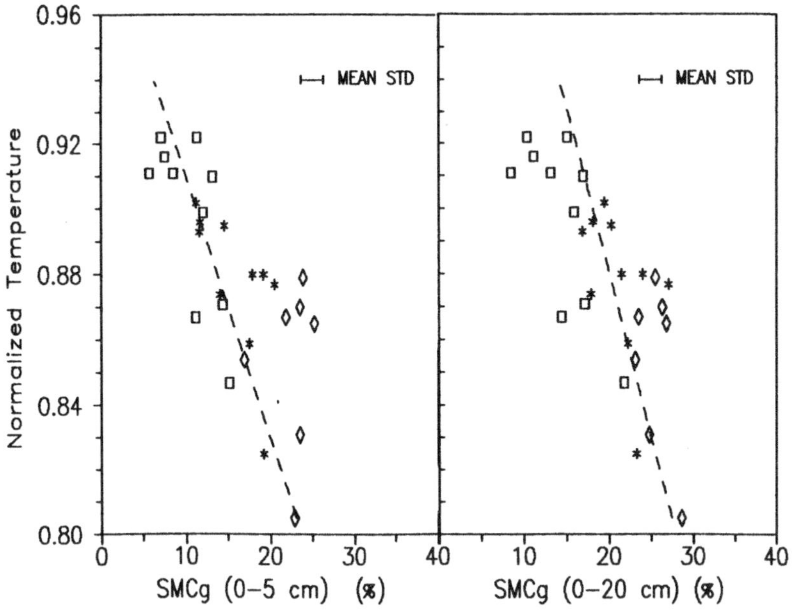

Figure 14.11: *T_n at 1.4 GHz (H pol., θ = 10°), collected on bare and vegetated fields during 3 flights (◇ May; + June and □ July 1988) as a function of the SMC_g of the first 0–5 cm (a) and 0–20 cm (b) layer. Bar represents the mean standard deviation of moisture spatial variability in the field. The slope (m) and the correlation coefficient (r) of the regression lines are for (a): m=-0.0073, r=0.72; for (b): m=-0.0074, r=0.74 (after Pampaloni et al., 1990).*

Date	T_b (K)	Predicted SMC (%)	Observed SMC (%)
July 10	263	12.9	13.4
July 15	228	22.7	23.1
July 17	231	25.8	21.3
July 18	246	18.8	18.9
July 19	245	19.2	17.6

Table 14.1: *Comparison between T_b and soil moisture.*

(Paul Farm Watershed), the soil moisture measured on ground and the soil moisture computed from MW data is shown. The total number of sample for soil moisture was around 40 and all these points have been averaged to obtain the mean values listed in the table.

Effects of soil roughness and vegetation

Soil roughness is known to play an important role in microwave emission from soils and many theoretical models based on various approximations have been developed to predict emission from rough surfaces in different frequency ranges (Fung et al., 1981; Tsang et al., 1982). On the other hand, experimental results have shown that the effect of roughness is usually an increase of the emissivity and a reduction of the sensitivity to SMC (Choudhury et al., 1979; Wang et al., 1983).

Since the effect of roughness is dependent on observation wavelength in that a surface can be rough at 3 cm wavelength and smooth at 21 cm, a multifrequency approach can be useful for estimating the surface roughness. In general, 10 GHz (3 cm) emission, mainly when measured at incidence angles higher than 30°, was found to be more sensitive to the soil roughness than 36 GHz (0.8 cm) emission, in that the latter saturates as soon as the roughness increases to a value in the order of one wavelength (< 1 cm). In the diagram of Fig. 14.12 we observe that, on bare soil with a SMC > 15%, the difference (δT_b) between the brightness temperatures at 36 GHz and 10 GHz, measured at θ=40° and $H_{pol.}$, decreases as the height standard deviation (H_{std}) increases. This correlation can be expressed by means of an exponential function $\delta T_b = 86.7 * \exp(-1.37 * H_{std})$ ($r^2 = 0.8$). Three zones can be roughly isolated in the diagram: smooth ($H_{std} < 0.4$ cm), medium rough (rough1) ($H_{std} = 0.4$–1.2 cm) and very rough (rough2) ($H_{std} = 1.2$–3 cm) surfaces, corresponding to three appreciable levels of δT_b (> 50 K, 20–50 K, < 20 K) (Paloscia et al., 1993).

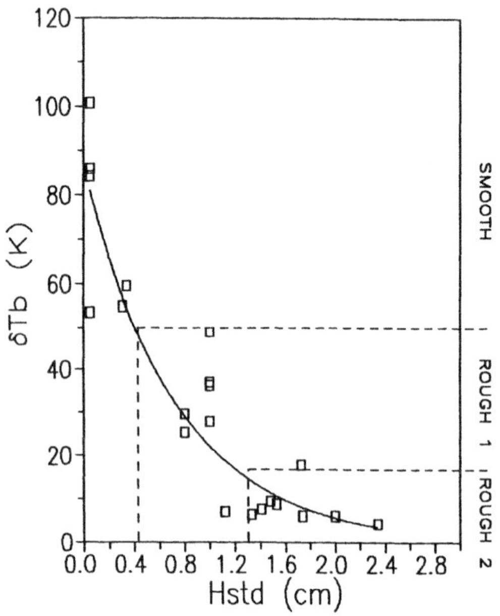

Figure 14.12: *δT_n (T_n 36 GHz - T_n 10 GHz) at H pol., $\theta = 40°$, as a function of the height standard deviation of the soil surface. The line represents the best fit of the experimental points: $\delta T_n = 86.7 \exp(-1.37 h_{std})$, $r^2 = 0.8$ (after Paloscia et al., 1993).*

The possibility of estimating surface roughness from radiometric measurements at 10 GHz and 36 GHz could allow some improvement in the precision of SMC measurement from L band data. An example of this approach is shown in the following. If we resume all data of T_n at 1.4 GHz collected during both the aircraft and the ground based experiments as a function of the volumetric SMC of the 0–5 cm layer, the points can be classified on the basis of their δT_b, computed using X and K_a band data,and separated into three groups which, according to Fig. 14.12, correspond to three different classes of roughness. A correlation line can then be computed for each of the three classes identified in the diagram, and used for estimating the soil moisture from T_n data at 1.4 GHz. The comparison between soil moisture computed by multifrequency radiometric measurements and "ground truth" is shown in Fig. 14.13. Here the SMC has been computed from L band measurements by using one of the three regression lines according to the δT_b value of each point.

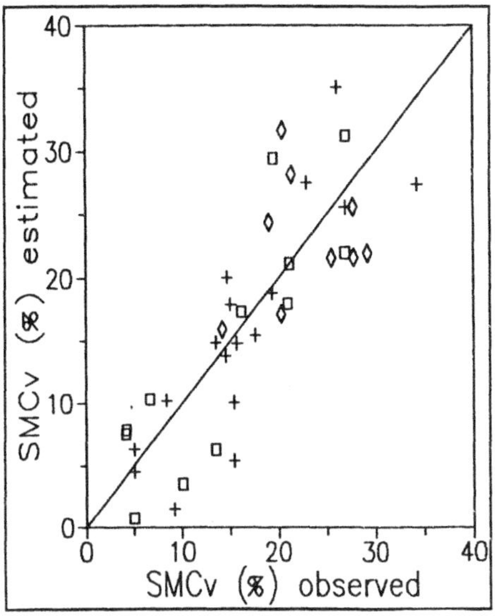

Figure 14.13: *SMC$_v$ (0–5 cm) computed using multi-frequency microwave data and corrected for the soil roughness effect as a function of SMC$_v$ measured on ground. (◇ = very rough, + = medium rough, and □ smooth soils) (after Paloscia et al., 1993).*

On canopy-covered fields the presence of vegetation reduces the sensitivity to soil moisture as well, in that plants absorb emission from soil and re-emit their own radiation. Emission from vegetation was first studied by Kirdiashev et al. (1979) who evaluated the sensitivity reduction to SMC measurement due to different types of vegetation, classified as small grains, broadleaf and forest. Except the case of forest, the maximum reduction in sensitivity was found in the presence of broadleaf crops and it appeared to be about 30% at 1 GHz and 90% at 10 GHz. At the latter frequency most of the radiation upwelling from a canopy covered soil is then emitted from plants. In general, the sensitivity reduction factor to SMC measurement depends on biomass quantity overlaying soil. Therefore, except for very light vegetation cover, a fairly accurate measurement of soil moisture under vegetated canopies requires some estimate of the biomass above ground. Observations of agricultural fields, at 10 GHz and 36 GHz, were able to detect quantitative relations between microwave emission and vegetation features (Pampaloni and Paloscia, 1985). It should be noted that monitoring of vegetation biomass and water conditions is very important in that these parameters can be used as inputs into soil moisture budget/water management models. As a matter of fact evapotranspiration does not depend entirely on SMC, but also depends on plant conditions. On the other hand plant water conditions may indicate SMC at root depth. Jackson et al. (1982) developed a parametric apprach based on the model proposed by Basharinov and Shutko (1975) which considers vegetation as an absorbing layer characterized by its water content. Using field measurements and a model based on the radiative transfer theory, Pampaloni and Paloscia (1986) established a semi-empirical relation between the optical depth of vegetation and the plant water content per unit area. The same authors pointed out that the normalized difference between vertical and horizontal polarization components (Polarization Index), of X band (10 GHz) emission, can be correlated to leaf area index (Pampaloni and Paloscia, 1985; Paloscia and Pampaloni, 1988; Paloscia and Pampaloni, 1992). Emission and scatter spectra, from L to X band, of several vegetated canopies, have been obtained by a team at the University of Bern, by means of a ground based sensor (Huppi et al., 1986).

Thanks to this intensive research work, the potential of microwave passive sensors in the SMC measurement has been demonstrated and operational systems have been tested in Russia, where maps with four levels of soil moisture have been obtained using a two frequency airborne sensor at 1.0 and 1.7 GHz (Fig. 14.14). The method is based on a model which considers a 'transition' layer over a uniform wetter or dryer layer and makes use of two frequency measurements to estimate the depth of the transition layer and moisture below it (Fig. 14.15) (Reutov et al., 1986). Research programmes based on this approach are being developed in Hungary (Ijjas et al., 1989; Ijjas et al., 1993) and Bulgaria (Mkrtchjan et al., 1988).

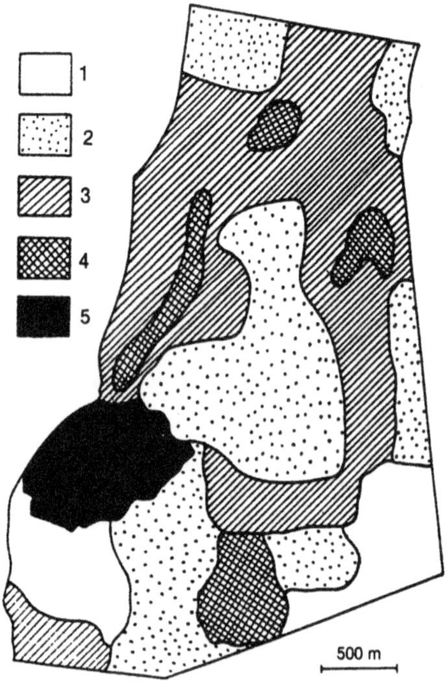

Figure 14.14: *Example of moisture content map obtained by radiometric measurements at 2.25 cm and 18 cm wavelength in Russia. Moisture ranges: 1<0.1 g/cm^3, 2=0.1-0.2 g/cm^3, 3=0.2-0.3 g/cm^3, 4=0.3-0.4 g/cm^3, 5>0.4 g/cm^3.*

The major problem in using passive microwave sensors for SMC measurements seems to be the ground resolution which is in the order of a few tens of meters for airborne sensors, and of many kilometers from space. However the use of synthetic aperture antennas could reduce this

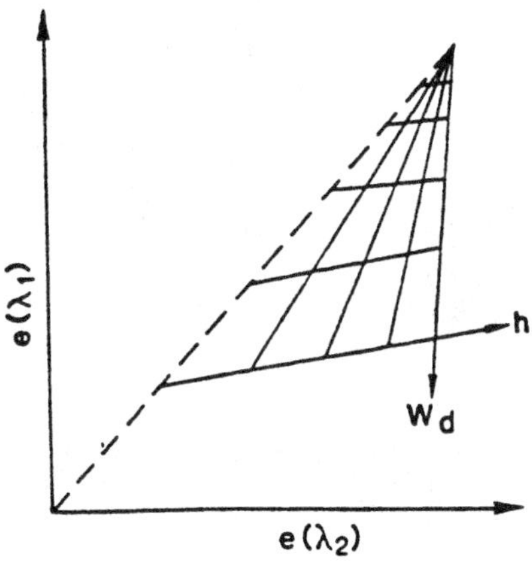

Figure 14.15: *The diagram used by Reutov and Shutko (1986) for estimating the depth of transition layer (h_d) and moisture below it (W_d) from the emissivity measurements at two wavelengths.*

drawback. In the past years the Electronically STeered Array Radiometer (ESTAR) technology has been developed and the application of an ESTAR L-band radiometer for soil moisture mapping has been investigated (Le Vine et al. 1990). This sensor has a ground resolution in the order of 20% of the altitude and therefore better than the PBMR. Results obtained from data collected during an experiment on the Walnut Gulch Watershed (Arizona) pointed out that this radiometer is able to provide soil moisture data with the same level of accuracy as existing systems and has the potential of satisfying application data requirements and spacecraft limitations (Jackson et al. 1993).

Another serious restriction in populated areas of Western Europe is the strong man-made interference, that is the strong noise introduced by TV and radio channels. In practice only the protected band for radioastronomy (1.42 GHz) can be used in industrialized countries.

Current research is now engaged in increasing the accuracy and reliability of the method for operational use. The main topics are the depth to which microwave radiometers sample SMC directly, and the quantification of the effects of spurious factors such as surface roughness, vegetation cover and salinity.

14.3.3 Active sensor measurements

Active microwave sensors transmit a beam of modulated electromagnetic energy and receive backscattered echoes which can be correlated, through the backscattering coefficient σ^o, to the morphological and physical characteristics of the target. When a wave is incident upon the soil, surface and volume scattering usually take place at the same time. Surface scattering is due to the dielectric discontinuity at the interface between air and soil while volume scattering is caused by spatial inhomogeneities in the illuminated volume at a scale comparable to that of the wavelength. Two types of radars are normally used in remote sensing: side looking imaging systems (with real or synthetic aperture) and scatterometers. The latter are calibrated systems specifically designed for backscattering measurement and in general do not operate as imagers. A simple scheme of an active system is shown in Fig. 14.16.

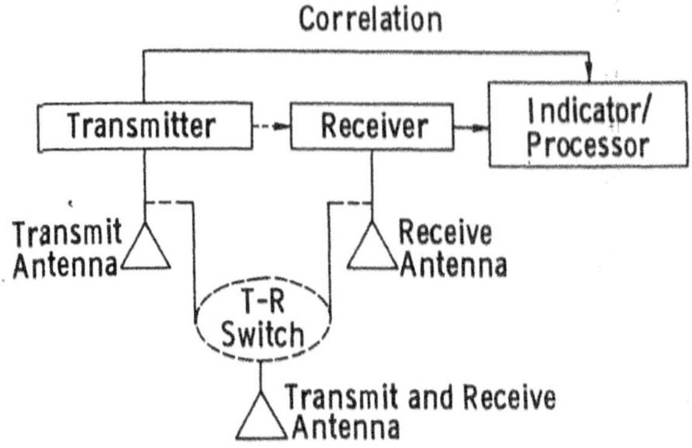

Figure 14.16: *Basic elements of a radar system.*

The measured radar backscatter σ° over a bare soil can be related to soil moisture according to this expression:

$$\sigma_s^\circ = f(R, a, SMC_v) \tag{14.5}$$

where R is the surface roughness term, a is a soil sensitivity term and SMC_v is the volumetric soil moisture. Many models have been developed to estimate these terms independently, but if semi-empirical models are easy to invert though not reliable enough, on the other hand, more sophisticated theoretical models require many input data which make the inversion rather difficult. If soils are vegetated the problem becomes more complicated and thus models which isolate the effects of vegetation and roughness are needed. Two types of models are currently used: continuous and discrete models. In the first case the dielectric constant of the medium is assumed to be a random process with means and correlation function known (e.g. Tsang and Kong, 1979). In the second case the medium is considered as an ensemble of dielectric scatterers (disks and cylinders) which represent leaves and stems of plants. Some of the last developed models are based on the discrete scatterers approach, such as the MIMICS (MIchigan MICrowave Scattering model) (Ulaby et al., 1990).

Experiments carried out by means of a multifrequency ground based scatterometer at the University of Kansas (USA) tried to define the best observation parameters for the measurement of SMC (Ulaby et al., 1982). According to these investigations, a high sensitivity to SMC of a top 5 cm layer was obtained at a frequency f = 4.5 GHz (C-band), HH polarization and incidence angle $\theta = 10^\circ$. As we can see in the diagrams of Fig. 14.17, where the backscattering coefficient at C-band, HH pol., is represented as a function of the soil moisture content of the first 5 cm layer for bare (a) and vegetated fields (b), the sensitivity values are about 0.15 dB (% of field capacity) on bare soils, and 0.13 dB (% of field capacity) on vegetation covered soils. The regression lines with the correlation coefficients are represented in the diagrams. For these observation parameters the sensitivity to surface roughness, which also influences radar response, seems to be minimized. Although these results have been partially reappraised by further experiments, the research work carried out by Ulaby and his group cast the basis for the development of the subsequent activities in this field.

The measurements taken in Toulouse (France) at CNES and based on the use of a four frequency (1.5, 3.0, 4.5, 9.0 GHz) ground based scatterometer (Le Toan 1982) confirmed the conclusions of Ulaby's group (1982) on optimum sensor parameters for detecting soil moisture. As an example, the backscattering coefficient σ° of soils with different surface roughness, measured

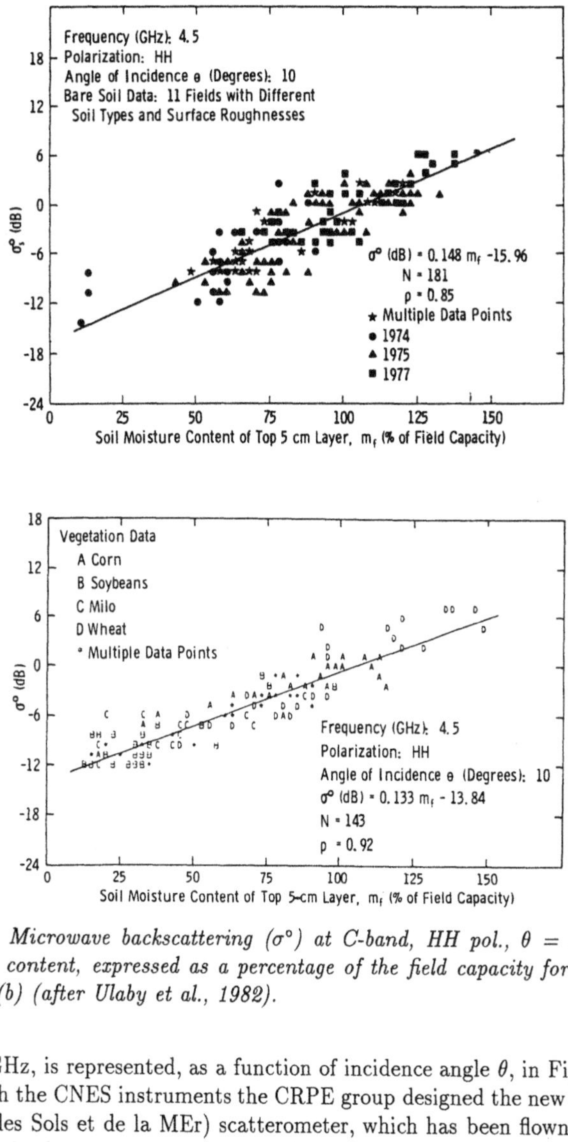

Figure 14.17: *Microwave backscattering ($\sigma°$) at C-band, HH pol., $\theta = 10°$, as a function of soil moisture content, expressed as a percentage of the field capacity for bare soils (a) and vegetated fields (b) (after Ulaby et al., 1982).*

at 1.5 and 4.5 GHz, is represented, as a function of incidence angle θ, in Fig. 14.18. After the experiments with the CNES instruments the CRPE group designed the new C-Band ERASME (Etude RAdar des Sols et de la MEr) scatterometer, which has been flown on helicopter and aircraft to collect backscattering data on several sites (Bernard et al., 1986). In the Netherlands the measurements made by the ROVE (Radar Observation of Vegetation) team proved that, at X band, an incidence angle of 15° is optimum for SMC determination, with the smallest dependency on roughness (de Loor, 1979).

Experimental campaigns with airborne sensors, organized in the last decade by the Joint Research Center (JRC) and the European Space Agency (ESA), have included many experiments devoted to SMC measurements. The JRC AGRISAR 86 Campaign made use of X band SAR VARAN S and was mainly devoted to vegetation monitoring, however one experiment for SMC

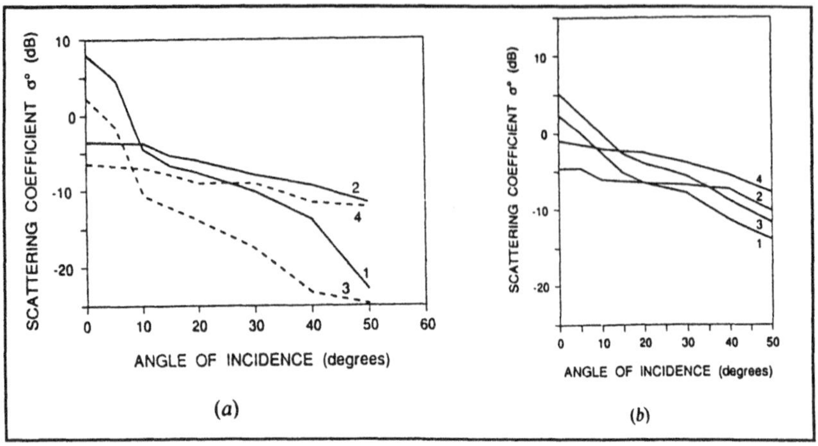

Figure 14.18: *Angular variations of σ° at 1.5 GHz (a) and 4.5 GHz (b) on four types of surface roughness (hstd, cm) and SMC(%): (a) 1-SMC=0.23, hstd=1.2; 2- SMC=0.19, hstd=5.2; 3-SMC=0.03, hstd=1.2; 4=SMC=0.03, hstd=5.2; (b) 1- SMC=0.18, hstd=1.2; 2-SMC=0.22, hstd=9.2; 3-SMC=0.24, hstd=1.2; 4=SMC=0.03, hstd=5.2 (sfter Le Toan, 1982).*

detection was successfully carried out in the Freiburg area. AGRISCATT experiments over many sites in Europe were organized by ESA in 1987 and 1988 making use of two scatterometers: the Delft University of Technology Scatterometer (DUTSCAT) in 1987–88 and the ERASME in 1988. An example of data collected at C band in 1988 on the Italian site "Montespertoli" (Florence), which confirms the sensitivity of scatterometric data to SMC, is shown in Fig. 14.19, where σ° of vegetated fields, measured by DUTSCAT (a) and ERASME (b), is represented as a function of volumetric SMC (Coppo et al., 1990).

A research carried out using C band airborne SAR images (Beaudoin et al., 1990) investigated the influence of random roughness and row pattern on radar backscatter and concluded that "for airborne SAR images, as far as a wide range of incidence angle is concerned, the inversion of SAR data into soil moisture data is unrealistic"; whereas, using the proposed European Remote Sensing Satellite (ERS-1) and RADARSAT instruments, with an incidence angle of about 20°, soil moisture inversion "appears highly probable" (Fig. 14.20). However, a theoretical study based on the scalar approximation, (Autret et al., 1989) has estimated the backscattering coefficient's sensitivity to a relative variation in soil parameter and has pointed out that the radiometric resolution of the ERS-1 SAR "is about the variation of the backscattering coefficient expected from a moisture and roughness variation".

Measurements of backscattering coefficients, performed with a truck mounted scatterometer operating from L to X band, indicated that the correlation between σ° and SMC is similar to the 4.6 GHz regression line for frequencies from 2.5 to 11 GHz (Fig. 14.21) (Huppi et al., 1985). Besides which, the measurements showed that freezing and thawing can cause immense changes in microwave signatures.

Satellite data

Two satellites boarding microwave radars have been launched in the past years: ERS-1/2 (at C-band, VV pol., $\theta = 22°$) and JERS-1 (at L-Band, HH pol., $\theta = 40°$). Although the ERS satellites were projected for sea studies, their data have been analyzed on solid earth's surface

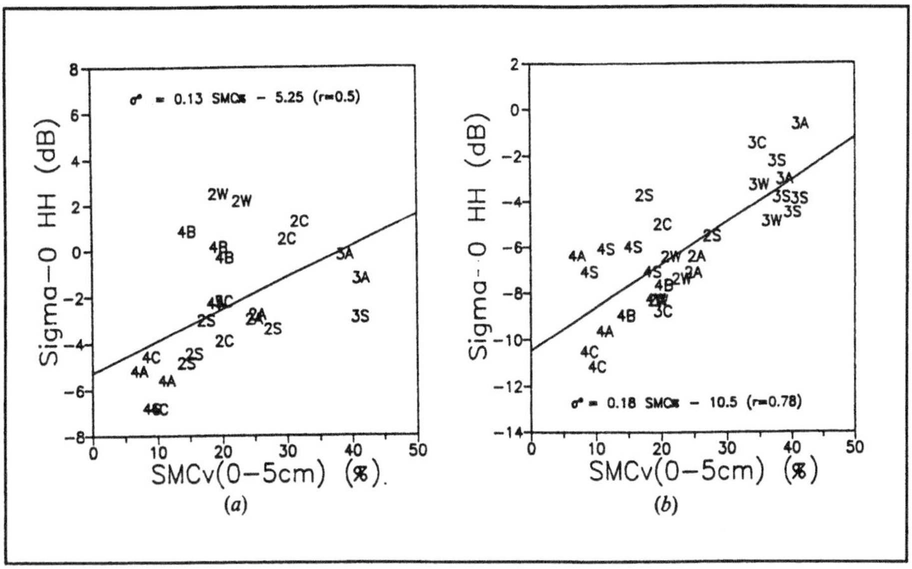

Figure 14.19: *σ^o as a function of SMC_v measured with (a) the DUTSCAT scatterometer at L-band (HH pol. and $\theta = 10^o$) and with (b) the ERASME at C-band (HH pol., $\theta = 15^o$). Letters refer to crop type (A = alfalfa, B = bare soil, C = corn, S = sunflower, W = wheat) and figures to overflight number (after Coppo et al., 1990).*

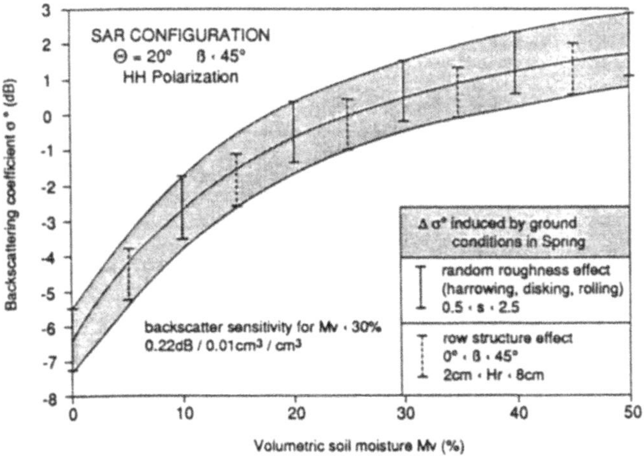

Figure 14.20: *Radar backscatter vs. volumetric SMC and simulated backscatter variations due to surface conditions in spring in RADARSAT configuration (after Beaudoin et al., 1990).*

as well, looking for a sensitivity to the soil moisture content. Some interesting results have been achieved in this sense by Dobson et al. (1992) using a first-order vector radiative transfer model (MIMICS) (Ulaby et al., 1990) with which the σ^o values over forest and grasslands has been predicted. In order to estimate the usefulness of ERS-1 SAR for estimating near-surface soil moisture, MIMICS simulations of σ^o have been compared to ERS-1 data on grasslands. The results showed that model outputs and ERS-1 measurements agreed to within 0.4 dB, whereas on forests the ERS-1 backscattering is dominated by properties of the crown layer and consequently none relationship with soil moisture can be obtained. The conclusion can be that the retrieval of soil moisture from ERS-1 data is feasible for surfaces covers characterized by low biomass (< 1 kg/m^2) such as short grasses.

JERS-1 is a satellite really conceived for soil and vegetation studies. Interesting results have been obtained in estimating biomass both of herbaceous and arboreous crops. On soil moisture some works have been carried out mainly by Japanese researchers and presented on the final report of JERS-1/ERS-1 system verification program (1995). Musiake et al. (1995) presented a comparison between ERS-1 and JERS-1 data in detecting soil moisture of agricultural fields. In general a fairly good relationship has been obtained between JERS-1 σ^o and the saturation degree S (which is the soil moisture content normalized to the saturation of soil), obtaining a regression equation of this type: $\sigma^o = 0.0458S - 14.219$, with a correlation coefficient of 0.93. This means that when saturated water content is 50% the sensitivity to soil water content is 0.1 dB/%. As far as ERS-1 data are concerned, a correlation with soil moisture variations is only obtained using multitemporal data sets, when the roughness and vegetation characteristics of the observed fields do not change in time.

MAC Europe campaign

A milestone in the improvement of SAR research has been represented by the NASA/JPL Aircraft Campaign named MAC-91 which has been conducted within the framework of SIR-C/X-SAR Project. It is well-known that, in general, multifrequency/multipolarization observations make it possible to obtain more information than single channel measurements do. For example: a combination of airborne active microwave (C band) and passive thermal infrared remote sensing (Soares et al., 1987), has shown that the spatial behaviour of surface temperature is correlated to the fields' canopy whilst the backscatter depends on soil parameters and mainly on SMC. Indeed, this experiment has been a systematic investigation, which aimed toward a better understanding of the information obtainable in agro-hydrology from the multi-frequency polarimetric SAR, and allowed the gathering of a large and useful data set which includes multi-temporal, multi-frequency and multi-incidence angle, polarimetric SAR measurements and ground truth.

In summer 1991, a DC-8, carrying a multifrequency, multipolarization SAR at P, L and C bands on board (AIRSAR), was flown over several European and American test-sites which were imaged on several dates and with different incidence angles. The SAR resolution in the multilook data was 6.66 m in slant range and 12 m in azimuth. In the framework of this experiment also the test site of Montespertoli (Italy), which was selected as a SIR-C / X-SAR 'supersite' for Hydrology, was imaged by the AIRSAR (Canuti et al., 1992). The landuse is represented by vineyards, oliveyards and some woods in the hills and agricultural crops (wheat, corn, sorghum, sunflower, alfalfa) in the flat areas along the rivers.

An example of recorded data is represented in Fig. 14.22, which shows the slant range images of the agricultural area at P-, L- and C-band in VH polarization. A simple photointerpretation of this composite image shows that the Pesa river which crosses the scene can easily be identified, as well as the village of Cerbaia on the left side which shows a very strong and diffused backscattering. Three corner reflectors are well recognizable as bright points on the right side of Cerbaia, south of the Pesa. Forests are also evident as bright areas and agricultural fields are clearly distinguishable: different levels of colour refer to different crops (Baronti et al., 1995).

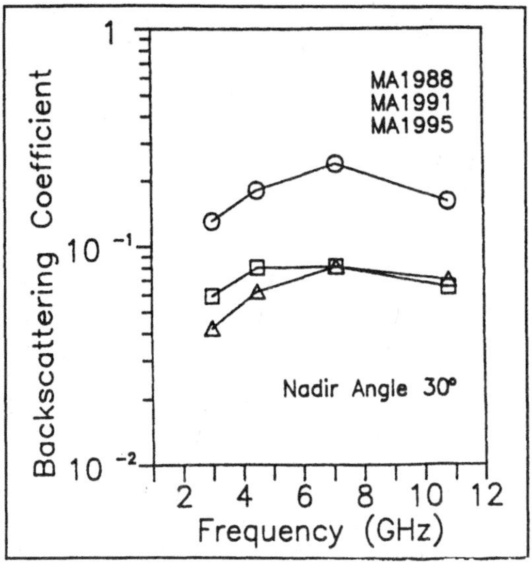

Figure 14.21: *Spectra of σ° of bare soil with 0.82 cm rms height and 2 cm correlation length. The variable parameter of 3 observations is the 0–4 cm layer SMC. (○ = 40%, □ =36%, △ =36%) (after Huppi et al., 1985).*

Despite the research effort carried out over the past years, a reliable extraction of soil moisture content (SMC) from SAR data is still questionable, especially since the signal is influenced by surface roughness and vegetation cover as well. In addition it is intrinsically difficult to obtain good data on SAR images at incidence angle close to 10° where the spurious effects are minimized. It has been rather difficult, even in our case too, as already noted, to identify fields and calibrate data even at $\theta = 20°$. Some preliminary results on the sensitivity to soil moisture (SMC) have been obtained at P band, at $\theta = 35°$ on bare or scarcely vegetated soils. Although such a high incidence angle is not yet the best configuration to enhance the sensitivity of radar to soil moisture and the range of variation of soil moisture is quite small, a clear increasing trend of σ° has been appreciated as the SMC increases. The obtained best fit regression line is: $\sigma^\circ = 1.07SMC - 34.7$. Analyzing L-band data, while at 35° the sensitivity and correlation to SMC were very poor, a fairly better result has been obtained, on agricultural fields, at 20° incidence angle and HH polarization, as shown in Fig. 14.23, where the σ°, for a certain number of bare and vegetation covered fields, is represented as a function of SMC. Data appear quite scattered, however, by separating bare from vegetated soils, the spread of data decreases and the correlation to soil moisture increases. As expected, the sensitivity is higher on bare smooth soils than on rough or vegetated terrains. The correlation coefficient, which is fairly good (r = 0.91) for bare soil, goes down to r = 0.43 on vegetated fields (Baronti et al., 1993; Baronti et al., 1995).

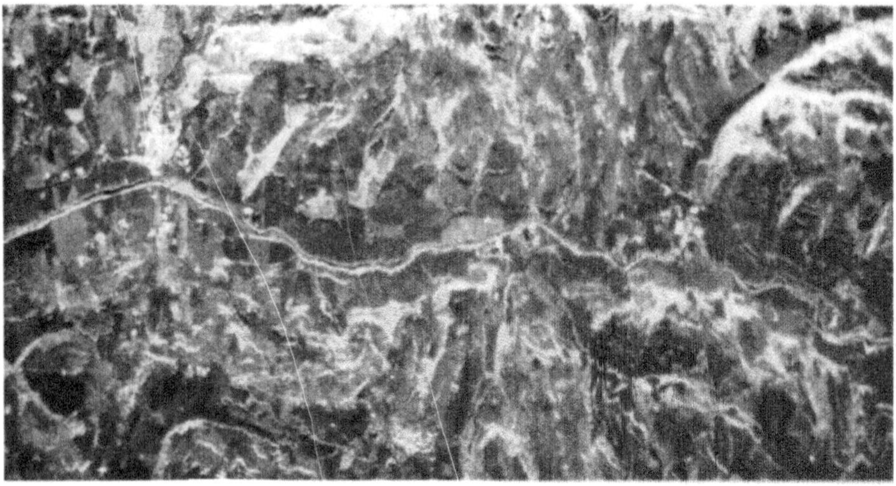

Figure 14.22: *SAR image at P-, L- and C-bands (HV pol.) of Montespertoli area (Tuscany) collected on June 1991 during the MAC '91 Campaign.*

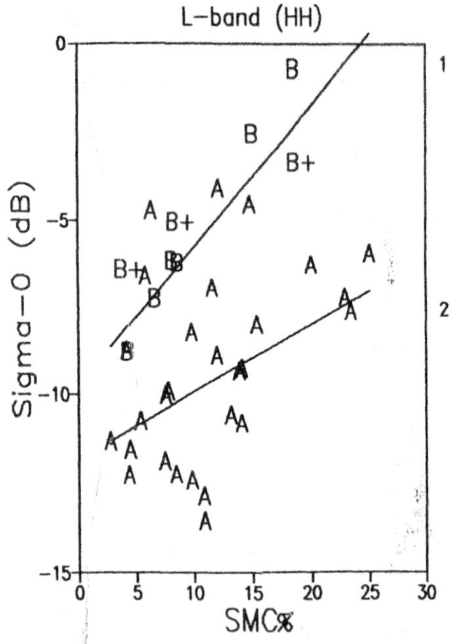

Figure 14.23: *$\sigma°$ at L-band (HH pol, $\theta = 20°$) as a function of SMC of bare (B) and vegetated fields (A). Lines represent the best fit for the two groups of data (after Baronti et al., 1995).*

SIR-C Measurements

The Shuttle Imaging Radar-C and X-Band Synthetic Aperture Radar (SIR-C/X-SAR) is a cooperative space shuttle experiment between NASA (National Aeronautics and Space Administration), DARA (German Space Agency) and ASI (Italian Space Agency). The experiment is the next evolutionary step in NASA's Spaceborne Imaging Radar (SIR) program that began with the Seasat Synthetic Aperture Radar (SAR) in 1978 and continues with SIR-A in 1981 and SIR-B in 1984. It is also a continuation of Germany's radar program which started with Microwave Remote Sensing Experiment (MRSE) flown aboard the Shuttle on the first SPACE-LAB mission in 1983. The SIR-C/X-SAR Mission benefits from synergism with aircraft sensors such as the JPL airborne SAR (AIRSAR) and the German Aerospace Establishment (DLR) E-SAR. The SIR-C, carrying on board a multifrequency SAR at X- (VV polarization), C-, and L-bands (fully polarimetric), was launched in April and in October 1994. This multifrequency, multipolarization capability can provide information about the Earth's surface over a wide range of scales for studying many land and ocean processes. Six research areas have been emphasized by this project: Ecology, Geology, Hydrology, Oceanography, Calibration/Validation Techniques and EM Theory. For each research area at least one supersite was selected among the other test sites, in order to concentrate the efforts of several groups on the same area and obtain as many data takes as possible on few sites. Montespertoli was the European test-site for Hydrology and 10 data takes were collected over it during the ten days of the mission, both in April and October. Ground truth measurements of soil, vegetation and meteo parameters were collected during the SIR-C passages. The objective of this experiment was mainly to analyze the radar response to many biophysical parameters, such as soil moisture, vegetation biomass, surface roughness and so on and test theoretical models for microwave propagation over natural surfaces. Till now only the data collected in April are available and researchers are working to analyze them and carry out the first results. An empirical algorithm for the retrieval of soil moisture and surface roughness, developed by Oh et al. (1992), has been tested by Dubois et al. (1995) using SIR-C and AIRSAR data collected on Washita Watershed site. This model was optimized for bare surfaces and requires two co-polarized channels at a frequency between 1.5 and 11 GHz. The result is shown in Fig. 14.24, where estimated and ground measured soil moisture values have been compared.

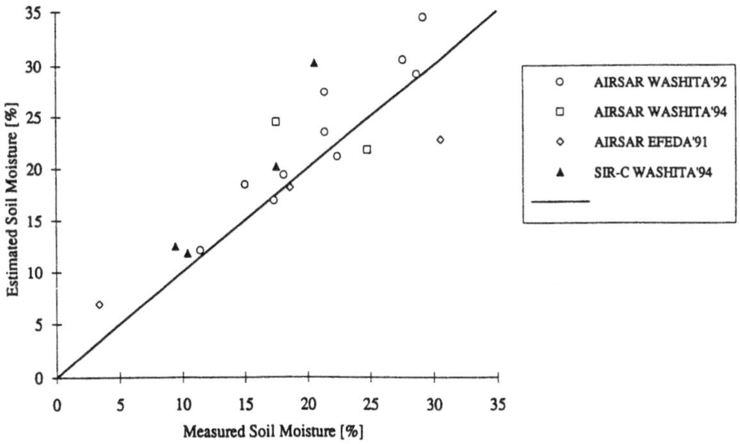

Figure 14.24: *Radar estimated and ground measured soil moisture content on Washita site. Both AIRSAR and SIR-C microwave polarimetric data sets have been used (after Dubois et al., 1995).*

14.4 Summary and Future Prospects

Remote sensing methods are of fundamental importance in the measurement of SMC on large areas. The most suitable sensors for this purpose are the ones operating in the microwave region of the electromagnetic spectrum. As a matter of fact these instruments, besides allowing all-weather, day/night operation, are very sensitive to the moisture content of a soil layer about 5 cm thick even if the terrain is covered by herbaceous crops.

Both active and passive systems showed good performances in such a type of measurements, although with some basic differences between them. Passive sensors in fact seem to be less affected by the geometrical features of the surface (i.e. surface roughness and vegetation) than active sensors and are therefore more accurate and sensitive, but their results are partially hampered by the poor spatial resolution, at least for an operational use from satellite. On the contrary, aircraft remote sensing seems to be very promising especially if the new generation of radiometers is used, such as the PBMR and the ESTAR. Active sensors have very high spatial resolutions, even from satellite platforms, but their sensitivity to moisture content is more attenuated by structural characteristics of the surface than passive sensors. These considerations suggest that the integrated use of multiband systems may allow still more precise and reliable estimates.

As far as passive sensors are concerned, we can say as conclusion that the L-band emission allows the measurement of the top-layer soil moisture content, while two or more frequency channels in the higher microwave bands (e.g. 10 GHz and 36 GHz) can be useful for estimating surface roughness on bare soils. On canopy covered fields, vegetation biomass can be estimated by using optical data or the X band microwave polarization index.

Multifrequency polarimetric SAR measurements, carried out over composite landscapes of forests and agricultural fields, indicate that SAR data are effective in discriminating among broad land surface categories. Although the first exciting results obtained with C-band data have been partially reappraised by later experiments, rather good correlation with soil moisture has been found on agricultural fields mainly at L-band (co-polar), a correlation which appreciably increases if bare soils are separated from vegetation surfaces. While a fairly good sensitivity and correlation to the Plant Water Content of some agricultural species has been observed at L-band especially for the cross-polarized σ°_{HV}.

The use of multifrequency, multipolarization remote sensors seems to be a promising approach both for evaluating the spurious factors which affect the relation between microwave signals and geophysical parameters and for improving the accuracy of measurements.

A valuable amount of experimental and theoretical research has been carried out worldwide, aimed at clarifying the physical principles of the measurement and to test and validate sensors and algorithms. For this purpose satellite data, as well as measurements performed by means of ground based and airborne sensors, have been used. Further research is in progress to implement the precision and reliability of the available methods according to users' requirements in different disciplines.

14.5 References

Autret M, Bernard R, Vidal-Madjar D (1989) Theoretical study of the sensitivity of the microwave backscattering coefficient to the soil surface parameters. *International Journal of Remote Sensing* 10: 171–179

Baronti S, Del Frate F, Ferrazzoli P, Paloscia S (1993) Interpretation of Polarimetric MAC-91 data over Montespertoli Agricultural area. *Proc. of the 25th Intern. Symposium, Remote Sensing and Global Change*, Graz (Austria), 4-8 April.

Baronti S, Del Frate F, Ferrazzoli P, Paloscia S, Pampaloni P, Schiavon G (1995) SAR Polarimetric features of agricultural areas. *International Journal of Remote Sensing* 16, 14, 2639–2656

Basharinov AE, Shutko AM (1975) Simulation studies of the SHF radiation characteristics of soils under moist conditions. NASA Tech. Trans. TTF-16, Greenbelt, MD

Basharinov AE, Krylova MS, Maslov AI, Shutko AM (1979) Remote sensing of subsurface soil moisture by means of microwave radiometers. *Water Res* 5: 538–542

Beaudoin A, Le Toan T, Gwyn Q (1990) SAR observations and modelling of the C-band backscatter variability, due to multi-scale geometry and soil moisture. *Institute of Electrical and Electronics Engineers Transactions on Geoscience and Remote Sensing* 28: 886–895

Becker F, Bolle H, Rowntree P (1988) The international satellite land-surface climatology project (ISLSCP). ISLSCP, Berlin Report no 10

Bernard R, Vidal-Madjar D, Baudin F, Laurent G (1986) Data Processing and Calibration for an Airborne Scatterometer. *Institute of Electrical and Electronics Engineers Transactions on Geoscience and Remote Sensing* 24: 709–716

Blyth K, Evans R (1985) Results in hydrology and soils. The European SAR-580 Experiment. Investigtors Final Report SAR-580 , edited by J.W.Trevett (Commission of European Communities, Joint Research Centre, Ispra Establishment) S.A./I.04.E2.85.12/2., 143–146

Campbell G, Norman J (1990) Estimation of plant water status from canopy temperature: an analysis of the inverse problem, in *Applications of Remote Sensing in Agriculture,* M.Steven and J.Clark editors, Butterworths, London, 255–271

Canuti P, D'Auria G, Pampaloni P, Solimini D (1992) MAC-91 on Montespertoli: an experiment for agro-hydrology. *Proc. of the International Geoscience and Remote Sensing Symposium - IGARSS 92*, Houston, Texas, pp. 1744–1746

Carlson T, Rose F, Perry E (1984) Regional-scale estimate of surface moisture availability from GOES Infrared satellite measurements. *Agronomy Journal* 76: 972–979

Coppo P, Ferrazzoli P, Paloscia S, Pampaloni P, Schiavon G, Solimini D (1990) Sensitivity of active and passive microwave sensors to soil moisture of vegetated fields. International Geoscience and Remote Sensing Symposium. Remote Sensing Science for the nineties, edited by R. Mills (IGARSS'90, Washington D.C.), 90CH2825-8, 1851–1854

Denmead OT (1976) Temperate Cereals, in "Vegetation and Atmosphere" Case study, vol. 2 (Ed. by J.L.Monteith), *Academic Press*, New York

De Loor GP (1979) Soil moisture determination at X band. European Association of Remote Sensing Laboratories (EARSeL). Workshop on Microwave Remote Sensing of Bare Soil, edited by N. Lannelongue (EARSeL,Paris), 198–202

De Loor GP (1982) The dielectric properties of wet materials. International Geoscience and Remote Sensing Symposium. Digest, edited by W.Keydel, (IGARSS'82, München), 82CH14723-6, TP-1, 1-5

Dobson C, Ulaby F, Hallikainen M, El-Rayes M (1985) Microwave Dielectric behaviour of wet soil - Part II: Four component dielectric mixing models. *IEEE Transactions on Geoscience and Remote Sensing* 23, 4, 35–46

Dobson C, Pierce L, Sarabandi K, Ulaby F, Sharik T (1992) Preliminary analysis of ERS-1 SAR for forest ecosystem studies. *IEEE Transactions on Geoscience and Remote Sensing* 30, 2, 203–211

Dubois P, van Zyl J, Engman T (1995) Measuring soil moisture with imaging radar. *IEEE Transactions on Geoscience and Remote Sensing* 33, 4, 915–926

Engman ET, Gurney RJ (1991) *Remote Sensing in Hydrology. Remote Sensing Applications*, Chapman and Hall, London

Hallikainen M,Ulaby F, Dobson M, El-rayes M, Wu L (1985) Microwave dielectric behavior of wet soil - Part I: Empirical models and experimental observations. *IEEE Transactions on Geoscience and Remote Sensing* 23, 25–34

Haefner H, Pampaloni P (1992) Water Resources. *International Journal of Remote Sensing* 13, 6-7, 1277–1303

Heilman J, Moore D (1982) Evaluating near-surface soil moisture using Heat Capacity Mapping Mission Data. *Remote Sensing of the Environment* 12: 117–121

Huppi R, Stotzer E, Schanda E (1985) Calibrated microwave signature measurements of soil and wheat. *Proceedings of the 3rd International Colloquium. Spectral Signatures of Objects in Remote Sensing*, edited by T.D. Guyenne (European Space Agency, Noordwijk) SP-247: 351–355

Huppi R, Schanda E (1986) L to X band scatter and emission measurements of vegetation. International Geoscience and Remote Sensing Symposium. Moving Towards the 21st Century, edited by T.D. Guyenne and J.J. Hunt (European Space Agency, Paris), SP - 284: 1113–1118

Idso S, Jackson R, Reginato R, Kimball B, Nakayama F (1975) The dependence of bare soil albedo on soil water content. *Journal of Applied Meteorology* 14: 109–113

Ijjas G, Juhasz I, Zilahy (1989) Airborne Microwave equipment for soil moisture investigation. Microwave Radiometry and Remote Sensing Applications, edited by P.Pampaloni (Utrecht, VSP), 321–328

Ijjas G, Rieger J, Aujesky L, Paloscia S, Susini C, Bechini C (1993) Microwave experiment on Montespertoli area. EARSeL Workshop on Agrometeorology, Florence, April 13-14 1992, EARSeL Advances in Remote sensing vol. 2 - VI: 40–45

Jackson RD (1982) Soil moisture inferences from thermal-infrared measurements of vegetation temperatures. *IEEE Transactions on Geoscience and Remote Sensing* 20: 282–286

Jackson T, Schmugge T, Wang J (1982) Passive microwave remote sensing of soil moisture under vegetation canopies. *Water Resources Research* 18: 1137–1142

Jackson T, Engman ET, Le Vine D, Schmugge T, Lang R, Wood E, Teng W (1994) Multitemporal passive microwave mapping in Machydro'90. *IEEE Transactions on Geoscience and Remote Sensing* 32,1, 201–206

Jackson T, Le Vine D, Griffis A, Goodrich D, Schmugge T, Swift C, O'Neill P (1993) Soil moisture and rainfall estimation over a semiarid environment with the ESTAR microwave radiometer. *IEEE Transactions on Geoscience and Remote Sensing* 31, 4, 836–841

Kirdiashev K, Chuklantsev A, Shutko A (1979) Microwave radiation of the earth's surface in the presence of vegetation cover. *Radio Engineering and Electronics* 24: 256–264 (English Translation)

Kramer PJ (1969) *Plant and Soil relationships. A modern synthesis.* McGraw - Hill Book Company, New York

Le Toan T (1982) Active microwave signatures of soil and crops: significant results of three years of experiments. International Geoscience and Remote Sensing Symposium. Digest, edited by W.Keydel, (IGARSS, Munchen), 82CH14723-6, TP2, 3.1-3.5

Le Vine D, Kao M, Tanner W, Swift C, Griffis A (1990) Initial results in the development of a synthetic aperture microwave radiometer. *IEEE Transactions on Geoscience and Remote Sensing* 28: 614–619

Mkrtchjan F, Reutov E, Shutko A, Kostov K, Michalev M, Nedeltchev N, Spasov A, Vichev B (1988) Experiments in Bulgaria for determination of soil moisture in the top one-meter layer using microwave radiometry and a priori information. International Geoscience and Remote Sensing Symposium. Moving Towards the 21st Century, edited by T.D. Guyenne and J.J. Hunt (European Space Agency, Paris), SP - 284, 665–666

Musiake K, Oki T, Nakaegawa T, Wakasa K (1995) Verification experiment of extraction of soil moisture information using SAR mounted on JERS-1/ERS-1. Final report of JERS-1/ERS-1 System Verification Program, MITI-NASDA, vol. II, 617–624

Oh Y, Sarabandi K, Ulaby F (1992) An empirical model and an inversion technique for radar scattering from bare soil surfaces. *IEEE Transactions on Geoscience and Remote Sensing* 30: 370–381

Paloscia S, Pampaloni P (1988) Microwave Polarization Index for Monitoring Vegetation Growth. *IEEE Transactions on Geoscience and Remote Sensing* 26: 617–621

Paloscia S, Pampaloni P, Chiarantini L, Coppo P, Gagliani S, Luzi G (1990) Multifrequency microwave radiometric measurements of soil moisture. International Geoscience and Remote Sensing Symposium. Remote Sensing Science for the Nineties, edited by R. Mills (IGARSS, Washington D.C.), 90CH2825-8, 1837–1840

Paloscia S, Pampaloni P (1992) "Microwave Vegetation Indexes for detecting biomass and water conditions of agricultural crops". *Remote Sensing of the Environment* 40: 15–26

Paloscia S, Pampaloni P, Chiarantini L, Coppo P, Gagliani S, Luzi G (1993) "Multifrequency passive microwave remote sensing of soil moisture and roughness". *International Journal of Remote Sensing* vol. 14, n.3, pp. 467–483

Pampaloni P (1981) Microwave Radiometry for Remote Sensing in agriculture. *Proceedings of 9th Conference of the Remote Sensing Society. Matching Remote Sensing Technologies and Their Applications*, edited by D. Lynn and J.Allan, (R.S.S., London), 273–283

Pampaloni P, Paloscia S (1985) Experimental relationships between microwave emission and vegetation features. *International Journal of Remote Sensing* 6: 315–323

Pampaloni P, Paloscia S (1986) Microwave emission and plant water content: a comparison between field measurement and theory. *IEEE Transactions on Geoscience and Remote Sensing* 24: 900–905

Pampaloni P, Chiarantini L, Coppo P, Gagliani G, Luzi G, Paloscia S (1990) Sampling Depth of Soil Moisture Content by Radiometric Measurements at 21 cm wavelength: Some Experimental Results. *International Journal of Remote Sensing* 11: 1085–1092

Reutov E, Shutko A (1986) Prior knowledge-based soil moisture determination by microwave radiometry. *Soviet Journal of Remote Sensing* 100–125

Rosema A, Bijleveld J, Reiniger P, Tassone G, Blyth K, Gurney R (1978) "Tellus", a combined surface temperature, soil moisture and evaporation mapping approach. 12th International Symposium on Remote sensing of Environment, Environmental Research Institute of Michigan, Manila

Schmugge T (1980) Effect of soil texture on the microwave emission from soils. *IEEE Transactions on Geoscience and Remote Sensing* 18: 353–361

Schmugge T, O'Neill P, Wang J (1986) Passive microwave soil moisture research. *IEEE Transactions on Geoscience and Remote Sensing* 24: 12–22

Schmugge T, Choudhury BJ A comparison of radiative transfer models for predicting the microwave emission fron soils. *Radio Science* 16, 5, 927–938

Schmugge T, Becker F (1991) Remote Sensing observations for the monitoring of Land-Surface fluxes and water budget, in *Land Surface Evaporation - Measurement and Parametrization*, Springer-Verlag, New York Inc., Ch.20, 337–348

Seguin B, Itier B (1983) Using midday surface temperature to estimate daily evaporation from satellite thermal IR data. *International Journal of Remote sensing* 4: 371–383

Shivola A (1989) Properties of dielectric mixtures with layered spherical inclusions. In *Microwave Radiometry and Remote Sensing Applications*, edited by P. Pampaloni, (Utrecht, VSP), 115–123

Shutko A (1982) Microwave radiometry of lands under natural and artificial moistening. *IEEE Transactions on Geoscience and Remote Sensing* 20: 18–26

Shutko A, Reutov E (1982) Mixture formulas applied in estimation of dielectric and radiative characteristics of soils and grounds at microwave frequencies. *IEEE Transactions on Geoscience and Remote Sensing* 20: 29–32

Soares J, Bernard R, Vidal-Madjar D (1987) Spatial and temporal behaviour of a large agricultural area as observed from airborne C-band scatterometer and thermal infrared radiometer. *International Journal of Remote Sensing* 8: 981–996

Simonett DS Ed. *Manual of Remote Sensing*, vol. I: Theory, Instruments and Techniques, American Society of Photogrammetry, Falls Church, Virginia, 1983

Stotzer E, Wegmüller U, Huppi R, Maetzler C (1986) Dielectric and surface parameters related to microwave scatter and emission properties. International Geoscience and Remote Sensing Symposium. Today's Solutions for Tomorrow's Information Needs, edited by T.D. Guyenne (European Space Agency, Nordwijk), SP - 254, 599–603

Tsang L, Kong J (1979) Wave theory for microwave remote sensing of a half- space random medium with three dimensional variations. *Radio Science* 14: 359–369

Ulaby F, Moore R, Fung A *Microwave Remote Sensing: Active and Passive*, Addison-Wesley Publication Co.. Advanced Book Program/World Science Division, Readings, Massachusetts, 1981

Ulaby F, Aslam A, Dobson C (1982) Effects of vegetation cover on the radar sensitivity to soil moisture. *IEEE Transactions on Geoscience and Remote Sensing* 20, 4, 476–481

Ulaby F, Razani M, Dobson C (1983) Effects of vegetation cover on the microwave radiometric sensitivity to soil moisture. *IEEE Transactions on Geoscience and Remote Sensing* 21: 51–61

Ulaby F, Sarabandi K, Mc DOnald K, Whitt M, Dobson C (1990) Michigan microwave canopy scattering model. *Int Journal of Remote Sensing* 11, 7, 1223–1253

Wang J, Schmugge T (1980) An empirical model for the complex dielectric permittivity of soils as a function of water content. *IEEE Transactions on Geoscience and Remote Sensing* 18: 288–295

Chapter 15

Remote Sensing Parameterization of Meso-scale Land Surface Evaporation

W.G.M. Bastiaanssen, M. Menenti, A.J. Dolman, R.A. Feddes and H. Pelgrum
DLO Winand Staring Centre for Integrated Land, Soil and Water Research
P.O. Box 25, NL-6700 AC Wageningen
The Netherlands

15.1 Introduction

Land surface fluxes vary spatially as a result of the spatial heterogeneity of soil physical properties, fractional soil cover, land use, rainfall and hydrological processes. The scale at which this variability of surface fluxes occurs may be as low as a fraction of a meter, and time variations over a period of hours may be significant. The spatial average land surface flux at regional scale is because of length scale and sampling size difficult to quantify from field measurements. Although considerable progress has been made in this area over the last decade (e.g. Michand and Shuttleworth, 1996), there is no general consensus at this moment how to extrapolate surface fluxes from patch scale to meso-scale.

Atmospheric Models are the only available tools to forecast climate trends and to assess the impact of reduced emissions strategies on global warming. The grid-lengths of these models at which the land surface fluxes need to be described are of the order of 100 km to 300 km; at this scale the land surface processes exhibit a considerable spatial variability. Hence, a serious problem exists in the application of large scale models: *How do we test the prediction of area-averaged land surface fluxes at meso-scale?* The number of land surface parameterization schemes which simulate meso- scale land surface fluxes by continuity of energy, momentum, heat and water is still growing (e.g. Sellers et al., 1986; Noilhan and Planton, 1989; Dolman, 1993) while research on techniques to quantify model input parameters and verify model output parameters at the scale of a grid-square is still lagging behind. Henderson-Sellers et al. (1995) for the PILPS project intercompared the performance of 24 land surface parameterization schemes and plans to use observational data of relative homogeneous grasslands to validate the fluxes. Tower-based flux measurements apply at scales smaller than the correlation length of the fluxes, and a direct link between model output at a horizontal scale manyfold larger than the correlation length of field measurements is not straightforward. In this respect, remote sensing measurements may be useful, since the land surface characteristics can be measured by satellite platforms with different spatial (as low as meters) and temporal resolutions (as low as half hours) over large areas.

NATO ASI Series, Vol. I 45
Radiation and Water in the Climate System:
Remote Measurements
Edited by Ehrhard Raschke

The sensitivity of results of Atmospheric Models to the representation of land surface processes was emphasized by Eagleson (1981). At that time, rather simple parameterizations of land surface hydrology were applied, for instance that developed by Manabe (1969) and based on earlier work by Thornthwaite and Mather (1955) and Budyko (1956). Land surface hydrology was basically reduced to a soil water reservoir, assumed to be of equal size everywhere in the world (i.e. water storage capacity was independent of soil type, depth, etc.). This land surface parameterization scheme, however, was sufficient to show that soil water content can have a significant effect on the allocation of net radiation into sensible, latent and soil heat flux and, therefore, on atmospheric circulation processes (Mintz, 1982; Rowntree and Bolton, 1983). More recent research with more advanced land surface hydrology parameterizations confirmed that surface albedo, soil water content and surface roughness are very critical for adequate simulations of the climate system (e.g. Sud et al., 1990; Milly and Dunne, 1994)

This paper describes some classes of remote sensing algorithms which cover different approaches to estimate large scale energy balances from remote sensing measurements. The key-problem associated with energy balance modelling on the basis of thermal infrared measurements is to adequately determine the sensible heat flux from surface radiation temperatures. The aim of this paper is to address the classical problem of inferring a vertical air temperature difference from spectral radiances and shows how several classes of existing remote sensing algorithms deal with this issue. Some new 'internal calibration' techniques to get meso-scale sensible heat fluxes in their correct physical range are the cornerstones of the Surface Energy Balance Algorithm for Land (SEBAL) and will be outlined (Bastiaanssen, 1995). Some examples of the mapping of meso- scale sensible and latent heat fluxes during the EFEDA experiment conducted in Castilla la Mancha, Spain (Bolle et al., 1993), will be described.

15.2 Land Surface Fluxes and their Relation to Remote Sensing Estimates of Surface Characteristics

Radiometers measure the radiances in selected bands throughout the visible, infrared and microwave region at the top of the atmosphere and no possibility exists to measure the turbulent fluxes directly from these platforms (Menenti, 1993). However, several land surface state variables and properties mentioned in Table 15.1 can be deduced from these spectral radiances and are essentially the key parameters marked by Garrat (1993) for improving the description of land surface and atmosphere boundary layer treatments.

Since only the parameters mentioned in Table 15.1 are measurable, regional latent and sensible heat fluxes with current satellite systems need to be estimated indirectly. To obtain estimates of meso- scale land surface evaporation, a remote sensing parameterization scheme has therefore to be developed, i.e. a schematization of the sensible (H) and latent heat (λE) fluxes based on the parameters listed in Table 15.1. The conventional micro-meteorological framework for turbulent fluxes needs to be briefly discussed prior to identify the data gaps between Table 15.1 and H and λE fluxes at composite land surfaces.

The relationship between mean turbulent momentum τ, sensible H and latent heat λ_E fluxes are usually described by:

$$\tau = \rho_a u_*^2 \quad (\mathrm{N\ m^{-2}}) \tag{15.1}$$

$$H = -\rho_a c_p T_* u_* \quad (\mathrm{W\ m^{-2}}) \tag{15.2}$$

$$\lambda E = -\lambda q_a q_* u_* \quad (\mathrm{W\ m^{-2}}) \tag{15.3}$$

where τ (N m^{-2}) is the momentum flux, ρ_a (kg m^{-3}) is the air density, u_* (m s^{-1}) is the friction velocity, c_p (J kg^{-1} K^{-1}) is the air specific heat at constant pressure, T_* (K) is the temperature

Remote sensing version	**Land surface attribute**	**Source**
Surface radiation temperature	Surface temperature, T_0	Wan and Dozier, 1989; Becker and Li, 1990
Spectral radiation temperature	Surface thermal emittance, ϵ_0	Kahle and Alley, 1992; Schmugge et al., 1995
Surface directional reflectance	Surface hemispherical reflectance, ρ_0	Pinty and Raymond, 1987; Menenti et al., 1989
Degree of cloudiness	Solar radiation, $K^{\downarrow}$	Dedieu et al., 1987; Gurney et al, 1994
Spectral reflected radiances	Vegetation indices, NDVI	Justice et al., 1985; Tucker, 1986
Laser beam profiles	Aerodynamic roughness length, z_{0m}	Menenti and Ritchie, 1994
Laser beam profiles	Leaf area index, LAI	Menenti and Ritchie, 1994
Spectral reflected radiances	Leaf area index, LAI	Asrar et al., 1992; Pierce et al. 1992
Backscatter coefficient	Soil moisture, Θ	Fung et al., 1992; Oh et al., 1992
Microwave emittance	Soil moisture, Θ	Schmugge et al., 1992; Van de Griend and Owe, 1994

Table 15.1: *Radiometric characteristics of the land surface and their interpretation into attributes being ultimately linked to sensible and latent heat fluxes.*

scale, λ (J kg^{-1}) is the latent heat of vaporization, q_a (-) is the air specific humidity and q_* (-) is the humidity scale. A full explanation of the symbols is presented in Appendix 1. It is useful to express the u_*, T_* and q_* scales into profiles of horizontal windspeed u, potential temperature T_p and air specific humidity q_a respectively:

$$u_*^{-1} = \frac{1}{u(z)k}\{\ln(\frac{z}{z_{0m}}) - \Psi_m(z, L)\} \quad (\mathrm{s\ m^{-1}}) \tag{15.4}$$

$$T_*^{-1} = \frac{1}{\{T_{z0h} - T_p(z)\}k}\{\ln(\frac{z}{z_{0h}}) - \Psi_h(z, L)\} \quad (\mathrm{K^{-1}}) \tag{15.5}$$

$$q_*^{-1} = \frac{1}{\{q_{z0h} - q_a(z)\}k}\{\ln(\frac{z}{z_{0v}}) - \Psi_v(z, L)\} \quad (\mathrm{Kg^{-1}m^{-3}}) \tag{15.6}$$

where k (-) is the von Karman constant, z (m) is the reference height, z_{0m} (m) is the roughness length for momentum transport, Ψ_m (-) is the stability correction for momentum transport, L (m) is the Monin Obukhov length, T_{z0h} (K) is the heat source temperature, z_{0h} (m) is the roughness length for heat transport, Ψ_h (-) is stability correction for heat, z_{0v} (m) is the roughness length for vapour transport and Ψ_v (-) is the stability correction for vapour transport. The Monin-Obukhov length is expressed as the ratio of mechanical production of kinetic energy divided by the theoretical convective production of turbulent kinetic energy and used to account for stability correction of the vertical profiles of horizontal wind speed $u(z)$, potential temperature $T_p(z)$ and humidity profiles $q_a(z)$ derivable from Eqs. (15.4) to (15.6):

$$L = -\frac{\rho_a c_p u_*^3 T_a}{kgH} \quad (\mathrm{m}) \tag{15.7}$$

The integrated stability correction functions $\Psi_m(z, L)$, $\Psi_h(z, L)$ and $\Psi_v(z, L)$ are related to the Monin-Obukhov length L by empirical functions (e.g. Paulson, 1970). Comparing Table 15.1 with Eqs. (15.4) and (15.5), shows that z_{0m} and T_{z0h} can be estimated from remote sensing data and that some simplification have to be introduced to quantify the many remaing model parameters. Alternatively, the sensible heat flux can be expressed in analogy to Ohm's law, as a vertical difference in entity and a resistance by combining Eqs. (15.2), (15.4) and (15.5):

$$H = \frac{\rho_a c_p}{r_{ah}}(T_{z0h} - T_p) \quad (\mathrm{W\ m^{-2}}) \tag{15.8}$$

where r_{ah} (s m-1) is the aerodynamic resistance to heat transport (see Eq. 15.10). Although not identical, the heat source temperature, T_{z0h}, controlled by turbulent exchange processes is theoretically related to the surface temperature, T_0, and it is customary to adjust z_{0h} as the 'radiometric surface roughness' (Sugita and Brutsaert, 1990) in a way that T_{z0h} and T_0 become identical ($T_{z0h} = T_0$). The same can be done for latent heat flux which after including a surface resistance (Monteith, 1965) and some manipulations becomes:

$$\lambda E = \frac{\rho_a c_p}{\gamma(r_{av} + r_s)}\{e_{sat}(T_0) - e\} \quad (\mathrm{W\ m^{-2}}) \tag{15.9}$$

The explicit expression for the aerodynamic resistance to heat transfer used in Eq. (15.8) is:

$$r_{ah} = \frac{1}{ku_*}\{\ln(\frac{z}{z_{0h}}) - \Psi_h(z, L)\} \quad (\mathrm{s\ m^{-1}}) \tag{15.10}$$

The roughness length for momentum (z_{0m}, Eq. (15.4) and heat (z_{0h}, Eqs. (15.5) and (15.10)) are not identical, because the transfer processes of heat and momentum in close proximity of obstacles are not similar (Brutsaert, 1982). A standard expression to account for z_{0m} and z_{0h} differences on homogeneous surfaces is the introduction of the kB^{-1} factor (Chamberlain, 1966):

$$kB^{-1} = \ln \frac{z_{0m}}{z_{0h}} \quad (-) \tag{15.11}$$

There is a slight indication that for 'permeable-rough' surfaces such as uniform canopies, kB^{-1} is approximately 2 to 3 (Choudhury 1989; Prevot et al., 1994). However, heterogeneous surfaces consisting of distinct components, for example bare soil and vegetation, generate different component momentum and heat fluxes and kB^{-1} can hardly be solved in a straightforward manner (Beljaars and Holtslag, 1991; Blyth and Dolman, 1995; Holtslag and Ek, 1996). The aerodynamic resistance to vapour transport necessary to solve Eq. (15.9) can be written as:

$$r_{av} = \frac{1}{ku_*}\{\ln(\frac{z}{z_{0v}}) - \Psi_v(z, L)\} \quad (\mathrm{s\ m^{-1}}) \tag{15.12}$$

The differences between z_{0m} and z_{0v} are poorly understood. The bulk surface resistance to evaporation r_s reflects the overall resistance a vapour molecule meets on its flow path from its source into the atmosphere. The resistance r_s for composite land use is somehow a function of canopy resistance r_c, soil resistance r_{soil} and fractional soil cover s_c:

$$r_s = f(r_c, r_{soil}, s_c) \quad (\mathrm{s\ m^{-1}}) \tag{15.13}$$

in which the canopy resistance is a non-linear function of the following parameters (Jarvis, 1976):

$$r_c = f\{r_c^{min}(LAI)g_1(\theta)g_2(K^\downarrow)g_3(\Delta e)g_4(T_a)\} \quad (\mathrm{s\ m^{-1}}) \tag{15.14}$$

and the soil resistance is a function of soil state variables and properties (Bastiaanssen, 1995):

$$r_{soil} = f\{\theta_0, \theta_z, \rho_{z0v}, \rho_v^{sat}(z_e), D_\theta^{eff}, D_v^{eff}\} \quad (\mathrm{s\ m^{-1}}) \tag{15.15}$$

Combination of the theories opens the possbility to relate the surface energy balance terms H and λE to the net available energy, $Q^* - G_0$ as a function of various Soil-Vegetation-Atmosphere-Transfer parameters (e.g. Stricker et al., 1993):

$$\begin{aligned} Q^*(r_0, K^\downarrow, L^\downarrow, \epsilon_0, T_0) &= G_0(T_0, T_s, \lambda_s) + H(u_*, T_0, T_a, z_{0m}, kB^{-1}, L) \\ &\quad + \lambda E(u_*, T_0, e, z_{0m}, kB^{-1}, L, \theta(z), \rho_v(z), K^\downarrow, LAI, \Delta e, D_\theta^{eff}, D_v^{eff}) \\ &\quad (\mathrm{W\ m^{-2}}) \end{aligned} \tag{15.16}$$

Considering the extensive number of Soil-Vegetation-Atmosphere-Transfer SVAT parameters mentioned in Eq. (15.16) needed to solve λ_E (or actually r_s defined in Eq. (15.13) and specified in Eqs. (15.14) and (15.15), as compared to Q^*, G_0 and H, it is common to compute land surface evaporation as the residue of the land surface energy budget (e.g. Jackson et al., 1977):

$$\lambda E = Q^* - G_0 - H \quad (\mathrm{W\ m^{-2}}) \tag{15.17}$$

Bastiaanssen et al. (1994) applied Eq. (15.17) to obtain λE. Afterwards, Eq. (15.9) was used to assess r_s by inversion of the transfer equation for latent heat flux. Such data assimilation of r_s was performed to test the consistancies between prognostic variables (e.g. θ in Eqs. (15.14) and (15.15)), fluxes (λE in Eq. (15.9)) and budgets (Q^*, G_0, H and λE in Eq. (15.17)). Hence, the skill of methods to obtain land surface evaporation using remote sensing data depends on H, if $Q^* - G_0$ can be solved with a reasonable accuracy. Choudhury (1991) and Kustas et al. (1994) provided examples of procedures to quantify $Q^* - G_0$ and this paper will not further elaborate this issue. The remaining concerns for solving H using Eq. (15.8) are then:

- Minimize the difference in remote sensing surface skin temperature T_0 and surface heat source temperature T_{z0h} by adjusting kB^{-1};
- Spatial variation of kB^{-1} at composite terrain;
- Spatial variation of T_a or $T_{z0h} - T_a$ at composite terrain;

These three items will now be elaborated upon.

15.3 Surface Temperature

Thermal radiometers measure spectrally emitted radiances which are related to temperature through Planck's law:

$$L_\lambda(T_0) = \frac{3.7427 \cdot 10^8}{\lambda^5}\{\exp(\frac{14388}{\lambda T_0}) - 1\}^{-1} \quad (\text{W m}^{-2}\mu\text{m}^{-1}) \tag{15.18}$$

where λ (μm) is the wavelength, L_λ (Wm^{-2} mm^{-1}) is the spectral radiance and T_0 (K) the surface temperature. The constants comprise the frequency, Plancks constant, the speed of ligth and the Boltzmann constant. Eq. (15.18) gives the radiation for a perfect black body emitter with the surface emittance one ($\epsilon_0 = 1.0$). On the contrary, grey body radiators reflect a small fraction of longwave radiation ($\epsilon_0 \neq 1.0$). Natural land surfaces behave usually as grey bodies and consequently a correction is required to interpretL_λ (T) measurements into temperatureT_0:

$$L_\lambda(T_0^R) = \epsilon_0(\lambda)L_\lambda(T_0) \quad (\text{W m}^{-2}\mu\text{m}^{-1}) \tag{15.19}$$

Salisbury and D'Aria (1992) provided tables on $\epsilon_0(\lambda)$ for terrestrial materials. Atmospheric disturbances may have a significant effect to convert the outgoing narrow band longwave radiation at the top of the atmosphere L_λ^{TOA} into radiation emitted and reflected at the land surface $L_\lambda(T_0^R)$. Atmospheric correction schemes are indispensible and may vary from highly sophisticated physically based radiative transfer models (e.g. Lowtran, Kneizys et al., 1988) to simple empirical methods. The split window technique (Price, 1984) can be applied if measurements of at least two thermal channels are available. Basically, the atmospheric correction consists of the following equation:

$$L_\lambda^{TOA} = \{\epsilon_0(\lambda)L_\lambda(T_0) + (1 - \epsilon_0(\lambda))L_\lambda^{\downarrow}\}\tau_\lambda + L_\lambda^{\uparrow} \quad (\text{W m}^{-2}\mu\text{m}^{-1}) \tag{15.20}$$

where $L_\lambda^{\uparrow}$ is the upwelling spectral radiance from atmospheric emission and scattering that reaches the sensor, $L_\lambda^{\downarrow}$ is the downwelling spectral radiance from atmospheric emission incident upon the land surface and τ_λ is the spectral atmospheric transmission. *In-situ* measurements of T_0 at homogeneous targets sufficiently large to cover at least one single pixel, can be used as input in Eq. (15.18) to compute $L_\lambda(T_0)$ at any desired wavelength interval. By comparing a set of at-surface $L_\lambda(T_0)$ values with L_λ^{TOA}-values for a range of cold and warm surface targets at the same spectral interval measurement by a remote thermal scanner, the unknown τ_λ and $L_\lambda^{\uparrow}$ values required to obtain the atmospheric correction specified by Eq. (15.20) can be obtained. Such anchor- station procedure to obtain T_0-values ensures realistic T_0-values and can be an attractive alternative if radiosoundings are not available.

Hence, T_0 can be obtained after atmospheric and emittance corrections, but it has to be recalled that Eq. (15.8) requires T_{z0h}. For large pixels with resolutions of 1 km or more (which are useful for a (sub-) continental coverage), it is likely that an individual pixel encompasses a composite land surface. The same applies to 120 m Thematic Mapper pixels overlaying a land surface with sparse canopies. Consequently, T_0 measurements may have a 'mixed' character because only the total radiance related to canopy temperature and bare soil temperature is measured. In that case, a transfer equation being similar to Eq. (15.8) but applying to heterogeneous land surfaces can be tentatively written:

$$H^{mix} = \rho_a c_p \frac{T_{z0h}^{mix} - T_a^{mix}}{r_{ah}^{mix}} \quad (\text{W m}^{-2}) \tag{15.21}$$

which defines T_{z0h}^{mix}, T_a^{mix}, r_{ah}^{mix} and H^{mix} and $T_a^{mix} = T_p^{mix}$ if the difference between the air layers is kept small.

15.4 Surface Roughness for Heat Transport

The intriguing question is whether a one-layer parameterization scheme, such as presented in Eq. (15.21), can provide a correct description of H^{mix} for a heterogeneous land surface. Basically two questions can be distinguished:

1. What is the relationship between T_{z0h}^{mix} required by Eq. (15.21) and T_0^{mix} derivable from space measurements after atmospheric and emittance corrections?

2. Does linear averaging of H_i from individual land surface elements having different surface temperatures into $< H >$ yield the same value H^{mix} obtained from T_0^{mix}, T_a^{mix} and r_{ah}^{mix} using Eq. (15.21)?

In mathematical terms, the statement to be verified is:

$$\{T_0^{mix} - T_a^{mix}\} \frac{\rho_a c_p}{r_{ah}^{mix}} = \frac{1}{n} \sum_{i=1,n} \{(T_{z0h} - T_a)(\frac{\rho_a c_p}{r_{ah}})\}_i \quad (\text{W m}^{-2}) \tag{15.22}$$

where n denotes the total number of elementary fluxes i being located within the area encompassed by a single T_0^{mix} measurement. The answer to question 1 will be dealt with first. Eq. (15.5) is verified by an infinite number of (z_{0h}, T_{z0h})-pairs at a constant T_* of the same $T_p(z)$ profile (see Eq. (15.5)). Since $z_{0m} \neq z_{0h}$, the definition of z_{0h} is somewhat loose. The height of z_{0h} can therefore be more pragmatically defined as the height z_{0h} so that $T_0 = T_{z0h}$ (Fig. 15.2). Kustas et al. (1989) optimized kB^{-1} in the same way by assuming $T_{z0h}^{mix} = T_0^{mix}$, as a starting point and gave by doing that, a precise definition of z_{0h}. The authors found an average value of $kB^{-1} = 5.6 \pm 3.2$ for natural vegetative surfaces in California. Beljaars and Holtslag (1991) advocated, applying the same philosophy, for complex sparse canopies and for $T_{z0h}^{mix} = T_0^{mix}$, z_{0m} to use 3 to 4 orders of magnitude larger than z_{0h} ($kB^{-1} = 8.8$). Brutsaert et al. (1993) also calibrated kB^{-1} in an inverse manner assuming that $T_0^{mix} = T_{z0h}^{mix}$, using measurements of H, and found that $kB^{-1} = 1$ to 2.3 for forests at a regional scale in HAPEX Mobilhy. Stewart et al. (1994) summarized kB^{-1} investigations in several large scale field experiments and gave values of $kB^{-1} = 4$ for grass and $kB^{-1} = 12.4$ for shrubs. Hence, T_{z0h}^{mix} may be equated to T_0^{mix} after adjusting the definition of by means of kB^{-1}, but the case studies mentioned reveal a wide range of kB^{-1} values which seems to change with land use and spatial scale. With the dual source model of Dolman (1993), Blyth and Dolman (1996) showed that z_{0h} for sparse canopies is affected by available energy, humidity deficit and vegetation cover and that general rules for kB^{-1} on heterogeneous land surfaces cannot be given *a priori*.

Dolman and Blyth (1996) used the concept of blending height, i.e. the elevation in the PBL at which fluxes are independent of surface features, to predict effective r_{ah}^{eff} values from which the

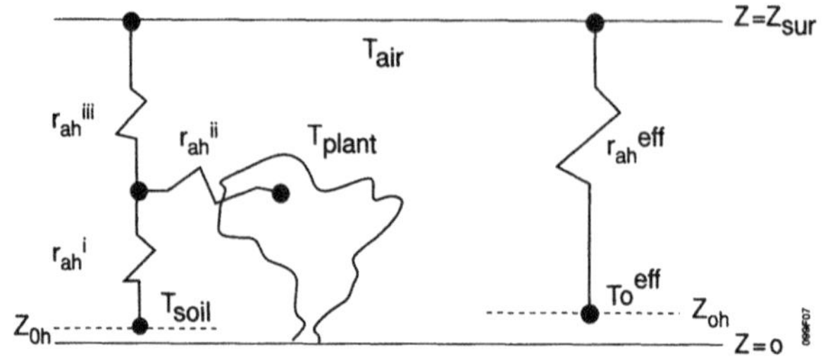

Figure 15.1: *Schematic representation of a bulk resistance formulation for sensible heat transfer taking into account an idealized surface plane having a surface temperature equal to the 'heat source' temperature.*

corresponding roughnesses could be inverted. They showed that the fractional soil cover and the contrast of r_s and z_{0m} between neighbouring patches have significant effects on kB^{-1} at meso-scale. Their recommendation is to apply two-layer resistance schemes which explicitly deal with different r, H and λE fluxes for soil and vegetation (Fig. 15.1A). Hence, the non- univocallity of kB^{-1} is a crucial problem for the application of one-layer resistance parameterization schemes in composite terrain (Fig. 15.1B) and although T_{z0h}^{mix} may be equated to T_0^{mix} if kB^{-1} is properly adjusted, it at the same time has to be concluded that guidelines for kB^{-1} at meso-scale are missing. In the following pages it will be shown that this z_{0h}-problem can be circumvented because of the observed correlation of the air-to-surface temperature difference with surface temperature. For a better appreciation of the background principles of the blending height, one is referred to Wieringa (1986) and Claussen (1990).

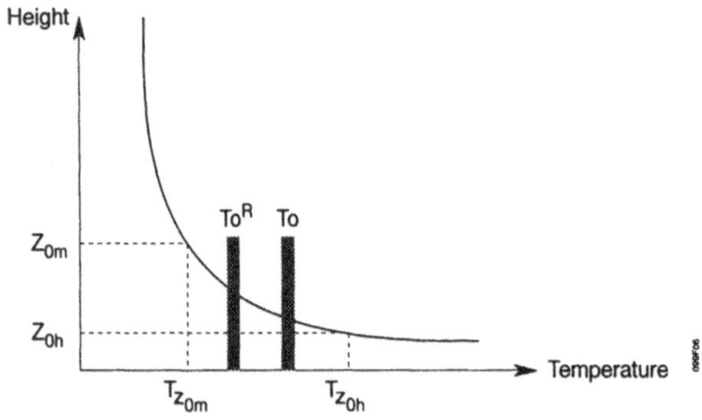

Figure 15.2: *Logarithmic temperature profile emphasizing the difference between aerodynamic surface temperature T_{z0m}, heat source temperature T_{z0h}, radiation temperature T_0^R and skin temperature T_0.*

15.5 Length Scale of Sensible Heat Flux

Surface energy balance studies with remote sensing data are traditionally based on an one-dimensional representation of the exchange rates between land and atmosphere. On heterogeneous land surfaces, horizontal advection between different land surface elements however becomes important and the vertical profiles of horizontal wind speed, heat and vapour may be perturbated by lateral effects (Fig. 15.3).

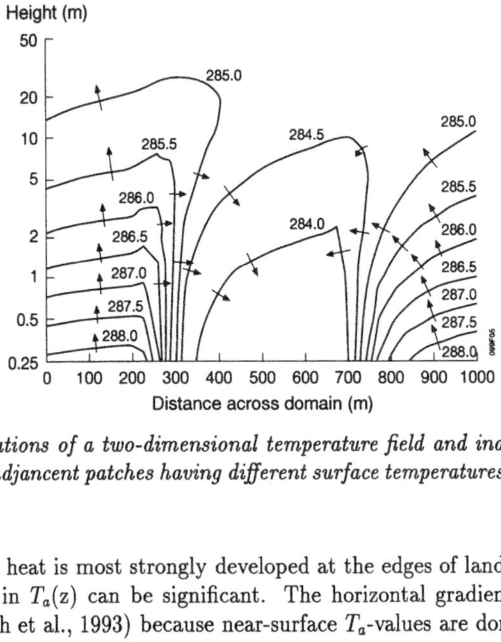

Figure 15.3: *Simulations of a two-dimensional temperature field and indications of heat flow directions above two adjancent patches having different surface temperatures (adapted after Blyth et al., 1993).*

Advection of sensible heat is most strongly developed at the edges of land elements where the horizontal difference in T_a(z) can be significant. The horizontal gradients are most evident at surface level (Blyth et al., 1993) because near-surface T_a-values are dominantly established by the underlaying hydro-meteorological processes. It may also be remarked from Fig. 15.3 that T_a at an elevation of 2 to 10 m differ in the horizontal domain which has consequences for the schematization of vertical near-surface H-fluxes. The vertical profiles at the leading edges at 250 and 750 m distance from the origin in Fig. 15.3 are distorted and the vertical fluxes should be calculated at some distance from this edge. The height of the upper air layer z_{sur} (Fig. 15.1) used to calculater_{ah} according to Eq. (15.10), should therefore be chosen following general laws for vertical and horizontal length scales. The horizontal scale of a fetch is defined by its surface roughness and friction velocity over which the surface flux becomes in equilibrium to that particular surface conditions. This scale depends primarily on the height above the terrain, i.e. the height of the internal adapated layer (De Bruin et al, 1991). For momentum transfer, horizontal L_x, and vertical L_y length scales, are generally related according to Townsend, (1965):

$$\frac{L_y}{L_x} \sim \frac{u_*}{u} \quad (-) \tag{15.23}$$

Using typical values suggests that $L_y/L_x \sim 0.01$ to 0.1 and the vertical profiles and fluxes are thus adapted at an altitude approximately one-tenth of the fetch length. Remote sensing measurements of T_0^{mix} on a pixel-by-pixel basis implies that spatial variation in T_0^{mix} cannot be detected at horizontal scales smaller than the size of a pixel. Hence, homogeneity must be assumed for each separated pixel and step-changes in surface conditions are schematized at the pixel edges (if the image comprises a composite land surface). If the size of a pixel is schematized to be 2 L_x (T_0^{mix} applies to the centre of the pixel), Eq. (15.23) implies that $z_{sur} \leq L_y \approx 0.1 L_x$ (Fig. 15.4).

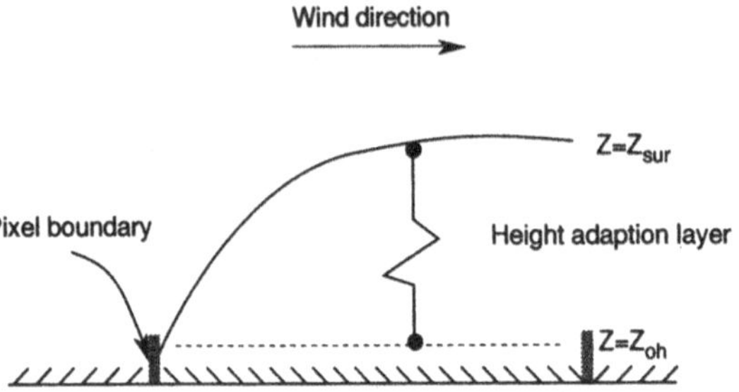

Figure 15.4: *Schematic representation of the height of the adapted air layer as a function of the distance from the leading edge.*

The selection of the vertical length scale z_{sur} has consequences for the quantification and spatial variability of T_a^{mix} and r_{ah}^{mix}. Different treatments of T_a^{mix} and r_{ah}^{mix} in published remote sensing studies can be distinguished. Table 15.2 presents an overview of the spatial treatments of $T_a^{mix}$and r_{ah}^{mix}, with r_{ah}^{mix} being further decomposed into z_{0h}^{mix}, u_*^{mix} and Ψ_h^{mix}. Table 15.3 provides some references dealing with the classes of remote sensing algorithms described in Table 15.2.

Obviously, most algorithms treat T_a^{mix} to be areally constant either at $z_{sur} \sim 2$ m. (e.g. class 2: Kustas et al., 1994) or at larger height inside the mixed layer (e.g. class 3: Brutsaert et al., 1993). Considering Eq. (15.23), a constant T_a^{mix}-value at larger elevation (e.g. 100 m) should be combined with large pixel sizes (NOAA-AVHRR and METEOSAT). Menenti and Choudhury (1994) stated ' the spatial resolution of infared temperature measurements must to be consistent with the aerodynamic characteristics associated with r_{ah}^{mix} '. Since the areal variation of T_a^{mix} above the blending height for heat is eliminated by advection, the sensible heat flux between the land surface and the blending height is not a strictly one-dimensional flow process. Pixelwise H-fluxes calculated with remote sensing algorithms schematizing sensible heat transport between the blending height and the land surface to be solely vertical contain therefore a bias per definition. Table 15.2 further shows that the momentum flux is often kept areally constant (i.e. u_{mix} is constant for all classes except class 5). This is far from trivial, and an issue which deserves considerable attention in future investigations. If the purpose of study is to assess the spatial variation of local near-surface fluxes, z_{sur} may be kept small (Figures 15.3 and 15.4) and advection may by ignored. Then T_a^{mix} should be taken close to the land surface and spatially variable. The basic question which now remains is:

How can the regional distribution of T_a^{mix} adjacent to the land surface be determined in the absence or incompleteness of synoptic stations?

15.6 Near-surface Vertical Air Temperature Difference

Derivation of T_a-patterns from satellite thermal infrared data was carried out by Chen et al. (1983) who investigated the capability of geostationary GOES thermal infrared data to infer

Classes of H-schematizations	T_a^{mix} ($z_{sur} \sim 2$ m)	T_a^{mix} ($z_{sur} \sim 100$ m)	z_{0h}^{mix}	u_*^{mix}	Ψ_h^{mix}
Class 1	c	n	c	c	n
Class 2	c	n	v	c	c
Class 3	n	c	c	c	c
Class 4	n	c	v	c	c
Class 5	v	c	v	v	v

Table 15.2: *Application studies on solving Eq. (15.21) using conceptually different solutions for T_a^{mix} and r_{ah}^{mix}: c = areally constant, v = areally variable, n = not considered.*

Class 1	Jackson et al., 1977; Seguin and Itier, 1983; Hall et al., 1992
Class 2	Kustas et al., 1994; Kalma and Jupp, 1990
Class 3	Rosema, 1990; Brustsaert et al., 1993
Class 4	Menenti and Choudhury, 1994; Diak and Whipple, 1994
Class 5	Klaassen and van den Berg, 1985; Carlson and Buffum, 1989; Bastiaanssen, 1995

Table 15.3: *Selected references on the classes of remote sensing algorithms listed in Table 15.2.*

T_a. A value of $R^2 = 0.76$ was found between linear regression of T_0 and 1.5 meter high shelter T_a- values. The standard deviation from the regression line was 1.3–2.0°C. Davis and Tarpley (1983) reported from NOAA temperature data also a standard linear trend between T_0 and T_a over a large area of North America with standard deviations of 1.6–2.6 K. Linear fitting between T_0 and T_a for a number of land surface elements implies a constant offset. The offset of a regional $T_a(T_0)$ relationship theoretically implies that the product of H and r_{ah} is constant, which is very unlikely to happen:

$$T_{a-i} = c_1 T_{0-i} - c_2 \quad (K) \tag{15.24}$$

where i represents a particular surface element. If $c_1 = 1.0$, c_2 will exactly represent the area-effective heat flux H^{eff} for a heterogeneous land surface and r_{ah}^{eff} the resistance of the same targets:

$$c_2 = \frac{r_{ah}^{eff} H^{eff}}{\rho_a c_p} \quad (K) \tag{15.25}$$

The linearity between T_{0-i} and T_{a-i} posed in Eq. (15.24) may however not be generally applied because the spatial variability of H and r_{ah} in heterogeneous land surfaces is evident, i.e. $H^{eff} \neq H_i$ and $r_{ah}^{eff} \neq r_{ah-i}$ and a constant offset in the form of c_2 is thus unlikely. Although T_{a-i} at near-surface level is certainly not constant, the dynamic range of T_{a-i} within a certain landscape is reduced as compared to the dynamic range of a large set of $r_{ah-i}$and H_i values within the same landscape. It is therefore preferred to consider the near-surface vertical air temperature T_{a-i} as the offset factor:

$$\delta T_{a-i} = c_3 T_{0-i} - c_4 \quad (°C) \tag{15.26}$$

Day	Location	Hour local time	n	R^2	Intercept c_4 (°C)	Slope c_3 (°C °C^{-1})
03/09/93	Degraded bush	9:00	5	0.93	33.9	1.24
		10:00	5	0.94	27.6	0.98
		11:00	5	0.98	30.3	1.01
		12:00	5	0.98	27.4	0.91
		13:00	5	0.98	29.8	0.95
05/09/93	Fallow bush/ grassland	9:00	4	0.89	30.4	1.08
		10:00	4	0.98	28.3	0.99
		11:00	4	0.99	30.4	1.01
		12:00	4	0.99	31.4	0.99
		13:00	4	0.98	34.9	1.04
		14:00	4	0.95	38.2	1.09
		15:00	4	0.99	40.9	1.14
		16:00	4	0.99	37.3	1.06
08/09/93	Fallow bush/ grassland	9:00	5	0.97	43.0	1.45
		10:00	5	0.97	32.7	1.08
		11:00	5	0.98	28.4	0.92
		12:00	5	0.99	28.5	0.89
		13:00	5	0.99	30.7	0.93
		14:00	5	0.99	33.0	0.98
		15:00	5	0.99	33.4	0.98
		16:00	5	0.98	35.6	0.98

Table 15.4: *Field observations performed during HAPEX-Sahel in Niger at the Central West Supersite to prove the linear relationship between near-surface vertical air temperature difference, δT_a, and surface temperature, T_0.*

If $c_3 = 1.0$, c_4 is an estimate of the area-representative air temperature T_a^{eff}. In a more general form, T_a^{eff} can be calculated from c_3 and c_4 as:

$$T_a^{eff} = (1 - c_3)T_0^{eff} + c_4 \quad (°C) \tag{15.27}$$

A deeper investigation towards the shape of the $\delta T_{a-i}(T_{0-i})$ relationship is required to check the generallity of Eq. (15.26). Data collected in Niger, Egypt and China has been used to study c_3 and c_4 in composite terrain.

Niger

During the Special Observation Period of HAPEX-Sahel (Goutorbe et al., 1993), T_{a-i} was recorded at 3 meter elevation at several sites in a Sahelian landscape with bushes and bare soil intermixed. δT_{a-i} was calculated for each surface type as the difference in $T_{0-i}$and T_{a-i}. The 4 to 5 different land units were located on an approximately 2 km long transect at the West Central Super-site. The results are presented in Table 15.4. The regression coefficients for the different days and moments of observation were rather similar with a mean intercept being $c_4 = 32.7$°C and a mean slope of $c_3 = 0.98$°C °C^{-1}. The slope is indeed very close to one which proves the correctness of Eq. (15.26) and $c_4 \sim T_a^{eff}$ may be applied as a first order estimate.

Egypt

During an evaporation study in the sub-sealevel Qattara Depression in the Western Desert of Egypt (Menenti et al., 1991), 3 (δT_{a-i}, T_{0-i}) parallel pairs were measured continuously with a 5' sampling interval during 7 days in September 1989. The observation height for T_{a-i} was 200 cm. The offset in the $\delta T_a(T_0)$ function shows a diurnal pattern (Fig. 15.5B) whereas the slope is fairly constant at $c_3 \sim 1.0$ (Fig. 15.5A). The diurnal variation of c_4 has a trend similar to $T_a(t)$ with warming ups in the morning and a plateau in the afternoon. The sudden decrease of $c_3 \sim 35°C$ at 20:00 hours to $c_3 \sim 20°C$ at 21:00 hours is most probably associated with the inversion of the $T_a(z)$ profiles and the collapse of the PBL. The average slope for the wet saline bare soils in the Qattara Depression is $c_3 = 1.09°C\ °C^{-1}$ and the intercept is $c_4 = 37.1°C$.

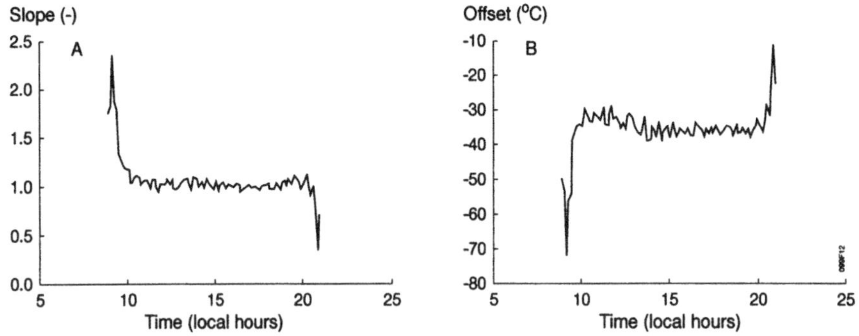

Figure 15.5: *Diurnal variation of the slope c_3 in the linear relationship between near-surface vertical air temperature differences and surface temperature (Part A). Intercept c_4 of the same relationship based on field measurements on September 15, 1989, Qattara Depression (Part B).*

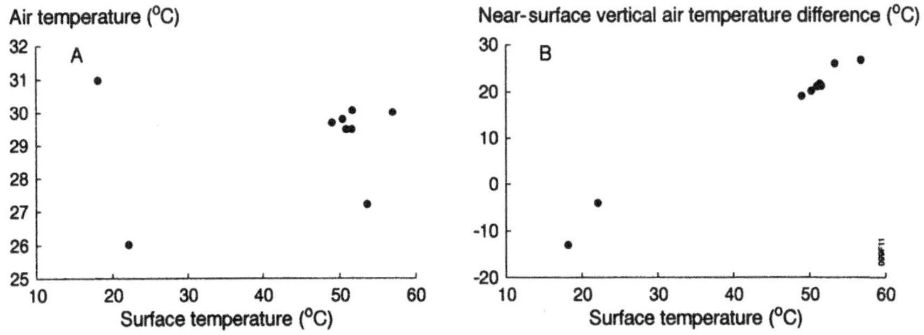

Figure 15.6: *Observed behaviour of near-surface vertical air temperature differences, δT_a, and surface temperature, T_0, in the HeiHe basin with distinct land surface types such as desert, oases and mountains along a 50 km transect, July 9, 1991: $T_a(T_0)$ part A; $\delta T_a(T_0)$ part B.*

China

Data from the HEIFE experiment in Central China (Wang et al., 1994) has been considered to analyse the $\delta T_a(T_0)$ relationship for landscapes where extremes in physiography and hydrology exists. The study area encompasses the Gobi desert, irrigated oases near the towns of Zhan Ye and Linzhe and mountainous hillslopes. The distance between the most far away measurement points was approximately 50 km. Fig. 15.6 shows the observed trends at 11:00 hours local time for T_0, T_a and δT_a. At this time, the lowest air temperature was $T_a = 26°$C while the hot spots gave already $T_a = 31°$C at the same elevation above the land surface. The depicted $T_a(T_0)$ relationship in Fig. 15.6A expresses much scatter with significant standard deviations from the regression line. On the contrary, the $\delta T_a(T_0)$ trend of Fig. 15.6B has a strongly reduced scatter. Hence, the expected linear trend for the $\delta T_a(T_0)$ relationship is also confirmed for the arid conditions of the HEIHE basin for several days with $c_3 = 0.97$°C °C^{-1} and $c_4 = 27.8$°C on average.

The experimental evidence of the $\delta T_a(T_0)$ relationship can be summarized by the following conclusions:

- The linear $\delta T_a(T_0)$ hypothesis for heterogeneous terrain conditions can be experimentally proven and the slope lies close to one (Niger 0.98, Egypt 1.09, China 0.97) which implies that c_4 is a good first order estimation of T_a^{eff};

- δT_a can be spatially interpolated using a T_0-image if c_3 and c_4 for that particular image and moment of acquisition are determined;

15.7 Implications of the Current kB^{-1} and δT$_a$ Findings for Meso-scale Sensible Heat Flux Estimations

Because T_{z0h}^{mix} and T_a^{mix} or T_{a-i} need to be sampled at exactly the same moment and place (remote radiometers: T_{z0h}^{mix}; *in-situ* observations: T_{a-i}, radiosoundings: T_a^{mix}) and with a very high precision, it is challenging to obtain T_{z0h}^{mix} and T_a^{mix} for Eq. (15.21) with the required accuracy. Furthermore, it was concluded previously that kB^{-1} needs to be calibrated against local τ and H-fluxes, which outside HAPEX experiments at meso-scale is hardly feasible. However, kB^{-1} and δT_a are related because T_0 increases with kB^{-1} (Fig. 15.1) which directly affects δT_a ($\delta T_a = T_0 - T_a$). In other words, an increased resistance r_{ah}^{mix} by an increase of kB^{-1} should according to the concept of logarithmic profiles in the constant flux layer result in an increased δT_a^{mix} value. On similar grounds, Kustas et al. (1989) found that kB^{-1} can be related to δT_a and u. They calibrated kB^{-1} and extrapolated $kB^{-1}(t)$ in the time domain using $T_0(t)$ (not in the space domain). Hence, either kB^{-1} or δT_a^{mix} can be calibrated in one-layer resistance schemes (Fig. 15.1) to obtain appropriate H-fluxes. Considering on one hand the results of the previous section showing the potential to interpolate δT_a regionally for a wide range of surface types and climate conditions and on the other hand the difficulties to predict the spatial distribution of kB^{-1} in complex terrain (Dolman and Blyth, 1996), it is preferred for heterogenoues meso-scale land surface to calibrate δT_a^{mix} rather than kB^{-1}. Then kB^{-1} should be areally fixed a priori and δT_a^{mix} will be adjusted to that kB^{-1} value in order to maintain adequate H-fluxes. Since H-fluxes are usually not available, it is further proposed to optimize δT_a^{mix} from sub-areas (groups of pixels) where H reaches its limits. For those sub-areas, an inversion of the transfer equation for sensible heat can be applied which excludes the necessity to collect δT_a^{mix} from scarse density observation networks in complex landscapes:

$$\delta T_a^{mix} = \frac{H^{mix} r_{ah}^{mix}}{\rho_a c_p} \quad (^\circ\mathrm{C}) \tag{15.28}$$

It should be stressed that r_{ah}^{mix} in Eq. (15.28) should be based on as best as possible estimates of kB^{-1}. For specific pixels where $\lambda E^{mix} \sim 0$, H^{mix} may be determined from $\{Q_* - G_0\}^{mix}$ without involving Eqs. (15.1) to (15.7). The areal maximum value for δT_a^{mix} will in this way be obtained. Also the other extreme, viz. $H^{mix} = 0$, provides information on the magnitude of $\delta T_a^{mix}(\delta T_a^{mix} \sim 0)$. A new approach by combining $(T_{0-j}^{mix}, \delta T_{a-j}^{mix})$ and $(T_{0-jj}^{mix}, \delta T_{a-jj}^{mix})$ where j represents a large group of pixels with dry land surfaces ($\lambda E^{mix} \sim 0$) and jj the class with wet surfaces ($H^{mix} \sim 0$), provides the opportunity to internally calibrate c_3 and c_4 on a regional scale for the moment of image acquisition without needing *in-situ* T_a^{mix} observations or radiosoundings. This approach should be conceived as an essential element in the improvement for estimating H from thermal remote sensing at data sparse terrain. In this way, the high absolute accuracy of T_0^{mix} required in sensible heat flux estimations (error ≤ 0.5 K: Brutsaert et al., 1993) is not longer needed: δT_a^{mix} is solved from H^{mix} of specific surfaces according to Eq. (15.28), rather than from T_0^{mix} and T_{a-i}^{mix} seperated. Although T_0^{mix} is still needed to spatially interpolate δT_a^{mix}, the standard error of T_0^{mix} becomes to a certain extent irrelevant (it only affect c_3 and c_4, not the dynamic range of H).

15.8 Classes of Remote Sensing Flux Algorithms

The impact of the selection of z_{sur} and spatial variability of T_a^{mix} for the different H-schematizations addressed at Table 15.2 is briefly examined by means of an example with EFEDA data. Five different land use types (A to E) which differ in surface radiative properties are selected from a Thematic Mapper image acquired on June 12, 1991 (Pelgrum and Bastiaanssen, 1996). Table 15.5 shows the primary remote sensing measurements on T_0^{mix}, r_0^{mix} and $NDVI^{mix}$. The remaining SVAT- parameters in Table 15.5, are derived from semi-empirical relationships, i.e. ϵ_0^{mix}, z_{0m}^{mix}, δT_a^{mix} and u_*^{mix} and empirically determined using the equations proposed by Bastiaanssen (1995).

The air temperature was in this case considered at 10 m above surface level. The air temperature at the blending height (in this case an arbitrary value of 100 m. has been considered) was 22.8°C while the wind speed at the same height is 5.7 m s^{-1}. The air heat capacity was taken as 1230 J m^{-3} K^{-1} and $kB^{-1} = 2.3$ has been taken as an arbitrary value. To compare the H-fluxes computed with the various remote sensing algorithm classes distinguished in Table 15.3, area-effective values of u_*^{mix}, z_{0m}^{mix}, T_0^{mix} and $T_a^{mix}(z_{sur})$ are required as input for algorithm class 1 to 4. The following aggregation rules have been applied to obtain u_*^{eff}, z_{0m}^{eff}, T_0^{eff} for class 1 to 4. The value for T_a^{eff} has been quantified from T_0^{eff} using Eq. (15.27).

$$u_*^{eff} = \{\frac{1}{n}\sum \rho_a u_*^{mix2}\}^{0.5}\rho_a^{-1} \quad (\mathrm{m\ s^{-1}}) \tag{15.29}$$

$$z_{0m}^{eff} = \exp\{\frac{1}{n}\sum \ln z_{0m}^{mix}\} \quad (\mathrm{m}) \tag{15.30}$$

$$T_0^{eff} = \{(\frac{1}{n}\sum \epsilon_0^{mix}\sigma T_0^{mix4})(<\epsilon_0>\sigma)\}^{0.25} \quad (\mathrm{K}) \tag{15.31}$$

The impact of the selection of the algorithm on local H^{mix}-fluxes for land use types A to E and on the area- averaged flux, $< H >$, is presented in Fig. 15.7. The area-averaged flux was

Land use type Parameter	A	B	C	D	E	Regional (A to E aggregated)
T_0^{eff} (°C)	23.6	29.7	35.0	42.8	46.8	30.3
r_0 (-)	0.17	0.17	0.20	0.35	0.23	0.22
NDVI (-)	0.71	0.24	0.41	0.17	0.18	0.34
ϵ_0 (-)	0.99	0.94	0.97	0.93	0.93	0.97
z_{0m} (m)	$1.2\ 10^{-1}$	$5.9\ 10^{-3}$	$1.8\ 10^{-2}$	$3.8\ 10^{-3}$	$4.1\ 10^{-3}$	0.011
$\delta T_a(10)$ (°C)	1.8	6.0	9.9	15.6	18.3	7.0
u_* (m s^{-1})	0.35	0.24	0.27	0.23	0.23	0.27
$T_a(10)$ (°C)	21.8	23.7	25.1	27.2	28.5	23.3

Table 15.5: *Remote sensing measurements of land surface and related hydro-meteorological parameters for five different land use types in EFEDA (A to E) having different radiative surface properties (A: Irrigated cropped pivots, B: Irrigated agricultural plots at crop emergence, C: natural vegetation with sparse canopies, D: bare soil (calcareous) and E: bare soil.*

computed as the arithmetic mean of the five local H^{mix}-fluxes. Considerable differences can be identified for the isolated land use types, especially for type E. The difference of algorithm classes 1 and 4 at land type E is 34%. Land type B seems to be rather insensitive to the schematization of H because u_*^{eff} (0.27 m s^{-1}), z_{0m}^{eff} (0.011 m), T_0^{eff} (30.3° C) and T_a^{eff} (23.3°C) are approximately similar to the local values at type B ($u = 0.24 ms^{-1}, z_{0m} = 0.006m, T_0 =$ 29.7°C, $T_a = 23.7$°C). At regional scale the differences are somewhat averaged out but are still significant (upto 30%) and thought to be a direct consequence of the length scale and areal variation of the various SVAT parameters considered. Since it describes most extensively the local micro-meteorological processes, algorithm class 5 is preferred. Whether its performance is better as compared to the other classes is something to be examined with field measurements in the near-future. Classes 3 and 4 are simpler to implement in regional studies as compared to class 5 (for instance Klaassen and van de Berg need a PBL model). Hence, regional evaporation from heterogeneous land surfaces can be reasonably obtained from class 3,4 and 5, while class 5 is from a theoretical viewpoint preferred to describe local fluxes and soil moisture conditions. SEBAL belongs to class 5 and its potential will be further demonstrated with EFEDA ground and aircraft fluxes in the next section.

15.9 A Case Study with SEBAL Using EFEDA Data

During June 1991, more than 30 scientific teams worked in Castilla-La Mancha (Spain), studying the energy and water transfer processes between soil, vegetation and the atmosphere in semi- arid conditions within the coordinated (Bolle et al., 1993) European research project EFEDA (European Field Experiment in Desertification-threatened Areas). Measurements were made from the micro-scale (e.g. measurements on single plants) up to a scale compatible with the grid size of Atmospheric Models. For this purpose, three super-sites were selected 70 km apart and heavily instrumented at the scale of the order of 30 km (Fig. 15.8): the Tomelloso, Barrax and Belmonte supersites. Aircraft missions, satellite data and movable equipment were deployed to provide a bridge to the larger scale.

An estimation of macro-scale surface energy balances using SEBAL including the whole Iberian Peninsula was made by using NDVI from NOAA-AVHRR, surface hemispherical reflectance r_0 from METEOSAT and skin temperature T_0 also from METEOSAT. Fig. 15.9 shows the sensible

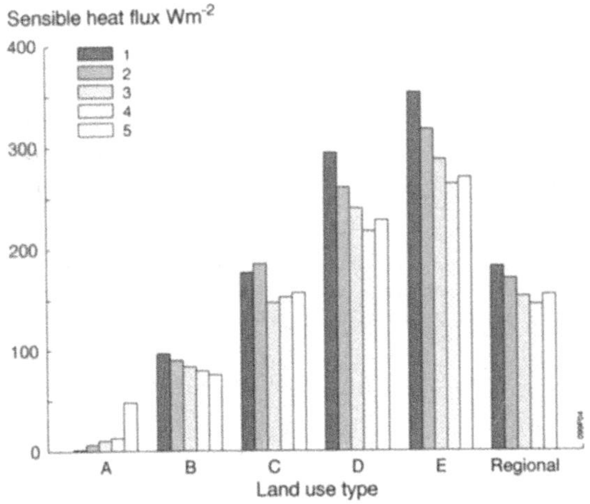

Figure 15.7: *Patch scale and regional scale sensible heat fluxes calculated with the classes of remote sensing algorithms summarized in Table 15.2 and using the data presented in Table 15.5.*

heat flux isolines on June 29, 1991 at 10:00 GMT for the entire EFEDA grid. Due to the limited possibility to measure the natural variation of fluxes over heterogeneous land surfaces, flux maps with a horizontal scale of 4 km such as Fig. 15.9 are usually difficult to validate. Field measurements of turbulent fluxes by means of instrumented towers are representative for relatively small source areas, i.e. the footprint in the upwind direction which contributes in the establishment of a certain flux (e.g. Gash, 1986). Fig. 15.10 demonstrates the range of in-situ evaporative fraction values as well as the area-aggregated value obtained from all flux towers which were installed within a given METEOSAT pixel. It seemed that only 3 out of the 595 METEOSAT pixels could be used for comparison. To overcome the differences in integration times of the differnent flux stations and the SEBAL fluxes (having an instantaneous time scale), relative fluxes by means of the evaporative fraction, Λ, were compared (Bastiaanssen et al., 1996) where Λ is defined as $\lambda E/(Q_* - G_0)$. Fig. 15.10 shows that the SEBAL evaporative fractions are at least within the range of ground measurements, but a validation in the real sence could not be accomplished because the estimated area-effective Λ-value derived from *in-situ* observations is not necessarily correct since all footprints of the flux towers together represent only a minor fraction of one single 16 km^2 large METEOSAT pixel.

Since the METEOSAT flux maps cannot be compared with field measurements directly, an intercomparison with the aircraft fluxes was made (Fig. 15.11). The DLR Falcon 20 aircraft was flown on June 19, 21 and 28 and measured the turbulent fluxes at flighing altitude (Jochum et al., 1993). Although these fluxes, due to elevation effects, cannot be compared directly with the at- surface fluxes and sea breeze induced lateral inflow of heat and moisture. A comparison at a scale of 85 km by 125 km is worked out. Fig. 15.11 shows that Λ on June 29 and calculated with SEBAL lies close to Λ-Falcon on June 28. The average evaporative fraction calculated according to Eq. (15.32) considering leg ML1, ML2, ML4, ML6 and ML7 gave $\Lambda^{eff}_{FALCON} = 0.24$. The June 29 image for an area of 85 km $\times$ 125 km covering almost the entire EFEDA grid gave an average of $\Lambda^{eff}_{SEBAL} = 0.21$. Actually, there are clear reasons

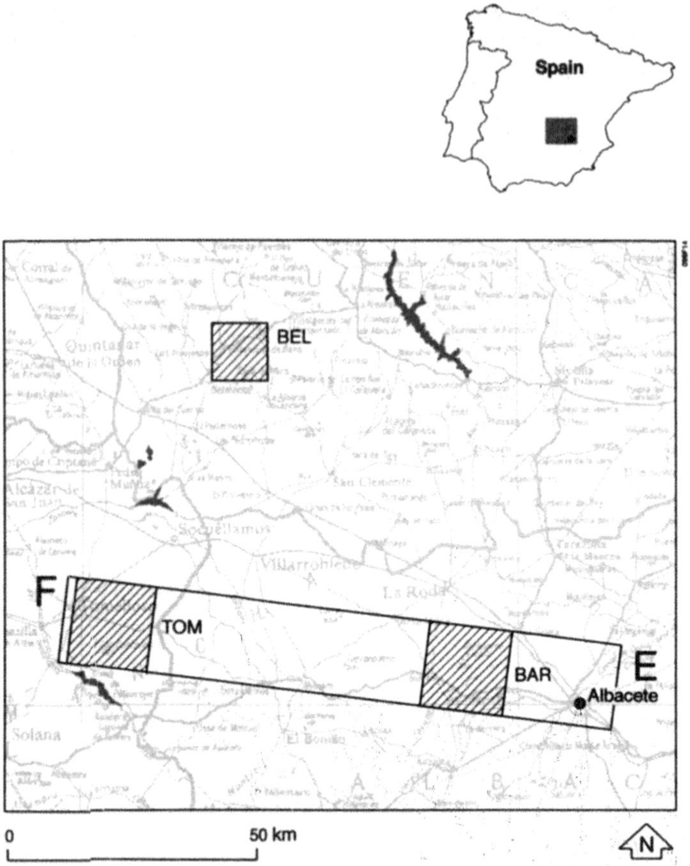

Figure 15.8: *Location of the EFEDA square grid in Castilla la Mancha, Central Spain. The three super-site, Tomelloso, Barrax and Belmonte are indicated as 'TOM', 'BAR' and 'BEL'.*

to assume that Λ^{eff}_{SEBAL}-value should be slightly higher approaching the $\Lambda^{eff}_{FALCON} = 0.24$ value rather well ($\Lambda^{eff}_{SEBAL} - \Lambda^{eff}_{FALCON} \sim 0.0$ to 0.02): the Northwest corner of Fig. 15.9, where the wettest spots are located around the Embalso de Alarcon lake, was overcast at 10:00 GMT and could therefore not be incorporated into Λ^{eff}_{SEBAL}.

$$\Lambda^{eff} = \frac{< \lambda E >}{< H > + < \lambda E >} \quad (-) \tag{15.32}$$

Figures 15.12 and 15.13 demonstrate the capability of METEOSAT thermal infrared measurements to describe the diurnal variation of the sensible and latent fluxes. One single pixel being located at the irrigated plains just North of Albacete has been selected comprising a mixture of agricultural crops (Fig. 15.12). The energy partioning between λE and H is almost similar and λE has a peak of approximately 250 Wm^{-2}. A daytime integration of λE (setting the nighttime evaporation at zero), yields a value of $E_{24} = 3.3$ mm d^{-1}. Fig. 15.13 shows the energy balance for an area West of Tomelloso, which on June 29 belongs to the driest sub-areas in the EFEDA

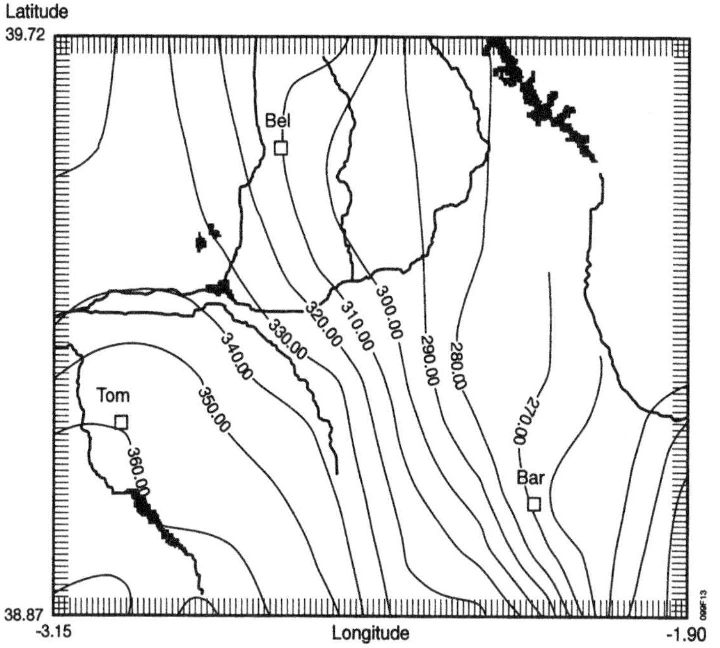

Figure 15.9: *SEBAL-based contour map of sensible heat flux interpreted from METEOSAT and NOAA-AVHRR measurements covering the entire EFEDA grid-square, June 29, 10:00 GMT, 1991.*

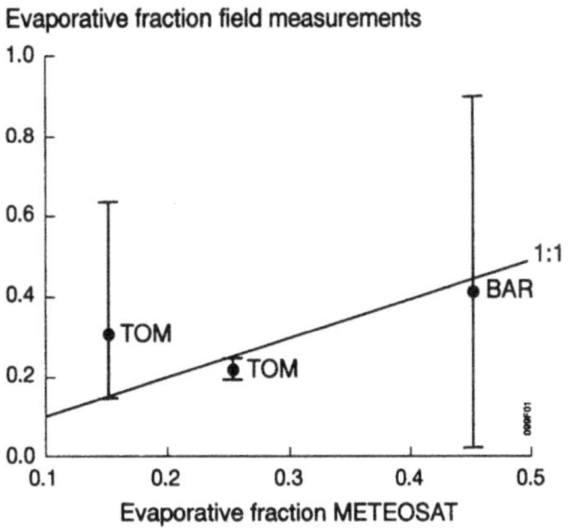

Figure 15.10: *Comparison of SEBAL based evaporative fraction obtained from METEOSAT and NOAA-AVHRR measurements vs. the dynamic range of evaporative fractions measured in-situ for all flux stations situated within the pixel boundaries of METEOSAT.*

grid-square. Fig. 15.9 showed for this sub-area the largest magnitudes in sensible heat flux. Except the early morning, the daytime evaporation is almost reduced to zero. We associate

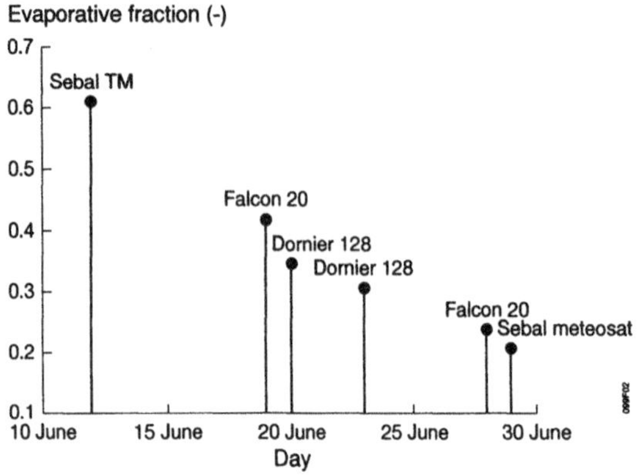

Figure 15.11: *Temporal behaviour of the aggregated evaporative fraction for the entire EFEDA grid according to SEBAL and aircraft measurements.*

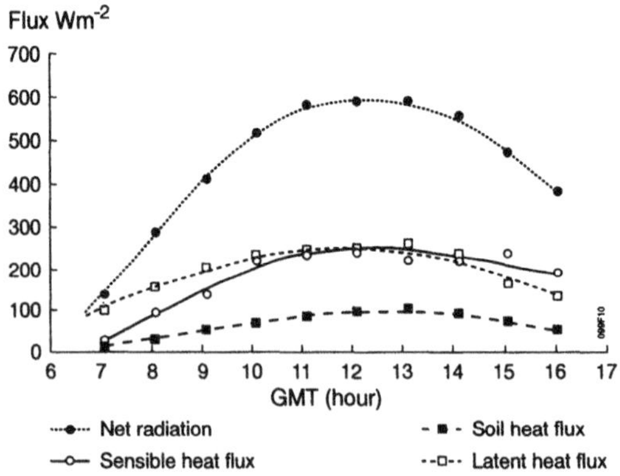

Figure 15.12: *Daytime surface energy balance of a heterogeneous and irrigated landscape with a 4 km horizontal scale near Albacte, June 29, 1991. UTM (x,y) 591168.0, 4334868.0.*

this morning evaporation with the presence of dew deposited on the bare soil, stones and rocks during the night. The daily evaporation adds after time integration up to a value of $E_{24} = 0.8$ mm d^{-1}.

For validation of the land surface parameterization schemes connected to Atmospheric Models, it is necessary to have access to the diurnal variation of the surface energy balance at grid-square scale, thus describing the diurnal tendencies of Figures 15.12 and 15.13 effectively. Fig. 15.14 shows the diurnal variation of the area-integrated fluxes for the entire EFEDA grid-square being based on all 595 METEOSAT pixels. On June 29, this will lead to a meso-scale land evaporation of 1.3 mm d^{-1}. The area integration of the fluxes for 595 METEOSAT pixels was established as:

$$\lambda E^{eff} = \frac{1}{595} \sum \lambda E^{mix} \quad (\text{W m}^{-2}) \tag{15.33}$$

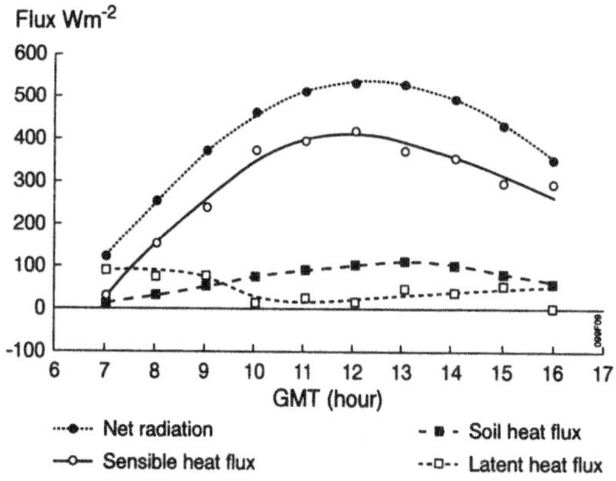

Figure 15.13: *Daytime surface energy balance of a heterogeneous and partially vegetated landscape with a 4 km horizontal scale near Tomelloso, June 29, 1991. UTM (x,y) 591168.0, 4334868.0.*

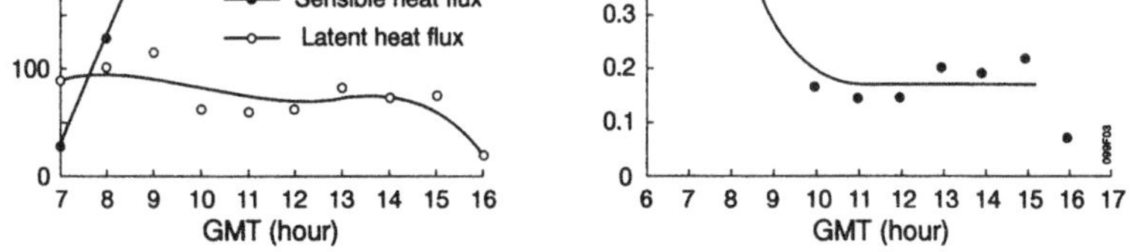

Figure 15.14: *Daytime surface energy balance for the entire EFEDA grid-square with a 85 km × 125 km horizontal scale, June 29, 1991.*

Part B of Fig. 15.14 shows the Bowen-ratio and evaporative fraction. Although Shuttleworth et al. (1989) found Λ to be temporally stable for homogeneous land surfaces, we conclude that the same holds true for heterogenoues land surfaces at a horizontal length scale of 100 km. Fig. 15.14 indicate that temporal stability especially emerges between 10:00 and 14:00 GMT.

15.10 Concluding Remarks

- Land surface fluxes for climate studies are simulated by means of land surface parameterization schemes coupled to Atmospheric Models. Although the validation of these schemes at micro- scale is useful (e.g. van den Hurk et al., 1995), studies on their peformance at meso-scale should be pursued using meso-scale flux estimates from remote sensing algorithms such as illustrated in Fig. 15.4.

- The key-issue in evaporation estimations based on remote sensing data is an adequate description of sensible heat flux: Latent heat flux will be obtained from energy budget residuals and any error in H will be directly conveyed into error in λE;

- Patch scale fluxes can be best estimated by a near-surface level for the upper air layer to which Eq. (15.21) applies because (i) the height of the internal adapted layer over homogeneous patches is generally small and effects from (ii) horizontal advection are reduced. The latter implies that the areal variation of near-surface T_a-values need great attention and that the surface energy balance can be regarded as a descriptor of vertical exchange processes;

- Regional scale fluxes can be well estimated by considering T_a of the mixed layer (remote sensing flux algorithm classes 3 and 4) and it is not strictly necessary to apply class 5 of algorithms with spatially variable near-surface air temperature for this purpose;

- The use of aerodynamic temperatures ($kB^{-1} = 0$) to determine sensible heat fluxes overestimates of the fluxes (Hall et al., 1992): kB^{-1} needs therefore to be calibrated locally to obtain realistic H^{mix}- fluxes from one-layer resistance schemes. At the same time it is concluded that kB^{-1} is almost impossible to extrapolate spatially;

- The linearity of $\delta T_a(T_0)$ relationships provides a solution to spatially interpolate δT_a in a simple manner for a wide variety of surface types and climate conditions. If δT_a is coupled to an a priori chosen and areal fixed kB^{-1} value (physics shows that there is a relationship among them), it forms a promising technique to calculate H^{mix} at composite terrain with a one-layer parameterization scheme;

- Inversion of the transport equation for sensible heat flow for land surface elements where $\lambda E^{mix} \sim 0$ and $H^{mix} \sim 0$ provides an opportunity to derive δT_a in an inverse mode which excludes the need to involve collateral T_{a-i} data. This 'internal calibration' procedure at meso-scale, may substitute kB^{-1} calibrations valid at micro scale only (e.g. Lhomme et al., 1994).

- The SEBAL parameterization is less sensitive to T_0^{mix} as compared to conventional schematizations of H because T_0^{mix} is used as a predictive variable to estimate δT_a^{mix} rather as a dependent parameter to compute δT_a^{mix}. Since in general, the error in deriving δT_0^{mix} from narrow band radiometers is ± 2 to 3°C, this may be conceived as an essential step forward.

- Validation of fluxes calculated with low resolution data against multiple tower-based fluxes can only be realized qualitatively. A better option may be to validate the regional evaporation against regional water balances, or against high resolution images being validated with *in-situ* fluxes, i.e. a nested calibration procedure.

15.11 References

Asrar G, Myneni RB, Choudhury BJ (1992) Spatial heterogeneity in vegetation canopies and remote sensing of absorbed photosynthetically active radiation: a modeling study. *Rem Sens Env* 41: 85–103

Bastiaanssen WGM, Hoekman DH, Roebeling RA (1994) A methodology for the assessment of surface resistance and soil water storage variability at mesoscale based on remote sensing measurements. IAHS Special Publication no. 2, *IAHS Press Wallingford*, UK: pp. 66

Bastiaanssen WGM (1995) Regionalization of surface fluxes and moisture indicators at composite terrain, Ph.D. Thesis, Wageningen Agricultural University, Wageningen, The Netherlands, appeared also as *Staring Centre Report* 109: pp. 273

Bastiaanssen WGM, Pelgrum H, Menenti M, and Feddes RA (1996) Estimation od surface resistance and Priestly-Taylor α-parameter at different scales, in (Eds.) JB Stewart, ET Engman, RA Feddes and Y Kerr, Scaling up in hydrology using remote sensing, Institute of Hydrology, Wallingford, UK: 93-111

Becker F, Li ZL (1990) Towards a local split window method over land surfaces. *Int J of Rem Sens*, vol. II, no. 3: 369–393

Beljaars ACM, Holtslag AAM (1991) Flux parameterization over land surfaces for atmospheric models. *J of Applied Met*, vol. 30, no. 3: 327–341

Bolle HJ, Andre JC, Arrue JL et al. (1993) EFEDA: the European Field Experiment in a Desertification threatened Area. *Ann Geophys* II:173–189

Blyth EM, Dolman AJ, Wood N (1993) Effective resistance to sensible and latent heat flux in heterogeneous terrain. *Q J R Met Soc* 119: 423–442

Blyth EM, Dolman AJ (1995) The roughness length for heat of sparse vegetation. *J Applied Met* (Feb. 95): 583–585

Bruin de HAR, Bink NJ, Kroon LJM (1991) Fluxes in the surface layer under advective conditions, in (Eds. Schmugge and Andre), *Land surface evaporation, measurement and parameterization*, Springer Verlag, pp. 157–169

Brutsaert W (1982) *Evaporation into the atmosphere, theory, history and applications*, Reidel Dordrecht, 299 pp

Brutsaert W, Hsu AY, Schmugge TJ (1993) Parameterization of surface heat fluxes above forest with satellite thermal sensing and boundary layer soundings. *J of Applied Met* vol. 32 (5): 909–917

Budyko MI (1956) Teplovoi balans zemnoi poverkhnosti. Gidrometeoizdat, Leningrad; Heat balance of the Earth's Surface; translated by N.A.Stepanova (1958). U.S. Weather Bureau, Washington, D.C.

Carlson TN, Buffum MJ (1989) On estimating total daily evapotranspiration from remote surface temperature measurements. *Rem Sens Env* 29: 197–207

Chamberlain AC (1966) *Proc Roy Soc London* A290: 236–265

Chen E, Allen LH, Bartholic JF, Gerber JF (1983) Comparison of winter-nocturnal geostationary satellite infared-surface temperature wityh shelter-height temperature in Florida. *Rem Sens of Env* 13: 313–327

Choudhury BJ (1989) Estimating evaporation and carbon assimilation using infared temperature data: vistas in modelling, in G. Asrar (ed.), *Theory and applications of optical remote sensing*, John Wiley, New York, pp. 628–690

Choudhury BJ (1991) Multispectral satellite data in the contex of land surface heat balance. *Review of Geophysics* 29: 217–236

Clauusen M (1990) Area-averaging of surface fluxes in a neutrally stratified, horizontally inhomogeneous atmospheric boundary layer. *Atmospheric Environment*, vol. 24A, NO. 6: 1349–1360

Davis PA, Tarpley JD (1983) Estimation of shelter temperatures from operational satellite sounder data. *J of Climate and Applied Meteorology*, vol. 22: 369–376

Dedieu G, Deschamps PY, Kerr YH (1987) Satellite estimation of solar irradiance at the surface of the earth and of surface albedo using a physical model appliesd to METEOSAT data, *J of Climate and Applied Meteorology*, vol. 26: 79–87

Diak G, Whipple MS (1993) Improvements to models and methods for evaluating the land-surface energy balance and 'effective' roughness using radiosonde reports and satellite-measured 'skin' temperature data. *Agr and For Met* 63: 189–218

Dolman AJ (1993) A multiple-source land surface energy balance model for use in general circulation models. *Agr and For Met* 65: 21–45

Dolman AJ, Blyth EM (1996) Patch scale aggregation of heterogeneous land surface cover for mesoscale meteorological models, submitted to *J of Hydr*, special issue Tucson aggregation workshop, in press

Eagleson PS (1981) (Ed.). Land surface processes in atmospheric general circulation models. JSC/WCP Study Conference, Greenbelt, Md. WCP 46: 56 pp

Fung AK, Zongqian L, Chen KS (1992) Backscattering from a randomly rough dielectric surface. *IEEE Transactions on Geoscience and Remote Sensing* vol. 30: 356–610

Gash JHC (1986) A note on estimating the effect of a limited fetch on micrometeorological measurements. *Bound Layer Met* 35: 409–413

Goutorbe JP, Lebel T, Tinga A et al. (1993) HAPEX-SAHEL: a large scale study of land-atmosphere interactions in the semi-arid tropics. *Ann Geophys* 12: 53–64

Griend van de AA, Owe M (1994) The influence of polarization on canopy transmission properties at 6.6 Ghz and implications for large scale soil moisture monitoring in semi-arid environments. *IEEE Transactions on geoscience and remote sensing* vol. 32, no. 2: 409–415

Gurney RJ, Foster JL, Parkinson CL (1994) *Atlas of satellite observations related to global change*, Cambridge University Press, 470 pp

Hall FG, Huemmrich KF, Goetz SJ, Sellers PJ, Nickeson JE (1992) Satellite remote sensing of surface energy balance: success, failures and unresolved issues in FIFE. *J of Geophysical Research* vol. 97, no. D17: 19061–19089

Henderson-Sellers A, Pitman AJ, Love PK, Irranejad P, Chen TH (1995) The Project for Intercomparison of Land Surface Parameterization Schemes (PILPS): Phases 2 and 3. *Bulletin of the American Meteorological Society* vol. 76, no. 4: 489–503

Holtslag AAM, Ek M (1996) Simulation of surface fluxes and boundary layer development over the pine forest in HAPEX-Mobilhy. *J Applied Met*, in press

Hurk van den BJJM, Verhoef A, van den Berg AR, de Bruin HAR (1995) An intercomparison of three vegetation/soil models for a sparse vineyard canopy. *Q J R Met Soc* 121: 1867–1889

Jackson RD, Reginato RJ, Idso SB (1977) Wheat canopy temperature: a practical tool for evaluating water requirements. *Water Resources Research* vol. 13, no. 3: 651–656

Jochum AM (1993) Evaporation and energy fluxes during EFEDA: horizontal variability and area averaging, in (eds.) Bolle, Feddes and Kalma, Exchange processes at the land surface for a range of space and time scales. *IAHS Publ* no. 212: 373–380

Justice CO, Townsend JRG, Holben BN, Tucker CJ (1985) Analysis of the phenology of global vegetation using meteorological satellite data. *Int J of Rem Sens* 6–8: 1271–1318

Kahle AB, Alley RE (1992) Separation of temperature and emittance in remotely sensed radiance measurements. *Rem Sens Env* vol 42: 107–112

Kalma JD, Jupp DLB (1990) Estimating evaporation from pasture using infrared thermometry: evaluation of a one-layer resistance model. *Agr and Forest Met* 51: 223–246

Klaassen W, van de Berg W (1985) Evapotranspiration derived from satellite observed surface temperatures. *J of Climate and Applied Met* vol. 24: 412–424

Kneizys FX, Shettle EP, Abreu LW (1988) User's guide to LOWTRTAN7, Env. Res. Papers no. 10101, Report AFGL-TR-88-0177, Air Force Geophysics Laboratory, Hanscom Air Force Base, MA 01737: pp 137

Kustas WP, Choudhury BJ, Moran MS, Reginato RJ, Jackson RD, Gay LW, Weaver HL (1989) Determination of sensible heat flux over sparse canopy using thermal infrared data. *Agr and Forest Met* 44: 197–216

Kustas WP, Moran MS, Humes KS, Stannard DI, Pinter PJ, Hipps LE, Swiatek E, Goodrich DC (1994) Surface energy balance estimates at local and regional scales using optical remote sensing from aircraft platform and atmospheric data collected over semiarid rangelands. *Water Resources Research* vol. 30, no. 5: 1241–1259

Lhomme JP, Monteny B, Amadou M (1994) Estimating sensible heat flux from radiometric temperature over sparse millet. *Agr and Forest Met* 68: 77–91

Manabe S (1969) Climate and ocean circulation: 1. The atmospheric circulation and the hydrology of the earth's surface. *Monthly Weather Review* vol. 97(11):739–774

Menenti M, Bastiaanssen WGM, van Eick D (1989) Determination of surface hemispherical reflectance with Thematic Mapper data. *Rem Sens Env* 28: 327–337

Menenti M, Bastiaanssen WGM, Hefny K, Abdel Karim MH (1991) Mapping of groundwater losses by evaporation in the Western Desert of Egypt, DLO-Winand Staring Centre Report 43, Wageningen, The Netherlands: pp. 116

Menenti M (1993) Understanding land surface evapotranspiration with satellite multispectral measurements. *Adv in Space Research* vol. 13, no. 5: 89–100

Menenti M, Ritchie JC (1994) Estimation of effective aerodynamic roughness of Walnut Gulch watershed with laser altimeter measurements. *Water Resources Research* vol. 30, no. 5: 1329–1337

Menenti M, Choudhury BJ (1994) Synergistic use of thermal infrared and other spectral measurements for land surface studies, *Proc. Workshop on Thermal remote sensing*, La Londe, September 1993: 167–176

Michand , Shuttleworth WJ (1996) (eds.) *J of Hydr*, special issue (Tucson aggregation workshop)

Milly PCD, Dunne KA (1994) Sensitivity of the global water cycle to the water- holding capacity of land. *J Climate* 506–526

Mintz Y (1982) The sensitivity of numerically simulated climate to land surface boundary conditions. *NASA Technical Memorandum* No. 83985: 81 pp

Monteith JL (1965) Evaporation and the environment, In: *The state and movement of water in living organisms*, XIXth Symp. Soc. for Exp. Biol., Swansea, Cambridge University Press, pp. 205–234

Noilhan J, Planton S (1989) A simple parameterization of land surface processes for meteorological models. *Monthly Weather Review* vol. 117, no. 3: 536–549

Oh Y, Sarabandi K, Ulaby FT (1992) An empirical model and an inversion technique for radar scattering from bare soil surfaces. *IEEE Transactions on Geoscience and Remote Sensing* vol. 30: 370–381

Paulson CA (1970) The mathematical representation of wind speed and temperature profiles in the unstable atmospheric surface layer. *J Applied Met* 9: 857–861

Pelgrum H, Bastiaanssen WGM (1996) An intercomparison of techinques to determine the area-averaged latent heat flux from individual in situ observations: A remote sensing approach using the EFEDA data. *Water Resources Research*, in press

Pinty B, Ramond D (1987) A method for the estimate of broadband directional surface albedo from a geostationary satellite. *J Clim Appl Met* 26: 1709–1722

Prevot L, Brunet Y, Paw KT, Seguin B (1994) Canopy modelling for estimating sensible heat flux from thermal infrared measurements, in *Proc. Workshop on Thermal remote sensing*, La Londe, September 1993: 17–22

Pierce LL, Walker J, Downling TI, Vicar TR, Hatton TJ, Running SW, Coughlan JC (1992) Ecological change in the Murray-Darling Basin III: A simulation of regional hydrological changes. *J of Applied Ecology* 30: 283–294

Price JC (1984) Land surface temperature measurements from the split window channels of the NOAA-7 Advanced Very High Resolution Radiometer. *J Geophys Res* 89: 7231–7237

Rosema A (1990) Comparison of Meteosat-based rainfall and evapotranspiration mapping in the Sahel region. *Int J Rem Sens* vol. 11, no. 12: 2299–2309

Rowntree PR, Bolton J (1983) Simulation of the atmospheric response to soil moisture anomalies over Europe. *Quart J R Met Soc* vol. 109: 501–526

Salisbury JW, D'Aria DM (1992) Emittance of terrestrial materials in the 8–14 mm atmospheric window. *Rem Sens Env* vol 42: 83–106

Schmugge TJ, Jackson TJ, Kustas WP, Wang JR (1992) Passive microwave remote sensing of soil moisture; Results from HAPEX, F11 and MONSOON90, ISPRS. *J Photogrammetry and Remote Sensing* 47: 127–143

Schmugge TJ, Hook S, Kahle A (1995) TIMS observation of surface emittance in HAPEX-Sahel, Proc. IGARSS 1995, Proc. Florence, Italy

Seguin B, Itier B (1983) Using midday surface temperature to estimate daily evaporation from satellite thermal IR data. *Int J Rem Sens* 4: 371–383

Sellers PJ, Mintz Y, Sud YC, Dalcher A (1986) A SImple Biosphere model (SiB) for use within general circulation models, J. of the Atm. Sciences, vol. 43, no. 6: 505–531

Shuttleworth WJ, Gurney RJ, Hsu AY and Ormsby JP (1989) FIFE: The variation in energy partitioning at surface flux sites, Rem. Sens. and large-scale global processes, Proc. Baltimore Symp, *IAHS Publ* no. 186: 67–74

Stewart JB, Kustas WP, Humes KS, Nichols WD, Moran MS, de Bruin HAR (1994) Sensible heat flux - radiometric surface temperature relationship for 8 semi-arid areas, in *Proc. Workshop on Thermal remote sensing*, La Londe, September 1993: 27– 30

Stricker JNM, Kim CP, Feddes RA, van Dam J, Droogers P, de Rooy GH (1993) Modelling and observing the terrestrial hydrological cycle, in (eds.) E. Raschke and D. Jacob, NATO-ASI Glücksburg: 419–444

Sud YC, Sellers PJ, Mintz Y, Chou MD, Walker GK, Smith WE (1990) Influence of biosphere on the global circulation and hydrologic cycle - a GCM simulation experiment. *Agr and For Met* 52: 133–180

Sugita M, Brutsaert W (1990) Regionalization surface fluxes from remotely sensed skin temperature and lower boundary layer measurements. *Water Resources Research* vol. 26, no. 12: 2937–2944

Thornthwaite CW, Mather JR (1955) The water balance. *Publications in Climatology*, Laboratory of Climatology, Centerton, New Jersey. vol.8(1): 86 pp

Townsend AA (1965) The response of a turbulent boundary layer to abrupt changes in surface conditions. *J Fluid Mech* 22: 799–822

Tucker CJ (ed.) (1986) Monitoring the grasslands of semi-arid Africa using NOAA AVHRR data. *Int J Rem Sens* (special issue), 7–11: 1383–1622

Wan Z, Dozier J (1989) Land-surface temperature measurement from space: physical principles and inverse modeling. *IEEE Transactions on Geoscience and Remote Sensing* vol. 27, no. 3: 268–278

Wang J, Hu Y, Sahashi K, Mitsuta Y (1994) Outline of HEIFE field observations. in: Y. Mitsuta (ed.): *Proceedings Int. Symp. on HEIFE.* Disaster Prevention Research Institute, Kyoto University: 22–29

Wieringa J (1986) Roughness-dependent geographical interpolation of surface wind speed averages. *Q J R Met Soc* 112: 867–889

15.12 Appendix: List of symbols

c_p	air specific heat at constant pressure	J kg^{-1} K^{-1}
D_v^{eff}	effective soil vapour diffusivity	m^2 s^{-1}
D_Θ^{eff}	total soil liquid diffusivity	kg s m^{-1}
E	actual evaporation	m s^{-1}
e	actual vapour pressure	mbar
e_{sat}	saturated vapour pressure	mbar
Δe	saturation vapour pressure deficit	mbar
G_0	at surface soil heat flux	Wm^{-2}
g	acceleration due to gravity	m s^{-2}
H	sensible heat flux	Wm^{-2}
$K^\downarrow$	incoming shortwave radiation	Wm^{-2}
k	Von Karman's constant	-
kB^{-1}	normalized rougness coefficient	-
L	Monin-Obukov length	m
LAI	leaf area index	m^2 m^{-2}
L_y	vertical length scale	m
L_x	horizonal length scale	m
$L^\downarrow$	incoming longwave radiation	W m^{-2}
L_λ^{bb}	black body spectral radiance	W m^{-2} mm^{-1}
L_λ^{TOA}	spectral radiance at sensor aperture	Wm^{-2}mm^{-1}
$L_\lambda^\downarrow$	downwelling spectral radiance emitted from atmosphere	Wm^{-2} mm^{-1}
$L_\lambda^\uparrow$	upwelling spectral radiance emitted from atmosphere	Wm^{-2} mm^{-1}
L_λ^{land}	totally emitted and reflected spectral radiance from land	Wm^{-2} mm^{-1}
NDVI	normalized difference vegetation index	-
q_a	air specific humidity	-
q_*	humidity scale	-
Q^*	net radiation	Wm^{-2}
r_{ah}	aerodynamic resistance to heat transport	s m^{-1}
r_{av}	aerodynamic resistance to vapour transport	s m^{-1}
r_c	canopy resistance	s m^{-1}
r_c^{min}	minimum canopy resistance	s m^{-1}
r_s	bulk surface resistance	s m^{-1}
r_{soil}	surface resistance to total bare soil evaporation	s m^{-1}
r_0	hemispherical surface reflectance of shortwave radiation	-
s_c	fractional soil cover	-
T_a	air temperature	K
T_p	potential air temperature	K
T_s	soil temperature	K
T_{z0m}	aerodynamic surface temperature	K
T_{z0h}	heat source surface temperature	K

T_0	surface temperature	K
T_0^R	radiation temperature at land surface	K
T_*	temperature scale	K
t	time	s
u	horizontal wind velocity	m s^{-1}
u_*	friction velocity	m s^{-1}
z	height	m
z_{sur}	height upper layer of air temperature for sensible heat flux	m
z_{0m}	aerodynamical roughness length for momentum	m
z_{0h}	aerodynamical roughness length for heat	m
z_{0v}	aerodynamical roughness length for vapour	m
γ	psychrometric constant	mbar K^{-1}
δT_a	near-surface vertical air temperature difference	K
ϵ_0	broad band surface emittance	-
$\epsilon_0(\lambda)$	small band surface emittance	-
λ	latent heat of vapourization	J kg^{-1}
λ	wavelength	mm
$\lambda_{s'}$	apparent macroscopic soil thermal conductivity	Wm^{-1} K^{-1}
Λ	evaporative fraction	-
ρ_a	moist air density	kg m^{-3}
ρ_v	actual vapour density	kg m^{-3}
ρ_v^{sat}	saturated vapour density	kg m^{-3}
σ	Stefan Boltzmann constant	Wm^{-2} K^{-4}
τ	momentum flux	N m^{-2}
τ_λ	spectral transmittance	-
Θ	soil water content	cm^3 cm^{-3}
_0	surface soil water content	cm^3 cm^{-3}
Ψ_h	stability correction for atmospheric heat transport	-
Ψ_m	stability correction for atmospheric momentum transport	-
Ψ_v	stability correction for atmospheric vapour transport	-

Chapter 16

Satellite Remote Sensing of Evaporation over Sea

Peter Schlüssel
University of Colorado at Boulder
Colorado Center for Astrodynamics Research
Boulder, CO 80309-0431
USA

Abstract

The remote sensing of evaporation over sea mainly requires to derive three quantities from satellite data: Sea surface temperature, surface wind speed and specific air humidity. The sea surface temperature is derived from infrared imagery of the Advanced Very High Resolution Radiometer (AVHRR), flown on the NOAA polar orbiters and from the Along Track Scanning Radiometer (ATSR) operated on the ERS-1 satellite. Split- and dual-window techniques are employed in order to correct the satellite-measured brightness temperatures for atmospheric effects, mainly due to water vapour. Recent techniques include non-linear terms and information from other radiometers improving the retrieval accuracy. Attention must be paid to the difference between bulk and skin surface temperatures which can be as great as ± 1K due to the skin cooling and the diurnal thermocline. This difference is of the order of the retrieval accuracy achieved nowadays and must be known when interpreting satellite-derived sea surface temperatures. Surface wind speed and near-surface humidity are obtained from passive microwave measurements of the Special Sensor Microwave/Imager (SSM/I) that is part of the DMSP satellite series. While the surface wind speed is derived from brightness-temperature variations induced by emissivity changes in resposne to surface roughness and foam coverage the determination of the surface air humidity is related to the water-vapour columns that can be retrieved from multi-channel retrieval schemes applied to the satellite-measured microwave data. Recent investigations show that the SSM/I not only allows to retrieve the total precipitable water but also the water-vapour column of the lower 500 m which is closer related to the surface air humidity than the total water-vapour column. The retrieval methods are developed by means of radiative transfer simulations and subsequent multivariate analysis. An application of the theoretically developed techniques to real satellite data and comparisons with in-situ measurements are performed to test the single parameter retrievals. A subsequent intercomparison of the latent heat fluxes derived from satellites with those obtained from from ship measurements

NATO ASI Series, Vol. I 45
Radiation and Water in the Climate System:
Remote Measurements
Edited by Ehrhard Raschke

sheds light on the retrieval accuracy and on the differences between various methods in use. For single match-ups the standard deviation between ship and satellite measurements is about 30 W/m^2 leading to monthly mean errors less than 10 W/m^2.

16.1 Introduction

Ocean and atmosphere are coupled by fluxes of energy, matter, and momentum through the air-sea interface. The evaporation over sea accounts for two of these fluxes, for the transfer of fresh water and for the related latent heat flux. Besides these fluxes the evaporation affects the salinity budget of the ocean. Passive infrared and microwave remote sensing bear the potential for improving the evaporation estimates in data-sparse areas of the oceans, thus supplying to the overall goal of GEWEX, the improved understanding of the role of water vapour in meteorological, hydrological and climatological processes. Direct observation of the evaporation is only possible by means of eddy correlation measurements which are not accessible by satellite methods. The determination of the vertical water-vapour flux density E, or water-vapour flux for short, requires to determine the correlation product

$$E = \rho \overline{w'q'} \tag{16.1}$$

where w' and q' are the fluctuations of the vertical wind velocity and the specific air humidity, respectively, and ρ is the air density. Fluctuations of at least 10 Hz have to be resolved over averaging periods of about an hour in order to get a reliable measurement of E. Equation (16.1) is usually parameterised by the so-called bulk aerodynamic formula that allows the use of readily accessible surface measurements from the routine meteorological observation network:

$$E = -C_E \rho u (q_0 - q) \tag{16.2}$$

where u is the average horizontal surface wind speed, q_0 and q are the average specific humidities of the surface and the air, respectively, and C_E is the bulk transfer coefficient for water vapour, also known as Dalton number. The transfer coefficient is normally referred to measurements at 10 m and averaging periods of 10 minutes (Hasse, 1993). The key parameters to be entered in (16.2) are surface temperature, T_0, which determines q_0, surface wind speed, u, and surface air humidity, q. The latent heat flux related to E is $Q_E = LE$ where L is the latent heat of evaporation.

The satellite instruments being employed for the retrieval of latent heat fluxes are mainly infrared and microwave radiometers, but also scatterometer winds have been used so far. While microwave radiometry is used for the measurement of surface wind speed and atmospheric water vapour infrared radiometers are needed for an accurate measurement of sea surface temperature. Instruments that are currently operational and utilized for the retrieval of the evaporation over sea are the Special Sensor Microwave/Imager (SSM/I), described by Simmer (this issue), and the infrared imagers AVHRR (Advanced Very High Resolution Radiometer, flown on the NOAA satellites) and the ATSR (Along Track Scanning Radiometer, that is part of the ERS-1 satellite). Both infrared radiometers are designed to measure radiation in atmospheric window regions. Their main characteristics are listed in tables 16.1 and 16.2. The spectral response functions of their thermal channels are shown together with the atmospheric transmission spectra for a tropical and a subarctic atmosphere in figure 16.1. The main innovation of ATSR is the active detector cooling which leads to a reduced noise behaviour when compared to the AVHRR instrument that allows for a more accurate retrieval of sea surface temperatures. While AVHRR has a radiometric noise equivalent to a temperature difference $NE\Delta T$ = 0.12 K the ATSR values are 0.02, 0.04, and 0.08 K for the 11, 12, and 3.7 μm channels, respectively (Schlüssel

Channel	Band [μm]	Resolution [mrad]	Purpose
1	0.58 - 0.68	1.39	clouds, vegetation
2	0.725 - 1.10	1.41	clouds, vegetation
3	3.55 - 3.93	1.51	clouds, surface temperature
4	10.30 - 11.80	1.41	surface temperature, water vapour
5	11.50 - 12.50	1.30	surface temperature, water vapour
S/N (channels 1, 2)			3 at 0.5% reflectivity
$NE\Delta T$ (channels 3 - 5)			0.12 K
scan angle			±55.4°
pixels per line			2048
scan rate			1 line per 1/6 s
integration time			20 μs
calibration			internal black body and space view

S/R = Signal to noise ratio
$NE\Delta T$ = noise equivalent temperature difference

Table 16.1: *AVHRR-Characteristics*

and Meywerk, 1993). More details of the AVHRR are given by Lauritson et al. (1979), the ATSR has been described by Prata et al. (1990).

16.2 Sea Surface Temperature

Since more than 20 years sea surface temperature measurements are derived from satellite-borne infrared imagery. Under cloud-free atmospheric conditions brightness temperature measurements at 11 μm as performed from satellites are closely correlated with sea surface temperatures showing correlation coefficents higher than 98%. However, the atmospheric impact on the measurements is not negligible and without further corrections the brightness temperatures are of limited value for the derivation of sea surface temperature. Effects perturbing the radiative transfer through the cloud-free atmosphere are mainly caused by water-vapour absorption, but are also due to extinction by aerosol particles and, increasingly, by man-made trace gases like the chlorofluorocarbons $CFCL_3$ (F11) and CF_2Cl_2 (F12). Further modifications of the signal are caused by variations of the temperature profile in the planetary boundary layer and by the non-blackness of the sea surface. Table 16.3 lists the range of brightness-temperature deficits as simulated by means of radiative transfer calculations for five different standard atmospheric conditions (McClatchey et al., 1972) as caused by the various atmospheric constituents in the thermal AVHRR and ATSR channels. The table shows that the temperature deficits are not always positive; in cases with warm lower atmospheres and cool sea surfaces the deficits can be negative if the boundary layer is enriched with water vapour or aerosol particles. It is also noted that the anthropogenic trace gases F11 and F12 can no longer be ignored when looking for an optimum retrieval technique regarding T_0. Aerosols that have been brought into the stratosphere by volcanic eruptions can cause temperature deficits as large as that by water vapour. They require special attention since standard retrieval schemes are designed to account for normal background aerosol only. Currently, all sea surface temperatures derived from satellite infrared imagery are less accurate than before the eruption of Mt. Pinatubo in 1991 (Reynolds, 1993; Rao, 1992).

Channel	Band [μm]	Purpose
1	1.55 - 1.66	clouds, aerosols
2	3.47 - 3.98	surface temperature
3	10.37 - 11.48	surface temperature
4	11.30 - 12.60	surface temperature
$NE\Delta T$		<0.1 K
scan angle		$\pm 23.45°$
pixels per scan		555 (nadir), 371 (forward)
scan rate		1 line per 150 ms
integration time		80 μs
calibration		two internal black bodies
resolution		1.29 mrad (nadir), 1.54 mrad (forward)
Off-nadir distance [km]	**Zenith angle [°]**	**Relative air mass**
0	0	1.000
	55.0	1.743
50	4.4	1.003
	54.9	1.740
100	8.1	1.010
	54.6	1.728
150	12.7	1.025
	54.2	1.709
200	16.9	1.045
	53.5	1.680
250	21.5	1.075
	52.4	1.640

Table 16.2: *ATSR-Characteristics*

Anding and Kauth (1970) have shown for the first time that utilizing two spectral measurements in the atmospheric infrared window between 8 and 13 μm will lead to a so-called "split-window" algorithm that enables an atmospheric correction to be applied to the brightness-temperature measurements at 11 μm in order to retrieve sea surface temperatures. The main atmospheric influence on the satellite-measured signal is due to water vapour which can reduce the brightness temperature at 11 μm by more than 10 K.

For the split-window technique additional measurements are performed near 12 μm where the water-vapour absorption is even higher than at 11 μm. The surface temperature is then retrieved by

$$T_0 = a_0 + a_1 T_{11} + a_2 (T_{11} - T_{12}) \quad (16.3)$$

where T_{11} and T_{12} are brightness temperatures as measured at 11 and 12 μm, respectively. The coefficients a_i, $i = 0, 1, 2$ are usually optimized for each instrument with respect to its spectral response function accounting not only for the water-vapour absorption but also give a mean correction for other attenuating constituents of the atmosphere and for the non-blackness of

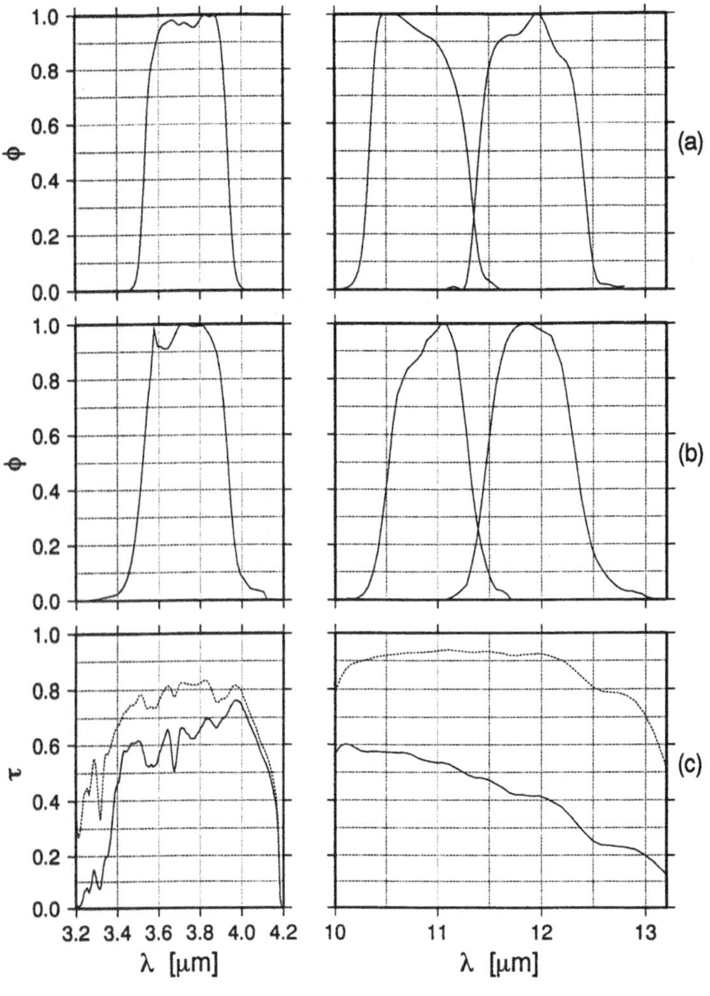

Figure 16.1: *Relative spectral filter functions for (a) AVHRR/NOAA-11 and (b) ATSR/ERS-1 as well as (c) the atmospheric transmission spectra in the atmospheric windows near 3.7, 11, and 12 μm for tropical (solid) and arctic conditions (dashed).*

the sea surface. McMillin (1975) and McMillin and Crosby (1984) have given the theoretical justifications for the use of (16.3) which has been successfully applied to AVHRR measurements (McClain et al., 1985). The accuracy reported is better than 1 K (e. g. Prabhakara et al., 1974; Deschamps und Phulpin, 1979; Llewellyn-Jones et al., 1984). The restriction of the split-window method to two channels being combined in a linear way limits the accuracy in cases with extremely high water-vapour loadings where only about 40% of the surface-leaving radiation is transmitted to the satellite. Also continental air masses advected over the near-coastal waters can prevent an accurate temperature correction by non-marine atmospheric temperature and humidity structures. Other anomalous cases are found in situations where strong temperature

	AVHRR					
	3.7 μm		11 μm		12 μm	
Absorber	Nadir	55°	Nadir	55°	Nadir	55°
CH_4	0.43-0.78	0.68-1.20	–	–	–	–
CO_2	0.02-0.04	0.04-0.06	0.13-0.50	0.20-0.78	0.04-0.15	0.06-0.22
F11	–	–	–	–	0.00-0.12	0.00-0.19
F12	–	–	0.18-0.32	0.29-0.53	0.01-0.36	0.00-0.58
H_2O	-0.03-1.78	-0.04-2.69	0.03-3.13	0.01-4.18	0.03-4.72	-0.08-5.54
HNO_3	–	–	0.01-0.02	0.02-0.04	0.00-0.01	0.01-0.02
N_2	0.10-0.21	0.18-0.35	–	–	–	–
N_2O	0.34-0.54	0.52-0.81	–	–	–	–
O_3	0.01-0.02	0.02-0.05	–	0.00-0.01	0.01	0.01-0.02
Aerosol (V23)	-0.04-0.16	-0.10-0.24	0.00-0.06	-0.03-0.09	-0.01-0.07	-0.09- 0.09
Aerosol (V5)	-0.14-0.74	-0.35-1.09	0.00-0.28	-0.11-0.39	-0.05-0.34	-0.37-0.42
Aerosol (SB)	0.00-0.02	0.02	0.00-0.01	0.01-0.02	0.01-0.02	0.01-0.02
Aerosol (SV)	0.71-1.03	1.21-1.75	0.42-0.71	0.71-1.20	0.59-1.00	0.98-1.68
	ATSR					
	3.7 μm		11 μm		12 μm	
Absorber	Nadir	55°	Nadir	55°	Nadir	55°
CH_4	0.43-0.81	0.68-1.24	–	–	–	–
CO_2	0.02-0.04	0.04-0.06	0.09-0.38	0.15-0.60	0.05-0.16	0.07-0.25
H_2O	-0.03-1.76	-0.06-2.63	0.03-3.26	0.01-4.35	-0.04-4.86	0.09-5.67
HNO_3	–	–	0.02-0.04	0.03-0.07	0.00-0.01	0.01-0.02
F11	–	–	–	–	0.00-0.12	0.00-0.19
F12	–	–	0.22-0.39	0.36-0.66	0.00-0.36	0.00-0.60
N_2	0.12-0.26	0.20-0.43	–	–	–	–
N_2O	0.33-0.52	0.49-0.77	–	0.00-0.01	–	–
O_3	0.01-0.03	0.02-0.05	–	0.00-0.01	0.00-0.02	0.02-0.03
Aerosol (V23)	-0.04-0.16	-0.11-0.24	0.00-0.06	-0.03-0.09	-0.01-0.05	-0.05- 0.06
Aerosol (V5)	-0.16-0.73	-0.38-1.08	0.00-0.29	-0.12-0.41	-0.03-0.22	-0.21-0.27
Aerosol (SB)	0.01-0.02	0.02	0.01	0.01-0.02	0.01-0.02	0.01-0.02
Aerosol (SV)	0.70-1.03	1.21-1.75	0.44-0.74	0.74-1.25	0.59-1.00	0.98-1.68

V23 marine tropospheric aerosol, 23 km visibility
V5 marine tropospheric aerosol, 5 km visibility
SB stratospheric background aerosol
SV volcanic aerosol shortly after outbreak

Table 16.3: *Brightness temperature deficits in the thermal channels of ATSR (NOAA-11) and ATSR (ERS-1)*

inversions occur in the planetary boundary layer or where deep water-vapour layers extend over the lower troposphere (Schlüssel et al., 1987).

Besides the split-window channels further spectral measurements that can be used for the remote sensing of sea surface temperature are available in the atmospheric window at 3.7 μm. For high water-vapour contents this window is far more transparent than the 8 to 13 μm window as seen in figure 16.1. However, the radiation field at 3.7 μm not only contains thermal radiation emitted from the earth and its atmosphere; during day it also contains a substantial part of scattered solar radiation which hampers the exploitation of radiance mesurements in this spectral region with respect to sea surface temperature. During night, however, the brightness temperatures measured in the 3.7 μm region can be used together with measurements at 11 or 12 μm in a similar way described by (16.3), rendering the split-window into a dual-window technique. Unfortunately, past measurements at 3.7 μm as taken by the AVHRR instrument were too noisy for a suitable utilization and the dual-window techniques were less frequently employed than the split-window method. The ATSR that is flown on the ERS-1 satellite carries a 3.7 μm channel with low noise amplitudes that allows its full utilization for the atmospheric correction of nocturnal infrared imagery when deriving sea surface temperature.

In some cases the coefficients being used in (16.3) are tuned to distinct regions in order to optimize the temperature retrieval for specific atmospheric situations (e.g. Llewellyn-Jones et al., 1984; Minnett, 1990). This regional tuning is only partly successful since even in small areas the variability of atmospheric water vapour can reach high values. For example, in the tropical Pacific Ocean day to day variations of the water-vapour columnn by more than 2 g/cm^2 can be observed in SSM/I water-vapour measurements. Besides this the regionalization of retrieval schemes can lead to discontinuities in the derived fields at the boundaries between adjacent regions so that in large-scale applications temperature gradients can be faked. However, with respect to studies of limited areas the regionalisation of retrieval schemes might be successful.

A further possibility for the improvement of the surface temperature retrieval is the knowledge of the actual state of the lower atmosphere. This could be obtained from radiosoundings or via satellite-derived temperature and moisture profiles. Even more advantageous seems the direct utilization of additional information from other instruments flying on the same satellites. Case studies for the synergetic use of different radiometers for the atmospheric correction have shown improved performances when compared to the use of a two- or three-channel technique (Schlüssel et al., 1987; Wick et al., 1992). A disadvantage of this combination of instruments is the difference of the horizontal resolution of the radiometers in use. While AVHRR resolves an area of 0.95 km^2 at nadir view an infrared sounder like the High-resolution Infrared Radiation Sounder (HIRS) covers 227 km^2 in a pixel near the sub-satellite track. Computationally expensive interpolations are necessary in order to match one instrument to the other.

16.2.1 Retrieval techniques

Usually, radiative transfer calculations are used for the development of retrieval techniques. Each retrieval scheme has to be optimized according to the intrument characteristics because of different noise behaviour and spectral response functions. It has been shown that the utilization of split-window coefficients developed for one radiometer (AVHRR/2 on board NOAA-7) in the retrieval of T_0 from another radiometer (AVHRR/2 on board NOAA-9) can introduce systematic errors up to 0.29 K (Minnett, 1990).

Examples of simulated brightness temperature deficits in the 12 μm channel at different viewing angles of the ATSR are shown in figure 16.2 for 289 globally distributed test cases. The corrections necessary broadly vary with the surface temperature which is roughly correlated with the atmospheric water-vapour amount. The slant view through the atmosphere strongly increases the differences. The values vary between -2 and 16 K at nadir view and up to 22 K at forward view of the radiometer. Similar but weaker relationships are found in the channels at 3.7 and 11 μm.

Different retrieval models have been formulated for the use with split-, dual- and triple-window techniques. The first one has been introduced above which was derived from the truncated radiative transfer equation:

$$L_\lambda(T_b) = B_\lambda(T_0)\tau_\lambda^* + B_\lambda(\overline{T_a})(1 - \tau_\lambda^*) \tag{16.4}$$

where L_λ is the radiance in a radiometer channel, T_b is the brightness temperature, τ_λ^* is the transmisison of the entire atmosphere, and $\overline{T_a}$ is the mean atmospheric temperature. The main assumptions made are vanishing surface reflectivity and atmospheric scattering. The formulation of (16.4) for two wavelengths, λ_1 and λ_2, together with further simplifying assumptions $\overline{T_a(\lambda_1)} = \overline{T_a(\lambda_2)}$ as well as $B(\lambda_1, T_0) = B(\lambda_2, T_0)$ lead to

$$B(\lambda, T_0) = L(\lambda_1) - (L(\lambda_1) - L(\lambda_2))\frac{1 - \tau_{\lambda_1}^*}{\tau_{\lambda_1}^* - \tau_{\lambda_2}^*} \tag{16.5}$$

For small arguments the exponential sum

$$\mathrm{e}^{-x} = 1 - \frac{x}{1!} + \frac{x^2}{2!} - \frac{x^3}{3!} + \ldots \tag{16.6}$$

can be approximated by $\mathrm{e}^{-x} = 1 - x$, hence, the transmission is approximated by $\tau^*(\lambda, \vartheta) = 1 - \sigma_{a,\lambda} z \sec\vartheta$ where $\sigma_{a,\lambda}$ is the spectral volume absorption coefficient, z is the vertical path length, and ϑ is the zenith angle. It follows

$$\frac{1 - \tau^*(\lambda_1, \vartheta)}{\tau^*(\lambda_1, \vartheta) - \tau^*(\lambda_2, \vartheta)} = \frac{\sigma_{a,\lambda_1}}{\sigma_{a,\lambda_1} - \sigma_{a,\lambda_2}} = \gamma, \tag{16.7}$$

from which a first form of the split-window formula is then given by

$$B(\lambda_1, T_0) = L(\lambda_1, \vartheta) + \gamma(L(\lambda_1, \vartheta) - L(\lambda_2, \vartheta)). \tag{16.8}$$

Introduction of (brightness) temperatures instead of radiances and expansion in a Taylor series with subsequent truncation leads to

$$T_0 = T_{\lambda_1}(\vartheta) + \gamma(T_{\lambda_1}(\vartheta) - T_{\lambda_2}(\vartheta)) \tag{16.9}$$

This form eliminates the non-linearity of the Planck's function and the split-window formula can also be used as dual-window formula using channels that are separated by wide spectral intervals. Similarly, one can derive a dual-angle method utilizing brightness temperature measurements at two viewing angles at a single wavelength:

$$T_0 = T_{\vartheta_1}(\lambda) + \gamma'(T_{\vartheta_1}(\lambda) - T_{\vartheta_2}(\lambda)) \tag{16.10}$$

where

$$\gamma' = \frac{\sec\vartheta_1}{\sec\vartheta_1 - \sec\vartheta_2} \tag{16.11}$$

The coefficients γ and γ' can theoretically be derived from spectroscopic data but the common use of the split-window technique requires the replacement of γ and γ' by statistically regularized coefficients in order to account for the non-blackness of the sea surface and for non-linear effects which are still included. The resulting retrieval model is given by (16.3). The coefficients can

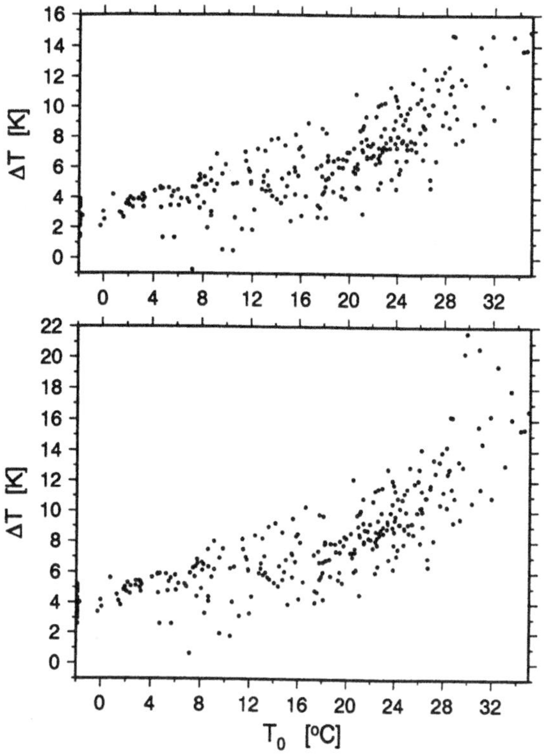

Figure 16.2: *Brightness temperature deficits in the ATSR channel at 12 μm simulated for 289 test cases, (top) at nadir view, (bottom) at forward view.*

be tuned to the viewing angle of the radiometer as shown in table 16.4 which gives a set of coefficients for the AVHRR flown on NOAA-11 and for the ATSR on board ERS-1. The discretization of the viewing angle in steps of 10° is accurate enough in order to avoid aditional errors greater than 0.1 K (Wick et al., 1992). Alternatively, the viewing-angle dependency can be included by the secans of the zenith angle. Barton et al. (1989) have used the form

$$T_0 = a_0 + a_1 T_{11} + a_2(T_{11} - T_{12}) \tag{16.12}$$

where

$$a_k = a_{0k} + a_{1k}(\sec\vartheta - 1) + a_{2k}(\sec\vartheta - 1)^2,\ k = 1, 2 \tag{16.13}$$

With respect to retrieval errors this method does not improve the accuracy in T_0 but is easy to use, although computationally more expensive. Table 16.4 also includes the theoretically estimated retrieval errors which demonstrate that the ATSR is superior to the AVHRR because of the smaller instrumental noise.

Effects due to the non-linear dependency of the brightness temperatures on the water-vapour amount have received increased attention during the last years. In view of the strong absorption by the water-vapour continuum the linear approach is hardly justified in the tropics, particularly, since the atmospheric transmission is far from unity the use of a linear truncation of (16.6)

ATSR					
θ [°]	Noise	a_0	a_1	a_2	ϵ
0-21	w/o	-1.72	1.0060	2.5663	0.34
0-21	with	-2.15	1.0077	2.5369	0.36
AVHRR					
θ [°]	Noise	a_0	a_1	a_2	ϵ
0	w/o	-3.48	1.0134	2.3999	0.28
0	with	-6.17	1.0235	2.1767	0.57
10	w/o	-3.50	1.0135	2.4163	0.28
10	with	-6.14	1.0234	2.1976	0.57
20	w/o	-3.69	1.0142	2.4667	0.31
20	with	-6.23	1.0238	2.2580	0.60
30	w/o	-4.53	1.0173	2.5676	0.37
30	with	-6.84	1.0260	2.3745	0.64
40	w/o	-8.12	1.0301	2.7470	0.49
40	with	-9.95	1.0372	2.5721	0.74
50	w/o	-22.71	1.0837	3.0201	0.78
50	with	-23.28	1.0863	2.8654	0.99

Table 16.4: *Coefficients and errors for equation (16.3).*

introduces great errors. A non-linear correction currently in use by NOAA is the so-called Cross Product Sea Surface Temperature (CPSST) algorithm according to Walton (1988):

$$T_0 = \frac{a_0 T_{12} - a_1}{a_2 T_{12} - a_3 T_{11} - a_4}(T_{11} - T_{12} + a_5) + a_6 T_{12} + a_7 (T_{11} - T_{12})(\sec\vartheta - 1) - a_8 \qquad (16.14)$$

However, comparisons with split-window retrievals do not reveal any advantage of the CPSST method but show that this method is rather sensible against undetected sub-pixel clouds and it enhances the radiometric noise more than the split-window method (Wick et al., 1992).

An improvement of the split-window technique can be obtained by including a quadratic term in (16.3) which better corrects for the non-linear impact of the water vapour because the optical depth in the window regions is mainly determined by the water-vapour amount. The quadratic approach leads to a term which is proportional to the water-vapour column w multiplied by the split-window difference, namely, $a_3(T_{\lambda_1} - T_{\lambda_2})w$, which results in $a_3(T_{\lambda_1} - T_{\lambda_2})^2$ when knowing that w is approximately proportional to w (Schlüssel, 1989; Harris and Mason, 1992). Emery et al. (1994) have used the water-vapour column as retrieved from SSM/I in order to yield

$$T_0 = a_0 + a_1 T_{11} + a_2(T_{11} - T_{12}) + a_3 w (T_{11} - T_{12}) \qquad (16.15)$$

with an accuracy better than obtained from (16.3). However, SSM/I and AVHRR are flown on different orbiters so that the temporal variability of w introduces unwanted errors. Therefore, the inclusion of the squared split-window differences should be as good as the use of (16.15) although the information about w is not as high as would be obtained from an SSM/I flown on the same orbiter carrying the AVHRR. For this reason the following retrieval scheme currently gives the same performance as (16.15):

$$T_0 = a_0 + a_1 T_{11} + a_2(T_{11} - T_{12}) + a_3(T_{11} - T_{12})^2 \qquad (16.16)$$

The coefficients a_i used in eqations (16.15) and (16.16) are different from those used in (16.3). An improvement can be expected from the Tropical Rainfall Monitoring Mission (TRMM) which is scheduled for flight in 1997 where AVHRR and SSM/I will fly on the same orbiter, allowing the simultaneous measurement of T_{11}, T_{12}, and w.

During night the use of the channels at 3.7 μm allows an even better retrieval than the foregoing models because of the smaller water-vapour influence. Corresponding retrieval schemes that have been developed include measurements made in all thermal channels. Again, the consideration of quadratic terms improve the expected accuracies:

$$T_0 = a_0 + a_1 T_{3.7} + a_2(T_{3.7} - T_{11}) + a_3(T_{11} - T_{12}) + a_4(T_{3.7} - T_{11})^2 + a_5(T_{11} - T_{12})^2 \quad (16.17)$$

The coefficients for retrieval models (16.16) and (16.17) designed for the use with AVHRR/NOAA-11 are listed in table 16.5.

AVHRR retrieval model (16.16)								
θ [°]	Noise	a_0	a_1	a_2	a_3	-	-	ϵ
0	w/o	-5.57	1.0221	1.7187	0.2000	-	-	0.23
0	with	-10.45	1.0411	1.0271	0.3369	-	-	0.51
10	w/o	-5.64	1.0224	1.7203	0.2037	-	-	0.23
10	with	-10.51	1.0414	1.0360	0.3385	-	-	0.51
20	w/o	-5.97	1.02381	1.7224	0.2152	-	-	0.25
20	with	-10.77	1.0425	1.0716	0.3400	-	-	0.54
30	w/o	-6.94	1.0276	1.7145	0.2419	-	-	0.30
30	with	-11.48	1.0453	1.1213	0.3501	-	-	0.58
40	w/o	-10.02	1.0394	1.6700	0.2979	-	-	0.39
40	with	-13.81	1.0544	1.1705	0.3794	-	-	0.67
50	w/o	-21.04	1.0816	1.5015	0.4227	-	-	0.66
50	with	-23.11	1.0903	1.2192	0.4472	-	-	0.91
AVHRR retrieval model (16.17)								
θ [°]	Noise	a_0	a_1	a_2	a_3	a_4	a_5	ϵ
0	w/o	-7.87	1.0299	-1.2631	1.7163	0.0848	0.2152	0.21
0	with	-10.96	1.0422	-0.7718	1.5016	0.3358	0.09737	0.46
10	w/o	-8.01	1.0304	-1.2694	1.7150	0.08407	0.2206	0.21
10	with	-11.09	1.0427	-0.7793	1.5088	0.3315	0.1015	0.47
20	w/o	-8.51	1.0324	-1.2701	1.7208	0.08815	0.2297	0.23
20	with	-11.54	1.0444	-0.7922	1.5487	0.3230	0.1070	0.49
30	w/o	-9.81	1.0373	-1.2698	1.7188	0.1000	0.2509	0.27
30	with	-12.70	1.0488	-0.8161	1.6082	0.3117	0.1221	0.53
40	w/o	-13.21	1.0502	-1.2620	1.6739	0.1121	0.3003	0.37
40	with	-15.78	1.0603	-0.8470	1.6972	0.3005	0.1503	0.62
50	w/o	-24.22	1.0926	-1.3086	1.5170	0.1609	0.40900	0.64
50	with	-26.42	1.1007	-0.9633	1.8015	0.3337	0.2072	0.86

Table 16.5: *Coefficients and errors for retrieval models (16.16) and (16.17).*

The two-angle view of ATSR can be included by the following retrieval models for day and night applications:

$$T_0 = a_0 + a_1 T_{11} + a_2(T_{11} - T_{12}) + a_3(T_{11} - T'_{11}) + a_4(T_{11} - T'_{12}) \quad (16.18)$$

$$T_0 = a_0 + a_1 T_{3.7} + a_2(T_{3.7} - T_{11}) + a_3(T_{3.7} - T_{12}) + a_4(T_{3.7} - T'_{3.7}) + a_5(T_{3.7} - T'_{11}) + a_6(T_{3.7} - T'_{12}) \qquad (16.19)$$

The brightness temperatures at forward view are indicated by primes. The errors achieved with these retrieval models are lower than for those described above. However, in order to account for the fact that the pixels at forward view are larger by a factor of about two when compared to the pixel size at nadir view additional noise has to be expected in areas with strong temperature gradients. The additional noise of the order 0.1 K propagates into the error of T_0; the expected accuracies for (16.18) and (16.19) are 0.34 K and 0.30 K, respectively. Further improvement can be expected again from including quadratic terms. An analysis of such retrieval schemes shows that the sixth term in equation (16.19) loses significance. Therefore, the optimized retrieval schemes read:

$$T_0 = a_0 + a_1 T_{11} + a_2(T_{11} - T_{12}) + a_3(T_{11} - T_{12})^2 + a_4(T_{11} - T'_{11}) + a_5(T_{11} - T'_{12}) \qquad (16.20)$$

$$T_0 = a_0 + a_1 T_{3.7} + a_2(T_{3.7} - T_{11}) + a_3(T_{3.7} - T_{11})^2 + a_4(T_{3.7} - T_{12}) + a_5(T_{3.7} - T_{12})^2 + a_6(T_{3.7} - T'_{3.7}) + a_7(T_{3.7} - T'_{12}) \qquad (16.21)$$

These equations deliver the retrieval schemes with least errors of 0.28 K and 0.24 K, respectively. However, the complex form of (16.20) and (16.21) might imply weak robustness of the schemes against undetected clouds in the infrared imagery. The coefficients for equations (16.18) to (16.21) are listed in table 16.6.

ATSR retrieval model (16.18)									
Noise	a_0	a_1	a_2	a_3	a_4	-	-	-	ϵ
w/o	4.70	0.9843	4.7233	3.0148	-2.4542	-	-	-	0.18
with	-0.48	1.0020	3.0831	0.8274	-0.6570	-	-	-	0.34
ATSR retrieval model (16.19)									
Noise	a_0	a_1	a_2	a_3	a_4	a_5	a_6	-	ϵ
w/o	-3.66	0.9883	-3.8879	4.2751	1.4770	0.7687	-1.8872	-	0.15
with	0.40	0.9992	-3.3514	3.0070	0.7148	0.2946	-0.7332	-	0.30
ATSR retrieval model (16.20)									
Noise	a_0	a_1	a_2	a_3	a_4	a_5	-	-	ϵ
w/o	3.12	0.9903	4.1508	0.0762	2.6796	-2.1512	-	-	0.18
with	-3.81	1.0153	1.7928	0.2539	0.5476	-0.3314	-	-	0.28
ATSR retrieval model (16.21)									
Noise	a_0	a_1	a_2	a_3	a_4	a_5	a_6	a_7	ϵ
w/o	-0.27	1.0026	-3.1714	-0.1105	3.5072	0.07055	1.5598	-1.3414	0.12
with	-5.84	1.0220	-3.3630	-0.1809	0.1289	2.3494	0.5564	-0.4423	0.24

Table 16.6: *Coefficients and errors for retrieval models (16.18) to (16.21).*

A prerequisite for the application of the retrieval algorithms is a careful cloud screening. Since clouds do not transmit enough surface leaving radiation to the satellite that can be analysed in terms of sea surface temperature any cloud that remains undetected will lead to increased errors in T_0. Cloud detection schemes using all spectral channels available have been widely described in literature (e. g. Olesen and Grassl, 1985; Saunders and Kriebel, 1988; Schlüssel and Grassl, 1990).

Before going into operational application each remote sensing method has to be assessed by independent means of comparison. For sea surface temperatures this is not an easy task since temperatures as measured in situ by the voluntary observing fleet are often inaccurate and they are restricted to bulk temperatures measured at depths of decimetres to metres whereas the temperature measured by infrared radiometers is representative of the upper 10 to 20 micrometres, i. e. the skin of the ocean. Ground measurements of the skin temperatures are only taken from research vessels or platforms during dedicated field campaigns. Therefore, it is almost impossible to get a statistically significant number of comparisons between satellite and surface skin measurements (Schlüssel et al., 1987). Hence, for a global validation of the retrieval schemes quality controlled data from drifting buoys are usually employed (Emery et al., 1994; Mutlow et al., 1994). Those validation studies confirm that the ATSR delivers superior accuracy when compared to the AVHRR retrievals and that the algorithms including additional water-vapour information from SSM/I or using quadratic terms for the water-vapour corrections show a better correlation with in-situ measurements.

A further augmentation of all retrieval schemes is necessary if volcanic aerosol reduces the atmospheric transparency as indicated in table 16.3. The additional correction that is needed can be performed by altering the retrieval coefficients just after a volcanic outbreak or by including additional information about the enhanced atmospheric opacity as retrieved from spectral measurements in the near-infrared at 1.6 μm. The latter is part of the ATSR measurements and will be utilized in future for correcting effects imposed by volcanic aerosol in the stratosphere but also by tropospheric aerosol particles.

16.2.2 The cool skin and the diurnal thermocline of the ocean

The sea surface temperature as measured with infrared sensors stems from the upper 10 to 20 μm of the ocean while the "surface temperature" defined by oceanographers is that measured by ships or buoys at depths of several decimetres or metres. The difference between bulk and skin temperature is determined by the thermal molecular sublayer of the ocean and, during daytime, also by the diurnal thermocline (figure 16.3). Usually, the surface is cooled by sensible, and latent heat fluxes as well as by longwave radiation fluxes causing a lower temperature at the surface when compared to the temperature at depths of millimetres. The solar radiation which is not acting directly at the surface penetrates to greater depths where it is absorbed depending on the actual turbidity of the ocean water. Consequently, it affects both the skin layer and the mixed layer below. Normally, the skin cooling is greater than the solar heating and the bulk versus skin temperature difference remains positive across the molecular sublayer. However, the formation of a diurnal thermocline during calm situations can alter the sign of the integrated effect. Under clear sky situations with a dry atmosphere aloft both effects can reach their extremes: During night the radiative cooling as well as the latent heat flux lead to a strong lowering of the surface skin temperature while during daytime the solar radiation is well transmitted by the atmosphere. Hence, the variablitity of the bulk versus skin temperature, ΔT, is maximized in situations where infrared remote sensing of sea surface temperature is possible. These situations can usually be found in the subtropical oceans and in the middle latitudes. The temperature differences caused by these two effects can range from -1 K to 1 K in the open ocean. Even larger values occur in coastal waters and in lakes. Developments of the parameterisation of ΔT in terms of energy and momentum fluxes through the sea surface

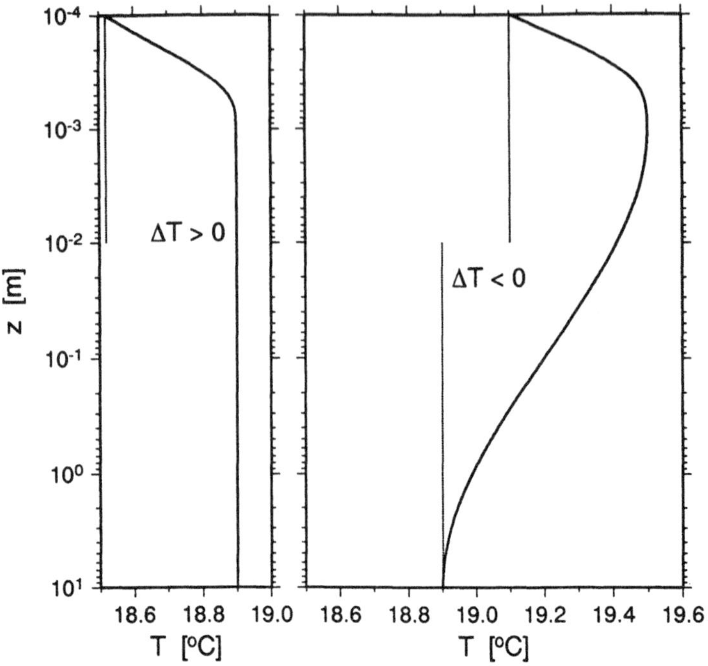

Figure 16.3: *Schematic presentation of possible temperature profiles in the upper ocean: cool skin (left) and diurnal thermocline with cool skin on top of it (right).*

are conducted in order to relate the satellite-measured skin temperatures to bulk temperatures as desired for the use in ocean models and for the interpretation of the satelite measurements in conjunction with past records of in-situ measurements of sea surface temperature (Robinson et al., 1984; Schlüssel et al., 1990).

A time series over two days showing the variability of hourly mean bulk versus skin temperature differences ΔT is shown in figure 16.4 together with surface heat and radiative fluxes as measured during the TOGA Coupled Ocean Atmophere Response Experiment (COARE) in the western equatorial Pacific in February 1993. Two time series referring to bulk temperatures at 0.1 m and 4 m are shown elucidating the high variablitity of ΔT under the influence of energy and momentum fluxes. In spite of the high solar net flux exceeding 1000 W/m^2 at the beginning of the second day the skin temperature as well as the bulk temperature show a pronounced diurnal cycle. At a depth of 0.1 m the cycle is delayed by about three to four hours. During the first day the clouds cover all tropospheric levels beginning with four to six oktas cumulus humilis and cumulus mediocris accompanied by cirrus clouds. The wind speed is close to 7 m/s causing a good mixing of the water which leads to ΔT values beginning near 0.2 K and increase with decreasing cloud amount and decreasing wind speed to about 0.5 K at the end of the night. The heat fluxes are between -200 and -300 W/m^2 during the first night. During the following morning the formation of cumulus begins and near the end of 3 February also altocumulus and opaque cirrus generated by high reaching cumulonimbus are present. Together with increasing solar irradiances ΔT reaches values near -0.1 K. Low

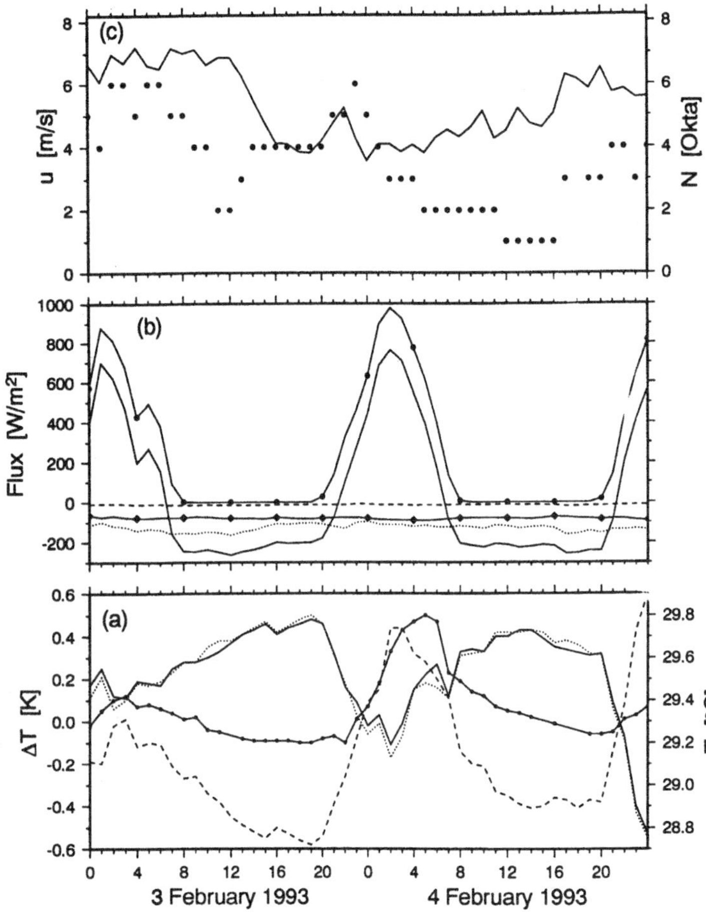

Figure 16.4: *Time series for 3 and 4 February 1993 in the western equatorial Pacific Ocean: (a) Hourly means of* $\Delta T_{0.1}$ *(—),* ΔT_4 *(...), skin temperature (- - -) and bulk temperature 0.1 m depth (○), (b) hourly means of shortwave net flux (○), longwave net flux (◇), latent (...), sensible (- - -) and total heat flux (—), (c) hourly means of surface wind speed (—) and hourly observations of cloud cover (○ ○○).*

cloud amounts and descreasing wind speeds at the beginning of 4 February cause again increasing temperature differences which occupy values near 0.4 K shortly after sun set. Near 6 UT pronounced differences have been observed between $\Delta T_{0.1}$ and ΔT_4 which are caused by heating due to strong solar radiation absorbed in the upper centi- and decimetres which are less pronounced at greater depth because of lacking wind mixing. Surface cooling by turbulent heat and longwave radiation transfer with flux densities below –200 W/m^2 takes place in the following night. In the second half of 4 February again deep cumulus convection is developed. At wind speeds of about 6 m/s ΔT is decreased to –0.5 K. The cause for the greater differences with respect to the previous day might be due to wind-speed fluctuations not seen in the hourly mean. On 3 February near 22 UT the wind speed reached peak values higher than 8 m/s for several minutes which lead to short but effectful mixing of the surface water which prevented the surface warming as near the end of the following day. Such short-term fluctuations might be partly responsible for the failure of simple parameterisations for ΔT when entering hourly means of heat and momentum fluxes.

Different processes prevail at different wind speeds which can be classified in the domains of free convection in calm situations, the forced convection by rollers on top of breaking wavelets at moderate wind speeds and the longwave breaking at gale force winds. For each of theses wind-speed classes a different parameterisation for ΔT is necessary. The mean temperature difference across the thermal boundary layer under moderate wind speeds at night has been described by Saunders (1967):

$$\Delta T = -\frac{\lambda(\nu/\kappa)Q_0}{u_* c_p \rho} \tag{16.22}$$

where $Q_0 = Q_T + Q_L + Q_H$ is the sum of longwave radiative flux (Q_T), latent heat flux (Q_L) and sensible heat flux (Q_H), u_* is the friction velocty, c_p is the specific heat capacity, ρ is the density, ν is the kinematic viscosity, and κ is the coefficient of thermal molecular diffusion, all referring to the upper ocean. The nondimensional coefficient λ was thought as constant of proportionality. However, later studies showed a variability depending on the wind speed (Grassl, 1976; Wu, 1985), especially, when applying (16.22) to other than the forced convection regime at moderate wind speeds. Therefore, Katsaros et al. (1977) worked out a description of ΔT for calm situations:

$$\Delta T = b(\alpha g \kappa^2/\nu)^{-1/4}(-Q_0/(c_p\rho))^{3/4}, \tag{16.23}$$

with another non-dimensional coefficient (presumably constant) b, the coefficient of thermal expansion α and the acceleration due to gravity g. Strictly, (16.23) is only valid at vanishing wind speeds, a smooth transition from equation (16.22) to (16.23) has been described by Kudryavtsev and Soloviev (1985) based on the surface Richardson number $Rf_0 = \alpha g Q_0 \nu/(c_p \rho u_*^4)$ which is the ratio between the energy supplied by buoyant forces due to thermal stratification and the energy flux furnished by the shear stress in the thermal boundary layer. A general formulation that covers all three wind speed regimes has been described by Soloviev and Schlüssel (1994):

$$\Delta T = -\Lambda_0 \frac{Q_0}{c_p \rho u_*} Pr^{1/2}(1 + Rf_0/Rf_{cr})^{-1/4}(1 + Ke/Ke_{cr})^{1/2} \tag{16.24}$$

where Λ_0 is a constant and $Ke = u_*^3/(g\nu)$ is the Keulegan number which controls the transition from forced convection to longwave breaking. $Rf_{cr} = -1.5 \times 10^{-4}$ and $Ke_{cr} = 0.18$ are critical values of the surface Richardson and the Keulegan numbers, respectively. At moderate and low wind speeds equation (16.24) can asymptotically be truncated to (16.22) and (16.23), respectively. The variation of ΔT with friction velocity according to (16.24) is shown for different heat fluxes in figure 16.5.

During daytime the solar heating and the possible formation of a diurnal thermocline have to be taken into account when describing the bulk versus skin temperature difference. Paulson and Simpson (1981) describe only the effect of the cool skin under solar illumination by

$$\Delta T = -\frac{\delta}{k}\{Q_0 + 0.93 Q_R^{\downarrow}[1 - \frac{1}{\delta}\sum_{i=1}^{9} F_i \zeta_i (1 - \exp(-\delta/\zeta_i))]\} \tag{16.25}$$

where δ is the depth of the thermal sublayer, k is the coefficient of thermal conductivity, $Q_R^{\downarrow}$ is the downwelling solar irradiance, the F_i represent the spectral distribution of $Q_R^{\downarrow}$ just below

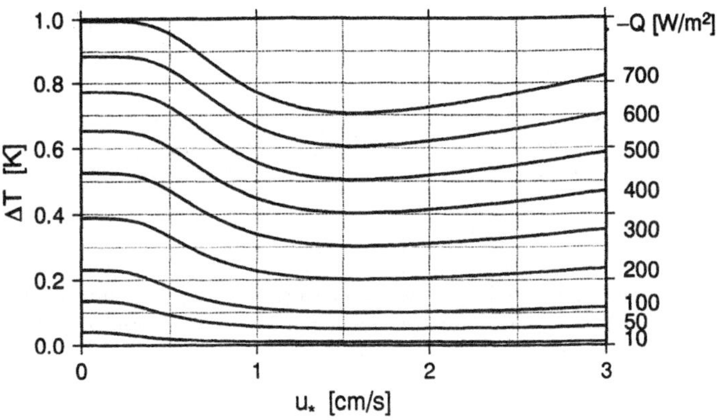

Figure 16.5: *Variation of mean temperature difference across the cool skin of the ocean with friction velocity at different heat fluxes.*

the sea surface in nine spectral bands, and ζ_i the distribution of penetration depths in these bands. Another coupled parameterisation has been given by Hasse (1971)

$$\Delta T = -C_1 Q_0 / u - C_2 Q_R / u, \tag{16.26}$$

where Q_R is the net solar radiative flux, C_1 and C_2 are coefficients depending on the depth where the bulk water temperature is measured. However, the daytime parameterisations are still inaccurate and require further research.

Global distributions of monthly mean surface temperature and nocturnal skin temperature deviation are shown in figure 16.6 for August 1987. The temperature field is derived from AVHRR according to the split-window technique with empirical scan-angle dependent coefficients. The field is an average from daily fields of the a.m. satellite orbits supplied by the Jet Propulsion Laboratory in the frame of the NOAA/NASA Pathfinder project. The distribution of the skin effect has been calculated from instantaneous surface fluxes and friction velocities according to equation (16.24) using remotely sensed surface wind speed, latent heat and longwave radiative fluxes retrieved from SSM/I and AVHRR data (Schlüssel et al., 1995), and a parameterised sensible heat flux where the surface air temperature is derived from the remotely sensed specific air humidity and the assumption of a constant relative humidity of 80%. The map reveals a high spatial variability of the skin effect even after averaging over an entire month. Lowest values are found in the northern Pacific Ocean and in the Arctic Ocean where warm and moist air is advected over cold water and concurrent deep cloud cover minimizes the the energy flux to the atmosphere. In those areas the bulk versus skin temperature differences vanish or even revert their sign showing slightly negative values. Maximum differences are found in subtropical areas and in the storm tracks of the southern oceans due to high energy fluxes leaving the ocean.

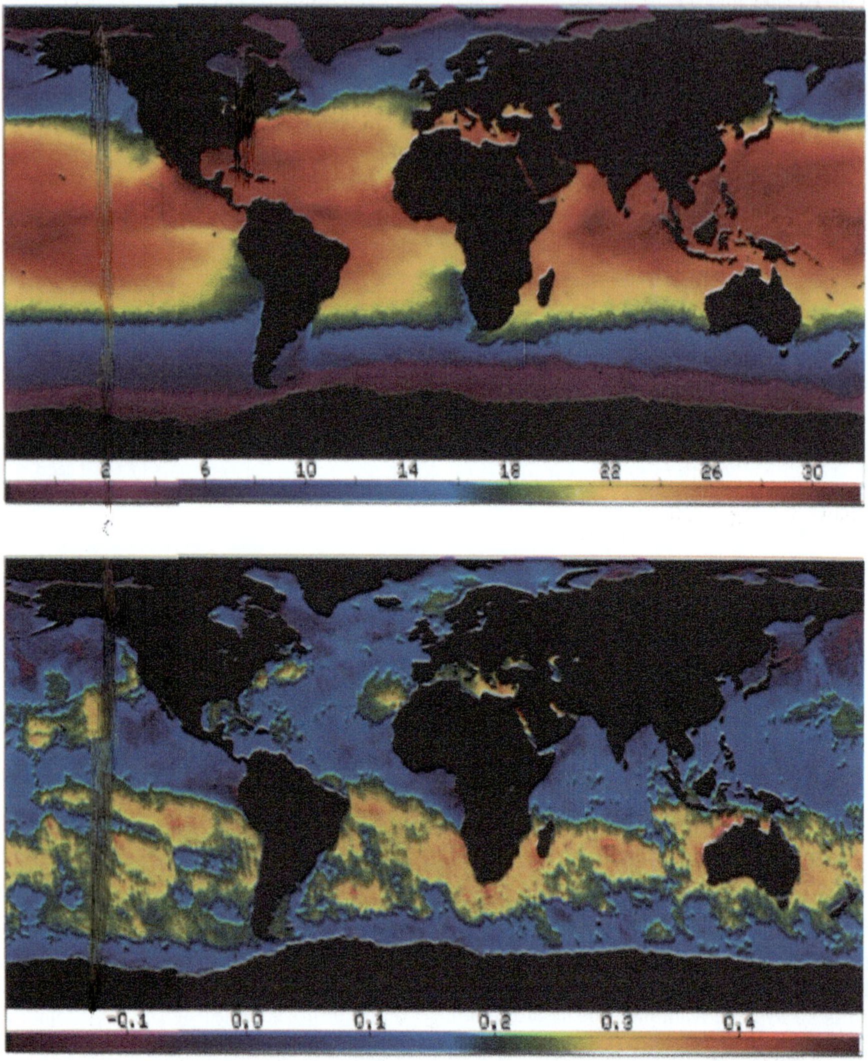

Figure 16.6: *Mean sea surface temperature in °C from AVHRR measurements (top) and nocturnal bulk versus skin temperature difference in K from AVHRR and SSM/I measurements (bottom) in August 1987; due to the limited sampling frequency of the SSM/I the boundaries of single swaths are visible in the lower image. The estimated error of ΔT is 0.05 K.*

16.3 Surface Wind Speed

Radiation reflected by a flat surface is highly polarized for incidence angles $0° < \vartheta < 90°$. This is also true for the smooth sea surface. When the surface is roughened by wind stress acting on it the polarization by the surface reflection decreases due to different mechanisms (Wentz, 1992):

- Surface waves with wavelengths greater than the wavelength of the radiation interacting with the surface change the local incidence angle due to tilted surface facets. This effects a mixture of horizontal and vertical polarizations by which the polarization of the reflected radiation is decreased (Stogryn, 1967).
- Whitecapping on the sea surface increases the reflectivity at visible wavelengths and decreases the reflectivity at infrared and microwave frequencies (Koepke, 1986; Stogryn, 1972).
- Surface ripples being small compared to the wavelength of the incident radiation cause diffraction of the radiation. Considering capillary waves as the shortest surface waves this effect is only relevant for radiation exceeding millimetre wavelengths (Wentz, 1975).

The three effects have been parameterised in terms of the rms slope of the large undulations, a fractional whitecap coverage and the correlation length of the small-scale ripples which all can in turn be described, more or less, as function of wind speed or, better, friction velocity (Cox and Munk, 1954; Monahan and O'Muircheartaigh, 1986; Mitsuyasu and Honda, 1982). The surface roughness especially modifies the reflectivity at microwave frequencies where the flat surface exhibits a rather high reflectivity ϱ as shown in figure 16.7 for 37 GHz where ϱ is shown for two temperatures according to the parameterisation of Klein and Swift (1977). The reflectivities increase monotonically at horizontal polarization; the vertically polarized component shows a minimum at Brewster's angles between 75° and 80°. By virtue of Kirchhoff's law and in view of the high opacity of sea water at microwave frequencies the surface emissivity[1] is related to the reflectivity by $\varepsilon = 1 - \varrho$. Several models have been developed describing the change of ϱ as function of wind speed or friction velocity. Pandey and Kakar (1982) proposed the form

$$\varrho_p(u) = [\varrho_{sp} - \Delta\varrho_{rp}(u)](1 - f(u)) + \varrho_{fp} f(u) \tag{16.27}$$

where ϱ_{sp} is the specular reflectivity of the smooth sea surface at polarization p, $\Delta\varrho_{rp}(u)$ is the change of the reflectivity due to tilted, not foam covered surface facets, ϱ_{fp} is the reflectivity of foam, and $f(u)$ is the fractional foam coverage depending on wind speed u. The wind speed dependence of $\Delta\varrho_{rp}(u)$ is a linear one while f varies with higher powers of u. Detailed descriptions of (16.27) in terms of temperature, salinity, incidence angle and frequency have been optimized for the use with radiometers operating at different scan angles and frequencies for the Scanning Multichannel Microwave Radiometer (SMMR) that was flown on the Nimbus-7 satellite (Francis et al., 1983; Wentz 1983) and for the Special Sensor Microwave/Imager (SSM/I) being flown as part of the DMSP satellites (Schlüssel and Luthardt, 1991; Wentz, 1992; Rosenkranz, 1992). With the knowledge about the wind dependent reflectivity of the sea surface it is straight forward to calculate the brightness of the sea surface for a given atmospheric/oceanic situation and subsequently the brightness temperature as measured by a satellite radiometer. Such calculations are used for the development of remote-sensing methods seeking for the wind-speed retrieval from a given set of spectrally distributed brightness tmperatures measured from satellites. Figure 16.8 shows the simulated brightness temperatures for the SSM/I channels at varying wind speed for a mid-latitude summmer situation. Especially, at horizontal polarization the brighness temperature can vary by 40 K within the wind speed range 0 to 30 m/s. However, the actual atmospheric situation induces further brightness-temperature changes that mask the relationship with the wind speed as shown for the brightness temperatures at 37 GHz, horizontal polarization (Figure 16.9). Consequently, multi-channel retrieval techniques have to

[1]instead of "emissivity" often the term "emittance" is used.

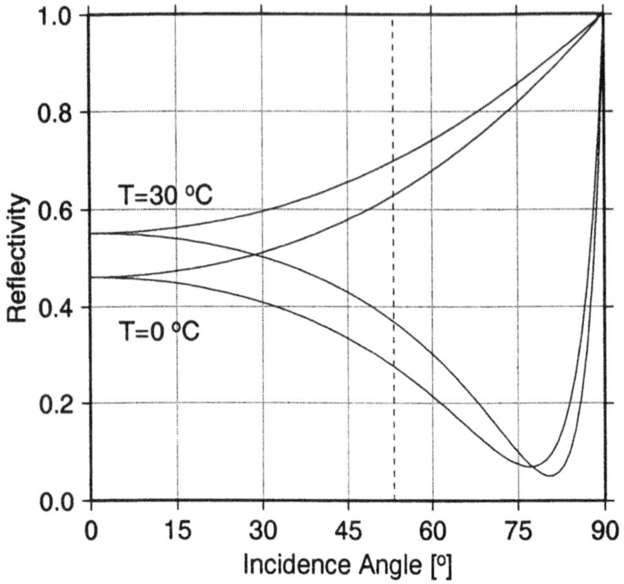

Figure 16.7: *Reflectivity of the flat sea surface at 37 GHz for $T = 0\,°C$ and $T = 30\,°C$ at horizontal and vertical polarization; the SSM/I zenith angle is indicated by the vertical dashed line.*

be developed that correct for the atmospheric impact due to water vapour, cloud and rain liquid water as well as varying atmospheric temperature structures. Corresponding schemes have been developed by means of radiative transfer simulations (Schlüssel and Luthardt, 1991) as well as empirically by matching SSM/I measurements with in-situ observations of wind speed (Goodberlet et al., 1989). Usually, dual polarized brightness-temperature measurements at several frequencies are necessary to perform an adequate correction. A simple retrieval model reads:

$$u = c_0 + c_1 T_{19v} + c_2(T_{19v} - T_{19h}) + c_3 T_{22v} + c_4 T_{37v} + c_5(T_{37v} - T_{37h}) \qquad (16.28)$$

where $c_i, i = 0, .., 5$ are coefficients which are optimized for different atmospheric liquid-water loadings separated by the thresholds given in table 16.7. In cases where T_{19h} and $(T_{37v} - T_{37h})$ exceed the given thresholds it is assumed that no wind-speed retrieval can be performed due to strong contamination by rain or cloud-liquid water. The wind speed derived from microwave radiometry is merely a neutral stability wind which is more closely related to the friction velocity at the sea surface because it is the wind stress that acts on the surface rather than the wind itself.

Higher accuracies can be achieved with more advanced retrieval models than given above. This requires to acknowledge the simultaneous impact of all parameters affecting the radiative transfer in the radiometer channels. The parameters retrieved with models like (16.28) are entered as first guess into a radiative transfer model in order to calculate brightness temperatures according to the first estimate. The differences between calculated and measured brightness temperatures are then used to alter the geophysical parameters. This procedure is repeated

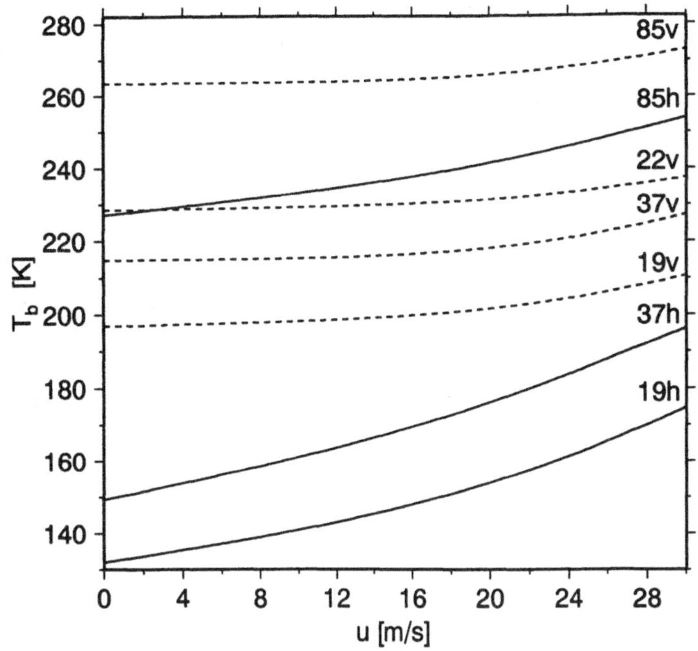

Figure 16.8: *Simulated variation of brightness temperatures with wind speed for a mid-latitude summer situation with clear sky.*

$T_{19h} \le 165\,K\ and\,(T_{37v} - T_{37h}) \ge 55\,K$						
c_0	c_1	c_2	c_3	c_4	c_5	Error
149.0	0.8800	-0.4887	-0.4642	-0.7131	-0.4668	1.4
$165\,K < T_{19h} \le 190\,K\ and\,55\,K < (T_{37v} - T_{37h}) \ge 20\,K$						
c_0	c_1	c_2	c_3	c_4	c_5	Error
205.6	-0.08349	0.07933	0.1066	-0.7346	-0.9132	1.6

Input: brightness temperatures in K
Output: wind speed in m/s

Table 16.7: *Coefficients and errors for retrieval model (16.28).*

until a final solution is found. An optimum method for such a solution, the minimum variance method, has been described by Rodgers (1976).

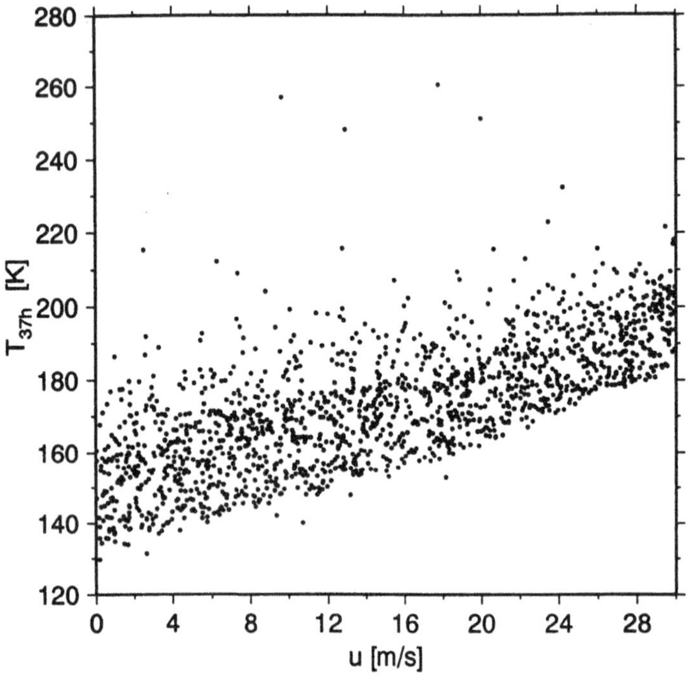

Figure 16.9: *Simulated SSM/I brightness temperatures at 37 GHz, horizontal polarization, for 546 globally distributed ocean-atmosphere situations.*

16.4 Water Vapour Over Sea

The near-surface specific humidity, q, that is desired for the parameterisation of the evaporation over sea cannot be remotely sensed because remote sensing techniques rely on the emission or attenuation of radiation from optically thick layers. Therefore, a proxy parameter must be found that can be remotely sensed and that is closely related to q. Liu and Niiler (1984) have chosen monthly mean total columnar water-vapour contents, w, as derived from SMMR measurements and related it to q by using a polynom of fifth degree. This method is restricted to average atmospheric situations and fails in single cases as well in monthly mean situations which deviate from normal cases. In particular, when the planetary boundary layer is decoupled from the higher troposphere by temperature inversions one can expect a lack of determinism between q and w. This is elucidated in figure 16.10 for about 9400 atmospheric situations collected in the time period July 1987 through June 1988. A detailed analysis of the data shows that in single situations q can be derived from w with an accuracy of 2.2 g/kg only when using a multilinear regression on powers of w up to five. Therefore, it is indicated to look for thinner layers in the lower troposphere that can be sensed by radiometry and that has a closer connection to the near-surface specific air humidity. Schulz et al. (1993) have shown that the water-vapour column of the lower 500 m, w_l, can be retrieved from SSM/I measurements independently from the total columnar water vapour. As seen in figure 16.10, q and w_l are indeed closely related to each other. However, outliers from this close relationship can be expected in coastal areas when dry air is advected over the ocean and the marine boundary layer is of shallow depth as indicated by the outliers in figure 16.10 that show high values of q and rather low water-vapour columns. From SSM/I measurements at 19, 22, and 37 GHz w_l can be retrieved by

$$w_l = -5.934 + 0.03697 T_{19v} - 0.02390 T_{19h} + 0.01559 T_{22v} - 0.00497 T_{37v} \tag{16.29}$$

where the brightness temperatures have to be entered in K and w_l results in g/cm^2 or cm. The main information about w_l comes from the channel at 19 GHz, vertical polarization, while T_{22} is used to suppress the effect of water vapour above the bottom layer. The information from T_{19h} and T_{37v} is used to account for wind speed and liquid-water effects, respectively. Equation (16.29) allows the retrieval of w_l with an accuracy of 0.07 g/cm^2 as long as no high liquid-water contents hamper the transmisison of radiation from the boundary layer to the satellite. Cases with $(T_{37v} - T_{37h}) > 50$ K or $T_{19h} > 165$ K are excluded from the computation of w_l. The specific humidity can then be obtained via

$$q = -0.17 + 20.00 w_l \tag{16.30}$$

where w_l is entered in g/cm^2 and q is retrieved in g/kg with an accuracy of 1.2 g/kg. The success of this two-step method in retrieving q has justified to go a step further in directly deriving q from microwave measurements of the SSM/I. This minimizes the propagation of retrieval errors and leads to lower systematic errors (Schlüssel et al., 1995):

$$q = -80.23 + 0.6295 T_{19v} - 0.1655 T_{19h} + 0.1495 T_{22v} - 0.1553 T_{37v} - 0.06695 T_{37h} \tag{16.31}$$

Here, information from the vertically polarized channel at 37 GHz is included in order to better account for the wind-induced surface roughness. This channel did not add significant information to the retrieval model (16.29) and was ignored in that case. The performance of equation (16.31) is slighly better than the combination (16.29)-(16.30), the estimated retrieval error is 1.1 g/kg. An intercomparison of the different methods for retrieving q with ship measurements in the time period July 1987 to June 1988 is shown in figure 16.11, covering 4245 globally distributed match-ups. Besides the methods (16.29)-(16.30) and (16.31) the

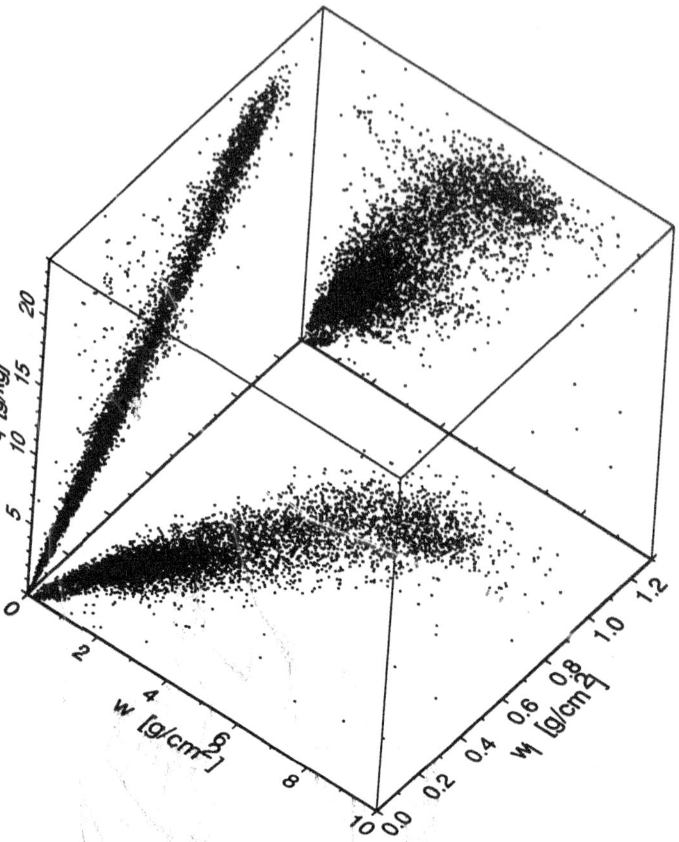

Figure 16.10: *Variation of near-surface specific air humidity with total columnar water vapour content and with the water-vapour column of the lower 500 m in the time period July 1987 to June 1988 over the global ocean.*

technique of Liu and Niiler (1984) is used with the water-vapour columns w as derived from SSM/I channels at 22 Ghz and 37 GHz, vertical polarization, according to Schlüssel and Emery (1990). Also, a technique is incorporated in determining q by fixing the relative air humidity to 80% and assuming the air temperature to be equal to the sea surface temperature. This method only requires the knowledge about the sea surface temperature for the retrieval of q. It is seen that the direct method according to (16.31) gives the lowest systematic and standard deviations with respect to the ship measurements of q. At high latitudes, the 80% - method gives similar results as the other techniques. In other latitudes great systematic errors must be expected when following this simple approach. However, it should be kept in mind that one of the main error sources in such intercomparisons is the temporal and spatial variability of water vapour between satellite measurements and ship observations that have been matched for compiling the statistics. Here, the time and space windows used cover differences of one hour and 25 km, respectively.

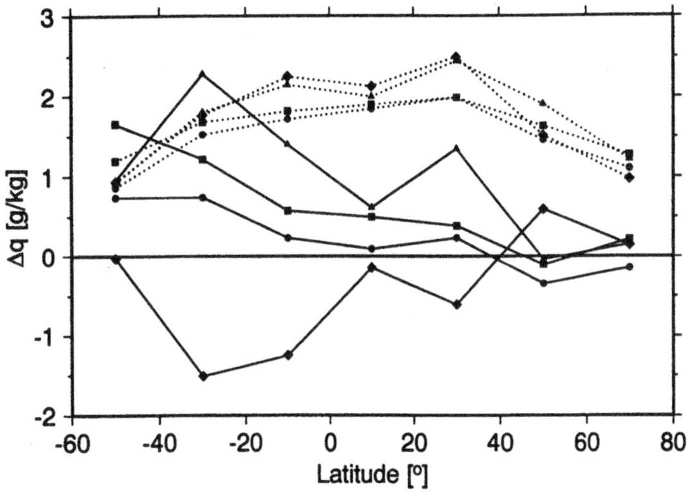

Figure 16.11: *Systematic and standard deviations between in-situ measurements and satellite retrievals of the specific air humidity in g/kg for latitude belts with a width of 20 degrees (○) using the direct method, (□) following models (16.29),(16.30), (△) according to the method of Liu and Niiler and (◇) with the 80%-method.*

16.5 Latent Heat Flux

The latent heat flux is usually parameterised with the bulk aerodynamic formula according to equation (16.2). Besides the three parameters, T_0, u, and q, discussed above the air density, the latent heat of evaporation and the Dalton number must be known. For the determination of ϱ a standard pressure can be used without loss of much accuracy and the air temperature is taken as $T_0 - 1$ K (Schlüssel et al., 1995) or calculated assuming a constant relative humidity of 80% and using the remotely sensed specific air humidity (Liu et al., 1994). The computation of C_E requires the knowledge of wind speed and the stability of the atmosphere close to the sea surface (e.g. Smith, 1989), L is a function of sea surface temperature. A problem that is currently of disadvantage is the fact that the surface temperature is not measured from the same satellite as the wind speed and surface air humidity. Although the sea surface temperature is a rather conservative parameter and a temporal shift of its measurement of several hours or even a day with respect to the observation of the other variables does not introduce great errors the sky is often obscured by clouds, thus, allowing a measurement of sea surface temperature only every two to four weeks in some regions. The performance statistics shown in figure 16.12 that intercompares latent heat fluxes derived from SSM/I and AVHRR with co-located ship measurements have been established with AVHRR measurements that have been taken up to one week apart from the corresponding SSM/I measurements. A total of 4014 ship measurements have been matched with satellite observations between July 1987 and June 1988. On the global average there are almost no systematic deviations between ship and satellite observations, the standard deviations range from 28 to 44 W/m^2, where the method labelled 'direct' shows the best performance while the 80%-method displays greatest deviations from the in-situ measurements. The zonally stratified comparisons of single match-ups show greatest deviations in the tropical regions for all methods. The results for the technique described by Liu and Niiler (1984) shows generally less accuracy than the methods utilizing the retrieval of boundary-layer water vapour.

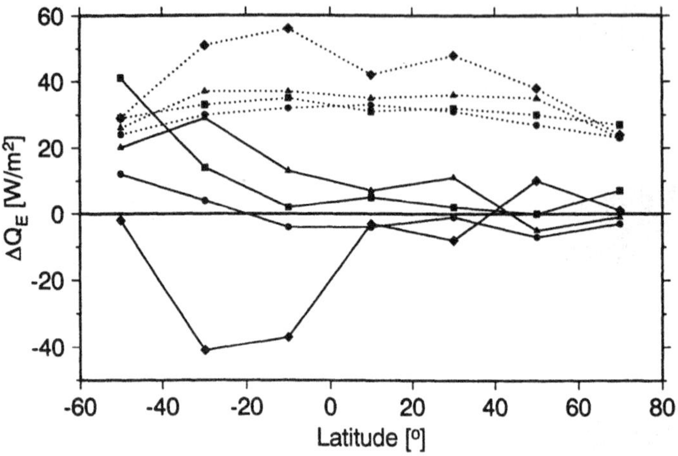

Figure 16.12: *Systematic and standard deviations between in-situ measurements and satellite retrievals of the latent heat flux in* W/m^2 *for latitude belts with a width of 20 degrees; the symbols refer to the different retrieval schemes for* q *as in figure 16.11.*

Global fields of latent heat fluxes over sea are shown in figure 16.13 for five months in the period August 1987 through August 1988. The fields have been derived from AVHRR sea surface temperatures as obtained from the NOAA/NASA Pathfinder project and from SSM/I antenna temperatures processed with the wind and humidity algorithms described above. High rates of evaporation are seen in areas where dry air is advected over warm water and steady winds produce a pronounced air-sea exchange. The highest values are found in the North Pacific Ocean and in the southern oceans during the corresponding seasons where the monthly evaporation exceeds 350 mm of water. Other areas with high evaporation are the trade wind regions. At the same time regions exist with no evaporation or even dew on the sea surface where warm humid air masses are advected over cool waters. Such situations can be found in polar regions of the northern hemisphere in summer.

16.6 Concluding Remarks

The combination of passive microwave measurements from the SSM/I with infrared imagery from the AVHRR allows the retrieval of sea surface temperature, surface wind speed and near-surface air humidity which in turn can be used to parameterise the latent heat flux at the sea surface whith rms errors near 30 W/m^2. Since the major part of the retrieval errors in the recently developed remote sensing methods is of random nature monthly averages of the fluxes can be obtained with an accuracy better than 10 W/m^2. Still higher accuracy in the retrieved parameters will be achieved from a new generation of intruments to be flown in the near future. Among the possible improving factors one can expect:

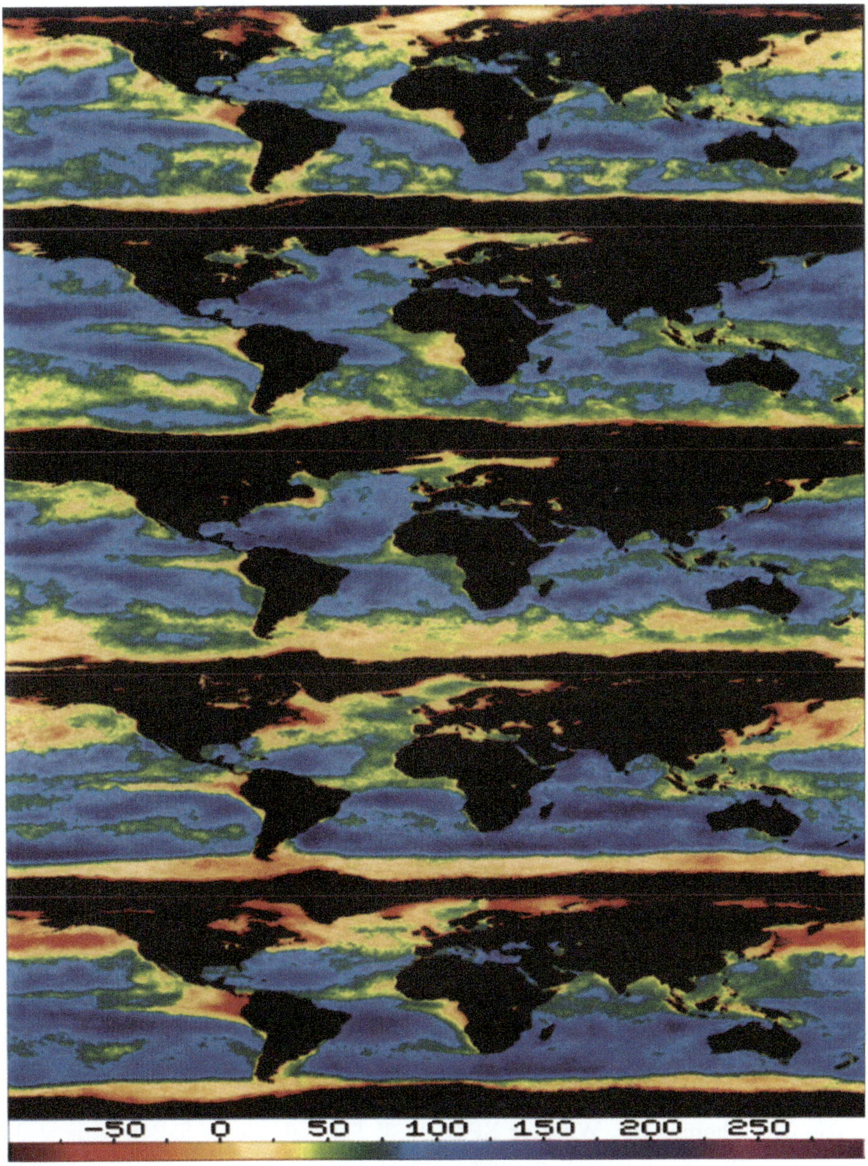

Figure 16.13: *Monthly mean latent heat flux density over oceans ($-Q_E$) in W/m^2 for August 87, November 87, February 88, May 88, and August 88 (top to bottom); the estimated error is 10 W/m^2.*

- Utilization of measurements at 1.6 μm for the correction of remotely sensed sea surface temperature with respect to atmopheric aerosol particles

- Measurement of the third and fourth components of the Stokes vector will give insight into the influence of the wind direction on the wind speed retrieval, hence, allow a more accurate wind-speed retrieval
- Measurement of all parameters entering the parameterisation of the latent heat flux from the same platform (AVHRR and SSM/I on TRMM, MODIS and MIMR on EOS-PM)
- Measurement of sea surface temperature with microwave radiometry allowing the observation under cloudy situations, too
- Correction of infrared measurements of sea surface temperature with water vapour columns derived from microwave radiometry on board the same satellite

Acknowledgements

This research was sponsored by Deutsche Forschungsgemeinschaft and by the Commission of the European Communities. The sea surface temperature data that have been used in this study have been provided by the NASA Physical Oceanography Distibuted Active Archive Center at the Jet Propulsion Laboratory, California Institute of Technology. All graphs have been plotted with free software supplied by Wessel and Smith (1991).

16.7 References

Anding D, Kauth R (1970) Estimation of sea-surface temperature from space. *Remote Sensing Environ* 1: 270

Barton IJ, Závody AM, Prata AJ, Llewellyn-Jones DT, Bailey P, Cechet RP, Gorman MR, Kent P, Lee DJ, Mutlow CT, Nilsson CS (1992) Intercomparison of satellite-derived sea surface temperatures. *Proceedings of the Central Symposium of the 'International Space Year' Conference*, held in Munich, Germany, 30 March - 4 April 1992, ESA SP-341: 367–372

Cox CS, Munk WH (1954) Measurement of the roughness of the sea surface from photographs of the sun's glitter. *J Opt Soc Am* 44: 838-850

Deschamps PY, Phulpin T (1979) Atmosphere correction of infrared measurements of sea surface temperature using channels at 3.7, 11 and 12 μm. *Bound Lay Met* 18: 131–143

Emery WJ, Yu Y, Wick GA, Schlüssel P, Reynolds RW (1994) Correcting infrared satellite estimates of sea surface temperature for atmospheric water vapor contamination. *J Geophys Res* 99: 5219–5236

Francis CR, Thomas DP, Windsor EPL (1983) The evaluation of SMMR retrieval algorithms, in: Allan TD (ed) *Satellite microwave remote sensing.* Ellis Horwood Ltd. Chichester, 481–498

Goodberlet MA, Swift CT, Wilkerson JC (1989) Remote sensing of ocean surface winds with the Special Sensor Microwave/Imager. *J Geophys Res* 94: 14547–14555

Grassl, H (1976) The dependence of the measured cool skin of the ocean on wind stress and total heat flux. *Bound Lay Met* 10: 465–474

Harris, AR, Mason IM (1992) An extension to the split-window technique giving improved atmospheric correction and total water vapour. *Int J Remote Sensing* 13: 881–892

Hasse, L (1971) The sea surface temperature deviation and the heat flow at the sea-air interface. *Bound Lay Met* 1: 368–379

Hasse, L (1993) Observations of air sea fluxes, in Raschke E, Jacob D (eds) *Energy and water cycles in the climate system*, NATO ASI Series I 5: 263–293

Katsaros KB, Liu TW, Businger JA, Tillman JA (1977) Heat transport and thermal structure in the interfacial boundary layer measured in an open tank of water in turbulent free convection. *J Fluid Mech* 83: 311–335

Koepke P (1986) Oceanic whitecaps: Their effective reflectance, in Monahan EC, MacNicoaill G (eds) *Oceanic whitecaps and their role in air-sea exchange processes.* Dordrecht, Reidel, pp 272–274

Kudryavtsev VN, Soloviev AV (1981) On thermal state of the ocean surface. *Izvestia Atmos Oceanic Phys* 17: 1065–1071

Lauritson L, Nelson GJ, Porto FW (1979) Data extraction and calibration of TIROS-N/NOAA radiometers. *NOAA Technical Memorandum* 107, U.S. Department of Commerce

Liu WT, Niiler PP (1984) Determination of monthly mean humidity in the atmospheric surface layer over ocean from satellite data. *J Phys Ocean* 14: 1451–1457

Liu WT, Zhang A, Bishop JKB (1994) Evaporation and solar irradiance as regulators of sea surface temperature in annual and interannual changes. *J Geophys Res* 99: 12623–12637

Llewellyn-Jones DT, Minnett PJ, Saunders RW, Závody AM (1984) Satellite multichannel infrared measurement of sea surface temperature of the N.E. Atlantic Ocean using AVHRR/2. *Quart J Roy Met Soc* 110: 613–631

McClain EP, Pichel WG, Walton CC, Ahmad Z, Sutton J (1983) Multi-channel improvements to satellite derived global sea surface temperatures. *Adv Space Res* 2: 43–47

McClatchey RA, Fenn RW, Selby JEA, Volz FE, Garing JS (1972) Optical properties of the atmosphere, *Environ Res Pap* 411 AFCRL-72-0497, Air Force Cambridge Research Laboratories Bedford Massachusetts

McMillin LM (1975) Estimation of sea surface temperature from two infrared window measurements with differential absorptions, *J Geophys Res* 80: 5113–5117

McMillin LM, Crosby DS (1984) Theory and validation of the multiple window sea surface temperature technique. *J Geophys Res* 89: 3655–3661

Minnett PJ (1990) The regional optimization of infrared measurements of sea surface temperature from space. *J Geophys Res* 95: 13497–13510

Mitsuyasu H, Honda H (1982) Wind-induced growth of of water waves. *J Fluid Mech* 123: 425–442

Monahan EC, O'Muircheartaigh IG (1986) Whitecaps and the passive remote sensing of the ocean surface. *Int J Remote Sens* 7: 627–642

Mutlow CT, Závody AM, Barton IJ, Llewellyn-Jones DT (1994) Sea surface temperature measurements by the along-track scanning radiometer on the ERS-1 satellite: Early results. *J Geophys Res* 99: 22575–22588

Olesen FS, Grassl H (1985) Cloud detection and classification over oceans at night with NOAA-7. *Int J Remote Sens* 6: 1435–1444

Pandey PC, Kakar RK (1982) An empirical microwaves emissivity model for a foam-covered sea. *IEEE J Ocean Eng* 7: 135–140

Paulson CA, Simpson JJ (1981) The temperature difference across the cool skin of the ocean. *J Geophys Res* 86: 11044–11054

Prabhakara C, Dalu G, Kunde VG (1974) Estimation of sea surface temperature from remote sensing in the 11 to 13 micron window region. *J Geophys Res* 79: 5039–5044

Prata AJF, Cechet RP, Barton IJ, Llewellyn-Jones DT (1990) The Along Track Scanning Radiometer for ERS-1 – scan geometry and data simulation. *IEEE Trans Geosci Remote Sens* 28: 3–13

Rao CRN (1992) Aerosol radiative corrections to the retrieval of sea surface temperature from infrared radiances measured by the Advanced Very High Resolution Radiometer (AVHR). *Int J Remote Sens* 13: 1757–1769

Reynolds RW (1993) Impact of Mount Pinatubo aerosols on satellite-derived sea surface temperatures. *J Climate* 6: 768–774

Robinson IS, Wells NC, Charnock H (1984) The sea surface boundary layer and its relevance to the measurement of sea surface temperature by airborne and spaceborne radiometers. *Int J Remote Sens* 5: 19–45

Rodgers CD (1976) Retrieval of atmospheric temperature and composition from remote measurements of thermal radiation. *Rev Geophys Space Phys* 14: 609–624

Rosenkranz PW (1992) Rough-sea microwave emissivities measured with the SSM/I. *IEEE Trans Geosci Remote Sens* 30: 1081–1085

Saunders PM (1967) The temperature at the ocean-air interface. *J Atmos Sci* 24: 269–273

Saunders RW, Kriebel KT (1988) An improved method for detecting clear sky and cloudy sky radiances from AVHRR data. *Int J Remote Sens* 9: 123–150

Schlüssel P (1989) Satellite-derived low-level atmospheric water vapour content from synergy of AVHRR with HIRS. *Int J Remote Sens* 10: 705–721

Schlüssel P, Emery WJ, Grassl H, Mammen T (1990) On the bulk-skin temperature difference and its impact on satellite remote sensing of sea surface temperature. *J Geophys Res* 95: 13341–13356

Schlüssel P, Emery WJ (1990) Atmospheric water vapour over oceans from SSM/I measurements. *Int J Remote Sens* 11: 753–766

Schlüssel P, Grassl H (1990) SST in polynias: A case study. *Int J Remote Sens* 11: 933–945

Schlüssel P, Luthard H (1991) Surface wind speeds over the North Sea from Special Sensor Microwave/Imager Observations. *J Geophys Res* 96: 4845–4853

Schlüssel P, Meywerk J (1993) ATSR correlative skin measurements of sea surface temperature. *Proceedings First ERS-1 Symposium - Space at the service of our environment*, Cannes France 4–6 November 1992, ESA SP-359: 779–784

Schlüssel P, Shin HY, Emery WJ, Grassl H (1987) Comparison of satellite-derived sea surface temperatures with in situ skin measurements. *J Geophys Res* 92: 2859–2874

Schlüssel P, Schanz L, Englisch G (1995) Retrieval of latent heat flux and longwave irradiance at the sea surface from SSM/I and AVHRR measurements. *Adv Space Res* 16: 107–116

Schulz J, Schlüssel P, Grassl H (1993) Water vapour in the atmospheric bounday layer over oceans from SSM/I measurements. *Int J Remote Sens* 14: 2773–2789

Smith SD (1989) Water vapor flux at the sea surface. *Bound Lay Met* 47: 277–293

Soloviev AV, Schlüssel P (1994) Parameterization of the cool skin of the ocean and of the air-ocean gas transfer on the basis of modelling surface renewal. *J Phys Ocean* 24: 1339–1346

Walton CC (1988) Nonlinear multichannel algorithms for estimating sea surface temperatures with AVHRR data. *J Appl Met* 27: 115–124

Wentz FJ (1975) A two-scale scattering model for foam-free sea microwave brightness temperatures. *J Geophys Res* 80: 3441–3446

Wentz FJ (1983) A model function for ocean microwave brightness temperature. *J Geophys Res* 88: 1892–1908

Wentz FJ (1992) Measurement of oceanic wind vector using satellite microwave radiometers. *IEEE Trans Geosci Remote Sens* 30: 960–972

Wessel P, Smith WHF (1991) Free software helps map and display data. *EOS Trans AGU* 72: 441, 445–446

Wick GA, Emery WJ, Schlüssel P (1992) A comprehensive comparison between satellite-measured skin and multichannel sea surface temperature. *J Geophys Res* 97: 5569–5595

Wu J (1985) On the cool skin of the ocean. *Bound Lay Met* 31: 203–207

Chapter 17

Applications of Passive and Active Microwave Sensors to the Study of Weather Systems over the Ocean

Kristina B. Katsaros
Département d'Océanographie Spatiale
IFREMER, B.P. 70,
29280 Plouzané
France

17.1 Introduction

Microwave remote sensing of storms over the ocean from satellites allows special features to be observed that are not obtainable with visible or infrared instruments. Most frequencies in use by both passive and active systems are not affected by the small ice particles in cirrus clouds, so the blocking by these large ice shields that occurs in the visible or infrared is avoided. Emitted microwave radiation penetrates the clouds and can reach the satellite sensor all the way from the sea surface at some frequencies. At other frequencies the brightness temperature observed at the satellite represents the emission by water vapor, cloud liquid water or precipitation in the column. The roughness of the sea surface affects the emissivity[1], and accordingly the intensity of the emitted radiation is an indicator of the surface wind speed (if there is no rain). Likewise the back-scattering of radar signals by a roughened sea surface corresponds to the surface wind speed. These features of passive and active microwave radiation have been used to study cloud systems and cyclonic storms from tropical regions to polar latitudes. Applications of data from the Special Sensor Microwave/Imager on the U.S. Defense Meteorological Satellites and from the scatterometer on the European Remote Sensing Satellite-1will be discussed. Resolution of the various parameters vary from 15 to 50 km, and are thus rather coarse compared to visible and infrared data. What they provide is different in nature. Features include: **Subtropical**

Stratus Clouds

* Estimates of cloud liquid water content can be obtained, albeit with rather large uncertainties.

[1]instead of "emissivity" often the term "emittance" is used.

NATO ASI Series, Vol. I 45
Radiation and Water in the Climate System:
Remote Measurements
Edited by Ehrhard Raschke

Midlatitude Cyclones

* In midlatitude cyclones the leading edge of frontal zones is related to strong gradients in integrated water vapor.

* Precipitation, flagged by an index, outlines the low and the fronts.

* Waves on the cold front are seen very well in water vapor content or precipitation indices.

* Rapidly deepening cyclones compared to standard storms show statistically larger amounts of anomalous water vapor content (anomalous with respect to the monthly mean for that location) and precipitation near the low center in the 6-hour period preceding the deepening.

Polar Lows

* These systems can be identified in the scattering index, and may identify such storms before classical methods do.

* The surface wind speeds are well observed with microwave radiometers in this type of storm, because of the low liquid water content in the clouds.

Tropical Cyclones

* The large amount of water in these storms leads to saturation of the microwave emission based precipitation algorithm (37 GHz polarization index) over most of the storm, but the 85 GHz scattering index gives valuable information about where the convective towers are located.

* Microwave radiometery is almost useless for observing surface winds because of the rain obscuring the surface, while scatterometry at C-band seems to have only weak interference by the rain.

* The scatterometer wind fields reveal long lines of convergence in the rear of Tropical Cyclones. These are typical features both in the southern and northern hemispheres.

* Amount of rain and organization of the rainbands as the storm develops and a convergence of the external rain bands to form a second eye-wall during a re-intensification phase have been noted in agreement with a suggestion by Willoughby et al. (1982).

This chapter starts with a description of the microwave instruments employed in these studies: the Special Sensor Microwave/Imager (SSM/I) flown on Defense Meteorological Satellites since 1987, and the Active Microwave Instrument (AMI) flown on the European Remote Sensing Satellites ERS-1 and 2 since 1992. The AMI can be operated as a scatterometer, providing surface vector winds, and in SAR (synthetic aperture radar) mode. Of special interest for this chapter is the SAR operated in Wave Mode, which provides 5 km x 10 km SAR 'imagettes' just inside the scatterometer swath.

After introduction of the instruments we outline the physics of radiometric measurements of the atmospheric moisture content (column integrated water vapor, liquid water, and precipitation size ice particles) and the surface wind speed, as well as scatterometer and SAR measurements of surface wind vectors and sea state. Several research projects concerned with use of microwave remote sensing to identify features in subtropical clouds and in storms are then discussed in some detail.

We begin with a multispectral study of stratus clouds which employs the combination of cloud liquid water estimates from microwave data with coincident visible and infrared measurements (Miletta and Katsaros, 1995). We then describe research on midlatitude cyclones which began before the launch of SSM/I emphasizing a recent study where the SSM/I signals were analyzed for their predictive value in identifying rapidly deepening cyclones off the U.S. east coast (McMurdie and Katsaros, 1996). This study is followed by a look at Polar Lows with numerous satellite systems. These cy̌clonic storms develop rapidly over oceanic regions where data other than those obtained by satellite are rare (Claud et al., 1993 and 1994).

Finally, we present the combined use of active and passive microwave systems to study tropical storms. Strong convergence lines in the surface wind field and very complex wave fields have been identified as well as the role of organized precipitation patterns surrounding the eye acting as precursors for storm intensification (Quilfen et al., 1993 and MS Thesis; Zhao, 1994). The chapter ends with conclusions and an outlook on future work.

17.2 The Instruments

17.2.1 The active microwave instrument : dcatterometer and SAR in wave mode

The European Remote Sensing Satellite 1, ERS-1, carries a complex radar system called the Active Microwave Instrument (AMI) operating at a frequency of 5.3 GHz (C- band). It functions in two mutually exclusive modes, either as a Wind Scatterometer, using three antennas or as a Synthetic Aperture Radar (SAR), using a very large (10m by 1m) antenna, as in a configuration illustrated by Figure 17.1. The SAR feature can be called upon during the scatterometer data collection for an instant, while the central scatterometer antenna would have been sampling. The SAR pixel is an area 5 km by 10 km just inside the scatterometer swath which is illuminated by the radar during these bursts.

The resulting SAR image is called an 'imagette' and the data collection operation is called 'SAR in wave mode'. (We do not discuss here regular SAR images 100 km by 100 km in size. These are obtained with the scatterometer mode turned off for reasons of power constraints.) The three antennas used for the AMI operated as a scatterometer generate radar beams looking 90° from the flight track and 45° fore and aft. Due to the variations in the power returned from the three looks at overlapping footprints, about 50 km by 50 km in size, wind direction as well as wind speed are determined. The three looks are obtained at different incidence angles and with a slight time delay. The backscattered power normalized by the emitted power form a ratio termed the normalized radar cross-section. These triplets of information for each footprint are entered into a mathematical algorithm to solve for the wind speed and direction. A certain ambiguity in the wind direction remains, since up and down-wind returns differ only slightly, often modified by signal noise, variability in the sea state or varibility in the wind over the footprint. To select one of the ambiguous wind vectors current methods use a 'first guess' approach based on numerical weather prediction models. The data products provide information on the multiple solutions to the users (ERS-1 System, 1992). The French Processing and Archiving Facility, CERSAT, distributes reprocessed AMI scatterometer data on CD-ROMS (CERSAT NEWS, Issue 5, 1995). The wind is perhaps the most important parameter controlling air-sea interactions. Scatterometry is one of four methods to observe the surface wind over the sea from space. Table 17.1 lists the methods, the characteristics of current systems and the principles behind the wind related signal.

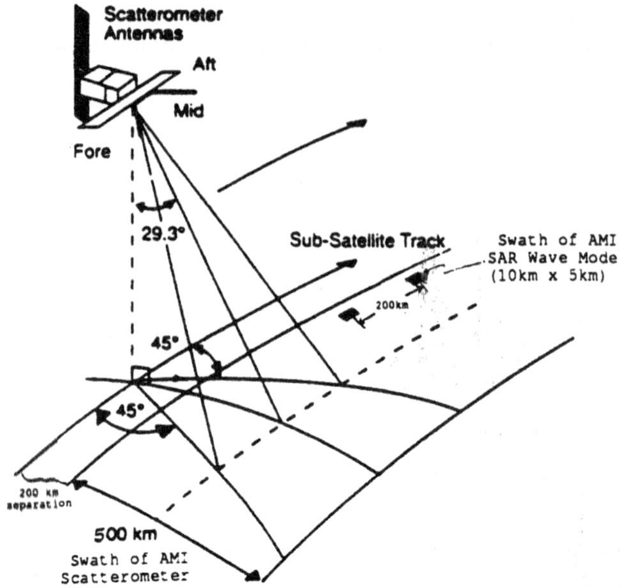

Figure 17.1: *Schematic of the Active Microwave Instrument on the ERS satellites used in Scatterometer Mode (three antennas) and as a SAR in Wave Mode (large antenna). The small rectangles, just inside the scatterometer swath (at an incidence angle of about 20 °), represent the so-called 'imagettes' of the SAR in wave mode.*

Instrument	Principle	Resolution	Swath width	Satellite
Scatterometer vector	Bragg resonance	50 km (25 km)	500 km	ERS-1,2 ADEOS
Altimeter speed only	Reflected power	10 km	10 km	GEOSAT ERS-1,2 TOPEX
Microwave radiometer speed + direction possible	Emissivity changes (roughness, foam)	50 km	1400 km	NIMBUS DMSP
SAR (imagette) vector	"Speckle" variance	50 m	5 km x 10 km	ERS-1,2

Table 17.1: *Four satellite methods for obtaining the surface wind over the sea.*

The AMI in Wave Mode provides two-dimensional spectra of the ocean surface waves over the area of the imagettes. The conversion of these SAR imagettes to two-dimensional wave spectra has been under development ever since the launch of the satellite, as pre-launch algorithms were not performing well. The fact that the SAR looks side-ways, with respect to the line of flight on a moving platform and attempts to deduce information about a moving surface, results in very complicated interpretation. Certain limitations exist on the observability of waves travelling parallel to the flight track. Currently, a first guess using the global WAM model (community Wave Model, Komen et al., 1995) is required to obtain wave spectra in pre-operational work with these data (e.g. Brüning et al., 1994). Wave statistical information and wind speed information using a so called 'speckle' algorithm can also be obtained from the imagettes (Chapron et al., 1995).

17.2.2 The passive microwave instruments

Microwave radiometers have operated on polar orbiting satellites since the 1960's. This technology is therefore well established. The major systems flown or scheduled for flight soon are listed in Table 17.2. The Scanning Multichannel Microwave Radiometer flown on Seasat and Nimbus 7 were considered experimental, while the Special Sensor Microwave/Imagers on DMSP satellites are operational systems. Even though the DMSP satellites are operated by the U.S. Defense Department, the data are now available for both research and for civilian operational uses.

The first SSM/I was launched in July 1987 with a total of 16 systems to be launched in the series as required. The series should continue well into the next millenium. The applications discussed below are based on data from these first two microwave radiometer systems, whose basic characteristics are given in Table 17.3.

Three of the SSM/I frequencies, 19, 37 and 85 GHz, are operated at horizontal and vertical polarization, while the 22 GHz channel measures vertical polarization only (Hollinger et al., 1990). The dual polarization and the addition of the 85 GHz channels, providing some information at 15 km resolution, make the SSM/I a particularly versatile instrument.

In general, the microwave radiometers are of low spatial resolution, of the order of 25 to 50 km, but the SSM/I covers a swath of 1400 km. With two functioning SSM/I instruments in space simultaneously (since fall 1991) several views per day of the same weather system can be obtained, providing valuable temporal sampling of evolution even in tropical regions. One valuable aspect of the DMSP satellites is that they have a visible and infrared high resolution system, the Operational Line-scan System (OLS), operating simultaneously.

17.2.3 The Operational Line-scan System

The Operational Line-scan System (OLS) is a two-channel radiometer flown on the DMSP satellites with the SSM/I. Its broad spectral bands are 0.4 to 1.1 m for the visible region and 10.2 to 12.8 m for the infrared, with a spatial resolution of 2.8 km for both channels. The OLS infrared data are calibrated and map linearly to the range of temperatures from 190 to 310 K (Heacock, 1985). The visible data are not calibrated, but may be used to compare relative brightnesses (Goyette et al., 1990). Since the DMSP are launched in sun-synchronous orbits near dawn and dusk and the visible channel therefore observes the earth at low light levels, with a gain control to obtain maximum contrast in the image.

Year	Platform	Sensor
1978	Seasat	SMMR
1978	Nimbus 7	SMMR
1987	MOS-1	MSR
1987	DMSP F8	SSM/I
1990	DMSP F10	SSM/I
1991	DMSP F11	SSM/I
1991	ERS-1	ATSR/M
1992	TOPEX/POSEIDON	TMR
1994	DMSP F12	SSM/I
1995	ERS-2	ATSR/M
1996	DMSP F13	SSM/I
1997	TRMM	TMI
1998	DMSP F14	SSM/I
1998	ADEOS II	ATSR/M
2000	EOS PM	MIMR
2000	METOP	MIMR

Acronyms	
ATSR/M	Along Track Scanning Radiometer and Microwave Sounder
EOS	Earth Observing System
ERS	European Remote Sensing Satellite
DMSP	Defense Meteorological Satellite Program
METOP	Meteorological Operational Satellite
MIMR	Multifrequency Imaging Microwave Radiometer
MSR	Microwave Scanning Radiometer
SMMR	Scanning Multichannel Microwave Radiometer
SSM/I	Special Sensor Microwave/Imager
TMI	TRMM Microwave Image
TMR	Tropospheric Moisture Radiometer
TOPEX	Topographic Experiment
TRMM	Tropical Raintall Measuring Mission

Table 17.2: *Major microwave radiometers flown in space since 1978 or planned for the near future.*

17.3 Physics of Microwave Remote Sensing

17.3.1 Microwave radiometry

Microwave radiometers are passive instruments responding to emitted, reflected, and scattered signals from the ocean and the atmosphere. The atmosphere is relatively transparent at microwave frequencies. The interpretation of the signal depends on the microwave frequency being used and the complexity of the scene viewed by the satellite. These two issues contribute to a certain ambiguity in interpreting microwave radiometer data. A comprehensive introduction to the interpretation of atmospheric and sea surface signals in the SSM/I frequency range is provided by Petty (1990, 1994a and b) and Petty and Katsaros (1992 and 1994). For microwave

SSMR		SSM/I	
Freq. Ghz	Approx. Resolution (km)	Freq. GHZ	Approx. Resolution (km)
6.6	150	–	–
10.7	100	–	–
18	65	19.35	55
21	60	22.235	50
37	35	37.	35
–	–	85.5	15
Swath Width Seasat : 650 km Nimbus -7 : 780 km		Swath Width 1400 km	
Period of Operation Seasat : July-October 1978 Nimbus -7 : October 1978 - Fall, 1987		Period of Operation July 1987-Present	

Table 17.3: *Characteristics of two satellite-borne microwave radiometers , the SMMR and the SSM/I. After Katsaros and Liu (1995).*

radiation, the Planck function for thermal emission can be simplified to the Rayleigh-Jeans approximation; radiance is simply proportional to the thermodynamic temperature of the radiating medium. This has given rise to use of the term "brightness temperature", T_B, for radiance which has units of degrees Kelvin,

$$T_B = R - \frac{\partial B(\lambda, T)}{\partial \bar{\omega}} \tag{17.1}$$

where R is the radiance, B is the black body radiance, λ is the wavelength of the radiation measured, T is the temperature of the radiating medium and $\bar{\omega}$ is the solid angle.

The ocean surface emits microwave radiation of feeble intensity, because its emissivity is only of the order of 0.5. Thus, the T_B of the sea is about 150K for a surface temperature of 300K. The microwave radiometers in use during the last decades operate in both horizontal and vertical polarization. The emission from the sea is strongly polarized, $T_{BV} - T_{BH}$ being of the order of 20K (where subscripts V and H refer to vertical and horizontal polarization; e.g. Petty and Katsaros, 1994). Petty (1990) and Petty and Katsaros (1994) employ the most recent electromagnetic constants for water and use Fresnel theory to calculate the polarized emittance from the sea at the SSM/I frequencies. The emissivity and the polarization are dependent on the roughness of the sea and therefore on the near surface wind speed (Nordberg et al., 1971; Wentz, 1983 and Petty and Katsaros, 1994).

Emission from water vapor and oxygen are responsible for the signals from the clear atmosphere (Figure 17.2). The weak absorption band at 22 GHz is used to measure the integrated atmospheric water vapor above the sea. Because this water vapor band is weak, even in the moist tropical atmosphere this band does not become opaque and the evaluation of water vapor content remains good (e.g. Katsaros, 1994 and Alishouse et al., 1990). However, thick precipitating clouds saturate the signals and make estimates of integrated water vapor erroneous.

Cloud liquid water and precipitation are determined from the normalized polarization difference. The physical principle behind these algorithms is the decrease of the polarization of the microwave signals at one frequency as more cloud and rain emissions are added to the original polarized signals emerging from the sea. Corrections for absorption by water vapor and for the degree of depolarization already present in the signals emerging from the sea due to wind roughening are required. These corrections are found from the clear areas in the pixels surrounding the clouds and precipitation. Cloud liquid water algorithms employing 37 and 85 GHz signals are described in more detail in section 17.4. For liquid precipitation, similar polarization based algorithms have been developed based on the 37 and 19 GHz signals. A second principle is used to infer precipitation from convective cloud systems that contain large wet ice particles, such as graupel and hail. This type of algorithm is referred to as a 'scattering index' and employs the higher frequencies, 37 and 85 GHz.

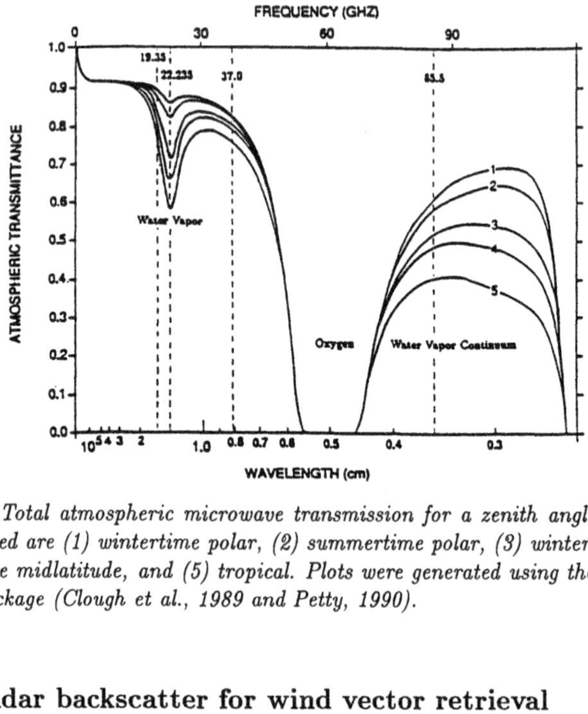

Figure 17.2: *Total atmospheric microwave transmission for a zenith angle of 53 °. Model atmospheres used are (1) wintertime polar, (2) summertime polar, (3) wintertime midlatitude, (4) summertime midlatitude, and (5) tropical. Plots were generated using the FASCOD3 line-by-line code package (Clough et al., 1989 and Petty, 1990).*

17.3.2 Radar backscatter for wind vector retrieval

The reflected radar signal from the sea surface received by the AMI, when operated in scatterometer mode, is due to Bragg scatter. This is the constructive resonant interference between signals returning from small roughness elements on the sea surface of a wavelength having the required relation to the wavelength of the radar for the incidence angle. For the C-band radar of the AMI operating at 18 to 45° for the mid-antenna and 25 to 59° for the fore and aft antennas, the water surface wavelengths are of the order of 5 to 25 cm. These short waves increase in amplitude with increasing wind speed, which results in increasing radar return for the scatterometer with increasing wind speed. (Wind speed retrieved from the power returned to an altimeter, which views the sea in the nadir direction only, is due to decreased radar return as the sea surface is roughened by increasing wind speed. No directional information is available for wind retrieved from an altimeter.) Details of the physics of AMI operated in SAR Wave Mode will not be included here.

17.4 Subtropical Stratus Cloud Study

17.4.1 Background

Subtropical marine stratocumulus clouds have been the subject of many climate studies because of their effect on the earth's radiative energy budget. Their high albedo reduces the flux of solar energy into the ocean, while their low altitude prevents a significant compensation by reduced thermal emission. Radiative properties of marine stratocumulus clouds are determined primarily by their water content, microphysical properties, and spatial distribution. Satellite data are a valuable tool for characterizing these cloud parameters. Several studies have used visible and infrared data (e.g. Coakley and Baldwin, 1984; Rossow et al., 1989; Minnis et al., 1992; Platnick and Twomey, 1994 and Bretherton and Pincus, 1994), but the addition of passive microwave data, from which we can derive the liquid water path, contributes to a more complete description of the observed cloud systems. Curry et al. (1990) and Liu and Curry (1993) combined microwave and ancillary data to characterize the cloud liquid water content of stratiform clouds, but the relatively low spatial resolution of the ancillary data precluded a sub-pixel-scale analysis. Lin and Rossow (1994) compared optical and microwave retrieval methods and examined the estimates on global and regional scales. In the study by Miletta and Katsaros (1995), a technique was developed employing observations from several instruments on the same satellite to examine the radiative properties of marine stratocumulus clouds. High resolution visible and infrared data are provided by the Operational Line-scan System (OLS), and passive microwave data are provided by the Special Sensor Microwave/Imager (SSM/I). The coincident visible and infrared data was used to characterize the cloud field within the larger microwave footprint. The objective was to explore the relationships between the SSM/I-derived and the OLS-derived cloud parameters, providing information about the reliability of the liquid water path estimates based on the Petty algorithms.

Identifying and eliminating sources of error in the liquid water path retrievals, rather than addressing the absolute accuracy of the measurements, were the objectives of the Miletta and Katsaros (1995) study. Sections 17.4.2 and 17.4.3 contain brief discussions of the retrieval algorithm and our method for incorporating the visible and infrared data. Techniques to identify the sources of error in the algorithm and a discussion of the application of these techniques on a set of sample images is found in section 17.4.4.

17.4.2 The SSM/I liquid water path algorithm

The SSM-I is a seven-channel radiometer (see section 17.2.2), with vertically and horizontally polarized channels at 19, 37 and 85 GHz, and a vertically polarized channel at 22 GHz. The 85 GHz channels are sampled every 12.5 km, and the lower frequency channels are sampled every 25 km. The footprints for each channel are centered on the same point, but they are not the same size as spatial resolution increases with frequency. We use Petty's algorithms to retrieve the liquid water path from the SSM-I brightness temperatures (Petty 1990; Petty 1994a and b). Polarization information provides a measure of the attenuation of the highly polarized emitted radiances from the open ocean by the intervening clouds. A parameter called the Normalized Polarization Difference (P) is the observed polarization difference normalized by an estimate of the polarization difference in the absence of clouds under otherwise similar meteorological conditions. For the 85 GHz channels, P is given by

$$P_{85} = \frac{P_{actual}}{P_{clear}} \cong \frac{T_{85V} - T_{85H}}{\exp[4.44 - (0.024.wind) - (0.027.vapor)]} \qquad (17.2)$$

For the 37 GHz channels, P is given by

$$P_{37} = \frac{P_{actual}}{P_{clear}} = \frac{T_{37V} - T_{37H}}{[77.0 - (0.970.wind) - (0.323.vapor)]} \quad (17.3)$$

The normalization factor (P_{clear}) includes a representation of the effects of water vapor and the wind-roughened sea surface on the polarization difference, so the only remaining effect is due to the liquid water in the column. Note that the expression for P_{clear} is exponential at 85 GHz and linear at 37 GHz.

In the absence of anisotropic extinction by precipitation particles, P is well approximated by the effective transmittance (τ) associated with cloud liquid water (L) raised to a power (Petty 1990). That is,

$$P \approx \tau^{\alpha} = \exp[\frac{(\alpha k_e L)}{\cos\theta}] \quad (17.4)$$

where k_e is the effective mass extinction coefficient of the liquid water in the cloud, and θ (= 53.1°) is the SSM/I viewing angle. A radiative transfer model was used to determine α (= 2.0) and k_e, which is a function of temperature and frequency (Petty and Katsaros, 1992). The mean cloud temperature, determined from the coincident OLS infrared data, is used to calculate k_e. For cloud temperatures ranging from 250 to 300K, k_e ranges from approximately 1.0 to 0.7 for 85 GHz channels, and 0.4 to 0.15 for 37 GHz channels. Equation 17.4 is then inverted to obtain two independent expressions for the liquid water path: one from the 37 GHz channels and the other from the 85 GHz channels (units are kg/m^2) :

$$L = -0.30(\frac{1}{k_e}\ln(P)) \quad (17.5)$$

For small values of liquid water path, the 85 GHz channels are the most sensitive. However, the non-linearity of the response may introduce a beam-filling bias when the field of view (~15 km) is not uniformly filled with clouds. Several studies have shown that a beam- filling bias exists for non-uniform coverage of precipitation (e.g. Short and North, 1990). A similar bias is likely to occur when measuring cloud liquid water in partly cloudy situations. The more linear response of the 37 GHz channels should provide a more accurate estimate of L in the case of broken clouds. (e.g. Dubroe, 1991, personal communication).

As L becomes large, the 85 GHz channels tend to saturate and lose their sensitivity. The 37 GHz estimate is more reliable in these situations, but unfortunately, the 37 GHz channels have poorer spatial resolution (~30 km) and are less sensitive to small cloud liquid water amounts. Therefore the 37 GHz channels may be less suitable for detecting thin stratocumulus clouds than the 85 GHz channels (Miletta, 1993).

17.4.3 Combining OLS and SSM/I observations

Our method to combine OLS and SSM/I observations accounts for the differences in spatial resolution of the two data sets. The method does not rely on the scanning geometries of the OLS and the SSM/I. Rather, collocation and footprint filling are based entirely on the latitude and longitude of individual pixels. The collocation step finds the OLS pixel that is closest (generally within 1 km) to the center of each SSM/I pixel. Then the footprint-filling step counts the OLS pixels surrounding the central pixel which fall within the boundaries of the SSM/I footprints. An average of 30 OLS pixels fall within an SSM/I 85 GHz footprint (150 within a 37 GHz footprint).

After counting the OLS footprints that fall within the boundaries of the SSM/I footprints, the mean infrared temperature, the mean visible count and a coefficient of variance (the standard deviation of the mean visible count divided by the mean) are calculated for all SSM/I 37 and 85 GHz footprints. Each OLS pixel is assumed to contribute equally to the mean (Miletta and Katsaros, 1995).

17.4.4 Liquid water path estimates for stratocumulus clouds in the azores region

The method described above is applied to a set of DMSP observations of stratocumulus clouds in the subtropical Atlantic Ocean near the Azores obtained in June 1992. In Figure 17.3 the relationship between the infrared temperature estimated within an SSM/I footprint and the brightness of the visible channel of the OLS (Mean Visible Count) is presented. In this figure a characteristic linear relationship for opaque, partly cloudy pixels and a non-linear relationship for semitransparent overcast pixels is shown. Figure 17.4 shows the much more variable liquid water path estimates for broken clouds as compared to semitransparent overcast clouds. It indicates that the classification of the scene by the visible and IR analysis allows an a priori error assessment of the liquid water path estimate. Similar curves relating the visible count within the pixel to the liquid water path estimates by the 37 and 85 GHz algorithms and further details of this study are found in Miletta and Katsaros (1995).

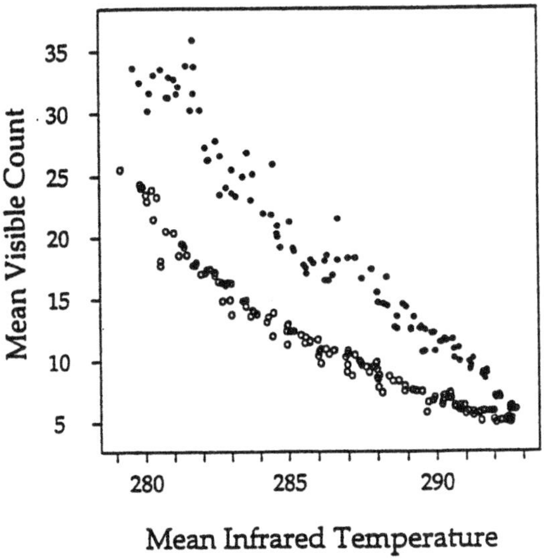

Figure 17.3: *Scatter diagram of OLS mean infrared temperature versus mean visible count within 85 GHz pixels for low-level clouds located southwest of the Azores Islands. Pixels determined to be overcast with semi-transparent clouds are show as open circles and pixels partly filled with opaque clouds as dark circles. After Miletta and Katsaros (1995).*

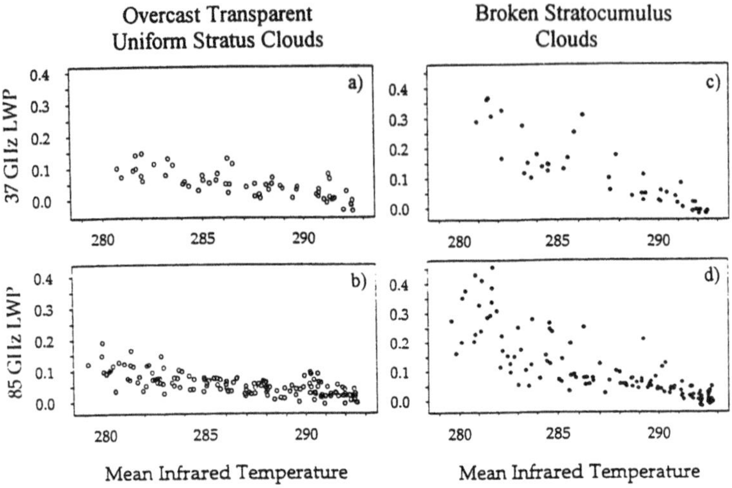

Figure 17.4: *Mean infrared temperature versus liquid water path (in kg/m^2) for overcast semi-transparent clouds on the left and for broken opaque clouds on the right. Figures on top in a) and c) are for 37 GHz liquid water path estimates, and figures at the bottom in b) and d) are for 85 GHz liquid water path estimates. After Miletta and Katsaros (1995).*

17.5 Midlatitude Cyclones

17.5.1 General features

Early looks at midlatitude cyclones by microwave radiometry were provided by the SMMR's on Seasat and Nimbus 7. The pattern in column integrated water vapor showed characteristic patterns for each stage of an evolving midlatitude cyclone. Examples of the mature and occluded stages of the frontal system around a depression are provided in Figure 17.5. A schematic of the typical water vapor distribution during the life-cycle of a frontal system is given by Katsaros and Brown (1991).

Statistical analysis by McMurdie and Katsaros (1991) based on SMMR data, showed that there existed systematic variability of the amount of water vapor in midlatitude cyclones with season and region. For instance, South Atlantic storms had much higher water vapor content than Southern Ocean storms. The differences were attributed to differences in ocean temperatures and moisture advection and convergence. Differences between the water vapor content in the regions surrounding warm fronts versus the regions surrounding cold fronts were found. Mature cyclones in both the North Atlantic and North Pacific oceans had significantly higher maximum water vapor content associated with the cold fronts as compared to the warm fronts, for example. Katsaros et al. (1987) and (1988) suggested that the water vapor gradient associated with fronts could be used to map frontal zones objectively from space. Even marine warm fronts that are often very weak and difficult to analyze by traditional methods were detected by this scheme 90% of the time. Mognard and Katsaros (1994) and Katsaros et al. (1996) demonstrated that surface wind observations from radars, such as the Geosat altimeter and SSM/I observations of cyclones are complementary. The radar signals are affected relatively little by precipitation

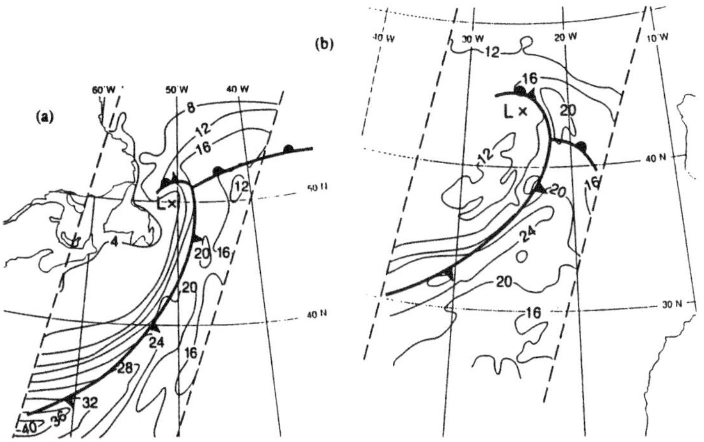

Figure 17.5: *Column integrated water vapor (in kg/m2) from the SSM/I showing the characteristic signatures in relation to the surface frontal positions interpolated to the time of satellite overpass for a) an intense storm off the eastern coast of America at 2212 UTC on 5 March, 1988, and b) and occluded depression in mid-Atlantic ocean at 2020 UTC on 25 January, 1988. Frontal analyses are provided by Bader et al. (1995). See this reference for other examples.*

(except when convective towers appear in the altimeter path), while the SSM/I does not provide surface wind data in regions of precipitation.

17.5.2 Rapidly deeping cyclones

Extratropical storms that deepen at a rate of 24 mb in 24 hours (Sanders and Gyakum 1980) or at rate of 10 mb in 6 hours (Hadlock and Kreitzberg 1988) are considered to be rapidly deepening cyclones. Storms of this strength most commonly occur in marine climates in the vicinity of warm ocean currents, such as the Gulf Stream or the Kuroshio, but some have occurred over land (Mass and Schulz 1993) or over the mid-ocean basins (Reed and Albright, 1986). These storms are often accompanied by hurricane force winds, extensive rain and cloud shields and have caused property damage and loss of life in near shore and offshore regions (Sanders and Gyakum, 1980 ; Bosart, 1981 and Kuo et al., 1992).

The important mechanisms contributing to rapid development in extra-tropical cyclones include upper level forcing and condensational heating due to latent heat realease.

Upper Level Forcing

The presence of an upper level trough preceding the development of the surface cyclone was found in several observational studies (e.g. Rogers and Bosart, 1986). Wash et al. (1992) compared rapidly deepening cases to non-rapidly deepening cases and found that strong upper-tropospheric features at the early stages of development produced stronger vertical motions and quicker development in the rapidly deepening cases. Several modeling and observational studies of rapid cyclogenesis demonstrated the presence of upper-level jet streaks, or upper-level potential vorticity anomalies during the incipient and deepening phases (e.g. Uccellini et al., 1985, Davis and Emanuel, 1991; Reed et al., 1992; Kuo et al., 1992; Reed et al., 1993 and Neiman and Shapiro, 1993).

Condensational Heating or Latent Heat Release

Several case studies have pointed to the presence of convection and extensive cloud shields associated with such storms (e.g. Reed and Albright, 1986; Neiman and Shapiro, 1993). Modeling studies have shown that moist processes must be accurately included to successfully simulate the deepening rate of the surface low, with deepening rates reduced from 30 to 70% when condensational heating is removed (e.g. Reed et al., 1993 and see further references in McMurdie and Katsaros, 1996). The details of how latent heat contributes to cyclone deepening has only been explored using model-derived fields (Kuo et al., 1991). The quantitative observational evidence of the role of moist processes is lacking.

Unfortunately, since these storms most frequently occur over oceanic regions, adequate in situ observations are unavailable. This has led forecasters and researchers to turn to remotely based platforms to observe rapid cyclogenesis. Forecasters often rely on infrared and visible satellite imagery to track their movement and rapid development. With the availability of remote sensing technology, other satellite indicators of rapid development are being sought. Satellite derived ozone maps and the 6.7 μm upper level moisture images gave evidence of strong upper tropospheric subsidence in two cases of rapid development (Uccellini et al., 1985 and Reed and Albright, 1986). Upper tropospheric thermal anomalies identified using the Microwave Sounding Unit (MSU) were shown to be greater in storms that deepened rapidly compared to those that did not (Velden ,1972).

With the integrated water vapor and precipitation information available from passive microwave radiometers such as the Special Sensor Microwave/Imager (SSM/I), it is now possible to document the quantitative observational evidence of the role of latent heat release in rapid cyclogenesis. Using rain rates derived from the brightness temperatures observed by the SSM/I, Petty and Miller (1995) found that average rain rates within a small region northeast of the surface low was strongly correlated with the rate of deepening pressure. With information derived from the earlier passive microwave instrument, the Scanning Multichannel Microwave Radiometer, McMurdie and Katsaros (1985) and Katsaros et al. (1987 and 1988) showed that the mesoscale structure in the integrated water vapor and rain distribution patterns are highly correlated with the location of the surface frontal zones. McMurdie and Katsaros (1996) searched for indicators of rapid development in the SSM/I derived water vapor and precipitation fields by analyzing storms that occurred in January through March 1988 and January and February 1989. The difference between ordinary and rapidly deepening storms is illustrated by Figure 17.6.

McMurdie and Katsaros (1996) found that the integrated water amount was too variable as a function of latitude to be a useful diagnostic parameter. However, the water vapor anomaly (WVA) with respect to the amounts typical for the season and latitude had predictive value over a 6 hours period. They also examined the correlation between rapid deepening and rain rate indicated by (1-P37) or (1-P19). P19 being less sensitive to precipitation and therefore less likely to saturate gave a slightly better correlation. This result is seen in Figure 17.7, which is representative of the resullts for the other predictors (WVA and 1-P37) as well.

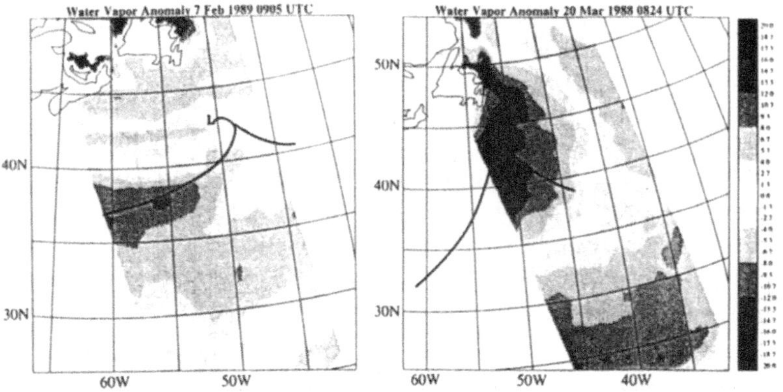

Figure 17.6: *Examples of the integrated water vapor anomaly pattern in an ordinary (left) and in a rapidly deepening cyclone (right). The anomaly is calculated with respect to the monthly average for the location of the observation. After McMurdie and Katsaros (1996).*

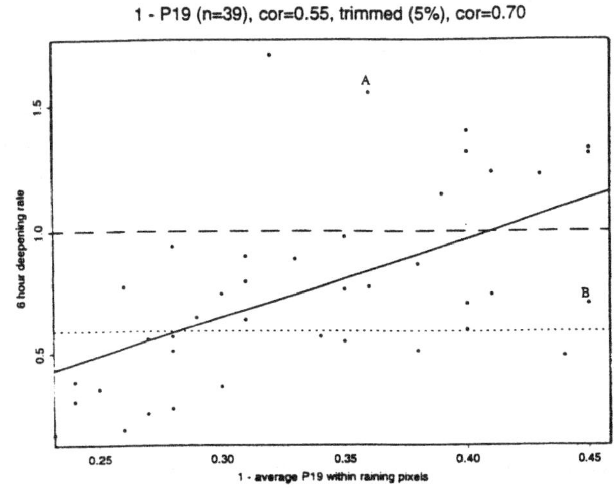

Figure 17.7: *Scatterplot of (1-P19) vs. 6-hour deepening rate for the 39 passes over rapidly deepening cyclones East of the U.S. Atlantic coast analyzed by McMudie and Katsaros (1996, their figure 7). The correlation coefficient is 0.55 and increases to 0.70 when the two extreme outliers, indicated by A and B, are eliminated. The horizontal long and short dashed lines delineate the rapidly deepening and the marginally rapidly deepening cases, respectively. The sloping solid line is the least squares regression line fitted to the full set of 39 cases. See further McMurdie and Katsaros (1996).*

17.6 Polar Lows

Polar lows are small-scale disturbances often generated at the boundary between sea ice and the ocean in regions of large sea surface temperature gradients. Those affecting Norway generally occur from September through April and occasionally develop in May and June. The horizontal scale of these lows is of the order of a few hundred kilometers, and they have a vertical depth of 1–5 km. Rather strong winds and snowfalls are associated with these phenomena, which can develop in only a few hours (Lystad, 1986). One major difficulty in studying these lows is that they develop in regions with very few meteorological stations. Therefore they sometimes go undetected until they affect coastal regions. In addition, the mechanisms involved in the formation, maintenance and decay of polar lows are still being discussed, i.e. low-level baroclinicity, convective forcing (CISK), or both, thermal instability, even barotropic instability (e.g. Rasmussen, 1979; Businger and Reed, 1989; Emanuel and Rotunno, 1989). A very complete description of polar lows may be obtained by research aircraft observations as shown by Shapiro et al. (1987) and Douglas et al. (1991). Data from polar-orbiting satellites could be of importance for monitoring and short-range forecasting since in polar regions, frequent sampling of any mesoscale weather system can be obtained due to overlapping swaths. In addition to using visible and infrared images from the Advanced Very High Resolution Radiometer (AVHRR) on NOAA satellites, the study by Claud et al. (1995) describes observations from three instruments. The Special Sensor Microwave/Imager (SSM/I) provides estimates of surface wind speed, scattering by precipitation size ice particles, integrated cloud liquid water, and water vapor contents over the ocean. The Geosat radar altimeter provides estimations of surface wind speed, significant wave height and swell height. The TIROS-N Operational Vertical Sounder (TOVS) provides temperature and humidity profiles in the atmosphere, as well as cloud parameters. Figure 17.8 illustrates the results obtainable with the SSM/I for a Polar Low that occurred off Norway on February 15, 1989.

17.7 Tropical Cyclones

These most violent of the cyclonic storms develop exclusively over oceanic areas. Their detection and monitoring, therefore, benefit especially from satellite observations. Since the precipitation is extreme and these storms are associated with widely spread cirrus cloud shields, the microwaves ability to penetrate clouds provides advantages over monitoring these storms in the visible and infrared (e.g. Quilfen et al., 1993 and 1994; Katsaros et al., 1993 and 1995). ERS-1 scatterometer and SSM/I observations have been found extremely useful in studying tropical cyclones.

17.7.1 Scatterometer and SAR observations

An example of the surface wind fields obtained by the ERS-1 scatterometer is seen in Figure 17.9. The pattern of convergence lines in the rear of the storm has been observed in numerous tropical cyclones since the launch of ERS-1, but a statistical and physical analysis has not yet been carried out to determine how frequent these features are, and whether they are directly correlated to the rain bands. The SSM/I and ERS-1 satellite sensors do not view such storms frequently enough to determine physical relationships.

The 500 km swath width and the small size of tropical cyclones (typically of radius less than 500 km), prevent frequent and dependable observations of these storms by the ERS-1 scatterometer. However, many such storms have been observed by the ERS-1 and an example is shown in Figure 17.9. Analysis of the directional wave field using the SAR in wave mode, shows rapid variations of the swell in space, and the persistence of swell, which can be traveling with or even perpendicular to the strong winds near the tropical cyclone center (Katsaros et al., 1995 and Biraud, 1995).

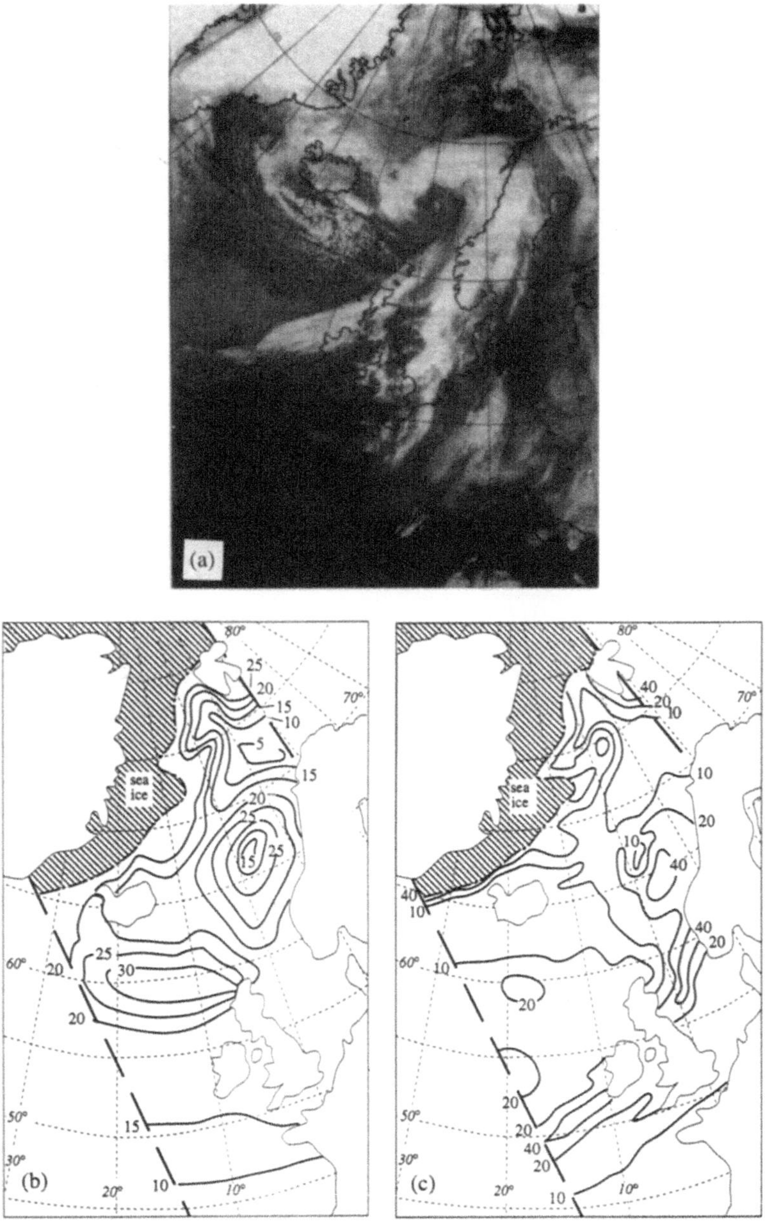

Figure 17.8: *Selected views of a Polar Low off the Norwegian Coast on February 15, 1989. a) AVHRR image of this small weather system. b) Wind speed obtained from the SSM/I a few hours later with low winds in the storm center and 24 m/s winds indicated in the southwest quadrant. (High values near the Norwegian coast can be attributed to interference in the signals by land.) c) Cloud liquid water content. After Claud et al. (1995).*

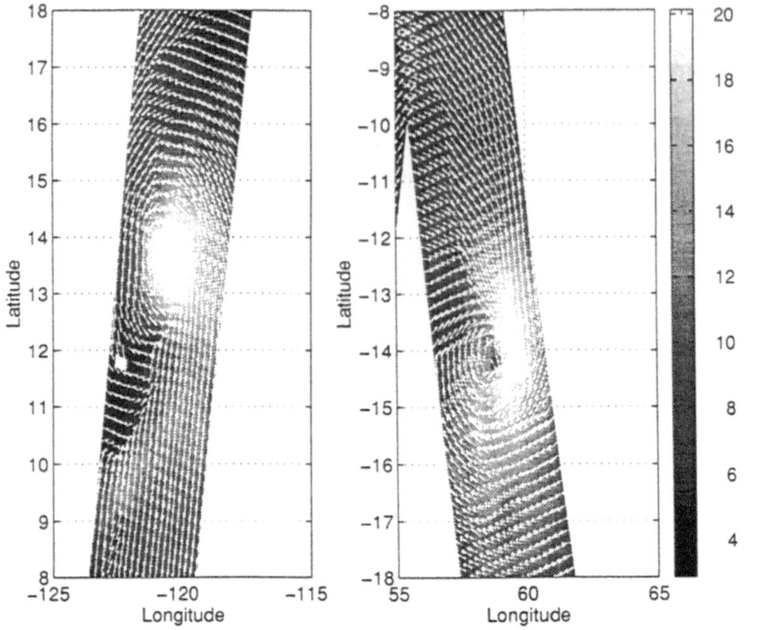

Figure 17.9: *ERS-1 scatterometer wind vectors in one northern hemisphere and one southern hemisphere tropical cyclone. In each case a pattern of convergence emerges, especially in the rear of the storms. (Courtesy Y. Quilfen).*

Knowledge of how well the existing scatterometer or SAR algorithms apply to these extreme wind conditions and confused sea states is not yet available, and will require much further study.

Initial retrievals however show intriguing possibilities for this new technology in reaching into the violent and dangerous regions of tropoical cyclones, where conventional measurements are limited to occasional penetrations by armored aircraft at some risk. Figure 17.10 shows an example of the surface wind vectors obtained by the ERS-1 scatterometer and the direction and magnitude of the dominant waves in the SAR imagettes, ocurring near the storm center.

17.7.2 SSM/I Observations of tropical cyclones

The precipitation algorithms based on SSM/I data are still far from proven, especially for extreme conditions that are found in tropical cyclones. But the delineation between widespread stratiform rain, obtained with the emission-based algorithms (P37) or (P19), and the convective rain in towering thunderstorms, based on the ice-scattering index, appears to give realistic patterns. These two features were used to study evolution of the precipitation in several tropical cyclones occurring in the Pacific Ocean and the Gulf of Bengal (Katsaros et al., 1993 and 1995; Zhao, 1994). Glass and Felde's idea (1992) that the SSM/I derived precipitation near the storm center was indicative of future accelerations, or even valid as a diagnostic of simultaneous storm intensity levels was examined. It is a physically sound since the release of latent heat due to precipitation is the major energy source for tropical cyclones. Katsaros et al. (1995) and Zhao (1994) used the published intensity estimate (the maximum surface wind speed) to correlate with the SSM/I derived estimate of precipitation in annuli of various radii surrounding the tropical cyclone center. Figure 17.11 shows a series of SSM/I precipitation

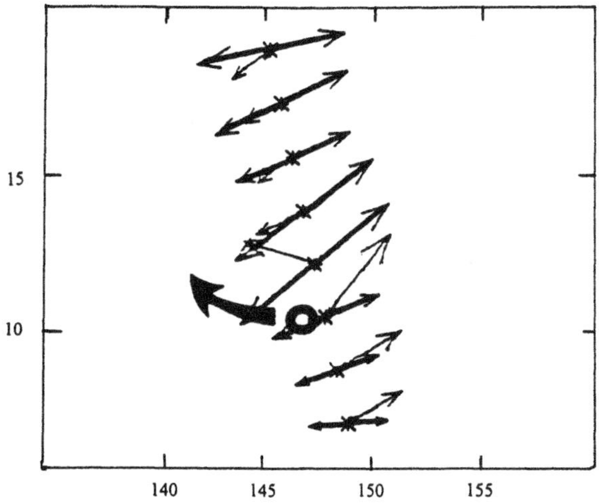

Figure 17.10: *Observations of surface wind vectors (thin lines) and dominant waves (thick lines) by the ERS-1 for 8 imagettes, 200 km apart, near the center of tropical cylone, ELSIE, November 1, 1992. The position of the center of the storm is marked with a black circle. The movement of ELSIE over time is illustrated with a large arrow. The magnitudes should be viewed as relative only in this illustration. The 180 ° ambiguity in the directional wave spectra has been retained pending further detailed study of the phenomenon. After Biraud (1995).*

maps during a period of re-intensification of typhoon Gay, in November, 1992. In Figure 17.12 the precipitation averaged around these annuli as a function of radius from the storm center is shown for 5 observations by SSM/I. (The precipitation values should be considered as indicative rather than as absolute values. The pattern shows a secondary maximum at a distance initially of 220 km (corresponding to 2° of latitude) that moves towards the center with time (marked by arrows on the figure). These new convective towers eventually replace the eyewall, which according to the Joint Typhoon Warning Center's annual report (JWTC 1992) corresponds to a redevelopment of the storm from a maximum wind speed of 90 kts to 159 kts. The inspiration to look for a relationship between maximum surface wind and motion of the tropical cyclone's rain bands to replace the eyewall was a suggestion in the literature by Willoughby et al. (1982) and Willoughby (1990).

17.8 Conclusions with Perspective on the Future

The use of microwave data for observing and predicting future development of weather systems particularly cyclonic storms over the ocean has been illustrated above. Operational weather forecast centers employ microwave data as auxiliary information and are beginning to include them in data assimilation schemes. The examples of application presented in this chapter show the promise of this type of data for monitoring remote oceanic regions and providing specialized information on atmospheric water content, latent heat release and surface wind patterns. One current drawback is the limited sampling by a single 500 km-wide scatterometer track and the rather slow transmission of SSM/I data via several centers to the forecast offices.

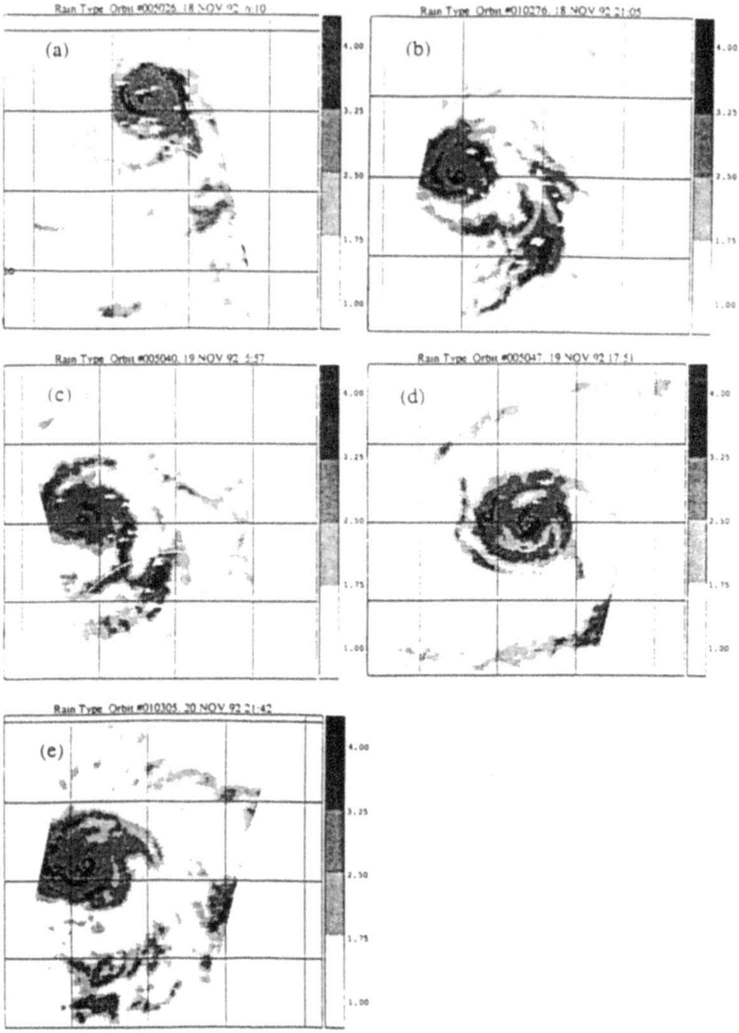

Figure 17.11: *Rain pattern in typhoon Gay as observed by SSM/I from early development to mature and re-intensifying stage. After Zhao (1994).*

Future systems could overcome these difficulties. The next scatterometer in space, NSCAT on the Japanese ADEOS satellite with launch date in August 1996, will have two swaths and thereby intercept most cyclones once per day (except at the Equator). Transfers of SSM/I data are improving over more efficient internet pathways. A general trend towards optimal use of several sensors on one or several satellites was also illustrated with the examples presented above. Combined use of visible, infrared and microwave data gives improved choice of liquid water path algorithms for strato-cumulus clouds. Multiple sensors allow more opportunities to

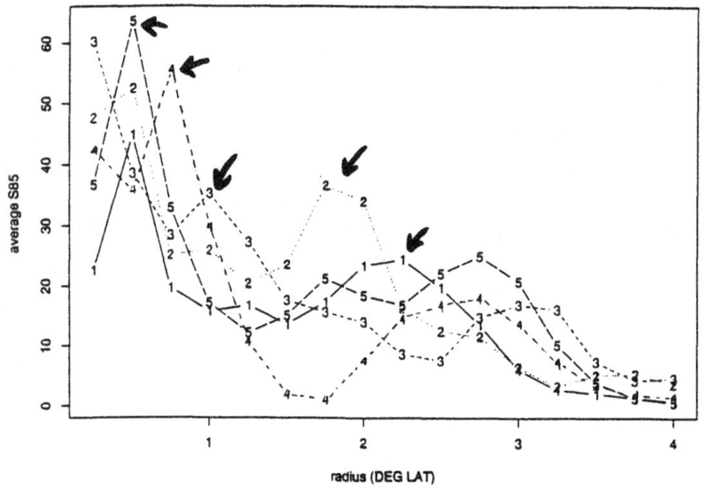

Figure 17.12: *Average S85 in different sizes of rings for the intensifying typhoon stage of Gay. Numbers 1 to 5 represent consecutive SSM/I passes. The arrows point to the outer ring of precipitation that migrates towards the center and replaces the eyewall at the time of maximum re-intensification of the typhoon.*

follow the time evolution of meteorological events such as Polar Lows. For tropical cyclones the use of scatterometer wind observations allows finding the center of the storm, and possibly a measure of the storms intensity, if calibration for the high wind speeds can be obtained. The maximum observed wind speed is also strongly dependent on the resolutions of the satellite sensors, since the regions of wind speed maximum may be only 10's of km in width. A few examples of ERS-1 wind fields at 25 km resolution have revealed details in the wind structure of these tropical cyclones missed by the 50 km standard resolution (the 25 km data are more noisy). Two maxima on opposite sides of the storm were revealed in one case, for instance, which may be very important for forecasting and warning. Multichannel microwave radiometers such as the SSM/I provides the structure of the precipitation field and can, therefore, identify the stage of the storm development from the initial unorganized stage, to the formation of the eyewall and even secondary developments with re-intensification.

The future scatterometer planned for a European operational satellite early in the next millenium, will have the 25 km resolution and dual swaths operating in the C-band, which is relatively uninhibited by precipitation, and NSCAT, on the Japanese satellite ADEOS will allow the possibility of higher resolution data as well. Improved microwave radiometers/imagers are also planned, (see Table 17.2). The challenge is to coordinate several space and meteorological agencies, so that the mix of operational and research instruments is optimal, the goal being to provide maximum benifits for both short-term forecasts and long-term, climatologic monitoring of the variability of the earth's weather systems.

Acknowledgements

This tutorial review is based on collaborations with several students and colleagues notably L. McMurdie, G. Petty, J. Miletta and H. Zhao of University of Washington, Y. Quilfen, A. Bentamy, B. Chapron and T. El Fouhaily at Institut Français de Recherche et d'Exploitation de la Mer, C. Claud at Laboratoire Dynamique Météorologique de l'Ecole Polytechnique Supérieure, and N. Mognard, CESBIO, Centre National des Etudes Spatiales. The material is largely drawn on published articles co-authored with them. The original ERS-1 data were provided by the European Space Agency and the SSM/I data were obtained from F. W. Wentz of Remote Sensing Systems, Inc. The support of these individuals and organizations is gratefully acknowledged. I am grateful for editorial assistance from M.-L. Quentel, S. Dickinson and F. Loubrieu.

17.9 References

Alishouse JD, Snyder JB, Vongsathorn J, Ferraro RR (1990) Determination of cloud liquid water content using the SSM/I. *IEEE Trans Geosci Rem Sens* 28: 817–821.

Bader MJ, Forbes JS, Grant JR, Lilley RBE, Waters AJ Eds (1995) Images in *Weather Forecasting, A Practical Guide for Interpreting Satellite and Radar Imagery*, Cambridge University Press, Cambridge, 499 pp

Biraud S (1995) Structures vent/vagues dans les cyclones tropicaux vues par des instruments micro-ondes embarqués. *Rapport de Stage* DRO/OS 95–11, 19–28

Bosart LF (1981) The President's Day snowtorm of 18-19 February 1979 : A subsynoptic scale event. *Mon Wea Rev* 109: 1542–1566

Bretherton CS, Pincus R (1994) Cloudiness and marine boundary layer dynamics in the ASTEX lagrangian experiments. Part I: Synoptic setting and vertical structure. *J Atmos Sci*, submitted

Brüning C, Hasselmann S, Hasselmann K, Lehner S, Gerling T (1994) A first evaluation of ERS-1 synthetic aperture radar wave mode data. *The Global Atmos and Ocean Sys* 2: 61–98

Businger S, Reed RJ (1989) Cyclogenesis in cold air masses. *Wea Forecast* 4: 133–156

CERSAT NEWS Issues 1–5, January 1993–1995, Available from CERSAT, IFREMER, BP70, 29280 Plouzané, France

Chapron B, El Fouhaily T, Quilfen Y (1995) A speckle wind algorithm from the ERS-1 SAR Wave-mode imagettes. *Extra Sheet of CERSAT NEWS* Issue 5, February, 2

Claud C, Mognard NM, Katsaros KB, Chedin A, Scott NA (1993) Satellite observations of a polar low over the Norwegian Sea by Special Sensor Microwave/Imager, Geosat, and TIROS-N Operational Vertical Sounder. *J Geophys Res* 98: 14487–14506

Claud C, Katsaros KB, Mognard NM, Scott NA (1995) Synergetic satellite study of a rapidly deepening cyclone over the Norwegian Sea. *The Global Atmos and Ocean Sys* 3: 1–34

Clough SA, Kneizys FX, Anderson GP, Shettle EP, Chetwynd JH, Abreu LW, Hall LA, Worsham RD (1989) FASCOD3: Spectral Simulation, IRS '88: *Current Problems in Atmospheric Radiation*, J. Lenoble and J.F. Gleyn Eds., A Deepak Publishing, 372–375

Coakley JA, Baldwin DG (1984) Towards the objective analysis of clouds from satellite imagery. *J Clim and Appl Met* 23: 1065–1099

Curry JA, Ardeel CD, Tian L (1990) Liquid water content and precipitation characteristics of stratiform clouds as inferred from satellite microwave measurements. *J Geophys Res* 95: 16659–16671

Davis CA, Emanuel KA (1991) Potential vorticity diagnostics of cyclogenesis. *Mon Wea Rev* 119: 1929–1953

Douglas MW, Fedor LS, Shapiro MA (1991) Polar low structure over the northern Gulf of Alaska based on research aircraft observations. *Mon Wea Rev* 119: 32–54

Emanuel KA, Rotunno R (1989) Polar lows as arctic hurricanes. *Tellus* 41A: 1–17

ERS-1 System (1992) Eds.: P.Vass and B. Battrick ESA Earthnet, ESRIN, Frascati, Italy, *Publ.: ESA Publications Division*, ESTEC, Noordwijk, The Netherlands, 87 pp

Goyette JA, Klein WD, Adams AL (1990) The Defense Meteorological Program Review, Preprint Volume, Fifth AMS International Conference on Satellite Meteorology and Oceanography, American Meteorological Society, London, England, 455–458

Glass M, Felde GW (1992) Intensity estimates of tropical cyclones using SSM/I brightness temperatures, Preprint Volume, Sixth AMS Conference on Satellite Meteorology and Oceanography, American Meteorological Society, 45 Beacon Street, Boston, MA, USA, J8–10

Hadlock R, Kreitzberg CW (1988) The Experiment on rapidly intensifying cyclones over the Atlantic (ERICA) field study: Objectives and plans. *Bull Amer Met Soc* 69: 1309–1320

Heacock EL Ed. (1985) Comparison of the Defense Meteorological Satellite Program (DMSP) and the NOAA Polar-Orbiting Operational Environmental Satellite Program (POES), NOAA Envirosat 2000 Report, p. 426

Hollinger JP, Peirce JL, Poe GA (1990) SSM/I instrument evaluation. *IEEE Trans Geosci Rem Sens* 28: 781–790

JTWC, 1992 Annual Tropical Cyclone Report, 1992, Joint Typhoon Warning Center, Guam, Mariana Islands

Katsaros KB (1994) Measurements of atmospheric water parameters and wind speed with the Special Sensor Microwave/Imager and the ERS-1 scatterometer, Proceedings of ECMWF/EUMETSAT Seminar on Developments in the Use of Satellite Data in Numerical Weather Prediction, September 6–10, 1993. Publication of ECMWF, Reading, U.K.

Katsaros KB, Bhatti I, McMurdie L, Petty G (1987) Passive microwave measurements of water vapor fields and rain for locating fronts in cyclonic storms, First Annual Report on Contract N00014-86-K-0453. Dept. of Atmos. Sci., University of Washington, Seattle, WA, USA

Katsaros KB, Brown RA (1991) Legacy of the Seasat mission for studies of the atmosphere and air-sea-ice interactions. *Bull Amer Met Soc* 72: 967–981

Katsaros KB, Liu TW (1995) Microwave Radiometers for Studies of the Ocean and Marine Atmosphere, Ch. 16, Oceanographic Applications of Remote Sensing, Eds., M. Ikeda and F. Dobson, CRC Press Inc., Boca Raton, Florida, USA, 261–274

Katsaros KB, Miletta J, Mognard NM (1996) Evolution of two oceanic extratropical cyclones as observed with the SSM/I and the GEOSAT and ERS-1 altimeters. *J Geophys Res*, in press

Katsaros KB, Petty GW, Bhatti I, Miller D (1988) Application of Special Sensor Microwave/Imager data for analysis of cyclonic storms in midlatitudes over the sea, Second Annual Report on Contract N00014-86-K-0453, Dept. of Atmos. Sci., University of Washington, Seattle, WA, USA

Katsaros KB, Zhao H, Miletta J, Quilfen Y (1993) Monitoring severe storms by passive and active satellite sensors in the microwave frequency range, Tropical Cyclone Disasters: *Proceedings of ICSU/WMO International Symposium*, October 12–16, 1992, Beijing, China,

J. Lighthill, Z. Zhemin, G. Holland and K. Emanuel, Eds., Peking University Press, Beijing, China, 69–78

Katsaros KB, Quilfen Y, Zhao H, Chapron B, El Fouhaily T (1995) Structure and development of Tropical Cyclones Hunt and Gay, revealed by satellite-borne active and passive microwave sensors, Preprint Volume, 21st Conference on Hurricanes and Tropical Meteorology, April 24–28, 1995, Miami, Florida, American Meteorological Society, 45 Beacon Street, Boston, MA, USA 460–462

Komen G, Cavalieri JL, Donelan M, Hasselmann K, Hasse S, Janssen PAEM (1994) *Dynamics and Modeling of Ocean Waves*, Cambridge University Press, Cambridge, 532 pp

Kuo YH, Reed RJ, Low-Nam S (1992) Thermal structure and airflow in a model simulation of an occluded marine cyclone. *Mon Wea Rev* 120: 2280–2297

Lin B, Rossow WB (1994) Observations of cloud liquid water path over oceans: optical and microwave remote sensing methods. *J Geophys Res* 99: 20907–20927

Liu G, Curry JA (1993) Determination of characteristic features of cloud liquid water from satellite microwave measurements. *J Geophys Res* 98: 5069–5092

Lystad M (1986) The Norwegian Polar Lows Project, International Conference on Polar Lows, DNMI, Oslo, Norway, May 20–23

Mass CF, Schultz DM (1993) The structure and evolution of a simulated midlatitude cyclone over land. *Mon Wea Rev* 121: 889–917

McMurdie LA, Katsaros KB (1985) Atmospheric water distribution in a midlatitude cyclone observed by the Seasat Scanning Multichannel Microwave Radiometer. *Mon Wea Rev* 113: 584–598

McMurdie LA, Katsaros KB (1991) Satellite derived integrated water vapor distribution in oceanic midlatitude storms; variation with region and season. *Mon Wea Rev* 119: 589–605

Miletta J (1993) Meteorological Applications of Coincident Visible, Infrared, and Microwave Observations from the Defense Meteorological Satellite Program, MS Thesis, Dept. of Atmos. Sci., University of Washington, Seattle, USA, 86 pp

Miletta J, Katsaros KB (1995) Using coincident multispectral satellite data to assess the accuracy of SSM/I liquid water path measurements. *J Geophys Res* 100: 16333–16339

Minnis P, Heck PW, Young DF, Fairall CW, Snider JB (1992) Stratocumulus cloud properties derived from simultaneous satellite and island-based instrumentation during FIRE. *J Appl Met* 31: 317–339

Mognard NM, Katsaros KB (1995) Weather patterns over the ocean observed with the Special Sensor Microwave/Imager and the Geosat altimeter. *The Global Atmos and Ocean Sys* 3: 301–323

Neiman PJ, Shapiro MA (1993) The life cycle of an extratropical marine cyclone. Part I: Frontal-cyclone evolution and thermodynamic air-sea interaction. *Mon Wea Rev* 121: 2153–2176

Nordberg W, Conaway J, Ross DB, Wilheit T (1971) Measurement of microwave emission from a foam-covered wind-driven sea. *J Atmos Sci* 28: 429–435

Nordeng TE (1990) A model-based diagnostic study of the development and maintenance mechanism of two polar lows. *Tellus* 42A: 92–108

Petty GW (1990) On the Response of the Special Sensor Microwave Imager to the Marine Environment – Implication for Atmospheric Parameter Retrievals, Ph.D. Dissertation, Dept. of

Atmos. Sci., University of Washington, Seattle, WA, USA, 291 pp. (Available from University Microfilm, 305 N. Zeeb Road, Ann Arbor, MI 48106.)

Petty GW, Katsaros KB (1992) The response of the SSM/I to the marine environment Part I. An analytic model for the atmospheric component of observed brightness temperatures. *J Atmos and Oceanic Tech* 9: 746–761

Petty GW, Katsaros KB (1994) The response of the SSM/I to the marine environment. Part II: A parameterization of the effect of the sea surface slope distribution on emission and reflection. *J Atmos and Oceanic Tech* 11, in press

Petty GW (1994a) Physical retrieval of over-ocean rain rate from multichannel microwave imagery. Part I: Theoretical characteristics of normalized polarization and scattering indices. *Met and Atmos Phys* 54: 79–99

Petty GW (1994b) Physical retrieval of over-ocean rain rate from multichannel microwave imagery. Part II: Algorithm implementation. *Met and Atmos Phys* 54: 101–121

Petty GW, Miller DK (1995) Satellite microwave observations of precipitation correlated with intensification rate in extratropical cyclones. *Mon Wea Rev* 123: 1904–1911

Platnick S, Twomey S (1994) Determining the susceptibility of cloud albedo to changes in droplet concentration with the Avanced Very High Resolution Radiometer. *J Appl Met* 33: 334–347

Quilfen Y, Katsaros KB, Chapron B (1993) Surface wind and precipitation patterns in tropical cyclones observed the ERS-1 scatterometer and with the Special Sensor Microwave/Imager. *Proceedings of the Second ERS-1 Symposium*, October, 11–14, Hamburg, Germany

Quilfen Y, Katsaros KB, Chapron B (1994) Structures des champs de vent de surface dans les cyclones tropicaux observés par le diffusiomètre d'ERS-1 et relation avec les vagues et les précipitations, Mémoires de l'Institut Océanographique, Monaco, no. 18, 13–22

Rasmussen E (1979) The polar low as an extra-tropical CISK disturbance. *Q J R Meteor Soc* 105: 531–549

Reed RJ, Albright MD (1986) A case study of explosive cyglogenesis in the eastern Pacific. *Mon Wea Rev* 114: 2297–2319

Reed RJ, Stoelinga MT, Kuo YH (1992) A model-aided study of the origin and evolution of the anomalously high potential vorticity in the inner region of a rapidly deepening marine cyclone. *Mon Wea Rev* 120: 893–913

Reed RJ, Grell GA, Kuo YH (1993) The ERICA IOP 5 storm. Part I: Analysis and simulation. *Mon Wea Rev* 121: 1577–1594

Rogers E, Bosart LF (1986) An investigation of explosively deepening oceanic cyclones. *Mon Wea Rev* 114: 702–718

Rossow WB, Gardner LC, Lacis AA (1989) Outlook for research on subtropical marine stratiform clouds. *Bull Amer Met Soc* 2, 419-458

Sanders F, Gyakum JR (1980) Synoptic-dynamic climatology of the "Bomb". *Mon Wea Rev* 108: 1589–1606

Shapiro MA, Fedor LS, Hampel T (1987) Research aircraft measurements of a polar low over the Norwegian Sea. *Tellus* 39A: 272–306

Shapiro MA, Keyser D (1990) Fronts, Jet Streams, and the Tropopause. In Extratropical Cyclones, The Erik Palmen Memorial Volume, C. Newton and E.O. Holopainen, Eds., American Meteorological Society, Boston, MA, USA, 167–191

Short DA, North GR (1990) The beam filling error in the Nimbus 5 Electronically Scanning Microwave Radiometer observations of Global Atlantic Tropical Experiment rainfall. *J Geophys Res* 95: 2187–2193

Uccellini LW, Keyser D, Brill KF, Wash CH (1985) The Presidents' Day cyclone of 18- 19 February, 1979: Influence of upstream trough amplification and associated tropopause folding on rapid cyclogenesis. *Mon Wea Rev* 113: 962–988

Velden CS (1992) Satellite-based microwave observations of tropopause-level thermal anomalies: Qualitative applications in extratropical cyclone events. *Wea Forecast* 7: 669–682

Wash CH, Hale RA, Dobos PH, Wright EJ (1992) Study of explosive and nonexplosive cyclogenesis during FGGE. *Mon Wea Rev* 120: 40–51

Wentz FJ (1983) A model function for ocean microwave brightness temperatures. *J Geophys Res* 88: 1892–1908

Willoughby HE (1990) Temporal changes of the primary circulation in tropical cyclones. *J Atmos Sci* 47: 242–264

Willoughby HE, Clos JA, Shoreibah MG (1982) Concentric eyewalls, secondary wind maxima, and the evolution of the hurricane vortex. *J Atmos Sci* 39: 395–411

Zhao H (1994) Analysis of Tropical Cyclones using the Microwave Data from the Special Sensor Microwave/Imager, MS Thesis, Dept. of Atmos. Sci., University of Washington, Seattle, WA, USA, 82 pp

Chapter 18

Physical Basis for Microwave Remote Sensing of Sea Ice and Snow

Martti Hallikainen
Laboratory of Space Technology
Helsinki University of Technology
Otakaari 5 A, 02150 Espoo
Finland

18.1 Introduction

About 10% of the world Ocean is covered by sea ice during some portion of the year. Sea ice participates in the key large-scale processes of the Earth's climate system, the absorption and emission of radiant energy, and the poleward flux of heat (Carsey et al., 1992). This participation comes about through processes involving the atmosphere, the ocean, and the radiation field. Some of these are outlined below:

Radiative balance: The snow-covered sea ice has a very high albedo relative to that of the open ocean. Therefore, changes in sea ice extent cause drastic changes in the surface albedo of the high-latitude seas.

Surface heat: Although the fully developed ice cover is an effective insulator between the cold air and the relatively warm ocean, areas of open water and thin ice lose heat rapidly during the cold seasons (Maykut, 1986).

Ice margin processes: The abrupt transition to open water gives rise to unique processes, including water mass formation, oceanic upwelling, eddy formation, and generation of atmospheric instability (Muench et al., 1987).

Operations: Navigation and trafficability on and below the surface, drill ship operations in the marginal seas and harbor operations are concerned with locating areas of thin ice, identifying hazards such as very thick and deformed ice, and forecasting ice conditions.

Snow is an important component of the Earth's climate and it impacts significantly global weather and is vital to the water supply in many areas. Snow serves as a water vapor input to the atmosphere through the processes of sublimation and evaporation and as a source of water to the soil and river system when melting occurs. It changes the albedo and the radiation balance and may reduce the solar energy available to heat the surface and the lower atmosphere by more than 50%. Therefore, snow cover represents a changing atmosphere output that is a function of variability in the Earth's climate and is also a boundary condition that affects meteorology and

NATO ASI Series, Vol. I 45
Radiation and Water in the Climate System:
Remote Measurements
Edited by Ehrhard Raschke

Variable	Accuracy requirement
Ice extent	
Thickness distribution	
Snow depth	5 cm
Snow wetness	1%
Summer melt pond coverage	3%
Ice motion and deformation on 5 km scale	2 cm/s
Wind, currents, clouds, air/ocean temperatures	

Table 18.1: *Primary sea ice variables and their accuracy requirements (Carsey et al., 1992).*

Variable	Accuracy requirement
Extent	5%
Water equivalent	10 mm
Water content	2%
Surface temperature	0.5° C
Albedo	5%

Table 18.2: *Observational accuracy requirements for snow surveys (Kuittinen, 1992).*

climate through the changes that it makes in the radiation balance and the water vapor input to the atmosphere (Foster and Chang, 1992). The increasing demand for water throughout the world imposes a challenging problem for water resources managers. Since precipitation deposited as snow represents an important source of the water supply, improved knowledge of the water equivalent of snow over large regions should result in a more accurate estimate of eventual runoff (Foster et al., 1984).

The sea ice variables of primary interest to both the operational and scientific data user are shown in Table 18.1. The observational requirements for snow are presented in Table 18.2.

The time table for spaceborne missions and instruments collecting data valuable for sea ice and snow research is depicted in Table 18.3. Many visible and infrared sensors with various spatial resolutions, the U.S. DMSP SSM/I microwave radiometer system, synthetic aperture radars onboard the ESA ERS-1 and Japanese JERS-1satellites collect these data presently. The ESA follow-up to ERS-1, called ERS-2, was launched in April 1995 and Canada plans to launch the Radarsat SAR satellite in 1995.

Operational retrieval of sea ice and snow characteristics from satellite data in civilian applications is still scarce. Research projects aiming on the development of advanced methods to use satellite data are well underway in several countries.

The main advantage of using microwave sensors for sea ice and snow research is their day/night and practically all-weather capability. Unlike optical sensors, microwave radiometers and radars can acquire data through cloud cover and independent of lighting conditions. This is important at high latitudes, where cloudy weather dominates in winter.

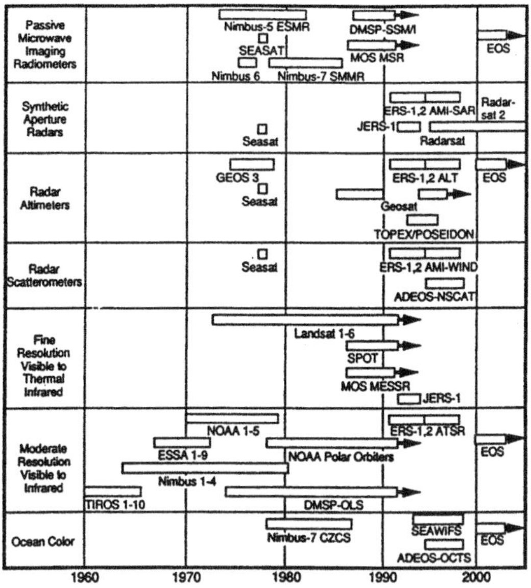

Table 18.3: *Planned and historical satellite missions with data valuable to sea ice and snow research (Carsey et al., 1992). Russian/Soviet satellites are not included.*

18.2 Basic Quantities in Microwave Remote Sensing

18.2.1 Brightness temperature

In microwave radiometry, the measured emission intensity is converted to an equivalent brightness temperature

$$T_B(f, \theta_0) = e(f, \theta_0)T \tag{18.1}$$

where e is the dimensionless emission coefficient ($0 \leq e \leq 1$) at this wavelength, T is the physical temperature of the target in degrees Kelvin (K), and is the local incidence angle off nadir. The emissivity[1] is determined by the dielectric properties and the surface roughness of the target. In radiometry of sea ice, radiation emitted by each layer of the water-ice-snow system contributes to the observed brightness temperature. Similarly, in radiometry of snow (ground-snow system, possibly with vegetation cover) emission from each layer contributes to the observed brightness temperature. Due to complexity within the snow and sea ice media, Equation (18.1) describes only approximately the observed brightness temperature.

[1]instead of "emissivity" often the term "emittance" is used.

18.2.2 Backscattering coefficient

The microwaves transmitted by a radar are reflected and scattered by a target. The backscattered intensity is measured by the receiver and the result is expressed as the differential backscattering coefficient σ^0 (usually referred to as the backscattering coefficient).

The backscattering coefficient is defined in terms of the incident and scattered electric fields E_t^i and E_r^s as follows:

$$\sigma_{tr}^0(\theta_i, \phi_i) = \frac{4\pi R^2 |E_r^s|^2}{A|E_t^i|^2} \tag{18.2}$$

where r and t are the polarizations of the transmitted and received fields, respectively, R is the distance from the radar to the target, and A is the area illuminated by the radar. Angles θ_i and ϕ_i define the incident direction of the transmitted power. The backscattering coefficient is usually expressed in decibels (dB),

$$\sigma^0(dB) = 10 \log_{10} \sigma^0 \tag{18.3}$$

18.2.3 Basic electromagnetic quantities describing natural media

In order to characterize the microwave interaction with the snow-ice-water system, we define several basic electromagnetic quantities (Hallikainen and Winebrenner, 1992).

18.2.3.1 Relative permittivity

The relative permittivity, denoted $\epsilon = \epsilon' - j\epsilon''$ (where $j = \sqrt{-1}$) is a complex number that characterizes the electrical properties of media. The real part of the relative permittivity (ϵ') gives the contrast with respect to free space ($\epsilon'_{air} = 1$), whereas the imaginary part of the permittivity (ϵ'') gives the electromagnetic loss of the material. ϵ' and ϵ'' are often referred to as the dielectric constant and the dielectric loss factor, respectively.

18.2.3.2 Propagation, absorption and phase constant

For an electromagnetic plane wave propagating in the z-direction, the intensity of the electric field at point z can be expressed as

$$E(z) = E_0 e^{-\gamma z} \tag{18.4}$$

where E_0 is the field intensity at $z = 0$. The complex propagation constant of the medium is denoted by γ and is given by

$$\gamma = \alpha + j\beta \tag{18.5}$$

where α is the absorption constant and β is the phase constant. The absorption constant describes transformation of energy into other forms of energy, such as heat. The phase constant is equal to the wave number $k = 2\pi/\lambda$ in a lossless medium. They are related to the complex permittivity by

$$\alpha = k_0 |Im\{\sqrt{\epsilon}\}| \tag{18.6}$$

$$\beta = k_0 Re\{\sqrt{\epsilon}\} \tag{18.7}$$

where k_0 is the wave number in free space. The power absorption coefficient κ_a is defined as

$$\kappa_a = 2\alpha \tag{18.8}$$

It is often expressed in dB/m through the relation

$$\kappa_a(dB/m) = 8.686\alpha(Np/m) \tag{18.9}$$

18.2.4 Extinction and scattering coefficient

The total electromagnetic loss in a scattering medium consists of absorption loss (electromagnetic power transformed into other forms of energy, such as heat) and scattering loss (energy is caused to travel in directions other than that of the incident radiation). Scattering loss is caused by particles of different ϵ embedded in a host medium. The extinction coefficient (total loss) is thus

$$\kappa_e = \kappa_a + \kappa_s \tag{18.10}$$

where κ_s denotes the scattering loss.

Penetration depth. Part of a wave incident upon the surface of a medium from the air in the direction z is transmitted across the boundary into the medium. The penetration depth is defined as the depth δ_p at which (Ulaby et al., 1986)

$$\frac{P(\delta_p)}{P(0+)} = \frac{1}{e} \tag{18.11}$$

where z is the direction normal to the surface, $P(\delta_p)$ is the transmitted power at δ_p, and $P(0+)$ is the transmitted power just beneath the surface. If scattering in the medium is ignored, $\kappa_e \equiv \kappa_a = 2\alpha$ and if, additionally, α does not depend on z,

$$\delta_p = \frac{1}{\kappa_a} = \frac{\sqrt{\epsilon'}}{k_0\epsilon''} \quad ; \quad \epsilon'' \ll \epsilon' \tag{18.12}$$

The penetration depth indicates the maximum depth of the medium that contributes to the backscattering coefficient and brightness temperature.

18.3 Physical Properties of Sea Ice and Snow

18.3.1 Sea ice

Sea ice is composed of fresh-water ice, inclusions of liquid brine and air bubbles. Fresh water ice forms from sea water at temperatures below the freezing point of sea water. Some of the salt liquid is entrapped within the ice forming microscopic pockets of brine. The salinity of brine is much higher than that of sea water and depends on the temperature of the surrounding ice with which it is in thermal equilibrium. The composition of sea ice is shown in Figure 18.1. At a temperature of −10°C brine forms only 0.7% of sea ice of 1 part per thousand salinity. At −2°C it is 2.7%. The air content by weight is negligible but by volume it may be up to 10% for first-year sea ice.

The brine pockets are usually elongated and their average length is of the order of millimeters. Their dimensions are determined by the temperature. The average salinity of sea water in the

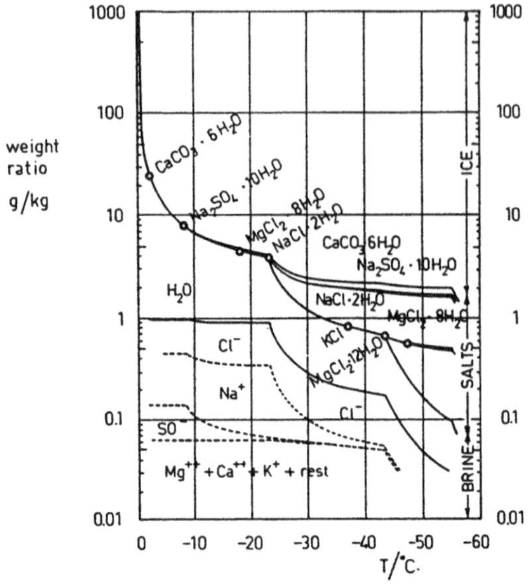

Figure 18.1: *Amount of different components in one kilogram of sea ice (Assur, 1960). A salinity of 1 $^0/_{00}$ is assumed. For a given sea ice the values for salt and brine must be changed in proportion to the salinities. Circles on the brine-salt line indicate temperatures at which the different salts precipitate.*

oceans is 32 $^0/_{00}$ and that of first-year sea ice from 5 to 10 $^0/_{00}$. The evolutionary sequence for thin sea ice is shown in Figure 18.2.

Initially, the water surface takes on an oily, opaque appearance, at which point grease ice is said to exist. Upon further freezing, grease ice develops into nilas or ice rinds, the former appearing as an elastic crust and the latter as a brittle, shiny crust. When the wind is blowing, the ice develops into masses as a result of water motion. Freezing occurs in the center of the masses and then spreads outward to form flat, circular ice disks up to 10 cm in diameter. When these disks grow together, they form pancake ice.

The international ice code is given in Table 18.4.

The concentration of ice is expressed as the ratio, in tenths, of the sea surface actually covered by ice to the total sea surface area over a defined area. It is expressed in the terms given in Table 18.5.

Wind, ocean currents, waves, and pressure cause deformation and drifting of ice. Pressure may cause the sheets of ice to pile up in ridges and hummocks. Moreover, one piece of ice may override another, a process known as rafting.

In columnar ice brine is trapped in inclusions within ice crystals in a platelet substructure, whereas in frazil ice the inclusions are located between the crystal boundaries. Brine drainage begins immediately after ice formation, occurring slowly during the growth season, but increasing considerably during summer (Tucker et al., 1992). In the Arctic, enhanced surface melting, coupled with increased interconnectivity of the brine inclusions, almost completely flushes the salt from the ice in the upper layers, leaving air voids and channels. In the Antarctic sea ice,

Ice type	Thickness (cm)
New ice, including frazil, grease, slush, shuga, and pancake ice	<10
Nilas	<10
Young ice	10–30
Gray ice	10–15
Gray-white ice	15–30
First-year ice	30–200
Thin first-year or white ice	30–70
Medium first-year ice	70–120
Thick first-year ice	>120
Old ice - second-year or multiyear ice	
Second-year ice	
Multiyear ice	

Table 18.4: *International ice code.*

Term	Concentration	Comments
Compact pack ice	10/10	No water visible
Consolidated pack ice	10/10	Floes frozen together
Very close pack ice	9/10 to less than 10/10	
Close pack ice	7/10 to 8/10	Floes mostly in contact
Open pack ice	4/10 to 6/10	Floes generally not in contact
Very open pack ice	1/10 to 3/10	

Table 18.5: *Concentration classes of sea ice.*

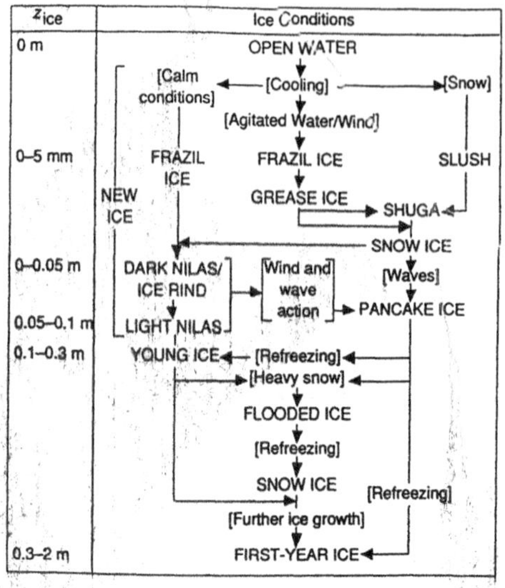

Figure 18.2: *Evolutionary sequence for thin sea ice (Grenfell et al., 1992). Ice types are shown in capital letters and related environmental processes are enclosed in square brackets.*

pronounced surface melting apparently does not occur, resulting in an absence of both surface melt ponds and the melt-pond-hummock relief so characteristic of similar floes in the Arctic.

Pressure ridges are evident in both polar regions, although they are larger and more numerous in the Arctic. Frost flowers and snow on recently formed ice absorb brine, affecting the surface dielectric properties. Excessive snow cover can depress both first-year and multiyear ice sufficiently to cause sea water to flood the surface.

Ice undergoes dramatic changes during the melt season. The most obvious change is the large decrease in salinity, Figure 18.3. The greatest desalination occurs in the upper layers of the ice. Associated with the brine volume increase, the brine pockets tend to coalesce into vertical channels that facilitate gravity drainage of the brine and desalination of the ice (Tucker et al., 1992).

The second major effect of summer in the Arctic is flushing of the ice with surface meltwater. Fresh water can percolate through the ice so that salinities of the upper 50 to 100 cm are reduced to less than one part per thousand, down from the previous 4 to 7 $^0/_{00}$. In Antarctic sea ice, the desalination is much less than in the Arctic, mainly because surface melting is uncommon.

In semi-enclosed seas, only first-year ice occurs. In the Baltic Sea, ice starts to form in November and it reaches its maximum thickness (usually less than one meter) in March. By the end of May, all ice is gone. The salinity of Baltic ice is low compared to Arctic ice, usually about 1.5 to 2 $^0/_{00}$ for new ice and less than one part per thousand for older ice. The maximum extent of ice varies substantially. During a severe winter the whole Baltic Sea is covered by ice, whereas in a mild winter only the northernmost 10% is covered by ice.

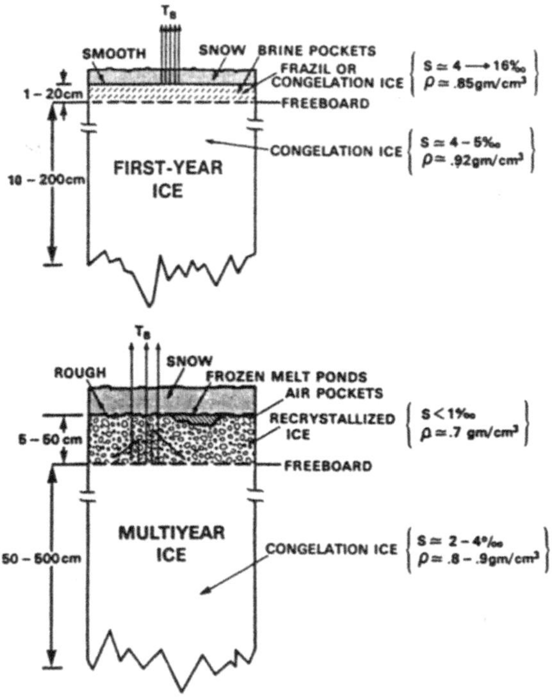

Figure 18.3: *Schematic diagram illustrating the basic difference between first- year and multiyear ice (Ulaby et al., 1986).*

18.3.2 Snow

Dry snow is a mixture of ice particles and air voids. After a snowfall, the shapes of the numerous ice particles in dry snow are modified by destructive metamorphism, resulting in smoothing the pointed angles of the crystals to rounded ice shapes. Thermodynamically, the ice crystals seek equilibrium, for which the ratio of surface area to volume is minimum (Gray and Male, 1981).

Surface effects cause another class of metamorphism. Melt-freeze cycles can form a very strong crust. The strength results from bonding of the ice grains. Another surface structure is surface hoar. Sublimation of water vapor from the air on the snow surface is the mechanism of growth. Surface hoar is ice dew which occurs on clear cold nights under high humidity conditions when the dewpoint is reached.

If large temperature gradients exist within the snowpack, differences in equilibrium vapor pressure will exist at different depths. The net effect is vapor migration towards the lower vapor pressure cold snow. This process, termed constructive metamorphism, can create very large crystals, on the order of 1 cm in diameter. The vapor pressure gradient is larger with respect to temperature gradient near 0°C. therefore, since the ground temperature is consistently near 0°C, formation of depth hoar is generally at the base of the snowpack. As soon as the temperature gradient is removed, destructive metamorphism begins.

Any water in the snowpack collects at points of contact between the grains. The geometry and porosity of snow at high and low liquid water contents have been examined by Colbeck (1981). He concluded that snow, like any porous material, has two distinct regimes of liquid saturation. In the lower range, called the pendular regime, air is continuous throughout the pore space, and liquid occurs in the forms of isolated inclusions. In the higher range of liquid saturation, the funicular regime, the liquid is continuous throughout the pore space and air occurs as distinct bubbles trapped by narrow constrictions in the pores. There is a sharp transition between the two regimes.

The metamosphism caused by melting and freezing changes the microstructure of snow; the grains become rounded during the melting process, and some of the smaller grains disappear completely. Snow that has undergone several melt-freeze cycles tends to form multiple clusters.

The density of dry newly fallen snow is of the order of 0.1 g/cm^3. Metamorphism increases the density so that it is of the order of 0.3 g/cm^3. Under extreme conditions the density may be as high as 0.5 g/cm^3. The snow grain size increases, due to metamorphism, from 0.2 mm for new snow to 1–2 mm for refrozen snow. The snowpack can include several layers with different physical properties, often with thin icy layers between them. The liquid water content of snow may be up to more than 10% by volume.

18.4 Dielectric Properties of Sea Ice and Snow

18.4.1 Sea ice

Sea ice consists of pure ice, brine inclusions and air bubbles. In order to understand the dielectric behavior of sea ice, the permittivity of water, brine, and pure ice are reviewed first.

18.4.1.1 Pure water and sea water

Compared with other natural media, the complex permittivity of water is exceptionally high. This is due to its polar molecular structure (Debye, 1929). Dissolved salts in sea water (salinity up to 35 $^0/_{00}$ by weight) increase the ionic conductivity. Consequently, the dielectric loss factor of sea water is higher than that of pure water at frequencies below 10 GHz. The salinity of liquid brine in sea ice is governed by its temperature and can exceed 200 $^0/_{00}$ at low temperatures.

The dielectric behavior of pure water follows the Debye equation (Debye, 1929). Sea water contains dissolved salts, including $NaCl$, $MgCl_2$, Na_2SO_4, $CaCl_2$, and KCl, and the average salinity in the oceans is 32.54 $^0/_{00}$ (Anderson, 1960). Salts increase the dielectric loss of water by adding free charge carriers. Hence, the loss factor of sea water includes an additional term due to ionic conductivity and the complex permittivity of sea water follows the modified Debye equation,

$$\epsilon_{sw} = \epsilon_{sw\infty} + \frac{\epsilon_{sw0} - \epsilon_{sw\infty}}{1 + j2\pi f \tau_{sw}} - j\frac{\sigma_{sw}}{\epsilon_0} \tag{18.13}$$

where $\epsilon_{sw\infty}$ is the high-frequency limit of ϵ_{sw}, ϵ_{sw0} is the static dielectric constant of sea water, f is the frequency, τ_{sw} is the relaxation time of sea water, σ_{sw} is the ionic conductivity of sea water, and ϵ_0 is the dielectric constant of vacuum.

Figure 18.4 shows the complex permittivity of pure and sea water (salinity 32.54 $^0/_{00}$) at 0°C and 20°C. The real part is practically independent of salinity. The imaginary part is governed by the Debye relaxation process at frequencies above 10 GHz and by the ionic conductivity at frequencies below 3 GHz.

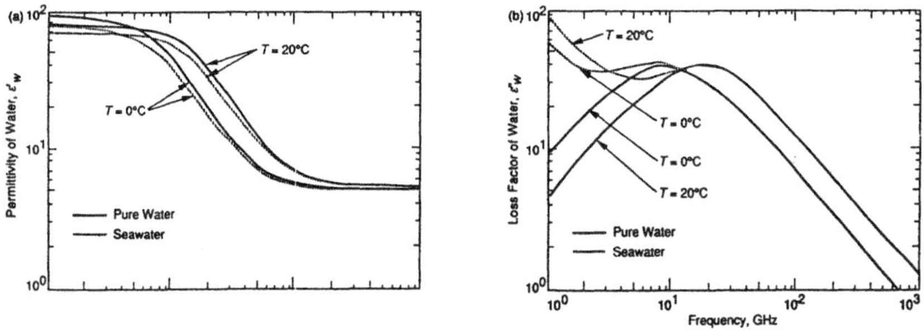

Figure 18.4: *Complex permittivity of pure and sea water at 0° C and 20° C (Ulaby et al., 1986).*

18.4.1.2 Brine

Liquid brine is concentrated in the form of elongated inclusions in sea ice (Assur, 1960). The permittivity of liquid brine follows basically Equation (18.13). However, the salinity of liquid brine is determined by its temperature; when the ice temperature decreases, additional water freezes out, thereby increasing the salt concentration in the brine. Empirical expressions for the brine salinity have been derived by Assur (1960).

Since the salinity of brine may be much higher than that of sea water, some of the empirical expressions derived for the terms of Equation (18.13) are not valid for liquid brine. Stogryn and Desargeant (1985) measured the dc-conductivity and the dielectric properties of sea water liquid brine in equilibrium with sea ice. Their results for the dielectric properties of liquid brine are shown in Figure 18.5. In general, ϵ_b' and ϵ_b'' decrease with decreasing temperature. The rapid increase of ϵ_b'' with decreasing frequency below 3 GHz is caused by the ionic conductivity. The Debye relaxation is just barely visible at $T = -5$°C; at lower temperatures, it is masked by the ionic conductivity.

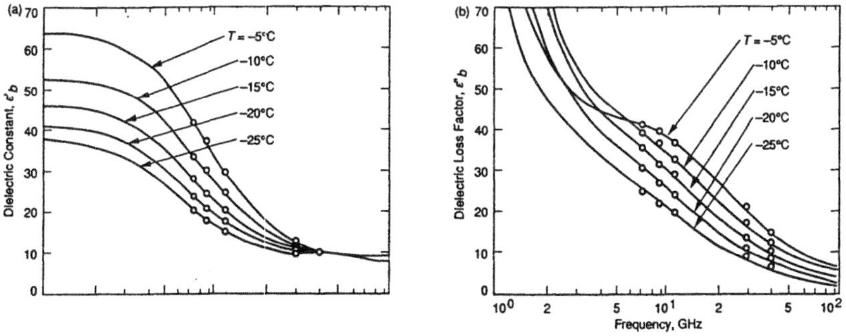

Figure 18.5: *The dielectric constant (a) and loss factor (b) of brine as a function of frequency with temperature as a parameter (Stogryn and Desargeant, 1985).*

18.4.1.3 Fresh-water ice

Fresh-water ice is free from salt but usually it includes air bubbles, impurities, and cracks. Pure ice is an idealization of fresh-water ice (frozen distilled water). The dielectric properties of fresh-water ice at microwave frequencies are determined by two physical processes, namely the high-frequency tail of a relaxation spectrum (relaxation frequency in the kilohertz range) and the low-frequency tail of far-infrared absorption bands (Evans, 1965), (Mätzler and Wegmüller, 1987). These processes result in a practically constant value for the dielectric constant ϵ_i' at microwave frequencies and in a minimum of the dielectric loss factor ϵ_i'' centered around 1 to 5 GHz.

Based on experimental data, the relative dielectric constant of pure and fresh-water ice may be assigned the constant value

$$\epsilon_i' = 3.17 \tag{18.14}$$

between 10 MHz and 1000 GHz (Mätzler and Wegmüller, 1987; Cumming, 1952).

For the imaginary part of the permittivity, results of various investigators show substantial scatter. The scatter is caused, in addition to measurement inaccuracies, partly due to small values of ϵ_i'', by varying amounts of impurities in the ice samples. Results from two empirical studies are shown in Figure 18.6.

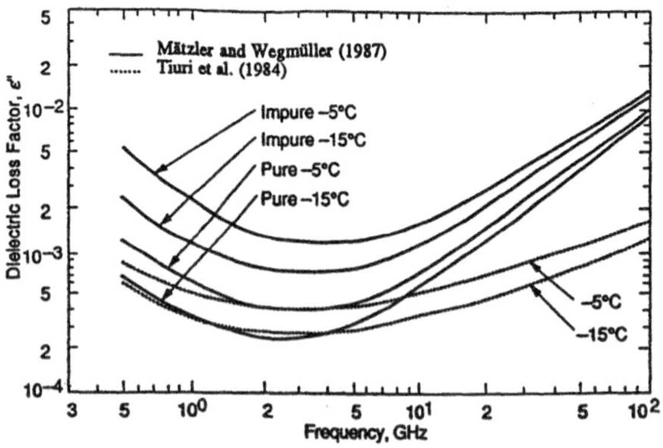

Figure 18.6: *Dielectric loss factor of freshwater ice by Mätzler and Wegmüller (1987), and by Tiuri et al. (1984).*

ϵ_i'' is between 0.001 and 0.01 in the 1 to 100 GHz range. Thus, fresh-water ice is a low- loss medium and its power absorption coefficient at 1 GHz (temperature −5°C) is 0.09 dB/m and penetration depth is 100 m. At 10 GHz, the corresponding numbers are 0.9 dB/m and 10.0 m. This means that contributions to the brightness temperature and the backscattering coefficient are not limited to the topmost ice layers.

18.4.1.4 Sea Ice

Sea ice consists of fresh-water ice, liquid brine and air. In addition to the dielectric properties of these constituents, several additional parameters influence the complex permittivity of sea ice, including the volume fraction of each constituent and geometry (shape, size and orientation) of brine pockets with respect to the propagation direction of the electromagnetic wave. Most of these parameters depend on ice temperature.

Experimental investigations of the microwave dielectric behavior of sea ice have been carried out at frequencies below 40 GHz. A review of these investigations is given in (Hallikainen and Winebrenner, 1992).

Figures 18.7 to 18.9 show experimental values of ϵ'_{si} and ϵ''_{si} values obtained by various investigators as a function of temperature. Data are shown for 1 GHz, 4 to 5 GHz, and 10 to 16 GHz, because these frequency bands are widely used for both active and passive remote sensing of sea ice. Note that the temperature scale is logarithmic in order to show the rapid changes occurring at temperatures above $-5°C$. For the permittivity of sea ice at frequencies above 16 GHz, the reader is referred to (Hallikainen and Winebrenner, 1992). Due to substantial scatter in most of the original data sets, only curves fitted to data points are shown in Figures 18.7 to 18.9.

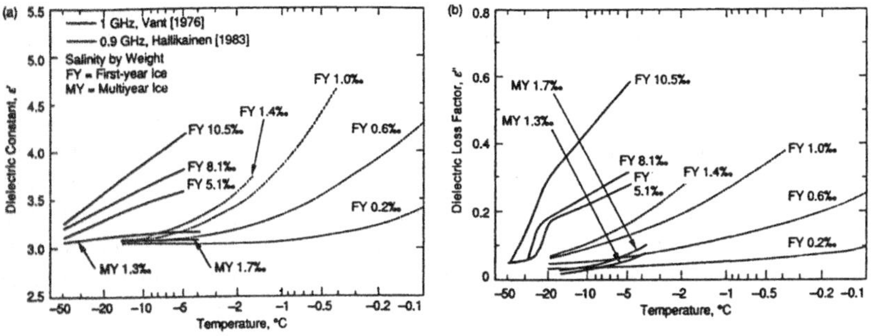

Figure 18.7: *Experimental permittivity values for sea ice at 1 GHz, obtained by various investigators (Hallikainen and Winebrenner, 1992).*

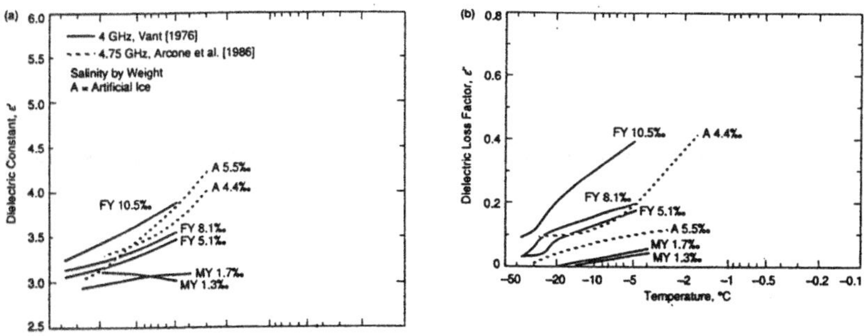

Figure 18.8: *Experimental permittivity values for sea ice in the 4 to 5 GHz range, obtained by various investigators (Hallikainen and Winebrenner, 1992).*

Experimental values by Vant et al. (1978) at 1 GHz and by Hallikainen (1983) at 0.9 GHz are shown in Figure 18.7. The values of Vant et al. (1978) show that ϵ'_{si} and $\epsilon''_{si}$for multiyear ice are lower than those for first-year ice; this is due to the lower salinity. Since the relative brine

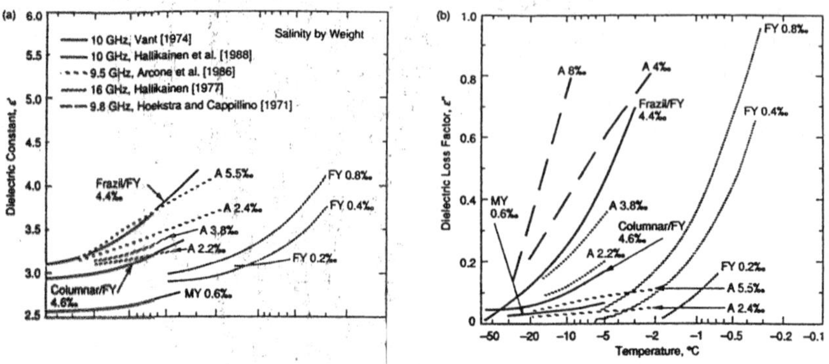

Figure 18.9: *Experimental permittivity values for sea ice in the 10 to 16 GHz range, obtained by various investigators (Hallikainen and Winebrenner, 1992).*

volume of low-salinity sea ice near 0°C may be higher than that of high-salinity sea ice at lower temperatures (the relative brine volume of sea ice of salinity 1 $^0/_{00}$ is 103 $^0/_{00}$ at -0.5°C, whereas the corresponding number for sea ice of salinity 4 $^0/_{00}$ is 40 $^0/_{00}$ at -5°C), the maximum values obtained by Hallikainen (1983) are of the same order of magnitude as those obtained by Vant et al. (1978) for high-salinity arctic sea ice.

At C-band, few measurements have been made at temperatures above −5°C. The results of Vant et al. (1978) for arctic sea ice at 4 GHz are compared against the values of Arcone et al. (1986) for simulated sea ice at 4.75 GHz in Figure 18.8. The abrupt changes in the slopes of ϵ''_{si} curves of Vant et al. (1978) may be due to crystallization of NaCl at −22.9°C. The unusual decrease of ϵ''_{si} with increasing salinity in the results for simulated sea ice may be due to differences in the size and length of brine pockets and thus, conductivity (Arcone et al., 1986).

Experimental values at 10 GHz cover frazil, columnar, multiyear, and artificial sea ice (see Figure 18.9). The dielectric loss is higher for frazil than for columnar sea ice with the electromagnetic wave propagating in the vertical direction. The values of ϵ''_{si} obtained by Hoekstra and Cappillino (1971) are higher than those by other investigators. This may be due to their sample preparation techniques (flashfreeze sea water in the sample holder) which results in an orientation of brine pockets different from that in natural sea ice.

The effect of the orientation of brine pockets with respect to the electric field of an electromagnetic wave propagating in the sea ice medium has been observed to be substantial (Sackinger and Byrd, 1972). The difference in ϵ''_{si} may be up to 300 % at −7°C, and decreases with decreasing temperature and is about 150% at −21.5°C. Hence, the dielectric loss of sea ice increases with increasing incidence angle.

Experimental values for the absorption coefficient κ_a of natural sea ice (incidence angle 0 degrees) are shown in Figure 18.10a. In general, κ_a increases with increasing frequency and temperature. The loss of multiyear ice is considerably lower than that of first-year ice, mainly due to much lower salinity. The values for first-year low-salinity sea ice show that the temperature dependence of the absorption coefficient is high at temperatures near 0°C. At 10 GHz, the differences between the results for columnar, frazil, and multiyear ice are substantial.

Figure 18.10b shows experimental values for the penetration depth of natural sea ice, computed from the results in Figure 18.10a. In the 1 to 10 GHz range, the penetration depth is between 100 and 5 cm for first-year ice and 500 and 30 cm for multiyear ice. Figure 18.10b indicates that sensors operating at X-band provide information on sea ice mainly from the topmost 5 to 80 cm, depending on ice type, salinity and temperature, whereas the corresponding numbers for L-band sensors are 40 to 500 cm.

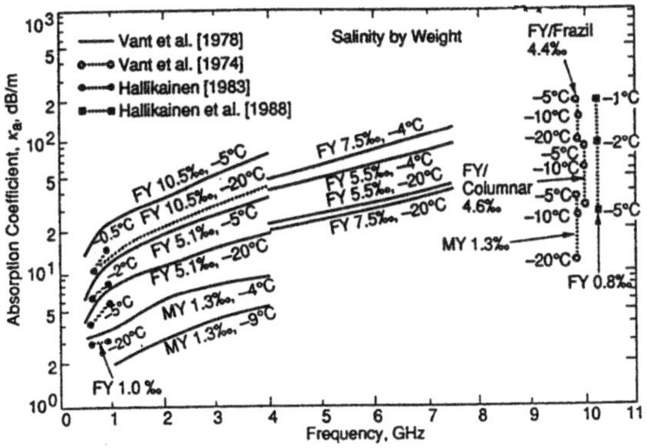

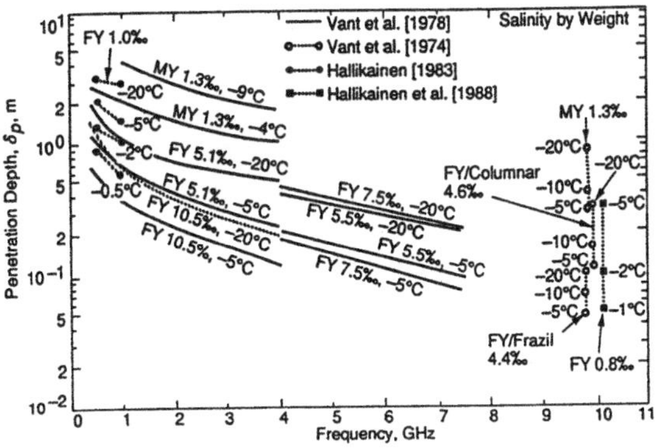

Figure 18.10: *Experimental (a) power absorption coefficient and (b) penetration depth of sea ice, obtained by various investigators (Hallikainen and Winebrenner, 1992).*

A number of attempts have been made to relate the dielectric properties of sea ice to the relative brine volume V_b. Numerical relations for 1 GHz, 4 GHz, and 10 GH are (Hallikainen and Winebrenner, 1992)

$$\epsilon'_{si} = 3.12 + 0.009V_b; 1GHz \tag{18.15}$$

$$\epsilon''_{si} = 0.04 + 0.005V_b; 1GHz \tag{18.16}$$

$$\epsilon'_{si} = 3.05 + 0.0072V_b; 4GHz \tag{18.17}$$

$$\epsilon''_{si} = 0.02 + 0.0033V_b; 4GHz \tag{18.18}$$

$$\epsilon'_{si} = 3.00 + 0.012V_b; 10GHz \tag{18.19}$$

$$\epsilon''_{si} = 0.0 + 0.010V_b; 10GHz \tag{18.20}$$

where $V_b \leq 70\ ^0/_{00}$. Equations (18.15) to (18.20) do not explicitly take into account the effect of ice density upon the real part of the permittivity. The correlation coefficient for Equations (18.15) to (18.20) is, in general, between 0.7 and 0.8.

Theoretical models are reviewed by Hallikainen and Winebrenner (1992).

18.4.2 Snow

18.4.2.1 Dry Snow

Dry snow consists of ice crystals and air voids. After a snowfall, the shapes of ice particles are naturally modified. Thermodynamically, the ice crystals seek equilibrium, for which the ratio of surface area to volume is minimum (Gray and Male, 1981). Any water in the snow medium collects at points of contact between the ice grains. The metamorphism caused by melting and freezing changes the microstructure of snow. The ice grains become rounded during the melting process and some of the smaller grains disappear completely. Snow that has undergone several melt-freeze cycles tends to form multiple clusters. In general, the density of snow slowly increases with time due to metamorphism and melt-freeze cycles. The density of dry snow varies from 0.1 g/cm^3 (newly fallen snow) to 0.5 g/cm^3 (refrozen snow). The interface between snow and sea ice may be gradual, consisting of a high-density now-ice layer. Due to flooding, this layer may be saline.

18.4.2.1.1 Dielectric properties of dry snow Electromagnetically, dry snow is a dielectric mixture of ice and air and, therefore, its complex permittivity is governed by the dielectric properties of ice, snow density and ice- particle shape. Since the real part of the permittivity of ice is $\epsilon_i' = 3.17$ at frequencies between 10 MHz and 1000 GHz (Equation (18.14)) and practically independent of temperature, the dielectric constant of dry snow, ϵ_{ds}' is only a function of density. For the most extensive data set used in dielectric measurements of snow, the result between 3 and 37 GHz is (Hallikainen et al., 1986)

$$\epsilon_{ds}' = 1 + 1.9\rho_{ds}; \quad \rho_{ds} \leq 0.5 \text{ g/cm}^3 \tag{18.21}$$

where ρ_{ds} is the density of dry snow in g/cm^3 . Based on Equation (18.21), $\epsilon_{ds}' = 1.57$ for snow density 0.3 g/cm^3 . The density of snow is mostly below 0.5 g/cm^3 .

As discussed in Section 18.4.1.3, fresh-water ice is a low-loss material and air is a lossless material. Since the density of dry snow is 0.1 to 0.5 g/cm^3 and that of pure ice is 0.916 g/cm^3 , the dielectric loss factor of dry snow is even lower than that of ice.

Experimental data for the loss factor of dry snow are limited to frequencies below 13 GHz (Cumming, 1952), (Tiuri et al., 1984), shown in Figure 18.11.

18.4.2.1.2 Extinction properties of dry snow Dry snow is a strongly scattering medium at high microwave frequencies due to its structure. Based on theoretical calculations using the Mie theory (assuming independent scattering), scattering dominates over absorption in dry snow (ice grain size 1 mm) at frequencies above 15 GHz (Ulaby et al., 1986). The following empirical expressions have been developed to relate the extinction coefficient of dry snow, denoted κ_{eds}, to the observed snow particle diameter d_s:

$$\kappa_{eds} = 1.5 + 7.4 d_s^{2.3} \quad \text{dB/m} \quad \text{at 18 GHz} \tag{18.22}$$

$$\kappa_{eds} = 30 d_s^{2.1} \quad \text{dB/m} \quad \text{at 35 GHz} \tag{18.23}$$

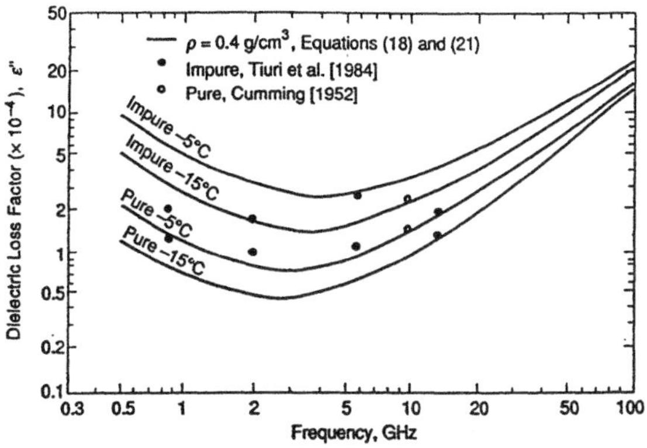

Figure 18.11: *Experimental data for the dielectric loss factor of dry snow by various investigators and comparison with an empirical expression (Hallikainen and Winebrenner, 1992). Equations (18.18) and (18.21) are from Ulaby et al., (1986).*

$$\kappa_{eds} = 180 d_s^{2.0} \quad \text{dB/m} \quad \text{at 60 GHz} \tag{18.24}$$

$$\kappa_{ds} = 300 d_s^{1.9} \quad \text{dB/m} \quad \text{at 90 Ghz} \tag{18.25}$$

The particle diameter (observed by photography) is in millimeters. Equations (18.22) to (18.25) hold for particle sizes below 1.6 mm.

The empirical expressions 18.22 to 18.25 are depicted in Figure 18.12. They indicate that the extinction coefficient at 90 GHz may exceed 500 dB/m for large grain sizes. At 18 GHz, the corresponding maximum value is about 20 dB/m. Figure 18.12 implies that at 35, 60, and 90 GHz, the scattering coefficient $\kappa_s = \kappa_e - \kappa_a$ is much larger than the absorption coefficient for all realistic grain sizes. Note that at 18 GHz, the absorption and scattering coefficient are comparable in magnitude for small-grained snow.

The strong fluctuation theory provides results that agree reasonably well with the empirical extinction coefficient values in the 18 to 60 GHz range for all realistic grain sizes, and also at 90 GHz for grain sizes smaller than 0.9 mm (Hallikainen et al., 1987).

18.4.2.2 Wet Snow

Wet snow is a mixture of ice crystals, liquid water, and air. For snow on top of sea ice layer, liquid water may be saline. The geometry and porosity of wet snow depend on its liquid water content. It has been concluded that snow has two distinct regimes of liquid saturation (Colbeck, 1982). In the lower range (pendular regime), air is continuous throughout the pore space, and liquid water occurs in the form of isolated inclusions. In the higher range of liquid saturation (funicular regime), liquid water is continuous throughout the pore space and air occurs as distinct bubbles trapped by narrow constrictions in the pores. There is a sharp transition between the two regimes.

Electromagnetically, wet snow is a three-component dielectric mixture consisting of ice particles, air, and liquid water. As discussed in Section 4.1, both water and ice exhibit Debye-type relaxation spectra. The relaxation frequency (frequency at which the maximum dielectric loss occurs) of ice is in the kilohertz range, whereas that for water at 0°C is 9 GHz. The

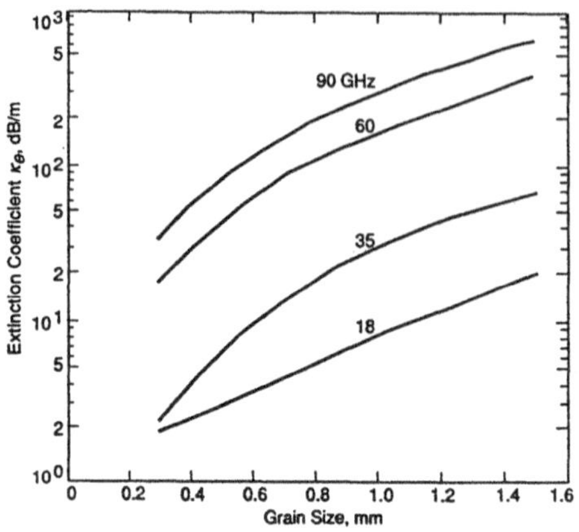

Figure 18.12: *Empirical extinction coefficient of dry snow as a function of snow particle diameter (Hallikainen et al., 1987).*

complex permittivities of ice and water depend on frequency and temperature. Consequently, the permittivity of wet snow is a function of frequency, temperature, volumetric water content, snow density, ice-particle shape and the shape of water inclusions.

Since the permittivity of water is substantially higher than that of ice and air, the dielectric behavior of wet snow is governed by the volume fraction of water. Basically, ϵ'_{ws} and ϵ''_{ws} are compressed versions of those for water at 0°C. The dielectric behavior of wet snow has been investigated by Hallikainen et al. (1986) between 3 and 37 GHz. The samples had densities ranging from 0.09 to 0.42 g/cm^3 and liquid water contents ranging from 0 to 12.3% by volume. The snow particle size varied between 0.5 and 1.5 mm and the sample temperatures ranged from -5°C to 0°C. Based on these data, the applicability of several theoretical and empirical dielectric models was evaluated.

The argument for a Debye-like model is the following: Since the water inclusions in snow have an ϵ_{ws} about 40 times larger than ϵ_{ds}, the spectral behavior of wet snow is dominated by the dispersion behavior of water. The developed Debye-like model is discussed in detail in (Hallikainen et al., 1986).

Figure 18.13 shows the effect of liquid water content on the real and imaginary part of the permittivity of wet snow, calculated using the Debye-like model. The results in Figure 18.13 confirm that the dielectric behavior of wet snow follows that of water at 0°C; the level of ϵ_{ws} is determined by the liquid water content of snow.

The modified Debye-like equation presented above does not take into account the geometry of the wet snow medium (shape of water inclusions, funicular and pendular regimes). The two-phase Polder-Van Santen dielectric mixing model (Polder and Van Santen, 1946) can be applied to wet snow by assuming that wet snow consists of dry snow as the host material with water inclusions embedded in it. The water droplets are randomly distributed and randomly oriented ellipsoids with depolarization factors Aw1, Aw2, and Aw3.

According to Colbeck (1982), liquid water inclusions in snow are approximately needle-shaped in the pendular regime (low values of water content), but they become approximately disk-shaped in the funicular regime (high values of water content). The best fit of the developed two-phase Polder-Van Santen mixing model to experimental data was obtained by assuming that their shapes depend on snow water content and, additionally, that the water inclusions are nonsymmetrical in shape (Hallikainen et al., 1986). Transition from the pendular regime to the funicular regime was observed to take place around $m_v = 3$ %.

Both the modified Debye-like model and the two-phase Polder-Van Santen model (with variable, nonsymmetrical shape factors) provide good accuracy (Hallikainen et al., 1986). As stated before, the simplified Debye-like model works well only at frequencies below 15 GHz. The two-phase Polder-Van Santen model with wetness-dependent shape factors provides a slightly better overall fit to the experimental observations of wet snow. The modified Debye-like model is easier to use and is likely to provide adequate accuracy for most applications.

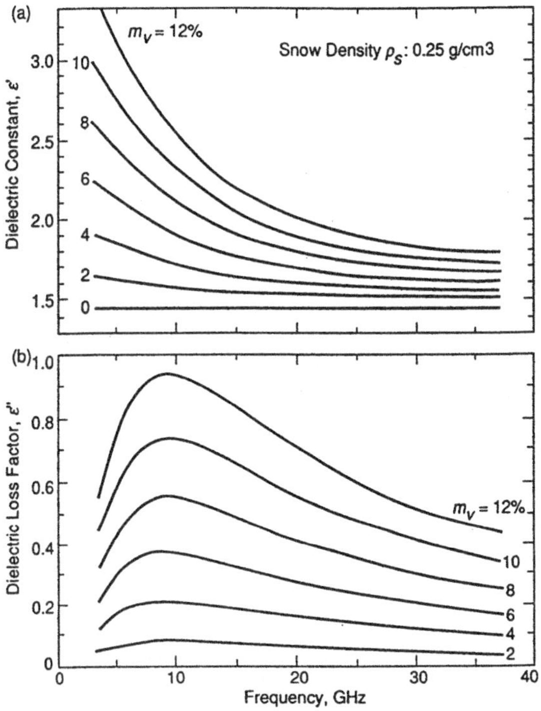

Figure 18.13: *Permittivity of snow according to the modified Debye-like model plotted as a function of frequency, with liquid-water content as a parameter (Hallikainen et al., 1986).*

Figure 18.14a depicts the power absorption coefficient of wet snow, calculated using the modified Debye-like model. Even a volumetric water content of 2% results in absorption of tens of decibels per meter, depending on frequency. Figure 18.14b shows the corresponding penetration depth, indicating that, in the case of wet snow, microwave radar and radiometer measurements of sea ice only provide information on the snowpack on top of the ice layer.

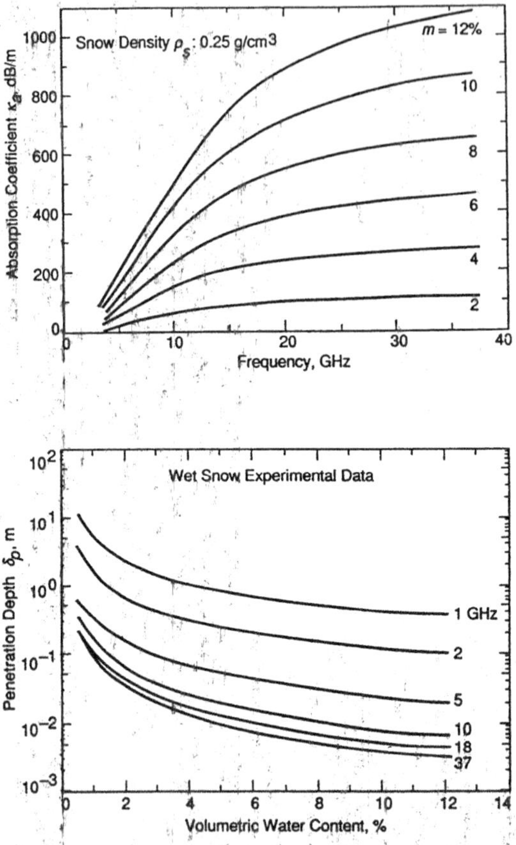

Figure 18.14: *Absorption coefficient (top) and penetration depth (bottom) for snow, calculated using the modified Debye-like model, as a function of frequency and volumetric liquid water content (Hallikainen and Winebrenner, 1992).*

18.5 The Physical Nasis for Remote Sensing of Sea Ice and Snow

18.5.1 Effect of atmosphere

The microwave emission received by a spaceborne radiometer consists of contributions from the Earth, atmosphere and space:

$$T_B = eT_s e^{-\tau} + T_{B,up} + (1-e)T_{B,down}e^{-\tau} + (1-e)T_{B,cosm}e^{-2\tau} \tag{18.26}$$

where e = surface emissivity
T_s = physical temperature of the radiating portion of the surface
τ = atmospheric opacity

Sensor Characteristics	Sea ice characteristics	Water characteristics
Frequency	Surface roughness	Temperature
Polarization	Temperature profile	Wave characteristics
Incidence angle	Salinity profile	Salinity
	Air bubble characteristics	
	Snow depth	
	Snow liquid water content	
	Snow density	
	Snow grain size (dry)	

Table 18.6: *Sensor and target characteristics that determine the brightness temperature and backscattering coefficient of sea ice and open water.*

$T_{B,up}$ = atmospheric upwelling radiation
$T_{B,down}$ = atmospheric downwelling radiation
$T_{B,cosm}$ = cosmic background radiation

The first term in Equation (18.26) is the surface emission attenuated by the atmosphere, the second term is the atmospheric upwelling radiation, the third term is the atmospheric downwelling radiation reflected from the surface and attenuated by the atmosphere, and the fourth term is the cosmic background radiation reflected by the surface and attenuated twice by the atmosphere.

At low microwave frequencies (below 5 GHz) the atmospheric effects can be ignored practically independent of weather conditions. Presence of rainfall in the sensor's field of view may introduce large errors at higher frequencies. Above 18 GHz, the varying amounts of atmospheric water vapor and cloud liquid water may cause errors. However, the atmosphere is usually very dry over consolidated ice and cloudiness generally consists of underlying stratus that contains mostly ice crystals. These clouds have low opacity in the microwave region, but they may affect the sea ice surface temperature. Cloud effects as high as 7 K have been observed on the physical temperature of sea ice (Gloersen et al., 1973).

For present spaceborne radars, the atmospheric effects are negligible under normal weather conditions. Precipitation may attenuate the signal and mask the target.

18.5.2 Remote sensing of sea ice

The brightness temperature and backscattering coefficient of sea ice depend on the sensor and target characteristics depicted in Table 18.6.

The brightness temperature (emissivity) and backscattering coefficient of sea ice vary with its dielectric properties (determined mainly by temperature, salinity, and liquid water content), surface roughness (both large-scale and small-scale) and the characteristics of snow cover.

18.5.2.1 Microwave radiometry

18.5.2.1.1 Brightness temperature of open water Since one of the most important applications in remote sensing of sea ice is discrimination of sea ice from open water, we examine briefly the behaviour of the brightness temperature of sea surface. Due to the exceptionally high permittivity of sea water compared to other natural media (Figure 18.4), the brightness temperature of calm sea water is low, Figure 18.15. The brightness temperature increases with increasing frequency, because the permittivity gradually decreases when approaching millimeter wave frequencies. Other sensor parameters affecting the brightness temperature of sea water are polarization and incidence angle. The main target parameter affecting the brightness temperature is wind speed. Figure 18.16 shows its effect in the 1.4 to 19.3 GHz frequency range. The effect is much larger for horizontal polarization, due to its Fresnel transmission coefficient at the air-ocean interface.

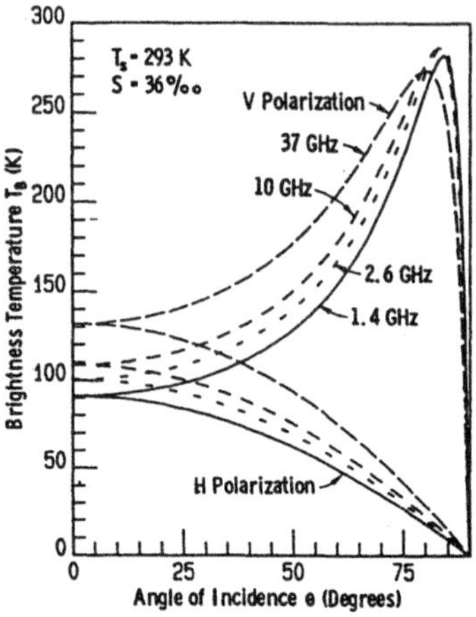

Figure 18.15: *Brightness temperature of calm sea water (salinity 36 $^0/_{00}$, temperature 20° C) as a function of incidence angle (Ulaby et al., 1986).*

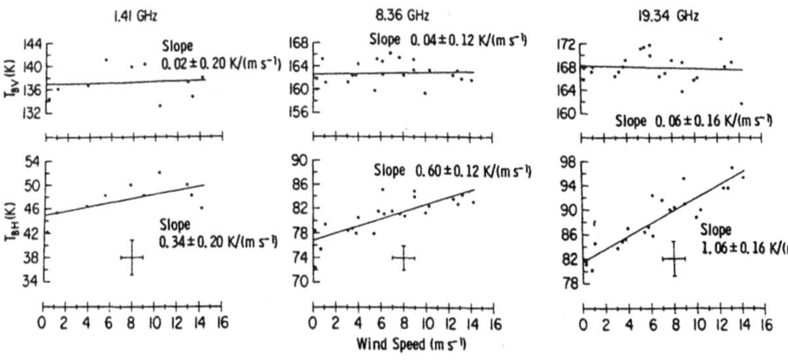

Figure 18.16: *Wind-speed dependence of sea surface brightness temperature at an incidence angle of 55 degrees (Hollinger, 1971).*

18.5.2.1.2 Evolution of sea ice signature The permittivity of sea ice is, depending on salinity, temperature and measurement frequency, of the order of 3 to 5. Hence, the emissivity of sea ice is substantially higher that that of sea water and the brightness temperature of sea ice is in most cases above 200 K. Figures 18.17 and 18.18 show the behaviour of the brightness temperature as a function of ice thickness at 0.6, 10, 20, and 30 GHz. The results indicate that the brightness temperature of sea ice is determined mostly by its surface properties and the snow cover.

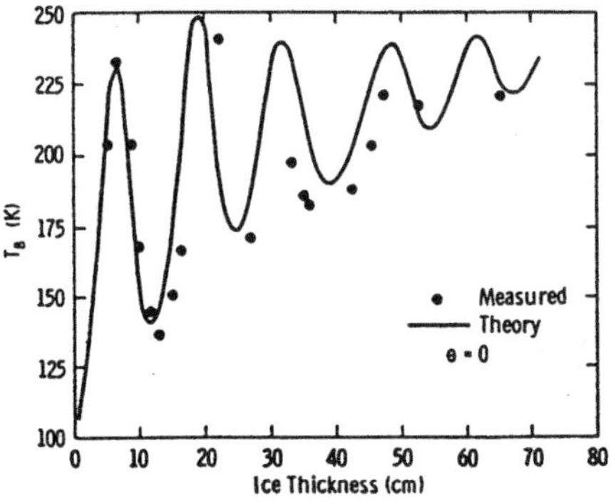

Figure 18.17: *Comparison of experimental low-salinity sea ice results and the layered theoretical model (Hallikainen, 1983).*

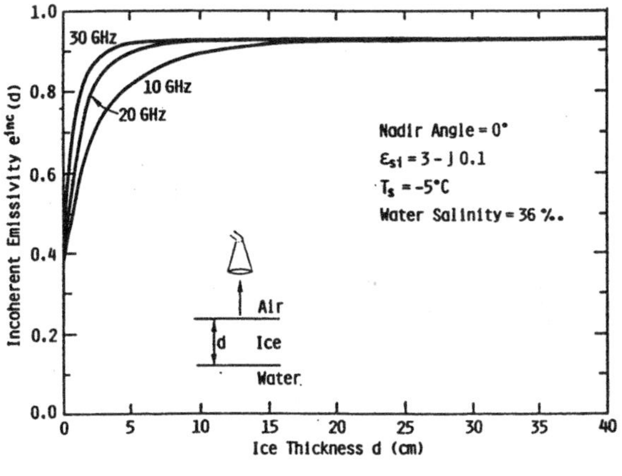

Figure 18.18: *Theoretical emissivity of sea ice as a function of ice thickness (Ulaby et al., 1986).*

18.5.2.1.3 Effect of snow cover The effect of snow cover on top of the ice layer to the brightness temperature depends on both weather history and present weather. Melt-freeze cycles gradually increase the snow grain size. Large-grained dry snow scatters emission from the ice layer and underlying water, decreasing the brightness temperature. Wet snow both absorbs emission from ice and water and emits radiation. In an extreme case, the brightness temperature of sea ice depends on the characteristics of the snow cover only. Since the permittivity of wet snow is lower (1.5 to 2.5) than that of sea ice (3 to 5), the brightness temperature of wet snow-covered sea ice is higher than that of snow-free sea ice.

18.5.2.1.4 Multiyear ice At high microwave and millimeterwave frequencies, the brightness temperature of multiyear ice is lower than that of first-year ice. This is due to the presence of air voids and flushed brine drainage channels that cause volume scattering in ice. The scattering increases with increasing frequency, as shown by the behavior of the brightness temperature in Figure 18.19.

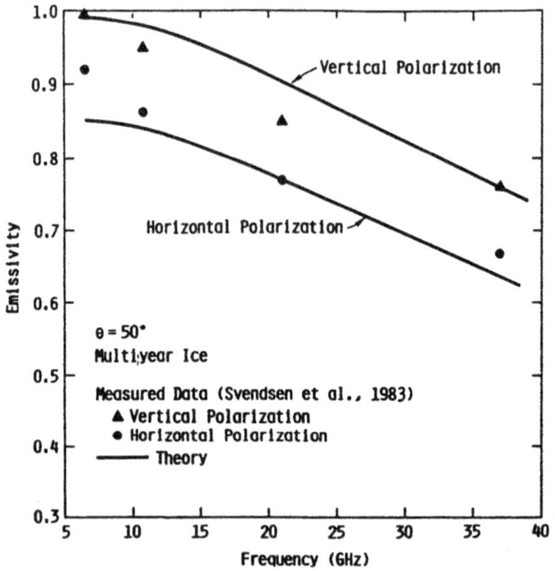

Figure 18.19: *Comparison of measured emissivity of multiyear sea ice at an incidence angle of 50 degrees with theoretical results (Ulaby et al., 1986).*

18.5.2.2 Radar

18.5.2.2.1 Backcattering coefficient of open water The effect of wind speed and direction relative to radar has been reported by Unal et al. (1990). Based on their values for C-band, VV polarization, 23° incidence angle, the value of the backscattering coefficient for upwind increases from −13 dB to −7 dB, when the wind speed increases from 4 m/s to 12 m/s. The maximum effect of wind direction is about 6 dB, resulting in a total variation of 12 dB for this wind speed range. Variation of the backscattering coefficient with wind speed is used to derive the near-surface wind field over oceans. For remote sensing of sea ice, this variation handicaps retrieval of sea ice parameters from radar data.

18.5.2.2.2 Backscattering coefficient of sea ice The backscattering coefficient of sea ice consists of contributions from air-snow interface, snow medium, snow-ice interface, ice medium and ice-water interface:

$$\sigma_{ice}^0 = \sigma_{as}^0 + t_{si}^2 \left\{ \sigma_s^0 + \frac{1}{L_s^2} \left(\sigma_{si}^0 + t_i^2 \sigma_i^0 + t_i^2 \frac{1}{L_i^2} \sigma_{iw}^0 \right) \right\} \tag{18.27}$$

where
σ_{as}^0 = backscattering coefficient of air-snow interface
t_{si} = Fresnel power transmission coefficient at air-snow interface
σ_s^0 = backscattering coefficient of snow medium
L_s = total attenuation of snow layer
σ_{si}^0 = backscattering coefficient of snow-ice interface
σ_i^0 = backscattering coefficient of ice medium
t_i = Fresnel power transmission coefficient for ice layer
L_i = total attenuation of ice layer
σ_{iw}^0 = backscattering coefficient of ice-water interface

It should be noted that all terms in Equation (18.27) are functions of the incidence angle. Figure 18.20 shows the predicted backscattering coefficient at 5.3 GHz, HH polarization for Baltic sea ice as a function of incidence angle(Ulander et al., 1991). For ice surface scattering, either the Kirchhof model with the scalar approximation or the perturbation model was used, depending on the observed values of the surface roughness parameters. The volume scattering contribution was computed according to (Kim et al., 1985). Contributions from snow-ice interface, snow volume, ice volume and air-snow interface are shown. The total backscattering coefficient is dominated by the snow-ice interface and has, additionally, a small contribution from the snow volume.

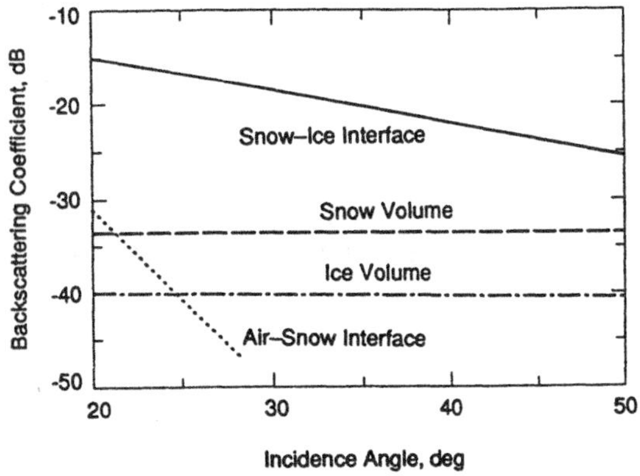

Figure 18.20: *Predicted backscattering coefficient at 5.3 GHz, HH polarization for Baltic sea ice as a function of incidence angle (Ulander et al., 1991). Volume and surface scattering contributions from snow and ice are shown. The snow-ice surface rms height is 1.81 mm and the correlation length 22.1 mm.*

18.5.2.2.3 Effect of ice roughness and type Figure 18.21 depicts the mean values of the backscattering coefficient for various low-salinity sea ice types in the Gulf of Bothnia at 5.4 GHz and 9.8 GHz, incidence angle 23 degrees off nadir. In general, the backscattering coefficient increases with increasing surface roughness. Due to the small wind speeds in the test site, the values for open water are low. However, in windy conditions the backscattering coefficient of open water may be close to those of different ice types.

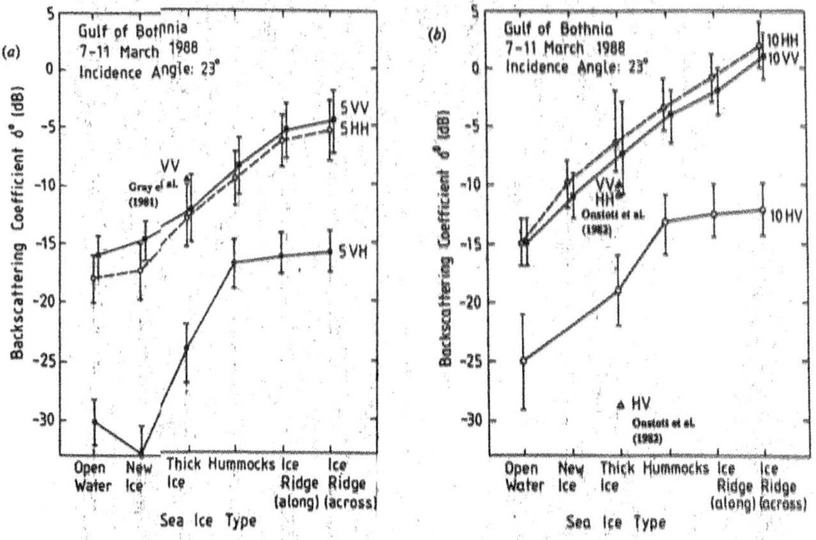

Figure 18.21: *Mean values of the backscattering coefficient for different low- salinity sea-ice types and open water in the Gulf of Bothnia, March 1988, at an incidence angle of 23 degrees and at 5.4 GHz (left) and 9.8 GHz (right). Ninety percent of the results fall within the maximum and minimum values indicated by the vertical bar. For comparison, results for Arctic sea ice are included.*

18.5.2.2.4 Effect of weather Under wet snow conditions, the air-snow interface and the snow volume contributions may dominate at high frequencies. Figure 18.22 shows the values of the backscattering coefficients measured for three ice types, new ice, level ice and deformed ice, under dry and wet snow/ice conditions. The general trend in Figure 18.22 is that in the case of wet snow/ice conditions the values of the three ice types get closer to each other. Hence, wet snow tends to mask the underlying sea ice layer.

18.5.2.2.5 Comparison of Arctic vs. Baltic ice Figure 18.23 shows a comparison between C-band sea ice signatures in the Arctic and low-salinity sea ice in the Baltic (Ulander, 1991). The difference between the like-polarization backscattering coefficients measured in the two areas for an ice type is usually less than 5 dB. For the cross-polarization cases, the difference may be larger.

18.5.2.2.6 Evolution of Arctic first-year ice signature Evolution of the backscattering coefficient of Arctic first-year ice is presented schematically in Figure 18.24. The changes in

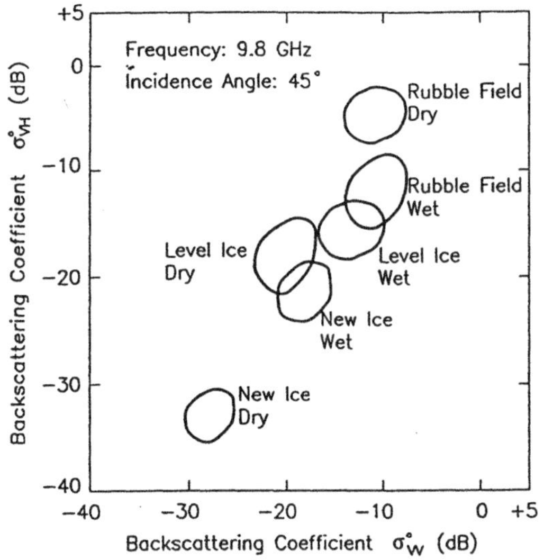

Figure 18.22: *Effect of dry and wet snow/ice conditions to the backscattering coefficients of three ice types at 9.8 GHz, VV and VH polarizations, incidence angle 45 degrees, in the Baltic Sea. Each contour line is based on the mean value and the standard deviation of data points measured for the two polarizations (Hallikainen and Toikka, 1992).*

the physical properties, dielectric properties, and surface roughness characteristics contribute to the evolution process. The highest backscattering coefficient values are obtained when the frost flower is fully developed. With time and the eventual infiltration of snow and wind, the structure of frost flowers becomes irreversibly altered. The snow and frost flower then undergoes a restructuring, which results in a steady reduction in the backscatter intensity until thicknesses near 2 m are reached.

18.5.2.2.7 Multiyear ice and seasonal effects Backscatter intensities for first-year ice, multiyear ice and open water are shown in Figures 18.25 and 18.26 (Onstott, 1992). During winter, the microwave signatures of multiyear ice are different from those of first-year ice. In summer, the situation is more complex due to the melting snow and ice, melt pool formation and desalination.

18.5.3 Remote sensing of snow

The sensor and target characteristics that determine the brightness temperature and backscattering coefficient of snow-covered and snow-free terrain are listed in Table 18.7.

In microwave radiometry (frequency range 6 to 90 GHz), the main parameters are the snow water equivalent (water layer thickness resulting from melt of snow, determined by snow depth and density), snow grain size and snow liquid water content. The backscattering coefficient in the frequency range of 1 to 10 GHz is mainly determined by the snow and ground surface roughnesses and snow and ground wetnesses.

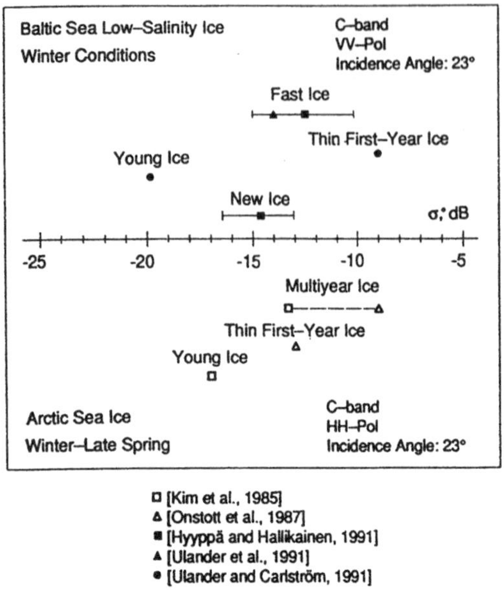

Figure 18.23: *Comparison of C-band sea-ice signatures in the Arctic and low- salinity sea ice in the Baltic Sea (Ulander, 1991).*

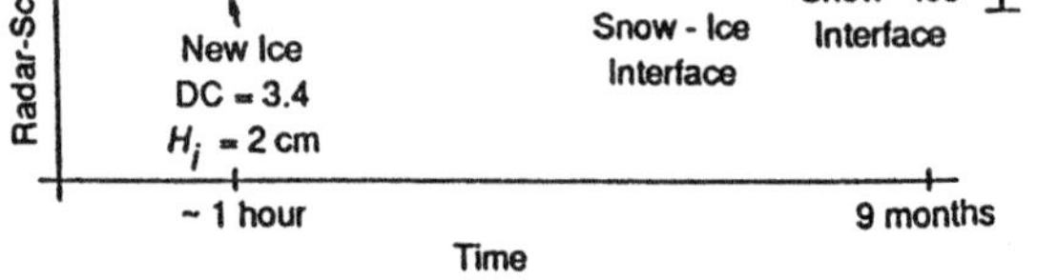

Figure 18.24: *Evolution of the microwave signature of first-year sea ice (Onstott, 1992). DC denotes dielectric constant and Hi is ice thickness.*

18.5.3.1 Microwave radiometry

Due to volume scattering by snow, its brightness temperature cannot be modelled in simple terms. The radiative transfer equation describes sufficiently the brightness temperature of snow-covered terrain, although it cannot explain all observations:

$$T_{Bsnow} = T_{Bg}e^{-\tau(0,d_s)} + \int_0^{d_s} \kappa_e(r')\{(1-a)T_s(r') + aT_{SC}(r')\}e^{-\tau(r',d_s)}\,dr' \qquad (18.28)$$

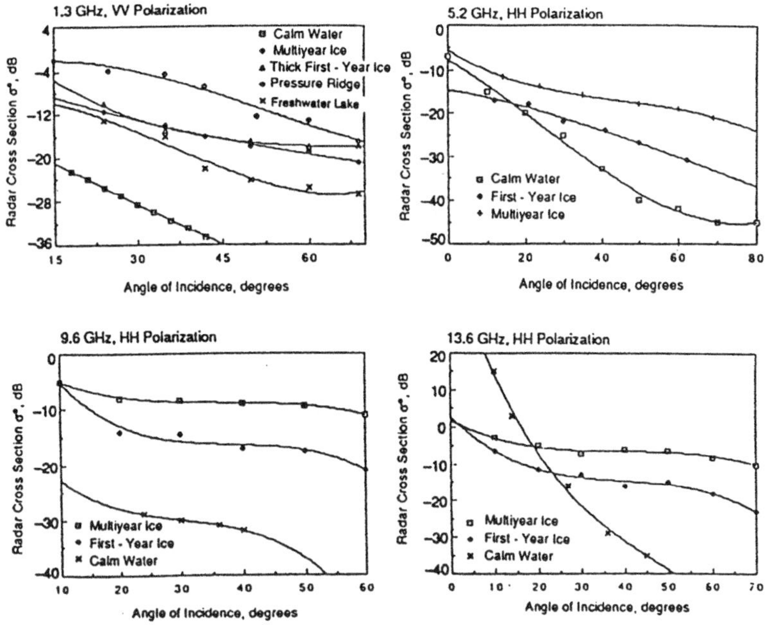

Figure 18.25: *Radar backscatter cross-sections at four frequencies during winter as a function of incidence angle (Onstott, 1992).*

Sensor Characteristics	Snow characteristics	Ground characteristics
Frequency	Depth	Surface roughness
Polarization	Liquid water content	Temperature
Incidence angle	Density	Liquid water content
	Grain size	Land use category
	Land use category	Vegetation characteristics
	Ground characteristics	
	Vegetation characteristics	

Table 18.7: *Sensor and target characteristics that determine the brightness temperature and backscattering coefficient of snow-covered terrain.*

where
T_{Bsnow} = brightness temperature of snow- covered terrain
T_{Bg} = brightness temperature contribution from underlying ground
$\tau(0, d_s)$ = optical depth of snowpack
d_s = snow depth
κ_e = extinction coefficient of snow
a = $\kappa_s - \kappa_e$ = albedo
κ_s = scattering coefficient of snow
T_s = physical temperature of snow
T_{SC} = scattered brightness temperature (from other directions)

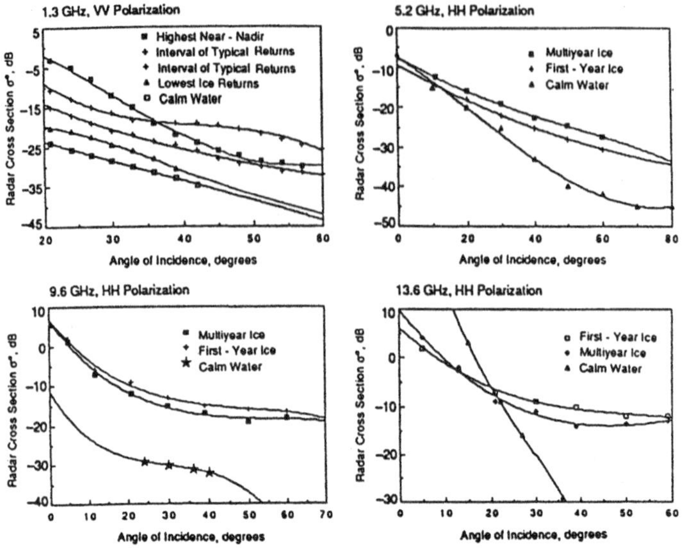

Figure 18.26: *Radar backscatter cross-sections at four frequencies during summer as a function of incidence angle (Onstott, 1992).*

$\tau(r', d_s)$ = optical depth from radiating point to snow surface

In Equation (18.28), the term $e^{-\tau(r_1,r_2)}$ describes the extinction (scattering plus absorption) between points r_1 and r_2. Calculation of the term T_{SC} is difficult and can be done reasonably well only by simplifying the physical snow model.

18.5.3.1.1 Dry snow Volume scattering becomes the dominant loss mechanism in dry snow at 15 GHz, assuming a grain size of 1 mm. Consequently, above 15 to 20 GHz (depending on grain size), the brightness temperature decreases with increasing snow depth and increasing frequency, as illustrated in Figure 18.27a. Several algorithms developed to retrieve the water equivalent of snow cover from radiometer data employ 18 GHz and 37 GHz data. Figure 18.27b shows that the effective grain size is inevitably an important parameter that has to be eliminated.

The average grain size of seasonal snow cover slowly increases with time. This is due to snow going through melt-freeze cycles. Consequently, the brightness temperature of snow- covered terrain does not follow a single curve shown in Figure 18.27b. Rather, it may start by following a curve of d = 0.5 mm for small values of the water equivalent and finally reach the curve for d = 1 mm after several melt-freeze cycles.

18.5.3.1.2 Wet snow With increasing liquid water content in the snow, absorption starts to dominate over scattering as the principal loss mechanism. As shown in Figure 18.14, the loss of wet snow is high even at low volumetric wetness levels. Consequently, brightness temperature

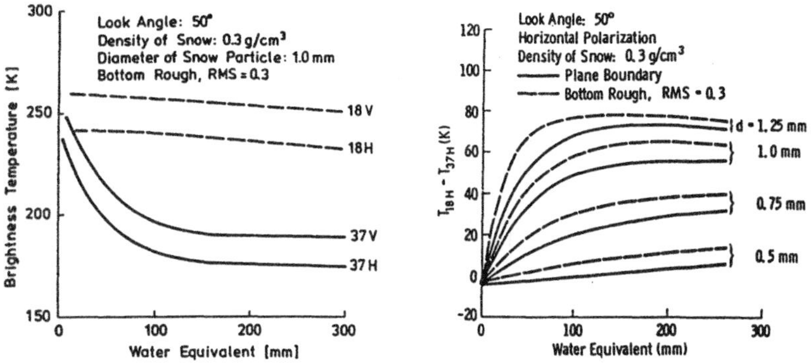

Figure 18.27: *(a) Theoretical brightness temperature of dry snow-covered terrain at 18 and 37 GHz as a function of snow water equivalent and (b) theoretical brightness temperature difference between 18 and 37 GHz, horizontal polarization as a function of snow grain size and water equivalent (Hallikainen, 1989).*

contributions only from the topmost snow layers reach the air and the brightness temperature does not depend on the thickness or the water equivalent of snow cover.

The effect of liquid water in snow is illustrated in Figure 18.28 for a volumetric content of 2% . The angular behavior of the brightness temperature of wet snow is the same as for dry snow both at 6 GHz and at 37 GHz. The difference in the absolute level between the dry and wet case is 10 to 20 K at 6 GHz, whereas it is about 70 K at 37 GHz. Figure 18.29 shows the variation of the brightness temperature as a function of the liquid water content. Even a wetness of 1% is enough to saturate the brightness temperature. The results indicate that it is difficult to determine the water equivalent of wet snow cover.

18.5.3.1.3 Effect of land-cover categories The spatial resolution of past and present spaceborne microwave radiometers is so modest that, for each pixel, the measured value is a contribution from several land-cover categories:

$$T_{Bsnow} = \sum_i f_i T_{Bsnow,i} \tag{18.29}$$

where $T_{Bsnow,i}$ and f_i are the brightness temperature contribution and the relative fraction of land-cover category i. Using land-use maps together with SMMR data, the microwave response to various land-cover categories for snow-covered terrain was determined by Hallikainen and Jolma (1986). The results show that boreal forests may reduce the response by up to 50% compared to farmland.

Recent helicopter-borne measurements with a multi-frequency microwave radiometer have provided more accurate information on the effect of land use categories to the brightness temperature of snow-covered terrain (Kurvonen and Hallikainen, 1995a). Figure 18.30 shows the effect at 35 and 94 GHz. The results suggest that, for dry snow, the effect is the larger the higher the biomass content is. Pine I and Pine II stand for stem volumes of 50 to 100 m^3/ha and 100 to 150 m^3/ha, respectively. For wet snow, the effect is very small, since the brightness temperature of snow-covered terrain is practically the same as that of forest canopies.

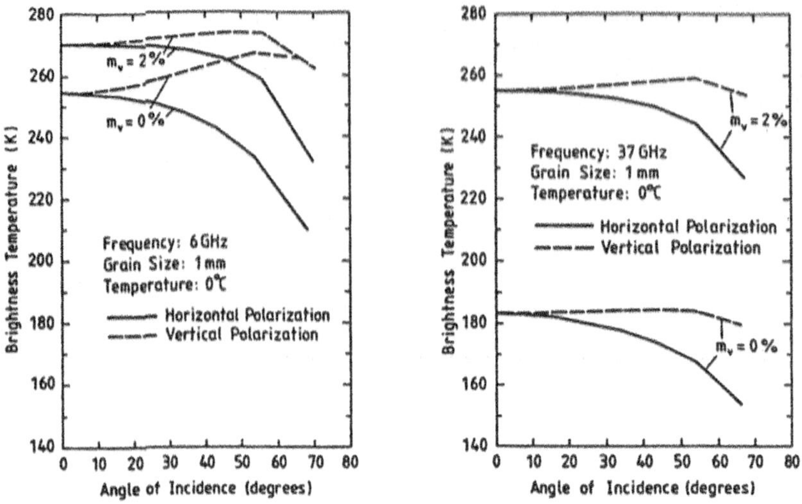

Figure 18.28: *Calculated brightness temperature of snow-covered terrain for water contents of 0% and 2% , at 6 and 37 GHz (Ulaby et al., 1986) . The thickness of snow cover is 50 cm.*

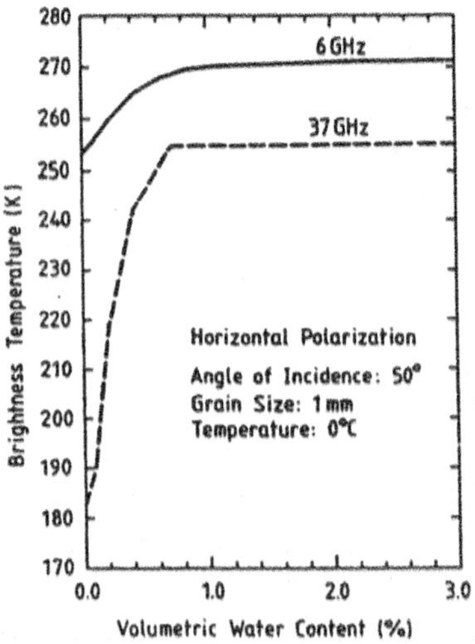

Figure 18.29: *Calculated brightness temperature of snow-covered terrain as a function of water content (Ulaby et al., 1986). The thickness of snow cover is 50 cm.*

18.5.3.2 Radar

The form of the expression for the backscattering coefficient of snow-covered terrain is similar to that of snow-covered sea ice. Contributions from the snow surface, snow volume, and snow-ground interface have to be accounted for:

$$\sigma^0_{snow} = \sigma^0_{as} + t^2_{si}\left(\sigma^0_s + \frac{1}{L^2_s}\sigma^0_{sg}\right) \tag{18.30}$$

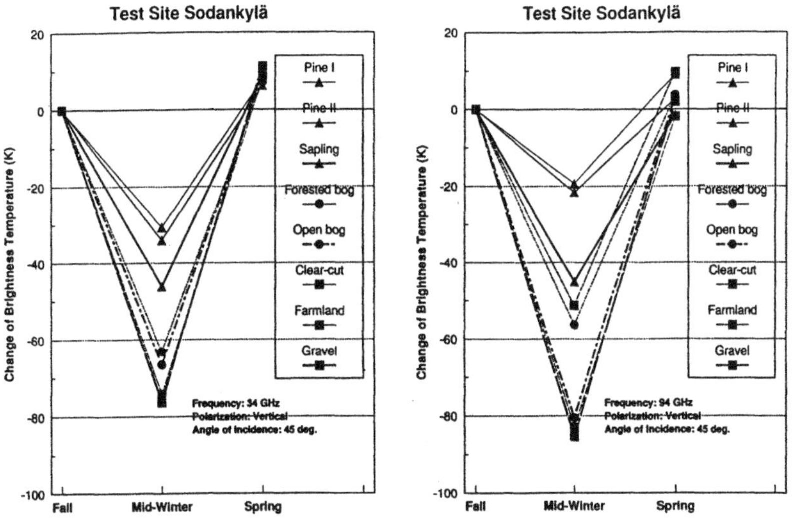

Figure 18.30: *Effect of snow situation and land-cover categories to the brightness temperature at 35 GHz (left) and 94 GHz (right) (Kurvonen and Hallikainen, 1995a). Results for snow-free ground are used as a reference.* **Fall**: *No snow, frozen ground, air temperature 269 K.* **Mid-winter**: *50 to 80 cm of dry snow, air temperature 266 K.* **Spring**: *25 to 80 cm of wet snow, air temperature 279 K.*

where σ^0_{as} = backscattering coefficient of air- snow interface
t_{si} = power transmission coefficient of air-snow interface
σ^0_s = backscattering coefficient of snow volume
L_s = total attenuation of snow layer
σ^0_{sg} = backscattering coefficient of snow-ground interface

Again it should be noted that all terms in Equation (18.30) are functions of the incidence angle.

18.5.3.2.1 Dry snow Usually, the real part of the permittivity is below 2 for dry snow. Hence, the backscattering contribution due to scattering by the snow surface, σ^0_{as}, may be ignored for angles away from nadir. The cloud model that assumes identical volume scatterers uniformly distributed throughout the volume can be applied to dry snow. Ignoring multiple scattering, the volume backscattering coefficient σ_{sv} (1/m) and the extinction coefficient κ_e (1/m) are given by

$$\sigma_v = N\sigma_b \tag{18.31}$$

$$\kappa_e = NQ_e \tag{18.32}$$

where N is the number of scattering particles per unit volume and σ_b and Q_e are the backscattering cross-section and extinction cross-section for a single particle, respectively. This approach leads to the following expression to the backscattering coefficient of dry snow volume:

$$\sigma^0_s = \frac{\sigma_{sv}\cos\theta'}{2\kappa_e}\left(1 - \frac{1}{L_s^2}\right) \tag{18.33}$$

In Equation (18.33) θ' is the propagation direction in snow. If the backscattering contribution from the snow surface can be neglected, σ_s^0 should be expected to be insensitive to the surface roughness. It has been shown that this is the case (Stiles and Ulaby, 1980).

18.5.3.2.2 Wet snow Stiles and Ulaby (1980) also noted that, for wet snow, σ_{as}^0 is sensitive to the snow surface roughness. In addition to effecting its overall level, it also affects the shape of its angular pattern. Hence, Equation (18.33) does not describe adequately the backscattering behavior of wet snow. This is demonstrated in Figure 18.31.

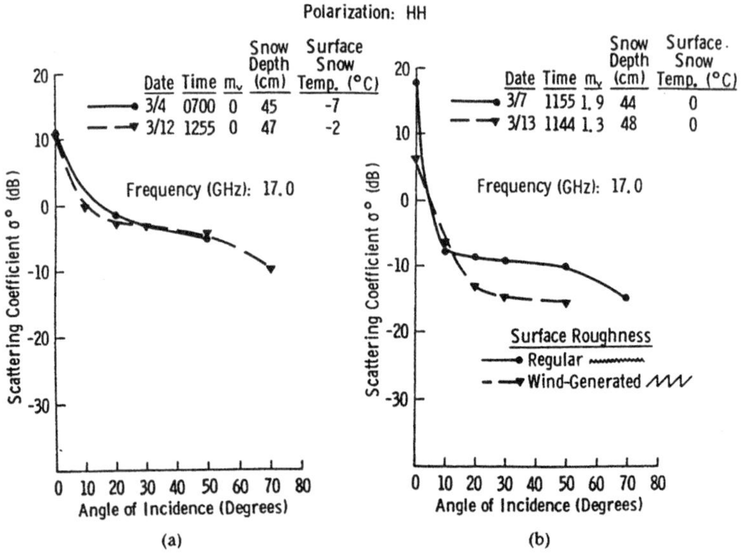

Figure 18.31: *Snow surface roughness appears to have no effect on the backscattering coefficient for (a) dry snow, but can change the level and shape of the angular pattern of that for (b) wet snow (Stiles and Ulaby, 1980).*

Due to the difference between the real parts of permittivity for dry and wet snow, the reflectivity of wet snow at normal incidence may be six times larger than that of dry snow. Hence, the snow surface must be treated as a rough surface. The presence of liquid water also causes a dramatic increase in the dielectric loss factor of the snow layer. This results in a large increase in the absorption coefficient, thereby reducing the relative importance of the ground contribution. Hence, the snow-ground surface can be assumed to be a plane boundary. Stiles and Ulaby (1980) found that the direct backscattering by the snow surface dominates for incidence angles below 20 degrees, and volume backscattering by snow is the dominant contribution at higher incidence angles.

18.5.3.2.3 Effect of forest canopies In the presence of forest canopies, Equation (18.30) has to be modified to account for the absorption and backscattering of biomass:

$$\sigma_{snow,forest}^0 = \sigma_f^0 + \frac{1}{L_f^2}\sigma_{snow}^0 \qquad (18.34)$$

where σ_f^0 = backscattering coefficient of forest canopy
L_f = total attenuation of forest canopy
σ_{snow}^0 = backscattering coefficient of snowpack without forest canopy.

The absorption of boreal forest canopies has been observed to be less than 0.5 dB at 24 GHz; hence it is even smaller in the frequency ranges used by present spaceborne radars (1 to 10 GHz). The forest canopy contribution increases with increasing biomass; however, the contribution from the snow-covered ground decreases almost by the same amount.

18.6 Future Aspects

ESA will launch its MIMR microwave radiometer after the year 2000. MIMR operates at six frequencies covering the 6.6 to 89 GHz range. Its spatial resolution will be better than that of previous spaceborne radiometers (4 to 5 km at 89 GHz). The recently launched ERS-2 guarantees availability of C-band SAR data for several years. The Canadian RADARSAT will be launched in September 1995. The SAR will provide C-band HH polarized data with a substantially wider swath (up to 500 km) than the ERS-1 SAR sensors (100 km).

18.7 References

Anderson DL (1960) The physical constants of sea ice. *Res Appl in Industry* 13: 310–318

Arcone SA, Gow AG, McGrew S (1986) Structure and dielectric properties at 4.8 and 9.5 GHz of saline ice. *J Geophys Res* 91: 14281–14303

Assur A (1960) Composition of sea ice and its tensile strength. U.S. Army Snow and Ice and Permafrost Res Establ, Wilmette, IL

Carsey FD, Barry RG, Weeks WF (1992) Introduction, Chapter 1 in: Carsey, F.D. (editor), Microwave remote sensing of sea ice, *American Geophysical Union Monograph* 68, Washington, DC, USA

Colbeck SC (1982) The geometry and permittivity of snow at high frequencies. *Journal of Applied Physics* 53: 4495–4500

Cumming W (1952) The dielectric properties of ice and snow at 3.2 cm. *Journal of Applied Physics* 23: 768–773

Debye P (1929) Polar Molecules, Dover, New York

Evans S (1965) Dielectric properties of ice and snow - a review. *Journal of Glaciology* 5: 773–792

Foster JL, Hall DK, Chang ATC, Rango A (1984) An overview of passive microwave snow research and results. *Reviews of Geophysics and Space Physics* 22, no. 2: 195–208

Foster JL, Chang ATC (1992) Snow cover, in: Guerney RJ, Foster JL, and Parkinson CL (editors), *Atlas of satellite observations related to global change*, University Press, Cambridge

Gloersen P, Nordberg W, Schmugge TJ, Wilheit TT, Campbell WJ (1973) Microwave signatures of first-year and multiyear sea ice. *Journal of Geophysical Research* 78: 3564–3572

Gray AL, Hawkins RK, Livingstone CE, Arsenault LD (1981) Seasonal effects on the microwave signatures of Beaufort Sea ice, Proc. 15th International Symposium on Remote Sensing of the Environment, pp. 239–257, Ann Arbor, Michigan, USA, May 11–15

Gray DM Male DH (1981) *Handbook of snow: Principles, processes, management and use*, Elmsford, NY, USA: Pergamon Press

Grenfell TC, Cavalieri DJ, Comiso JC, Drinkwater MR, Onstott RG, Rubinstein I, Steffen K, Winebrenner DP (1992) Considerations for microwave remote sensing of thin sea ice, Chapter 14 in: Carsey, F.D. (editor), Microwave remote sensing of sea ice, *AGU Geophysical Monograph Series* No. 68: 291–301, Washington DC, USA

Hallikainen MT (1977) Dielectric properties of NaCl ice at 16 GHz, Report S 107, Helsinki University of Technology, Radio Laboratory, 37 pp., Espoo, Finland

Hallikainen MT (1983) A new low-salinity sea ice model for UHF radiometry. *International Journal of Remote Sensing* 4: 655–681

Hallikainen MT, Jolma P (1986) Retrieval of snow water equivalent in Finland by satellite microwave radiometry. *IEEE Transactions on Geoscience and Remote Sensing* 24: 855–862

Hallikainen MT, Ulaby FT, Abdelrazik M (1986) Dielectric properties of snow in the 3 to 37 GHz range. *IEEE Trans Antennas Propagation* AP-34: 1329–1340

Hallikainen MT, Ulaby FT, Van Deventer TE (1987) Extinction behavior of dry snow in the 18 to 90 GHz range. *IEEE Trans. Geoscience and Remote Sensing* GE-25, No. 6: 737–745

Hallikainen MT, Toikka M, Hyyppä J (1988) Microwave dielectric properties of low-salinity sea ice, Proc. IGARSS'88 Symposium, pp. 419–420

Hallikainen MT (1989) Microwave radiometry of snow. *Advances in Space Research* 9 : (1)267–(1)275

Hallikainen MT, Winebrenner D (1992) The physical basis for sea ice remote sensing, Chapter 3 in: Carsey, F.D. (editor), Microwave remote sensing of sea ice, *AGU Geophysical Monograph Series* No. 68: 291–301, Washington DC, USA

Hoekstra P, Cappillino P (1971) Dielectric properties of sea and sodium chloride ice at UHF and microwave frequencies. *J Geophys Res* 76: 4922–4931

Hollinger JP (1971) Remote passive microwave sensing of the ocean surface. *Proc 7th International Symposium on Remote Sensing of the Environment*, pp. 1807–1817, Ann Arbor, Michigan, USA, 17-21 May.

Hyyppä J, Hallikainen MT (1991) Classification of low-salinity sea ice types by ranging scatterometer. *International Journal of Remote Sensing* 13: 2399–2413

Kim YS, Moore RK, Onstott RG, Gogineni S (1985) Towards identification of optimum radar parameters for sea ice monitoring. *Journal of Glaciology* 31(109): 214–219

Kuittinen R (1992) Remote sensing for hydrology: progress and prospects, World Meteorological Organization, Operational Hydrology Report No. 36, 62 pp., Geneva, Switzerland

Kurvonen L, Hallikainen MT (1995a) Classification of Baltic Sea ice types by airborne multifrequency microwave radiometer. *IEEE Transactions on Geoscience and Remote Sensing*, in press

Maykut GA (1986) The surface heat and mass flux, in: Untersteiner, N., *The geophysics of sea ice*, pp. 395–464, NATO ASI Series B: Physics Vol. 146, Plenum Press, New York

Muench RD, Martin S, Overland J (1987) Preface: Second marginal ice zone research collection. *Journal of Geophysical Research* 92, entire issue

Mätzler C, Wegmüller U (1987) Dielectric properties of fresh-water ice at microwave frequencies. *J Phys D: Appl Phys* 20: 1623–1630

Onstott RG, Moore RK, Gogineni S, Delker C (1982) Four years of low-altitude sea ice broad-band backscatter measurements. *IEEE Journal of Oceanic Engineering* OE-7: 44–50

Onstott RG, Grenfell TC, Mätzler C, Luther CA, Svendsen EA (1987) Evolution of microwave sea ice signatures during early summer and midsummer in the marginal ice zone. *Journal of Geophysical Research* 92(C7): 6825–6835

Onstott RG (1992) SAR and scatterometer signatures of sea ice, Chapter 5 in: Carsey, F.D. (editor), Microwave remote sensing of sea ice. *AGU Geophysical Monograph Series* No. 68, Washington DC, USA

Sackinger WM, Byrd RC (1972) Refection of millimeter waves from snow and sea ice, *IAEE Report* 7203, Institute of Arctic Environmental Engineering, University of Alaska, Fairbanks, Alaska, USA

Stiles WH, Ulaby FT (1980) Microwave remote sensing of snowpacks, RSL Technical Report 340-3, University of Kansas, Lawrence, KS, USA

Stogryn A, Desargeant GJ (1985) The dielectric properties of brine in sea ice at microwave frequencies. *IEEE Trans Antennas Propagation* AP-33: 523–532

Tiuri M, Sihvola A, Nyfors E, Hallikainen MT (1984) The complex dielectric constant of snow at microwave frequencies. *IEEE Journal of Oceanic Engineering* OE-9:377–382

Tucker WB, III, Perovich DK, Gow AJ, Weeks WF, Drinkwater MR (1992) Physical properties of sea ice relevant to remote sensing, Chapter 2 in: Carsey, F.D. (editor), Microwave remote sensing of sea ice. *AGU Geophysical Monograph Series* No. 68, Washington DC, USA

Ulaby FT, Moore RK, Fung AK (1986) *Microwave Remote Sensing - Active and Passive*, Vol. III: From Theory to Applications, Artech House, Inc., Dedham, MA, USA

Ulander L, Johansson R, Askne J (1991) C-band radar backscatter of Baltic ice: Theoretical predictions compared with calibrated SAR measurements. *International Journal of Remote Sensing*

Ulander L (1991) Radar remote sensing of sea ice: Measurements and theory, Technical report 212, Chalmers University of Technology, Göteborg, Sweden

Ulander L, Carlström A (1991) Radar backscatter signatures of Baltic ice, *Proc. 1991 IEEE IGARSS Symposium*, Vol. 3: 1215–1218, Espoo, Finland, 3–6 June 1991

Unal CMH, Snoeij P, Swart PJF (1990) The polarization dependent relation between radar backscatter from the ocean surface and surface wind vector at frequencies between 1 and 18 GHz, Proc. IEEE IGARSS'90 Symposium, pp. 2165–2167, College Park, Maryland, USA, 20–24 May 1990

Vant MR, Gray R, Ramseier R, Makios V (1974) Dielectric properties of fresh and sea ice at 10 and 35 Ghz. *Journal of Applied Physics* 45: 4712-4717

Vant MR (1976) A combined empirical and theoretical study of the dielectric properties of sea ice over the frequency range 100 MHz to 40 GHz, Technical Report, Carleton University, Ontario, Ottawa, Canada

Vant MR, Ramseier RO, Makios V (1978) The complex-dielectric constant of sea ice at frequencies in the range 0.1–40 GHz. *Journal of Applied Physics* 49: 1264–1280

Chapter 19

Retrieval of Sea Ice and Snow Parameters from Microwave Radiometer and Radar Data

Martti Hallikainen
Laboratory of Space Technology
Helsinki University of Technology
Otakaari 5 A, 02150 Espoo
Finland

19.1 Available Data Sets

In microwave radiometry the U.S. Nimbus-7 SMMR instrument (frequency range 6.6 to 37 GHz) provided data in 1978-1986 and these data have been used for developing snow and sea ice algorithms. Presently microwave radiometer data are available from the U.S. DMSP SSM/I sensors (frequency range 19 to 85 GHz). Airborne microwave radiometers include the recently constructed Helsinki University of Technology system that operates in the 6.6 to 94 GHz range. Ground-based sensors are an important tool in algorithm development.

In radar remote sensing of sea ice and snow, ERS-1 SAR (5.3 GHz, VV polarization) data are widely used. Additional data sets include NASA/JPL airborne SAR system (0.4 , 1.2, 5.3 GHz, fully polarimetric) data and Japanese JERS-1 SAR data. The recently completed EMISAR airborne sensor (5.3 and 1.2 GHz, fully polarimetric) has produced its first sea ice and snow data sets in the ESA EMAC campaign in Finland. Ground-based sensors provide useful data.

19.2 Sea Ice

19.2.1 Retrieval of ice concentration and ice type

19.2.1.1 Radiometer

19.2.1.1.1 Single-channel radiometry Retrieval of ice concentration by microwave radiometry is based on the contrast between the brightness temperatures of sea ice and open water. Single-channel radiometry can be used to determine ice concentration in areas, where

NATO ASI Series, Vol. I 45
Radiation and Water in the Climate System:
Remote Measurements
Edited by Ehrhard Raschke

a single ice type dominates. It cannot be used successfully to determine ice type. For an area of sea ice and water with brightness temperatures of T_{Bi} and T_{Bw}, respectively, and with atmospheric effects ignored, the observed brightness temperature is

$$T_B = CT_{Bi} + (1 - C)T_{Bw} \tag{19.1}$$

where C denotes the sea-ice concentration, and the fractional area covered by open water, therefore, is (1-C). If brightness temperatures T_{Bi} and T_{Bw} are known or can be estimated with sufficient accuracy, sea ice concentration can be determined from Equation (19.2):

$$C = \frac{T_B - T_{Bw}}{T_{Bi} - T_{Bw}} \tag{19.2}$$

Equation (19.2) can be used in areas where a single ice type dominates. In the Baltic Sea, only first-year ice exists. SMMR data from the Nimbus-7 satellite were used to map the ice extent in the Gulf of Bothnia (northern Baltic Sea) using 18 GHz and 37 GHz horizontally polarized data. However, in spring conditions involving dry snow on top of the ice layer in north and wet snow in south, reasonably good accuracy was obtained only by assuming that T_{Bi} is equal to the brightness temperature of snow-covered terrain in coastal areas, Figure 19.1b. Attempts to estimate T_{Bi} from local sea ice data did not produce a good estimate for the ice concentration, Figure 19.1a. For Antarctic areas, the sea-ice concentration can be estimated with an accuracy of about 15% using Equation (19.2) together with estimated ice physical temperatures (Comiso and Zwally, 1982; Zwally et al., 1983). The physical temperature of sea ice is usually somewhat warmer than the air temperature, except near the melting point (Steffen et al., 1992).

In areas, where open water, first-year ice and multiyear ice are present, the observed brightness temperature is

$$T_B = C_w T_{Bw} + C_{FY} T_{BFY} + C_{MY} T_{BMY} \tag{19.3}$$

where C_w, C_{FY} and C_{MY} are the fractions of water, first-year ice (FY) and multiyear ice (MY); they add to unity. The values of both first-year ice and multiyear ice fractions cannot be obtained from the brightness temperatures measured by a single-channel sensor.

19.2.1.1.2 Multichannel radiometry Several algorithms have been developed for estimating the sea-ice concentration and multiyear ice fraction from multifrequency, multipolarization microwave radiometer data (Steffen et al., 1992). As an example of these, the NASA Team Algorithm is presented here. It is originally based on the algorithm presented by Cavalieri et al. (1984).The algorithm employs two concepts defined below for the 18 GHz vertically (T_{18V}) and horizontally (T_{18H}) and 37 GHz vertically (T_{37V}) polarized brightness temperatures, namely the polarization ratio (PR) and the spectral gradient ratio (GR):

$$PR(18) = \frac{T_{18V} - T_{18H}}{T_{18V} + T_{18H}} \tag{19.4}$$

$$GR(37V/18V) = \frac{T_{37V} - T_{18V}}{T_{37V} + T_{18V}} \tag{19.5}$$

The main advantage of using ratios of radiances is that they reduce the dependence of ice-temperature variations. Equations (19.6) to (19.8) are used together to solve the first-year and multiyear ice concentrations:

$$C_{FY} = \frac{F_0 + F_1 PR + F_2 GR + F_3 (PR)(GR)}{D} \tag{19.6}$$

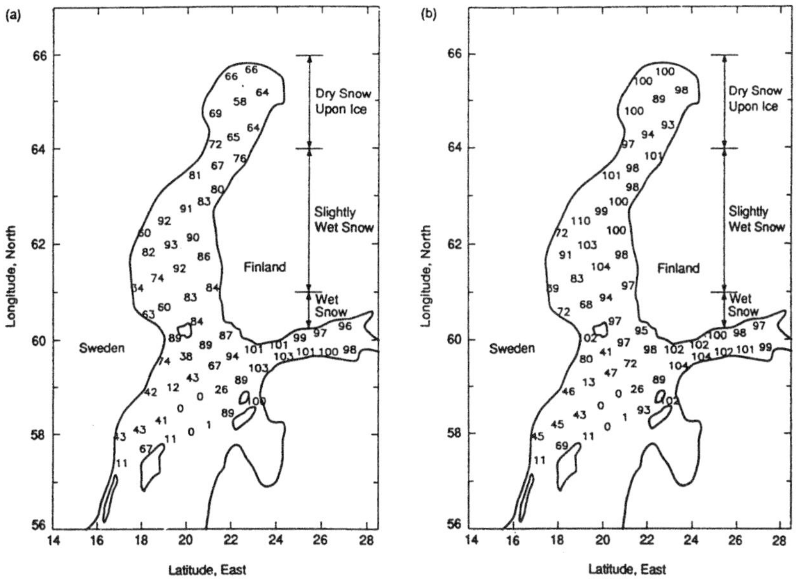

Figure 19.1: *Satellite-derived sea ice concentration in the Baltic Sea for March 1, 1979, using SMMR 37-GHz data (Hallikainen, 1992). The brightness temperature of water is assumed to be 140 K. The whole Gulf of Bothnia (northern Baltic Sea) and Gulf of Finland (eastern Baltic Sea) are covered by ice. a) The brightness temperature of sea ice is assumed to be 242 K. b) The brightness temperature of ice is assumed to be equal to that of coastal snow-covered terrain at the same latitude in Finland.*

$$C_{MY} = \frac{M_0 + M_1 PR + M_2 GR + M_3 (PR)(GR)}{D} \tag{19.7}$$

where

$$D = D_0 + D_1 PR + D_2 GR + D_3 (PR)(GR). \tag{19.8}$$

The total ice concentration is the sum of the first-year and multiyear ice concentrations,

$$C = C_{FY} + C_{MY} \tag{19.9}$$

The coefficients of Equations (19.6) to (19.8) are functions of the brightness temperatures T_{18V}, T_{18H}, and T_{37V} measured for open water, first-year ice and multiyear ice. These values are known as tie points and they represent the normal behavior of each category. They also include the atmospheric effects. The results for the Northern Hemisphere are (Gloersen et al., 1992):

$$C = \frac{2091.92 - 5054PR - 8744GR - 1600(PR)(GR)}{1648.2 + 7735.6PR - 4112.2GR - 10200(PR)(GR)} \tag{19.10}$$

$$C_{MY} = \frac{-690.38 + 14990PR - 27579GR - 43260(PR)(GR)}{1648.2 + 7735.6PR - 4112.2GR - 10200(PR)(GR)} \tag{19.11}$$

Multiyear ice calculations require the surface of multiyear ice to be dry. Consequently, multiyear ice concentration values cannot be obtained during summer. In the Southern Hemisphere, the amount of ice experiencing summer melt is limited. The same polarization ratio/spectral gradient ratio approach is used, but the coefficients in the two equations are slightly different from those in Equations (19.10) and (19.11). In the absence of reliable in situ data, the two ice types in the Southern hemisphere are called Type A and Type B ice (Parkinson and Gloersen, 1992). Figure 19.2 depicts the ice concentration nomograms for the Northern and Southern hemisphere.

If hemispherically chosen brightness temperatures are used as tie points, the variations of T_{Bw} (wind speed, foam, atmosphere), T_{BFY} (surface roughness, temperature, snow cover, atmosphere), and T_{BMY} (surface roughness, temperature, snow cover, atmosphere) over time and space are ignored. To account for these factors, ice concentrations based on local tie points determined in the area of interest can be derived. Local-ocean tie point values are usually higher than the hemispheric ones, due to e.g. wind speed. Differences as high as 18 K between the hemispheric and local tie points for water have been observed (Steffen and Schweiger, 1991).

Other ice concentration algorithms include those by Swift et al. (1985), Svendsen et al. (1992), Comiso (1986). The algorithm discussed above has been widely applied. Accuracy estimates for the calculated sea ice concentrations are 5 to 9% for the Nimbus-7 SMMR data (Cavalieri et al., 1984; Steffen and Maslanik, 1988).

The algorithm used for calculating sea ice concentrations from SSM/I data has the same formulation as the SMMR algorithm presented above; only the coefficients are different due to slightly different channel frequencies.

A basic problem in developing ice type discrimination algorithms for the polar regions is poor availability of in situ data. Validation of microwave observations by Landsat optical sensors can be done only in clear weather conditions, even this may give a bias to the results.

In the Baltic Sea winter shipping traffic is handicapped by deformed ice (ice ridges, hummocks). In the winter of 1992 classification of ice types and water with a microwave radiometer was investigated. Especially, measurements were made on compact pack ice and close pack ice that are difficult for ship navigation. Compact pack ice is formed from crushed fast or new ice, which has been packed and refrozen. This ice type is common in the open sea where waves and wind crush existing fast or new ice. The ice surface is rough and, hence, existing snow is spatially unevenly distributed. Close pack ice is formed from crushed ice and slush and it is not solid. This ice type exists between a solid ice field and open sea. It absorbs waves which keep it fractured and wet, without any snow cover.

The microwave radiometer system operates at 24, 35, 48, and 94 GHz, vertical polarization, incidence angle 45 degrees. Measurements were made under three different ice/snow conditions: dry, wet, and refrozen. Figure 19.3 shows the observed signatures for three ice categories and open water.

In the presence of dry snow cover on top of the ice layer, the brightness temperature decreases with increasing frequency. This can be used as a rough indicator of the ice type. Open water can be discriminated easily in all measurements.

The results suggest that a high-frequency receiver (90 GHz) should be combined with a lower-frequency (35 GHz) receiver for ice type classification. The high frequency is sensitive to dry and refrozen snow; hence, it is useful for discrimination of old first-year ice (compact pack ice and fast ice) which is often covered by snow. The brightness temperature of water may, under suitable wind conditions, be nearly the same as that of old snow-covered ice. Using the lower frequency, water can be discriminated under all weather conditions. A limiting factor in the Baltic Sea is the spatial resolution; information with resolution worse than 10 to 20 km is not useful.

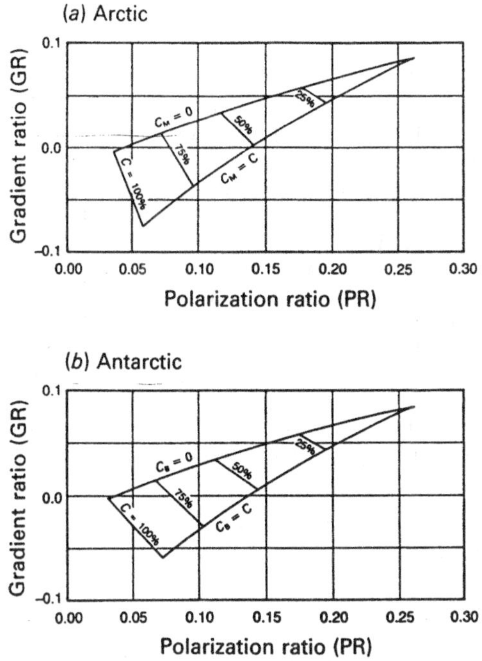

Figure 19.2: *Relationship between (a) the gradient ratio, polarization ratio, sea ice concentration, C, and multiyear ice concentration, C_{MY}, in the Arctic, based on Equations (19.9) and (19.10) and (b) the gradient ratio, polarization ratio, sea ice concentration, C, and Type B ice concentration, C_B, in the Antarctic (Parkinson and Gloersen, 1992).*

19.2.1.2 Radar

The capability of ERS-1 SAR to discriminate the three main categories, open water, first- year ice and multiyear ice is depicted in Figure 19.4. Data from the Beaufort Sea in November 1991 are used. The average value of the backscattering coefficient is −22 dB for open water/new ice, −14 dB for first-year ice, and −9 dB for multiyear ice.

The reason for using a single class for open water and new ice is the similarity of these surfaces in calm weather conditions: both targets have a specular surface. The good results shown in Figure 19.4 may be partly due to using data from a single image. The effect of weather conditions (especially wind) to the discrimination capability may be substantial, as demonstrated by the use of ERS-1 SAR data in the Baltic Sea by various groups. It turned out to be very difficult to discriminate open water from sea ice.

The capability of radar to discriminate sea ice categories at 0.44, 1.25, and 5.3 GHz has been reported by Rignot and Drinkwater (1992). Results obtained with the NASA/JPL airborne SAR are shown in Figure 19.5 and Tables 19.1 and 19.2.

C-band, VV polarization (C/VV) alone achieves 68% overall relative classification accuracy, as multiyear ice and compressed first-year ice are not separated. The corresponding figure for C/HH is 71%. At L-band, the results for HH and HH+VV are 75% and 84%, respectively. Performance at P-band is lower with poor separation of multiyear and first-year ice. Combining two frequencies at a single polarization results in an improvement of 10 to 20% in classification

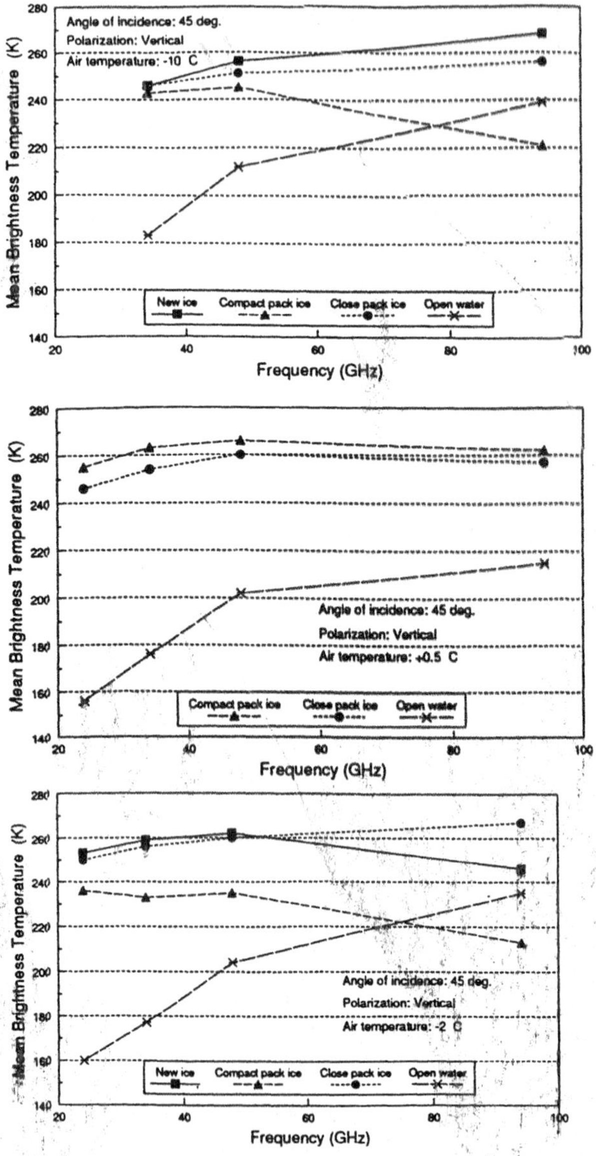

Figure 19.3: *Mean brightness temperatures of ice types and open water measured in the Baltic Sea using a helicopter-borne radiometer (Kurvonen and Hallikainen, 1995a). (a) Dry ice/snow conditions (air temperature* $-10°C$*), 13 February 1992 (b) Wet ice/snow conditions (air temperature* $+0.5\ °C$*), 13 March 1992 (c) Refrozen ice/snow conditions (air temperature* $-2°C$*), 17 March 1992.*

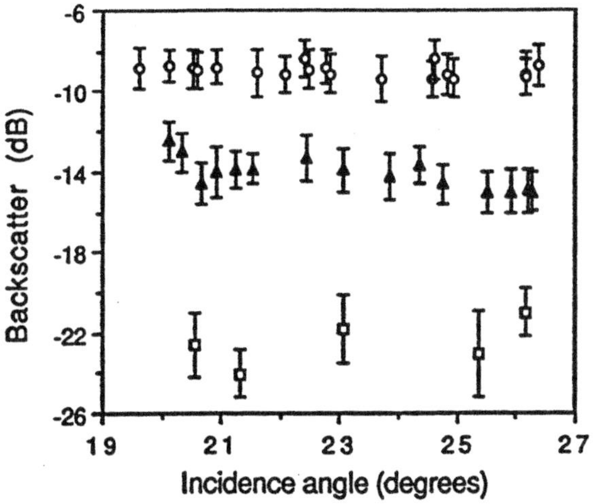

Figure 19.4: *Mean backscatter values with one standard deviation bars plotted versus ERS-1 SAR incidence angle from one image for multiyear ice (circles), first-year ice (triangles) and open water/new ice (squares) (Cunningham et al., 1992).*

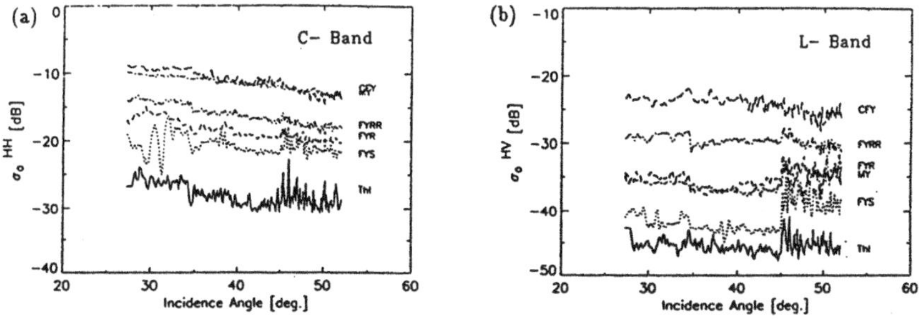

Figure 19.5: *Experimental results from NASA/JPL airborne SAR measurements of Beaufort sea ice in November 1988 as a function of incidence angle at (a) 5.3 GHz, HH polarization and (b) 1.25 GHz, VH polarization (Rignot and Drinkwater, 1992). ThI = thin ice FYS = smooth first-year ice FYR = rough first-year ice CFY = compressed first-year ice FYRR = first-year rubble and ridges MY = multiyear ice.*

accuracy. A 10% improvement results from the use of two polarizations instead of a single polarization.

In the Baltic Sea, helicopter-borne scatterometer studies using an incidence angle of 23 degrees show that the backscattering coefficient depends on sea ice type more at 9.8 GHz than at 5.4 GHz (Hallikainen, 1995). Based on these values, broad categories of open water/new ice, level ice and deformed ice can be discriminated.

	Ice Type	σ_{hh}	σ_{hv}	σ_{vv}	σ_{hhvv^*}	ϕ_{hhvv^*}	$\lvert\rho_{hhvv^*}\rvert$	E	vv/hh	hv/hh
P-	ThI	-38.7	-47.6	-34.7	-39.8	0.0	0.49	0.57	4.0	-8.9
	FYS	-30.4	-42.4	-26.5	-29.8	0.1	0.73	0.39	3.9	-12.0
	FYR	-24.8	-36.5	-21.7	-24.6	0.1	0.74	0.41	3.0	-11.7
	MY	-28.0	-42.0	-24.9	-27.2	-0.1	0.85	0.28	3.1	-14.0
	FYRR	-20.5	-31.6	-18.1	-20.6	0.0	0.74	0.44	2.4	-11.1
	CFY	-19.2	-29.4	-16.9	-19.3	-0.0	0.75	0.45	2.2	-10.2
L-	ThI	-35.9	-44.7	-30.6	-35.9	0.0	0.55	0.48	5.4	-8.8
	FYS	-26.2	-37.8	-22.5	-25.6	0.1	0.74	0.39	3.6	-11.6
	FYR	-23.2	-33.9	-19.6	-22.7	0.1	0.74	0.41	3.6	-10.7
	MY	-23.0	-34.6	-18.8	-21.6	-0.0	0.85	0.28	4.2	-11.5
	FYRR	-20.1	-30.0	-16.7	-19.7	0.0	0.74	0.43	3.4	-9.8
	CFY	-17.7	-25.4	-14.2	-17.5	-0.0	0.71	0.50	3.5	-7.8
C-	ThI	-27.9	-35.8	-26.0	-30.8	-0.2	0.41	0.71	1.9	-7.9
	FYS	-20.8	-33.5	-20.6	-22.4	-0.2	0.68	0.50	0.2	-12.6
	FYR	-19.5	-31.7	-19.1	-20.8	-0.2	0.71	0.49	0.4	-12.1
	MY	-12.8	-22.2	-11.8	-14.1	-0.0	0.67	0.56	1.0	-9.3
	FYRR	-17.6	-28.9	-17.2	-19.0	-0.2	0.70	0.51	0.5	-11.2
	CFY	-12.9	-21.2	-12.1	-14.9	-0.1	0.58	0.65	0.8	-8.3

Table 19.1: *Multifrequency multipolarization summary of SAR image class characteristics for various sea ice types at 45 degree incidence angle (see Figure 19.5 for category abbreviations) (Rignot and Drinkwater, 1992).*

a

Ice Type	ThI	FYS	FYR	MY	CFY
ThI	96	4	0	0	0
FYS	4	72	19	0	0
FYR	0	24	79	1	15
MY	0	0	1	55	32
CFY	0	0	0	43	52

b

Ice Type	ThI	FYS	FYR	MY	CFY
ThI	98	5	0	0	0
FYS	2	70	15	0	0
FYR	0	25	84	1	12
MY	0	0	1	77	40
CFY	0	0	0	22	47

c

Ice Type	ThI	FYS	FYR	MY	CFY
ThI	98	6	0	0	0
FYS	2	74	12	0	0
FYR	0	20	87	1	9
MY	0	0	1	81	44
CFY	0	0	0	18	47

d

Ice Type	ThI	FYS	FYR	MY	CFY
ThI	98	13	0	0	0
FYS	2	83	8	1	0
FYR	0	3	65	32	31
MY	0	1	21	68	4
CFY	0	0	6	0	64

e

Ice Type	ThI	FYS	FYR	MY	CFY
ThI	96	12	0	0	0
FYS	4	86	6	0	0
FYR	0	1	87	7	34
MY	0	1	3	93	1
CFY	0	0	4	0	65

f

Ice Type	ThI	FYS	FYR	MY	CFY
ThI	95	14	0	0	0
FYS	5	84	5	1	0
FYR	0	1	87	11	32
MY	0	2	5	88	0
CFY	0	0	3	0	68

g

Ice Type	ThI	FYS	FYR	MY	CFY
ThI	69	27	4	0	1
FYS	11	34	14	2	1
FYR	17	33	65	13	40
MY	1	3	2	82	17
CFY	2	2	14	3	41

h

Ice Type	ThI	FYS	FYR	MY	CFY
ThI	99	11	0	0	0
FYS	1	84	7	0	0
FYR	0	5	91	0	12
MY	0	0	1	98	10
CFY	0	0	1	1	78

i

Ice Type	ThI	FYS	FYR	MY	CFY
ThI	98	5	0	0	0
FYS	2	86	9	0	0
FYR	0	9	90	1	13
MY	0	0	1	98	2
CFY	0	0	1	1	85

Table 19.2: *Confusion matrices for (a) C/HH, (b) C/HH+VV, (c) C/full polarimetry, (d) L/HH, (e) L/HH+VV, (f) L/full polarimetry, (g) P/full polarimetry, (h) L/HH+C/VV, (i) L/HV+C/HH.*

Additional measurements performed using various incidence angles show that an angle of 45 degrees seems to provide a much better classification capability than 23 degrees. Figure 19.6 shows a comparison between the results obtained with an ERS-1 SAR-type radar configuration (5.4 GHz, incidence angle 23 degrees) and the optimal configuration (9.8 GHz, incidence angle 45 degrees).

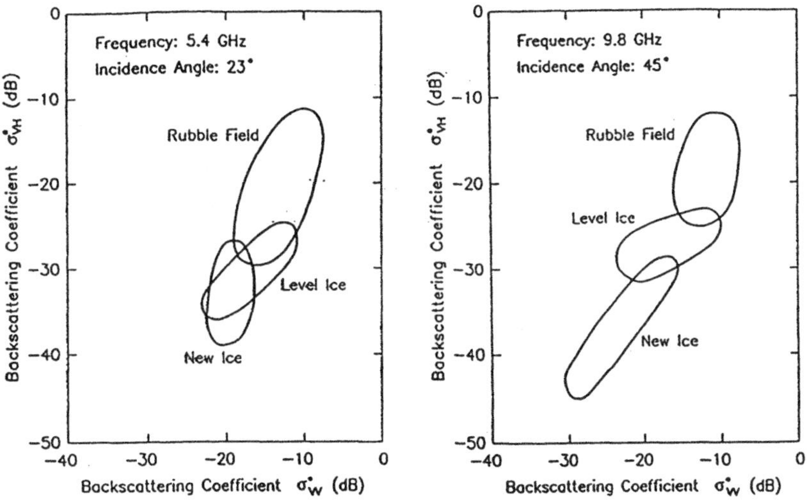

Figure 19.6: *Classification of three main sea ice types in the Baltic Sea using two combinations of frequency and incidence angle, with VV and VH polarization results shown in each case (Hallikainen and Toikka, 1992).*

19.2.2 Ice temperature

19.2.2.1 Microwave radiometry

The algorithm developed by the NASA Team for retrieving the physical temperature of the radiating portion of the ice from the SMMR 6.6 GHz vertically polarized brightness temperatures is:

$$T_i = \frac{T_{6.6V} - (1 - C)T_{w,6.6V}}{e_{i,6.6V}C} \tag{19.12}$$

where $T_{6.6V}$ is the observed brightness temperature and $T_{w,6.6V}$ is the brightness temperature of open water. Sea-ice concentration is denoted by C and $e_{i,6.6V}$ is the emissivity of ice. For ice concentrations less than 80%, T_i is set to 271 K. Comiso (1983) determined a value of 0.96 for the ice emissivity. The uncertainty in T_i was determined to be 9.7 K for 50% ice concentration to 6.6 K at 100% concentration (Steffen et al., 1992).

The NASA Team SMMR algorithm has been modified for SSM/I data. This modification involves estimates of ice emissivity as a function of frequency and ice type, and a model of the Arctic atmosphere as a function of surface temperature.

19.2.2.2 Radar

Due to the sensitivity of radar backscatter to surface roughness rather than surface temperature, determination of the ice temperature is difficult with radar.

19.2.3 Monitoring of ice movement

19.2.3.1 Microwave radiometry

The spatial resolution of present spaceborne microwave radiometers is modest, so drifting of only very large ice floes can be done.

19.2.3.2 Radar

The main advantages of spaceborne SAR in ice motion monitoring are its fine spatial resolution and all-weather, day/night operational capability. The most common approach to automatic ice tracking is cross-correlation between two images. A common feature is likely when a small region ($n \times n$ pixels) in one image has a peak correlation with a small region in another image. The cross-correlation between the brightness a_{ij} of the first image and the brightness b_{kl} of the second image is (Holt et al., 1992)

$$R(i,j:k,l) = \sum_{p=-t}^{+t} \sum_{q=-t}^{+t} \frac{(a_{(i+p,j+q)} - \overline{a})(b_{(k+p,l+q)} - \overline{b})}{n^2 \sigma_a \sigma_b} \tag{19.13}$$

where t is $(n\text{-}1)/2$ and $\overline{a}\sigma_a$ are the mean and the standard deviations, respectively, of a_{ij} over the $n \times n$ array. The original full-resolution images are first degraded by averaging pixels and then some initial common features are found. More highly correlated features are then sought as the image resolution is increased through successive stages, until the displacement grid is filled in as completely as possible. Zhang (1991) used a global motion predictor model initialized with a small set of visually identified features to guide the correlation matching.

19.3 Snow

19.3.1 Snow extent

19.3.1.1 Dry snow

19.3.1.1.1 Radiometer For non-scattering media containing liquid water, the brightness temperature decreases with increasing frequency, due to the spectral behavior of the permittivity of water. Dry snow is one of few exceptions in this regard and this exceptional behavior is used to discriminate bare land from snow-covered terrain.

At frequencies above 15 GHz (snow grain size 1 mm), the brightness temperature of dry-snow covered terrain decreases with increasing frequency and increasing snow thickness. Consequently, high microwave frequencies should be used to discriminate dry-snow covered terrain from snow-free terrain.

The following expression is used to discriminate dry snow:

$$T_{18V} - T_{37V} \geq 5K \tag{19.14}$$

A similar expression was used by (Künzi et al., 1982). Recent experiments with SSM/I data suggest that the 85 GHz channel is very sensitive to even shallow snow: however, it is sensitive to atmospheric effects as well.

19.3.1.1.2 Radar Discrimination of dry snow from snow-free ground is difficult, because the dielectric properties of dry snow do not provide a strong contrast with those of air. Hence, the backscattering coefficients of dry snow-covered terrain and snow-free terrain are close to each other. Studies carried out in Sodankylä, northern Finland using a helicopter-borne radar operating at 5.4 and 9.8 GHz, 4 linear polarizations, incidence angle 23 degrees (Hallikainen et al., 1994), suggest that the best radar channel to discriminate dry snow from snow-free ground is 9.8 GHz, VV polarization. In the case of wet snow-free ground the contrast is 2.5 to 5 dB, depending on land use category. For frozen snow-free ground the contrast is about 6 dB, Figure 19.7.

Ground-based studies in Switzerland show practically the same values for dry snow covered terrain and snow-free terrain (Mätzler and Schanda, 1984).

19.3.1.2 Wet snow

19.3.1.2.1 Radiometer Discrimination of wet snow from snow-free frozen terrain is not straightforward, due to the similar dielectric properties of wet snow and frozen soil (Hallikainen et al., 1986). This is demonstrated in Figure 19.8.

Discrimination of wet snow from wet soil can be done best by using a low microwave frequency; then the permittivity of wet snow is much smaller than that of wet soil. However, forest canopies may mask the brightness temperature difference.

19.3.1.2.2 Radar Using ERS-1 SAR, the backscattering coefficient has been observed to be about 3 dB smaller for wet snow than for other snow/soil categories, Figure 19.9 (Koskinen et al., 1994). This difference is rather independent of the biomass of boreal forests and can be used to discriminate wet snow from other snow/soil categories. By averaging ERS-1 SAR data to eliminate speckle, classification accuracies of the order of 90% were achieved.

The physical basis for the discrimination is the following: wet snow absorbs incident radar power on its way to the ground surface and again on its way into air. Absorption by dry snow is negligible at presently used radar frequencies.

Similar results have been obtained by Rott and Nagler (1994) for Alpine test sites.

19.3.2 Snow water equivalent

Since wet snow absorbs microwave power, determination of the water equivalent of snow cover seems to be possible only for dry snow.

19.3.2.1 Microwave radiometry

Most algorithms used to derive the water equivalent of dry snow cover rely on observations made at frequencies 18 and 37 GHz. Experience about the SSM/I 85 GHz data is still limited, although it seems to be useful. The algorithms are based on linear correlations observed between the water equivalent and satellite-derived quantities.

The quantities are used in the algorithms include the spectral and polarization difference (SPD) (Aschacher, 1989), spectral and temporal difference (STD) (Hallikainen and Jolma, 1992), spectral gradient (SG) (Künzi et al., 1982) and spectral difference (SD) (Chang et al., 1987):

$$SPD = (T_{18V} - T_{37V}) - (T_{18V} - T_{18H}) \tag{19.15}$$

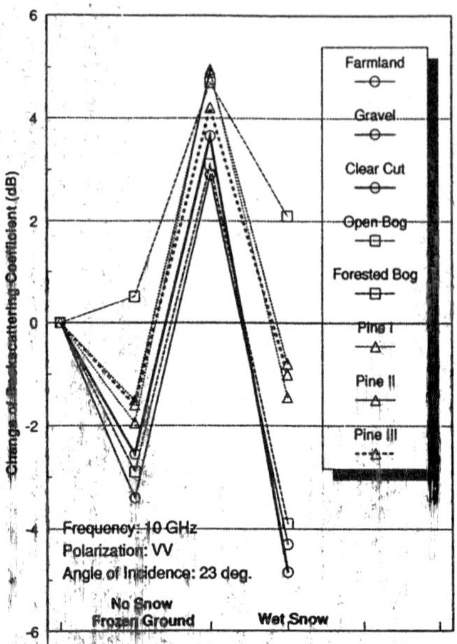

Figure 19.7: *Mean values of the backscattering coefficient for various snow situations and land use categories at 9.8 GHz, VV polarization, incidence angle 23 degrees (Hallikainen et al., 1994).*

$$STD = (T_{18V} - T_{37V})_{W_{eq}>0} - (T_{18V} - T_{37V})_{W_{eq}=0} \qquad (19.16)$$

$$SG = \frac{T_{18H} - T_{37H}}{18 - 37} \qquad (19.17)$$

$$SD = T_{18H} - T_{37H} \qquad (19.18)$$

STD takes into account the effect of land use category by comparing the results for snow-covered terrain with those for the same area before the first snowfall.

Using the defined quantities, the water equivalent (or the snow depth, d_s) can be calculated from the following equations (test site locations and amount of data pointed out):

$$W_{eq} = 2.02SPD - 7.42 \quad \text{(mm); Hungary, Germany: 4 years} \qquad (19.19)$$

$$W_{eq} = 10.1STD - 98 \quad \text{(mm); Southern Finland / forested: 4 years} \qquad (19.20)$$

$$W_{eq} = 8.7STD - 108.7 \quad \text{(mm); Northern Finland / sparse forest: 4 years} \qquad (19.21)$$

$$W_{eq} = -2.7\frac{SG - 0.085}{0.036} \quad \text{(mm); Finland, Russia, Canada: 1 year} \qquad (19.22)$$

$$d_s = 1.59SD \quad \text{(cm);} \quad \rho_s = 0.3\text{g/cm}^3\text{, grain size 0.3 mm} \qquad (19.23)$$

Equation (19.22) assumes the snow density to be $0.27 g/cm^3$ The three last equations can be directly compared by assuming in Equation (19.23) the snow density to be 0.27 g/cm^3 instead of 0.3 g/cm^3 . The slope of the water equivalent vs. $T_{18} - T_{37}$, horizontal or vertical polarization) is

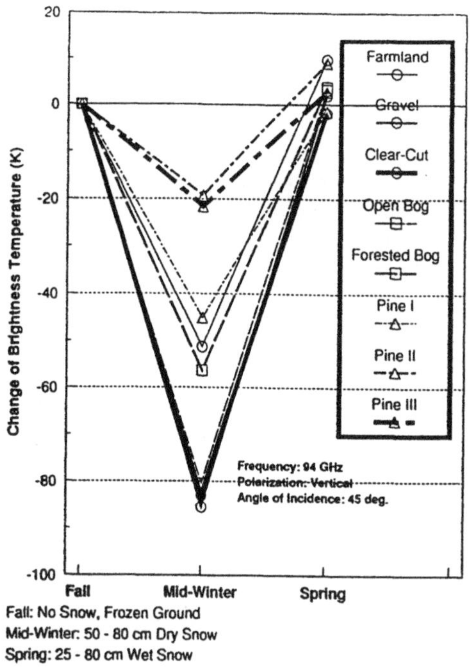

Figure 19.8: *Experimental behavior of 94-GHz brightness temperature for various land-use categories under different snow conditions (Hallikainen et al., 1994).*

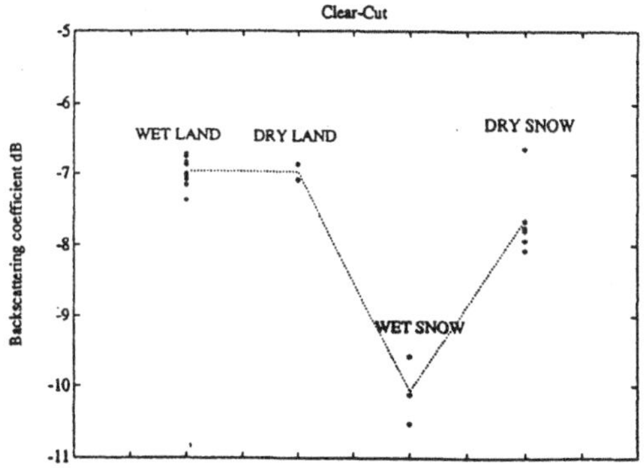

Figure 19.9: *ERS-1 SAR-derived backscattering coefficient for various snow/soil surfaces in the Sodankylä test site, northern Finland (Koskinen et al., 1994). Each point represents the average backscattering coefficient for clear-cut areas in an ERS-1 SAR image.*

S [mm/K]	= 10.1	;	*STD* for southern Finland
	= 8.7	;	*STD* for northern Finland
	= 3.84	;	*SG* for Finland, Canada, Russia

$$= 4.29 \quad ; \quad SD$$

The comparison suggests that local variations (land-cover category, forest canopies, terrain relief) need to be accounted for when water equivalent algorithms are used. A digital land use map and a digital elevation model are relevant tools for future algorithm development. Another method is to use "anchor stations" to correct for errors produced by the varying snow surface properties. The anchor stations have tower-based radiometers having identical technical parameters (frequency, polarization, incidence angle) as the spaceborne radiometers, and, additionally, the water equivalent is measured manually over a snow course (2 to 3 km long path with varying land use categories).

Figure 19.10 shows observed $T_{18V} - T_{37V}$ for northern Finland during the entire winter of 1980–1981. The scatter in the satellite data is due to differences in snow particle size, land use categories, distribution of forested/open areas and atmospheric conditions. Additional scatter is caused by inaccuracy of ground-based values of the water equivalent.

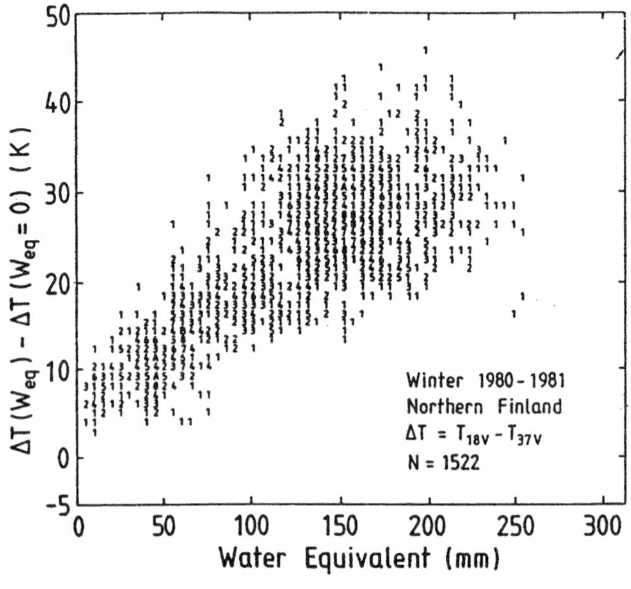

Figure 19.10: *Observed brightness temperature difference between 18 and 37 GHz, vertical polarization, as a function of ground-based manually measured values of the water equivalent (Hallikainen, 1989). Nimbus-7 SMMR data for northern Finland, covering the winter of 1980–1981, are shown. Each number shows the number of identical data points (A = 10, B = 11, etc.).*

19.3.2.2 Radar

The presently used spaceborne radar frequencies (1 to 10 GHz) are too low for determining the water equivalent of dry snow. The process is based on volume scattering by dry snow at frequencies between 18 and 90 GHz.

19.3.3 Onset of melting

19.3.3.1 Microwave radiometry

The onset of melting for seasonal snow cover is characterized by the presence of wet snow in the daytime and dry snow (at least the snow surface is dry) at night. This effect shows up in both the vertically and horizontally polarized brightness temperature as a diurnal variation, the amplitude (difference between the maximum and minimum values) being the larger the higher the frequency is. Figure 19.11 shows an example of the diurnal variation observed in tower-based measurements near Helsinki, Finland.

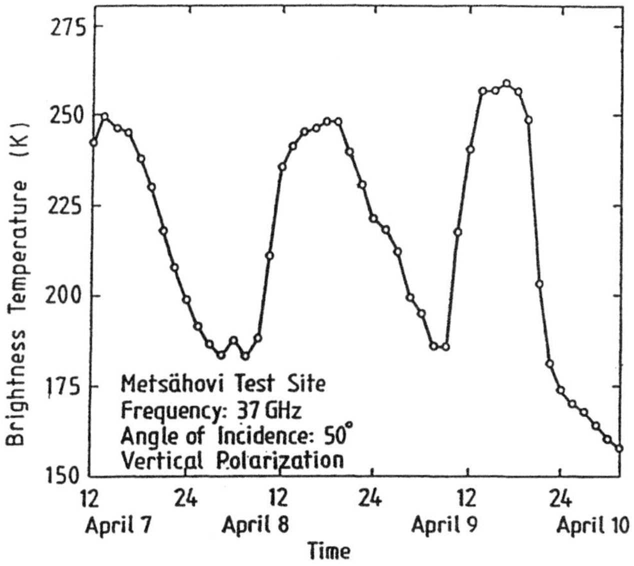

Figure 19.11: *Diurnal variation of the 37-GHz vertically polarized brightness temperature in April 1985 near Helsinki. The bottom of the snowpack is wet at all times.*

19.3.3.2 Radar

Based on the results of Ulaby and Stiles (1981), the difference in the backscattering coefficient between dry and wet snow increases with increasing frequency. It is about 3 dB at 5 GHz, 5 to 6 dB at 15 GHz, and 15 dB at 36 GHz. Hence, a frequency of 36 GHz (or perhaps even higher) should be used to discriminate dry snow from wet snow. However, at 36 GHz forest canopies may mask effectively the snow cover. Figure 19.12 shows the diurnal patterns in radar results at several frequencies, caused by a maximum daytime volumetric liquid water content of 1.3%.

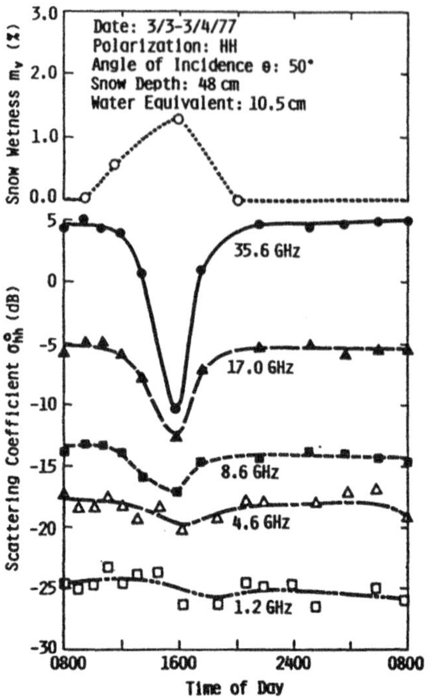

Figure 19.12: *Diurnal radar backscatter behavior between 1.2 and 36 GHz, caused by a maximum volumetric liquid water content of 1.3% (Stiles and Ulaby, 1980).*

19.3.4 Monitoring of snow melt

19.3.4.1 Microwave radiometry

When seasonal snow has started to melt, snow-free patches appear first in open areas. In forested areas, the snow cover remains much longer. Using microwave radiometry, snow- covered areas can be discriminated using a millimeterwave radiometer which is very sensitive to dry snow. Figure 19.13 shows results from helicopter-borne measurements in northern Finland, obtained early in the morning (dry snow) and in the afternoon (wet snow) using a frequency range of 24 to 94 GHz.

The results suggest that the 94 GHz channel can discriminate the remaining snow patches well even in forested areas; the change in the brightness temperature is 27 K for snow-covered areas and 14 K for snow-free areas (change of air temperature was 12 K).

19.3.4.2 Radar

Monitoring of snow melt by radar is based on the comparison of radar images acquired during the snow melt period to images acquired under dry snow conditions. The 3 dB difference in the ERS-1 SAR-derived backscattering coefficient between wet snow and dry snow/snow-free land can be used to derive the relative fraction of snow-free land (Koskinen et al., 1994). These results are based on a limited set of ERS-1 SAR images.

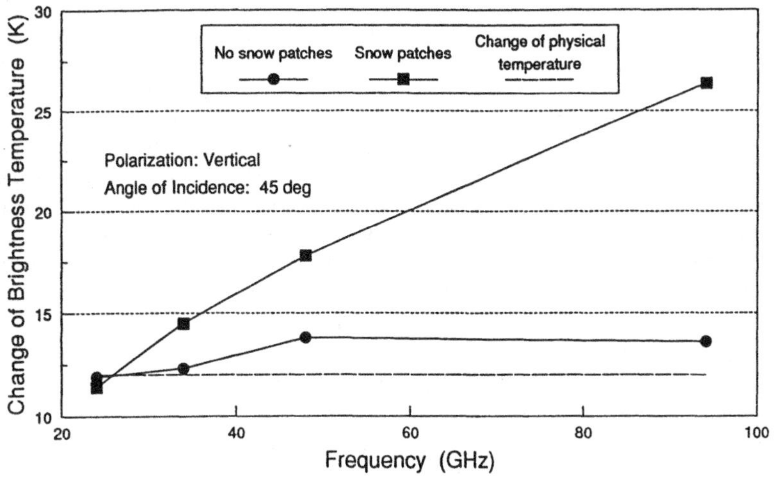

Figure 19.13: *Change in the observed brightness temperature between early morning values (air temperature −2° C) and afternoon values (air temperature 10° C) for forested snow-covered and snow-free areas (Kurvonen and Hallikainen, 1995b).*

19.4 References

Aschbacher J (1989) Land surface studies and atmospheric effects by satellite microwave radiometry, Doctoral dissertation, University of Innsbruck

Cavalieri DJ, Gloersen P, Campbell WJ Determination of sea ice parameters with Nimbus-7 SMMR. *Journal of Geophysical Research* 89: 5355–5369

Chang ATC, Foster JL, Hall DK (1987) Nimbus 7 derived global snow cover parameters. *Annals of Glaciology* 9: 39–44

Colbeck SC (1982) The geometry and permittivity of snow at high frequencies. *Journal of Applied Physics* 53: 4495–4500

Comiso JC, Zwally HJ (1982) Antarctic sea ice concentrations inferred from Nimbus 5 ESMR and Landsat imagery. *Journal of Geophysical Research* 87: 5836–5844

Comiso JC (1983) Sea ice effective microwave emissivities from satellite passive microwave and infrared observations. *Journal of Geophysical Research* 88: 7686–7704

Comiso JC (1986) Characteristics of Arctic winter sea ice from satellite multispctral microwave observations. *Journal of Geophysical Research* 91: 975–994

Cunningham G, Kwok R, Holt B (1992) Preliminary results from the ASF/GPS ice classification algorithm, *Proc. IEEE IGARSS'92 Symposium*, pp. 573–575, Houston, TX, USA, May 26–29, 1992

Gloersen P, Campbell WJ, Cavalieri DJ, Comiso JC, Parkinson CL, Zwally HJ (1992) Arctic and Anarctic Sea ice, 1978–1987: *Satellite Passive-Microwave Observations*, NASA SP-511, National Aeronautics and Space Administration, Washington, DC, USA

Gray AL, Hawkins RK, Livingstone CE , Arsenault LD (1981) Seasonal effects on the microwave signatures of Beaufort Sea Ice, *Proc. 15th International Symposium on Remote Sensing of the Environment*, pp. 239–257, Ann Arbor, Michigan, USA, May 11–15, 1981

Hallikainen MT, Ulaby FT, Abdelrazik M (1986) Dielectric properties of snow in the 3 to 37 GHz range. *IEEE Trans Antennas Propagation* AP-34: 1329–1340

Hallikainen MT (1989) Microwave radiometry of snow. *Advances in Space Research* 9: (1)267–(1)275

Hallikainen MT, Jolma P (1992) Comparison of algorithms for retrieval of snow water equivalent from Nimbus-7 SMMR data in Finland. *IEEE Transactions on Geoscience and Remote Sensing* 30: 124–131

Hallikainen MT (1992) Microwave remote sensing of low-salinity sea ice, Chapter 20 in: Carsey, F.D. (editor), Microwave remote sensing of sea ice, *AGU Geophysical Monograph Series* No. 68, Washington DC, USA

Hallikainen MT, Toikka M (1992) Classification of sea ice types radar radar, *Proc 22nd European Microwave Conference*, pp. 967–972, Espoo, Finland, 24–27 August 1992

Hallikainen MT, Kurvonen L, Jääskeläinen V, Koskinen J, Herland E-A, Perälä J (1993) Microwave remote sensing of snow using radar and radiometer, *Proc Second ERS-1 Symposium*, ESA SP-361, pp. 977–982, 1994, Hamburg, Germany, 11–14 October 1993

Hallikainen MT (1996) Physical basis for microwave remote sensing of sea ice and snow, NATO ASI Summer School on *Remote Sensing of Processes Governing Energy and Water Cycles in the Climate System*, Plön, Germany, 1-12 May, 1995

Holt B, Rothrock DA, Kwok R (1992) Determination of sea ice motion from satellite images, Chapter 18 in: Carsey, F.D. (editor), Microwave remote sensing of sea ice. *AGU Geophysical Monograph Series* No. 68, Washington DC, USA

Koskinen J, Kurvonen L, Jääskeläinen V, Hallikainen M (1994) Capability of radar and microwave radiometer to classify snow types in forested areas, *Proc IEEE IGARSS'94 Symposium*, pp. 1283–1286, Pasadena, California, USA, 8–12 August 1994

Kurvonen L, Hallikainen M Classification of Baltic Sea ice types by airborne multifrequency microwave radiometer. *IEEE Transactions on Geoscience and Remote Sensing* in print, 1995a

Kurvonen L, Hallikainen M Influence of land-cover category on brightness temperature of snow. *IEEE Transactions on Geoscience and Remote Sensing* in print, 1995b

Künzi K, Patil S, Rott H (1982) Snow-cover parameters retrieved from Nimbus-7 Scanning Multichannel Microwave Radiometer (SMMR) data. *IEEE Transactions on Geoscience and Remote Sensing* 20: 452–467

Mätzler C, Schanda E (1984) Snow mapping with active microwave sensors. *International Journal of Remote Sensing* 5: 409–422

Parkinson CL, Gloersen P Global sea ice coverage, in: Gurney, R.J., J. L. Foster, C.L. Parkinson, *Atlas of satellite observations related to global change*, Cambridge University Press, 1992

Rigot E, Drinkwater M On the application of multifrequency polarimetric radar observations to sea-ice classification, *Proc IEEE IGARSS'92 Symposium*, pp. 576– 578, Houston, TX, USA, May 26–29, 1992

Rott H, Nagler T Capabilities of ERS-1 SAR for snow and glacier monitoring in Alpine areas, *Proc of the Second ERS-1 Symposium*, Hamburg, Germany, 11–14 October 1993, ESA SP-361, pp. 965–970

Steffen K, Maslanik JA (1988) Comparison of Nimbus-7 Scanning Multichannel Microwave radiometer radiance and derived sea ice concentrations with Landsat imagery for the North Water area of Baffin Bay. *Journal of Geophysical Research* 93: 10769–10781

Steffen K, Schweiger A (1991) DMSP-SSM/I NASA algorithm validation using Landsat satellite imagery. *Journal of Geophysical Research* 96(C12): 21971–21987

Steffen K, Key J, Cavalieri DJ, Comiso J, Gloersen P, St. Germain K, Rubinstein I (1992) The estimation of geophysical parameters using passive microwave algorithms. Chapter 10 in Carsey, F.D. (editor), Microwave remote sensing of sea ice. *AGU Geophysical Monograph Series* No. 68, Washington DC, USA

Svendsen E, Kloster K, Farrelly B, Johannessen OM, Johannessen JA, Campbell WJ, Gloersen P, Cavalieri D, Mätzler C (1983) Norwegian Remote Sensing Experiment: Evaluation of the Nimbus-7 Scanning Multichannel Microwave Radiometer for sea ice research. *Journal of Geophysical Research* 88: 2781–2791

Stiles WH, Ulaby FT (1980) Microwave remote sensing of snowpacks. *RSL Technical Report* 340-3, University of Kansas, Lawrence, KS, USA

Swift CT, Fedor LS, Ramseier RO (1985) An algorithm to measure sea ice concentration with microwave radiometers. *Journal of Geophysical Research* 90: 1087–1099

Ulaby FT, Stiles WH (1981) Microwave response of snow. *Advances in Space Research* 1: 131–149

Zhang H (1991) Parameter retrieval algorithms and data analysis system for sea ice remote sensing, Ph.D. dissertation, Publication LD 85, 231 pp., Technical University of Denmark, Lyngby, Denmark

Zwally HJ, Comiso JC, Parkinson CL, Campbell WJ, Carsey FD, Gloersen P (1983) Antarctic sea ice, 1973–1976: Satellite passive-microwave observations, 206 pp, NASA SP–459. National Aeronautics and Space Administration, Washington, DC, USA

PART D

NEW DEVELOPMENTS

Chapter 20

TRMM: The Tropical Rainfall Measuring Mission

Anthony D. Del Genio
NASA Goddard Institute for Space Studies
2880 Broadway
New York, NY 10025
USA

20.1 Introduction

Precipitation is one of the most important parts of both the global water and energy cycles, yet it is one of the most poorly observed climate parameters. Globally, the net latent heating associated with precipitation accounts for about 80% of the total diabatic heating required to balance radiative cooling to space. The spatial pattern and temporal variability of this heating serves as the primary driver for the tropical general circulation. The local balance between precipitation and evaporation over ocean helps determine the thermohaline component of the ocean circulation and may influence both tropical El Niño-Southern Oscillation (ENSO) events and high latitude deep water formation. Over land, the precipitation-evaporation balance controls the tendency for drought to occur and directly influences the surface temperature by altering the relative contributions of sensible and latent heat fluxes to the surface energy budget. Within clouds, precipitation formation is thought to be the major sink of cloud water and thus affects the reflectance of bright tropical cumulus anvils. Drizzle formation in low-level marine stratocumulus may help detach those clouds from surface moisture fluxes as the drizzle evaporates in the subcloud layer, thereby contributing to the stratocumulus-trade cumulus transition. Land- ocean differences in precipitation efficiency, associated with different concentrations of nucleating aerosols, may cause cloud feedback in a changing climate to differ for marine and continental clouds.

Despite its importance to the climate system and its practical consequences in our everyday lives, current precipitation data sets are woefully inadequate. Global precipitation rates are uncertain to at least 20%, and monthly precipitation estimates over oceans can disagree by 50% or more depending on the estimation technique (cf. Arkin and Meisner, 1987). The main problem over oceans is the lack of direct sampling of precipitation. Existing ocean climatologies historically were simple extrapolations from coastal measurements or were heavily weighted by a few island stations which may give biased estimates due to the different atmospheric stability structure and diurnal cycle over land and ocean. More recently, ocean rainfall has

NATO ASI Series, Vol. I 45
Radiation and Water in the Climate System:
Remote Measurements
Edited by Ehrhard Raschke

been estimated indirectly from satellite using simple infrared or microwave indices, but often based on detailed observations for only a single region. Over land, precipitation is more well defined climatologically, but errors are nonetheless significant - precipitation averages over large areas are notoriously difficult to measure with a network of local rain gages because much of it occurs in high rain rate events of small spatial scale and short time duration (cf. Dai, 1995).

In a companion paper (Del Genio et al., 1996) I have suggested that a useful accuracy standard for data sets used to validate climate models is that the uncertainty in observed monthly mean values of a climate parameter over spatial scales resolved by the model (hundreds of kilometers) be significantly less than the differences among various models. For many climate parameters, this criterion is met by existing data sets, but not so for precipitation. For example, the global mean boreal winter precipitation simulated by 8 coupled atmosphere-ocean general circulation models (GCMs) for the most recent IPCC Scientific Assessment (IPCC, 1995) varies from 2.4 to 3.8 mm/d. (Since 1 mm/d $\approx$ 30 W/m^2 if converted to equivalent energy flux units, this difference is comparable to model errors in simulation of the planetary radiation budget.) But existing climatologies give estimates as low as 2.7 mm/d (Jaeger, 1976) and as high as 3.5 mm/d (Legates and Willmott, 1990), almost as large as the range of model disagreements and much larger than equivalent observational errors in radiation fluxes. Figure 20.1 shows differences in January rainfall between two data sets, and between one of them and the most recent version of the GISS GCM (cf. Del Genio et al., 1996). Although model-data discrepancies are somewhat larger than differences between data sets, in many places the differences in both cases are 1–2 mm/d and $>$ 4 mm/d in certain parts of the tropics. Model disagreements are also particularly large in the tropics: in an intercomparison of 19 GCMs, Randall et al. (1992) reported zonal and monthly mean rainfall differences of as much as 6–7 mm/d between individual GCMs in the Intertropical Convergence Zone (ITCZ). It is also worth noting that existing precipitation data exist only in climatological form: Annual cycle information is available, but neither interannual variations nor diurnal cycle information exists.

To address the absence of suitable precipitation data sets for climate and tropical dynamics applications, the Tropical Rainfall Measuring Mission (TRMM) satellite will be launched in August 1997. This paper outlines the TRMM philosophy and approach to obtaining meaningful rainfall estimates, the challenges that must be overcome to achieve the desired accuracy standard, the instruments and retrieval techniques to be used, and supporting observations and modeling that will enhance the impact of the data set. Section 20.2 describes the TRMM mission and instrumentation and Section 20.3 the challenges of measuring rainfall that produced the final mission configuration. Section 4 discusses rainfall retrieval approaches with the primary TRMM instruments, while Section 20.5 describes some of the complementary science that can be done with other TRMM instruments and models. Section 20.6 outlines the validation approach to be used by TRMM. We conclude in Section 20.7 with some thoughts about extending the TRMM results to other data sets to create a longer-term record of precipitation variations.

20.2 TRMM Mission and Instrumentation

TRMM is a joint mission of the United States and Japan. The Science Team is led by Project Scientist Dr. Joanne Simpson of the NASA Goddard Space Flight Center. The TRMM spacecraft will be launched in August 1997 on a Japanese H-II rocket into a rapidly precessing 35° inclination, 350 km altitude circular Earth orbit. The spacecraft will complete 16 orbits per day, resulting in the ground track pattern indicated in Figure 20.2. TRMM is part of the Earth Probes series of NASA launches, designed as relatively low-cost missions with targeted science

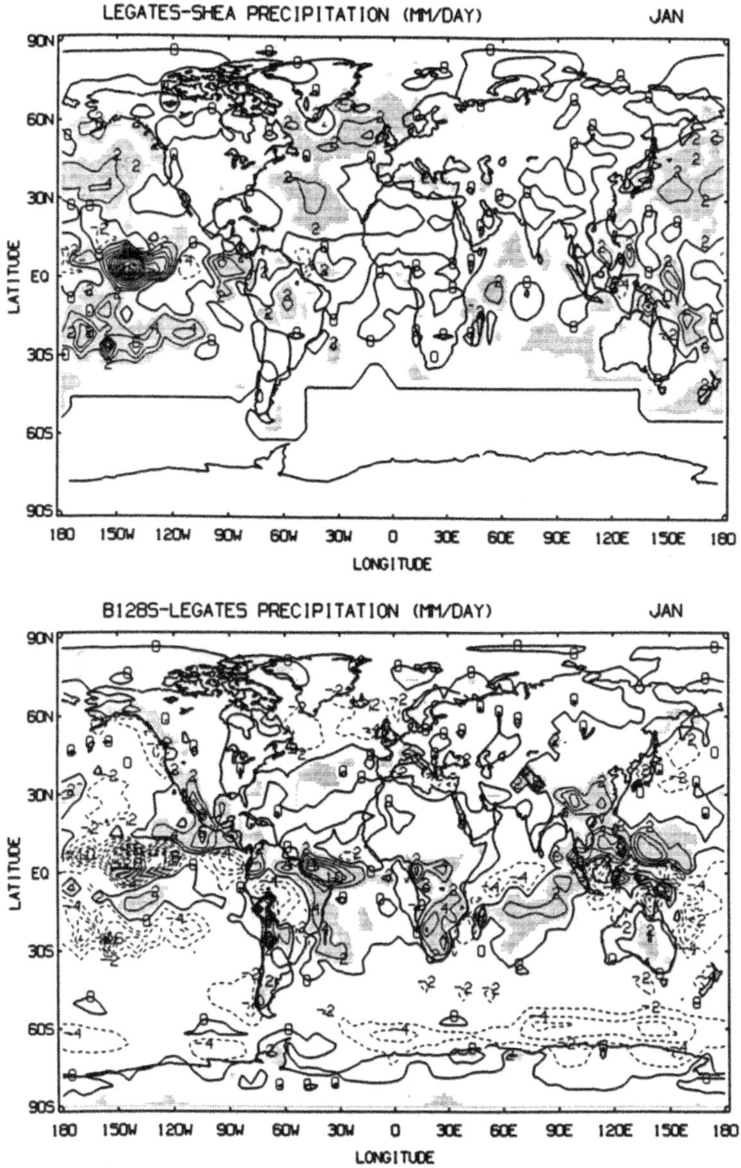

Figure 20.1: *Differences in January precipitation between the climatologies of Legates and Willmott (1990) and Shea (1986) (upper) and between the GISS GCM and the Legates and Willmott climatology (lower). Hatched areas represent differences > 1 mm/d.*

objectives; the TRMM spacecraft will weigh 3620 kg, of which 550 kg will consist of scientific

Instrument	Frequencies/wavelengths	Resolution	Swath
Precipitation Radar (PR)	14 GHz	250 m (range) 4 km (footprint)	220 km
TRMM Microwave Imager (TMI)	10, 19, 21, 37, 85.5 GHz (dual polarized except 21)	45,21,17,11, 5 km	760 km
Visible/Infrared Sensor (VIRS)	0.63,1.61,3.75,10.8, 12.0 μm	2.2 km	720 km

Table 20.1: *Summary of characteristics of the three remote sensing instruments making up the primary rain package for TRMM.*

instrument mass. The TRMM spacecraft is being built at the NASA Goddard Space Flight Center.

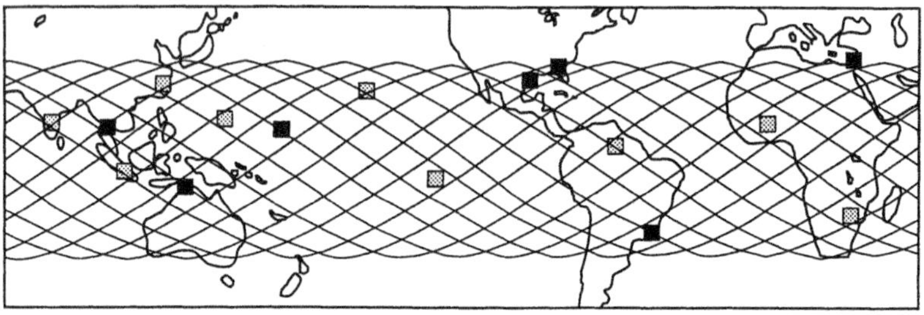

Figure 20.2: *TRMM satellite ground tracks for 1 day (16 orbits). The squares denote current (solid) and potential future (shaded) ground truth validation sites.*

The primary objective of TRMM is to produce a 3-year data set of monthly mean tropical precipitation with an accuracy of 10% at a spatial resolution of about 500 km, including diurnal cycle statistics for larger regions (Simpson, 1988). To achieve this goal, the TRMM spacecraft will carry a rain package of three complementary instruments (Table 20.1). The innovative feature of TRMM is its precipitation radar (PR), the first rain radar to ever fly in space. The radar, being built by Japan, will have a footprint of 4 km and 250 m range resolution and will scan cross-track over a 220 km swath. The PR represents a "flying rain gauge" that is expected to provide the most accurate measurements of rainfall ever achieved from space.

Because of its small swath, the PR by itself cannot provide the large-scale coverage required for a complete survey of tropical rain rates. The mass and power requirements of a rain radar also make it impractical as a candidate for routine rain observations on future operational meteorological satellites. Thus, the TRMM rain package includes two other instruments, a passive microwave radiometer (TMI) and a visible/infrared radiometer (VIRS), both supplied by the U.S. In keeping with the low-cost philosophy of the mission, both of these facility instruments are modifications of existing instruments that have already flown on other missions: TMI is a derivative of SSM/I, which has been used extensively for space retrievals of rainfall in recent years, while VIRS has 5 channels in common with AVHRR and provides a link to the long record of geostationary and polar-orbiter Earth imagery used to identify clouds associated with weather systems. Both TMI and VIRS have wider swaths than the PR and can potentially

Instrument	Wavelengths	Resolution	Swath
Lightning Imaging Sensor (LIS)	0.7774 μm	5 km	537 km
Clouds and Earth's Radiant Energy System (CERES)	0.3 - > 50, 0.3-5, 8-12 μm	10 km	3293 km + 360° azimuthal

Table 20.2: *Characteristics of the complementary EOS instruments flying on TRMM.*

be used as operational rain sensors once satisfactory retrieval algorithms have been developed. The TRMM strategy is to use two or all three of the instruments in combination to produce the most definitive estimate of tropical rainfall possible over the three-year mission life. Beyond the lifetime of TRMM, though, it is hoped that the PR will serve as a validation standard that can be used to "train" more accurate rainfall algorithms that utilize only data from the passive microwave, infrared and visible channels and can thereby extend the time period of useful climatological rain data.

In 1990 the Earth Observing System (EOS) Program decided to add two other instruments to TRMM: Clouds and the Earth's Radiant Energy System (CERES), a successor to the ERBE instrument for estimating top-of-the-atmosphere radiation fluxes, and a Lightning Imaging Sensor (LIS). The characteristics of these instruments are listed in Table 20.2. Neither CERES nor LIS is a rain-measuring instrument per se, but both produce complementary information on the convective and mesoscale processes associated with rain production. As such they provide a link to other aspects of the climate system and will aid in interpreting the rain data in terms of the larger questions about the role of precipitation in climate and climate change.

20.3 Challenges to Accurate Measurement of Rain from Space

There are a variety of problems involved in measuring precipitation remotely. Some involve the structure and microphysics of rain systems and the nature of the interaction between radiation and condensed water particles of various sizes. Perhaps the most important aspect of remote rain estimation is that passive measurement techniques often do not directly sense the rain they are attempting to estimate, but rather detect some other part of the rain system that is indicative of the intensity of rain falling below. As a result, different types of rain regimes behave differently, and no single technique or wavelength is effective at measuring rain in all situations. Different approaches must be used to distinguish light from heavy rain, to measure rain over ocean vs. land, and to detect warm vs. cold rain. Each of these difficulties will be discussed in more detail in the following section in connection with individual TRMM instruments.

Aside from the difficulty of converting radiances to a physical parameter such as precipitation, TRMM faces a severe sampling problem in obtaining monthly means to 10% accuracy. Unlike many other meteorological quantities, precipitation occurs primarily in high intensity, short duration events that occupy a very small fractional area, i.e., most of the time it is not raining in any particular location. Figure 20.3 shows the log- normal probability density function (pdf), which appears to be typical of rainfall in a variety of convective environments (Lopez, 1977). The log-normal distribution is Gaussian when plotted on a logarithmic scale. For the distribution in the figure, about 50% of the rain falls in a few events at rates > 10 mm/hr.

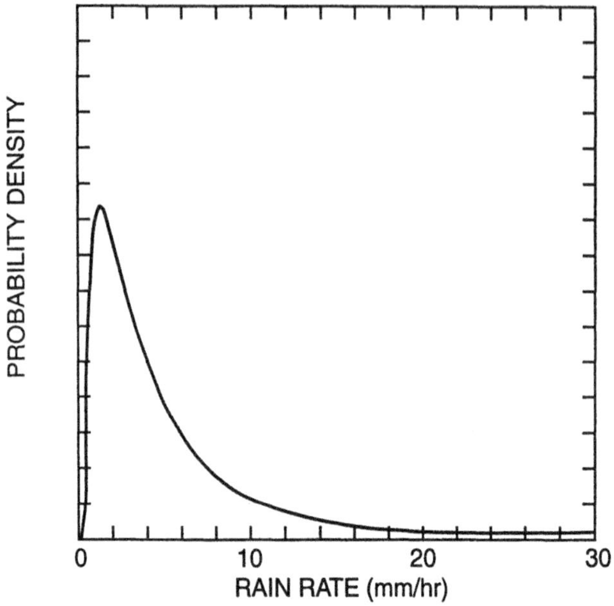

Figure 20.3: *An example of a log-normal probability density function, which is characteristic of rainfall and other cloud statistics in a variety of tropical convective regions.*

A geostationary satellite, which flies at high altitude, views essentially all of one face of the Earth almost continuously, and so can sample a given location adequately. But TRMM must fly at very low altitude to have acceptable footprint sizes for the PR, whose antenna size is limited by mass and power constraints to be significantly smaller than a typical surface-based rain radar. Thus, TRMM views only a small fraction of the Earth at a given instant and returns to the same location only every several days. Within a particular (500 km)^2 region, some satellite passes may cover all of the area, while others will only partly sample the raining region. The sampling error that results is a function of the spacecraft altitude and orbit inclination. Thus, the realized sampling error involves several tradeoffs: (1) Increasing orbit altitude provides better spatial coverage, but at the price of reduced resolution and accuracy for individual rain retrievals; (2) Decreasing orbit inclination provides better coverage of the diurnal cycle and thus reduces sampling error, but also reduces the fraction of the globe over which rain information will be acquired. The diurnal cycle issue is an especially important one for TRMM: Precipitation has an especially strong diurnal variation over land (cf. McGarry and Reed, 1978), with maxima occurring in afternoon over many locations but in the evening at others. Over oceans, the diurnal cycle is much weaker but noticeable, with early morning maxima typical over many locations (cf. Gray and Jacobson, 1977). The geographic variability of diurnal cycle strength and phase mandates that TRMM fly at a fairly low inclination to adequately sample these variations, not only so that its monthly means are sufficiently accurate, but because the diurnal cycle itself contains information about the physical processes that determine rain occurrence. A variety of sampling studies have been conducted in support of TRMM (cf. Shin and North, 1988); these suggest that a 10% monthly mean accuracy is achieveable at 500 km scales for a 350 km altitude orbit and 35° inclination. It is largely these sampling considerations that have determined the orbit configuration for TRMM.

A second difficulty associated with the small spatial scale of rain variations is the so-called "beam-filling" problem. The horizontal extent of a typical vigorous thunderstorm cell is of order 5 km, and the area of active updrafts producing most of the rain within this cell may be smaller, perhaps 1 km in extent. Since the TRMM instruments all have footprints larger than this, only part of the area covered by a TRMM pixel will in general be raining at any time. The radiance measured by a TRMM instrument will thus be a combination of signals from raining and non-raining areas, weighted in an unknown fashion. For a given instrument and wavelength, the retrieval algorithm must take this subpixel variability into account in order not to make a biased precipitation estimate (cf. Rosenfeld et al., 1992). The problem becomes worse when multichannel algorithms for a single instrument such as TMI, or multi-instrument (TMI-PR, TMI-VIRS) algorithms, are considered, because different instruments and different TMI channels have different size footprints, and so no two observe exactly the same raining scene.

There is no completely satisfactory way to eliminate the beam-filling problem. The most useful strategy is to use models of precipitating cloud systems to simulate the radiance seen by the spacecraft at various resolutions under different raining conditions (Smith et al., 1991; Adler et al., 1992), and to validate algorithms developed in this way against ground-based estimates of precipitation on smaller spatial scales. This requirement thus dictates the need for TRMM to include a substantial "ground-truth" validation program. We will return to this subject in Section 20.6.

20.4 Rain Retrieval Using TRMM Instruments

In-depth discussions of passive microwave (Simmer, 1996) and radar (Joe, 1996) retrieval techniques for precipitation are presented elsewhere in this volume. In this section I simply present an overview of the principles underlying active and passive remote sensing of rain as they apply to instruments on TRMM (see also Simpson, 1988). Additional information on recent developments in remote rainfall estimation algorithms can be found in the special February 1993 issue of *Journal of Applied Meteorology*.

20.4.1 Precipitation Radar (PR)

The estimation of rainfall by surface-based radars is one of the most widespread applications of remote sensing in existence. The requirement for surface rain radars is that they operate at a sufficiently long wavelength (typically > 5 cm) so that attenuation of the beam by rain droplets is insignificant. Under such circumstances a relationship between the radar reflectance Z and the rainfall rate R exists, taking the form $Z = aR^b$, where a and b are empirically determined constants. One immediate problem for a spaceborne radar algorithm is that the constants vary regionally, because Z is a strong function of the droplet size distribution (proportional to D^6, where D is the droplet diameter). Thus, a single Z-R relationship valid for rain in different parts of the tropics may not be feasible.

A larger problem is that to get satisfactory resolution from space, a long wavelength radar would require unacceptably large antenna dimensions. For space applications, then, a shorter wavelength (2.16 cm) radar is being used by TRMM. Z-R relationships can thus only be used by TRMM in light rain situations, where attenuation is small. At higher rain rates, the TRMM PR will instead make use of the information contained in the attenuation of the beam to derive rain rates. The general concept is as follows: If the attenuation coefficient k (attenuation per unit rain rate and distance through the raining layer) is approximately constant, then the observed attenuation A is given by $A \sim kRH$, where R is the path-averaged rain rate and H the path length through the raining layer. If we assume that H is equal to the slant range between the ground and the freezing level (identified by the radar bright band), then R can be determined from a measurement of H. In practice, though, supercooled liquid water exists above the freezing level, limiting the accuracy of this simple estimate.

A variety of other algorithms have been developed for space rain estimation. The surface reference method (Meneghini et al., 1983) compares the surface power return in raining and non- raining conditions to isolate the attenuation due to rain. This procedure is effective unless the nature of the surface return changes with time or is different underneath a rain system. The utility of such techniques is of course limited by the sensitivity of the radar to different rain rates and its ability to "see" the surface through the rain. For the PR, Okamoto and Kozu (1987) estimate that rain rates < 1 mm/hr are detectable, with maximum observable rain rates of 40 mm/hr and the surface return detectable through rain rates as high as 90 mm/hr. The mirror image method (Meneghini and Atlas, 1986) uses the surface in a different way: Energy reflected from the surface to the rain layer and then back again to the surface produces a virtual image of the precipitation below the surface in the radar return. The weaker return from the backscatter of the mirror image relative to that directly scattered back to the radar by the rain contains information about the rain rate. Since this method requires data from an even further attenuated portion of the signal, it fails at lower rain rates than does the surface reference technique.

The Height Area Rainfall Threshold technique takes a completely different approach to rain estimation (Atlas et al., 1990). Since rain distributions seem to be fairly well described by the log normal pdf, the fraction of a given area that is observed to be raining above some threshold detectable rain rate (i.e., above a threshold value of Z) is correlated with the total area of rain (and thus total rain volume, the integral of the rain rate over area and time). The rain volume in turn is related to rain area by the average rain rate in areas that are raining; this depends on the height of the storm, which can be measured also by the radar (or by other TRMM instruments). The strength of this approach is that it potentially provides a "snapshot" of the area-wide rain, rather than requiring multiple TRMM visits to a region over a month to build up suitable accuracy.

The weakness is its dependence on the existence of a universal rain pdf; a priori this is not known to be the case, although TRMM data should constrain the degree of spatial and temporal variability of the pdf.

20.4.2 TRMM Microwave Imager (TMI)

Figure 20.4 illustrates schematically the nature of the sources of radiation seen by a passive microwave sensor. Note first that the Planck function is approximately linear in the microwave region of the spectrum, so that microwave radiances can conveniently be discussed in terms of the brightness temperature of the emitting object. Microwave rain detection depends heavily on the fact that the ocean surface has low microwave emittance (typically $e \leq 0.5$). The brightness temperature of the ocean surface (T_{bs}) is thus much lower than the actual sea surface temperature (T_s). Microwave emission from the top of a rain- producing system ($e \sim 1$) consequently appears as a warm anomaly above a cold surface (just the opposite of infrared detection of cumulonimbus, which appear as cold cloud tops against a warm surface).

A number of complicating factors arise, though, in uniquely estimating rain rate from passive microwave data. Land surfaces have much higher microwave emissivities than the ocean, and the emittance varies with vegetation type, soil moisture, etc., so the contrast between rain

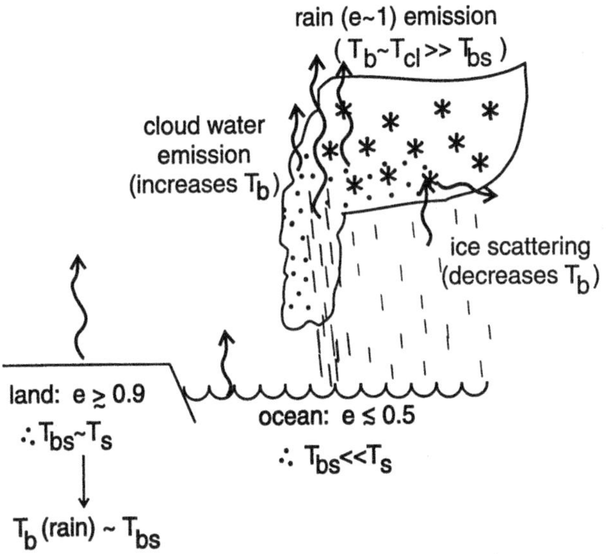

Figure 20.4: *Schematic illustration of the physical factors contributing to the passive microwave radiation observed over a raining scene.*

and an underlying land surface is both smaller and less predictable than in the ocean case. This means that microwave retrievals are significantly more accurate over ocean, although efforts to construct successful land algorithms are being made. Even over ocean, much of the microwave signal comes from particles other than the rain droplets themselves; emission from cloud liquid water increases the observed brightness temperature, while scattering by large ice crystals above the rain layer (as well as by the large rain droplets themselves) decreases the brightness temperature. The scattering effect increases with frequency, so microwave algorithms are described in terms of a low-frequency "emission regime" and a high-frequency "scattering regime ".

At low frequencies, emission initially increases linearly with rain rate, but begins to saturate at about 40 mm/hr; emission techniques are thus utilized mostly in light rain situations. Heavier rain rates are usually accompanied by a greater concentration of large ice particles. At the lowest frequencies typical of previous microwave instruments (19-22 GHz), this causes brightness temperature to decrease with rain rate slightly at high rain intensities. Thus, for a given brightness temperature, the solution may be double-valued. The 10 GHz channel on TRMM is at a sufficiently low frequency that scattering effects are negligible, thus permitting a unique solution for low to moderate rain rates. For the higher intensities, the high-frequency (85 GHz) channel should exhibit decreasing T_b with increasing rain rate due to increasing ice scattering, implying that scattering-based algorithms are preferred for intensely raining systems (cf. Adler et al., 1992 for model-based simulations of the dependence of microwave brightness temperatures on the different parts of a mesoscale convective system).

A wide variety of approaches to rainfall estimation using passive microwave data exist, but all can be considered as variants of five basic methods: (1) Single channel emission- based empirical techniques (cf. Wilheit et al., 1984); (2) Single channel scattering-based empirical techniques (cf. Spencer, 1986); (3) Algorithms exploiting the effect of attenuation by rain on the polarization of observed microwave brightness temperatures (cf. Petty and Katsaros, 1990); (4) Multichannel statistical techniques (cf. Adler et al., 1993); (5) Multichannel physically-based radiative transfer algorithms (cf. Wu and Weinman, 1984; Smith and Mugnai, 1988; Kummerow et al., 1989; Liu and Curry, 1992).

A particular problem for the microwave is warm (coalescence-dominated) rain over land. Emission techniques do not work here because of the confusion of the land surface seen through the rain, while ice scattering, which hides the land surface and permits microwave rain retrieval for strong systems over land, is nonexistent. Algorithm intercomparisons indicate that shallow, orographic warm rain is simply missed by typical microwave algorithms (cf. Adler et al., 1993).

20.4.3 Visible/Infrared Sensor (VIRS)

By far the most widely available data on clouds consist of the visible and infrared window channel imagery from geostationary and polar orbiter operational satellites. The obvious appearance of precipitating systems in such images as (visibly) bright and (thermally) cold regions suggests that a quantitative measure of precipitation might be derived from such data, which are global in coverage and several decades in length. The problem is that rainfall itself has little direct signature in either the visible or infrared. Optical thickness is inversely proportional to particle size, so large rain droplets contribute little to the visible reflectance of precipitating systems relative to smaller cloud droplets. Infrared emission depends on the temperature of the cloud top for an optically thick cloud, and in general higher intensity rain is associated with higher altitude (and therefore colder) cloud tops. But moderate optical thickness non-raining cirrus can appear just as cold as heavily precipitating cumulonimbus, and spatial/temporal variations in atmospheric temperature and water vapor prevent the association of a unique rainfall rate with a given infrared brightness temperature.

The standard infrared approach to rain estimation is the GOES precipitation index (GPI; Arkin, 1979), which relates rain rate over a 2.5° x 2.5° area to the fraction of the area whose IR brightness temperature is colder than 235 K via a simple linear regression. This algorithm, based on GATE data for the tropical Atlantic, not surprisingly works well over that region and is used to produce an operational rainfall product for the entire tropics. As a general indicator of rainfall patterns and their seasonal or interannual variability, the GPI is a useful index. But its quantitative accuracy is limited, it does not work well outside the tropics, where non-convective rain is more common, it does not work well for shallow rain systems in general, and it can give spurious information on the diurnal cycle of rain (cf. the discussion in Fu et al., 1990).

To address the problem of distinguishing precipitating cumulonimbus from equally cold non-precipitating cirrus, Inoue (1987) developed the "split-window technique ". This method relies on the different wavelength dependences of the infrared emissivities of ice and liquid water. By utilizing differences between the brightness temperatures at 11 and 12 μm, Inoue separates the cirrus (ice-only) cloud field from the cumulonimbus (ice and liquid) area, and bases the derived rainfall on the latter. VIRS contains channels at 10.8 and 12 μm to test this approach.

A different philosophy has been adopted recently by Rosenfeld (personal communication). A common feature of current operational satellites is the existence of near infrared channels that are sensitive to cloud droplet effective radius near cloud top (Han et al., 1994). While such channels also do not see the precipitating particles themselves, the possibility exists that the droplet size distributions as a whole are shifted to larger sizes in precipitating systems. If so, precipitation might be inferred indirectly in the near-IR as a function of cloud-top droplet radii. This approach may provide a means of detecting the shallow rain systems over land that currently elude both conventional thermal infrared and passive microwave techniques. VIRS contains channels at 1.61 and 3.75 μm to test the utility of this approach.

For modelers, it is important to know not only the total rainfall, but the partitioning between convective and stratiform components, for several reasons: (1) Convective and stratiform rain are produced by different physical processes (instability vs. forced uplift), even though both can occur in a single mesoscale convective cluster, and thus are predicted by different physics subroutines in GCMs. (2) Budget studies (Gamache and Houze, 1983) of convective clusters suggest that both stratiform and convective precipitation contribute significantly to the total rain, but the former consists of modest rain rates over large areas and many hours while the latter occurs as high-intensity events of short duration on small scales; thus, impacts of rain on soil moisture and runoff in land-surface models depend on a knowledge of the spatial and temporal distribution of the rain, not just the rain total over a given area. (3) Cumulus anvil radiative properties are perhaps the biggest uncertainty in estimates of cloud feedback in GCMs (Del Genio et al., 1996), and stratiform precipitation is the biggest sink of anvil cloud water while detrainment of the non- precipitating component of convective condensate is a major source (Figure 20.5).

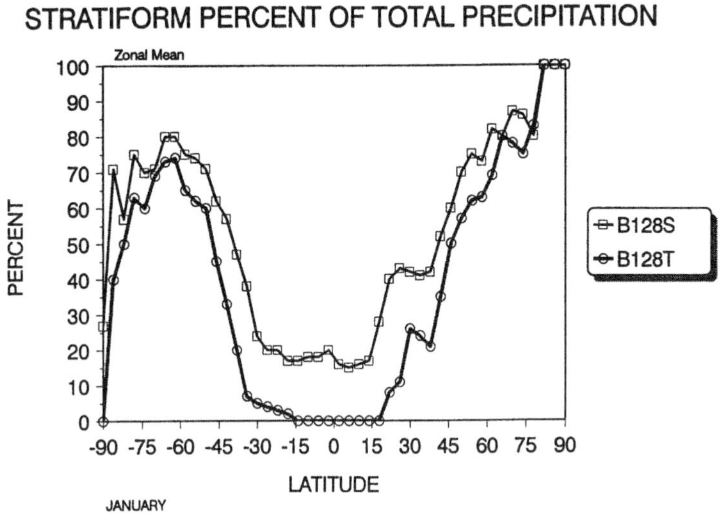

Figure 20.5: *Zonal mean percentage of total January precipitation that falls as stratiform precipitation as simulated by two different versions of the GISS GCM. Squares: Detrainment of ice from cumulus updrafts into stratiform anvils is included, leading to significant stratiform tropical rain. Circles: All cumulus condensate immediately precipitates or evaporates, leading to virtually no tropical stratiform rain.*

To address this issue, Adler and Negri (1988) developed an infrared algorithm called the "convective-stratiform technique" (CST). CST searches for possible thunderstorms by identifying minima in the IR brightness temperature (235 K or colder), and identifies the convective rain pixels as those for which the temperature difference between the minimum and the average of those for the surrounding pixels exceeds a certain threshold (1.5 K for the coldest cloud tops, increasing as the minimum temperature increases). The stratiform component is defined using the remaining area below a certain threshold temperature.

Surprisingly, little effort has been made to utilize visible channel information to discriminate precipitating from non- precipitating cold clouds. Fu et al. (1990) have shown that a combined visible-infrared threshold method, applied either to ISCCP B3 radiances or C1 optical thicknesses and cloud top temperatures, improves upon standard outgoing longwave radiation identifications of locations of deep convection by isolating the convecting region from other, less reflective high clouds. The visible channel appears to be particularly useful in identifying convection in different regions with different temperatures (e.g., east vs. west tropical Pacific): IR-only techniques validated in warm, heavily precipitating regions sometimes miss convection in regions with cooler temperatures and hence shallower convective cloud tops. Indications that their deep convective cloud (DCC) index might bear some relationship to convective precipitation come from the demonstration that the diurnal cycle of DCC over the tropical Pacific agrees in phase with that from rain gage data, and from point correlations of enhanced DCC cover with radiosonde-derived positive values of convective available potential energy (Fu et al., 1994). But no quantitative relationship of DCC to rainfall has been demonstrated. TRMM data will provide a conclusive test of whether visible reflectances can enhance the quantitative determination of rain.

20.4.4 Combined algorithms

The real advantage of TRMM lies in its potential to use information from several types of instruments to derive the best absolute monthly rainfall product. At this time, it has not yet been determined what form such an algorithm will take in practice once TRMM starts returning data. Since each instrument, different channels in a given instrument, and different retrieval philosophies all have strengths and weaknesses that are understood to some extent, an initial TRMM "combined" algorithm might simply be a "decision tree", in which different individual algorithms are applied in specific situations for which they are judged most reliable. Of more interest, though, is the simultaneous application of data from two or more instruments to exploit the advantages of each (rain sensitivity, vertical profile information vs. spatial coverage).

One class of algorithms combines single-frequency radar data with passive microwave data to derive rain rate. For example, the vertical profile of hydrometeors can be estimated from the radar and used as an input to a passive microwave cloud radiative transfer model to refine the passive microwave rain estimate (Fujita et al., 1985). Alternatively, the path-integrated extinction derived from passive microwave measurements can be used as a constraint on the radar equation to derive rainfall ate (Weinman et al., 1990). Another class of approaches is using passive microwave data to refine infrared estimates of rainfall. Adler et al. (1993), for example, compute separate rainfall totals from passive microwave and infrared algorithms, the latter sampled to coincide with the times and locations of the microwave data. The ratio of the two estimates is then used to calibrate a fully-sampled infrared estimate of the rainfall. Finally, Atlas and Bell (1991) relate radar-derived area-time integrals to analogous infrared integrals based on the GPI to improve the accuracy of large-scale rain estimates.

20.5 Complementary TRMM Science

20.5.1 Lightning Imaging Sensor (LIS)

Lightning occurrence is not by itself an indicator of rainfall, but recent evidence suggests that its presence or absence is indicative of the dynamical setting in which convective rain is generated. Optical detectors on the DMSP satellite reveal an instantaneous global distribution of lightning that matches the preferred locations of deep convection such as the ITCZ and the summer midlatitude continents (Orville and Henderson, 1986). But while convection occurs frequently over both land and ocean, lightning occurrence is an order of magnitude more frequent over continental convection regions than over maritime locations of persistent convection.

The most plausible explanation for this discrepancy is the different strength of convection over land vs. ocean. The microphysical factors leading to lightning production in thunderstorms are still not totally understood, but appear to be dependent on aspects of mixed phase microphysics (Williams et al., 1991). This in turn implies that lightning is favored by vertical advection of liquid above the freezing level into the ice-forming regions of a convective storm. Thus, the more vigorous the cumulus updrafts, the more likely the occurrence of lightning.

Observations support the idea of systematic differences between continental and maritime convection. Jorgensen and LeMone (1989) show that updraft strengths are significantly weaker in oceanic deep convection than in midlatitude continental thunderstorms. The likely reason is that the ocean is a consistent source of low-level moisture that allows the tropical atmosphere to convect in response to even a small amount of conditional instability, thereby remaining close to neutral stability. Over land, on the other hand, boundary layer relative humidity may at times be too low to trigger convection despite conditionally unstable lapse rates in the free troposphere, allowing large amounts of convective available potential energy (CAPE) to build up until sufficient low-level moistening occurs. Since in an undilute buoyant rising parcel, vertical velocity is proportional to $(\mathrm{CAPE})^{1/2}$, oceanic convection should be characterized by small CAPE, weak updrafts, and infrequent lightning, while continental convection should be more vigorous and thus more prone to lightning formation. TRMM LIS observations may thus provide us with an indirect near-global climatology of CAPE, a fundamental quantity in theories of deep convection. Because of the implications for upward advection of liquid water, LIS data may also serve as an indirect source of validation for the vertical structure of hydrometeors derived by other TRMM instruments and assumed in TRMM physically-based retrieval algorithms.

20.5.2 Clouds and the Earth's Radiant Energy System (CERES)

TRMM is the first of several platforms on which the CERES instrument will fly to provide sufficient geographical, diurnal and temporal sampling to permit accurate estimates of the shortwave and longwave components of the top-of-the-atmosphere (TOA) radiation budget. The science goals for CERES are more general than simply studies of tropical precipitating systems, encompassing the radiative effects of all clouds as well as the Earth's surface.

But CERES plays a central role in the connection between TRMM science and questions about global climate change. The radiation budget signature of tropical convective systems is large and unique, with simultaneous large negative TOA shortwave cloud forcing and large positive TOA longwave cloud forcing. These are correlated so as to produce small net TOA cloud forcing (cf. Figure 3 of Del Genio, 1996, chapter 2), but this does not imply a small net radiative effect because the vertical profiles of the shortwave and longwave effects differ dramatically. In the GISS GCM, the cloud feedback and climate sensitivity in a climate change appear to depend more on changes in the radiative properties of tropical anvils than on any other aspect of clouds in any part of the globe (Del Genio et al., 1996).

The radiative properties in turn depend on the sources and sinks of anvil water, e.g., the condensate detrained from cumulus updrafts and the stratiform precipitation produced by the anvil itself. The former depends on the strength of updrafts, an index of which should be provided by LIS, while the latter will be measured by the TRMM rain package instruments. By combining these data with CERES information on anvil reflectance variations, it should be possible to determine the role of convective cluster rain processes in determining the radiative signature of tropical anvils. This information in turn will lead to improved parameterizations of the microphysical aspects of anvils and the coupling between anvil and convective clouds in global climate models.

20.5.3 Cumulus Ensemble Model (CEM)

The precipitation distribution measured by TRMM is an indicator of the net vertically integrated latent heating of the atmosphere and as such provides a valuable constraint for model assessments of the realism of simulated forcing of the general circulation. But the vertical integral of heating by itself is not sufficient for several reasons. The vertical profile of heating determines the effect of moist convection on static stability and thus the vigor of large-scale vertical motions as well as the profile of radiative cooling to space. Furthermore, vertical variations in heating affect cloud formation both directly by altering relative humidity and indirectly by forcing or suppressing the upward motion that forms clouds. Interactions between convection and the large-scale circulation are expected to be sensitive to the vertical profile of heating.

Several field studies have indirectly derived vertical profiles of convective heating in a few locations as large-scale residuals using radiosonde-observed temperature tendencies and advection rates (cf. Reed and Recker, 1971; Thompson et al., 1979). But a large-scale depiction of variations in heating profile does not yet exist. Furthermore, the heating profile of a convective cluster differs between the convective and stratiform anvil regions (Johnson, 1984) and hence changes with time over the life cycle of a convective system. The heating profile may also be different for different types of convective structures (isolated cells, squall lines, non-squall clusters, etc.), but information is lacking.

Several TRMM instruments (PR, TMI) will produce information on the vertical profile of hydrometeors of different types. Tao et al. (1990, 1993) have developed a procedure for estimating the vertical profile of convective heating from this information. The procedure makes use of a cumulus ensemble model (CEM), i.e., a model that explicitly resolves individual cumulus clouds over a limited spatial domain. The CEM is an anelastic model with parameterized Kessler-type two-category liquid phase microphysics (cloud water, rain) and parameterized three-category ice microphysics (cloud ice, snow, hail/graupel). The CEM runs in either 2 or 3 dimensions with typically 0.5–1.5 km horizontal resolution and 0.2–1.05 km vertical resolution.

The CEM produces its own simulation of both heating profiles and hydrometeor profiles. Based on these simulations, Tao et al. have developed a parameterization of the former using the latter as inputs. The scheme ignores convective-scale heat transport and the vertical and horizontal transport of hydrometeors. With these approximations, the apparent heat source due to convection Q_1 can be written in terms of the microphysical sources and sinks as

$$Q_1 = C_r + C_c + D_s + D_c + E + M$$

where the terms on the right hand side represent condensation of rain, condensation of cloud liquid water, deposition of snow and hail/graupel, deposition of cloud ice, evaporation of rain, and melting of snow and hail/graupel, respectively. Each of these is parameterized on the basis of features of observed hydrometeor profiles (e.g., the vertical gradient of rainwater is related to rain evaporation, while the surface precipitation rate constrains the net condensation and deposition).

An alternative latent heating algorithm makes no use of observed hydrometeor profiles. Instead, it assumes that a limited number of heating profiles characterize different types of convective systems in different geographic regions; for each type a convective and stratiform heating profile are defined, normalized by surface precipitation rate. If the type of system can be defined by its IR characteristics and the surface precipitation rate and fraction of stratiform precipitation are determined from space by PR, TMI and VIRS, the heating profile can be inferred from a look-up table.

20.6 TRMM Ground Validation Strategy

Since a rain radar has never flown in space, a concerted effort must be made to acquire surface-based "ground truth" rain data to use as validation standards for the space-based instruments. This is in itself a major undertaking, because there is no universally accepted standard for ground-based large-scale rain estimation, given the highly variable nature of rain in space and time.

To deal with this problem, TRMM is setting up a network of ground stations with a variety of rain-measuring instruments in different tropical climate regimes to provide the best possible long-term record of rainfall at selected locations for comparison to the satellite data. Figure 20.2 shows the distribution of currently operating and planned ground validation sites. The network consists of four primary sites (Florida, Texas, Darwin, and Kwajalein) equipped with various combinations of conventional rain gages, disdrometers, and rain radars (zenith pointing or scanning). The Primary Rainfall Test Facility in Florida is intended as a testbed for surface rain measurement techniques. It consists of a polarized multi-frequency microwave attenuation link with a number of state-of-the-art rain gages underneath for calibration, as well as several experimental rain gages (optical, Doppler). The four primary sites will be supplemented by a number of secondary "Special Climatology" sites (Brazil, Taiwan, Thailand, Israel, Japan, Hawaii, Guam, and perhaps India) whose mission is to provide several months of uninterrupted, high quality rainfall data per year to verify that the physics used by the TRMM satellite algorithms properly translates to different geographical locations (Kummerow, personal communication).

The validation strategy has several components. Since the most accurate estimate of rainfall on small scales comes from a dense network of gages, instantaneous space-based rainfall estimates at individual locations will be calibrated against gage data at the ground truth sites for spacecraft overpasses of the sites. The gages in turn will be used to refine Z-R algorithms for individual ground-based radars, and the larger-scale radar-derived rainfall patterns, averaged over time to form a climatology, will be used to validate the 500 km x 500 km TRMM monthly averages. The long time series obtained at the ground truth sites will also permit rainfall statistics to be accumulated to provide a better understanding of the sampling issue and to generate rainfall pdfs for use in some of the space-based instrument algorithms. Additional data on vertical profiles of hydrometeors, drop size distributions, convective/stratiform separation, temperature and humidity soundings, and vertical velocities will supplement the basic rain validation product. Finally, a number of airborne and shipborne Doppler radar data sets will be acquired over the oceans adjacent to the ground truth sites, up to 200 km away from the coast, to provide a comparison with the land-based estimates and an independent validation standard for TRMM estimates of rainfall over ocean.

20.7 Conclusion

It must be emphasized that TRMM is a research mission to determine the optimum techniques for measuring global rainfall on climatic time scales, rather than an operational rain mapping mission. Although TRMM will produce a 3-year tropical data set with accuracies useful to climate modelers, this is far from a climatology. Figure 20.6 shows a Hovmöller diagram of equatorial Pacific precipitation simulated by the GISS GCM (Del Genio et al., 1996). Although the basic pattern of ENSO development and decay is obvious in this figure and can be validated against satellite visible and infrared data, it is worth noting that the quantitative predictions of precipitation anomalies in this figure cannot be validated with current data sets, nor will such a product be produced directly by TRMM itself.

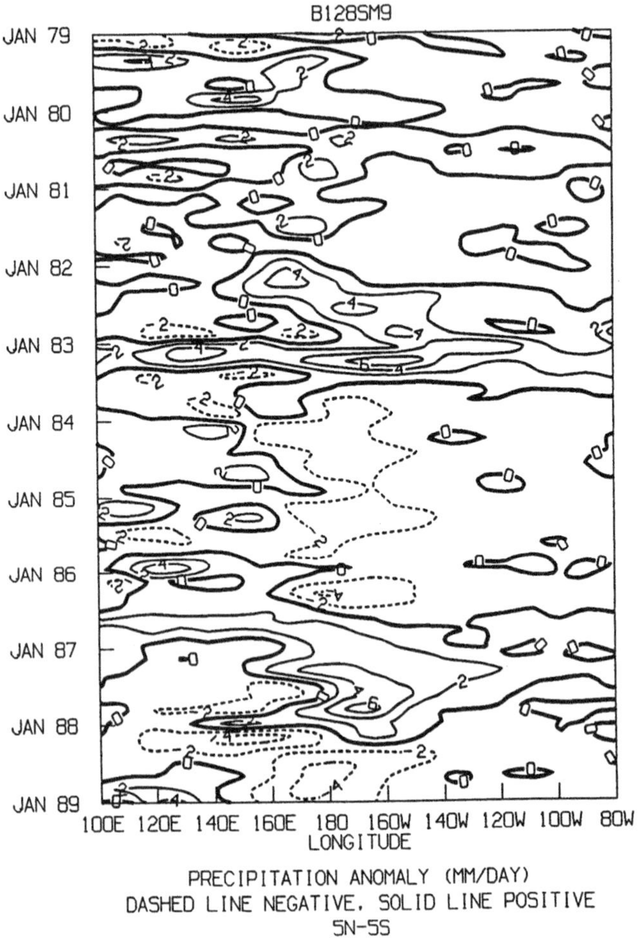

Figure 20.6: *Hovmöller diagram of equatorial Pacific precipitation anomalies relative to the 1979-1988 mean, simulated by the GISS GCM (Del Genio et al., 1996).*

In order for TRMM to have a maximum impact, it will be necessary for a hierarchy of "training" exercises to occur. The ground truth data will first validate the basic concept of radar rain measurement from space. Colocated radar and passive microwave data will then be used to train an optimum microwave-only rain algorithm that can be applied operationally to current and future satellite microwave instruments. Most ambitious is the final step, that of training a useful infrared (or visible-infrared) algorithm that can quantitatively estimate rainfall and replace the simple GPI. It is this last step that will allow the record of rainfall variations to extend backward in time over more than a decade and produce a true climatology for comparison to GCM estimates such as those in Figure 20.6. There is also a need to extend the TRMM experience to higher latitudes to make such climatologies truly global. This is a difficult problem because of the fundamentally different nature of stratiform precipitation in baroclinic wave systems from the convective systems that dominate tropical rainfall. Planning is currently taking place for a possible TRMM follow-on mission to address these concerns.

Acknowledgments

I thank William Kovari for assistance with the figures used in this paper. This research was supported by the NASA Tropical Rainfall Measuring Mission.

20.8 References

Adler RF, Negri AJ (1988) A satellite infrared technique to estimate tropical convective and stratiform rainfall. *J Appl Meteor* 27: 30–51

Adler RF, Yeh H-YM, Prasad N, Tao W-K, Simpson J (1992) Microwave simulations of a tropical rainfall system with a three-dimensional cloud model. *J Appl Meteor* 30: 924–953

Adler RF, Negri AJ, Keehn PR, Hakkarinen IM (1993) Estimation of monthly rainfall over Japan and surrounding waters from a combination of low-orbit microwave and geosynchronous IR data. *J Appl Meteor* 32: 335–356

Arkin PA (1979) The relationship between the fractional coverage of high cloud and rainfall accumulations during GATE over the B-scale array. *Mon Wea Rev* 107: 1382–1387

Arkin PA, Meisner BN (1987) The relationship between large-scale convective rainfall and cold cloud over the Western Hemisphere during 1982-84. *Mon Wea Rev* 115: 51–74

Atlas D, Rosenfeld D, Short DA (1990) The estimation of convective rainfall by area integrals, Part I: The theoretical and empirical basis. *J Geophys Res* 95: 2153–2160

Atlas D, Bell TL (1991) The relation of radar to cloud area-time integrals and implications for rain measurements from space. *Mon Wea Rev* 120: 1997–2008

Dai A (1995) Global precipitation variability and its relationship to other climate changes. Ph.D. dissertation, Columbia University, New York

Del Genio AD (1996) Observational requirements for modeling of global and regional climate change. This volume, chapter 2

Del Genio AD, Yao M-S, Kovari W, Lo KK-W (1996) A prognostic cloud water parameterization for global climate models. *J Climate* 9: 270–304

Fu R, Del Genio AD, Rossow WB (1990) Behavior of deep convective clouds in the tropical Pacific deduced from ISCCP radiances. *J Climate* 3: 1129–1152

Fu R, Del Genio AD, Rossow WB (1994) Influence of ocean surface conditions on atmospheric vertical thermodynamic structure and deep convection. *J Climate* 7: 1092–1108

Fujita M, Okamoto K, Yoshikado S, Nakamura K (1985) Inference of rain rate profile and path-integrated rain rate by an airborne microwave rain scatterometer. *Radio Sci* 20: 631–642

Gamache JF, Houze RA Jr (1983) Water budget of a mesoscale convective system in the tropics. *J Atmos Sci* 40: 1835–1850

Gray WM, Jacobson RW Jr (1977) Diurnal variation of deep cumulus convection. *Mon Wea Rev* 105: 1171–1188

Han Q, Rossow WB, Lacis AA (1994) Near-global survey of effective cloud droplet radii in liquid water clouds using ISCCP data. *J Climate* 7: 465–497

Inoue T (1987) An instantaneous delineation of convective rainfall areas using split window data of NOAA-7 AVHRR. *J Meteor Soc Japan* 65: 469–481

IPCC (1995) Intergovernmental Panel on Climate Change: Second Scientific Assessment of Climate Change (Houghton J, Meira Filho LG, eds): in press

Jaeger L (1976) Monatskarten des Niederschlags für die ganze Erde. Berichte des Deutschen Wetterdienstes 18, No. 139. Im Selbstverlag des Deutschen Wetterdienstes, Offenbach, W. Germany

Joe P (1996) Precipitation at the ground: Radar techniques. This volume, chapter 12

Johnson RH (1984) Partitioning tropical heat and moisture budgets into cumulus and mesoscale components: implication for cumulus parameterization. *Mon Wea Rev* 112: 1656–1665

Jorgensen DP, LeMone MA (1989) Vertical velocity characteristics of oceanic convection. *J Atmos Sci* 46: 621–640

Kummerow C, Mack RA, Hakkarinen IM (1989) A self-consistency approach to improve microwave rainfall estimates from space. *J Appl Meteor* 28: 869–884

Legates D, Willmott C (1990) Mean seasonal and spatial variation in gauge-corrected, global precipitation. *J Climatology* 3: 111–128

Liu G, Curry JA (1992) Retrieval of precipitation from satellite microwave measurement using both emission and scattering. *J Geophys Res* 97: 9959–9974

Lopez RE (1977) The log-normal distribution and cumulus cloud populations. *Mon Wea Rev* 105: 865–872

McGarry MM, Reed RJ (1978) Diurnal variations in convective activity and precipitation during phases II and III of GATE. *Mon Wea Rev* 106: 101–113

Meneghini R, Eckerman J, Atlas D (1983) Determination of rain rate from spaceborne radar using measurements of total attenuation. *IEEE Trans on Geoscience and Remote Sensing* GE-21: 34–43

Meneghini R, Atlas D (1986) Simultaneous ocean cross section and rainfall measurements from space with a nadir-looking radar. *J Atmos and Ocean Tech* 3: 400–413

Okamoto K, Kozu T (1987) Conceptual design of TRMM radar. Unpublished report, Radio Res. Lab., Japan, April 3, 1987

Orville RE, Henderson RW (1986) Global distribution of midnight lightning: September 1977 to August 1978. *Mon Wea Rev* 114: 2640–2653

Petty GW, Katsaros KB (1990) Precipitation observed over the South China Sea by the Nimbus-7 Scanning Multichannel Microwave Radiometer during winter MONEX. *J Appl Meteor* 29: 273–287

Randall DA, Cess RD, Blanchet JP, Boer GJ, Dazlich DA, Del Genio AD, Deque M, Dymnikov V, Galin V, Ghan SJ, Lacis AA, LeTreut H, Li Z-X, Liang X-Z, McAvaney BJ, Meleshko VP, Mitchell JFB, Morcrette J-J, Potter GL, Rikus L, Roeckner E, Royer JF, Schlese U, Sheinen DA, Slingo J, Sokolov AP, Taylor KE,

Washington WM, Wetherald RT, Yagai I, Zhang M-H (1992) Intercomparison and interpretation of surface energy fluxes in atmospheric general circulation models. *J Geophys Res* 97: 3711–3724

Reed RJ, Recker EE (1971) Structure and properties of synoptic- scale wave disturbances in the equatorial western Pacific. *J Atmos Sci* 28: 1117–1133

Rosenfeld D, Atlas D, Wolff DB, Amitai E (1992) Beamwidth effects on *Z-R* relations and area-integrated rainfall. *J Appl Meteor* 31: 454–464

Shea DJ (1986) Climatological atlas: 1950-79. Surface air temperature, precipitation, sea-level pressure and sea surface temperature (45S-90N). NCAR/TN-269+STR, 153 pp

Shin K-S, North GR (1988) Sampling error study for rainfall estimate by satellite using a stochastic model. *J Appl Meteor* 28: 1218–1231

Simmer C (1996) Precipitation from satellites. This volume, chapter 11

Simpson J (1988) Tropical Rainfall Measuring Mission (TRMM): A satellite mission to measure tropical rainfall. Report of the Science Steering Group (Simpson J, ed), NASA Publ., U.S. Govt. Printing Office, Washington, DC

Smith EA, Mugnai A (1988) Radiative transfer to space through a precipitating cloud at multiple microwave frequencies. Part II: Results and analysis. *J Appl Meteor 27*: 1074–1091

Smith EA, Mugnai A, Cooper HJ, Tripoli GJ, Xiang X (1991) Foundations for statistical-physical precipitation retrieval from passive microwave satellite measurements. Part I: Brightness temperature properties of a time-dependent cloud- radiation model. *J Appl Meteor* 31: 506–531

Spencer RW (1986) A satellite passive 37 GHz scattering based method for measuring oceanic rain rates. *J Clim Appl Meteor* 25: 754–766

Tao W-K, Simpson J, Lang S, McCumber M, Adler R, Penc R (1990) An algorithm to estimate the heating budget from vertical hydrometeor profiles. *J Appl Meteor* 29: 1232–1244

Tao W-K, Lang S, Simpson J, Adler R (1993) Retrieval algorithms for estimating the vertical profiles of latent heat release: Their applications for TRMM. *J Meteor Soc Japan* 71: 685–700

Thompson RM Jr, Payne SW, Recker EE, Reed RJ (1979) Structure and properties of synoptic-scale wave disturbances in the intertropical convergence zone of the eastern Atlantic. *J Atmos Sci* 36: 53–72

Weinman JA, Meneghini R, Nakamura K (1990) Retrieval of precipitation profiles from airborne radar and passive radiometer measurements: Comparison with dual-frequency radar measurements. *J Appl Meteor* 29: 981–993

Wilheit TT, Greaves JR, Gatlin JA, Han D, Krupp BM, Milman AS, Chang ES (1984) Retrieval of ocean surface parameters from the scanning multichannel microwave radiometer (SMMR) on the Nimbus 7 satellite. *IEEE Trans Geosci Remote Sens* 22: 133–143

Williams ER, Zhang R, Rydock J (1991) Mixed-phase microphysics and cloud electrification. *J Atmos Sci* 48: 2195–2203

Wu R, Weinman JA (1984) Microwave radiances from precipitating clouds containing aspherical ice, combined phase, and liquid hydrometeors. *J Geophys Res* 89: 7170–7178

Chapter 21

Model Development for Passive Microwave Remote Sensing at ESA ESTEC

J. Noll, J.P.V. Poiares Baptista, M. Borgeaud
European Space Agency, ESA ESTEC
Postbox 299, NL 2200 AG
Noordwijk, The Netherlands

21.1 Introduction

In Europe, the information gathered by METEOSAT as well as ESA's remote sensing satellite ERS-1 have demonstrated the benefit of space-borne instruments for observation of the Earth's environment. ESA is preparing for future Earth-observation missions. In particular, these are the development of Meteosat Second Generation (MSG), METOP and ENVISAT as well as studies for future Earth Explorer and Earth Watch missions in the next century. A more detailed information of ESA's Earth observation programme is attached as an annex.

Microwave measurements from space can provide all-weather monitoring of the Earth's surface and atmosphere on a global, day and night basis. This kind of observations plays today a vital role in monitoring the Earth environment.

The paper will focus on user requirements and model development for the Multi-frequency Imaging Microwave Radiometer (MIMR), acknowledging the wide range of applications for passive microwave radiometers. MIMR is scheduled to fly on the European METOP and the American EOS-PM satellites in about 2000. The paper will discuss direct models and inverse algorithms, and relate instrument features to the geophysical products that can be obtained. Snow monitoring is presented as one example application for modelling and geophysical parameter retrieval.

21.2 Geophysical Products as Derived from MIMR Data

In the following ESA's multi-frequency Imaging Microwave Radiometer (MIMR) will be briefly explained, Due to the multi-frequency capabilities of MIMR, it's data will be used for geophy-

NATO ASI Series, Vol. I 45
Radiation and Water in the Climate System:
Remote Measurements
Edited by Ehrhard Raschke

sical parameter retrieval for cryosphere, land, ocean and atmosphere communities. Geophysical products related to the instrument parameters will be addressed.

21.2.1 Short history of microwave radiometers

Starting with the Cosmos 243 satellite in 1968, microwave radiometers have been carried aboard several spacecrafts (Table 21.1). To fulfill multiple requirements, multi–channel radiometers

Sensor, Spacecraft, Country	year	freq. & pol.	inc. ∠ [°]	resolution [km]
Cosmos 243, USSR	1968	3.5, 8.8	0	37
Cosmos 384, USSR	1970	22.2, 37	0	
ESMR, Nimbus 5, USA	1972	19.35	var	25
NEMS, Nimbus 5, USA	1972	22.2, 31.4, 53.6, 54.9, 58.8	0	200
S–193, Skylab, USA	1973	13.9	var.	16
S–194, Skylab, USA	1973	1.4	0	115
Meteor, USSR	1974	37 V H	35	
ESMR, Nimbus 6, USA	1975	37 V H	var	20 × 43
SCAMS, Nimbus 6, USA	1975	22.2, 31.6, 52.8, 53.8, 55.4		150
SMMR, Seasat, USA	1978	6.6, 10.7, 18.0, 21.0, 37.0 V H	50	22 × 35
SMMR, Nimbus 7, USA	1978	6.6, 10.7, 18.0, 21.0, 37.0 V H	49	18 × 27
SAMIR, Bhaskara–II, India	1981	19.4, 22.2, 31.4		125
SSM/I, DMSP F08 (- F11), USA	1987 (- 1991)	19.3 VH, 22.2 V, 37.0 VH, 85.5 VH	53	70x45 – 16x14
MIMR, ESA	future	6.8, 10.7, 18.7, 23.8, 36.5, 90 all VH	50	39x60 – 3x5
AMSR, NASDA	future	6.9, 10.7, 18.7, 23.8, 36.5, 50.3, 52.8, 89	50	5 - 60 km

Table 21.1: *Space-borne microwave radiometers for earth observation.*

have been designed, starting with the SMMR (Scanning Multi–Channel Microwave Radiometer) on Nimbus–7 and SEASAT. The SMMR was able to measure the sea surface temperature and wind speed together with the atmospheric water vapour, the liquid water and rain rate, the sea ice concentration and snow coverage. At present, the Special Sensor Microwave/Imager (SSM/I) is monitoring the Earth's surface (Hollinger et al., 1990). It provides high–performance microwave measurements of geophysical parameters such as wind speed, water vapour, ice type and ice extent. As the number of frequencies is no longer a challenging requirement for instrument designers, modern multi-frequency radiometers can satisfy the needs of various user communities.

21.2.2 MIMR description

MIMR, scheduled to fly on both the European METOP and the American EOS-PM platform around 2000, is a 12–channel six frequency linearly polarized, passive microwave radiometric system which measures brightness temperatures at 6.8, 10.6, 18.7, 23.8, 36.5 and 89 GHz (Table 21.2). It receives both vertically (V) and horizontally (H) linearly polarized radiation for all channels except for the only vertically receiving 23.8 GHz channel. The advantages in terms of power and mass consumption for passive instruments compared to active instruments is seen also for this sensor consuming 190 W of power for a weight of 200 kg.

frequency [GHz]	6.8	10.65	18.7	23.8	36.5	90
pixel [km]	60	38	22	20	11.6	4.9
sampling interval [km]	15.5	15.5	15.5	15.5	7.75	3.87
radiometric sensitivity [°K]	0.2	0.4	0.5	0.5	0.5	0.7
radiometric stability [°K]	0.2	0.4	0.5	0.5	0.5	0.7
radiometric accuracy [°K]	1	1	1.5	1.5	1.5	1.5

Table 21.2: *MIMR system performance.*

MIMR characteristics can be explained using the specifications of the existing Special Sensor Microwave Imager (SSM/I) onboard the satellites of the Defense Meteorological Satellite Program (DMSP) (Fig. 21.1).

The DMSP F08 spacecraft flies a circular sun–synchronous orbit with an inclination of 98.8° and an altitude of 835 km. At this height the orbit period is 102.0 minutes, so that approximately 14.1 orbits per day are performed. With a swath width of 1400 km an almost global coverage is obtained within one day. Only the grey shaded areas of Fig. 21.1b) are not covered within one day. Due to the 98.8° inclined orbit a small circular area with a radius of 280 km on the north and south pole is not covered at all. The instrument is mounted on top of the spacecraft and consists of an offset parabolic reflector (61 × 66 cm), a corrugated, broad–band seven port horn antenna and the radiometric subsystem.

Compared with SSM/I, MIMR will consist of an offset parabolic reflector of dimension 1.6 m × 1.4 m. MIMR rotates continuously about an axis parallel to the local spacecraft vertical, resulting in a conical scan with an incidence angle of 50° on earth. The spin rate of 26 rpm provides a period of 2.3 s during which the spacecrafts sub-satellite point travels 15.5 km. The concept followed by MIMR considerably improves the performances compared to previous space-borne radiometric systems (Table 21.2).

21.2.3 Geophysical products and user requirements

The MIMR mission objectives are the measurement and full-time monitoring of geophysical parameters related to atmosphere, ocean, cryosphere and land (Table 21.3). The subset summarized in Table 21.4 reflects the channel sensitivity of the used 6.8 to 90 GHz channels in terms of geophysical parameters.

A short overview is presented reflecting the User Requirements and Data Products for the different user communities for MIMR.

Ocean and marine atmosphere

The key geophysical parameters that will directly be retrieved by MIMR over the ocean includes: sea surface temperature, surface wind speed (or friction velocity), total water vapour of the atmosphere; and liquid water column. These parameters are used both operationally (the first three) as well as in oceanographic and meteorological climate and process studies. The resolution requirements and measurements accuracy required for the retrieval of the geophysical parameters are summarized in Table 21.5.

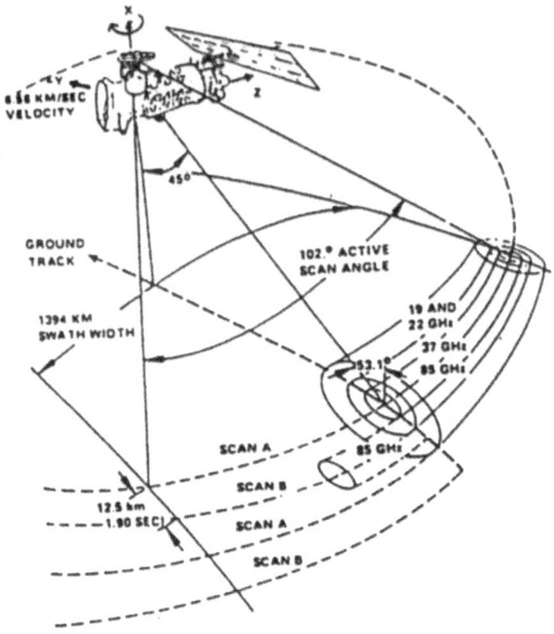

a) Scan geometry

b) 24 h coverage

Figure 21.1: *Scan geometry and global coverage of the SSM/I aboard the DMSP F08 spacecraft (Hollinger et al., 1990).*

Parameter	Sampling distance [km]	Range of values	Quantization levels	Absolute accuracy
Water vapour	15.5	0 to 1 kg/m^2	0.05 kg/m^2	±0.1 kg/m^2
Cloud liquid water	15.5	0 to 60 kg/m^2	0.1 kg/m^2	±0.2 kg/m^2
Precipitation over ocean	15.5	0 to 25 mm/h	0, 5, 10, 15 20, $\leq$ 25 mm/h	±5mm/h
Cloud amount	15.5	0 to 100%	1%	±20%
Ocean temperature	15.5	0 to 30°C	0.2°C	±1°C
Ocean Surface wind speed	15.5	3 to 25 m/s	1 m/s	±2 m/s
Ice	15.5			
- area covered		0 to 100%	5%	±10%
- age		First year, multi-year		
- edge location				±2 km
Soil moisture	15.5	0 to 60%	1%	
Snow water equivalent	15.5	0 to 30 cm	1 cm	±3 cm
Surface type	15.5	12 types		

Table 21.3: *MIMR Environmental Products Performance.*

		MIMR frequency [GHz]					
Mission	Product	6.8	10.65	18.7	23.8	36.5	89
Atmosphere	Water vapour			x	x	x	
	Cloud liquid water				x	x	x
	Precipitation over ocean				x	x	x
Ocean	Temperature	x	x	x	x		
	Wind speed	x	x	x	x		
Sea-ice	Concentration			x		x	
	Type (age)			x		x	x
	Edge location				x	x	x
Land	Soil moisture	x	x	x			
	Snow water equivalent			x		x	x

Table 21.4: *Geophysical products and related microwave frequencies for MIMR.*

Other parameters which are related to and will benefit from the availability of the above mentioned set of parameters but might need additional information from other sources are parameters related to air-sea interaction such as: latent and sensible heat fluxes, longwave radiative heat flux, momentum flux, surface rain rate, sea ice extent, type and coverage, oceanic particle flux (biogenic and lithogenic), oceanic white-cap coverage and ocean-atmosphere gas exchanges.

Parameter	primary frequency [GHz]	spatial resolution [km]	range of values	expected RMS accuracy
SST	6.8/10.65	85	-2 to 35 C	0.5K
Wind speed	36.5/10.65	15/55 *1	0 - 40 m/s	1.5 m/s
Water vapour	23.8	25	0 to 70 mm	1 mm
Water vapour column (500m) in PBL	18.7/23.8/36.5	25	0 to 15 mm	0.5 mm
Liquid Water Column	36.5/89	15	0 to 1 mm	0.05 mm

Table 21.5: *Measurement accuracy at specified resolutions for primary MIMR ocean products (*1 non-precipitation/precipitation)*

Precipitation

There is a wide variety of applications for rainfall data. The spatial resolution at which the rainfall measurements are needed depends upon the application. For many of the applications, the spatial resolution needed for the rainfall is that of the models which typically runs at about 250 km currently and could improve to the order of 100 km in the foreseeable future. However, this is not the same as the needed instrumental resolution. Typical rain features are of the order of 10 km or less. The relationships between the observed brightness temperature and the rainfall intensity are highly non-linear in such a manner that the average rain rate over a partially filled beam will always be underestimated. The degree of underestimate depends on the actual distribution of the rainfall. Thus, any correction for this beam filling error can, at best, only be valid in an average sense over a statistical ensemble of measurements. Even at this level, determination of the beam filling correction is problematic. The closer the beam size comes to the scale of the rain features the less the correction will be and thereby the less the impact of its uncertainty and variability. At the 70 km resolution of the SSM/I, the correction is between 1.5 and 2 (an instrument with a delta function for an antenna pattern would have a beam filling correction of 1). In the Tropical Rainfall Measurement Mission (TRMM) the resolution will improve by slightly better than a factor of 2 over that of SSM/I. This is expected to keep the beam filling correction at acceptable levels. Thus, for many rainfall measurements a resolution of the order of 25 km at 18 GHz is acceptable.

The accuracy of rainfall totals, in order to be useful for most purposes must be of the order of 20% or better. From a low orbiting spacecraft one cannot measure all the rainfall at a point. Rather one gets a series of instantaneous views spaced rather widely in time; i.e. we have only a sampled subset of the rainfall time series. This sampling introduces serious errors. From a single polar orbiting satellite, even with a perfect rain measuring instrument, we cannot achieve this accuracy for space-time volumes less than about 500 km x 500 km x 1 month. The use of additional satellite observations to complement the MIMR will be essential. A second MIMR in orbit would be quite useful. It would reduce the dominant term in the error budget for rainfall totals by a factor of at least $\sqrt{2}$ enabling meaningful totals to be generated for smaller space time volumes. The errors introduced by undersampling coherent rainfall signals such as the diurnal cycle or the propagation of storm systems would be larger.

Data Products for Precipitation

Data products at pixel level concerning precipitation can be found in Table 21.6, with the resolution of the 36.5 GHz channel as the driver of the pixel level products.

Geophysical parameter	Application	Coverage Spatial / Temporal	Resol-ution / grid	Delivery time	Algorithm Maturity	Ac-curacy
Instantaneous precipitation over land	Operational Meteorology / Hydrology / crop forecast	Regional	17 km	< 15 min	Low	50%
Instantaneous precipitation over land	Operational Meteorology / Hydrology / crop forecast	Global	17 km	< 3 hours	Low	50%
Instantaneous precipitation over ocean	Operational Meteorology	Regional	17 km	< 15 min	Medium	50%
Instantaneous precipitation over ocean	Operational Meteorology	Global	17 km	< 3 hours	Medium	50%
Rain / No Rain Flag	Other instruments	Regional	17 km	< 15 min	Medium to High	
Rain / No Rain Flag	Other instruments	Global	17 km	< 3 hours	Medium to High	
Stratiform / Convective Flag	Operational Meteorology	Regional	17 km	< 15 min	Low to Medium	
Stratiform / Convective Flag	Operational Meteorology	Global	17 km	< 3 hours	Low to Medium	

Table 21.6: *Level 2a products for precipitation monitoring with MIMR (resolution from 36.5 GHz channel).*

Table 21.7 shows the foreseen Level 2b products. These are on a monthly basis and the accuracies shown assume that morning and afternoon instruments are used. If only one instrument is used the accuracies of these products will be severely degraded since the diurnal cycle of precipitation will be sampled only once. Results recently published indicate that there may be two daily peaks for the probability of occurrence of rain in the tropics and these occur at different times over the ocean or over land. The use of a AM and PM platforms has also an impact on the Level 2a products (pixel level) in Operational Meteorology.

Geophysical parameter	Application	Coverage Spatial / Temporal	Resol-ution / grid	Delivery time	Algorithm Maturity	Ac-curacy
Monthly precipitation total over land	Climate & crop forecast	Global	2.5°	Month	Medium	20%
Monthly precipitation total over ocean	Climate	Global	2.5°	Month	Medium	20%

Table 21.7: *Level 2b products (AM & PM Platforms assumed).*

The resolution indicated in Table 21.7 is the one presently used in GPCP (Global Precipitation Climate Project). Some communities (hydrology) would be interested in higher resolutions however these are only achievable with a degradation of the accuracy.

Cryosphere

The key objectives of the MIMR mission as envisioned by the cryosphere group are: (a) to generate data useful for global climate modeling and related studies, (b) to improve our understanding of geophysical processes in ice covered and adjacent regions, and (c) to monitor ice cover changes for e.g. ship routing.

Frequency [GHz]:	6.8	10.7	19	23	36.5	89	how well	req. accur.
SEA ICE:								
ice concentration	+	o	o	o	o	+	9	< 5%
ice type	o	o	o	o	o	o	8	15%
ice edge location	*	o	o	o	o	o	8	TBD
area ice extent	-	-	o	o	o	o	9	4%
actual ice area	-	o	o	o	o	+	9	5%
snow cover	+	o	o	o	o	o	5	10%
melt cover	+	o	o	o	o	o	6	TBD
polynya areas	+	o	o	o	o	o	7	TBD
large scale vel.	-	-	o	o	o	o	6	TBD
melt water prod.	-	o	o	o	o	o	6	TBD
ice regimes	o	o	o	o	o	o	6	TBD
ice temperature	o	o	o	o	o	-	6	TBD
ICE SHEETS:								
accumulation rates	+	o	o	o	o	o	6	TBD
melt area	-	-	o	-	o	o	8	TBD
surface type	-	o	o	o	o	o	6	TBD
temperature	o	+	o	o	o	o	5	TBD
icebergs stats.	-	-	-	o	o	o	5	TBD

Table 21.8: *Geophysical Parameters and Requirements for cryosphere monitoring (required channels: o is a must; + is helpful; and - is desirable how well are user req. met: rating from 1 to 10, with 10 being excellent)*

MIMR meets the basic resolution requirements for most of the sea ice and ice sheet applications (Table 21.8). The improved resolution of MIMR at all channels over current systems will enable higher accuracies of the derived products and improved characterization of sea ice in important economic zones like the Gulf of Bothnia. Assuming a 12.5 by 12.5 km pixel size (comparable to the 37 GHz resolution), leads or polynyas with a total open water area of around 8 km^2 within the pixel can be detected by the sensor (appearing in the ice concentration maps as 5% open water). While even better resolution would be desirable, the current system as designed is a good compromise considering limitations in terms of antenna size and data volume.

The swath width of 1400 km (or greater) is also adequate for cryosphere applications. In less than a day, because of overlapping coverage near the poles, the entire polar regions would be almost completely covered. Also, large spatial coverage that includes a study region is made with each orbit and a temporal resolution as good as an hour and a half, using consecutive orbits is possible.

Land and Snow

The land applications of MIMR fall into four classes, i.e. soil moisture monitoring using the lower frequencies (6.8 and 10.65 GHz), large scale vegetation monitoring with polarization difference and multi-frequency techniques, land surface temperature monitoring (18 GHz H&V), snow extent monitoring using both spectral and polarisation information and flood monitoring. The main land parameters with desired and expected performances are summarized in Table 21.9.

Parameter	Radiometric Sens. [K]		Spatial Resolution [km]		
	Ideal	MIMR Perf.	Ideal	Adequate	MIMR Perf.
Vegetation extent	0.5	0.4	5-10	25	38
Soil moisture	0.5	0.15	5-10	25 (*1)	60
Snow cover extent	0.1-1 0.7 <10	10	22-11		
Surf. temp.	0.2	0.4-0.5	1	25-50	30-20
Vegetation biomass	0.2	0.2-0.6	1	10	50-20
Flooding	0.2	0.4-0.5	1		
Frozen soil extent	0.2	0.2-0.6	1	20	50-20
Snow liq. water content	0.2	0.2-0.4	1	5-10	50-30
Snow charac.	0.5-1	0.7	<10	10-20	22-11

Table 21.9: *Main land parameters together with desired and expected performances (*1 in arid zones)*

On land surfaces temporal variations are especially large. These variations are expressed by brightness temperature changes and they contain important geophysical information on the state of the surface. Important parameters are snow state, effects of thermal inertia (related to soil moisture and soil type/surface cover) and surface fluxes. MIMR will have a better temporal coverage as compared to optical and IR sensors which depend on cloud-free - and therefore biased - observations. Nevertheless all types of sensors can contribute to the complex information needed for the heterogeneous surfaces (e.g. bare soil, vegetation, land water, snow). In case of MIMR a large fraction of the data analysis will be based on multi-temporal data. In order to make such data reliable users will need:

- high pointing accuracy and stability (1 km or better)
- separate datasets for ascending and descending passes
- high temporal coverage. For optimum coverage of the diurnal cycle at least one observation in the morning and one in the afternoon are required.

21.2.4 Model development for MIMR

The development of land applications for passive microwave radiometers has been poor so far, as the resolution of the SSM/I with 70 × 45 km for 19 GHz and 16×14 km for 85 GHz is not very high. Due to the increase of the antenna size of 1.6 m compared to 0.65 m for the SSM/I the spatial resolution of the corresponding channels is improved by about a factor three for MIMR. The radiometric accuracy is, despite the antenna size, kept very similar to the SSM/I. With a sampling distance of 15.5 km for the 6.8 to 23.8 GHz channels and even smaller for the 36.5 and 90 GHz channels (Table 21.2), the use of MIMR data will open new fields of applications.

Modelling is an important part of the data analysis as it combines the measured T_B data with geophysical parameters and vice versa. In principle modelling uses two types of algorithms: the direct, and the inverse algorithms (Fig. 21.2).

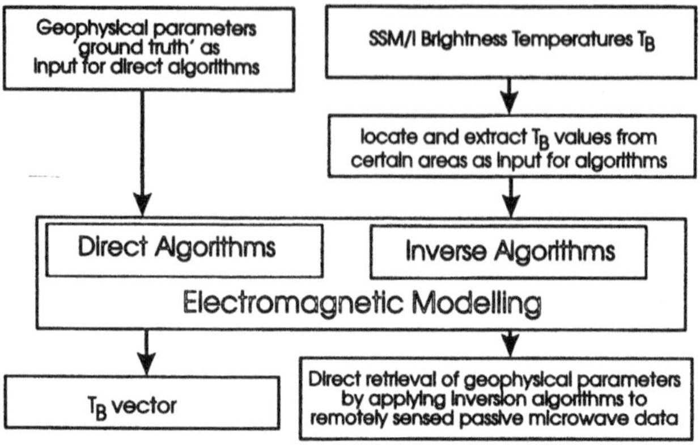

Figure 21.2: *Connection of SSM/I-T_B data and geophysical parameters for modelling.*

Direct algorithms are the ones that from a given geophysical environment allow for the calculation of T_B for a given frequency and polarization. These algorithms are for example used by instrument designers when it is fundamental to know what can be expected, in terms of measurement using a particular instrument, from a certain environment.

On the other hand users need to retrieve geophysical parameters (e.g. wind speed, sea surface temperature) out of the observed brightness temperatures. This relation is expressed by inversion algorithms. Modelling is complete, if both kind of algorithms are available and allow to close the loop from geophysical parameters to temperatures and back. In such a case, sensitivity analysis can be performed to give the accuracy of the observed results.

Radiative transfer theory can be used to model the influence of the physical scene that includes the solid earth and the atmosphere. The latter can be defined by a vertical temperature profile, absorptance due to gas and water vapour as well as scattering due to water droplets and ice crystals are also taken into account by the radiative transfer approach. However a general inversion of this theory is, in most cases, very difficult to perform. Instead most inversion algorithms used in an operational environment are based on empirical or statistical formulae that are very efficient in terms of computing time. Several have been proposed in the literature to retrieve geophysical parameters from the measured brightness temperatures (Pulliainen et al., 1993).

Radiative transfer models give, however a better insight into the physical processes and for this reason are used in R&D environments.

21.3 Geophysical Parameter Retrieval

21.3.1 Surface - Atmosphere model

In order to model the MIMR brightness temperature, an emission simulation and inversion tool has been developed under ESA supervision (Pulliainen et al., 1990, Pulliainen et al., 1993).

21.3.1.1 Emission simulation

In this section the emission simulation as part of the MIMR-software is briefly described (Pulliainen et al., 1990). Emittance models are implemented to simulate the surface emittance ϵ_s from

- ocean surface using Pandey's or Wilheit's model (Pandey et al., 1982, Wilheit, 1979),
- snow-covered land using HUT's model (Hallikainen et al., 1988),
- vegetation covered land using Kerr's model (Pulliainen et al., 1990), and
- different ice types using constant emittance values.

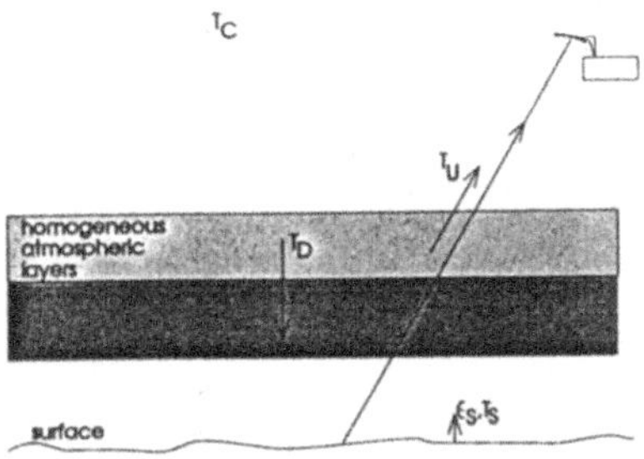

Figure 21.3: *Simulation model of a non scattering atmosphere.*

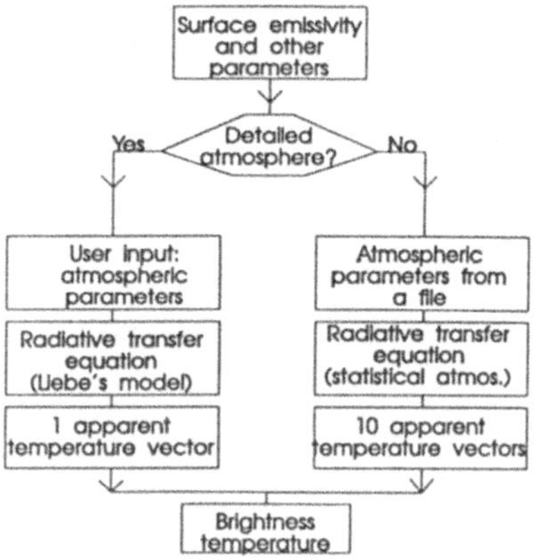

Figure 21.4: *Modelling with statistical or detailed atmosphere.*

Knowing the emittance of the surface the apparent temperature observed from space (Fig. 21.3) is calculated using

$$\begin{aligned} T_B &= T_U(1 - t_{trans}) + \epsilon_S t_{trans} T_S + (T_D(1 - t_{trans}) + \\ &\quad + t_{trans} T_C)(1 - \epsilon_S) t_{trans} \end{aligned} \tag{21.1}$$

where ϵ_s is the surface emittance, t_{trans} is the transmittance of the atmosphere, T_U and T_D are the atmospheric apparent radiative temperatures for up-welling and down-welling radiation. T_C and T_S are respectively cosmic and surface temperatures.

Two options are implemented to consider the influence of the atmosphere (Fig. 21.4). For a detailed atmosphere specific input parameters are requested. Liebe's MPM model is then used for calculating the upwelling and downwelling contributions of the atmosphere (Liebe, 1989).

If no a priori information about the atmosphere is available, the space observed brightness temperature is simulated by using statistical transmittance values. Using

$$T_{U,D} = \alpha_{U,D} T_s (1 - t_{trans}), \tag{21.2}$$

the up- and downwelling temperatures T_U, T_D are based on the atmospheric profile factors $\alpha_{U,D}$ and the transmittance t_{trans}. For a full explanation of this approach see Pulliainen et. al. (Pulliainen et al., 1993). Equation (21.2) is an approximation that has shown very good results for climates similar to Norway's.

21.3.1.2 Inversion algorithm

Inversion for retrieving geophysical parameters are implemented following the two different approaches shown in Fig. 21.5:

(A) inversion algorithms as (linear) combinations of MIMR channels or

(B) search for an inverse solution by using a model with input parameters, which represents the actual measurements.

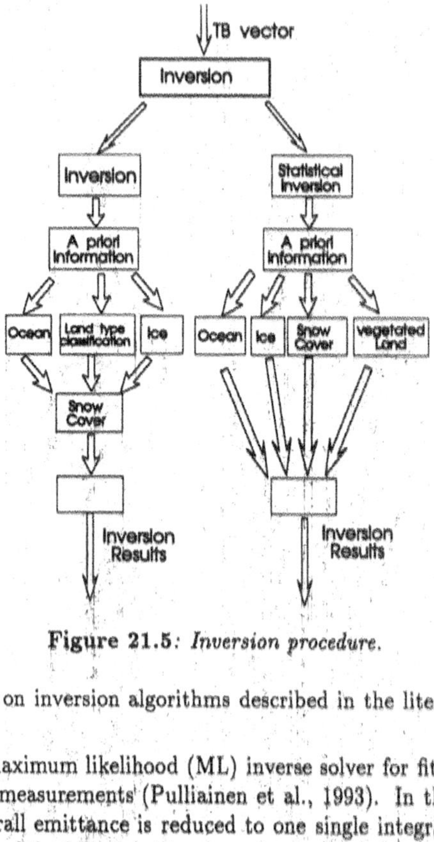

Figure 21.5: *Inversion procedure.*

Procedure (A) is based on inversion algorithms described in the literature (Pulliainen et al., 1993).

Procedure (B) uses a maximum likelihood (ML) inverse solver for fitting the model with the results of multichannel measurements (Pulliainen et al., 1993). In this case the atmospheric contribution to the overall emittance is reduced to one single integral parameter in order to simplify the application of the ML algorithm.

After discussion some of the in-build models in the next section, an application is performed using both the direct models and inversion algorithms for properties of land covered snow.

21.3.1.3 Examples of in-build models and algorithms

Most algorithms are quite complex and their detailed descriptions goes beyond the scope of this chapter. Only some example empirical or statistical algorithms are presented in the following to demonstrate the use of direct and inverse modelling.

Water vapour over sea

International programmes have been initiated to understand the dynamic ocean/atmosphere system. The Tropical Ocean and Global Atmosphere (TOGA) Programme started in 1985 and focussed on the tropical ocean/atmosphere system and its effect on the climate at higher latitudes. The goal of the World Ocean Circulation Experiment (WOCE) is to develop a data set of unprecedented magnitude and quality for use in developing numerical models of ocean circulation and physical processes and to couple them with atmospheric models. Passive remote sensing with ESMR, SMMR, SSM/I and MIMR can provide global information about sea surface temperature, sea surface wind speed and other ocean parameters (Goodberlet et al., 1990).

For water vapour (WV) Lojou (1990) has proposed algorithms for SMMR, SSM/I and MIMR (Sciré Scappuzzo, 1992). Using the MIMR channels (see Table 21.1) WV can be expressed as a function of the 18.7, 23.8 and 37 GHz channels as follows

$$\begin{aligned} WV[\mathrm{g/cm^2}] &= 8.45 + 3.71 ln(280 - T_{H19}) + 1.05 ln(280 - T_{V18}) - 6.52 ln(280 - T_{H23}) \\ &\quad -2.69 ln(280 - T_{V23}) + 3.32 ln(280 - T_{H37}) - 0.417 ln(280 - T_{V37}). \end{aligned} \quad (21.3)$$

This algorithm (Eq. 21.3) is valid only for $T_B < 280$ K and assumes an emittance with an averaged homogeneous distribution that can only be found when observing ocean surfaces. After applying this formula to measured data the water vapour is obtained and its accuracy proved by comparing the results with measurements taken from other sensors. Similar formulae can be applied for other parameters like rain rate and liquid water content.

Most inversion algorithms that are used in an operational environment are based on formulations as simple as the one given above. These type of formulae are obtained through parameter fitting to either existing measurements or to calculated values obtained from previously tested direct models.

Wind speed and sea surface temperature

Brightness temperature as a function of ocean parameters is presented as an example for direct algorithms. Three major physical properties influence the emission characteristics, namely the dielectric constant, the surface geometry and the foam cover (or indirectly the wind speed).

The permittivity ε_w of sea water is a function of frequency, salinity and water temperature and can be calculated by the modified Debye equation (Ulaby et al., 1986). At frequencies higher than 6 GHz (all SSM/I channels) the effect of salinity is negligible. The surface geometry is regarded as a rough surface with partial foam cover depending on the wind speed W. Wilheit (1979) has proposed an algorithm to model this behaviour where the reflectance $\Gamma_{h,v}$, given by the Fresnel reflections coefficients with $\Gamma_{h,v} = R^2_{h,v}(f, \Theta)$, is averaged over the surface slope distribution of a wind roughened sea and it uses a Geometrical Optics (GO) approach of the sea surface. A partial foam cover is taken into account for $W > 7$ m/s, as it reduces the surface reflectance. An empirical expression for the foam reduction is given as

$$K_{foam} = \begin{cases} \alpha(1 - e^{-f/f_0})(W[\mathrm{m/s}] - 7) & \text{for} \quad W \geq 7\mathrm{m/s} \\ 0 & \text{for} \quad W < 7\mathrm{m/s} \end{cases}, \quad (21.4)$$

where $\alpha = 0.006$ and $f_0 = 7.5$ GHz.

Finally the emittance e_{sea} is defined by the foam reduced averaged reflectance as

$$e_{sea}(f, \Theta, W, SST) = 1 - [\Gamma_{h,v;ave}(f, \Theta, W, SST)\,(1 - K_{foam}(f, W))] \quad (21.5)$$

and has a typical behaviour shown in Fig. 21.6 as a function wind speed and in Fig. 21.7 as a function of SST.

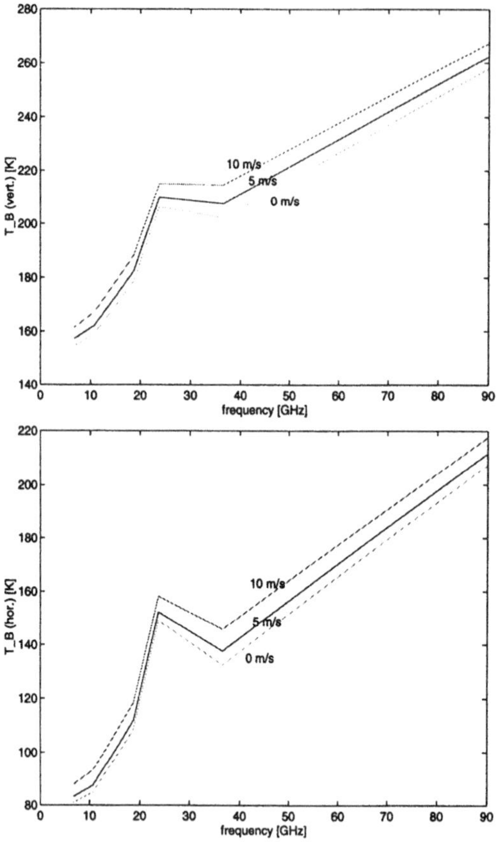

Figure 21.6: *Vertical (top) and horizontal (bottom) brightness temperature as function of wind speed using Wilheit's model for partial foam covered sea (SST = 20°).*

In Fig. 21.6 the effect of the Brewster angle is clearly shown, leading to higher vertical than horizontal brightness temperatures. T_B increases with wind speed due to the increased sea surface roughness.

T_B also increases with SST and frequency. The 23 GHz channels shows the highest sensitivity that can be explained by the influence of the water vapour absorptance around 22.2 GHz.

It should be noted that the model was tested only for frequencies lower than 37 GHz. However, due to the theoretical basis of the model the results at 90 GHz seem to be acceptable. Further investigations have to be undertaken to describe the polarization dependence of foam.

Land surface type classification

This section will show a completely empirical algorithm for the classification of land surface types. The algorithm can be considered as a combined physical/statistical method, where frequency and polarization differences are used to set up the classification rules.

The classification is in terms of topography, natural vegetation, land use, climate soils and water resources. A sizable number of formulae/relations with all the channel measurements

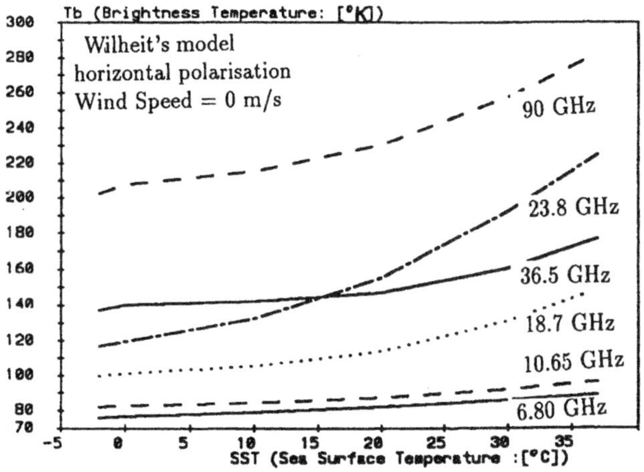

Figure 21.7: *Brightness temperature as function of sea surface temperature (SST) using Wilheit's model for partial foam covered sea ($W = 0m/s$, $\Theta = 0°$).*

of SSM/I are used to perform this classification. An example using the set of formulae for identification of natural high density forests, as found in the Amazon, Congo jungles or in the Appalachian forest *Dense Vegetation* is presented below.

$$T_{22V} - T_{19V} \leq 4 \text{ K} \tag{21.6}$$

$$0.5\,(T_{19V} + T_{37V} - T_{19H} - T_{37H}) \leq 1.9 \text{ K} \tag{21.7}$$

$$T_{85V} - T_{37V} < 0.5 \text{ K} \tag{21.8}$$

$$T_{85H} - T_{37H} < 6.0 \text{ K} \tag{21.9}$$

$$T_{19V} > 262 \text{ K} \tag{21.10}$$

Each of these equations is related to the influence of a different physical parameter. Eq. 21.7 represents the polarization behaviour of the vegetation covered soil. In case of a lossy dense canopy layer emission is isotropic showing very small polarimetric differences. As the density decreases (*jungle* $\rightarrow$ *desert*) polarization differences caused by the soil become significant, thus classifying the density of vegetation (Fig. 21.8).

Condition 21.8 is a precipitation flag showing precipitation clouds over vegetation. Rule 21.10 ensures that no ice or snow are present and that what the instrument is seeing is vegetation for this class of land type. The two remaining conditions 21.6 and 21.9 check the presence of large quantities of water on the surface or, on the other hand, differentiate between dry and moist soil, depending on the vegetation coverage. This check is based on the influence of water in an essentially bare soil as it decreases the brightness temperature and increases the polarization difference.

All this rules can be explained by the physical behaviour of the soil types. However classification is useful only for relatively large areas, with sizes greater than at least a couple of footprints (pixels).

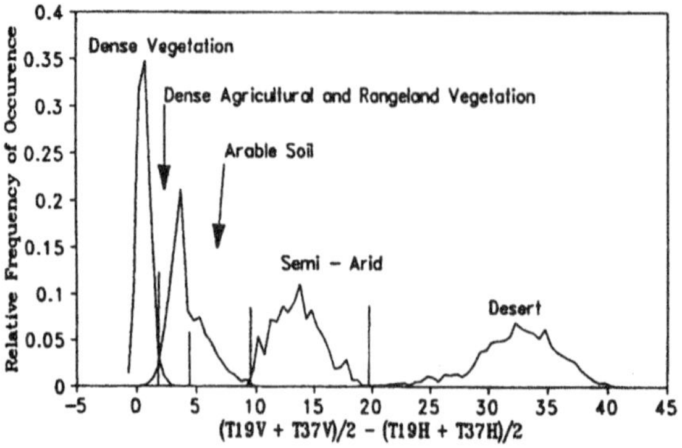

Figure 21.8: *Vegetation classification using the polarimetric difference of the 19 and 37 GHz channels (Neale et al., 1990).*

21.3.2 Snow monitoring using snow/atmosphere models

Simulations of satellite observed brightness temperatures requires an integrated model, combining the emittance of the surface with the characteristics of the atmosphere. This section will give an example of monitoring snow cover extent from space, using the surface - atmosphere model described before.

The test site selected for snow retrieval is the Hardangervidda in southern Norway. The area is a plateau at an altitude of 1200 - 1400 m with some tops up to 1600 m. The basin of this plateau is controlled by dams and used for hydro electric power. Knowledge of the amount of water in the snow in the basin is of vital interest for the Norwegian state power board (Statkraft). So far this snow information is gathered by in situ measurements and estimates from AVHRR images. Meteorological information is provided by the Norwegian meteorological service (DNMI).

Simulation of the emittance by the snow covered land surface is done following the model described in section 21.3.1.1. Snow data measured by Statkraft on 11th April 1988 at the station Dyratjönn in the north of the Hardangervidda are used as inputs for the emittance model. Meteorological data from local observations at the same time are used to describe the atmosphere. The input parameters as used in this emission simulation are summarized in Table 21.10.

The results of the simulation are shown in Fig. 21.9 and compared with SSM/I observations. It can be noted that the observed brightness temperature polarization difference for the 19 GHz channel (PD_{19}) is in the order of 30 to 35°K, whereas the simulated temperatures have a PD_{19} of less than 20°K. Second, the gradient for the spectral difference of the 37 and 85 GHz channels ($SD_{37/85}$) has a negative value for the emission simulation which is slightly higher than the observed vertical gradient and opposite to the positive gradient for the observed horizontal T_B.

A more detailed analysis was carried out to quantify the contributions from the snow parameters and those related to the atmosphere.

Meteorological data from February to April 1988 measured by the DNMI station on the Hardangervidda (Table 21.11) are used as input parameters for a detailed atmosphere.

The simulated brightness temperature as shown in Fig. 21.10 leads to the following conclusions: If no precipitation occurs, the 19 GHz channel is only slightly affected (<5°K). For the gra-

Ground emittance	from Pulliainen et al. (1990)
Ground temperature	$-1°C$
Grain size	0.8 mm
Snow density	364 kg/m^3
Snow depth	147 cm
Snow water equivalent	535 mm
Forest coverage	0%
Atmosphere	using (Liebe, 1989)
- pressure	1013 mbar
- temperature	$-6.5°C$
- humidity	64%
- standard atmosphere	southern Finland
	from Pulliainen et al. (1993), Salonen et al. (1992)

Table 21.10: *Geophysical parameters used for the emission simulation, station Dyratjönn, 11 Apr 1988.*

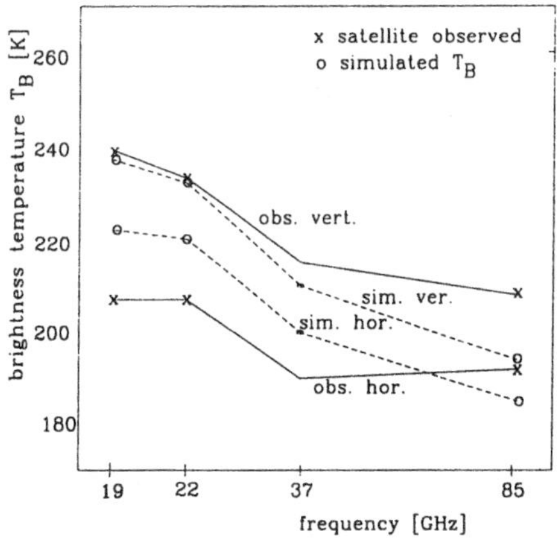

Figure 21.9: *Comparison between space observed ('x') and ground based simulated ('o') brightness temperature T_B. Parameters in Table 21.10 were used*

Table 21.11: *Detailed atmosphere information from stations Dyratjönn (D), Haukeliseter (H1, H2), and Finse (F1, F2). Cloud types are stratus (2), stratocumulus (3) and cumulus (7) clouds.*

station	D	H1	H2	F1	F2
pressure [mbar]			1013		
max. cloud thickness [km]	1	2	0.2	4	1
cloud type	7	7	2	7	3
rain rate [mm/h]	0	0	0	5	0
air temp.[°C]	-1	2.5	-7.7	0.8	-11
humidity [%]	64	86	64	99	77

dient of the $SD_{37/85}$ no general trend is noticeable. However, both negative gradients as for the observed vertical $SD_{37/85}$ and positive gradients for the observed horizontal $SD_{37/85}$ are represented within the simulation. As the simulation of the atmospheric contribution to the observed brightness temperature shows, investigations have to be undertaken in order to better describe the emission from the snow surface.

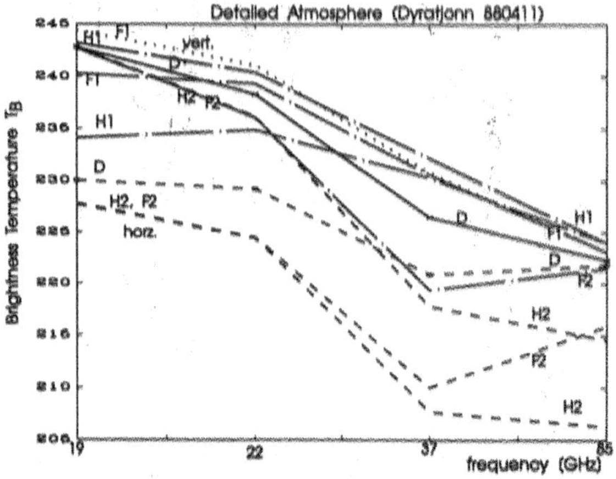

Figure 21.10: *Detailed atmosphere information taken from stations Dyratjönn (D), Haukeliseter (H1, H2), and Finse (F1, F2) (Table 21.11)*

Inversion for retrieving geophysical parameters are implemented as a (linear) combinations of MIMR channels.

Three major snow parameters are required for this type of application: snow extent, snow height, and snow water equivalent. Inversion algorithms based on linear combinations of the observed channels are briefly reviewed and applied to SSM/I observed T_B.

Snow detection algorithms

Amongst all snow parameters dry versus wet snow shows the highest spectral contrast. As a consequence retrieval algorithms have been developed for dry and wet snow surfaces.

In this paper the relationships from Mätzler (1987) are used for dry snow coverage and from Pulliainen et. al. (1990) for wet snow coverage.

For **dry snow**, mainly the spectral and polarization differences for the 19 and 37 GHz channels are used. The footprint is assumed to be completely covered by snow, if at least one of the following relations is true

$$T37V - T19V < -5\text{ K}, \quad (21.11)$$

$$(T37V - T19V) * (T19V - T19H) < -60\text{ K}^2, \quad (21.12)$$

$$(T37V - T19V) * (T37V - T37H) < -50\text{ K}^2, \quad (21.13)$$

$$(T37V - T19V) * (T90V - T90H) < -30\text{ K}^2. \quad (21.14)$$

Even though the algorithms presented are developed for MIMR frequencies, validation is performed with SSM/I data. Due to missing 85 GHz data in 1988 Eq. 21.14 is left out of the inversion performed here.

Mapping of **wet snow** is more difficult, as slightly wet soils, wet snow and open water have similar emissivities. The relations proposed by Pulliainen et. al. (1990)

$$T37V > 263\text{ K}, \tag{21.15}$$

$$|T37V - T19V| < 3\text{ K}, \tag{21.16}$$

$$12K < T19V - T19H < 33\text{ K}, \tag{21.17}$$

$$|T19V - T19H - 1.4 * (T37V - T37H)| < 3\text{ K}, \tag{21.18}$$

have all to be true in order to exclude false detections.

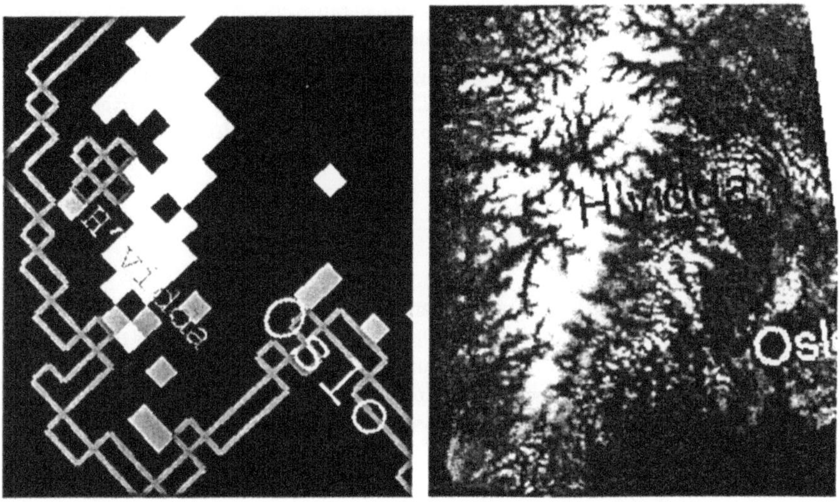

Figure 21.11: *Comparison of SSM/I retrieved snow coverage (left) and AVHRR image (right) from 13th June 1990. On the SSM/I image, dry snow is indicated by the white pixels, wet snow by gray pixels, the coast by the double lines.*

The algorithms are applied for multi-temporal SSM/I data of the Hardangervidda (Fig. 21.11). The test site can be identified by the outlined cross, which indicates the Sognefjorden. Both AVHRR image and SSM/I snow cover image are oriented north to ease comparison. The images show a quite similar snow coverage, indicating the possibility of retrieval from space of snow extent. It has to be mentioned however that the spatial resolution is rather unsatisfactory.

Conclusions

ESA's future microwave radiometer MIMR has been presented, showing the possibility of geophysical parameter retrieval for ocean, land, cryosphere and atmosphere. Direct and inversion algorithms have been described for atmospheric (water vapour), oceanic (sea surface temperature) and land (type classification) retrieval applications.

Modeling and retrieval was investigated for snow monitoring. It was shown, that passive spaceborne microwave remote sensing of snow properties is a suitable task for future space-borne microwave radiometers. The order of magnitude can be correctly retrieved for snow height and snow water content, the differences to ground observed data are in the order of a factor 2-3. The difficulty in the validation of the retrieved snow properties are mainly due to the comparison of 25 × 25 km SSM/I data with ground truth at selected points. The major limitation here is the footprint size, which for land applications should be in the order of 10-15 km.

With respect to MIMR, the increase of the antenna size of 1.6 m compared to 0.65 m for the SSM/I will improve the spatial resolution of the corresponding channels by about the factor three for MIMR. The radiometric accuracy is, despite the antenna size, kept very similar to the SSM/I. With a sampling distance of 15.5 km for the 6.8 to 23.8 GHz channels and even smaller for the 36.5 and 90 GHz channels, the use of MIMR data will provide not only snow extent and melt-onset, but also snow water equivalent and snow depth.

The simulations have shown also, that atmospheric parameters affect significantly the measured emittance, thus combined models for atmosphere and surface parameter retrieval are required.

The advantages of microwave remote sensing compared to optical remote sensing is the all weather, day and night remote data retrieval. This advantage has to be further evaluated in models describing the change of geophysical parameters (*change detection*) rather than performing singular event calculations.

Acknowledgment

The authors would like to thank Frank Wentz (Remote Sensing Systems in California), the University of Colorado in Boulder (World Data Center A) and ESA–ESRIN in Frascati for providing us with SSM/I data. Thanks to our colleagues Johnny Johannessen and Paul Ingmann at ESA ESTEC for information and comments. All those, who were involved in ESA contracts during the last years and delivered various direct and inverse algorithm are gratefully acknowledged.

21.4 References

Alishouse JC, Snider JB, Westwater ER, Swift CT, Ruf CS, Snyder SA, Vongsathorn J, Ferraro RR (1990) Determination of Cloud Liquid Water Content Using the SSM/I. *IEEE Trans Geosci Rem Sens* 28(5): 817–822

CEOS (1994) CEOS Dossier; Volume A: Satellite Missions. *ESA ESTEC*, Noordwijk (NL)

ESA (1995a) New Views of the Earth, Scientific Achievements of ERS-1. *ESA SP-1176/I* ESA ESTEC, Noordwijk (NL)

ESA (1995b) Land and Sea, ERS-1 Applications. *ESA BR-109* ESA ESTEC, Noordwijk (NL)

Goodberlet MA, Swift CT, Wilkerson JC (1990) Ocean Surface Wind Speed Measurements of the SSM/I. *IEEE Trans Geosci Rem Sens* 28(5): 823–828

Gudmandsen P, Skou N (1991) User Wishlist for future Spaceborne Microwave Radiometry. *ESTEC Contract No 8968/90/NL/US(SC) Vol 2*, ESA ESTEC, Noordwijk (NL)

Hallikainen MT, Jääskeläinen VI (1988) Microwave Emission behaviour of Snow. *IGARSS'88* Edinburgh, Scotland

Hollinger JP (1990) Introduction. *IEEE Trans Geosci Rem Sens* 28(5): 779–780

Hollinger JP, Peirce JL, Poe GA (1990) SSM/I Instrument Evaluation, *IEEE Trans Geosci Rem Sens* 28(5): 781–790

Liebe H (1989) MPM - an atmospheric millimeter-wave propagation model. *Int J Infrared and Millimeter Waves* 10: 631–650

Lojou J-Y (1990) Algorithmie et methodes de validation des instruments en radiometrie hyperfrequence. *Thesis*, Universite de Paris

Menard Y, Reynolds, M (1991) The design of the ESA multiband imaging microwave radiometer MIMR. *IGARSS'91* Espoo, Finland, 2359–2363

Mätzler C (1987) Applications of the interaction of microwaves with the natural snow cover. *Rem Sens Reviews* 2: 259–387

Mätzler C (1994) Passive Microwave Signatures of Landscapes in Winter. *Meteorol Atmos Phys* 54: 241–260

Neale CMU, McFarland MJ, Chang K, (1990) Land–Surface–Type Classification Using Microwave Brightness Temperatures from the Special Sensor Microwave/Imager. *IEEE Trans Geosci Rem Sens* 28(5): 829–838

Noll J, Poiares Baptista JPV, Buonomo S, Rognes A (1994) Special Sensor Microwave/Imager (SSM/I): Instrument, Data Analysis and Applications, in Remote Sensing and Global Climate Change. (RA Vaughan and AP Cracknell) *NATO ASI Series* I 24: 119–133

Noll J, Poiares Baptista JPV, Borgeaud M, Rognes A (1994) Multi-Temporal Aspects in Snow Retrieval using Spaceborne Radiometers, in μrad94: Specialist Meeting on Microwave Radiometry and Remote Sensing Applications. *VSP*, Zeist (NL)

NSIDC (1992) User's Guide: DMSP SSM/I Brightness Temperature and Sea Ice Concentration Grids for the Polar Regions on CD–ROM. *National Snow and Ice Data Center, NSIDC Spec Rep 1* Boulder (USA)

Pandey P, Kakar R (1982) An Empirical Microwave Emissivity Model for a Foam-Covered Sea. *IEEE Journ of Oceanic Engin* OE-1(3): 135–140

Pulliainen J, Hallikainen M, Somersalo E, Kärnä J-P, Jääskeläinen V, Hyyppä J, Tavela J, Luntama J-P, Manninen T (1990) Study of Microwave Interaction with the Earth's Surface. *Final report ESA ESTEC contract 8447/89/NL/PB*

Pulliainen J, Kärnä J-P, Hallikainen M (1993) Development of Geophysical Retrieval Algorithms for the MIMR. *IEEE Trans on Geosc Rem Sens* 31(1): 268–277

Readings CJ, Dubock PA (1993) Europe's major contribution to Earth observation for the late nineties. *ESA bulletin* 76: 15–29

Readings CJ (1995) The Post ENVISAT Strategy: The ESA Earth Explorer and Earth Watch Missions. *IGARSS'95* Florence, Italy

Salonen E, Karhu S, Jokela P, Zhang W, Uppala S, Aulamo H, Sarkkula S, Poiares Baptista JPV (1992) Modelling and Calculation of Atmospheric Attenuation for Low-Fade Margin Satellite Communications. *ESA Journal* (16)3: 299–317

Sciré Scappuzzo F (1992) Study of Passive Remote Sensing Emission/Inversion Models and Use of SSM/I Data for MIMR Analysis. *ESTEC working paper no 1639*, ESA ESTEC, Noordwijk (NL)

Ulaby FT, Moore RK, Fung AK (1986) Microwave Remote Sensing, Vol III: From Theory to Applications. *Artech House* Dedham, 2018ff

Van de Berg L, Schmetz L (1991) Meteosat Monitors Humidity in the Earth's Troposphere. *ESA Bulletin* 67: 84–88

Wentz FJ (1991) User's Manual SSM/I Antenna Temperature Tapes. *RSS Technical Report 120191*, Remote Sensing Systems, Santa Rosa

Wilheit T (1979) A Model for the Microwave Emissivity of the Ocean's Surface as a function of Wind Speed. *IEEE Trans on Geosci Electron* GE-17(4): 244–249

21.5 Appendix

ESA's Earth Observation Programme

This annex will give a short overview of ESA's existing and planned Earth observation satellites. It will briefly describe the sensors onboard the satellites and highlight the application areas. In addition, the concept of Earth Explorer and Earth Watch missions for the next century is explained.

A.1 ERS-1 and ERS-2

The European Earth Remote Sensing Satellite ERS-1, launched in July 1991, has during the last successful years of operation demonstrated the promising use of radar satellite technology for Earth observation. Originally designed as an ice and oceanic mission, ERS-1 has proven to be also useful for land applications. As the full achievements of the use of ERS-1 data goes beyond this study, only some applications are mentioned (Table A-1). More detailed information can be found in the related literature (ESA, 1995a, ESA, 1995b).

Application area	Keywords
sea bottom topography	navigation, offshore construction, pipelines
sea-ice monitoring	maritime activities for coast guards, regional ice-breaking services, oil exploration
sea state climatology	wave height, wind speed, wind direction
sea state forecasting	fishing and offshore safety, ship routeing
ocean topography	global ocean circulation and sea level
ship routeing	real-time knowledge of weather, sea state and ice cover
oil spill detection, ship traffic	monitoring and control
mapping	environmental changes, erosion
mineral and hydrocarbon exploration	trace of streamlines of former glaciers, subsurface sounding
crop monitoring	planning harvests, harvest forecasts, crop inventory
flooding surveillance	area monitoring, flood forecast
deforestation	global ecosystem
SAR interferometry	digital elevation map

Table A-1: *Applications of ERS-1 for land and sea monitoring.*

With the successful launch of ERS-2 on 20 April 1995 a new era of Earth observation by radar is opened. ERS-2 will operate with ERS-1 in a tandem mission, allowing SAR interferometry of the the whole Earth's surface (with exception of areas not covered by ground stations). SAR interferometry is already used for observing earthquakes, monitoring volcanic areas as well as estimating glacier motions. The tandem mission will establish a database of a highly precise digital, three-dimensional map of the Earth's continental surface.

In addition to the instruments of ERS-1, ERS-2 carries an advanced PRARE for the accurate orbit determination and GOME, the Global Ozone Monitoring Experiment.

A.2 Research oriented missions, ENVISAT

ENVISAT, the follow-on to ERS-1 and ERS-2, will cater the future needs of the international community of both science- and operational application of remote sensing data. ENVISAT will continue and improve meteorological observations such as 'greenhouse effects', acid rain, depletion of the ozone layer, and improve observations for ocean, sea ice and land applications, study the Earth structure and the dynamics of the Earth's crust and interior.

A.2.1 Mission overview

Scheduled to be launched at the end of 1998, ENVISAT will collect remote sensing data, which in combination with retrieval algorithms will allow the observation of geophysical parameters for the atmosphere, land, ocean and the cryosphere (Readings et al, 1993). ENVISAT will operate in a polar sun synchronous orbit at 98.55° at an altitude of 800 km, leading to $14\frac{11}{35}$ orbits/day with an orbit cycle of 100.6 min.

A.2.2 Instrumentation of ENVISAT

ENVISAT will carry eleven instruments (Table A-2), which will allow in the first place to monitor geophysical parameters on land, ocean, cryosphere and the atmosphere (Table A-3).

AATSR	Advanced Along Track Scanning Radiometer	Optical/IR radiometer
ASAR	Advanced Synthetic Aperture Radar	C-band SAR
DORIS	Doppler Orbitography and Radiopositioning Integrated by Satellite	RF Orbitography
GOMOS	Global Ozone Monitoring by Occultation of Stars	UV/Optical spectrometer
LRR	Laser Retro Reflector	Passive Optical Reflector
MERIS	Medium Resolution Imaging Spectrometer	VIS/near IR Spectrometer
MIPAS	Michelson Interferometer for Passive Atmospheric Sounding	Limb-view IR Interferometer
MWR	Microwave Radiometer	Two-Channel Nadir View Radiometer
RA-2	Radar Altimeter	Pulse Radar
SCARAB	Scanner for Radiation Budget	4 chan. VIS/IR radiometer
SCIAMACHY	Scanning Imaging absorptance spectrometer for Atmospheric Cartography	Nadir + Limb UV/VIS/IR spectrometer

Table A-2: *ENVISAT-1 instruments (Readings et al., 1993) (the highlighted instruments are sensors operating in the microwaves).*

As monitoring of the ecosystem requires detailed knowledge of not only the environmental characteristics, but also of the associated physical/biological processes, the instruments on ENVISAT are selected in order to be complementary. For land surfaces it is expected that the payload will provide the data to allow the estimation of surface albedos, vegetation cover, vegetation characteristics, etc. All of these data correlate with the surface energy and water budgets and the transfer of trace species between the land surface and the atmosphere. In detail, the AATSR is foreseen to measure the land surface temperature and the vegetation characteristics,

Atmosphere	Land	Ocean	Ice/Snow
Trace gases Clouds Humidity Fluxes Temperature	Surface Temp. Vegetation char.	Ocean colour Sea Surf. Temp. topography wave charac. wind speed	Extent Snow cover topography temperature

Table A-3: *Primary Mission objectives for ENVISAT.*

supported by ASAR and MERIS. MERIS and AATSR will provide also atmospheric information about clouds, radiative fluxes and aerosols, which are essential for the modelling of the interaction between the surface and the atmosphere. In order to evaluate surface observations, RA-2 and ASAR will provide information about the surface topography, leading to an overall view of the vegetation state.

A.3 Meteorological oriented missions, METOP, MSG

METOP and MSG, the next generation of operational meteorological satellites, are developed by ESA and will be operated by EUMETSAT. These missions are planned in order to continue the successful operation of the current METEOSAT programme. One of the major tasks of the meteorological satellites is to monitor the distribution of water in the Earth's atmosphere. The knowledge of trace gases and as such also atmospheric water vapour is essential to understand and model the Earth's climate system. It is believed that a doubling of the carbon-dioxide content will lead to a mean global warming at the surface of between 1.5° and 4.5°C (v.d. Berg et. al., 1991). These estimates for the magnitude of the global warming have been obtained from model calculations that try to simulate the relevant physical mechanisms operating in the Earth's atmosphere and ocean. MSG will continue and improve the observation from a geostationary orbit, whereas METOP is designed as a low altitude orbiting satellite on a polar, sun-synchronous orbit.

A.3.1 Mission overview

The **MSG** satellites are to be launched in 2000, 2002 and 2006 into a geostationary orbit. The main objective of the MSG mission is to serve the operational meteorological community through a series of geostationary observing satellites, thus allowing the monitoring of atmospheric dynamics as well as water and energy cycles. In addition, the - with respect to METEOSAT - higher spatial resolution of 1 km (VIS channel) to 3 km (all other channels) will provide more details about cloud properties, thus giving the inputs for improved three-dimensional atmosphere models. The duty cycle of 15min will also allow the monitoring of cloud system evolution and movements over the ocean.

METOP will complement MSG, extending the range of observations by including the polar regions and use a sun synchronous polar orbit with an altitude of 825 km.

A.3.2 Instrumentation of MSG and Metop

MSG will carry two types of instruments (Table A-4) operating in the ultra violet (UV), visible (VIS) and infra red (IR) spectra. GERB, having a resolution of 48 km will focus on atmospheric dynamics and in particular on water and energy cycles in order to determine the total radiation. SEVIRI, offering high resolution (1 km) visible broad band channel as well as specific channels in the IR window and in the water vapour, CO_2 and ozone absorptance bands, will provide specific information about cloud properties as well as sea surface temperature and total ozone concentration.

This information may help to improve geophysical models in order to correctly describe feedback mechanisms of global warming. Changes in the distribution and the amount of clouds can affect both the influx of solar radiation to the Earth-atmosphere system and the loss of heat through radiation to space.

Parameter	Meteosat	MSG
	SEVIRI	
coverage	full Earth	full Earth
repeat	30 min	15 min
sample	2.5 km VIS; 5 km IR	3 km
channels	VIS, WV, IR	3 VIS, 5 IR
	High Res. VIS	
coverage	not carried	1/2 of 1/4 of full Earth
repeat	not carried	5 min
sample	not carried	0.5 km
	GERB - Geost. Earth Rad. Budget	
coverage	not carried	full Earth
repeat	not carried	15 min
sample	not carried	48 km
channels	not carried	0.2 - 4.0 μm (SW), 4 - 50 μm (LW)

Table A-4: *Comparison of Meteosat and Meteosat Second Generation (MSG) instruments (CEOS, 1994) (full Earth as seen from a geostationary orbit)*

METOP will carry eleven instruments to provide the meteorological community with operational atmosphere and ocean data (Table A-5). In particular, the operational needs are covered as follows:

- improved high resolution temperature and humidity profiles (AMSU-A, MHS, IASI, HIRS/3);
- global ocean surface wind information (scatterometer);
- improved observation of precipitation and sea ice (MIMR, AVHRR);
- routine monitoring of the Earth's radiation budget (ScaRaB) and ozone (GOMI).

The second objective is to provide global observation of land surfaces needed for research and climate monitoring, in particular of vegetation cover, soil moisture and snow cover.

METOP will clearly support remote sensing data for monitoring and modelling of the Earth's global energy and water cycle, as identified by the World Climate Research Programme (WCRP) to be one of the most important tasks for research.

A.4 Future Earth Explorer and Earth Watch Missions

Even though the near future generation of ESA's Earth Observation satellites (MSG, METOP, ENVISAT) will provide the operational and scientific community with remote sensing data

AVHRR/3	Advanced Very High Resolution Radiometer
HIRS/3	High resolution Infra-Red Sounder
AMSU-A	Advanced Microwave Sounding Unit-A
MHS	Microwave Humidity Sounder
IASI	Infrared Atmospheric Sounder Interferometer
ASCAT	Advanced SCATterometer
MIMR	Multi-frequency Imaging Microwave Radiometer
AATSR	Advanced Along Track Scanning Radiometer
ScaRaB	Scanner for Radiation Budget
GOMI	Ozone Monitoring Instrument
SEM	Space Environment Monitor

Table A-5: *METOP instruments (the highlighted instruments are sensors operating in the microwaves).*

to improve modelling of climate processes, other parts of the ecosystem are not appropriately covered by these missions.

Two general classes of Earth Observation Missions have been identified (Readings, 1995):

Earth Explorer Missions are research or demonstration missions with the emphasis to better understand different Earth processes. Missions under this header are also intended to demonstrate new observation techniques, leading to (pre)operational services.

Earth Watch Missions are pre-operational missions covering the needs of specific Earth observation application areas. It is foreseen to eventually transfer the responsibility of these missions to operational entities, e.g. EUMETSAT. To ensure data continuity a mission period of at least ten years is foreseen.

The user community has identified areas for future Earth Explorer Missions covering the following interests:

Radiation - Determination of radiation budget components at the surface and throughout the atmosphere and the derivation of flux divergences;
Determination of cloud/aerosol characteristics.

Precipitation - Contribution to the study of the hydrological cycle in tropical regions;
Study of interactions between precipitation and atmospheric circulation in tropical regions;
Sea surface salinity variations and their impact on the evaporation-precipitation ratio and the hydrological cycle as well as thermohaline circulation.

Atmospheric Dynamics - Contribution to the study of the Earth's global energy budget by measuring 3-dimensional wind fields globally in cloud free areas
Contributions to the study of specific phenomena such as precipitation systems, tropical jet streams etc.

Stratos. Profiling - Contributions to better derive temperature (T), pressure (p) and humidity, derived from the p-T retrieval, in the stratosphere.

Chemistry - Advance understanding of the processes responsible for the ozone balance
Quantify the ozone budget;
Contribute to the study of the budgets of greenhouse gases and the processes controlling their distribution.

Gravity Field and Ocean Sea Level - Measurement the Earth's gravity field and derive the reference geoid with an accuracy of better than 3 cm
Derive the stationary dynamic ocean topography everywhere and at any time
Achieve these objectives with a spatial resolution of about 100 km.

Magnetometry - Determine the Earth's Anomaly field to better understand the behaviour or the Main field and the Earth's interior.

Land Surface Processes and Interaction - Monitor change detection and enhance understanding of geophysical and biological interactions;
Contribution to the development of land/hydrosphere/cryosphere/atmosphere interaction models
Determine the climate impacts on the biosphere and the water cycle.

Topography - Satisfy the needs for up-to-date global topographic data for agricultural, hydrological, ecological and geophysical sciences as well as polar ice sheets.
Extend knowledge on processes determined by ERS-1.

Out of these possible Earth Explorer Missions, the user community has given priority to three missions, the *Gravity-Field Mission*, the *Radiation Mission* and the *Land Surface Processes Mission*, foreseen to continue after ENVISAT-1.

Earth Watch Missions have been identified covering the following subjects:

Coastal Zones - Monitoring of sea ice and current, oil spill, flooding, river discharge, coastal erosion, etc. and forecast of sea state.

Ice Monitoring - Operational monitoring of sea ice fields and forecast of evolutions.

Land Surface Covers aspects such as crop forecasting, crop damage assessment, forestry, land use, cartography and hydrology.

Atmospheric Chemistry - Monitor wide range of atmospheric molecules.

Open Oceans - Addresses ship routeing, ocean wave climatology, sea state monitoring and forecasting.

As some of these aspects are already addressed in the coming up MSG and Metop missions, the *Coastal Zone Mission* is a strong candidate to address purely observed applications.

It should be noted, that the above mentioned plans are preliminary. Further extensive consultations with the users and the representatives of the Agency's member states as well as cost benefit analyses have to be undertaken to finalize the future Earth Watch and Earth Explorer Missions.

PART E

POSTER SESSIONS

Chapter 22

Poster Summaries

1 Group A: Trace Gas Characteristics

An important goal of atmospheric remote sensing deals with the detection and modelling of atmospheric trace gases. It is well known that they directly and indirectly influence the climate of the earth due to their radiative and chemical properties. However, the net effect and the global variability of these gases is not well known. The modelling of the earths climate requires long term global datasets of trace gas concentrations.

Three studies related to the subject of trace gas characteristics were presented in student posters. They dealt with new remote sensing technique of water vapor and carbon monoxide. Water vapor is the most important greenhouse gas in the atmosphere . Also it is the origin for atmospheric energy fluxes, clouds and precipitation. CO is an important trace gas as it is the largest sink of tropospheric hydroxyl.

In the poster of **Servazi et al.** the measurements of the atmospheric water vapor using the Raman lidar technique is described. A short technical description of the instrument and a summary of how to derive the water vapor mixing ratio from the detected, incoherently backscattered light intensities is given. The comparison of the water vapor mixing ratio, measured simultaneously from radiosonde and lidar, just at the start time of the radio-sonde, shows very good correspondence between 1500 and 8000 m. Additionally, the strong variability of the atmospheric water vapor during the transit time of the radiosonde (1h), inside of the vertical column between 1500 and 8000 m above the lidar, is reported.

Bartsch et al. are involved in the development of a new algorithm for the column water vapor content land surfaces with relative error less than 10% using backscattered solar radiation. This involved the calculation of more than 1600 high resolution spectra. These spectra were calculated in the wavelength range between 880 and 1000 nm - a strong water vapor absorption band in the near IR - as a function of the following parameteres: various real measured atmospheric profiles, 22 different surfaces, gaussian distributed aersosol parameters and 5 solar zenith angles.

Using a principal component analysis, it can be concluded that only 3 to 4 wavelenght channels, instead of the original 210 channels, are necessary to obtain the atmospheric column water vapor content. Using the original 210 channels, to which a 1% noise is added, the developed algorithm yields a rms error of 5%.

The validation of the algorithm with the airborne high resolution spectrometer OVID will be done in future. Also the reduction of channel number with respect to the MERIS spectrometer onboard the European satellite ENVISAT is one further goal of her study.

In the poster by **Tolton** and **Drummond** a new form of correlation radiometer known as a length modulated radiometer (LMR) is presented. The technique of correlation radiometry has been used in satellite-based measurements of atmospheric temperature and composition for over two decades. The LMR will be an integral component of the satellite instrument known as MOPITT (Measurements Of Pollution In The Troposphere), which has been selected to be launched on the EOS-AM1 platform of the Earth Observing System in 1998. The primary objective of MOPITT will be to enhance our knowledge of the lower atmosphere by measuring atmospheric profiles of CO and CH_4.

The development of the LMR for MOPITT has led to the building of a ground-based solar absorption instrument for measuring atmospheric columns of CO. This is the first remote sounding instrument to utilise a LMR. A short decription of the LMR technique and a summary of the sources of imbalance in the current design of the LMR (Mk III) is provided. The results of a measurments campaign in the Toronto region in the late summer of 1994 is presented.

Bartsch B Meteorologisches Institut, Universität Hamburg, Germany
Bakan S Max-Planck-Institut für Meteorologie, Hamburg, Germany
Fischer J Institut für Weltraumwissenschaften, Freie Universität Berlin, Germany
Water vapour detection above land surface using backscattered solar radiation.

Serwazi M, Reichardt J, Weitkamp C GKSS Forschungszentrum Geesthacht GmbH, Postfach 1160, D-21494 Geesthacht, Germany
Wandlinger U, Ansmann A IfT-Institut für Atmosphärenforschung, Permoserstrasse 15, D-04303 Leipzig , Germany
Water vapour profiling using the Raman lidar techniques

Tolton BT, Drummond JR Department of Physics, University of Toronto, Toronto, Ontario, Canada
Ground based measurements of atmospheric CO using length modulated radiometer

2 Group B: Cloud-Radiative Modelling / Parameterizations

Gaining a clear understanding of the physical processes that occur in the earths atmosphere due to the presence of condensed water is a fundamental goal of many aspects of atmospheric research. While clouds represent an important link in the earths hydrologic cycle, they also impact other facets of human activity. This is demonstrated by **Seker** who considers the propagation of microwave radio signals in a cloudy atmosphere. He finds that mixed phase clouds can significantly impact radio communication both by attenuation and phase shifting of the transmitted signal using his stochastic multi-component propagation model of clouds. Clouds are also an important regulator of the earths climate. The coupling of clouds with the climate system stems from the well known effect of clouds to increase the planetary albedo leading to a net decrease in the globally averaged downward solar flux while simultaneously warming the earth system by decreasing the upward-directed top of atmosphere (TOA) terrestrial flux. Because the interaction of clouds with solar and terrestrial radiatiation is poorly characterized in models of the earths climate, climate change predictions from General Circulation Models (GCMs) are somewhat ambiguous. Before general circulation models (GCMs) can realistically predict changes in the climate due to anthropogenic modification of radiatively active trace gases, the nonlinear feedback mechanisms induced by clouds must be sucessfully parameterized in these models. The complexity of this problem is illustrated by **Mace** and **Ackerman** who, using data, show that the decrease in surface insolation can vary naturally from 10% to more than 80% of the clear sky value for identical cloud types. The difficulty is further illustrated by **Doutriaux** and **Yu** who show that simply validating the results of a GCM against ISCCP C1 cloud climatologies can lead to significant biases owing to sampling ambiguities in the satellite retrieval algorithms. When validating GCM results in this way, the GCM cloud predictions must be carefully procesed so that the resulting modeled cloud prediction matches the sampling characteristics of the satellite platform. Beyond this, simply determing the proper vertical distribution of parameterized cloudiness is shown to be a significant difficulty, especially with low level clouds.

The influence of clouds on the current climate was sucessfully demonstrated by the Earth Radiation Budget Experiment (ERBE) where clouds were shown, on average, to lead to a slight cooling of the earth system. SCARAB, the follow-on to the ERBE, was launched in 1993 and collected one year of radiation budget data. Before these data can be analyzed, the sensors must be carefully calibrated. **Raison** finds that it is possible to calibrate the short wave channel of the SCARAB with a geophysical technique and that biases in the data arising from a thermal leak associated with the silica filter can be corrected using statistical methods. After this calibration is implemented, TOA radiation budgets will be calculated and comparison to the ERBE can be conducted.

Not only do tropospheric clouds influence the earths climate, but condensed water in the polar stratosphere has been shown to influence the chemistry of this atmospheric region leading to significant decreases in ozone concentrations. The Japanese Space Agency will launch ADEOS in 1996 and attempt to quantify the aerosol concentrations in the stratosphere. However, before this can be accomplished, the retrieval of aerosol characteristics from the data must be understood theoretically. **Okamato et al.**, therefore, consider this important procedure and show that it will be possible to infer the chemical composition of stratospheric aerosol particles, as well as their size distribution and shape from the ADEOS data.

Determination of cirrus cloud radiative and microphysical characteristics is a primary goal of much of the cloud research currently being conducted. The global coverage of cirrus, the nonsphericity of the ice particles and their remoteness in the upper troposphere make understanding their influence on climate very difficult. Using data from EUCREX 94, **Chepfer** and **Brogniez** illustrate both the strengths and weaknesses of multi-aircraft cirrus observational experiments. By carefully coordinating aircraft flight tracks, visible and infrared optical depths associated with an extensive cirrus layer are considered and meso gamma scale variability in the cirrus radiative properties are clearly demonstrated. However, side-by-side flights of the three aircraft involved in the experiment revealed nontrivial biases in the flux radiometer measurements making quantitative interpretation of these data a significant challenge. As a special case of the cirrus cloud-climate problem, **Drummond et al.** examine cirrus modification by jet aircraft condensation trails. They show that the ice particles generated in the aircraft effluent are likely composed of high concentrations of radiatively active particles with characteristic effective radii of less than 10 microns. Their data also suggest modification of the ambient cirrus by interaction of the condensation trail with the natural cirrus generation process.

In summary, these results represent significant progress in understanding the physical mechanisms that make clouds an important component of the earth's atmosphere. However, these results clearly demonstrate that considerable research in all aspects of cloud-climate coupling must be conducted before adequate characterization of clouds in GCMs can be accomplished.

Chepfer H, Brogniez G: First analysis of radiative measurements for cirrus clouds during EUCREX94

Doutriaux M, Yu W: Validation of cloud cover simulated by a GCM using ISCCP C1 cloud climatology

Drummond C, Spinhirne JD, Hart WD: Case study: contrail microphysics from aircraft remote sensing analysis

Mace GG, Ackerman T: Examination of observed radiative forcing during the 94–95 winter

Okamoto H, Mukai S, Sano I: Aerosol retrieval algorithms from ADEOS/ILAS measurements

Seker S: Stochastic propagation of model clouds

Raison F: ScaRaB Scanner for radiation budget

3 Group C: Cloud Precipitation Characterisations

Our studies concern liquid water, solid ice and snow in the atmosphere. The projects were organised in terms of their scale, to show that they (almost!) covered the full range of cloud and precipitation phenomena, from the molecular to the global.

Okamoto explains in his poster the modellng of 94 Ghz radar backscatter from hexagonal ice crystals, using the Discrete Dipole Aproximation (DDA). **Lemke** and **Oldag's** poster analyses 94 Ghz radar observatons of cirrus, in which the DDA computed backscatter from hexagonal column and plate crystals eliminates the errors arising from the spherical particle assumption of Mie theory.

Füg's poster is concerned with the problem of different spatial resolutions at different frequencies used in microwave radiometry. Using the high correlation between the different frequencies, a method is developed on the basis of multiple regression to increase the poorer resolution of the lower SSM/I frequency channels to that of the 85 Ghz channel. Taking into acount spatial and temporal restrictions with this method, it is possible to use the SSM/I data unaffected by land up to 15 km off the coast.

The following group members presented their work to the audience at a poster session, in ascending order of atmospheric scale.

Simpson: Analysis of 3 Ghz and 35 Ghz radar signals can measure particle size distributions in stratiform cloud, by exploiting Mie scatter from the larger particles at 35 Ghz. Results demonstrated truly astonishing agreement with simultaneous sampling by aircraft. The size distributions obtained can be used to compute essential cloud parameters such as Effective Radius, Liquid Water Content, and Ice Water Content.

Oury et al.: X band radar beams are deeply attenuated when passing through rain cells. The "Stereoradar Analysis" allows us to retrieve non-attenuated reflectivity fields and the corresponding attenuation fields from the apparent reflectivities. Obtaining boundary conditions and a sampling of the cells from two viewing angles, the "Stereoradar Analysis" is currently able to retrieve the cells in coplan and cartesian co-ordinates.

Bennartz: In the 1990s, satellites with improved remote sensing capabilities will be launched. In preparation for these, a method to derive semi-statistical algorithms based on a systematic use of data from different sensors was developed and tested for combined SSM/I and SSM/T2 datasets. First intercomparisons with classical algorithms showed the reliability and improvement of this method.

Dioszeghy and **Fejes:** For operational purposes, satellite based cloud classifications should be done on datasets with high time resolution, such as Meteosat. This method determines the main cloud classes on the two-dimensional histogram of the infrared and visible brightness values by a dynamic cluster analysis, according to cloud heights and thicknesses. The cumuliform and layer clouds were seperated by a threshold in the standard deviation field of the IR image. Comparisons with surface observations and visual analysis were encouraging.

Michelson et al.: Image processing methods were applied to identify and suppress undesired effects in C-band weather radar imagery. The first method removes echoes that are static in time and space, using a time series of images. The second method uses a radial wind velocity image from Doppler radar to remove precipitation echoes that are not matched with wind information.

Booysen and **Truter:** Ground radar and Meteosat image data were used to develop a procedure that produces daily rainfall maps over South Africa. This system is based on the assumption that rain rate is proportional to cloud depth. Cloud depth is determined from satellite image data in terms of the difference in temperature pixels between cloud top and cloud base. Rain-rates are determined from ground radar data on a 2000 m Constant Altitude Plan Position Indicator (CAPPI). A cloud classification scheme is used to assign the satellite imagery to rainfall criteria. Statistical relationships between cloud depth and rain rates are then determined which are used to compile rainfall maps by using Meteosat image data alone.

Jones: A new precipitation climatology for the tropical oceans has been developed, using available observations of rainfall from atoll stations, and mean Outgoing Longwave Radiation (OLR) flux measurements. The climatology spans the period June 1974 to November 1994, and is a gridded dataset of monthly precipitation accumulations, at a spatial resolution of 2.5° of latitude and longitude.

To close our presentation, we illustrated the scope of our projects on a chart of remote sensing measurements required for continuing studies, and invited our friends in the audience to fill in the blanks during the remainder of the week.

Bennartz R
A semi-statistical approach for cloud and rain identification using SSM/I and SSM/T2 data

Booysen RG, Truter MT
A real time rainfall remote sensing system

Dioszeghy M, Fejes E
Cloud classification derived from METEOSAT data involving the standard deviation fields of the brightness values

Füg C
Resolution enhancement of microwave observations over the baltic sea

Jones C
Development of a tropical ocean precipitation climatology from OLR flux measurements

Lemke H, Oldag J
Theoretical studies for remote sensing with a 95 GHz radar

Michelson DB, Anderson T, Häggemark L
Precipitation analysis with radar effects of anomalous propagation echoes

Oury S, Testud J, Kabeche A
The stereoradar analysis

Simpson P
Dual-wavelength radar estimate of ice particle size

4 Group D: Surface Processes

Participants of that group examine different land surface types and processes using various techniques. The first group of posters are related to the computation of the turbulent fluxes of sensible heat, water vapour and momentum at the land surfaces (SVAT- models). The LMD GCM contains a complex land surface scheme, SECHIBA, which explicitly takes into account the sub-grid scale variability of vegetation. The study of **de Rosnay** and **Polcher** shows the impact of the introduction of subgrid scale variability of the soil hydrology on the surface fluxes computed by SECHIBA.

The next poster presents a SVAT scheme, extended by a long- and a short wave radiative transfer model, which will be used to simulate the dependency between the radiometric surface temperature and NDVI. This dependency represents an approach to infer the regional scale surface soil moisture (**Thonfeld et al.**). There was also a presentation of the main processes represented in the UCL/SVAT model, as well as a few validation results. This model is now a part of a mesoscale atmospheric model (UCL/MAR) and will be used to study the influence of the land surface on mesoscale circulations (**de Ridder** and **Schayes**).

Further posters present results from experiments. The analysis of aircraft boundary layer flux measurements from NOPEX 1994 over a heterogeneous land surface show that surface induced mesoscale scale divergences cannot be neglected in the boundary layer moisture budget on a regional scale (**Frech et al.**) The mesoscale fluxes are organized into fairly stationary patterns along the analysed flight legs. Measurements from ground by simultaneous use of different remote sensing techniques are presented by **Abbate**, namely: kinematic eddy sensible heat flux by Doppler SODAR and cryogenic radiometer, characteristics of sea breeze circulation and PBL structure by a network of Doppler SODARs, kinematic water vapour flux and latent heat by Doppler SODAR and Raman LIDAR. Results are in accordance with values obtained during other experiments. **Kowalewska et al.** have analysed and compared Landsat TM (band 1–4) optical and ERS-1 SAR radar data to obtain the surface details of the urban and countryside areas of Rome. Preliminary classification of water, urban and various vegetated areas has been shown. Vegetation pattern in the urban and countryside areas at Northeastern Blacksea Region in Turkey using NOAA AVHRR data recorded in 24 April 1995 are presented in poster by **Okcu et al.** The time series of temperature, relative humidity and pressure for long term data at each stations have been analysed by using Fourier Analysis to study how the vegetation pattern affects the climatology of the region.

Abbate G ENEA - Ambiente, Rome, Italy
Lower atmosphere study of urban areas by means of remote sensing techniques: observations and measurements in the area of Rome

Frech M, Jochum AM DLR, Oberpfaffenhofen, Germany
Samuelsson P, Bergström H Uppsala University, Sweden. A first step towards aggregating energy fluxes over the NOPEX area

Kowalewska H Institute of Marine Sciences, University of Szczecin, Szczecin, Poland
Abbate G, Borfecchia F ENEA, Rome, Italy
Land cover characteristics of Roman urban area and its countryside from Landsat TM and ERS- 1 SAR data

Okcu D, Sögüt AS Bogazici University, Meteorology Laboratory, Istanbul, Turkey
Aslan Z Kafkas University, Faculty of Forestry, Artvin, Turkey
Thermal and vegetation map of Northeastern Blacksea Region in Turkey

de Ridder K, Schayes G Institut d'Astronomie et de Geophysique, Louvain-la-Neuve, Belgium

Modelling soil-vegetation-atmosphere transfer processes and their influence on mesoscale atmospheric circulations

de Rosnay P, Polcher J Laboratorie Meteorologie dynamique, LMD, CNRS, Paris, France
A parameterization of subgrid-scale variability of hydrology linked to vegetation

Thonfeld H, v.Schonermark M, Tornow C, Weimann A, Vogt P, Witt H DLR, Berlin, Germany
A simulation scheme for testing relationships between vegetation index and surface radiometric temperature to estimate surface soil moisture

5 Group E: Applications to Dynamical-Thermodynamical Processes

Remote sensing offers many possibilities in increasing our understanding of dynamical and thermodynamical processes in the atmosphere. Opportunities exist on a wide variety of scales, ranging from the synoptic to the mesoscale, and also including the turbulent scales in the atmospheric boundary layer (ABL).

Klein Baltink presented a poster on recent developments in ground-based remote sensing techniques. These allow continuous, unmanned measurement of vertical profiles of wind and temperature using a windprofiler/RASS. For studying boundary layer processes in order to improve parametrizations this system offers data of high temporal and spatial resolution, of which some examples were presented. The data are collected within the framework of the TEBEX project which addresses the problem of parameterization of sub-grid scale processes above land in climate models.

Descriptions of the stability conditions in the ABL are also of increasing importance to climate studies and scientific studies of local thermal and orographic produced wind-systems. To study such systems, **Kerschgens et al.** worked with a Phased Array Doppler SODAR. He presented in his poster results of an intercomparison of SODAR and tower measurements at two different heights, as well as of a comparison between horizontal windspeed measured with SODAR and computed by a prognostic windfield model.

The complexity of the retrieval of geophysical parameters from space often makes it necessary to use simultaneous remote sensing systems working in different spectral regions and at different spatial scales. For instance, more accurate information on cloudiness derived from AVHRR may improve the quality of the TOVS temperature and humidity retrieval products. **Csiszár** presented a poster in which the mean fractional cloud cover, cloud top temperature, cloud emissivity and droplet size of semi-transparent single-layered clouds for the HIRS spots were derived from collocated AVHRR data. The information on droplet size was used to derive the cloud emissivities at the HIRS wavelengths from pre-calculated emissivity look-up tables.

Lieder and **Heinemann** presented in their poster the use of passive microwave remote sensing in polar regions. He showed the need for additional, remotely sensed meteorological information in data-scarce areas for monitoring synoptic as well as meso-scale systems. For a case study of a polar low in the Antarctic region he examined the fields of SSM/I-retrieved integrated water vapour, cloud liquid water path, surface wind speed and rain rates. Presenting corresponding ECMWF model analyses for a comparison, he showed that the mesocyclone is not analyzed by the model. Already on the synoptic scale numerical weather prediction models have difficulties describing the antarctic meteorology. An international project addressing this problem was described (FROST, First Regional Observing Study of the Troposphere in the Antarctic) and the use of SSM/I retrievals for the detection of fronts and other synoptic features in the FROST project was outlined.

Apart from the monitoring approach desribed above, numerical simulations allow a further examination of the (thermo -) dynamical processes responsible for their formation. **Köpken** and **Heinemann** presented work in this domain, concentrating on the simulation of polar mesocyclones in the Antarctic. In order to obtain a realistic description of the synoptic environment and the mesoscale systems, observations are assimilated using a 4-dimensional intermittent scheme. In a case study, the positive impact of conventional observations and measurement campaign data was shown and the technique used for assimilating integrated water vapour content and near-surface windspeeds derived from SSM/I measurements was presented. Due to the extreme scarcity of conventional observations in this area, these satellite data as well as TOVS temperature and humidity profiles and ERS-1 wind vectors provide an important source of large scale and mesoscale information. These will be used in further simulation studies.

Remote sensing also has applications to larger, synoptic scale phenomena. **Clerk** presented an observational study of an upper-level jet-front system using actual in-situ flight data, as well as data from the TOVS instrument and ECMWF analyzed fields. The results of the study indicated that contrary to conventional thinking, water vapour may not be a simple passive tracer of stratospheric intrusion events associated with such systems. This suggests the technique of tracing these intrusions by following water-vapour anomalies revealed by satellite data should be used cautiously.

Clerk A Department of Physics, University of Toronto, Canada
Stratospheric intrusion associated with a jet front system

Csiszár I Hungarian Meteorological Service, Satellite Research Laboratory, Budapest, Hungary
Use of AVHRR data in the TOVS cloud clearing process

Kerschgens M, Jasbinschek D, Mole FJ Meteorological and Geophysical Institute, University of Cologne, Germany
Windfield measurements with a REMTECH phased array (PA-2) Doppler Sodar

Klein Baltink H Royal Netherlands Meteorological Institute, De Bilt, the Netherlands
Tropospheric Energy Budget Experiment: TEBEX

Köpken Ch, Heinemann G Meteorological Institute, University of Bonn, Germany
Assimilation of satellite data into a mesoscale model for polar low studies

Lieder M, Heinemann G Meteorological Institute, University of Bonn, Germany Remote sensing of polar mesocylones with passive microwave radiometry

6 Group F: Earth System Modelling/Validation

Modelling the climate and Earth System is a very complex task. Processes, surfaces and interfaces in nature have, in general, complicated morphologies, often exhibit statistical self-affine properties and are characterized by finite-domain power-law power spectra. **Ivanova et al.** have developed an approximate self-affine model which has been applied to rough metal deposits and provided their reasonably accurate description. This implies that the model might also prove useful for characterization of the interfaces and processes in the atmosphere.

A simulation of the end-Cretaceous (65 Myr. ago) climate is carried out by **Pierazzo** using the latest version of the NCAR General Circulation Model (GCM) CCM2, with input data set relative to the end-Cretaceous paleogeography and topography, sea surface temperature, vegetation, and CO_2 content. Preliminary results indicate a more equable climate, with a somewhat different precipitation pattern than today's and a similar planetary albedo, except for the polar regions (that lack a permanent ice sheet).

Development of reliable models for microwave emission and scattering from terrain, i.e. soil, vegetation, snow, forest, etc., is one of the important problems of microwave remote sensing. The simple case of scattering by a thin dielectric strip was examined by **Arslan**. Extinction cross sections were conducted for certain dielectric constants and strip dimensions corresponding to the parameters of vegetation elements.

Schröder and **Hense** applied inverse modelling, concerning different parameterizations of the horizontal energy transports in atmosphere and ocean, to validate simple Energy Balance Models. Up to now two different diffusive approaches for the transports and one including some transport velocity were examined by estimating coarse patterns of the free parameters of each parameterization from net radiation data at the top of the atmosphere. First results suggest the parameterizations (models) are of similar quality.

Models can be used to interpret remote sensing images. In the case study by **Li et al.**, the two-layer finite depth model is used to simulate the oceanic internal wave propagation over the continental shelf of United States imaged by Airborne High Resolution Synthetic Aperture Radar in 1991. The results agree well with coincident oceanographic in-situ measurements. Today a big effort is devoted to calibrate models with available measured data or to use models to help explain processes in the atmosphere-ocean-cryosphere.

Pereira et al. have used a modified version of the physical model for solar radiation calculation by Möser and Raschke (1983) to map the solar radiation in Brazil by use of METEOSAT data. Validation is being made with the network of 22 pyranometers of the Brazilian meteorological service. Results show standard errors of the order of 13%. A microwave radiative transfer model is used by **Mehrtens** to derive weighting functions for precipitating clouds. An adjoint formulation allows to solve the full equation of transfer including the phase function to include scattering processes in the weighting functions. The results show that even below a precipitating cloud (RR=10 mm/h) there is a contribution from the lowest layers and the ground at 85 Ghz.

Karlsson used monthly cloud cover statistics, derived from NOAA-AVHRR data, to validate cloud cover information generated from the HIRLAM forecasting model both with and without initial cloudiness states. When initial model cloudiness is set to zero, a considerable spin-up time is required in order to generate realistic cloudiness fields. When 6-hour HIRLAM cloud parameters are used as initial states, the model spin-up is drastically reduced; modelled and observed cloudiness amounts are similar whereas some spatial discrepancies remain.

Chen et al. study water vapor distribution obtained from ECHAM atmospheric GCM and compare the simulation with satellite observations of total precipitable water from SSM/I and upper tropospheric humidity from TOVS. In general, the model simulation agrees well with satellite observed climatological mean, seasonal variation and interannual variation of moisture. Cloud liquid water and water vapor are the two important atmospheric error sources for the retrieval of sea-ice concentration and ice type/age with passive microwaves from satellites. Ch. Oelke modelled these effects using a radiative transfer model and a high number of radiosonde ascents from Antarctica and compared them with observed concentration changes in SSM/I data.

Burford created a 1-D model to investigate sublimation of snowfall profiles and particle spectral evolution in the atmosphere. By assuming a Marshall Palmer type distribution and using the diffusion equation, the effects of varying the temperature and humidity fields over a 1 km fall depth were investigated. Most surprising results included large mass losses at relatively high temperatures and low subsaturation values. The Sahara dust intrusion in the atmosphere and the deposition in the mixed-layer of the Mediterranean Sea is studied by N. Kubilay et al.

A dust concentration equation was built into the existing ETA/NMC regional forecast model for the dynamics of the atmosphere. The model can give integrated dust concentration in the atmosphere along with wet/dry depositional areas.

Arslan AN Cukurova University, Adana, Turkey
Diffraction problems in microwave remote sensing

Burford J York University, North York, Canada
Sublimation in Arctic environments

Chen C-T, Roeckner E, Soden BJ MPI für Meteorologie, Hamburg, Germany
A comparison of satellite observations and model simulations of column integrated moisture and upper tropospheric humidity

Ivanova K, Jordanov OI, Michalev MA Institute of Electronics, BAS, Sofia, Bulgaria
Characterization and identification of fractal-like interfaces in the climate formations and the environment

Karlsson KG SMHI, Norrköping, Sweden
Comparison of modelled and satellite-observed cloudiness over northern Europe

Kubilay N, Nickovic S, Dobricic S Institute of Marine Sciences, Erdemli, Turkey
Climate processes related to Sahara dust in the Mediterranean

Li X, Morrison JM, Pietrafesa LJ North Carolina State University, Raleigh, USA
Dynamic analysis of oceanic internal waves from airborne synthetic aperture radar images

Mehrtens H Institut für Meereskunde, Kiel, Germany
Adjoint calculations of weighting functions

Oelke Ch Alfred-Wegener Inst. for Polar and Marine Research, Bremerhaven, Germany
Atmospheric signatures in passive microwave sea-ice algorithms

Pereira EB, Abreu SL, Stuhlmann R, Rieland M, Colle S University of Santa Catarina, Santa Catarina, Brazil
Incident solar radiation in Brazil using the IGMK model (validation and first results)

Pierazzo E University of Arizona, Tucson, USA
Cretaceous climate simulation using NCAR general circulation model CCM2-BATS

Schröder C, Hense A Meteorologisches Institut der Universität Bonn, Bonn, Germany
Statistical validation of simple energy balance model with satellite data

Index

The ASI Series Books Published as a Result of Activities of the Special Programme on Global Environmental Change

This book contains the proceedings of a NATO Advanced Research Workshop held within the activities of the NATO Special Programme on Global Environmental Change, which started in 1991 under the auspices of the NATO Science Committee.

The volumes published as a result of the activities of the Special Programme are:

Vol. 1: **Global Environmental Change.** Edited by R. W. Corell and P. A. Anderson. 1991.
Vol. 2: **The Last Deglaciation: Absolute and Radiocarbon Chronologies.** Edited by E. Bard and W. S. Broecker. 1992.
Vol. 3: **Start of a Glacial.** Edited by G. J. Kukla and E. Went. 1992.
Vol. 4: **Interactions of C, N, P and S Biogeochemical Cycles and Global Change.** Edited by R. Wollast, F. T. Mackenzie and L. Chou. 1993.
Vol. 5: **Energy and Water Cycles in the Climate System.** Edited by E. Raschke and D. Jacob. 1993.
Vol. 6: **Prediction of Interannual Climate Variations.** Edited by J. Shukla. 1993.
Vol. 7: **The Tropospheric Chemistry of Ozone in the Polar Regions.** Edited by H. Niki and K. H. Becker. 1993.
Vol. 8: **The Role of the Stratosphere in Global Change.** Edited by M.-L. Chanin. 1993.
Vol. 9: **High Spectral Resolution Infrared Remote Sensing for Earth's Weather and Climate Studies.** Edited by A. Chedin, M.T. Chahine and N.A. Scott. 1993.
Vol. 10: **Towards a Model of Ocean Biogeochemical Processes.** Edited by G. T. Evans and M. J. R. Fasham. 1993.
Vol. 11: **Modelling Oceanic Climate Interactions.** Edited by J. Willebrand and D. L. T. Anderson. 1993.
Vol. 12: **Ice in the Climate System.** Edited by W. Richard Peltier. 1993.
Vol. 13: **Atmospheric Methane: Sources, Sinks, and Role in Global Change.** Edited by M. A. K. Khalil. 1993.
Vol. 14: **The Role of Regional Organizations in the Context of Climate Change.** Edited by M. H. Glantz. 1993.
Vol. 15: **The Global Carbon Cycle.** Edited by M. Heimann. 1993.
Vol. 16: **Interacting Stresses on Plants in a Changing Climate.** Edited by M. B. Jackson and C. R. Black. 1993.
Vol. 17: **Carbon Cycling in the Glacial Ocean: Constraints on the Ocean's Role in Global Change.** Edited by R. Zahn, T. F. Pedersen, M. A. Kaminski and L. Labeyrie. 1994.
Vol. 18: **Stratospheric Ozone Depletion/UV-B Radiation in the Biosphere.** Edited by R. H. Biggs and M. E. B. Joyner. 1994.
Vol. 19: **Data Assimilation: Tools for Modelling the Ocean in a Global Change Perspective.** Edited by P. O. Brasseur and J. Nihoul. 1994.

Vol. 20: Biodiversity, Temperate Ecosystems, and Global Change.
Edited by T. J. B. Boyle and C. E. B. Boyle. 1994.

Vol. 21: Low-Temperature Chemistry of the Atmosphere.
Edited by G. K. Moortgat, A. J. Barnes, G. Le Bras and J. R. Sodeau. 1994.

Vol. 22: Long-Term Climatic Variations – Data and Modelling.
Edited by J.-C. Duplessy and M.-T. Spyridakis. 1994.

Vol. 23: Soil Responses to Climate Change.
Edited by M. D. A. Rounsevell and P. J. Loveland. 1994.

Vol. 24: Remote Sensing and Global Climate Change.
Edited by R. A. Vaughan and A. P. Cracknell. 1994.

Vol. 25: The Solar Engine and Its Influence on Terrestrial Atmosphere and Climate.
Edited by E. Nesme-Ribes. 1994.

Vol. 26: Global Precipitations and Climate Change.
Edited by M. Desbois and F. Désalmand. 1994.

Vol. 27: Cenozoic Plants and Climates of the Arctic.
Edited by M. C. Boulter and H. C. Fisher. 1994.

Vol. 28: Evaluating and Monitoring the Health of Large-Scale Ecosystems.
Edited by D. J. Rapport, C. L. Gaudet and P. Calow. 1995.

Vol. 29: Global Environmental Change Science: Education and Training.
Edited by D. J. Waddington. 1995.

Vol. 30: Ice Core Studies of Global Biogeochemical Cycles.
Edited by R. J. Delmas. 1995.

Vol. 31: The Role of Water and the Hydrological Cycle in Global Change.
Edited by H. R. Oliver and S. A. Oliver. 1995.

Vol. 32: Atmospheric Ozone as a Climate Gas.
Edited by W.-C. Wang and I. S. A. Isaksen. 1995.

Vol. 33: Carbon Sequestration in the Biosphere.
Edited by M. A. Beran. 1995.

Vol. 34: Climate Sensitivity to Radiative Perturbations: Physical Mechanisms and Their Validation.
Edited by H. Le Treut. 1996

Vol. 35: Clouds, Chemistry and Climate.
Edited by P. J. Crutzen and V. Ramanathan. 1996.

Vol. 36: Diachronic Climatic Impacts on Water Resources.
Edited by A. N. Angelakis and A. S. Issar. 1996.

Vol. 37: Climate Change and World Food Security.
Edited by T. E. Downing. 1996.

Vol. 38: Evaluation of Soil Organic Matter Models.
Edited by D. S. Powlson, P. Smith and J. U. Smith. 1996.

Vol. 39: Microbiology of Atmospheric Trace Gases.
Edited by J. C. Murrell and D. P. Kelly. 1996.

Vol. 40: Forest Ecosystems, Forest Management and the Global Carbon Cycle.
Edited by M. J. Apps and D. T. Price. 1996.

Vol. 41: Climate Variations and Forcing Mechanisms of the Last 2000 Years.
Edited by P. D. Jones, R. S. Bradley and J. Jouzel. 1996.

Vol. 42: The Mount Pinatubo Eruption: Effects on the Atmosphere and Climate.
Edited by G. Fiocco, D. Fuà and G. Visconti. 1996.

Vol. 43: Chemical Exchange Between the Atmosphere and Polar Snow.
Edited by E. W. Wolff and R. C. Bales. 1996.

Vol. 44: **Decadal Climate Variability: Dynamics and Predictability**
Edited by D. L.T. Anderson and J. Willebrand. 1996.

Vol. 45: **Radiation and Water in the Climate System: Remote Measurements**
Edited by E. Raschke. 1996.

Vol. 46: **Land Surface Processes in Hydrology: Trials and Tribulations of Modeling and Measuring.**
Edited by S. Soroosh and H. Gupta. 1996.

GPSR Compliance

The European Union's (EU) General Product Safety Regulation (GPSR) is a set of rules that requires consumer products to be safe and our obligations to ensure this.

If you have any concerns about our products, you can contact us on ProductSafety@springernature.com

In case Publisher is established outside the EU, the EU authorized representative is:

Springer Nature Customer Service Center GmbH
Europaplatz 3
69115 Heidelberg, Germany

Batch number: 10593112

Printed by Printforce, the Netherlands